Lecture Notes in Computer Science 8416

Commenced Publication in 1973
Founding and Former Series Editors:
Gerhard Goos, Juris Hartmanis, and Jan van Leeuwen

Maarten de Rijke   Tom Kenter
Arjen P. de Vries   ChengXiang Zhai
Franciska de Jong   Kira Radinsky
Katja Hofmann (Eds.)

Advances in Information Retrieval

36th European Conference on IR Research, ECIR 2014
Amsterdam, The Netherlands, April 13-16, 2014
Proceedings

 Springer

Volume Editors

Maarten de Rijke
Tom Kenter
University of Amsterdam, The Netherlands
E-mail: {derijke, t.m.kenter}@uva.nl

Arjen P. de Vries
Centrum Wiskunde en Informatica, Amsterdam, The Netherlands
E-mail: arjen@acm.org

ChengXiang Zhai
University of Illinois at Urbana-Champaign, Urbana, IL, USA
E-mail: czhai@cs.uiuc.edu

Franciska de Jong
University of Twente, Enschede, The Netherlands
E-mail: f.m.g.dejong@utwente.nl

Kira Radinsky
SalesPredict, Haifa, Israel
E-mail: kiraradinsky@gmail.com

Katja Hofmann
Microsoft Research, Cambridge, UK
E-mail: katja.hofmann@microsoft.com

ISSN 0302-9743   e-ISSN 1611-3349
ISBN 978-3-319-06027-9   e-ISBN 978-3-319-06028-6
DOI 10.1007/978-3-319-06028-6
Springer Cham Heidelberg New York Dordrecht London

Library of Congress Control Number: 2014934424

LNCS Sublibrary: SL 3 – Information Systems and Application, incl. Internet/Web
and HCI

Typesetting: Camera-ready by author, data conversion by Scientific Publishing Services, Chennai, India

Printed on acid-free paper

Springer is part of Springer Science+Business Media (www.springer.com)

Preface

These proceedings contain the full papers, short papers, and demonstrations selected for presentation at the 36th European Conference on Information Retrieval (ECIR 2014). The event was organized by the University of Amsterdam and supported by the Intelligent Systems Lab Amsterdam (ISLA), the Information Retrieval Specialist Group at the British Computer Society (BCS–IRSG), and Girl Geek Dinner NL. The conference was held April 13–16, 2014, in Amsterdam, The Netherlands.

ECIR 2014 received a total of 288 submissions in three categories: 142 full papers, 120 short papers, 26 demonstrations. The geographical distribution of the submissions is as follows: 58% were from Europe (including 8% from The Netherlands), 18% from Asia, 19% from North and South America, and 6% from North Africa and the Middle East. All submissions were reviewed by at least three members of an international two-tier Program Committee. Of the full papers submitted to the conference, 33 were accepted for oral presentation (23%). Of the short papers submitted to the conference, 50 were accepted for poster presentation (42%). In addition, 15 demonstrations (58%) were accepted.

The accepted contributions represent the state of the art in information retrieval, cover a diverse range of topics, propose novel applications, and indicate promising directions for future research. Out of the accepted full paper contributions, 73% have a student as the primary author. We gratefully thank all Program Committee members for their time and efforts ensuring a high-quality level of the ECIR 2014 program.

Additionally, ECIR 2014 hosted three tutorials and four workshops covering a range of information retrieval topics. We are extremely grateful to Leif Azzopardi (University of Glasgow) and his committee for selecting the ECIR 2014 workshops:

- Bibliometric-Enhanced Information Retrieval
- 4th Workshop on Context-awareness in Retrieval and Recommendation (CaRR)
- Gamification for Information Retrieval (GamifIR 2014)
- Information Access in Smart Cities (i-ASC)

We are equally grateful to Edgar Meij (Yahoo! Research) and his committee for selecting the ECIR 2014 tutorials:

- Designing Search Usability
- Text Quantification
- The Cluster Hypothesis in Information Retrieval

We would like to thank our invited speakers for their contributions to the program: Eugene Agichtein (Emory University) and Gilad Mishne (Twitter). We are very grateful to a committee lead by Ayse Goker for selecting the winner of the 2013 Karen Spärck-Jones Award.

ECIR 2014 featured a panel on the relation between academic and industrial research in information retrieval, led by Mark Sanderson as moderator.

During the final day of the conference, an industry session took place. It featured a diverse and attractive range of presentations by researchers and practitioners. We are grateful to David Carmel (Yahoo! Labs) and Thijs Westerveld (WizeNoze) for organizing the program of the industry session.

Finally, ECIR 2014 would not have been possible without the generous financial support from our sponsors: the City of Amsterdam (gold level); Google, Microsoft Research, The Netherlands Organization for Scientific Research, Text-kernel, Yahoo! Labs and Yandex (silver level); CRC Press, The Nederlands Instituut voor Beeld en Geluid, NOW Publishers and School voor Informatie- en Kennissystemen (SIKS) (bronze level). The conference was supported by the BCS-IRSG, the ELIAS Research Network Programme of the European Science Foundation, Girl Geek Dinner NL, the Intelligent Systems Lab Amsterdam, and through the Friends of SIGIR program. Thank you all!

February 2014   Maarten de Rijke
                                                            Tom Kenter
                                                         Arjen P. de Vries
                                                        ChengXiang Zhai
                                                        Franciska de Jong
                                                          Kira Radinsky
                                                         Katja Hofmann

Organization

ECIR 2014 was organized by the University of Amsterdam. It was supported by the Intelligent Systems Lab Amsterdam (ISLA), the Information Retrieval Specialist Group at the British Computer Society (BCS–IRSG), Girl Geek Dinner NL, and the Friends of SIGIR program.

General Chair

Maarten de Rijke   University of Amsterdam

Local Chair

Tom Kenter   University of Amsterdam

Program Committee Chairs

Arjen P. de Vries   Centrum Wiskunde en Informatica and
     Technical University of Delft
ChengXiang Zhai   University of Illinois at Urbana-Champaign

Short Paper Chairs

Franciska de Jong   University of Twente and Erasmus University
     Rotterdam
Kira Radinsky   SalesPredict

Demonstrations Chair

Katja Hofmann   Microsoft Research Cambridge

Workshop Chair

Leif Azzopardi   University of Glasgow

Tutorial Chair

Edgar Meij   Yahoo! Research

Industry Session Chairs

David Carmel   Yahoo! Labs
Thijs Westerveld   WizeNoze

Best Paper Award Chair

John Tait   johntait.net Ltd.

Local Organizers

Richard Berendsen   University of Amsterdam
David Graus   University of Amsterdam
Caroline van Impelen   University of Amsterdam
Daan Odijk   University of Amsterdam
Maria-Hendrike Peetz   University of Amsterdam
Ridho Reinanda   Royal Netherlands Academy of Arts and
                                   Sciences and University of Amsterdam
Zhaochun Ren   University of Amsterdam
Anne Schuth   University of Amsterdam
Wouter Weerkamp   904Labs

Program Committee

Senior Program Committee

Eugene Agichtein   Emory University
Giambattista Amati   Fondazione Ugo Bordoni
Jaime Arguello   University of North Carolina at Chapel Hill
Leif Azzopardi   University of Glasgow
Krisztian Balog   University of Stavanger
Nicholas Belkin   Rutgers University
Bettina Berendt   Katholieke Universiteit Leuven
David Carmel   Yahoo! Labs
Paul Clough   University of Sheffield
Bruce Croft   University of Massachusetts
David Craig Elsweiler   University of Regensburg
Juan M. Fernandez-Luna   University of Granada
Norbert Fuhr   University of Duisburg-Essen
Eric Gaussier   University of J. Fourier/Grenoble 1
Ayse Goker   Robert Gordon University
Julio Gonzalo   UNED
Cathal Gurrin   Dublin City University

David A. Hull	Google
Gareth Jones	Dublin City University
Joemon M. Jose	University of Glasgow
Jaap Kamps	University of Amsterdam
Noriko Kando	National Institute of Informatics
Evangelos Kanoulas	Google
Gabriella Kazai	Microsoft Research Cambridge
Birger Larsen	Aalborg University
Jimmy Lin	University of Maryland
David E. Losada	University of Santiago de Compostela
Stefano Mizzaro	University of Udine
Josiane Mothe	Institut de Recherche en Informatique de Toulouse
Vanessa Murdock	Bing
Fabrizio Sebastiani	Consiglio Nazionale delle Ricerche
Pavel Serdyukov	Yandex
Fabrizio Silvestri	Yahoo! Labs
Jun Wang	University College London
Emine Yilmaz	Microsoft Research Cambridge

Additional Reviewers

Mikhail Ageev	Moscow State University
Dirk Ahlers	Trondheim University
Ahmet Aker	University of Sheffield
Elif Aktolga	University of Massachusetts Amherst
Dyaa Albakour	University of Glasgow
Omar Alonso	Microsoft
Ismail Sengor Altingovde	L3S
Robin Aly	University of Twente
Avi Arampatzis	Democritus University of Thrace
Javed A. Aslam	Northeastern University
Ricardo Baeza-Yates	Yahoo! Research
Alvaro Barreiro	University of A Coruña
Barry Smyth	University College Dublin
Roberto Basili	University of Rome, Tor Vergata
Srikanta Bedathur	IIIT-Delhi
Michel Beigbeder	Ecole des Mines de Saint-Etienne
Alejandro Bellogín	Universidad Autónoma de Madrid
Patrice Bellot	Aix-Marseille Université (AMU) - CNRS
Klaus Berberich	Max Planck Institute for Informatics
Toine Bogers	Aalborg University Copenhagen
Gloria Bordogna	CNR
Mohand Boughanem	University of Toulouse
Marc Bron	University of Amsterdam
Peter Bruza	Queensland University of Technology

Katriina Byström	University of Boras
Fidel Cacheda	University of A Coruña
Fazli Can	Bilkent University
Mark Carman	Monash University
Claudio Carpineto	Fondazione Ugo Bordoni
Marc-Allen Cartright	Google, Inc.
Carlos Alberto Alejandro Castillo Ocaranza	Qatar Computing Research Institute
James Caverlee	Texas A&M University
Max Chevalier	University of Toulouse
Paul-Alexandru Chirita	Adobe Systems Inc.
Ingemar Cox	Copenhagen University
Fabio Crestani	University of Lugano
Bin Cui	Peking University
Ronan Cummins	University of Greenwich
Alfredo Cuzzocrea	ICAR-CNR and University of Calabria
Na Dai	Yahoo
Franciska de Jong	Universiteit Twente
Pablo de la Fuente	Universidad de Valladolid
Adriel Dean-Hall	University of Waterloo
Thomas Demeester	University of Ghent
Romain Deveaud	Glasgow University
Giorgio Maria Di Nunzio	University of Padua
Shuai Ding	Polytechnic Institute of New York University
Vladimir Dobrynin	St. Petersburg State University
Huizhong Duan	WalmartLabs
Pavlos Efraimidis	Democritus University of Thrace
Carsten Eickhoff	Delft University of Technology
Liana Ermakova	Institut de Recherche en Informatique de Toulouse (IRIT) / Perm State National Research University
Hui Fang	University of Delaware
Yi Fang	Santa Clara University
Nicola Ferro	University of Padua
Luanne Freund	University of British Columbia
Karin Friberg Heppin	University of Gothenburg
Ingo Frommholz	University of Bedfordshire
Patrick Gallinari	Université Pierre et Marie Curie, Paris 6
Kavita Ganesan	3M
Shima Gerani	University of British Columbia
Giorgos Giannopoulos	National Technical University of Athens
Richard Glassey	Robert Gordon University
Ayse Goker	Robert Gordon University
David Adam Grossman	IIT

Antonio Gulli   Microsoft Bing
Matthias Hagen   University of Weimar
Allan Hanbury   Vienna University of Technology
Preben Hansen   Stockholm University
Donna Harman   NIST
Morgan Harvey   University of Lugano (USI)
Claudia Hauff   Delft University of Technology
Jer Hayes   IBM Research Lab - Ireland
Ben He   University of Chinese Academy of Sciences
Daqing He   University of Pittsburg
Jiyin He   Centrum Wiskunde en Informatica
Yulan He   Aston University
Nathalie Hernandez   IRIT
Katja Hofmann   Microsoft Research
Andreas Hotho   University of Würzburg
Yuexian Hou   Tianjin University
Gilles Hubert   University of Toulouse
Dmitry Ignatov   National Research University Higher School
                                 of Economics
Jiepu Jiang   University of Massachusetts Amherst
Richard Johansson   University of Gothenburg
Frances Johnson   Manchester Metropolitan University
Hideo Joho   University of Tsukuba
Kristiina Jokinen   University of Helsinki
Maryam Karimzadehgan   University of Illinois at Urbana-Champaign
Mostafa Keikha   University of Massachusetts Amherst
Diane Kelly   University of North Carolina
Liadh Kelly   Dublin City University
Yiannis Kompatsiaris   Information Technologies Institute - CERTH
Marijn Koolen   University of Amsterdam
Alexander Kotov   Wayne State University
Manolis Koubarakis   National and Kapodistrian University
                                 of Athens
Udo Kruschwitz   University of Essex
Oren Kurland   Technion University
Dmitry Lagun   Emory University
James Lanagan   Facebook
Monica Angela Landoni   University of Lugano
Birger Larsen   Aalborg University
Fotis Lazarinis   University of Western Greece
Kyumin Lee   Utah State University
Wang-Chien Lee   Pennsylvania State University
Hyowon Lee   Singapore University of Technology and Design
Johannes Leveling   Dublin City University
Yanen Li   University of Illinois at Urbana-Champaign

Elizabeth Liddy	Syracuse University
Christina Lioma	University of Copenhagen
Xiaozhong Liu	Indiana University
Elena Lloret	University of Alicante
Yue Lu	Twitter
Bernd Ludwig	University of Regensburg
Yuanhua Lv	Microsoft Research
Craig Macdonald	University of Glasgow
Andrew MacFarlane	City University London
Marco Maggini	University of Siena
Thomas Mandl	University of Hildesheim
Stephane Marchand-Maillet	University of Geneva
Miguel Martinez-Alvarez	Queen Mary, University of London
Yosi Mass	IBM Research
Qiaozhu Mei	University of Michigan
Wagner Meira Jr.	Universidade Federal de Minas Gerais
Marcelo Mendoza	Yahoo! Research
Alessandro Micarelli	Roma Tre University
Dunja Mladenic	J. Stefan Institute
Marie-Francine Moens	Katholieke Universiteit Leuven
Hannes Mühleisen	CWI
Henning Müller	University of Applied Sciences Western Switzerland
Wolfgang Nejdl	University of Hannover
Boris Novikov	St. Petersburg University
Andreas Nuernberger	Otto-von-Guericke-Universität Magdeburg
Neil O'Hare	Yahoo! Research
Michael O'Mahony	University College Dublin
Michael Philip Oakes	University of Wolverhampton
Douglas Oard	University of Maryland
Iadh Ounis	University of Glasgow
Monica Paramita	University of Sheffield
Gabriella Pasi	Università degli studi di Milano Bicocca
Virgil Pavlu	Northeastern University
Karen Pinel-Sauvagnat	Toulouse University
Vassilis Plachouras	ATHENA Research Center
Barbara Poblete	University of Chile
Tamara Polajnar	Cambridge University
Andreas Rauber	Vienna University of Technology
Paolo Rosso	Universitat Politècnica de Valencia
Dmitri Roussinov	University of Strathclyde
Alan Said	CWI
Michail Salampasis	Alexander Technology Educational Institute (ATEI) of Thessaloniki

Rodrygo Santos	Universidade Federal de Minas Gerais
Markus Schedl	Johannes Kepler University (JKU)
Ralf Schenkel	University of Passau
Falk Scholer	RMIT University
Florence Sedes	Université Paul Sabatier
Giovanni Semeraro	University of Bari "Aldo Moro"
Jangwon Seo	Google Inc.
Azadeh Shakery	University of Tehran
Jialie Shen	Singapore Management University
Mario J. Silva	Instituto Superior Tecnico / INESC-ID
Alan Smeaton	Dublin City University
Mark D. Smucker	University of Waterloo
Parikshit Sondhi	WalmartLabs
Yang Song	Microsoft Research
Herman Stehouwer	Max Plack Institute for Psycholinguistics, Nijmegen
Benno Stein	Bauhaus-Universität Weimar
Simone Stumpf	City University London
L. Venkata Subramaniam	IBM Research India
Lynda Tamine-Lechani	IRIT
Martin Theobald	Max Planck Institute for Informatics
Bart Thomee	Yahoo Research
Marko Tkalcic	Johannes Kepler University
Anastasios Tombros	Queen Mary University of London
Dolf Trieschnigg	University of Twente
Ming-Feng Tsai	National Chengchi University
Theodora Tsikrika	Information Technologies Institute, CERTH
Denis Turdakov	The Institute for System Programming of the Russian Academy of Sciences (ISPRAS)
Ata Turk	Yahoo Labs
Yannis Tzitzikas	University of Crete
Marieke van Erp	VU University Amsterdam
Jacco van Ossenbruggen	CWI
Natalia Vassilieva	HP Labs
Olga Vechtomova	University of Waterloo
Sumithra Velupillai	Stockholm University
Suzan Verberne	Radboud University Nijmegen
Robert Villa	University of Sheffield
Stefanos Vrochidis	Centre for Research and Technology Hellas
Jeroen Vuurens	Delft University of Technology
V.G. Vinod Vydiswaran	University of Michigan
Xiaojun Wan	Peking University
Hongning Wang	University of Illinois at Urbana-Champaign
Fan Wang	Microsoft Bing
Lidan Wang	University of Maryland

Wouter Weerkamp	904Labs
Ryen William White	Microsoft Research
Jun Xu	Huawei Technologies
Tao Yang	UCSB & Ask.com
David Zellhöfer	BTU Cottbus
Peng Zhang	Tianjin University
Dan Zhang	Facebook Inc.
Dell Zhang	Birkbeck, University of London
Duo Zhang	Twitter
Lanbo Zhang	UC Santa Cruz
Peng Zhang	Tianjin University
Zhaohui Zheng	Yahoo! Labs Beijing
Ke Zhou	Glasgow University
Guido Zuccon	CSIRO

Demonstration Selection Committee

Alejandro Bellogín	Universidad Autónoma de Madrid
Rafael Berlanga-Llavori	Universitat Jaume I
Sebastian Blohm	Microsoft Research Ltd.
Toine Bogers	Aalborg University Copenhagen
Marc Bron	University of Amsterdam
Thomas Demeester	University of Ghent
Thomas Gottron	Universität Koblenz Landau
Morgan Harvey	University of Lugano
Jiyin He	Centrum Wiskunde en Informatica
Udo Kruschwitz	University of Essex
Dong Nguyen	University of Twente
Filip Radlinski	Microsoft
Markus Schedl	Johannes Kepler University
Anne Schuth	University of Amsterdam
Milad Shokouhi	Microsoft
Benno Stein	Bauhaus-Universität Weimar
Dolf Trieschnigg	University of Twente
Manos Tsagkias	University of Amsterdam
Özgür Ulusoy	Bilkent University
Suzan Verberne	Radboud University Nijmegen
Wouter Weerkamp	904Labs

Tutorial Selection Committee

Ricardo Baeza-Yates	Yahoo! Research
Krisztian Balog	University of Stavanger
Berkant Barla Cambazoglu	Yahoo! Research
Fabio Crestani	University of Lugano
Gianluca Demartini	University of Fribourg
Peter Knees	Johannes Kepler University Linz
Mounia Lalmas	Yahoo Labs London

Ilya Markov	University of Lugano
Massimo Melucci	University of Padua
Marie-Francine Moens	Katholieke Universiteit Leuven
Jian-Yun Nie	University of Montreal
Daan Odijk	University of Amsterdam
Benjamin Piwowarski	CNRS
Filip Radlinski	Microsoft
Tony Russell-Rose	UXLabs
Ian Ruthven	University of Strathclyde

Workshop Selection Committee

Hideo Joho	University of Tsukuba
Stefano Mizzaro	University of Udine
Paul Ogilvie	LinkedIn
Rodrygo Santos	Universidade Federal de Minas Gerais
Falk Scholer	RMIT University
Emine Yilmaz	Microsoft

Sponsoring Institutions

Gold sponsors	City of Amsterdam
	ELIAS Research Network Programme
Silver sponsors	Google
	Microsoft Research
	Netherlands Organization for Scientific Research (NWO)
	Textkernel
	Yahoo! Labs
	Yandex
Bronze sponsors	CRC Press
	Nederlands Instituut voor Beeld en Geluid
	NOW Publishers
	Taylor & Francis Group
	School voor Kennis- en Informatiesystemen (SIKS)

Inferring Searcher Attention and Intention by Mining Behavior Data (Keynote)

Eugene Agichtein

Emory University
eugene@mathcs.emory.edu

Abstract. This is a summary of the keynote talk delivered by the winner of the Karen Spärck-Jones Award at ECIR 2014.

A long standing challenge in Web search is how to accurately determine the intention behind a searchers query, which is needed to rank, organize, and present results most effectively. The difficulty is that users often do not (or cannot) provide sufficient information about their goals. As this talk with show, it is nevertheless possible to read their intentions through clues revealed by behavior, such as the amount of attention paid to a document or a text fragment. I will overview the approaches that have emerged for acquiring and mining behavioral data for inferring search intent, ranging from robust models of click data in the aggregate, to modeling fine-grained user interactions such as mouse cursor movements in the searchers browser. The latter can also be used to measure the searchers attention "in the wild, with granularity approaching that of using eye tracking equipment in the laboratory. The resulting techniques and models have already shown noteworthy improvements for search tasks such as ranking, relevance estimation, and result summary generation, and have applications to other domains, such as psychology, neurology, and online education.

Real-Time Search at Twitter
(Keynote)

Gilad Mishne

Twitter
gilad@twitter.com

Abstract. This is a summary of the keynote talk delivered at ECIR 2014.

Twitter's search engine faces some of the most unique challenges in information retrieval and distributed systems today. On the scaling front, it's a relatively young system with a massive user base, billions of queries daily, and many billions of indexed documents - with thousands being added every second. On the ranking side, the combination of realtime and social requires new solutions to relevance estimation for newly-created documents, blending different types of live content, evaluation in the absence of direct user feedback, and more. On top of this, the dynamic nature of a nascent company and product leads to a multitude of operational challenges and opportunities.

In this talk, I'll cover some of these challenges and how we approach them at Twitter.

Panel on Academic and Industrial Research in Information Retrieval

Abstract. This is a brief description of panel that was held at ECIR 2014.

The relation between academic and industrial research in information retrieval has been a frequent topic of discussion.

In discussions about this relation, the use of proprietary data in IR research is one issue that gives rise to a range of positions. For instance, it was recently decided to turn ICTIR into a more 'foundational' conference, with an emphasis on experiments on data that is openly available to the research community. This move probably reflects the position that for scientific results to be reproducible, the data being used should be open. The inclusion of industry days or industry sessions at recent editions of SIGIR and ECIR represents a different position, namely that valuable insights can be gained from proprietary data.

But there's more. As the community's research agenda develops, a strong emphasis on online algorithms and online evaluation is emerging. Is this shift a source of divergence between academic and industrial research in IR? Work on online evaluation may require access to live systems. And it already seems that expertise on online algorithms and time-aware ranking methods can more easily be found in industrial research environments than in academia.

The ECIR 2014 organizers organized a panel to discuss the IR research ecosystem and the roles of academic and industrial research in this ecosystem. Our goal was to stimulate discussion in the IR community that may inform future strategy development.

The panel consisted of Mark Sanderson as moderator and Behshad Behzadi, Norbert Fuhr, Birger Larsen, and Peter Mika as panelists.

Panel on Academic and Industrial Research in Information Retrieval

Abstract. This is a brief description of a panel that was held at ECIR 2014.

The relation between academic and industrial research in information retrieval has been a frequent topic of discussion.

In discussions about this relation, the use of proprietary data in IR research is one issue that gives rise to a range of positions. For instance, it was recently decided in that ECIR has a new foundational confidence with emphasis on experiments on data that is openly available to the research community. This raises probably reflects the position that for scientific results to be reproducible, the data being used should be open. The inclusion of industry at conferences such as those of SIGIR and ECIR presents a different position, namely that valuable insights can be gained from proprietary data.

But there's more. As the community's research agenda develops, a strong emphasis on online algorithms and online evaluation is emerging. In this shift a source of divergence between academic and industrial research is felt. Work on online evaluation may require access to live systems. And it already seems that expertise on online algorithms and techniques are readily and more easily be found in industrial research environments than in academia.

The ECIR 2014 organizers organized a panel to discuss the different research on and the roles of academic and industrial research in this congress. Our goal was to stimulate discussion in the IR community in this much needed but in many ways neglected research topic.

The panel consisted of Maarten de Rijke as moderator and Richard Boulton, Arjen P. de Vries, Jimmy Lin, and Peter Mika as panelists.

Table of Contents

Session 4B: Mining Social Media

Session 5: Digital Libraries

Session 6: Efficiency

Session 7: Information Retrieval Theory

Short Paper Session 1

Short Paper Session 2

Demonstration Session 1

Demonstration Session 2

Workshops

Tutorials

Reducing Reliance on Relevance Judgments for System Comparison by Using Expectation-Maximization

Ning Gao[1], William Webber[2], and Douglas W. Oard[1]

[1] College of Information Studies/UMIACS, University of Maryland, College Park
{ninggao,oard}@umd.edu
[2] William Webber Consulting
william@williamwebber.com

Abstract. Relevance judgments are often the most expensive part of information retrieval evaluation, and techniques for comparing retrieval systems using fewer relevance judgments have received significant attention in recent years. This paper proposes a novel system comparison method using an expectation-maximization algorithm. In the expectation step, real-valued pseudo-judgments are estimated from a set of system results. In the maximization step, new system weights are learned from a combination of a limited number of actual human judgments and system pseudo-judgments for the other documents. The method can work without any human judgments, and is able to improve its accuracy by incrementally adding human judgments. Experiments using TREC Ad Hoc collections demonstrate strong correlations with system rankings using pooled human judgments, and comparison with existing baselines indicates that the new method achieves the same comparison reliability with fewer human judgments.

1 Introduction

Information retrieval systems are generally evaluated on the relevance of the documents they retrieve. Relevance judgments must be made by humans, and obtaining these judgments can be the most expensive part of retrieval evaluation. The expense is greatest when building a reusable test collection, since judgments are required not only for documents returned by a particular system or set of systems, but also for those that might be returned by future systems on the same collection. The traditional method of managing the judgment task for reusable collections is through pooling, where only the top-ranked documents (such as the first 100 documents) from participating systems are judged. Even pooling can lead to a heavy judgment burden, however, with depth 100 pooling on Text Retrieval Conference (TREC) Ad Hoc collections requiring thousands of document judgments (133,681 for TREC 5, for example).

In this paper, we introduce a novel Expectation-Maximization (EM) algorithm for comparing system effectiveness. The expectation step predicts the relevance of documents based on which systems return them; then the maximization step estimates system effectiveness from the predicted relevance of the documents the system returns. This process is repeated iteratively until the algorithm converges. A particular feature of our method is that it can work with no human judgments at all, but can also incrementally incorporate human judgments to improve accuracy.

M. de Rijke et al. (Eds.): ECIR 2014, LNCS 8416, pp. 1–12, 2014.
© Springer International Publishing Switzerland 2014

We conduct experiments comparing our EM algorithm with existing state-of-the-art methods for reduced-effort evaluation, both without and with human judgments, on four TREC Ad Hoc collections. The task is to estimate a ranking of systems by retrieval effectiveness. We find that our method generally outperforms existing judgment-free methods. Our method improves still further as human relevance judgments are added, outperforming existing limited-judgment methods for most judgment set sizes.

The remainder of the paper is laid out as follows. In Section 2, we survey related work on the reduced-effort evaluation of retrieval systems. Section 3 describes our expectation-maximization method for retrieval system ranking, along with the various methods of calculating system votes for document relevance and, where human judgments are available, of selecting documents for judgment. In Section 4, we evaluate the different voting and document selection methods against each other, and against established methods of judgment-free and limited-judgment system evaluation. Finally, Section 5 presents our conclusions.

2 Related Work

The use of system pools to select documents for relevance judgment was first proposed by Spärck-Jones and van Rijsbergen [13]. Zobel examined the reliability of pools in early TREC Ad Hoc collections, finding that although they only captured half or fewer of the relevant documents in the collection, the resulting judgment sets were not significantly biased against unpooled systems [17]. Cormack et al. proposed Move-To-Front (MTF) pooling [6]. Under MTF, the documents from each system are judged in order of rank, and the systems themselves are also prioritized according to the previously judged documents. Various approaches have been proposed for evaluation with incomplete judgments. One is to use an evaluation metric that handles missing judgments. Buckley and Voorhees introduced this idea with the BPref metric, which evaluates only by judged documents [3].

A second approach to a limited judgment budget is to sample documents for judgment, then statistically estimate the evaluation measure. Aslam et al. described Hedge, a greedy algorithm for creating pools likely to contain large numbers of relevant documents [1]. Yilmaz and Aslam introduced a uniform random sample of the pooled documents and infer the pooled Average Precision (AP) metric to create infAP [14]. Aslam et al. also proposed an unequal sampling method based upon AP weights [2]. A simpler and more general stratified-sampling approach, xinfAP, is described in Yilmaz et al. [16]. Carterette and Allan suggested that the documents to (incrementally) prioritize for judgment are those which, if found relevant, would maximize the difference in scores between evaluated systems [5]. Carterette's Robust Test Collection (RTC) treats the retrieval systems as experts, and their retrieval of documents as votes for relevance [4].

All the methods in this paper are item-based, estimating pseudo-judgments of relevance for each document, and then evaluating the systems from these pseudo-judgments. There are also distribution-based methods. In Dai et al., the distribution of scores over documents is modeled by a mixture of two distributions, a Gaussian distribution for relevant documents and an exponential distribution for non-relevant documents [7].

An alternative approach to system evaluation is to do without human relevance judgments, ranking systems based solely on the system document rankings themselves.

The first to propose this approach was Soboroff et al., who used voting based on random assignment of relevance to documents in the document pool [12]. Nuray and Can's Condorcet method used system document rankings to rank documents by predicted relevance, and then assigned binary relevance to documents up to some cutoff [11]. Hauff et al. compared all the above methods across 16 different test collections (from TREC and elsewhere), finding Soboroff et al's random-voting method best on nine collections, and Nuray and Can's Condorcet method best on six [8]. Hosseini et al. used the EM framework to solve the problem of acquiring relevance judgements for Book Search tasks through crowdsourcing when no true relevance labels are available [9].

3   System Comparison by EM

In this section, we introduce our method for performing system comparisons with limited or no human judgments. We begin with a description of our fully automatic EM method, using no human judgments at all, in Section 3.1. We then describe how we adapt the technique to make use of incomplete human judgments in Section 3.2.

3.1   Zero Relevance Judgments

To illustrate the operation of our EM algorithm, we start from the simplest case, which is when no human judgments are available. In the expectation step (E-step), real-valued pseudo-judgments are estimated using the document rankings produced by a diverse range of systems. The loss between the result vector of a system and the pseudo-judgment vector is then calculated for each system. In the maximization step (M-step), the weights of systems are tuned to minimize the loss with respect to the pseudo-judgments. The E-step and M-step alternate until the values converge to a fixed point. The final real valued pseudo-judgments of the documents are then taken as estimates of document relevance, and used to evaluate retrieval systems, for instance by binarizing relevance at a prediction threshold and then calculating an effectiveness metric.

The process can be more precisely explained with reference to the "learning matrix" shown in Table 1. Suppose that there are p systems $\{S_1, ..., S_p\}$ to be compared using a measure that averages over k topics $\{T_1, ..., T_k\}$, and that for each topic T_m there are $t(m)$ documents $\{D_{m,1}, ..., D_{m,t(m)}\}$ over whose relevance the effectiveness measure will be computed. These documents could be the whole collection, but in our experiments they are the judgment pool that was created for TREC. Because different topics have different judgment pools, we subscript the documents separately for each topic. Suppose further that $V_{S_j,D_{m,n}}$ is the score that system S_j assigns to document n for topic m. We refer to this score as the *retrieval status value* of the respective document.

The EM algorithm iteratively estimates two sets of values: the hidden variables, which are the pseudo-judgment scores for each document on each topic; and the parameters, which are the loss-minimizing weights for each system. For each document $D_{m,n}$, a real-valued pseudo-judgment $J_{m,n}$ reflects the EM algorithm's current degree of belief that document n is relevant to topic m. Similarly, the system weight w_j represents the algorithm's present degree of belief that the results produced by system j are correct. The better a system is at contributing to the estimation of accurate pseudo-judgments,

Table 1. EM Learning Matrix: $\mathbf{S}$ for systems, $\mathbf{w}$ for system weights, $\mathbf{T}$ for topics, $\mathbf{D}$ for documents, $\mathbf{J}$ for pseudo-judgments, and $\mathbf{V}$ for predicted scores

		$\mathbf{w_1}$	...	$\mathbf{w_p}$	
		S_1	...	S_p	
T_1	$D_{1,1}$	$V_{S_1,D_{1,1}}$		$V_{S_p,D_{1,1}}$	$J_{1,1}$
	...	...		...	...
	$D_{1,t(1)}$	$V_{S_1,D_{1,t(1)}}$		$V_{S_p,D_{1,t(1)}}$	$J_{1,t(1)}$
...	...	...		...	...
T_k	$D_{k,1}$	$V_{S_1,D_{k,1}}$		$V_{S_p,D_{k,1}}$	$J_{k,1}$
	...	...		...	...
	$D_{k,t(k)}$	$V_{S_1,D_{k,t(k)}}$		$V_{S_p,D_{k,t(k)}}$	$J_{k,t(k)}$

the higher its weight should be. We require that the p system weights be bounded to $[0,1]$ and that they always sum to 1, $(||w_i||_1 = 1)$; thus they represent a distribution.

The iteration begins by setting the initial system weights $w^1 = \{w_1^1, ..., w_p^1\}$ to the uniform distribution, as $1/p$, and then performing the first E-step. In the t-th E-step, suppose we have system weights $w^t = \{w_1^t, ..., w_p^t\}$. We then compute the pseudo-judgment $J_{m,n}$ for each document $D_{m,n}$ as:

$$J_{m,n}^t = \sum_{j=1}^{p} w_j^t \cdot f(V_{S_j,D_{m,n}}) . \tag{1}$$

The score transformation function $f(\cdot)$ allows us to implement a range of estimators for the E-stage by transforming the system's retrieval status value $V_{S_j,D_{m,n}}$ prior to performing the weighted linear combination across systems. In this paper, we try three score transformation functions:

$f_{\text{Score}}(v)$. A scaled version of the system-produced document score, $f_{\text{Score}}(v) = v/v_{\max}$, where $v_{\max}$ is the maximum score assigned by system S_j to any document for topic T_m. This scaling treats the retrieval status value as being measured on an interval scale, while avoiding giving one system (or one topic) more emphasis than another at the outset.

$f_{\text{BordaCount}}(v)$. The (single-ranking) Borda count of the score, computed as $f_{\text{BordaCount}}(V_{S_j,D_{m,n}}) = R_{j,m} - r(D_{m,n})$, where $R_{j,m}$ is the number of results that system S_j returns for topic T_m, and $r(D_{m,n})$ is the rank of document $D_{m,n}$ when retrieval status values $V_{S_j,D_{m,n}}$ are sorted decreasing, breaking ties arbitrarily.

$f_{\text{Vote}}(v)$. A binarized version of the score, in which if the document $D_{m,n}$ is in the top 1000 results returned by system S_j, then $V_{S_j,D_{m,n}}$ is set to 1; otherwise, $V_{S_j,D_{m,n}}$ is set to 0. The pseudo-judgment score for document $D_{m,n}$ at iteration t is therefore the w^t-weighted average of the number of systems returning that document.

The t-th M-step is then initiated by computing the loss function L_j for each system S_j in iteration t as the square of the Euclidean distance between the (scaled) vector of system scores and the corresponding vector of pseudo-judgments from the t-th E-step:

$$L^t_j = \sum_{m\in[1,k],n\in[1,t(m)]} \left(w^t_j \cdot V_{S_j,D_{m,n}} - J^t_{m,n}\right)^2 . \tag{2}$$

L^t_j will be small when a system's results exhibit little (scaled) disagreement with the pseudo-judgments; to the extent that the pseudo-judgments are reasonable estimates of the actual judgments, L^t_j can be interpreted as an inverse estimate of the effectiveness of system S_j. We therefore update the system weights by first computing the *inverse loss* I^t_j of system j, by inverting the sign of the loss and then applying an additive offset to guarantee that the result is non-negative:

$$I^t_j = \sum_{m,n,j} \left(w^t_j \cdot V_{S_j,D_{m,n}}\right)^2 - \sum_{m,n} \left(w^t_j \cdot V_{S_j,D_{m,n}} - J^t_{m,n}\right)^2 . \tag{3}$$

We can regard the inverse loss as an improved estimate for the relative system weight; so all that remains to get the actual system weight for the $(t+1)$-th step is to normalize the values to sum to 1:

$$w^{t+1}_j = \frac{I^t_j}{\sum^p_{j=1} I^t_j} . \tag{4}$$

The iteration stops when the algorithm converges to a point where the weights and pseudo-judgment scores no longer change.[1] The final pseudo-judgment vector J can then be used to estimate the relevance judgments. When binary relevance judgments are required for computation of system effectiveness measures (as in our experiments), we treat the λ_m documents with the highest pseudo-judgment values for topic m as relevant and the others as not relevant. Because we use this value only for reporting results, it has no effect on our EM iteration. In this paper, we set λ_m to the true number of relevant documents in the NIST TREC judgment pools. In experiments, we also give λ_m to the baseline techniques against which we compare. We leave parameter estimation for λ_m (or the design of parameter-free measures for system comparison directly from pseudo-judgments) for future work.

3.2 Incomplete Relevance Judgments

The dependence of the M-step in our algorithm on the prior generation of informative pseudo-judgments in the immediately previous E-step suggests that we might be able to improve our results by substituting some actual judgments for selected pseudo-judgments. This leads naturally to two questions. If some incomplete set of actual judgments were available, how should they be used? And if those actual judgments were to be created on demand, which documents should be judged? In this section, we address these two questions in turn.

Using Human Judgments. Suppose that after the E-step in the t-th iteration (but before the M-step) we obtain a human judgment for document $D_{m,n}$, denoted $H_{m,n} \in \{0,1\}$. Then we can simply set $J^t_{m,n} = H_{m,n}$. Indeed, since once we learn $H_{m,n}$ we expect that there would be no value to forgetting it, we perform the same substitution after each

---

[1] Convergence Proof: http://www.umiacs.umd.edu/~ninggao/publications

subsequent E-step. Moreover, because the human judgments represent ground truth, we can reasonably give them greater influence when we estimate system weights in the M-step. We therefore modify the computation of the loss function to implement this idea, replacing Equation (2) with:

$$L_j^t = \sum_{m\in[1,k],n\in[1,t(m)]} T_{m,n} \cdot (w_j^t \cdot f(V_{S_j,D_{m,n}}) - J_{m,n}^t)^2 , \qquad (5)$$

where

$$T_{m,n} = \begin{cases} \gamma & \text{if human judgment} \\ 1 & \text{if pseudo-judgment.} \end{cases} \qquad (6)$$

Here $T_{m,n}$ encodes the weight to place on the (human or pseudo) judgment for $D_{m,n}$. When we have a human judgment $H_{m,n}$, we set $T_{m,n}$ to γ ($\gamma > 1$). For this paper we have arbitrarily set $\gamma = 2$, leaving the question of optimizing γ for future work. With this one small change, our EM algorithm proceeds as before.

Requesting Human Judgments. In general, we can request some number N_m^t of new judgments for topic m after each E-step t (and before the subsequent M-step), selecting those judgments using selection policy P_i, where a selection policy is a rule for selecting previously unjudged documents for human judgment. In this paper we set N_m to be 1% of the available TREC relevance judgments[2] at each iteration, and we consider four selection policies, consistently selecting using the same policy at each iteration. Note that after 100 steps, all available judgments will be used, and thus we will produce the same results as reported at TREC. We then compare policies based on which most rapidly approach those TREC results. The four policies are:

P_1 Select the N_m as-yet unjudged documents with highest pseudo-judgment scores $J_{m,n}$, breaking ties arbitrarily.

P_2 For each document $D_{m,n}$ compute (across systems) the arithmetic mean $M_{m,n}$ and the standard deviation $SD_{m,n}$ for $(f(V_{S_1,D_{m,n}}),...,f(V_{S_p,D_{m,n}}))$ and then select the as-yet unjudged documents for which $M_{m,n} + \beta SD_{m,n}$ is largest. Here β is a tunable parameter, arbitrarily set to 2 in our experiments.

P_3 Randomly select N_m as-yet unjudged documents by simple random sampling.

P_4 Select the documents which would maximize the weighted average system loss if they were not relevant. This is the Hedge loss function defined by Aslam et al. [1].

In our experiments, we select documents to be judged only from the TREC judgment pool, as we do not know the correct judgment for documents outside the pool. Both P_1 and P_2 reflect the fact that the distinction between positive judgments and unknown judgments is consequential for Mean Average Precision, our evaluation measure, while the distinction between negative judgments and unknown judgments is inconsequential for that measure. Our use of standard deviation in P_2 reflects our intuition that a greater diversity of system responses for the same document might be an indication of poorly estimated pseudo-judgments, and also of documents likely to be discriminative between systems.

---

[2] The mean of N_m across all topics and test collections is 22.5 (min 11, max 46).

4   Experiments

We start with a description of our test collections and evaluation measures in Section 4.1. We then present zero-judgment results in Section 4.2, followed by incomplete-judgment results for each judgment selection policy in Section 4.3.

4.1   Test Collections and Evaluation Measures

We use four TREC Ad Hoc test collections, from TREC 5, 6, 7 and 8. In each case, we use all topics, and all the Ad Hoc runs officially submitted to NIST that searched the full collection (known as Category A runs). For each test collection, the top 100 documents returned by each system (for TREC 5), or for selected systems (for TREC 6 through 8), form the pool for which human judgments, created by NIST, are available. For comparability with prior work, we focus on the degree to which system comparisons made on the basis of numerical differences in Mean Average Precision (MAP) reflect those which would have been made had the full TREC judgment pool been available. We use *trec_eval* version 9.0 to compute MAP truncated at 1,000 documents ($MAP@1k$) as the measure of effectiveness, treating unjudged documents as not relevant, and binarizing pseudo-judgments as described above.

Our first measure of reliability is Kendall's τ [10] between the system ranking using MAP@1k and the ranking produced by the method under analysis. When the purpose of our evaluation is to measure relatively small improvements over already-good systems, we would prefer to have a reliability metric that is more influenced by reversals among the best systems than among relatively poor ones. Kendall's τ weights both equally; the τ_{ap} [15] measure, however, places greater weight on more highly ranked systems. The τ_{ap} measure is asymmetric; we designate the full-pool TREC rankings as the objective and our EM algorithm's system ranking as the comparison.

4.2   Zero-Judgment Results

In this section, we present the reliability of our EM algorithm for system comparisons without human judgments. Various voting and learning methods are employed. Vote, Score, and BordaCount denote the results after the first E-step using f_{Vote}, f_{Score} and $f_{BordaCount}$, respectively, while Vote.EM, Score.EM, and BordaCount.EM denote the corresponding results after convergence. BordaCount is equivalent to the Borda count method of Nuray and Can [11]. Additionally, we reimplement two other previously published baselines for the zero-judgment task, the Condorcet method [11] and Soboroff's random-voting method [12].

Table 2 indicates the effect of EM, with (+) when EM numerically improves the reliability measure τ or τ_{ap} over the first E-step alone, and (-) to indicate a numerically adverse affect. Over 24 such comparisons (two measures, three score transformation functions, four test collections), 17 favored EM and the remaining 7 favored the first E-step, a significant improvement under a paired Wilcoxon test at $p < 0.05$.

Taking TREC MAP@1k as ground truth scores, Figure 1 shows the difference between the estimated and ground truth MAP@1k score. Systems are ordered by TREC MAP@1k on the X-axis with best systems on the left. For each system, the value on

Table 2. Zero-judgment (highest value in **bold**), training and testing on all systems

	τ				τ_{ap}			
	TREC 5	TREC 6	TREC 7	TREC 8	TREC 5	TREC 6	TREC 7	TREC 8
Score	0.469	0.551	0.473	0.471	**0.337**	0.289	0.227	0.156
Score.EM	**0.479** (+)	**0.563** (+)	0.496 (+)	0.490 (+)	0.326 (-)	**0.314** (+)	**0.257** (+)	0.169 (+)
BordaCount	0.464	0.537	0.471	0.462	0.322	0.276	0.229	0.147
BordaCount.EM	0.453(-)	0.555 (+)	0.487 (+)	0.461 (-)	0.321 (-)	0.294 (+)	0.237 (+)	0.146 (-)
Vote	0.468	0.537	0.473	0.459	0.336	0.291	0.227	0.143
Vote.EM	0.452 (-)	0.554 (+)	**0.497** (+)	0.467 (+)	0.322 (-)	0.292 (+)	0.253 (+)	0.152 (+)
Condorcet	0.428	0.491	0.485	0.486	0.287	0.245	0.248	0.165
Soboroff	0.416	0.480	0.477	**0.533**	0.285	0.233	0.245	**0.227**

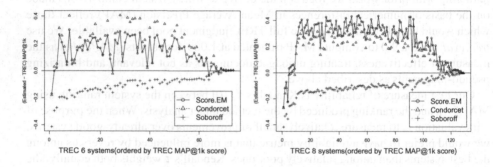

Fig. 1. Difference between the Estimated and TREC MAP@1k, Zero-judgment, Systems ordered by TREC MAP, best systems on the left and worst systems on the right

the Y-axis is calculated by (estimated MAP@1k - TREC MAP@1k). Perhaps the most easily noted effect is that our Score.EM method and Condorcet tend to substantially overestimate the value of MAP@1k, while Soboroff's random-voting method tends to substantially underestimate MAP@1k. However, because our focus is on system comparison, systematic tendencies towards overestimation or underestimation are not consequential. The methods tend to give higher estimated MAP scores to the systems that are in the middle range, but lower scores to the best and worst systems at the two ends. This may be because the systems in the middle use similar searching strategies and return similar results, while the best and worst systems may return results that are not retrieved by other systems. The effect of majority voting causes underestimation for systems that behave distinctively, and all the zero-judgment methods tend to make similar errors.

4.3 Incomplete Judgment Results

Reliability of System Comparisons. The difficulty that zero-judgment techniques exhibit with even detecting, much less distinguishing between, the best systems, motivates our interest in approaches that can use partial, but still incomplete, human judgments. Figure 2 shows how τ_{ap} grows as we add increasing numbers of human judgments. We start from the Score.EM method, which was the best overall choice in the zero-judgment case, and we consider four ways of selecting documents to be judged at each step, policies P_1, P_2, P_3 and P_4 from Section 3.2. We compare to the following four state-of-the-art baselines: xinfAP [16], BPref-10 [3], RTC [4] and Hedge [1].[3]

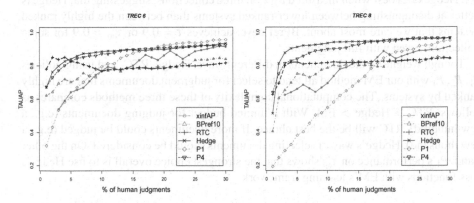

Fig. 2. τ_{ap} by percent of pooled judgment used for different TREC Ad Hoc collections

Table 3. Human judgments (%) required to achieve $\tau = 0.9$ or $\tau_{ap} = 0.9$ (* means unachievable)

	τ				τ_{ap}			
	TREC 5	TREC6	TREC 7	TREC 8	TREC 5	TREC 6	TREC 7	TREC 8
P_1	21	14	21	16	31	19	31	25
P_2	22	**12**	22	16	28	20	33	24
P_3	78	72	75	79	89	80	84	86
P_4	**5**	17	5	3	**5**	**16**	**12**	11
Hedge	**5**	17	**3**	**2**	13	19	21	**4**
RTC	16	80	2	26	29	*	77	*
BPref	*	68	26	37	*	*	*	78
xinfAP	23	21	18	19	26	27	32	29

For our RTC results, we used the code package *mtc-eval.tgz* from the RTC developers.[4] The computational complexity of our proposed EM-based method is $O(CSTn)$, where S is the number of systems, T the number of topics, n the number of unique documents for a topic in the pool, and C the number of iterations before the convergence. Empirically, C is smaller than 40. The time complexity of RTC is $O(S^2 T n^3)$.

---

[3] In our implementation of Hedge, the tunable parameter β is set constantly as 0.9; the results are reported on MAP@1k; the unjudged documents are considered as not relevant; losses are transformed linearly into the range of [0,1]; for each topic, there will be one document selected to be judged in each iteration.

[4] http://ir.cis.udel.edu/~carteret/downloads.html

Running MAP@1000 on TREC 5 would take more than two days on a four-core 3.10GHz PC. Therefore, we only return RTC results for MAP@100.

As can be seen from Figure 2, Hedge and P_4 achieve better τ_{ap} with fewer human judgments. On TREC 6, RTC briefly dominates other methods until more than 5% of the judgments are available. Table 3 shows the percentage of human judgments needed for the methods to achieve $\tau = 0.9$ or $\tau_{ap} = 0.9$. As can be seen, on three of the four collections P_4 needs fewer human judgments than the other methods when measured by τ_{ap}. Hedge does best when measured by τ on three collections, suggesting that Hedge is better at distinguishing between lower-ranked systems than between the highly ranked systems that we care most about. BPref never achieves $\tau = 0.9$ or $\tau_{ap} = 0.9$ for some collections, even with full judgments.

Though weighting the documents in different ways, Hedge, RTC, and using policies P_1, P_2, P_4 with our EM method all tend to select for judgment documents that are highly ranked by systems. The computational complexity of these three methods compares as follows: RTC > Hedge > EM. With a limited budget for judging documents (e.g., a few hundred), RTC will be the best choice. If more documents could be judged (e.g., a few thousand), Hedge's way of selecting documents should be considered. On the other hand, P_4's performance on τ_{ap} shows that the strongest choice overall is to use Hedge's loss functions with EM's learning framework.

Effect of EM Learning. Figure 3 shows the improvement on four collections of P_1 with human judgments, over a non-learning baseline, with available judgments increases from 1% to 20%, measured by τ_{ap}. In the non-learning baseline, the pseudo-judgments of the as-yet unjudged documents are taken from the zero-judgment iteration of the EM algorithm; the effect of additional human judgments is only to assign true relevance

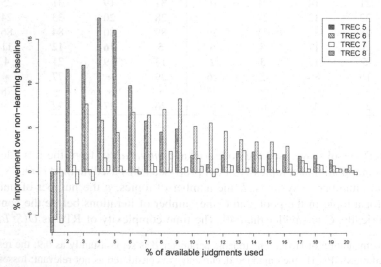

Fig. 3. % improvement by using EM learning, measured by τ_{ap}, with human judgments from 1% to 20%

judgments to documents for evaluation, not to train the learner. In other words, there is no E-step or M-step in the non-learning baseline. The values shown are the differences from this baseline of the τ_{ap} of the full EM P_1 method (Section 3.2), which refines its pseudo-judgements with each iteration of new data.

A notable result is that learning with P_1 actually hurts performance with 1% judgments. The reasons for this are unclear. Possibly, a small number of human judgments are insufficient to allow the the model to distinguish the better (often, "manual") systems from the less effective (often "automatic") systems. As more human judgments are added, the P_1 method seems to quickly overcome the spurious effect of this mistuning. The improvements are substantial with 2% to 9% of the judgments, and decrease progressively from 10% to 20%. In part, this is because there is less room for improvement; as more documents are assessed, there are fewer for which pseudo-judgments are required, and the effectiveness of the pseudo-judgment method therefore becomes progressively less important.

5 Conclusion

In this paper, we have introduced a novel method for automated or semi-automated system evaluation, based on the expectation-maximization (EM) algorithm. Our EM method predicts document relevance by system rankings, and system quality by the predicted relevance of the documents they return, iterating till convergence.

For zero-judgment evaluation, we have compared our method with the Condorcet method of Nuray and Can [11] and the random-voting method of Soboroff et al. [12]. Our methods beat the Condorcet method on all four of the test collections that we tried, and beat Soboroff's random-voting method on three of four. For incomplete judgments, we find that with a limited judging budget, a few hundred documents for example, RTC will be the better choice. However, the high computational complexity of RTC may make it impractical for interactive use with large collections. If more documents are judged, a few thousand for example, our results indicate that using EM's learning framework with Hedge's loss function would be a good choice.

Acknowledgments. This material is based in part on work supported by the National Science Foundation under Grant No. 1065250 and by the Human Language Technology Center of Excellence at Johns Hopkins University. Any opinions, findings, and conclusions or recommendations expressed in this material are those of the authors and do not necessarily reflect the views of the National Science Foundation.

References

[1] Aslam, J., Pavlu, V., Savell, R.: A unified model for metasearch and the efficient evaluation of retrieval systems via the hedge algorithm. In: Proc. 26th Annual International ACM SIGIR, pp. 393–394 (2003)
[2] Aslam, J., Pavlu, V., Yilmaz, E.: A statistical method for system evaluation using incomplete judgments. In: Proc. 29th Annual International ACM SIGIR, pp. 541–548 (2006)

[3] Buckley, C., Voorhees, E.: Retrieval evaluation with incomplete information. In: Proc. 27th Annual International ACM SIGIR, pp. 25–32 (2004)

[4] Carterette, B.: Robust test collections for retrieval evaluation. In: Proc. 30th Annual International ACM SIGIR, pp. 55–62 (2007)

[5] Carterette, B., Allan, J.: Incremental test collections. In: Proc. 14th ACM International Conference on Information and Knowledge Management, pp. 680–687 (2005)

[6] Cormack, G.V., Palmer, C.R., Clarke, C.L.: Efficient construction of large test collections. In: Proc. 21st Annual International ACM SIGIR, pp. 282–289 (1998)

[7] Dai, K., Pavlu, V., Kanoulas, E., Aslam, J.A.: Extended expectation maximization for inferring score distributions. In: Baeza-Yates, R., de Vries, A.P., Zaragoza, H., Cambazoglu, B.B., Murdock, V., Lempel, R., Silvestri, F. (eds.) ECIR 2012. LNCS, vol. 7224, pp. 293–304. Springer, Heidelberg (2012)

[8] Hauff, C., Hiemstra, D., Azzopardi, L., de Jong, F.: A case for automatic system evaluation. In: Gurrin, C., He, Y., Kazai, G., Kruschwitz, U., Little, S., Roelleke, T., Rüger, S., van Rijsbergen, K. (eds.) ECIR 2010. LNCS, vol. 5993, pp. 153–165. Springer, Heidelberg (2010)

[9] Hosseini, M., Cox, I.J., Milić-Frayling, N., Kazai, G., Vinay, V.: On aggregating labels from multiple crowd workers to infer relevance of documents. In: Baeza-Yates, R., de Vries, A.P., Zaragoza, H., Cambazoglu, B.B., Murdock, V., Lempel, R., Silvestri, F. (eds.) ECIR 2012. LNCS, vol. 7224, pp. 182–194. Springer, Heidelberg (2012)

[10] Kendall, M.G.: Rank Correlation Methods, 1st edn. Charles Griffin, London (1948)

[11] Nuray, R., Can, F.: Automatic ranking of information retrieval systems using data fusion. Information Processing & Management 42(3), 595–614 (2006)

[12] Soboroff, I., Nicholas, C., Cahan, P.: Ranking retrieval systems without relevance judgments. In: Proc. 24th Annual International ACM SIGIR, pp. 66–73 (2001)

[13] Spärck Jones, K., van Rijsbergen, C.J.: Report on the need for and provision of an 'ideal' test collection. Tech. rep., University Computer Laboratory, Cambridge (1975)

[14] Yilmaz, E., Aslam, J.: Estimating average precision with incomplete and imperfect judgments. In: Proc. 15th ACM International Conference on Information and Knowledge Management, pp. 102–111 (2006)

[15] Yilmaz, E., Aslam, J., Robertson, S.: A new rank correlation coefficient for information retrieval. In: Proc. 31st Annual International ACM SIGIR, pp. 587–594 (2008a)

[16] Yilmaz, E., Kanoulas, E., Aslam, J.A.: A simple and efficient sampling method for estimating ap and ndcg. In: Proc. 31st Annual International ACM SIGIR, pp. 603–610 (2008b)

[17] Zobel, J.: How reliable are the results of large-scale information retrieval experiments? In: Proc. 21st Annual International ACM SIGIR, pp. 307–314 (1998)

Best and Fairest: An Empirical Analysis
of Retrieval System Bias

Colin Wilkie and Leif Azzopardi

School of Computing Science, University of Glasgow,
Scotland, UK
{colin.wilkie,leif.azzopardi}@glasgow.ac.uk

Abstract. In this paper, we explore the bias of term weighting schemes used by retrieval models. Here, we consider bias as the extent to which a retrieval model unduly favours certain documents over others because of characteristics within and about the document. We set out to find the least biased retrieval model/weighting. This is largely motivated by the recent proposal of a new suite of retrieval models based on the Divergence From Independence (DFI) framework. The claim is that such models provide the fairest term weighting because they do not make assumptions about the term distribution (unlike most other retrieval models). In this paper, we empirically examine whether fairness is linked to performance and answer the question; is fairer better?

1 Introduction

Retrieval bias in retrieval systems and models has been a long standing problem when developing term weighting schemes [3, 4, 21]. It occurs when a document or group of documents are overly or unduly favoured due to some document feature (such as length, term distribution, etc). Bias is largely seen as problematic, especially, when a retrieval model's weighting function places too much emphasis on certain features. This leads to documents being ranked highly, not because they are relevant, but because of the bias present in the retrieval model's term weighting scheme. For example in [21], pivoted length normalisation was introduced to avoid overly favouring longer documents in the vector space model [19]. In [8], a study of IR heuristics showed how various models violated common sense principles relating to term weighting schemes. By addressing these violations, improvements in performance were obtained. Again linking bias with performance. In this work, instead of directly focusing on performance or looking at whether particular heuristics are violated, we will directly measure the bias associated with different retrieval models and then examine the relationship with performance.

To this end, we conduct a comprehensive empirical analysis across four TREC test collections using seventeen term weighting functions from six families of retrieval models (vector space models, best match models, language models, DFR, Log-Logistic and DFI models) and issue over 80 million queries in the process. We show that there is generally quite a strong correlation between fairness and performance, although this is collection dependent. In the next

M. de Rijke et al. (Eds.): ECIR 2014, LNCS 8416, pp. 13–25, 2014.
© Springer International Publishing Switzerland 2014

section, we shall provide an historical overview of the development of retrieval
models. Following this, we detail the experiment methodology we employed. We
then report our results and discuss the implications of our findings.

2   Retrieval Models and Term Weighting Schemes

The development of retrieval models has been largely driven by the focus on
improving the effectiveness of the retrieval system. Over the years numerous
retrieval models have been developed, each leading to subtle and not so subtle differences in the term weighting schemes. For example in [20] Salton tried
over 1400 term weighting schemes before arriving at TF.IDF. While reasonably
effective, it was later shown that TF.IDF overly favoured the retrieval of long
documents and that this bias is detrimental to retrieval performance [21]. Since
then, document length normalisation has been a key component in most modern
term weighting schemes as a way to mitigate such bias.

Conversely, probabilistic approaches to the term weighting problem have looked
towards modelling different aspects of documents, including their term distributions and the distributions of relevance depending on the model in order to obtain
a more accurate fit of the data [1, 6, 9, 11, 13, 16, 17]. For example, Harter [11] proposed the 2-Poisson model to represent speciality words and non-speciality words,
where speciality words are those which occur densely in elite documents. These
speciality words contribute to how informative a document is, while non-speciality
words essentially occur at random and do not contribute to how informative a document is. While the 2-Poisson model worked well for particular samples of text, it
did not perform very well in practice as it was problematic to estimate the distributions accurately. However, it did lead to the development of more generalised probabilistic models, which became known as Best Match (BM) models [13, 18], from
which BM25 has become the most commonly used. Again, document length normalisation was a key feature in ensuring that BM25 did not overly favour shorter
or longer documents.

In [17], Ponte and Croft proposed the Language Modelling approach which took
an alternative view on estimating the relevance of a document. Language Models
are based on generative probabilistic models, and look to estimate the probability
of a query, given a document [17]. Under this framework, a mutlinomial distribution is typically used to model the document [23]. This free-er form means that
the document data can be fitted more accurately. When the document model is
estimated, document statistics are combined/smoothed by the background collection's statistics. Terms appearing in the document are akin to speciality words,
while terms in the collection are akin to the non-speciality words that occur at
random. Depending on how the document model is smoothed determines what
features of a document dominate the term weighting function, and whether any
document length normalisation is applied [23]. For instance, Jelinek-Mercer (JM)
smoothing does not include any document length normalisation, focusing mainly

on the information content of the document to rank documents[1], whereas Bayes smoothing includes document length normalisation.

In [1], Amati and van Rijsbergen proposed the Divergence from Randomness (DFR) framework to construct probabilistic term weighting schemes. The framework was also inspired by Harter's 2-Poisson model [11] and further refined the notion of eliteness using semantic information theory and Popper's notion of information content. A DFR model is typically composed of two divergence functions (referred to as $Prob_1$ to characterise the randomness of a term, and $Prob_2$ to model the risk of using the term as a document descriptor [1]) and a normalisation function. Rather than using two Poisson functions, in [1] the authors explored various distributions to characterise the randomness and eliteness of terms. Two of the best performing models they found were the Poisson-Laplace with their second normalisation of term frequency (PL2) and a hyper-geometric model that used Popper's normalisation (DPH). It was argued that the better fit to the underlying term distributions with the document collection, the use of information content and the normalisation of term frequency resulted in superior term weighting schemes (these models have been shown to perform empirically well at subsequent TRECs). It is also worth noting that under the DFR framework it is also possible to instantiate BM25.

In [5], another statistical model was proposed by Clinchant and Gaussier. This model attempted to account for the burstiness of terms within text (and thus obtain a better fit of the data). The term weighting function used a log-logistic distribution to provide a simpler information-based weighting as an alternative way to represent the eliteness of terms. The log-logistic distribution (LGD) model bridges Language Models and the Divergence from Randomness models.

More recently, the Divergence from Independence (DFI) framework has been proposed by Kocabas et al [14]. The DFI framework is most related to the DFR framework where the statistical independence takes place of the randomness. Rather than treating non-speciality words as random, DFI characterises them by their independence i.e. does this term occur independently of this document, or not? This alternative viewpoint means that DFI is essentially the non-parametric counterpart of DFR. No assumptions are made a priori about the distribution of underlying data. Therefore, if the data does not fit the prescribed distribution (i.e. Poisson or Laplace) then the non-parametric approach should work better, but even if it does, the non-parametric approach should work just as well. These models are also parameter free - and do not require tuning. In [14], the authors proposed three variants based on different measures of independence (the saturated model of independence, the standardisation model, and normalised Chi-Square distance). In [7], it was shown that the term weighting scheme derived from these different methods performed empirically well (and often the best at TREC). Since these schemes make no assumptions about the distribution of the data, it is contested that they are fairer than other models, and thus better. In this work, we shall test this claim.

---

[1] In [12], it was shown that Language Modelling with Jelinek-Mercer smoothing provides a probabilistic justification for TF.IDF.

To summarise, we have described a range of models that are related but make different assumptions about how to model a term's relevance. Most models are composed of two main parts: one to estimate the value of the information content of the term, and one to regulate the influence of the document's length. Overly focusing on one part or another, or ignoring one part invariably leads to some form of bias creeping into the retrieval model.

3   Experimental Method

The focus of this study is to assess the level of bias exhibited by each retrieval model/weighting, and then to determine whether there is any relationship between the level of bias and retrieval performance. Specifically, we wish to determine whether the recently proposed DFI models are fairer and better, as they claim to be. To this end, we employed the following methodology. On each test collection, and for each retrieval model/weighting, we measured the bias using retrievability measures, and then measured the corresponding performance using the topics and relevance judgements associated with the collection. Below we shall first outline the collections and topics used, before describing how we estimated the bias and set of retrieval models/weightings we used.

3.1   Data and Materials

For our experiments we report results from four TREC collections: TREC 123 (T123), Aquaint (AQ), WT10G (WT) and DotGov (DG)[2]. These collections are typical of the collections used to test the models outlined in Section 2, where

Table 1. Summary of each Collection Statistics. * denotes whether the difference from the whole collection is significant at $p < 0.05$.

		Collection			
		AQ	TREC 123	DG	WT
TREC Topics		301-400	1-200	551-600	451-550
Number of Documents		1,033,461	1,078,166	1,247,753	1,692,096
Number of Queries		273,245	237,810	337,275	212,201
Avg. Doc. Length	All	439	420	1108	617
	Pool	623*	3913*	2056*	6737*
	Relevant	583*	1280*	2175*	2903*
Avg. Doc. Info. Cont.	All	2.114	2.194	2.498	2.721
	Pool	2.059*	2.115*	2.183*	2.626*
	Relevant	2.00*	2.120*	2.082*	2.475*

---

[2] We also conducted these experiments on AP and TREC678 where we found similar results and trends.

the focus is on creating the best term weighting scheme. These collections, while sufficient, are not so large that it is not possible to estimate the bias associated with each collection given each term weighting scheme (and parameter setting). Table 1 shows the collections used along with their statistics. To report performance we used a number of standard TREC measures: MAP, P@10, NDCG@100 and Mean Reciprocal Rank.

3.2 Measuring Bias

The retrievability of a document provides an indication of how easily or likely a document is to be retrieved, given a particular configuration of a retrieval system [3]. Intuitively, if a retrieval system has a bias towards longer documents, then we would expect that the retrievability of longer documents to be higher than shorter documents, and vice versa. If a retrieval system tends to retrieve documents with a higher information content, then we would expect to see that the retrievability of such documents would be higher, and vice versa. Formally, the retrievability $r(d)$ of a document d is defined by: $r(d) \propto \sum_{q \in Q} f(k_{dq}, c)$ where q is a query from a large set of queries Q, and k_{dq} is the rank of document d for query q. While there are various measures of retrievability, the simplest is the cumulative based measure which defines $f(k_{dq}, c)$ to be equal to 1 if the document d is retrieved in the top c results for the query q. The retrievability of a document is essentially the count of how many times the document is retrieved in the top c results. The bias that the retrieval model exhibits on the collection can be summarised by using the Gini Coefficient [10], which is commonly used to measure the inequality with a population (usually the wealth of people in a population). In the context of retrievability, if all documents were equally retrievable (i.e the retrieval function fairly retrieve all documents) then the Gini Coefficient would be 0 (denoting equality within the population). Conversely if only one document was retrievable and the rest were not, then the Gini Coefficient would be 1 (denoting total inequality). Usually, documents have some level of retrievability for any given retrieval function, and thus the Gini Coefficient is somewhere between 1 and 0. In [2], it was shown that for a given retrieval model, if a retrieval function was tuned to minimise bias (represented by the lowest Gini Coefficient) then this led to near optimal retrieval performance. Similar relationships were shown in [22] for P@10 and MAP and in [4] for recall based measures. However, here we examine the relationship between retrieval bias and performance across the spectrum of retrieval models and term weighting schemes as opposed to how bias changes when tuning a particular retrieval model (as done in [2, 4, 22]).

Estimating the retrievability of the documents within a collection requires a large number of queries. This is usually done by extracting all the bigrams from the collection and then selecting the most frequent [2–4, 22]. We employed the same method, but only selected those bigrams that appeared at least 20 times. Table 1 shows the total number of queries used on each collection (approximately 250,000 queries per collection). These queries were then issued to the retrieval system with a particular configuration (retrieval model/weighting/parameter setting). Using the cumulative based retrievability with a cut-off $c = 100$, we computed the retrievability of each document and subsequently the Gini Coefficient for each collection given the term weighting scheme.

3.3   Term Weighting Schemes

For our experiments, we used 17 term-weighting schemes stemming from the 6 model/frameworks presented in Section 2. The score assigned to a document given a query is determined by: $s(q,d) = \sum_{t \in q} s(t,d)$ where $s(t,d)$ is the term weighting assigned to term t in document d and q is the query which is composed of sequence of terms. The following schemes were implemented within the Lemur/Indri Framework[3].

The first four term weighting schemes we used were: (i) Term Frequency (TF) where $s(t,d) = n(t,d)$, (ii) Normalised TF (NTF) where $s(t,d) = \frac{n(t,d)}{n(d)}$, (iii) Term Frequency Inverse Document Frequency (TF.IDF) where $s(t,d) = n(t,d).idf(t)$, and (iv) Normalised TF.IDF (NTF.IDF) where $s(t,d) = \frac{n(t,d)}{n(d)}.idf(t)$. Here, $n(t,d)$ is the number of times t occurs in a document d, $n(d)$ is the total number of terms in a document, $idf(t) = \log \frac{N}{df(t)}$ where N is the number of documents in the collection, and $df(t)$ is the number of documents in which t appears. The fifth weighting scheme employed was Pivoted Length Normalisation (PTF.IDF) [21] where $a(d)$ is the average number of terms in d in the collection, and b controls the level of normalisation ($0 < b < 1$):

$$s(t,d) = \frac{n(t,d)}{(1-b) + b.\frac{n(d)}{a(d)}}.idf(t)$$

From the series of Best Match models [13] we employed BM25 ($0 > b > 1$), BM11($b = 0$) and BM15 ($b = 1$):

$$s(t,d) = \frac{(k_1 + 1).n(t,d)}{\left(k_1.(1-b) + b.\frac{n(d)}{a(d)}\right) + n(t,d)}.idf(t) \tag{1}$$

From the space of Language Models we implemented three different smoothing methods: Laplace smoothing (LP), which doesn't take into consideration document length normalisation and is similar to TF/NTF models (see Equation 2, where α is the level of smoothing parameter, and V is the number of unique terms in the collection); Jelinek Mercer Smoothing (JM) which is similar to the TF.IDF/NTF.IDF models, again without any explicit document length normalisation (see Equation 3), where $p(t,d)$ is the maximum likelihood estimate of the probability of a term appearing in a document, i.e. $p(t,d) = \frac{n(t,d)}{n(d)}$, and $p(t)$ is the the maximum likelihood estimate of the probability of a term appearing in the collection; and Bayes Smoothing with Dirichlet Prior (BS) which is often

---

[3] See www.projectlemur.org/. Code is freely available on GitHub.

the best performing Language Model [23] and has a β parameter which controls the amount of length normalisation implicitly (see Equation 4).

$$s(t,d) = \frac{n(t,d) + \alpha}{n(d) + V.\alpha} \tag{2}$$

$$s(t,d) = \lambda.p(t,d) + (1-\lambda).p(t) \tag{3}$$

$$s(t,d) = \frac{n(t,d) + \beta.p(t)}{n(d) + \beta} \tag{4}$$

From the Divergence from Randomness framework [1], we choose two of the best performing models, DPH and PL2. DPH is parameter-free model:

$$s(t,d) = \frac{\left(1 - p(t,d)\right)^2}{n(t,d)+1}.\left(\frac{n(t,d).N}{n(t)}.\log\left(\frac{n(t,d).a(d)}{n(d)}\right)\right.$$
$$\left. + \frac{1}{2}\log\left(2\pi.n(t,d).\left(1 - p(t,d)\right)\right)\right) \tag{5}$$

where $n(t)$ is the number of times term t appears in the collection. While PL2 has a parameter c to control document length normalisation:

$$s(t,d) = \frac{1}{n_2(t,d,c)+1}.\left(n_2(t,d,c).log_2\left(\frac{n_2(t,d,c)}{\Lambda(t)}\right.\right. +$$
$$\left(\Lambda(t) + \frac{1}{12.n_2(t,d,c)} - n_2(t,d,c)\right).\log_2 e + \frac{1}{2}log_2(2\pi.n_2(t,d,c)) \tag{6}$$

where the second normalisation component $n_2(t,d,c)$ is equal to $n(t,d).\log_2\left(1 + \frac{c.a(d)}{n(d)}\right)$ and the mean of the Poisson distribution is defined by $\Lambda(t) = \frac{\sum_d n(t,d)}{N}$. As previously mentioned the Log Logistic Distribution (LGD) model [5] bridges DFR and Language models and is defined as:

$$s(t,d) = \log_2\left(\frac{\frac{d(t)}{N} + \log_2\left(n(t,d).(1 + \frac{c.a(d)}{n(d)})\right)}{\log_2\left(n(t,d).(1 + \frac{c.a(d)}{n(d)})\right)}\right) \tag{7}$$

where the c parameter controls the amount of smoothing ($c > 0$). Finally, we explored three of the best Divergence from Independence models [7]): DFIa referred to as irra12a in [7] based on the saturated model of independence (see Equation 8), and DFIb referred to as irra12b in [7] which is based on the standardisation model (see Equation 9):

$$s(t,d) = \log_2(1 + \frac{(n(t,d) - e(t,d))^2}{e(t,d)}) \tag{8}$$

$$s(t,d) = \log_2(1 + \frac{(n(t,d) - e(t,d))}{\sqrt{e(t,d)}}) \tag{9}$$

where $e(t,d) = \frac{n(t).n(d)}{N.a(d)}$. The third model DFIc based on the normalised Chi-Square measure of independence (referred to as irra12c in [7]):

$$s(t,d) = \left((n(t,d)+1).\log_2\left(\frac{n(t,d)+1}{\sqrt{e_p(t,d)}}\right) - n(t,d).\log_2\left(\frac{n(t,d)}{\sqrt{e(t,d)}}\right) \right).\Delta(t,d)$$

where:
$$\text{(10)}$$

$$\Delta(t,d) = \left(\frac{n(d)-n(t,d)}{n(d)}\right)^{\frac{3}{4}} \times \left(\frac{n(t,d)+1}{n(t,d)}\right)^{\frac{1}{4}}$$

and:

$$e_p(t,d) = \frac{\big(n(t)+1\big).\big(n(d)+1\big)}{N.a(d)} + 1$$

Table 2. Performance values along with Gini scores for each model. The final rows report the correlation between each performance measure and Gini. * denote whether the correlation is significant at $p < 0.05$.

Model	AQ Gini	MAP	P@10	NDCG	MRR	DG Gini	MAP	P@10	NDCG	MRR
TF	0.979	0.054	0.165	0.132	0.264	0.987	0.011	0.018	0.021	0.074
TF.IDF	0.977	0.071	0.180	0.148	0.296	0.987	0.014	0.026	0.029	0.094
NTF	0.971	0.034	0.057	0.039	0.122	0.975	0.017	0.032	0.031	0.067
NTF.IDF	0.967	0.048	0.086	0.061	0.175	0.971	0.024	0.038	0.040	0.099
PTF.IDF	0.956	0.063	0.122	0.094	0.227	0.970	0.049	0.088	0.087	0.160
BM25	**0.544**	0.162	0.316	0.263	0.478	**0.614**	0.167	**0.222**	**0.277**	**0.479**
BS	0.581	0.140	0.290	0.240	0.457	0.637	0.151	0.210	0.257	0.455
JM	0.669	0.125	0.253	0.206	0.436	0.666	0.108	0.172	0.213	0.410
LP	0.572	0.127	0.253	0.204	0.436	0.632	0.109	0.166	0.208	0.406
LGD	0.576	0.145	0.310	0.249	0.448	0.715	0.130	0.188	0.228	0.427
PL2	0.605	0.169	0.331	0.281	0.503	0.803	**0.182**	0.220	0.271	0.440
DPH	0.548	**0.181**	**0.369**	**0.315**	**0.564**	0.869	0.149	0.196	0.244	0.425
DFIa	0.612	0.173	0.349	0.297	0.519	0.870	0.153	0.194	0.243	0.426
DFIb	0.607	0.173	0.349	0.297	0.530	0.868	0.154	0.196	0.245	0.426
DFIc	0.610	0.135	0.274	0.226	0.439	0.807	0.139	0.196	0.235	0.410
Correlation	-0.941*	-0.909*	-0.897*	-0.925*	-	-0.686*	-0.759*	-0.765*	-0.808*

Model	T123 Gini	MAP	P@10	NDCG	MRR	WT10G Gini	MAP	P@10	NDCG	MRR
TF	0.997	0.020	0.040	0.042	0.107	0.996	0.023	0.059	0.064	0.140
TF.IDF	0.997	0.025	0.048	0.050	0.122	0.996	0.030	0.085	0.084	0.169
NTF	0.977	0.017	0.031	0.030	0.060	0.979	0.010	0.006	0.008	0.020
NTF.IDF	0.974	0.024	0.048	0.045	0.085	0.975	0.015	0.022	0.017	0.051
PTF.IDF	0.957	0.047	0.165	0.164	0.277	0.972	0.031	0.083	0.069	0.115
BM25	**0.575**	0.122	0.301	0.307	0.478	0.670	0.094	0.183	0.176	0.313
BS	0.593	0.111	0.285	0.292	0.461	0.657	0.098	0.167	0.160	0.265
JM	0.664	0.049	0.253	0.253	0.401	0.680	0.058	0.128	0.127	0.260
LP	0.586	0.048	0.256	0.253	0.387	**0.653**	0.056	0.126	0.124	0.250
LGD	0.787	0.105	0.228	0.229	0.370	0.816	0.096	0.120	0.107	0.218
PL2	0.656	0.122	0.311	0.312	0.474	0.776	0.098	0.187	0.189	0.318
DPH	0.837	**0.128**	0.319	0.327	**0.490**	0.872	**0.121**	**0.226**	**0.228**	**0.409**
DFIa	0.781	0.126	0.319	0.326	0.484	0.859	0.116	0.194	0.198	0.379
DFIb	0.770	**0.128**	**0.323**	**0.330**	0.489	0.854	0.116	0.194	0.198	0.379
DFIc	0.690	0.111	0.278	0.285	0.449	0.754	0.118	0.194	0.201	0.385
Correlation	-0.624*	-0.807*	-0.801*	-0.811*	-	-0.554*	-0.597*	-0.570*	-0.563*

Using this selection of retrieval models and term weighting-schemes we hypothesised that we would observe a reduction in bias as the models evolved over time from TF and TF.IDF to the more sophisticated DFR/DFI models. While some of the models are parameter-free (i.e. TF, NTF, TF.IDF, NTF.IDF, DPH, DFIa,DFIb and DFIc), we were required to estimate the free parameter for the other models. To estimate the parameters, we explored a parameter sweep, and selected the setting that resulted in the fairest or least biased model according to the Gini Coefficient. Setting the model this way requires no recourse to relevance judgements, and means that we can determine whether a model is capable of being the fairest. Furthermore, prior work has shown that setting the model this way is close to optimal [22], we therefore believe that this is an appropriate way to determine which model is the fairest, and to see if it also is the best.

With BM25 we used 11 parameter settings for b between 0.0 and 1.0 increasing in steps of 0.1 (where BM11 is when $b = 0$ and BM15 is when $b = 1$). For PL2 and LGD we set parameter c to values between 1 and 10 but also included 0.1 and 100 to test the extremes. For Bayes (BS) and Laplace (LP), we set their respective smoothing parameters (β and α) to: 1, 10, 100, 500, 1000, 2000, 3000, 5000 and 10000 while on Jelinek Mercer and PTF.IDF we set their respective parameters (λ and c) between 0.1 and 0.9 increasing in steps of 0.1.

Overall, this resulted in 81 different configurations given the 17 term weighting schemes. With four collections and approximately 250,000 queries per collection, this amounted to well over 80 million queries being issued to generated the data for the results reported here.

4   Results and Analysis

Table 2 shows the bias (expressed by the Gini Coefficient, where lower is fairer) along with the performance associated with each model for each of the collections. For each measure we calculated the Pearson's correlation with Gini (where an * denotes whether the result was statistically significant at $p < 0.005$). For all the collections and measures we see that a moderate to high negative correlation exists between bias and performance (and is significant in most cases). To provide an impartial presentation of the results, Figure 1 shows a plot of Gini versus MAP for AQ (where the strongest correlations were observed) and WT (where the weakest correlations were observed). On AQ, we see that the fairest model also obtains the best MAP, while on WT the fairest model is mediocre at best. From the plots and tables, we can also see that the DFI models were not the fairest models, but they are all reasonably effective (and the best performing model on T123, and second best on AQ and WT) with the DFIb term weighting scheme performing the best out of all DFI models. It is also apparent from the plots that there are two main groups: the TF/TF.IDF based models which don't explicitly perform document length normalisation and the other models, which do. Within this second group, however, the relationship is bit more complicated, as can be seen from the WT plot in Figure 1. On WT, without the TF/TF.IDF models the correlation would appear to be positive.

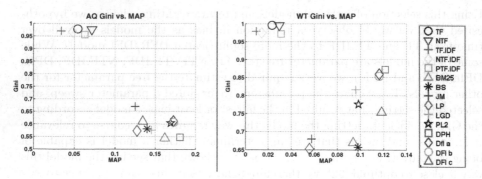

Fig. 1. AQ (Left) & WT (Right): Gini vs. MAP across the range of retrieval models

To explore the relationship between models further and to see what document features they make more or less retrievable, we plotted the retrievability of documents versus document length (and versus average information content of a document[4]). This was done by sorting the documents according to the length or information content, then grouping documents into buckets we calculated the average length (information content) and the average retrievability. The results are plotted in Figure 2 for both AQ and WT.

The first observation we can make here is to see that on TF.IDF (blue triangles) longer documents are much more retrievable than shorter documents. However, when TF is normalised (NTF.IDF shown as yellow triangles), the trend is reversed and shorter documents become highly retrievable. It is clear that most of the bias associated with these models stems from the lack of length normalisation, similar and corresponding patterns were observed for TF and NTF. Of note, is the erratic shape of the plot of information content versus retrievability for TF.IDF (blue triangles), suggesting that the term weighting is not particular robust or consistent when compared to other models.

As previously mentioned, BM25 is consistently the fairest model. If we examine the length plots then we can see that across the different lengths BM25 (purple triangles) tends not to overly favour longer documents when compared to the other retrieval models (though has a tendency to favour average length documents). In terms of information content, BM25 provides about the same level of retrievability. Given these plots, it is clear why BM25 is the fairest.

On AQ, we see that DPH (red squares) tends to be very similar to BM25 and very fair in terms of length. However, on WT, DPH clearly favours documents proportional to their length such that longer documents are much more retrievable. If we examine Table 1, we can see that relevant documents in WT are also much longer - so this bias seems to improve the performance. PL2 (black stars) shows a similar bias toward longer documents, however it is somewhat mitigated when compared to DPH because it has a parameter that can be tuned.

---

[4] This was calculated by summing the TF.IDF scores of the all the terms in the document and then dividing by the document's length.

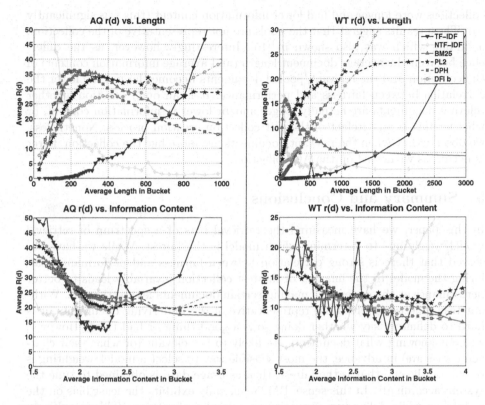

Fig. 2. AQ (Left) & WT (Right) $R(d)$ vs length (top), $R(d)$ vs average information content (bottom)

With respect to the Divergence From Independence Models, we have plotted the best performing model, DFIb (green circles). Much like the DFR models, on AQ the model is quite fair across document lengths. However, when applied to the web collections which have much more varied lengths, the DFIb (and the other DFI models) tend to make longer documents more retrievable. On the information content plots, DFIb also favours documents which have lower information content. This perhaps, is to be expected as longer documents tend to have lots of non-informative words which reduces their overall average information content.

On WT, in particular, models that favour documents that are longer and that contain less information content on average, tend to perform much better than the fairer models (i.e. DPH, PL2 and DFI all outperform BM25). To examine why this is the case we examined the length and information content of documents in the collection, the pool of judged documents the relevant documents. Table 1 reports the mean values for each collection. To determine whether relevant documents and pooled documents were different, either in terms of length or information content, compared to other documents in the collection, we performed un-paired t-tests. We found that relevant/pooled documents in these test collections were longer and had lower information content (this was significantly so). These results suggest that the pools are not representative of the collection, a finding which was also shown in [15]. Interestingly however, as the difference between the average document length (and average information content) of the collection and the relevant/pooled documents becomes larger, the lower the correlation between fairness and performance. It is an open question whether relevant documents are actually longer and/or lower in information content, or whether this is an artefact of the test collection creation process. Nonetheless, we observe that when the relevant documents are more like the collection, fairer is better, as witnessed on the AQ collection.

5   Summary and Conclusions

In this paper, we have measured the retrieval bias of a spectrum of retrieval model/weightings to determine which model is the fairest. While we have observed that there is strong correlation between fairness and performance, tailoring the model to the nuances of the test collection invariably leads to better performance at the expense of making certain documents less retrievable. Without test collections which are representative of the underlying documents, it is hard to definitely say whether doing so is a good thing or bad thing. However, without knowing what documents are likely to be relevant (or what their characteristics are) in advance, the most sensible way to select a model/weighting is to choose the one that is the fairest, then as usage data is obtained to tune the system accordingly. In this sense, BM25 generally exhibits the least bias on the collection, while delivering competitive retrieval performance. This is quite remarkable given all the subsequent models developed. This work prompts further research questions: (i) how do we optimise performance given such biases, (ii) how do we make more representative test collections, (iii) what is the impact of such biases on future sets of queries (i.e. what if shorter and more informative documents were more likely to be relevant), and (iv) if the performance measures took into account document length and utility (such as Time Biased Gain and the U-Measure), would fairer lead to better?

Acknowledgements. This work is supported by the EPSRC Project, *Models and Measures of Findability* (EP/K000330/1).

References

1. Amati, G., Van Rijsbergen, C.J.: Probabilistic models of ir based on measuring the divergence from randomness. ACM Trans. on Info. Sys., 357–389 (2002)
2. Azzopardi, L., Bache, R.: On the relationship between effectiveness and accessibility. In: Proc. of the 33rd international ACM SIGIR, pp. 889–890 (2010)
3. Azzopardi, L., Vinay, V.: Retrievability: An evaluation measure for higher order information access tasks. In: Proc. of the 17th ACM CIKM, pp. 561–570 (2008)
4. Bashir, S., Rauber, A.: On the relationship bw query characteristics and ir functions retrieval bias. J. Am. Soc. Inf. Sci. Technol. 62(8), 1515–1532 (2011)

5. Clinchant, S., Gaussier, E.: Bridging language modeling and divergence from randomness models: A log-logistic model for ir. In: Azzopardi, L., Kazai, G., Robertson, S., Rüger, S., Shokouhi, M., Song, D., Yilmaz, E. (eds.) ICTIR 2009. LNCS, vol. 5766, pp. 54–65. Springer, Heidelberg (2009)
6. Crestani, F., Lalmas, M., Van Rijsbergen, C.J., Campbell, I.: Is this document relevant? probably: a survey of probabilistic models in information retrieval. ACM Computing Survey 30(4), 528–552 (1998)
7. Dinçer, B.T., Kocabas, I., Karaoglan, B.: Irra at trec 2010: Index term weighting by divergence from independence model. In: TREC (2010)
8. Fang, H., Tao, T., Zhai, C.: A formal study of information retrieval heuristics. In: Proc. of the 27th ACM SIGIR Conference, SIGIR 2004, pp. 49–56 (2004)
9. Fuhr, N.: Probabilistic models in ir. Computer Journal 35(3), 243–255 (1992)
10. Gastwirth, J.: The estimation of the lorenz curve and gini index. The Review of Economics and Statistics 54, 306–316 (1972)
11. Harter, S.P.: A probabilistic approach to automatic keyword indexing. part i. on the distribution of specialty words in a technical literature. Journal of the American Society for Information Science 26(4), 197–206 (1975)
12. Hiemstra, D.: A probabilistic justification for using tf.idf term weighting in information retrieval. International Journal on Digital Libraries 3(2), 131–139 (2000)
13. Jones, K.S., Walker, S., Robertson, S.E.: A probabilistic model of information retrieval: development and comparative experiments (parts 1 and 2). Information Processing and Management 36(6), 779–808 (2000)
14. Kocabas, I., Dinçer, B.T., Karaoglan, B.: A nonparametric term weighting method for information retrieval based on measuring the divergence from independence. Information Retrieval, 1–24 (2013)
15. Losada, D.E., Azzopardi, L., Baillie, M.: Revisiting the relationship between doc. length and relevance. In: Proc. of the 17th ACM CIKM 2008, pp. 419–428 (2008)
16. Maron, M.E., Kuhns, J.L.: On relevance, probabilistic indexing and information retrieval. Journal of the ACM 7(3), 216–244 (1960)
17. Ponte, J.M., Croft, W.B.: A language modeling approach to information retrieval. In: Proc. of the 21st ACM SIGIR Conference, SIGIR 1998, pp. 275–281 (1998)
18. Robertson, S.E., Walker, S.: Some simple effective approx. to the 2-poisson model for probabilistic weighted retrieval. In: Proc. of ACM SIGIR 1994, pp. 232–241 (1994)
19. Salton, G., Wong, A., Yang, C.S.: A vector space model for automatic indexing. Communications of the ACM 18(11), 613–620 (1975)
20. Salton, G.: Automatic Information Organization and Retrieval (1968)
21. Singhal, A., Buckley, C., Mitra, M.: Pivoted document length normalization. In: Proce. of the 19th ACM SIGIR Conference, SIGIR 1996, pp. 21–29 (1996)
22. Wilkie, C., Azzopardi, L.: Relating retrievability, performance and length. In: Proc. of the 36th ACM SIGIR Conference, SIGIR 2013, pp. 937–940 (2013)
23. Zhai, C., Lafferty, J.: A study of smoothing methods for language models applied to ad hoc ir. In: Proc. of the 24th ACM SIGIR, pp. 334–342 (2001)

Tackling Biased Baselines
in the Risk-Sensitive Evaluation of Retrieval Systems

B. Taner Dinçer[1], Iadh Ounis[2], and Craig Macdonald[2]

[1] Department of Statistics & Computer Engineering, Muğla University,
48000, Muğla, Turkey
dtaner@mu.edu.tr
[2] School of Computing Science, University of Glasgow,
Glasgow G12 8QQ, UK
{iadh.ounis,craig.macdonald}@glasgow.ac.uk

Abstract. The aim of optimising information retrieval (IR) systems using a risk-sensitive evaluation methodology is to minimise the *risk* of performing any particular topic less effectively than a given baseline system. Baseline systems in this context determine the reference effectiveness for topics, relative to which the effectiveness of a given IR system in minimising the risk will be measured. However, the comparative risk-sensitive evaluation of a set of diverse IR systems – as attempted by the TREC 2013 Web track – is challenging, as the different systems under evaluation may be based upon a variety of different (base) retrieval models, such as learning to rank or language models. Hence, a question arises about how to properly measure the risk exhibited by each system. In this paper, we argue that no model of information retrieval alone is representative enough in this respect to be a true reference for the models available in the current state-of-the-art, and demonstrate, using the TREC 2012 Web track data, that as the baseline system changes, the resulting risk-based ranking of the systems changes significantly. Instead of using a particular system's effectiveness as the reference effectiveness for topics, we propose several remedies including the use of mean within-topic system effectiveness as a baseline, which is shown to enable unbiased measurements of the risk-sensitive effectiveness of IR systems.

1 Introduction

Different approaches in information retrieval (IR) such as query expansion [1, 2] and learning to rank [3] behave differently across topics, often improving the effectiveness for some of the topics while degrading performance for others. This results in a high variation in effectiveness across the topics. To address such variation, there has been an increasing focus on the effective tackling of difficult topics in particular (e.g. through the TREC Robust track [4]), or more recently, on the risk-sensitive evaluation of systems across many topics [5].

In general, the evaluation of risk is performed on the variation of a particular system against a baseline. In all the previous works, the baseline is taken as a predefined configuration of the system under consideration [5–7]. The TREC Web track has recently introduced the risk-sensitive task, to achieve a comparative evaluation of the risk across many systems [8]. In the proposed TREC risk-sensitive evaluation, the baseline is not

M. de Rijke et al. (Eds.): ECIR 2014, LNCS 8416, pp. 26–38, 2014.

necessarily a variation of the system deployed by an individual participating group. In this paper, we argue that this makes the unbiased evaluation of risk challenging, as we show that using a baseline that is not a variation of the same system under consideration has implications on the validity of the risk-sensitive measurements.

Indeed, this paper shows that the choice of an appropriate baseline is of paramount importance in ensuring an unbiased risk-sensitive measurement of the performance of individual systems. We show that the higher the correlation between any given system and the baseline system across queries, the higher the measured risk-sensitive scores of that system on average, leading to a *de facto* bias in the estimation of the systems' risks.

More precisely, in all the previous works, experiments are performed on the variation of the same system, taking a particular configuration as the baseline. Although using a particular system configuration as the baseline is valid in the context of an experiment concerning a single IR system, we show that if this experimental setup is applied to multiple systems, it results in biased measurements for those systems that are not a variation of this baseline. Systems with performance scores that are highly correlated in a positive direction with the baseline scores will get their risk-sensitive performances overestimated. In contrast, systems with performance scores that are highly correlated in a negative direction will have their risk-sensitive performances under-estimated.

To address this bias in risk measurements, this paper argues for and contributes a number of definitions for alternative baselines, such as the per-topic "average system performance" that gives equal weight to every system under consideration in determining the baseline performance for each topic, and the per-topic "maximum system performance", which is akin to the achievable retrieval performance on every topic in the current state-of-the-art (SOTA). Using the TREC 2012 Web track participating systems, as well as a TREC-provided baseline system, we show that for a world-wide experimental evaluation effort such as TREC, our alternative baselines lead to unbiased evaluation of the risk-sensitive performance of the participating systems. As demonstrated in the study presented in this paper, by using the per-topic maximum baseline, the risk-sensitive evaluation of IR systems can be turned into a loss-in-SOTA evaluation where the systems are compared with each other based on measuring to what degree their observed performances diverge from the performance achievable in SOTA.

The remainder of this paper is structured as follows: In Section 2, we empirically show, through a study on the TREC 2012 Web track, the inherent bias obtained using a single baseline for multi-system risk-sensitive evaluation; Section 3 proposes several alternative baselines and shows through the same methodology their unbiasedness within a risk-sensitive evaluation. We discuss the unbiasedness property and the limitations of the proposed baselines in Section 4. We provide concluding remarks in Section 5.

2   The Bias in Risk-Sensitive Evaluation

This section first discusses the evaluation of single IR systems in a risk-sensitive fashion (Section 2.1), then their comparative evaluation within a TREC setting (Section 2.2). Later, we show that the choice of baseline for a multi-system risk-sensitive evaluation can favour some systems over others (Section 2.3), which is explained through the use of a Principal Components Analysis (Section 2.4).

28   B.T. Dinçer, I. Ounis, and C. Macdonald

2.1 Measuring Risk

The risk-sensitive performance of a retrieval system (a *run* in TREC terminology) is typically measured as the risk-reward tradeoff between the system itself and a baseline configuration of that system. In particular, given a topic set Q with c topics, the risk-reward tradeoff between a run r and a baseline run br is given by:

$$U_{Risk}(r|br,Q) = \frac{1}{c}\left[\sum_{q \in Q_+}(r_q - br_q) - (\alpha+1)\sum_{q \in Q_-}(br_q - r_q)\right], \qquad (1)$$

where r_q and br_q are respectively the score of the run and the score of the baseline run on q measured by a retrieval effectiveness measure [7] (e.g. NDCG@20, ERR@20 [9]). The left summand in the square brackets gives a total win (or upside-risk) with respect to the baseline and the right summand the total loss (or downside-risk). The risk-reward tradeoff score of a run refers to the average difference of the total win from a weighted total loss with weight $(1 + \alpha)$ over c topics. For higher α, the penalty for under-performing with respect to the baseline is increased: typically $\alpha = 1, 5, 10$ [8].

2.2 Comparative Risk-Sensitive Evaluation of Systems

The TREC 2013 Web track aims to make a comparative evaluation of the risk of retrieval systems, in order to identify the systems that are able to consistently outperform a "provided baseline run"[1]. The provided baseline run for the TREC 2013 Web track risk-sensitive task is based on the *Indri* retrieval platform, which is developed under the Lemur project[2]. However, as the TREC 2013 campaign has yet to conclude at the time of writing, in the following we perform an empirical study based on runs submitted to the TREC 2012 Web track. Indeed, the 2013 track coordinators have made available a set of Indri runs on the TREC 2012 Web track topics[3] that correspond directly to the TREC 2013 baseline runs.

Table 1 lists the risk-sensitive scores calculated for the top 8 TREC 2012 adhoc runs and the corresponding performance ranks of the runs, for varying values of risk-sensitive parameter $\alpha = 1, 5, 10$. Here, the baseline is *IndriCASP*, an Indri run on the ClueWeb09 Category A document collection, with the Waterloo spam-page filter [10] applied - in effect, this equates to the TREC provided Indri baseline. The retrieval effectiveness measure used for the calculation of the U_{Risk} scores in Table 1 is the Expected Reciprocal Rank at 20 documents, $ERR@20$.

As can be seen from Table 1, increasing the risk-sensitive parameter α changes the risk-based performance ranking of the runs. For example, at $\alpha = 5$, run *uogTrA44xi* is demoted from rank 1 to rank 4, while *srchvrs12c09* is promoted from rank 2 to rank 1. According to the notion of risk-sensitive evaluation, as defined in [5–7], a system is promoted over another, if it minimises the risk over all topics better than the other system, i.e. a risk in the sense of showing a performance worse than that of the baseline

---

[1] http://research.microsoft.com/en-us/projects/trec-web-2013
[2] http://www.lemurproject.org
[3] https://github.com/trec-web/trec-web-2013

Table 1. U_{Risk} scores calculated for the top 8 TREC 2012 runs, based on ERR@20 measure, and the corresponding ranks of the runs (R), for varying values of risk-sensitive parameter α, where IndriCASP is the baseline

Run	ERR@20	R	U_{Risk} $\alpha = 1$	R	$\alpha = 5$	R	$\alpha = 10$	R
uogTrA44xi	0.313	1	0.0556	2	-0.1959	4	-0.5104	4
srchvrs12c09	0.305	2	0.0679	1	-0.1015	1	-0.3133	1
DFalah121A	0.292	3	0.0467	3	-0.1558	2	-0.4089	2
QUTparaBline	0.290	4	0.0385	4	-0.1893	3	-0.4740	3
utw2012fc1	0.219	5	-0.0558	6	-0.3782	6	-0.7813	6
ICTNET12ADR2	0.215	6	-0.0495	5	-0.3286	5	-0.6774	5
indriCASP	0.195	*	0	*	0	*	0	*
irra12c	0.172	7	-0.1182	7	-0.5014	7	-0.9805	7
qutwb	0.166	8	-0.1342	8	-0.5560	8	-1.0832	8

system for any particular topic. Thus, the results given in Table 1 suggest, in theory, a conclusion that can be stated roughly as *srchvrs12c09* is the best run in minimising the risk amongst the top 8 TREC 2012 adhoc runs, followed by *DFalah121A*, which is better than *QUTparaBline*, which in turn is better than *uogTrA44xi*, and so on. In addition, a risk-sensitive evaluation is meant to provide information on the robustness/stability of IR systems in terms of retrieval effectiveness across topics, such that *srchvrs12c09* is more robust than *DFalah121A*, which is in turn more robust than *QUTparaBline*, and so on.

In measuring the risk-sensitive performance of IR systems, we will show in the next section that the baseline system is a major factor in depicting the final risk-sensitive performance of a given system. Indeed, one could argue that, except for spam-page filtering, IndriCASP is a plain, out-of-the-box run of the Indri system that employs no advanced retrieval technology other than its core term weighting and ranking methods. Moreover, it has a moderate ERR@20 score, 0.195, compared to the top TREC 2012 runs. Thus, one can claim that it is a *fair* baseline, fair in the sense of being a true reference IR system relative to which the risk-sensitive performance of other IR systems can objectively be measured and compared with each other. There is no doubt that, to be a fair baseline, an IR system must satisfy certain conditions including but not limited to the ones considered thus far. However, the question that arises is whether this choice of a particular baseline system is unbiased with respect to the other systems under evaluation. We study this issue in details in Section 2.3.

2.3 Bias in the TREC Baseline

We argue that a baseline system for risk-sensitive evaluation must not only be fair, but must also be *unbiased*, in the sense of favouring no system over another in a systematic fashion. Risk-sensitive evaluation is originally proposed for those IR experiments that involve only the variants of a single retrieval strategy, where a supplementary method, such as query expansion, is applied to the system of base retrieval methods particular to that strategy, e.g., term weighting model and ranking function. This kind of experiment follows an experiment design called the *before and after* design in statistics, where measurements are made on the response variable before and after the treatment's exposure in order to decide whether the treatment has an effect on the response of the subject of

Table 2. U_{Risk} scores of top 8 TREC 2012 runs at $\alpha = 10$, when each run in turn is chosen as the baseline

Run	U_{Risk}							
	$\alpha = 10$	R	$\alpha = 10$	R	$\alpha = 10$	R	$\alpha = 10$	R
uogTrA44xi	0	*	-0.928	1	-0.882	2	-1.072	3
srchvrs12c09	-1.027	1	0	*	-0.944	3	-0.830	1
DFalah121A	-1.135	2	-1.098	3	0	*	-0.835	2
QUTparaBline	-1.349	3	-1.008	2	-0.858	1	0	*
utw2012fc1	-1.443	5	-1.412	4	-1.291	4	-1.387	4
ICTNET12ADR2	-1.434	4	-1.577	5	-1.542	5	-1.513	5
irra12c	-1.758	6	-1.941	7	-1.766	7	-1.706	6
qutwb	-1.814	7	-1.826	6	-1.715	6	-1.788	7

Run	U_{Risk}							
	$\alpha = 10$	R	$\alpha = 10$	R	$\alpha = 10$	R	$\alpha = 10$	R
uogTrA44xi	-0.319	1	-0.255	1	-0.068	1	-0.047	1
srchvrs12c09	-0.387	2	-0.497	2	-0.351	4	-0.159	2
DFalah121A	-0.421	3	-0.617	4	-0.329	3	-0.202	3
QUTparaBline	-0.540	4	-0.611	3	-0.293	2	-0.298	5
utw2012fc1	0	*	-0.843	5	-0.432	5	-0.224	4
ICTNET12ADR2	-0.897	6	0	*	-0.581	6	-0.613	7
irra12c	-0.997	7	-1.092	6	0	*	-0.610	6
qutwb	-0.866	5	-1.201	7	-0.687	7	0	*

interest. In relation a to risk-sensitive evaluation, treatments correspond to the supplementary methods to be applied to the system of base retrieval methods. For this kind of risk-sensitive evaluation, the baseline system is, by default, composed of the base methods determining the retrieval strategy under consideration, so that, after the exposure to the treatment, any measured gain or loss in the effectiveness of the system under evaluation can be attributed only to the supplementary method applied.

On the other hand, as opposed to the classical before and after experimental design setup, for a TREC-like large experiment involving multiple retrieval strategies (most of which could not be necessarily considered as variants of each others) there would arguably be no single reference system that is composed of the base retrieval methods in common to every retrieval strategy under evaluation. Table 2 shows the calculated U_{Risk} scores and the corresponding systems' rankings for each of the 8 possible selections of baseline system (i.e. each run is in turn chosen as the baseline). From the table, the observed systems' rankings markedly and significantly vary as the baseline changes ($p = 0.0003$ according to the nonparametric Friedman's test).

The variance in the effectiveness of IR systems across topics explains why different baseline systems yield different risk-based system rankings. IR systems with different retrieval strategies would in general show different performance profiles over a given set of topics. Two IR systems with similar/parallel performance profiles may be considered to be variants of each other, in analogy to before and after design experiments. However, having similar performance profiles over topics basically implies a positive correlation between the observed performance scores across the topics, while having discordant performance profiles implies a negative correlation. Using the classical risk-sensitive evaluation setup, given a set of IR systems, any particular baseline system promotes in a systematic fashion those IR systems whose performance profiles are similar to its performance profile, while demoting the systems with discordant performance profiles. In the next section, we use Principal Components Analysis (PCA) [11] to demonstrate

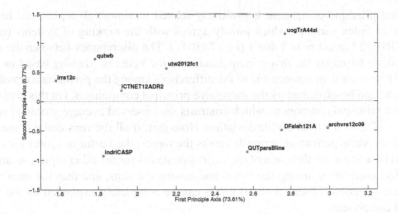

Fig. 1. PCA plot of the top 8 TREC 2012 Web track adhoc runs, based on the first and second principal axes

the correlation of the effectiveness of the TREC submitted runs across topics, thereby illustrating how particular choices of baselines can be correlated with particular runs.

2.4 Principal Components Analysis of Per-Topic Effectiveness

PCA – a dimension reduction technique – provides an intuitive way for visually exploring the performance correlations among IR systems across topics [12]. Since higher dimensional spaces are difficult to inspect, it becomes necessary to reduce them into lower dimensions. For a test collection with many topics, groups of topics would in general be performed similarly by IR systems. In PCA, each principal component is a linear combination of the original topics. The first principal component is a single axis in space. When you project the measured system performance scores for each topic on that axis, the resulting values form, in one sense, a meta topic. The variance of scores in this meta topic is the maximum among all possible choices of the first axis. The second principal component is another axis in space, perpendicular to the first. The projection of the measured system performance scores on this axis generates another meta topic. The score variance associated with this meta topic is the maximum among all possible choices of this second axis. All the principal components are orthogonal to each other, and hence, as opposed to the original topics, there is no redundant information on the performance relationships among IR systems across topics. Hence, the first two principal components together can be used to define two orthogonal dimensions, and therefore to visualise the performance relationships among IR systems across topics by means of a scatter plot that accounts for the major part of the total performance variations observed on the original topics. The PCA plot of the top 8 TREC 2012 Web track adhoc runs using the first and the second principal axes is given in Figure 1. For this PCA plot, the first principal component accounts for 73.61% of the total sum of squared deviations observed on 50 topics, and the second principal component accounts for 6.77% of the same total i.e., together accounting for 80% of the total variation in performance among the top 8 TREC 2012 Web track adhoc runs across the topics.

The first principal component is positively related to almost all topics, and hence it acts as an index variable, which mostly agrees with the ranking of systems based on the ERR@20 measure in Table 1 (i.e., 73.61%). The discrepancy between the systems' ranking based on the first principal axis and the systems' ranking based on the ERR@20 measure is in essence due to the differences among the performance profiles of runs, and can be explained by the successive principal components. On this account, the second principal component, which contrasts the observed average scores of runs, explains 6.77% of the total profile deviation. Note that, if all the runs under consideration had the same performance profile across the topics, the vector of scores for each topic will be a linear combination of the vectors of scores for the other topics, resulting in a perfect correlation among the topics and among the runs, and thus the total sum of squared deviations observed on 50 topics can be completely explained by the first principal component.

The interpretation of a PCA plot is simple. Runs that are clustered around the same location in the plot are runs that have both comparable average ERR@20 scores and similar performance profiles over all (50) topics, such as *srchvrs12c09*, *DFalah121A*, and *QUTparaBline*. For the PCA plot in Figure 1, since the first principal component is positively related to all topics, a low component score for any run implies a low average score over all topics, and a high component score implies a high average score. On the other hand, the second principal component contrasts the observed scores of runs on two subsets of topics. For any two runs that are contrasted by the second principal component, say *srchvrs12c09* and *uogTrA44xi*, a high component score in a positive direction implies a high average score for one of them (i.e., *uogTrA44xi*) and simultaneously a low average score for the other (i.e., *srchvrs12c09*) on the topic subset to which the second principal component is positively related. Conversely, on the topic subset to which the second principal component is negatively related, a high component score in a negative direction implies a high average score for the latter run (*srchvrs12c09*) and a low average score for the former (i.e., *uogTrA44xi*).

In summary, comparing the *uogTrA44xi* and *IndriCASP* runs, the PCA plot in Figure 1 reveals that *uogTrA44xi* performs most of the 50 topics better than *IndriCASP* (c.f. the position of the runs along the first principal axis) but on a particular subset of topics, to which the second principal axis is negatively related, *IndriCASP* has relatively high scores while *uogTrA44xi* has relatively low scores, compared to the within-topic average system scores. This is actually the case for all the runs having positive component scores on the second principal axis, such as *utw2012fc1* and *ICTNET12ADR2*. In contrast, the opposite case is true for the runs with negative component scores. For instance, for the runs *srchvrs12c09* and *DFalah121A*, which are comparable to *uogTrA44xi* in terms of average ERR@20 score, the PCA plot reveals that they show their low and high scores on every topic in synchronisation with *IndriCASP*. Due to the nature of a risk-sensitive evaluation, the use of *IndriCASP* as a baseline will favour, in a systematic fashion, those runs that have negative component scores on the second principal axis over the runs with positive component scores. This is the basic reason why one of the two comparable runs in terms of average ERR@20 score, *srchvrs12c09*, is promoted to rank 1 while *uogTrA44xi* is demoted to rank 4 in the risk-based systems' ranking obtained using *IndriCASP* as the baseline. Simultaneously, this is also why *uogTrA44xi* is prompted to rank 1 and *srchvrs12c09* is demoted to rank 4 of the ranking obtained when *irra12c* is used as the baseline (both the *uogTrA44xi* and *irra12c* runs can be considered as variants of each other with respect to their base retrieval strategies, *Divergence From Randomness* [13] and *Divergence From Independence* [14], as well as both being based upon the Terrier retrieval platform [15][4]). Hence, within a *comparative* risk-sensitive evaluation of different IR systems, we conclude that the choice of *IndriCASP* as the baseline run benefits systems similar to that run, and hinders the risk-sensitive performance of other runs. In the next section, we propose several alternatives to define the baseline performance for comparative risk-sensitive evaluations, that are both fair and unbiased.

3 Unbiased Comparative Risk-Sensitive Evaluation

As shown above, no IR system's retrieval strategy alone is representative enough to be used as a baseline for the risk-sensitive evaluation of different retrieval strategies. In this respect, every retrieval strategy biases towards its variants. To fairly measure and compare the risk-sensitive performance of multiple retrieval strategies that are stemmed from different base models of information retrieval (e.g., vector space vs. probabilistic) in an unbiased manner, there needs to be a baseline that is generalisable enough to be applied to each retrieval strategy under evaluation. In classical statistics, a remedy for such issues is the use of an estimate of the parameter of interest. Here, the parameter of interest is the performance of an unbiased baseline system for any given topic. Given a particular topic q and a set of r runs, the arithmetic mean of the r performance scores according to an evaluation measure observed on q is one of the possible estimates of the unbiased baseline score UBS_q:

$$\text{UBS}_q = \frac{1}{r} \sum_{i=1}^{r} s_i(q),\tag{2}$$

where $s_i(q)$ is the performance score of run i on topic q for $i = 1, 2, \ldots, r$ for a given evaluation measure (e.g. ERR@20). Since the arithmetic mean gives equal weight to every retrieval strategy in determining the UBS_q, a baseline system that is determined by the UBS_q scores, MEAN for short, will be unbiased with respect to the retrieval strategies yielding the r run scores.

We note that the notion of risk defined here by MEAN is different from the original notion of risk, in that the U_{Risk} measure now measures the risk-sensitive performance of an IR system in terms of the risk of a given topic being less effective than the mean system effectiveness *expected* on that topic, instead of a single system's effectiveness.

Next, the *median* of the within-topic scores, MEDIAN, can also be used as the unbiased baseline score for each topic. In addition, it is also possible to take the *maximum* topic scores as the baseline, MAX. Since such a baseline will represent the achievable system performance for each topic in the current state-of-the-art, it turns the risk-sensitive evaluation of IR systems into a loss-in-SOTA evaluation where the systems are compared with each others based on measuring to what degree their observed performances diverge from the performance achievable in SOTA. Table 3 shows the

---

[4] http://terrier.org

Table 3. U_{Risk} scores of the top 8 TREC 2012 runs, using MEAN, MEDIAN and MAX as the baseline, and the corresponding ranks of the runs (R), for $\alpha = 5, 10$

| | | | U_{Risk} | | | | | |
| | | MEAN | | MEDIAN | | MAX (SOTA) | |
Run	ERR@20 R	$\alpha = 5$ R	$\alpha = 10$ R	$\alpha = 5$ R	$\alpha = 10$ R	$\alpha = 5$ R	$\alpha = 10$ R
uogTrA44xi	0.313 1	-0.1061 1	-0.2845 1	0.0059 1	-0.0969 1	-1.3901 1	-2.5486 1
srchvrs12c09	0.305 2	-0.1668 2	-0.3848 2	-0.0067 2	-0.1140 2	-1.4398 2	-2.6396 2
DFalah121A	0.292 3	-0.1760 3	-0.4161 3	-0.0469 3	-0.1815 3	-1.5168 3	-2.7809 3
QUTparaBline	0.290 4	-0.1976 4	-0.4444 4	-0.0559 4	-0.1976 4	-1.5286 4	-2.8024 4
utw2012fc1	0.219 5	-0.3574 5	-0.6935 5	-0.1235 5	-0.2621 5	-1.9523 5	-3.5792 5
ICTNET12ADR2	0.215 6	-0.4204 6	-0.8149 6	-0.2757 6	-0.5620 6	-1.9796 6	-3.6293 6
indriCASP	0.195 7	-0.5343 7	-1.0224 7	-0.3500 7	-0.6616 7	-2.1011 7	-3.8521 7
irra12c	0.172 8	-0.5555 8	-1.0425 8	-0.3921 9	-0.7522 8	-2.2351 8	-4.0977 8
qutwb	0.166 9	-0.5908 9	-1.1067 9	-0.3825 8	-0.7553 9	-2.2736 9	-4.1682 9

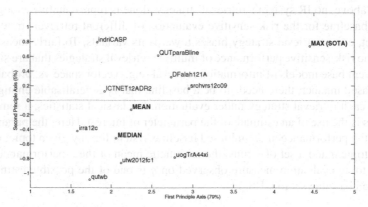

Fig. 2. PCA plot showing the performance relationships among the top TREC 2012 Web track runs, *indriCASP*, and the mean and median topic scores across 50 Web track topics

risk-based ranking of the top 8 TREC 2012 Web track adhoc runs, using MEAN, MEDIAN and MAX as the baseline, respectively.

In general, given a set of IR systems with different retrieval strategies for a comparative risk-sensitive evaluation, an unbiased baseline system can be thought of as the system that is jointly determined by the given systems, such that being selected as the baseline is equally likely for each system. This definition of unbiasedness applies to the three baselines that we have proposed. To demonstrate the unbiasedness of our proposed baselines, we again turn to PCA.

Figure 2 provides the PCA plot showing the MEAN, MEDIAN and MAX baselines as three virtual runs. In the PCA plot of Figure 2, a run with scores equal to the mean scores for each topic will be shown at the origin with respect to all contrasting principal axes. This is the reason why the MEAN baseline is close to the origin with respect to the second principal axis in Figure 2[5]. Here, based on the given definition of unbiasedness, one

---

[5] The MEAN baseline is not shown at the origin because the mean scores, median scores, and maximum scores for the corresponding baseline runs are calculated using the scores of 9 actual runs and then PCA is applied to a (run×topic) matrix of 12 runs (9 actual + 3 virtual) and 50 topics. Removing the median and max scores from the matrix would fix this issue.

may argue that the closer a baseline is to the origin, the higher its degree of unbiasedness w.r.t. the systems under evaluation. However, this is not the case. As can be seen from Figure 2, the MEDIAN and MAX baselines diverge from the origin, although they are unbiased. As statistical estimates, both MEDIAN and MAX have the same property as MEAN in giving equal weight to every system in determining the baseline scores for each topic, and hence, as a baseline, they have the property of unbiasedness. However, for the MEDIAN baseline, the PCA plot shows in essence that 1) the median scores for most of the topics are slightly below the corresponding mean scores (i.e., the positions of MEAN and MEDIAN with respect to the first axis), and that 2) the median scores tend to be higher in magnitude than the corresponding mean scores for those topics to which the second principal axis is negatively related. Similarly, the MAX baseline exhibits the same issue. Overall, the PCA plot of Figure 2 shows that our three proposed baselines are unbiased, but each serves a different purpose in minimising the risk attached to the IR systems under evaluation.

Our introduced definition of unbiasedness actually exposes a problem about the validity of the *comparative* risk-sensitive evaluation of IR systems, which we will discuss in detail in the next section. This issue is related to the rankings of risk-sensitive systems obtained using the unbiased baselines. Indeed, we will show that such a comparison of the risk-sensitive performances of different IR systems in an objective manner actually implies the comparison of the retrieval effectiveness of the individual systems based on the underlying measure, i.e. ERR@20.

4   Discussion

As can be seen in Table 3, the risk-based ranking of systems remains concordant with the original systems' ranking and also with each others for all values of the risk parameter α, as opposed to the risk-based systems' rankings given in Table 1 where *indriCASP* was used as the baseline. Indeed, this simply suggests that the use of *indriCASP* as a baseline is not unbiased. Note that, for each value of α, a risk-sensitive evaluation is applied to the same set of systems, where in fact the (true) risk associated with each system that the U_{Risk} measure is intended to measure remains constant for all α values. Under this particular condition, unless the baseline favours some systems over the others in a systematic fashion, it is expected that the calculated U_{Risk} scores for each system will increase in magnitude evenly, proportional to the α values, because of the magnification of the downside-risk, while the rankings of the top 8 TREC 2012 runs remain unchanged relative to each other. This is actually the case in Table 3 but not in Table 1.

Figure 3 shows (a) the observed per-topic ERR@20 scores of *uogTrA44xi* and *srchvrs12c09*, (b) the U_{Risk} scores for *uogTrA44xi* when *srchvrs12c09* is the baseline, and (c) the U_{Risk} scores for *srchvrs12c09* when *uogTrA44xi* is the baseline. *srchvrs-12c09* has 11 topic scores greater than 0.8, while *uogTrA44xi* has 8 topic scores greater than 0.8 (a). However, the per-topic loss of *srchvrs12c09* with respect to *uogTrA44xi* (a) is greater in number than the per-topic loss of *uogTrA44xi* with respect to *srchvrs12c09* (c). Thus, it can be argued that the risk of showing a performance worse than that of a fair baseline is higher for *srchvrs12c09* than for *uogTrA44xi*.

36   B.T. Dinçer, I. Ounis, and C. Macdonald

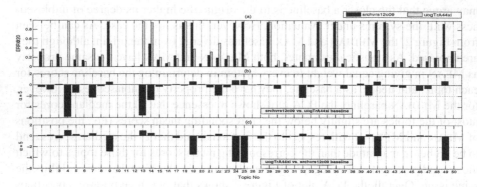

Fig. 3. Comparison of *uogTrA44xi* and *srchvrs12c09*: a) based on per topic ERR@20 scores, b) per topic U_{Risk} scores of *srchvrs12c09* for $\alpha = 5$ when *uogTrA44xi* is the baseline, and c) per topic U_{Risk} scores of *uogTrA44xi* for $\alpha = 5$ when *srchvrs12c09* is the baseline

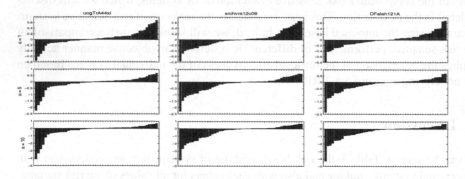

Fig. 4. Distribution of the U_{Risk} scores on 50 Web track topics for *uogTrA44xi*, *srchvrs12c09*, and *DFalah121A* runs, measured using topic mean scores as the baseline for $\alpha = 1, 5, 10$. U_{Risk} scores are shown in ascending order for each run of the ease in visual comparison.

For *uogTrA44xi*, *srchvrs12c09* and *DFalah121A*, Figure 4 shows the per-topic U_{Risk} scores when MEAN is used as the baseline. From the figure, it can be observed that the loss distribution on the topics for *uogTrA44xi* is steeper than that of *srchvrs12c09*, which in turn is steeper than *DFalah121A*, suggesting that the risk attached to *uogTrA44xi* is less than that of *srchvrs12c09*, which in turn is less than that of *DFalah121A*. As a result, we argue that the comparison of the risk-sensitive performances of different IR systems in an objective manner actually implies comparing the retrieval effectiveness of individual systems based on the underlying effectiveness measure, i.e. ERR@20.

5 Conclusions

Following the proper practice of risk-sensitive evaluation in before and after design experiments concerning single IR systems, it would appear, for the experimental evaluation of multiple IR systems, that every IR system employing a particular retrieval strategy requires a baseline system that is composed of the base retrieval methods particular to that strategy, for the purpose of the proper measurement of its risk-sensitive performance. However, since this experiment design asks for a different baseline for every IR system having a particular retrieval strategy that could not be considered the variant of the retrieval strategies represented by the baseline systems at hand, it would cause existing test collections not to be reusable for new IR systems. On the other hand, selecting a particular IR system as the baseline – as attempted by the TREC 2013 Web track – is also not a viable remedy for the issue of test collection reusability. Moreover, we show in this study that selecting a particular IR system as the baseline will result in biased performance measurements for all systems that are not a variation of the provided baseline. Finally, we also demonstrate that a comparative risk-sensitive evaluation of multiple IR systems using unbiased baselines actually implies the typical adhoc type evaluation of the systems based on a retrieval effectiveness measure like ERR@20.

Nevertheless, the benefit of the proposed baselines is that individual IR systems can be optimised for risk minimisation, using MEAN or MEDIAN as the baseline, with respect to the retrieval effectiveness expected for every topic on average by current SOTA IR technology, and, using MAX as the baseline, with respect to the retrieval effectiveness achievable in SOTA.

In summary, we question whether it is possible to conduct a comparative evaluation campaign for risk-sensitive approaches within a TREC-like setting, as it is impossible to derive a common, unbiased baseline that can measure risk-sensitivity separately from average effectiveness. Alternatively, a comparative evaluation for risk-sensitivity could be formulated with each participating risk-sensitive run being measured with respect to the effectiveness of its own declared baseline. However, such an operationalisation would make it difficult to combine measures of baseline effectiveness and risk-sensitivity into a theoretically defined final measure that is comparable across baselines/participating groups. A similar problem has been faced by some previous tasks, e.g., the TREC 2004 Terabyte track efficiency task used tradeoff graphs for comparing the efficiency and effectiveness of participating runs on different axes [16]. We leave the theoretical combination of baseline effectiveness and risk-sensitivity to future work.

References

1. Amati, G., Carpineto, C., Romano, G.: Query difficulty, robustness, and selective application of query expansion. In: McDonald, S., Tait, J.I. (eds.) ECIR 2004. LNCS, vol. 2997, pp. 127–137. Springer, Heidelberg (2004)
2. Carmel, D., Farchi, E., Petruschka, Y., Soffer, A.: Automatic query refinement using lexical affinities with maximal information gain. In: Proc. SIGIR, pp. 283–290 (2002)
3. Macdonald, C., Santos, R., Ounis, I.: The whens and hows of learning to rank for web search. Information Retrieval 16(5), 584–628 (2013)
4. Voorhees, E.M.: Overview of the TREC 2003 robust retrieval track. In: Proc. TREC (2003)
5. Collins-Thompson, K.: Accounting for stability of retrieval algorithms using risk-reward curves. In: Proceedings of SIGIR Workshop on the Future of Evaluation in Information Retrieval (2009)
6. Collins-Thompson, K.: Reducing the risk of query expansion via robust constrained optimization. In: Proc. CIKM, pp. 837–846 (2009)

7. Wang, L., Bennett, P.N., Collins-Thompson, K.: Robust ranking models via risk-sensitive optimization. In: Proc. SIGIR, pp. 761–770 (2012)
8. Collins-Thompson, K., Bennett, P.N., Diaz, F., Clarke, C., Voorhees, E.: TREC 2013 Web Track Guidelines,
   http://research.microsoft.com/en-us/projects/trec-web-2013/
9. Chapelle, O., Metlzer, D., Zhang, Y., Grinspan, P.: Expected reciprocal rank for graded relevance. In: Proc. CIKM, pp. 621–630 (2009)
10. Cormack, G., Smucker, M., Clarke, C.: Efficient and effective spam filtering and re-ranking for large web datasets. Information Retrieval 14(5), 441–465 (2011)
11. Jackson, J.E.: A users guide to principal components. John Wiley & Sons (1990)
12. Dinçer, B.T.: Statistical principal components analysis for retrieval experiments. Journal of the American Society for Information Science and Technology 58(4), 560–574 (2007)
13. Amati, G., van Rijsbergen, C.: Probabilistic models of information retrieval based on measuring the divergence from randomness. Transactions on Information Systems 20(4), 357–389 (2002)
14. Dinçer, B.T.: IRRA at TREC 2012: Index term weighting based on divergence from independence model. In: Proc. TREC (2012)
15. Macdonald, C., McCreadie, R., Santos, R., Ounis, I.: From puppy to maturity: experiences in developing Terrier. In: Proc. OSIR at SIGIR (2012)
16. Clarke, C.L.A., Craswell, N., Soboroff, I.: Overview of the TREC 2004 Terabyte track. In: Proc. TREC (2004)

Content + Attributes: A Latent Factor Model for Recommending Scientific Papers in Heterogeneous Academic Networks

Chenyi Zhang[1,3], Xueyi Zhao[2], Ke Wang[3,*], and Jianling Sun[1]

[1] College of Computer Science, Zhejiang University, China
[2] Dept. of Information Science and Electronic Engineering,
Zhejiang University, China
[3] School of Computing Science, Simon Fraser University, Canada
{chenyizhang,xueyizhao,sunjl}@zju.edu.cn, wangk@cs.sfu.ca

Abstract. This paper focuses on the precise recommendation of scientific papers in academic networks where users' social structure, items' content and attributes exist and have to be profoundly exploited. Different from conventional collaborative filtering cases with only a user-item utility matrix, we study the standard latent factor model and extend it to a heterogeneous one, which models the interaction of different kinds of information. This latent model is called "Content + Attributes", which incorporates latent topics and descriptive attributes using probabilistic matrix factorization and topic modeling to figure out the final recommendation results in heterogeneous scenarios. Moreover, we further propose a solution to handle the cold start problem of new users by adopting social structures. We conduct extensive experiments on the DBLP dataset and the experimental results show that our proposed model outperforms the baseline methods.

1 Introduction

Collaborative filtering (CF) has attracted a lot of attention during the past few years with emerge of popular social network services and recommender systems. However, the existing methods simply incorporate user profiles or social relationship into recommendation systems, e.g., many social network services provide user recommendation based on similar ratings of the same items or a friend of friend relation. Such trivial attempts hardly make precise recommendation in heterogeneous networks. In the general framework of latent factor models [1–4] that perform successfully in collaborative filtering, the single utility matrix is factorized into the latent user vectors and latent item vectors in the low dimensional latent space. Latent user vector p_i represents user i's personal interests while latent item vector q_j express item j's features; the predicted rating $\hat{r}_{ij}$ is computed by the inner product $(p_i^T q_j)$. Following this framework, much work

---

* Ke Wang's work is partially supported by a Discovery Grant from Natural Sciences and Engineering Research Council of Canada.

M. de Rijke et al. (Eds.): ECIR 2014, LNCS 8416, pp. 39–50, 2014.

further involved social networks of users [5–7] or latent topics of items [8] to improve the recommendation performance compared to that based on traditional utility matrix. Although these auxiliary information added to the recommender systems would help improve the performance, the previous works neglect the heterogeneous scenarios where usually more than single type of information exists.

This paper mainly focuses on the recommendation in academic networks, which is far more complex than the conventional cases. It is a heterogeneous scenario to recommend scientific articles in academic networks: each author has several co-authors and may write and cite some papers; this co-authorship forms social networks and such publishing or citation reflects implicit ratings for items; each paper has *content* (e.g., title and abstract in plain text) and *attributes* (e.g., author, venue, publish year), as shown in Fig. 1. However, conventional collaborative filtering methods cannot deal with the paper recommendation well if they still follow the pattern of friend's friend or item similarity. We have to recognize the diversity in the academic networks and admit that users prefer not all the papers of their co-authors. Moreover, due to large amount of publication, the item similarity based methods would return too many (even not convincing) papers. At this time, the attributes remain a good indicator of user's interests as the users may follow the big guy or prefer top conferences in research areas. This motivates us to develop a novel model to recommend scientific articles by taking advantage of all the heterogeneous information. Next, we present some discussions on how these heterogeneous information matter.

1.1 Discussions

According to the toy example exhibited in Fig. 1, there are 5 items binding with content and attributes. Assume that there are two topics "database" and "information retrieval" and three authors. According to the content, suppose that v_1, v_3, v_4 and v_5 are in one topic (e.g., information retrieval) and v_2 is in the other (e.g., database). v_1, v_2 and v_4 are written by the same author A_1, v_3 and v_5 are written by A_2 and A_3 respectively.

Item v_1 and v_2 have the same attributes but different topics in content; Item v_1 and v_3 have similar content but different attributes. The underlying pattern which involves both content and attributes should be that u_1 prefers the "data mining" topic except that of author A_3, u_2 prefers the "data mining" topic except that of author A_1, u_4 prefers both topics from author A_1. If we treat content alone, it is difficult to explain why u_2 likes v_3 but dislikes v_1 although they share the same topic and why u_4 likes both v_1 and v_2 although they have different topics. If we treat attributes alone, it is difficult to explain why u_1 likes v_1 but dislikes v_2 although they are written by the same author. In fact, both content and attributes information should be collaborated to determine a user's interest.

One possible solution is to assign certain weights to content and attributes (i.e., w_1 and w_2), build one recommender using content and another recommender using attributes, and then integrate them according to different weights.

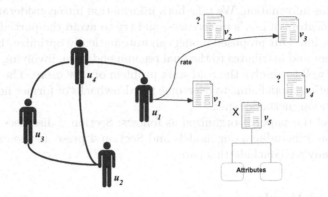

		User-item Utility Matrix			
	v_1	v_2	v_3	v_4	v_5
u_1	√	?	√	?	×
u_2	×	×	√	?	√
u_3	√	×	√	?	?
u_4	√	√	?	?	?

	CONTENT / WORDS				ATTRIBUTES						
	W_1	W_2	W_3	...	W_n	A_1 A_2 A_3	C_1 C_2 ... C_k	Year
v_1	6	5	0	...	0	√	√	2008
v_2	0	0	8	...	6	√	√	2008
v_3	3	6	0	...	1	√	√	2010
v_4	5	7	1	...	0	√	√	2012
v_5	5	9	2	...	1	√	√	2010

Fig. 1. Heterogeneous recommendation scenario: users have social networks and each item has content and attribute information. Each user marks the likes(√) and dislikes(×) for some items and the rest are unknown(?).

However, this approach does not consider correlation of content and attributes into a uniform model, and it is difficult to find right weights from separate recommenders. For example, the recommender using attributes would highly recommend v_2 to u_3, which lead to lower the value of w_2 to balance the final result. In that case, the recommender using content becomes dominant and would recommend v_5 to u_1 and v_1 to u_2 by mistake. The problem of contradicted results by two recommenders is hard to resolve.

Another solution is simply to combine content and attribute matrix together and treat each attribute value as a word. At this time, the effect of attributes may be overwhelmed by content because the number of attributes is much less than that of words. For instance, v_3 and v_5 should have been very similar in this case but u_1 likes v_3 and dislikes v_5.

Moreover, for out of matrix prediction, e.g., item v_4 has never been rated or a new user joins the network, traditional collaborative filtering methods cannot deal with this cold start problem, but we can address it through item's information and user's social network.

1.2 Contributions

From above discussions, in order to make precise recommendation in academic networks, we propose a latent factor model especially incorporating the content and attributes information. We take both information into consideration as they are good indicators of user's preferences, and try to avoid the partial and biased recommendation. Our proposed model can automatically optimize the contribution of content and attributes to the final recommendation, involving users social networks as well to resolve the cold start problem of new users. The generalized recommendation model aims to overcome the drawbacks of former homogeneous recommendation methodologies.

The rest of the paper is organized as follows: Section 2 discusses the related work. Section 3 introduces our models and Section 4 presents the experimental studies. Finally we conclude this paper.

2   Related Works

Traditional recommender system usually adopted two techniques: collaborative filtering (CF) and content based filtering, with explicit data. Later, CF used classic matrix factorization methods such as singular value decomposition (SVD) to achieve low-rank approximation on minimizing the sum-squared distance. In [4], probabilistic matrix factorization (PMF) model was proposed to handle the large, sparse, and very imbalanced utility matrix with good scalability and performance, which can overcome the overfitting problem of SVD.

Based on the framework of matrix factorization, there are some extensions to improve the recommendation results. Among them, one direction is the model completion such as fully Bayesian treatment [3], which automatically controls all the model parameters and hyperparameters. Others [9, 10] further resort to Bayesian nonparametric to bypass the model selection. But these methods still cannot resolve the cold start problem as no auxiliary information is considered, so we do not discuss or compare them in the following sections.

Another direction is to adopt auxiliary information such as social networks or item content. [5–7] incorporate social networks for social recommendation, based on the assumption that users should have close interests with their friends in the social network. [6] introduces the social regulation to conform user i to the friend f, through an individual based regularization term $\|p_i - p_f\|^2$. [7] proposes the friendship-interest propagation (FIP) mode that utilizes the inner product $(p_i^T p_f)$ between two users i and f. As indicated in [11], existing approaches have largely ignored the heterogeneity and diversity of the networks. It is not equivalent to say that preferences between friends should be similar even though they share certain interests.

[12] proposes fLDA to combine matrix factorization with topic modeling, which lets latent item vector and topic assignments, as well as latent user factor, contribute the rating prediction. [8] introduces collaborative topic regression (CTR) which captures item's content in latent space and derives the latent item vector through the topic proportion vector with Gaussian noise. Both models provide a good solution incorporating item's content in latent space, but cannot satisfy the requirements in dealing with the complex situations discussed in section 1.1. The major improvement in this paper is to further leverage the

descriptive attributes into the latent factor model and automatically tune their contributions to the latent item vectors, not with fixed settings in CTR.

3   Model

The Formulation of Recommendation Task in Academic Networks.
The recommender system has several input: (1) I users and J items with the user-item utility matrix R, in which each element $r_{ij} \in \{0, 1\}$ indicates user i's preference to item j. $r_{ij} = 1$ means the user rates the item (publish or cite it) and $r_{ij} = 0$ means the user does not rate the item (dislikes or unknown); (2) items' content and attributes, which would be summarized into K topics and L aspects respectively using topic modeling; (3) the social network $G = (V, E)$, where $(i, f) \in E$, indicate user i and f are linked. For each user, the task is to recommend scientific papers that are not rated by this user before. We assume that the latent topics and latent aspects for each item are obtained by topic modeling and we can represent users and items in the latent low-dimensional space of dimension $D = K + L$, with latent user vector $p_i \in \mathbb{R}^D$ and latent item vector $q_j \in \mathbb{R}^D$ through matrix factorization. The prediction $\hat{r}_{ij}$ represents that user i likes item j or not with inner product in their latent space $\hat{r}_{ij} = p_i^T q_j$.

Observed ratings in utility matrix are involved in a supervised approach to minimize the regularized squared error loss respect to $P = (p_i)_{i=1}^I$ and $Q = (q_j)_{j=1}^J$:

$$min_{P,Q} \sum_{i,j} \frac{c_{ij}}{2} (r_{ij} - p_i^T q_j)^2 + \frac{\lambda_p}{2} \sum_i \|p_i\|^2 + \frac{\lambda_q}{2} \sum_j \|q_j\|^2 \qquad (1)$$

where c_{ij} is a binary indicator that is equal to 1 if user i rated item j and equal to 0 otherwise.

In PMF [4], the matrix factorization is generalized as a probabilistic model, where latent user vector $p_i \sim \mathcal{N}(0, \lambda_p^{-1} I_D)$, latent item vector $q_j \sim \mathcal{N}(0, \lambda_q^{-1} I_D)$ and user-item rating $r_{ij} \sim \mathcal{N}(p_i^T q_j, c_{ij}^{-1})$. A local minimum of Eq. (1) can be achieved by applying gradient descent algorithm in P and Q. The final results can be used to predict user i's preference on item j by $\hat{r}_{ij}$. The disadvantage of PMF is that it cannot deal with the cold start problem.

In CTR [8], the latent topic vector θ_j is incorporated into PMF framework and the latent item vector is further confined by setting $q_j \sim \mathcal{N}(\theta_j, \lambda_q^{-1} I_D)$. CTR can address the cold start problem of new items, but is not fit for the feature information on items.

In this paper, we focus on the heterogeneity of content and attribute information of items and build a recommendation system to combine these two sides into a uniform model. We first propose a probabilistic topic model to process the item information, then introduce two naive solutions to import attribute information and later propose our Content + Attributes model to achieve better performances.

3.1   Item Information Processing

Each item has its unstructured information (i.e., content) and structured information (i.e., attributes). Both of them are useful in recommending items to users as discussed in the introduction. Usually, every item may contain hundreds of words in content and several attributes, and the item set may overall contain tens of thousands of words in vocabulary and attribute values. So we have to seek for an unsupervised method to reduce them to a low dimension.

For content information, the topic modeling methods such as LDA [13] can be used to achieve the goal of dimension reduction. The intuition behind LDA is that documents exhibit multiple topics which are represented by distributions over words. For paper dataset, the content of each item consists of title and abstract, which is a probabilistic mixture of latent topics. Refer to [13] for more details.

An attribute can be numerical (i.e., year) or categorical (i.e., venue and author). The values of categorical attributes can be treated as words in content, while the values of numerical attributes need to be processed in order to avoid the sparse problem. For example, the attribute "year" in a paper dataset can be ranged from late 20th century to now and each paper only has one value for this attribute. If we take a separated view on it, the connection of papers sharing the same topic but publishing in adjacent years would be concealed. It is also probable that the number of published papers is skewed on "year", much more in some years while much less in other years. So we need to partition the numerical attributes to ensure the effectiveness. The values in the attribute "year" are divided into four classes: (2005-now), (2000-2004), (1995-1999) and (before 1995) from newly published to long before. After such transforming, we can adopt the LDA method on attributes to obtain the interpretable low dimensional representation - latent aspect vectors for attributes.

After item information processing, we can reduce content and attributes into a low dimensional space and apply them into the matrix factorization discussed in the following sections.

3.2   First Cut Solutions

This section introduces two first cut solutions to deal with the additional attribute information in the recommendation systems. Both methods leverage the attributes of items to complement the only content based method.

The first method is **weighted attributes method**, which builds two different recommenders on content and attributes separately. Based on the content vector and attribute vector learned by LDA, weighted attributes method applies CTR to obtain two predicted ratings from content part and attribute part, denoted as $\hat{r}_{ij}^{(c)}$ and $\hat{r}_{ij}^{(a)}$. A weight $w \in [0,1]$ is chosen to adjust the contribution of content and attributes, so the final rating of user is preference on item j is:

$$\hat{r}_{ij} = (1-w)\hat{r}_{ij}^{(c)} + w\hat{r}_{ij}^{(a)} \tag{2}$$

The second method is **overwhelming attributes method**, which treats the attribute values as single word in content. This method merges the attribute values and words, and then adopts LDA to obtain the latent topic vectors mixed with attribute information. CTR is applied based on these latent vectors to figure out the final ratings. Compared to the words in content, the number of attribute values is very little, so the attribute values are largely overwhelmed by words, leading to a limited improvement.

3.3   Content + Attributes Model

This section introduces our Content + Attributes model to address the heterogeneity issue in recommendation systems. The model has loose coupling and integrality on content and attributes, as well as self-adaptation to latent item vectors. This model can offer more precious recommendations to users by considering both content and attributes.

Note that we have obtained the K-dimensional latent topic vectors for content and the L-dimensional latent aspect vectors for attributes, we should extend these vectors into a D-dimensional space ($D = K + L$). In our model, θ_j represents the latent topic vector and γ_j represents the latent aspect vector for item j where only first K elements of θ_j keep the topic information and only last L elements of θ_j keep the aspect information defined as follows:

$$\theta_j =< \theta_j(1), ..., \theta_j(K), 0(1), ..., 0(L) > \tag{3}$$

$$\gamma_j =< 0(1), ..., 0(K), \gamma_j(1), ..., \gamma_j(L) > \tag{4}$$

After the above transformations, we now describe the generative process for observed ratings of the utility matrix in Content + Attributes model:

1. For each user i, draw latent user vector $p_i \sim \mathcal{N}(0, \lambda_p^{-1} I_D)$;
2. For each item j, draw latent item offset $\varepsilon_j \sim \mathcal{N}(0, \lambda_q^{-1} I_D)$, and set latent item vector $q_j = \varepsilon_j + \tau(\theta_j + \gamma_j)$;
3. For each user-item (i, j), draw the rating $r_{ij} \sim \mathcal{N}(p_i^T q_j, 1)$.

Note that the latent item vector $q_j = \varepsilon_j + \tau(\theta_j + \gamma_j)$ indicates q_j should be close to scaled $(\theta_j + \gamma_j)$, and τ is the adjustment factor to adapt the importance of $(\theta_j + \gamma_j)$ to q_j

So after combining adjusted latent topic vector for content and latent aspect vector for attribute, we modify Eq. (1) to follow our generative process and set our goal of object function that is to minimize E given such variables and parameters as follows:

$$E = \sum_{i,j} \frac{c_{ij}}{2}(r_{ij} - p_i^T q_j)^2 + \frac{\lambda_p}{2} \sum_i p_i^T p_i + \frac{\lambda_q}{2} \sum_j (q_j - \tau(\theta_j + \gamma_j))^T (q_j - \tau(\theta_j + \gamma_j))$$

$$\tag{5}$$

A local minimum can be achieved by iteratively applying gradient descent method. In each iteration, first take the gradient of respect to variable p_i, q_j and τ:

$$\frac{\partial E}{\partial p_i} = \lambda_p p_i - \sum_j (r_{ij} - p_i^T q_j) q_j \tag{6}$$

$$\frac{\partial E}{\partial q_j} = \lambda_q q_j - \lambda_q \tau (\theta_j + \gamma_j) - \sum_i (r_{ij} - p_i^T q_j) p_i \tag{7}$$

$$\frac{\partial E}{\partial \tau} = -\lambda_q \sum_j (\theta_j^T + \gamma_j^T)(q_j - \tau(\theta_j + \gamma_j)) \tag{8}$$

Then update each variable by taking steps proportional to the negative of the gradient based on recent values:

$$p_i^{t+1} = p_i^t - \eta \frac{\partial E}{\partial p_i^t} \quad q_j^{t+1} = q_j^t - \eta \frac{\partial E}{\partial q_j^t} \quad \tau^{t+1} = \tau^t - \eta \frac{\partial E}{\partial \tau^t} \tag{9}$$

where η is a parameter called learning rate and p_i^t, q_j^t, τ^t stand for the value of p_i, q_j and τ at iteration t.

We can obtain the final results after enough iterations until the above equations (9) reach the convergence. Similar to the previous methods, we can use the latent user vector p_i and latent item vector q_j to predict user is rating value on item j: $\hat{r}_{ij} = p_i^T q_j$.

The Content + Attributes model introduces adjustment factor τ to automatically optimize the contribution of content and attributes to latent item vectors, because the closeness of $(\theta_j + \gamma_j)$ to q_j is unknown. Unlike CTR that fixes $q_j \approx \theta_j$, we use τ to control the scale of vectors and achieve better predictions.

3.4   Cold Start Resolution

As the cold start problem for new items is easily addressed using content or attribute information mentioned in [8]. This section mainly emphasizes the cold start problem for new users. In social networks, users are not isolated; in other words, they are connected according to their common interests. So we can incorporate social relationships to resolve the cold start problem of new users.

From the above model, we have learnt the latent user vectors for the existed users who have ratings for some items. Now if the recommender system has to predict the preference for a new user i, although he has no available ratings or other information, we can adopt his friends (i.e., co-authors in academic networks) who have ratings to substitute for the latent user vector of i as follows:

$$p_i = \frac{1}{|\mathcal{F}(i)|} \sum_{k \in \mathcal{F}(i)} p_k \tag{10}$$

where $\mathcal{F}(i)$ denotes the set of user i's friends. The above conclusions rely on the fact that, the interest of a user should be close to his friends. In other words, latent user vector of a new user can be regularized by those of friends. This trend is often detected as homophily phenomena in social networks.

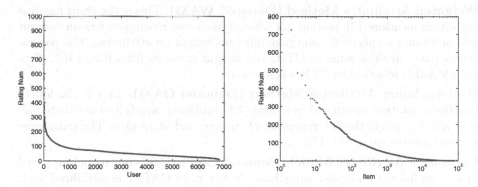

Fig. 2. Rating distributions of the dataset. The left subfigure shows the number of ratings by each user and the right subfigures shows the the number of ratings on each item (logistic scaled), all desecnt sorted.

4   Experimental Evaluation

4.1   Datasets

The original dataset[1] is from DBLP, containing over 1.5 million item (papers) and around 700 thousand users (authors). We preprocess this dataset to select those items with complete content (i.e., title and abstract) and attributes (i.e., author, venue and year), to remove the users with fewer than 10 papers. The final dataset consists of 6815 users and 78475 items with 436704 user-item ratings. The rating distribution of the processed dataset is exhibited in Fig. 2, which is following the power law distribution. We randomly select 10% percent of ratings from 1600 users who have more than 80 ratings, i.e., 26328 user-item ratings (approximately 6% of total ratings) as testing set withheld in model learning and use others as training set for learning the latent vectors. We conduct validations on the testing set and report the averaged results of 5 repeated experiments to measure the performance of different methods.

4.2   Methods

Followings are five methods compared in the experiments all with the same learning rate $\eta = 0.001$:

Probabilistic Matrix Factorization (Denoted PMF): This is the first baseline method only adopting matrix factorization on utility matrix. PMF is widely used in the collaborative filtering community. Following [4], we set the parameters $\lambda_p = \lambda_q = 0.01$.

Collaborative Topic Regression (Denoted CTR): This is the second baseline method incorporating topic model into matrix factorization. Following [8], we set the parameters $\lambda_p = \lambda_q = 0.01, \alpha = {}^{50}/\kappa$ and $\beta = 0.01$.

---

[1] http://arnetminer.org/citation

Weighted Attributes Method (Denoted WAM): This is the third baseline algorithm mentioned in section 3.2 which builds two recommenders on content and attributes respectively, and take different weight on attributes. The parameter settings are the same as CTR, and weight w varies from 0.1 to 0.5. Note that WAM is identical to CTR when $w = 0$.

Overwhelming Attributes Method (Denoted OAM): This is the fourth baseline algorithm mentioned in section 3.2 combining words and attribute values, which neglects the heterogeneity of content and attributes. The parameter settings are the same as CTR.

Content + Attributes Model (Denoted CAT): This is our model proposed in section 3.3 with adapted adjustment factor τ. In CAT, τ is initialized as 1. Other parameter settings are the same as CTR.

4.3 Metrics

We adopt the following metrics to evaluate the performance of different methods.

RMSE (root mean squared error) quantifies the difference between rating values implied by a recommender and the true values in the training/testing set. This metric is defined as follows: $RMSE = \sqrt{\frac{1}{N} \sum_{i,j} (r_{ij} - \hat{r}_{ij})^2}$ where r_{ij} is the true rating value, $\hat{r}_{ij}$ is the predicted rating value and N is the number of ratings in the testing set.

Coverage indicates the retrieval ratio from total ratings in the testing set using a transforming function $\sigma(x)$, which classifies the estimated rating values into "likes" and "dislikes" by a threshold: $Coverage = 1 - \frac{1}{N} \sum_{i,j} (r_{ij} - \sigma(\hat{r}_{ij}))$ where $\sigma(x) = \begin{cases} 1 & \text{if } x \geq 0.5 \\ 0 & \text{if } x < 0.5 \end{cases}$.

Recall@k quantifies the fraction of rated items that are in the top-k of the ranking list sorted by their estimated ratings, denoted by $N(k; u)$, from among all rated items $N(u)$ in the test set. For each user u: $Recall@k = \frac{N(k; u)}{N(u)}$. The metric is designed in the condition that a high coverage may be biased due to excessive estimated rating values.

We report average RMSE, coverage and recall over the whole testing set.

4.4 Results

We vary the number of dimension $D \in \{10, 20, 40, 80\}$, when the dimension is low or high, the matrix factorization cannot converge well on RMSE of training set due to less features or overfitting problem. So we report the results of best performance at $D = 20$. The mean rating values of the predicted utility matrix is 0.2-0.3, which would not lead to a biased coverage.

The experiment results for RMSE and coverage are shown in Table 1. Obviously, the conventional PMF without considering any auxiliary information performs worst. Two naive solutions combining attributes information work slightly

Table 1. RMSE and coverage results of different methods. Lower values on RMSE and higher values on coverage are better. RMSE on training set and testing set are shown respectively.

Models	RMSE train	RMSE test	Coverage
PMF	0.1554 ± 0.0013	0.3077 ± 0.0019	0.8810 ± 0.0034
CTR	0.1214 ± 0.0016	0.2775 ± 0.0020	0.9090 ± 0.0042
WAM	0.1193 ± 0.0020	0.2691 ± 0.0014	0.9158 ± 0.0033
OAM	$\mathbf{0.1190} \pm 0.0019$	0.2700 ± 0.0022	0.9157 ± 0.0031
CAT	$\mathbf{0.1190} \pm 0.0014$	$\mathbf{0.2580} \pm 0.0017$	$\mathbf{0.9237} \pm 0.0038$

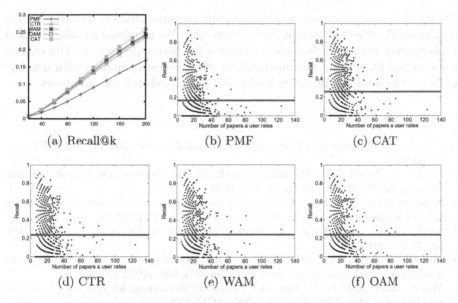

(a) Recall@k   (b) PMF   (c) CAT

(d) CTR   (e) WAM   (f) OAM

Fig. 3. Recall performance of different methods. The left top subfigure shows overall performance of different methods while varying k. The rest subfigures exhibit the individual level of performance each method while fixing k = 200.

better than CTR, which only considers the content information. As discussed in Section 1.1, these solutions lost specific patterns represented by attributes. In contrast, CAT incorporates both content and attributes into a uniform model which automatically adapts the contribution to the final latent item vectors. As pointed in [14], achievable RMSE values lie in a quite compressed range and small improvements in RMSE terms can have a significant impact on the quality of the top few presented recommendations. Our CAT model achieves a 16% and 7% improvement over PMF and CTR on RMSE.

The experiment results for recall are shown in Fig. 3, with an overall subfigure and five individual level subfigures (each dot represents a user and the red line reports the average) for each method. We observe that CTR, WAM and OAM perform very close, still due to the inappropriate usage of attributes. CAT model performs best on recall, with a 53% and 10% improvement over PMF and CTR, as well as more dense points above the average line.

For cold start testing, we randomly pick up 500 users as cold users and withhold their ratings as testing set. Following Eq. (10), we use the normalized latent vector according to the friends of the cold user for prediction. Still, CAT achieves the best coverage of 0.9051, followed by OAM (0.8733), WAM (0.8619), CTR (0.8613) and PMF (0.8348). This result further demonstrates the effectiveness of latent vectors learnt by the uniform model CAT.

5   Conclusion

This paper mainly discussed a heterogeneous recommendation scenario in academic networks where abundant information exist. We proposed a uniform model to incorporate item's content and attributes for better prediction. The experiments on real-life dataset demonstrated its effectiveness compared with baseline methods. The model parameters learnt also performed well for cold users.

References

1. Marlin, B.: Modeling user rating profiles for collaborative filtering. In: NIPS, vol. 16 (2003)
2. Marlin, B., Zemel, R.S.: The multiple multiplicative factor model for collaborative filtering. In: ICML, pp. 73–80. ACM (2004)
3. Salakhutdinov, R., Mnih, A.: Bayesian probabilistic matrix factorization using markov chain monte carlo. In: ICML, pp. 880–887. ACM (2008)
4. Salakhutdinov, R., Mnih, A.: Probabilistic matrix factorization. In: NIPS, pp. 1257–1264 (2008)
5. Ma, H., Yang, H., Lyu, M.R., King, I.: Sorec: social recommendation using probabilistic matrix factorization. In: CIKM, pp. 931–940. ACM (2008)
6. Ma, H., Zhou, D., Liu, C., Lyu, M.R., King, I.: Recommender systems with social regularization. In: WSDM, pp. 287–296. ACM (2011)
7. Yang, S.H., Long, B., Smola, A., Sadagopan, N., Zheng, Z., Zha, H.: Like like alike: joint friendship and interest propagation in social networks. In: WWW, pp. 537–546. ACM (2011)
8. Wang, C., Blei, D.M.: Collaborative topic modeling for recommending scientific articles. In: KDD, pp. 448–456. ACM (2011)
9. Hoffman, M.D., Blei, D.M., Cook, P.R.: Bayesian nonparametric matrix factorization for recorded music. In: Proc. ICML, pp. 439–446 (2010)
10. Wood, F., Griffiths, T.L.: Particle filtering for nonparametric bayesian matrix factorization. In: NIPS, vol. 19, pp. 1513–1520 (2007)
11. Shen, Y., Jin, R.: Learning personal+ social latent factor model for social recommendation. In: KDD, pp. 1303–1311. ACM (2012)
12. Agarwal, D., Chen, B.C.: flda: matrix factorization through latent dirichlet allocation. In: WSDM, pp. 91–100. ACM (2010)
13. Blei, D.M., Ng, A.Y., Jordan, M.I.: Latent dirichlet allocation. The Journal of Machine Learning Research 3, 993–1022 (2003)
14. Koren, Y.: Factorization meets the neighborhood: a multifaceted collaborative filtering model. In: KDD, pp. 426–434. ACM (2008)

Real-Time News Recommendation
Using Context-Aware Ensembles

Andreas Lommatzsch

Technische Universität Berlin, DAI-Lab
Ernst-Reuter-Platz 7, 10587 Berlin, Germany
andreas.lommatzsch@dai-labor.de

Abstract. With the rapidly growing amount of items and news articles on the internet, recommender systems are one of the key technologies to cope with the information overload and to assist users in finding information matching the their individual preferences. News and domain-specific information portals are important knowledge sources on the Web frequently accessed by millions of users. In contrast to product recommender systems, news recommender systems must address additional challenges, e.g. short news article lifecycles, heterogonous user interests, strict time constraints, and context-dependent article relevance. Since news articles have only a short time to live, recommender models have to be continuously adapted, ensuring that the recommendations are always up-to-date, hampering the pre-computations of suggestions. In this paper we present our framework for providing real-time news recommendations. We discuss the implemented algorithms optimized for the news domain and present an approach for estimating the recommender performance. Based on our analysis we implement an agent-based recommender system, aggregation several different recommender strategies. We learn a context-aware delegation strategy, allowing us to select the best recommender algorithm for each request. The evaluation shows that the implemented framework outperforms traditional recommender approaches and allows us to adapt to the specific properties of the considered news portals and recommendation requests.

Keywords: real-time recommendations, online evaluation, context-aware ensemble.

1 Introduction

Recommender systems support users in finding information or objects potentially relevant according to the user's preferences. Recommender systems are usually tailored to scenarios characterized by a rarely changing set of items (e.g., movies or books) and users, clearly identified by userIDs and user profiles. In this paper we develop a system optimized for recommending news articles. Our system differs from traditional product recommender systems due to the special requirements of the news domain: News articles have a short time-to-live; it is difficult to identify the users (the users usually have neither an account nor a profile); the user's interests strongly depend on the respective news portal and the information domain (politics, sports, etc.). In addition, the relevance of news articles depends on the context, e.g. time, day of week (working day/weekend),

M. de Rijke et al. (Eds.): ECIR 2014, LNCS 8416, pp. 51–62, 2014.

or the used device. Thus, a flexible recommender framework is required, able to adapt to the continuously changing news and able to consider changing user's preferences and the respective contexts. The news recommendations should be provided in real-time.

The remaining paper is organized as follows: Sec. 2 describes the analyzed scenario, the benchmark data, and the evaluation measures in detail. In Sec. 3, we present our approach and explain the implemented recommender algorithms and ensemble strategies. The evaluation results are discussed in Sec. 4. Subsequently, Sec. 5 presents related research; finally, a summary and an outlook on future work are given in Sec. 6.

2 Problem Description

We analyze the task of recommending news articles. In order to evaluate our framework in real-world conditions, we cooperate with a big German company, providing recommendations for many German news portals. PLISTA is the leading recommendation and advertising network in the German-speaking market. PLISTA collects user interactions from websites and provides recommendations that are integrated into the website. The cooperation allows us to perform an online as well as an offline evaluation, testing our algorithms in a real-world environment.

2.1 The PLISTA-Contest

The PLISTA contest[1] offers researchers the chance to test algorithms in a real-world news recommendation scenario.

Fig. 1. The figure visualizes the placement of news recommendation around a new article

Scenario Description. The task in the PLISTA contest is to provide suggestions that are placed around a news article on a web page.

Fig. 1 schematically shows the presentation of a news article. Below the article text and on the right side there are typically slots that should be filled with recommendations. Dependent from the design of the respective web page one to six articles must be provided. The median number of requested suggestions is four. Since the number of requested suggestions may influence the placement on the web page, a recommender should consider the number of requested recommendations.

The Recommendation Process. The course of interaction in the contest is visualized in Fig. 2. When a user visits a website participating in the contest, the PLISTA server randomly selects a recommender team and forwards the request to the respective recommender algorithm. The recommender algorithm must provide up to six suggestions (dependent from the request). A request must be answered within 100ms including the communication time. The PLISTA server renders the recommendations and embeds them in the website requested by the user. The performance of

---

[1] http://orp.plista.com

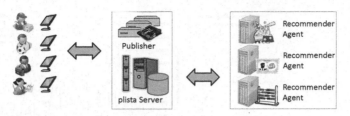

Fig. 2. The figure visualizes the communication between the users, the PLISTA server, and the recommender algorithms

the teams participating in the contest is measured by the number of suggestions clicked by the user.

There are several challenges for the participating teams. On the one hand each team receives only a certain percentage of requests. On the other hand, the news portals do not force the users to login but rely on web tracking techniques (e.g. cookies) for identifying users. Thus, the knowledge about the user's behavior is limited and noisy forcing the recommender algorithms to cope with uncertainty about already seen articles. Other challenges are the strict time limits for answering recommendation requests, the short lifecycle of news articles, and the heterogeneity of the portals for that recommendations must be provided.

2.2 The Dataset Properties

In the PLISTA contest there are three main types of messages used in the communication between the PLISTA server and the participating teams.

Impression events describe that a user visits a news article. Impression events are annotated with a userID, an articleID, the domainID (identifying the news portal), the news article text (title and the abstract), a timestamp, and several other features provided by the tracking component (e.g., browser or operating system).

Recommendation requests contain information about the article for which recommendations must be provided (similar to impression events). In addition, the request defines how many recommendations must be delivered.

Click Events describe which recommendations are clicked by users. Click events are provided to all teams even if a team was not responsible for the clicked recommendation.

Analyzing the stream of impression events, we find characteristic pattern. The left part of Fig. 3 visualizes the amount of impressions for a working day compared with the number of impressions for a Sunday. The graphs show that there is a variance in the number of impressions over one day. On working days the heaviest load is observed in the morning. On weekends the heavy use of the portals starts two hours later and the number of impressions is much steadier than on a working day.

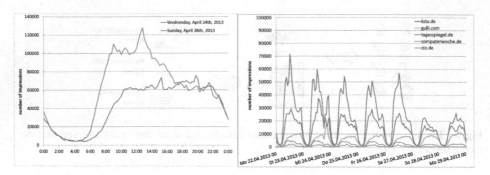

Fig. 3. The figure visualized the number of impressions in the news recommendation scenario

2.3   Evaluation Measures

The performance of the recommender algorithms are measured in an online and an *near-to-online* evaluation.

The **online** ("live") evaluation is realized by providing recommendations in real-time. When the user requests an article from the news portal, a recommendation request is created and the suggestions are embedded into the news article web page. The performance of the recommender is computed by counting the number of clicks on the suggestions. The click-through-rate (CTR) defines the proportion of clicked suggestions to the total number of presented suggestions. The click-through rate is usually very low ($\approx 0.5\%$), due to the fact that most users are only interested in the news and do not pay any attentions to the presented recommendations. The CTR highly depends on the portal ("domain"), since this affects the placement of the recommendations on the web page, the habits of the user groups and the time [7]. Thus, the context of the CTR measurement must be taken into account when interpreting the observed click through rates. In general, there are several "hidden" factors that result in "noise". Thus, the CTR should be seen as a probabilistic measure that should be aggregated over an adequate time span to get statistically significant results.

In the **near-to-online** evaluation we analyze how precise a recommender algorithm can predict the clicks of a user. Therefor we analyze the stream of click events provided by the PLISTA server. When a click event is received, we send a request to one (or more) recommender algorithms and ask for predicting what articles the user has clicked. The predicted article lists (of one to six articles) are checked against the click observed in the online evaluation. We compute the prediction precision [4] by counting an answer as correct if the list of suggestions contained the correct article. Based on this definition, the expected precision grows with the number of suggestions in the predicted list. In contrast to the online evaluation, the *near-to-online* evaluation allows us to benchmark several different algorithms in parallel enabling us to efficiently measure the performance of different algorithms and parameter settings.

Summary. The main goal in the PLISTA contest is improving the recommendation performance in the online evaluation. The *near-to-online* evaluation allows us to efficiently analyze the behavior of different recommender approaches in parallel. Our observations can be used for improving the online recommendations.

2.4 Challenges

Recommending online news articles raises several interesting research challenges. In contrast to most traditional recommender systems the lifecycle of items is short and the users are difficult to identify. The set of recommendable items is continuously changing forcing the recommender system to constantly adapt to the current changes. Dependent from the season, popular events, and the respective domain the number of articles varies greatly. Moreover, the suggestions must be provided in real-time (strict time constraints).

The objective is to create a recommender system able to meet the strict time constrains and providing highly relevant suggestions. The system must be robust and able to adapt to the specific properties of different domains. In addition, the recommender system should be able to consider the context and be able to learn from user feedback.

2.5 Discussion

In this section we have introduced the PLISTA contest. We discuss the different types of request and messages used by the recommender clients to communicate with the contest server. An important requirement in the scenario is that the recommendation requests have to be answered in less than 100ms. The statistical analysis of the dataset shows that the service usage (measure by the number of impressions) highly depends on the context (time, type of the portal). Thus, our recommender framework must be able to consider the diversity in the relevant contexts.

3 A Real-Time News Recommender Framework

We develop an agent-based recommender framework that integrates different recommender algorithms. The intelligent combination of different recommender algorithms allows us to cope with the heterogeneity of news portals and the changes in user preferences during a day. A manager agent analyzes the incoming requests, forwards the requests to appropriate agents and decides what results should be returned to the user. Therefore, the manager agent continuously evaluates the performance of the recommender algorithms (*near-to-online*) and learns a strategy for selecting the best-suited recommender algorithm for a new request. We implement different recommender algorithms and combine these recommender algorithms in ensembles.

3.1 Algorithms Optimized for Recommending News

At first we discuss different recommender algorithms optimized for the news recommendation scenario. Subsequently, we discuss the approaches for creating ensembles based on implemented recommender algorithms.

Recommending the Most Popular Articles. Items liked by most users, but still unknown to the current user are potentially a good recommendation. In order to implement a "most-popular" recommender strategy, we count for each article, how often each article is read by the users. We define a maximal "age" for the recommendable articles in order to ensure that the recommendations are not too old. When recommendations are requested, the algorithm computes the most frequently read articles and filters out the articles already read by the user.

The advantage of a most popular recommender approach is that the algorithm can be efficiently implemented (using sorted skip-lists). The suggestions do not consider the individual user preferences but relying on the "taste of the crowd".

Recommendations based on the Most Popular Sequence. The most-popular recommender discussed in the previous paragraph does not consider articles currently read by the user. In order to overcome this problem, we count for each article how often other articles are requested. This means, we create a "most-popular navigation sequence" statistic allowing us to predict the most popular article taking into account the article currently requested by the user.

In our implementation we limit the statistic to 10,000 news articles (using a LRU cache) and to 1,000 articles that can be reached from each of these 10,000 articles. The algorithm computes recommendations taking into account the currently requested article and the "knowledge of the crowd". It does not consider individual user preferences.

User-based Collaborative Filtering. One of the most used recommender algorithms is user-based collaborative filtering [10]. The algorithm follows the idea that users who interacted with the same items in the past, tend to have similar interests. An item that user U_1 liked is potentially interesting for user U_2 if user U_1 and U_2 showed similar interests in the past.

In order to implement user-based collaborative filtering, we store all user-item interactions in a LRU-cache limiting the number of considered interactions to 50,000. When a request is received, the recommender computes the must similar users (based on the stored user-item interactions) and determines the most popular items of these users. From these potentially relevant items the items already known to the user are filtered out. The algorithm provides personalized recommendations taking into account the taste of similar users. Unfortunately, the provided recommendations are not strongly related to the news article currently requested by the user.

Item-based collaborative filtering. An alternative approach to user-based collaborative filtering is item-based collaborative filtering. The algorithm follows the idea, that items are similar if they have been read by the same users. The score describing the similarity of two articles A and B is computed by the number of users that have accessed both articles A and B.

Analogue to the user-based collaborative recommender implementation, the item-based CF recommender stores the user-item interactions in a LRU-cache. The number of considered interactions is limited to most recent 50,000 interactions. When a request is received, the recommender computes the news articles most similar to the current article (based on the stored user-item interactions). From these articles the items already known to the user are filtered out. The algorithm does not provide personalized recommendations, but suggests articles relevant to the currently read article.

Content-based collaborative filtering. In the analyzed news recommendation scenario the shortened version of the news article text is part of the request. This allows us to compute the similarity of articles based on the article text. In order to reduce to complexity and to improve the reliability of the computed similarity score, linguistic techniques (stemming, stop word removal) are applied. The similarity of the news articles is computed based on the stem overlap between two articles.

When the content-based recommender is asked to provide recommendations for a request, the algorithm determines the articles most similar to the current article based on the content-based similarity ("stem overlap"). The articles with the highest similarity, still unknown to the user and not older than a predefined date (typically one day) are returned to answer the recommendation request.

Discussion. In this section we introduced the basic concepts of the different recommender algorithms used by our news recommender system. These algorithms are optimized to the news scenario, by means of taking into account, that news articles are relevant only for a short time span. In addition, the algorithms can handle new articles efficiently. This is achieved by focusing on memory-based algorithms that do not require batch updates of the internal recommender models.

As discussed, all the algorithms have specific strengths and weaknesses. That is the reason why we analyze approaches for combining different recommender algorithms (working based on different algorithms or parameter configurations). Ensembles enable us selecting the best recommender agent based on the context and the respective request properties.

3.2 Ensemble-Based Approaches

In our scenario, one challenge is the heterogeneity in the data and the diversity of news portals. That is why we do not rely on one single algorithm but analyze ensemble strategies enabling the context-aware delegation of requests and the combination of different algorithms. In this section we introduce our system architecture and explain the strategies for delegating requests to the best-suited recommender algorithm.

System Architecture. Instead of a single recommender we use an agent-ensemble. In order to take into account the differences between the news portals in the PLISTA contest, we define for each portal a group of agents. Each group is controlled by a manager agent and handles the requests for one news portal ("domain-based delegation").

Each group consists of several recommender agent using different algorithms or parameter settings. The manager agent observes the performance of each recommender agent and aggregates the recommendation precision according to context-specific criteria (e.g., hour of day, number of demanded recommendations in a request). Based on the aggregated performance data the manager agent selects the best recommender agent for the incoming requests considering the respective contexts. A simplified visualization of the system architecture is shown in Fig. 4.

Benchmarking the Recommender Agents. The success of the delegation strategy strongly depends on the question how the performance of the recommender agents can be measured. We decided to use the *near-to-online* precision (Sec. 2.3). This approach has several advantages: Many different recommender agents can be evaluated based on one PLISTA contest feedback message. Thus, new algorithms can be added and benchmarked together with already existing agents enabling the investigation of new recommender strategies. In addition, feedback message from all teams in the competition can be used for benchmarking speeding up the collection of a significant number of

58   A. Lommatzsch

feedback messages. The disadvantage of the *near-to-online* evaluation is that the evaluation is biased towards the popular algorithms: Since positive feedback is always the result of a recommendation, the algorithms selected for placing the results on a news article web page define what set of articles the user can click on. That means, that the algorithms used by other teams in the contest, influence the measure *near-to-online* performance. Due to the fact, there is nothing known about the algorithms used by other teams, the feedback (caused by recommendations from other teams) helps exploring new aspects. $\approx 95\%$ of the feedback messages are based on recommendations from other teams.

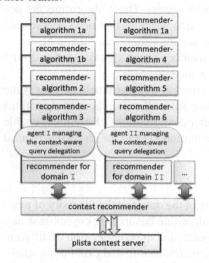

Fig. 4. The system architecture for the context-aware query delegation

Context-aware delegation. The manager agent for each domain aggregates the measured performance of the recommender agents. We analyze the following criteria for classifying (aggregating) the request: hour of the day, day of the week, number of requested recommendations, how often does the user visit the news portal. The aggregation must ensure that a significant number of requests is assigned to every context criteria preferring more general context properties. On the other hand, the criteria must be fine-grained enough to reflect the different demands of the respective requests.

Discussion. The developed framework for the context-aware request delegation provides an open platform for integrating different recommender algorithms. The framework benchmarks the recommender algorithms using the *near-to-online* recommendation precision, aggregates the results according to different context criteria, and learns which algorithms performs best in what context. The main advantages are that many different algorithms can be benchmarked in parallel, different criteria for delegating the requests can be analyzed simultaneously, and the ensemble-based system continuously adapts to the changes in the user's behavior.

4   Evaluation

We analyze the performance of the developed framework in an online as well as in a *near-to-online* evaluation.

Online Evaluation. Fig. 5 shows the Click-Through-Rate for the teams participated in the contest. The analysis shows that our teams were only outperformed by a team from PLISTA. This shows that the context-aware combination of standard recommender algorithms enables us to provide highly relevant recommendations.

Our three teams are implemented based on the same recommender algorithms but differ in the strategy used for delegating the requests.

Team A considers the day of the week and the number of requested recommendations. Internally the delegation distinguishes between requests asking for one suggestion and

CTR	week 38	week 39	week 40	week 41	Average
team name	Sep 16 - Sep 22 2013	Sep 23 - Sep 29 2013	Sep 30 - Oct 06 2013	Oct 07 - Oct 13 2013	CTR
plista GmbH	0.92%	1.64%	1.94%	1.78%	1.57%
MyTeam A	1.22%	1.32%	1.50%	1.26%	1.33%
MyTeam B	1.19%	1.32%	1.44%	1.32%	1.32%
MyTeam C	1.18%	1.25%	1.39%	1.20%	1.26%
recommenders.net	1.00%	0.97%	1.08%	0.92%	0.99%
inbeat	0.85%	0.91%	1.06%	1.06%	0.97%
riemannzeta	0.90%	0.91%	1.07%	0.97%	0.96%
comit-rainbow	0.88%	0.87%	1.11%	0.95%	0.95%

Fig. 5. The figure shows the click-through rate of the teams in the PLISTA-contest. The performance was been measured online, based on real user interactions. The data are taken from the contest's leaderboard. (http://orp.plista.com/leaderboard)

requests asking for more than one suggestion. Thus $2 \times 7 = 14$ states for describing the context are used by each manager agent (responsible for one domain).

Team B uses a trend-based delegation strategy. We analyze the *near-to-online* performance of the recommenders for the last 25 feedback messages and select the most successful recommender algorithm. This is based on the idea that the environment in the news recommendation scenario changes quickly. A trend-based delegation strategy seems to be promising allowing us to adapt to new events and trends.

Team C delegates the requests based on the popularity of the article that should be enriched with news recommendation. This is based on the idea that popular news article address another audience that rarely read news articles. We count the access frequency of the items for one hour and classified the news articles into 5 level of popularity (using a logarithmic scale).

Discussion. Comparing the analyzed delegation strategies, we found that the time-based delegation performed best in the contest. In previous experiments, we found that instead of day of week hour of the day leads to slightly better results; but in the weeks 38–41 we decided to investigate an alternative time-based delegation criterion. The success of the time-based delegation sounds plausible due to the observation that the habits in using news portal strongly depend on the time. Even though the domains may highly differ, the time has a strong influence what types of recommendation are most frequently clicked by the users.

The trend-based recommendation performed successfully, but did not outperform the time-based delegation. The strength of the trend-based delegation is that untypical events (e.g. public holiday in the middle of the week) can be taken into account without using additional knowledge. A challenge consists of efficiently detecting trends, but avoiding over-fitting caused by noise. The delegation based on the item popularity performed worse than the trend-based delegation. This shows that the item popularity does not tell much about the user preferences and the expected suggestions.

In a nutshell, the context-aware delegation of queries performed well in the PLISTA contest. A good recommendation quality is reached even by using simple criteria for selecting the recommender algorithm for a request. The use of more complex delegation criteria seems promising allowing us to consider a comprehensive set of properties of a request.

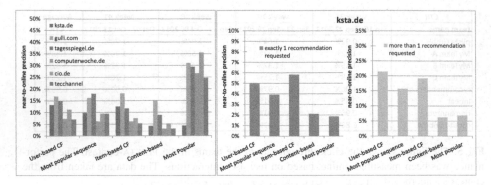

Fig. 6. The figure shows the precision of correctly predicted feedback messages. The left graph shows that the performance of the algorithms strongly depends on the news portal. The right figure shows the recommender performance depedencies on the number of requested suggestions (for the portal ksta.de). Item-based recommenders perform best for one article requests, whereas user-based recommendation works best if more than one suggestion is expected.

Near-to-Online Evaluation. We analyze the *Near-to-Online* precision of different recommender algorithms. The evaluation allows us to identify contexts in that the different algorithms perform best.

The left diagram in Fig. 6 shows the prediction precision of the implemented algorithms for different domains in the news recommendation challenge. We count a prediction as correct, if the article clicked by the user is contained in the list of six suggested articles. The results show that there are big differences in the observed precision. The implemented most popular recommender works best for the most portals, but fails for the portal ksta.de.

The right diagram in Fig. 6 shows the performance for the domain ksta.de dependent from the number of requested suggestions. We find that the item-based CF recommender performs best with requests asking for exactly one suggestion, whereas the requests asking for two or more results are answered best by the user-based recommender. This can be explained by the observation, that suggestions coming from a one-result request a presented on the news portal in a way letting the users expect a strong relation to the currently viewed article. The users expect from larger recommendation panels more generally interesting suggestions, not necessarily highly related to the currently read article. Thus, user-based collaborative filtering outperforms the item-based algorithms for requests asking for 2 – 6 results.

5   Related Work

In this section we review existing news recommender systems as well as state-of-the-art recommender approaches.

News Recommender Systems. Liu et al. [5] introduce a recommender system for Google news articles. The system computes content-based recommendations based on user profiles consisting of keyword describing the user preferences. The user behavior is analyzed in order to extract key phrases from interesting documents. These key phrases are added to the user profile to reflect the changes in the user's interests. Tan et al. [11] describe the system *PIN* providing a keyword-based news recommender system. The users define initially keyword describing their topics of interests; the profiles are incrementally updated based on the user's feedback. These news recommender systems focus on content-based recommender approaches assuming that the user interests can be modeled using keywords. The weaknesses of this approach are that a large set of keywords are required, new trends (topics of interests) are not considered and collaborative knowledge is not taken into account. These systems are focused on searching keywords than on suggestion new potentially interesting articles. In our research we use a more complex approach combining content-based as well as collaborative knowledge.

Recommender Approaches. The research in recommender algorithm traditionally focusses on finding items matching the user's preference based on a huge (but sparse) user-item matrix [1]. The *MovieLens* system [8] recommends movies based on user profiles containing explicit user ratings for movies. A similar scenario is the basis of the *Netflix* challenge: Based on ratings users assigned to movies in the past, ratings for new movies must be predicted [3]. Both systems focus on Collaborative Filtering methods predicting the relevance of movies for a user based on the ratings of similar users. Due to the huge size of the dataset, the challenges in these systems are efficient complexity reduction and weighting algorithms enabling the reduction of noise and the data sparsity. In contrast to the news recommendation scenario the set of users and movies does not change much over time allowing these movie recommender system to use complex models (e.g., low rank approximations of the user-movie matrix [3]). The time-constraints are not strict in most movie recommender systems enabling the use of computational complex algorithms.

Most existing evaluation scenarios use an offline evaluation applying complex data mining approaches on log data, e.g., Liu et al. [5]. In our research we focus on stream-based recommender algorithms able to adapt quickly to changes in the stream of news articles. We limit the number of considered items to the most recent news articles in order to reduce the computational complexity and to hold the strict time limits. Ensemble approaches combine different algorithms to benefit from the strength of different algorithms. Polikar [6] discusses strategies for aggregating results in order to improve the recommendation precision as well as the evidence that results are relevant. Strategies for selecting the most appropriate algorithm based on the respective request are discussed in [9] and [2]. In our research we use a more complex context description and optimize the delegation strategy to hold the strict time constraints.

Discussion. Most traditional news recommender systems focus on content-based algorithms searching for articles containing user-defined keywords. These systems deliver results as offline newsletters or personalized news portals avoiding problems with real-time constraints. In contrast to the analyzed news recommender systems tailored to exactly one news portal, our approach supports a variety of news portals. Thus, we focus on a flexible framework able to adapt to heterogeneous systems. That is why we do not deploy *one* optimized recommender algorithms; we focus on ensembles of different agents (using different algorithms and parameter settings) allowing us to incrementally learn what algorithm performs best based on the respective portal and the context.

6 Summary and Conclusion

We developed and evaluated a framework for providing high-quality news recommendation. We started with the analysis of different types of recommender algorithms. For this purpose, we adapted the state-of-the-art recommendation algorithms to be able to cope with the quickly changing amount of news articles and users. We analyze the correlation between the algorithms and evaluated the performance dependent from the respective context. The results show that there is not an optimal algorithm: The performance strongly dependent on the context and the respective publisher. Based on this observation, we implemented and evaluated three different ensemble approaches. The evaluation results show that strategies combining the suggestions from different agents slightly improved the recommendation precision. The context-aware delegation improved the evaluation quality significantly. This delegation strategy allows us the adaptation to the specific demands of the context and the respective portal. Thus, having trained the delegation agent for one week, we won the PLISTA contest by outperforming all other teams in the contest.

As future work we further plan to optimize our recommender framework. We want to test additional recommender algorithms, analyze approaches for incrementally improving the recommender strategies, and investigate strategies for combining results from different recommender agents.

References

1. Adomavicius, G., Tuzhilin, A.: Toward the next generation of recommender systems: A survey of the state-of-the-art and possible extentions. IEEE Transactions on Knowledge and Data Engineering 17(6) (2005)
2. Albayrak, S., Wollny, S., Lommatzsch, A., Milosevic, D.: Agent technology for personalized information filtering: The pia-system. Scalable Computing: Practice and Experience 8 (2007)
3. Bell, R., Koren, Y.: Lessons from the netflix prize challange. ACM SIGKDD Explorations 9(2), 75–79 (2007)
4. Herlocker, J.L., Konstan, J.A., Terveen, L.G., Riedl, J.T.: Evaluating collaborative filtering recommender systems. ACM Trans. Inf. Syst. 22(1), 5–53 (2004)
5. Liu, J., Dolan, P., Pedersen, E.R.: Personalized news recommendation based on click behavior. In: Proceedings of the 15th International Conference on Intelligent User Interfaces, IUI 2010, pp. 31–40. ACM, New York (2010)
6. Polikar, R.: Ensemble based systems in decision making. IEEE Circuits and Systems Magazine 6(3), 21–45 (2006)
7. Pu, P., Chen, L., Hu, R.: A user-centric evaluation framework for recommender systems. In: Proc. of the 5th ACM Conf. on Recommender Systems, RecSys 2011. ACM, NY (2011)
8. Research, G.: Movielens data sets (October 2006),
   http://www.grouplens.org/node/73
9. Scheel, C., Neubauer, N., Lommatzsch, A., Obermayer, K., Albayrak, S.: Efficient query delegation by detecting redundant retrieval strategies. In: Proceedings of SIGIR 2007 Workshop: Learning to Rank for Information Retrieval (2007)
10. Su, X., Khoshgoftaar, T.M.: A survey of collaborative filtering techniques. In: Advances in Artificial Intelligence (January 2009)
11. Tan, A.-H., Teo, C.: Learning user profiles for personalized information dissemination. In: Proc. of the IEEE World Congress on Comp. Intelligence, vol. 1, pp. 183–188 (1998)

A Personalised Recommendation System for Context-Aware Suggestions

Andrei Rikitianskii, Morgan Harvey, and Fabio Crestani

Faculty of Informatics, University of Lugano (USI), Switzerland
{andrei.rikitianskii,morgan.harvey,fabio.crestani}@usi.ch

Abstract. The recently introduced TREC Contextual Suggestion track proposes the problem of suggesting contextually relevant places to a user visiting a new city based on his/her preferences and the location of the new city. In this paper we introduce a more sophisticated approach to this problem which very carefully constructs user profiles in order to provide more accurate and relevant recommendations. Based on the track evaluations we demonstrate that our system not only significantly outperforms a baseline method but also performs very well in comparison to other runs submitted to the track, managing to achieve the best results in nearly half of all test contexts.

1 Introduction

Modern web services such as search engines and online mapping tools provide a means for people to find interesting things to do and places to go when visiting a new city. However in many cases, due to the sheer number of possibilities and perhaps a lack of time for planning, it is not possible or desirable to manually search through all of the available options. In such instances, recommender systems can provide users with a personalised set of suggestions of places to visit that they are likely to enjoy, recommendations which are often made based on the user's previous interactions with the system. Recommender systems are becoming an ever more frequently investigated area of research and are used by many thousands of web sites and services - including the likes of Amazon, Netflix and Trip Advisor - to improve the experience for their users and are especially useful on mobile devices [9].

The vast majority of work in the area has considered only 2 dimensions of interest - the users and the items to be recommended - however there has been a recent surge of interest in the inclusion of a third dimension: context [1]. Rather than simply recommending items that the user will like based on a profile, context-sensitive recommender systems attempt to also include various other sources of information to make the resulting recommendations more accurate and relevant. Examples of useful contextual information are, for example, the time of day or current location of the user. Use of context is clearly of significant benefit when suggesting places to visit since factors such as location will dictate which places can feasibly be visited. The increased awareness of the importance

M. de Rijke et al. (Eds.): ECIR 2014, LNCS 8416, pp. 63–74, 2014.
© Springer International Publishing Switzerland 2014

of context when making recommendations is demonstrated by the introduction of a new TREC track dedicated to this problem in 2012 which has continued into 2013 [4].

In this work we present a new approach to recommending places to users incorporating geographical information as context and exploiting data from multiple sources. Via analysis of results from TREC evaluations performed by a large group of users we demonstrate the high level of performance delivered by this method, showing that it is able to significantly outperform the Contextual Track baseline and all other track entrants in nearly half of all cases. Over all metrics our system performs considerably better than the median result. We conclude the paper with a more detailed analysis of the results, indicating potential avenues for future work.

2 Related Work

Recommendation and personalisation are 2 very common themes in modern Information Retrieval research. Much early work was conducted in the 90s and the field has seen a resurgence of interest lately, partially due to the Netflix prize [7] and also due to the realisation that context can play an important part in generating truly accurate and useful results [1,12]. In this work we focus solely on the problem of recommendation in which a potentially massive set of candidate items are filtered down such that the subset of remaining items will be of interest to the user based on their profile information. Ideally these final suggestions should be ranked such that the first item has the greatest probability of being liked by the user with probability of interest decreasing as rank increases [8].

Personalisation has been frequently applied to mobile search and recommendation problems such as tourist guides [2] and event navigation tools [9]. In many of these approaches various forms of context are considered. Ardissono et al. [2] present a system designed to help tourists navigate their way around the Italian city of Turin and make recommendations to users by assigning each to one of several tourist "groups" with certain requirements. Recommendations are made by taking various contextual factors into account such as opening hours, entrance prices and determine user preferences based on features such as the historical period of the attractions.

Schaller at al. [9] attempt to provide recommendations to visitors of a large-scale event composed of many smaller sub-events. Their system uses a variety of forms of profile information and constructs tours around sets of recommended events in an effort to maximise the amount of time visitors spend at attractions, rather than travelling between them. They demonstrate the importance of considering contextual information when constructing tours and recommendations such as visitor and event locations, start and end times and durations of events.

Many of the approaches submitted to last year's track approached the problem in a similar manner [4]. They first obtain a list of suitable venues by querying a search API using the geo context data, then construct profiles using the terms in the descriptions of places already given positive ratings by users and finally rank the list of potential candidates based on their descriptions' similarities to the positive user profile. In this work we expand on this simple framework in a number of ways. We consider both positive and negative ratings separately, are far more selective when choosing terms to build profiles, make use of term expansions techniques to mitigate matching issues caused by varied vocabulary use and use machine learning methods to build a classifier and ranker.

3   Dataset and Tasks

The TREC Contextual Suggestion Track investigates search techniques for complex information needs that are highly dependent on context and user interests. In this track the goal is to suggest personalised attractions to an individual, given a specific geographic context. The track imagines a traveler in a new city. Given a set of the traveler's preferences for places and activities in their home city, the system should suggest places and activities in a new city that the person may enjoy. In this paper we use the terms "attraction," "place" and "venue" interchangeably.

As input to the task, participants were provided with a set of 635 profiles, a set of 50 example suggestions, and a set of 50 geo contexts in CSV/JSON format. Example suggestions can represent attractions of different types, for example: bars, restaurants, museums, etc. All the attractions are from the Philadelphia area. Each profile corresponds to a single user, and indicates that user's preference with respect to each example suggestion. Each training suggestion includes a title, description, and an associated URL. Each context corresponds to the GPS coordinate of the centre of a number of cities in the United States. The set of cities is quite diverse in terms of population: starting from small cities such as Beckley, WV (with a population of 17,606) up to much larger cities such as Atlanta, GA and Wichita, KS (with populations in the hundreds of thousands or even millions). Profiles consist of two ratings for a series of attractions, one rating for the attraction's title and description and another for the attraction's website. The ratings are given on a five-point scale, ranging from "strongly disinterested" to "strongly interested", based on how interested the user would be in going to the venue if they were visiting the particular city it is located in.

As output to the task, for each profile/context pairing, the participant should return a ranked list of up to 50 ranked suggestions. Each suggestion should be appropriate to the profile (based on the user's preferences) and the context (according to the location), contains a title, description and attraction's URL. The description of the suggestion may be tailored to reflect the preferences of that user. Profiles correspond to the stated preferences of real individuals, who will return to judge proposed suggestions. Users were recruited through crowdsourcing sites or are university undergraduate and graduate students. For the purposes of this experiment, you can assume users are of legal drinking age at the location specified by the context.

4   A New Approach for Context Suggestion

To generate ranked lists of appropriate recommendations for each user profile
and geographical context we developed a geo context-aware system. The system
can be broken down into the following 4 steps:

1. processing geo contexts;
2. inferring user term preferences;
3. building a personal ranking model;
4. ranking suggestions

In the following section we describe these individual steps in more detail.

4.1   Processing Geographical Contexts

Before we can apply any user profile-based personalisation we first need a set
of appropriate candidate attractions located within a small radius of the geo
context specified. We used the Google Places API[1] to obtain a list of potential
suggestions, retrieved based on a query consisting of GPS coordinates and attraction types. We considered only types of venues, as defined by the Google
Places API, which were present within the training set. In doing so we retrieved
27 different types, such as: night clubs, amusement parks, libraries, movie the-
atres, shopping malls, etc. On average, for each geo context, we collected about
350 suggestions.

Google Places only provides a short *title* and a web site *URL* for each suggestion. So that users can evaluate the quality of each suggestion a description
of each venue should be provided. To generate these brief descriptions we first
queried the Yandex Rich Content API[2] which, given a URL as a query, returns
a short static textual description of the page's content. While the Yandex API
has generally quite good coverage, there were instances where it was unable to
return any information and in these cases we instead queried the Google Custom
Search API and used the web site snippet it returned.

4.2   Inferring User Term Preferences

In order to make personalised suggestions for each user we need to be able to
compare a new venue with that user's profile to determine how likely it is that the
user will like it. We therefore need to have some representation of the user's likes
and dislikes based on the training data made available to us, i.e. the venues each
user has already rated. In line with previous work on recommender systems [6],
we chose to maintain separation between positive and negative preferences using
descriptive terms from already rated venues. We used Natural Language Toolkit
(NLTK) software[3] to extract only nouns, adjectives, adverbs and verbs from each

---

[1] Google Places API - https://developers.google.com/places/documentation/
[2] Yandex Rich Content API - http://api.yandex.com/rca/
[3] Toolkit version 1.0 used, available from http://nltk.org/

description and represented these as binary vectors, indicating the presence or absence of each term. For each user, we extracted positive and negative terms, using them to build separate positive and negative profiles. A positive term is one derived from a positively rated venue, a negative term is one from a venue that was given a negative rating. Venues with a *title and description* rating of more than 2 was considered to be positive, while a venue was considered to be negatively rated when it was allocated a rating of less than 2. Terms from neutral suggestions were ignored.

Due to the relative brevity of the descriptions, this approach of using only the terms present is unlikely to result in many exact term matches and will therefore deliver quite unreliable similarity scores. Consider, for example, if the negative profile for a user contains the word "sushi" and this is matched against a description containing the terms "raw" and "fish." Without performing any kind of term expansion these concepts, despite their obvious similarity, would not make any contribution to the similarity score. However, by expanding existing raw terms using similar words we can (at least partially) overcome this vocabulary mismatch. When comparing profiles with venue descriptions, instead of simply using the raw terms, we checked for matches between the synonym lists returned for each term in WordNet[4]. Given a list of synonyms for a term a and those for a term b, we consider the terms to be matching if the two lists share at least one component (i.e. if there is some overlap).

Using both the positive and negative models, we can estimate what the user's opinion might be about a potential suggestion based on its description. To estimate how positive the description is for user u, we can calculate the cosine distance between vector $\vec{D_i}$ representing the description of venue i and positive user profile $\vec{M_u^+}$ as follows:

$$cos^+(\vec{D_i}, \vec{M_u^+}) = \frac{\vec{D_i} \cdot \vec{M_u^+}}{\|\vec{D_i}\| \cdot \|\vec{M_u^+}\|}$$

The same formula was applied to estimate how negative the description is, by using the negative user profile. cos^+ and cos^- scores were used in the final ranking model as described in section 4.3.

4.3   Building a Personal Ranking Model

After obtaining a set of potential suggestions for a given geographical context (described in section 4.1), we need to rank these potential candidates according to the user's preferences. Because of the variability of individual preferences among users it is nearly impossible to build an accurate global ranking model and given that the task is to provide *personalised* suggestions this would not be suitable anyway. To investigate how varied user preferences were, we measured the level of agreement between all judgments from user profiles by using a standard statistical *overlap* metric [10] where the overlap between two sets of items A and B is defined as:

---

[4] WordNet - http://wordnet.princeton.edu

$$Overlap(A, B) = \frac{|A \cap B|}{min(|A|, |B|)} \qquad (1)$$

The mean pairwise overlap between *title and description* ratings is 0.38 and between *website* ratings is 0.39. Both of these overlaps are quite small, suggesting that users have different preferences. Therefore, we decided to build a personal ranking model for each user in order to more precisely adapt suggestion to their own preferences.

Model and Training Data. We consider the choice of suitable candidates as a binary classification problem. We separate relevant and non-relevant suggestions for each individual user, and then rank those classed as relevant based on confidence score estimated by the classifier. To generate training set data for each profile we used example suggestion weight as a linear combination of *title and description* and *website* ratings:

$$Weight(S) = \lambda R_{desc,s} + (1 - \lambda)R_{url,s}$$

with the condition $\lambda \in [0, 1]$ and $Weight \in [0, 4]$. In the formula , S indicates the example suggestion from a particular profile, $R_{desc,s} \in [0, 4]$ is the suggestion *title and description* rating, $R_{url,s} \in [0, 4]$ is the suggestion *website* rating. We then assigned a positive label to suggestions with a combined weight of more than a threshold T^+ and negative label to the suggestion with weight less than threshold T^-.

These thresholds T^+ and T^- were tuned to try to balance the number of positive and negative samples in the training set. The degree of imbalance is represented by the ratio of sample size of the small class to that of the large class. We considered a training set to be imbalanced if the ratio was less than 1:5, i.e. if the largest class was more than 4 times greater than the smallest. T^+ and T^- were turned for each profile by using a simple iterative algorithm. If the algorithm was unable to converge (i.e. sufficiently balance the 2 classes) after 3 iterations we consider this particular profile to be unsuitable for classification and use an alternate approach for making recommendations which we outline later.

By default we set uniform weights to the 2 different ratings for each example ($\lambda = 0.5$), meaning that the importance of *title and description* is the same as *website*. It is possible that the true influence of these two factors may not be equal and this may depend on the kind of venue under consideration. For example, for many cafs the description may provide sufficient information upon which to base a decision, whereas for a restaurant the user may wish to browse the website first, perhaps to look at the menu before making a decision. These type-dependent values for λ could perhaps be learnt from the training data, however we leave this for future work.

Learning Algorithm and Features. We chose a Naïve Bayes classifier as our learning algorithm. This is a simple probabilistic classifier based on applying

Bayes' theorem and making the assumption that each feature's weight (or in the binary case, presence or absence) is independent of the weights of other features, given the class variable. Although this assumption is unlikely to be entirely true in many cases, it greatly simplifies the model - making it tractable - and does not significantly degrade performance in practice. The motivation for choosing such a simple classifier was that it generally performs better on small data sets than more sophisticated machine learning techniques [3] and does not require any complex parameter tuning or additional learning. In our case, the size of the training data set is never greater than 50 examples; the number of examples for each profile lies in the range 30-49. We use the Weka implementation of the classifier [11] for all of our experiments.

Each suggestion in the training set is represented by a feature vector, consisting of two different types of features: boolean and real-valued. The boolean features were derived based on attraction types, representing the user's preferences with regard to the kind of venue suggested. As described in section 4.1, the Google Places API returns a simple type for each place, indicating what kind of venue it is. Each place can be assigned to multiple types and as such our binary feature vector encodes the types each suggestion has been assigned to: 1 if it is assigned to that type, 0 otherwise.

The 3 real-valued features were based on the cosine distance between the suggestion description and both user profiles (positive and negative), reflecting user term preferences, and the description length. For a given user u and suggestion i, we calculated two features $cos^+(\overrightarrow{D_i}, \overrightarrow{M_u^+})$ and $cos^-(\overrightarrow{D_i}, \overrightarrow{M_u^-})$, where D_i is a description of suggestion S and M_u^+ and M_u^- are the positive and negative profiles for user u. The description length feature is simply the length of the description in characters. We believe that the length of description may be an important factor when the user explores the suggestions for an attraction as a longer description may provide more detailed information, allowing the user to be more sure of their rating.

4.4   Ranking Suggestions

To rank the potential suggestions for each user, we use their individual personal ranking model as described in Section 4.3. The personal ranking model was first used to determine the 50 most relevant suggestions for each geographical context, in descending order of confidence score as estimated by the classifier. The confidence score, in the case of a Naïve Bayes classifier, is simply the posterior probability that the suggestion belongs to class "relevant" and therefore encodes, in some sense, how likely it is that the user will like the candidate venue. This approach has been demonstrated to work well for ranking problems [13].

As mentioned in the previous section, there were a few profiles for which it wasn't sensible to build a classifier due to the level of imbalance between the 2 classes in the training data. In this case, potential suggestions were ranked by using only the user term preferences. We ordered the suggestions in descending order of their scores, which were calculated as the difference between cos^+ and

cos^-. This is a reasonable, if slightly simplified approach, since it will return a positive value if the similarity between the candidate venue and the positive profile is greater than its similarity compared with the negative profile and vice-versa.

5 Results

In this section we present an overview of the performance of our system. Output suggestions were judged both by the original group of users who supplied the training data and NIST assessors. The user corresponding to each profile judged suggestions in the same manner as for the training examples, assigning a value of 0-4 for each *title and description* and *url*. NIST assessors judged suggestions in terms of geographical appropriateness. In total, 223 of the potential 31750 profile/context pairs were judged, i.e. not all pairs were used for evaluation. The top 5 suggestions for each profile/context pair were taken into account for evaluation.

To evaluate the performance of the model for the problem of Contextual Suggestion, three different measures were used: Precision at Rank 5 (P@5), Mean Reciprocal Rank (MRR) and Time-Biased Gain (TBG). P@5 and MRR are traditional evaluation metrics to measure the overall effectiveness of an IR system in terms of its ability to return a good ranked list. The TBG metric, on the other hand, was developed specially for the contextual suggestion task [5]. As the basis for evaluation, a suggestion was counted as "relevant" if the user liked both the description and the geographically appropriate document. All other suggestions were counted as "non-relevant". P@5 and MRR are calculated by using these definitions for relevant and non-relevant. The TBG metric is more complex and takes into account the impact of descriptions and disliked suggestions, which are ignored by P@5 and MRR. All the metrics were computed for each profile/context pair, and then averaged across all pairs.

Since we submitted our system as an entrant to the TREC Contextual Suggestion track we can compare our results with those of two baselines other competing systems from other institutions. The baselines use the Google Places API to retrieve context-relevant candidates and rank them based on their overall popularity (a common baseline measure in recommender systems). BaselineA uses all candidates whereas baselineB uses only candidate places which are present in the ClueWeb12 collection. For the description, Google Places provided a description, review, or a blurb from the meta-description tag on the website. Personalisation was not attempted for either baseline runs. BaselineB is used as a baseline for participants who gathered suggestions from ClueWeb12 datasets only. Table 1 shows evaluation results for our proposed model, baselineA system and two best systems. It also shows the median score, which is calculated based on the results from all 34 systems submitted to the TREC track. The column *P@5 Rank* presents a rank of system result among all 34 systems based on P@5 metric.

The results show that our system greatly outperforms baselineA, however the baseline result does appear quite weak in comparison to the submitted runs

Table 1. Results for our method compared with baselineA run, two best runs and median scores

	P@5 Rank	P@5	MRR	TBG
UDInfoCS1	1	0.5094	0.6320	2.4474
UDInfoCS2	2	0.4969	0.6300	2.4310
our method	**3**	**0.4332**	**0.5871**	**1.8374**
baselineA	25	0.1372	0.2316	0.5234
median		0.2368	0.3415	0.8593

(based on the median score). This means that most participants overcame baselineA result. Nevertheless, our run performs significantly better than median score: P@5 +45%, MRR +41%, TBG +53%. More detailed analysis of our results shows that, according to the MRR metric, our system was able to return the best result over all entrants for 48.43% of user/context pairs. When considering P@5, our system returned the best result in 23% of cases and was better than the median score 61% of the time. Moreover, according to P@5 metric, our system placed 3rd among all 34 systems and 2nd among all 19 groups that participated in the track. We found that about 1.5% of all suggestions had a website which could not be loaded during the assessment procedure. Removing these suggestions from the top 5 leads to performance improvements of: P@5: +0.5%, MRR: +1% and TBG: +1.5%.

6   Analysis

Besides final/max/min/median results for each profile/context pair, the organisers also provided all judgments (description, website, geographical relevance) for each suggestion from the 223 profile/context pairs which were judged. Using an evaluation script provided for the track and these judgments we performed a more detailed analysis of our results. Table 2 shows how the geographical relevance(G), description(D) and the website(W) ratings contributed to the P@5 and MRR scores. According to this statistic, almost all documents retrieved by our system were geographically appropriate to the context, suggesting that the approach of pre-filtering candidates is effective. The website of each suggestion and its description appear to contribute equally to final result quality. We found that there was a small correlation (0.271) between the number of candidates returned by the Google Places API for a city and the performance of the system, suggesting that it is easier to make good recommendations when there is a wide variety of possible candidates. The 4 context cities with the smallest population also have the worst performance in terms of P@5 and, unsurprisingly, there is a strong correlation between the population of a city and the number of candidates returned for it by the API (0.693).

In general, assessors judged 1115 suggestions from the Top 5 suggestions for 136 different profiles . These suggestions represent 772 unique venues, i.e. some venues were recommended for different profiles at the same time. We explored

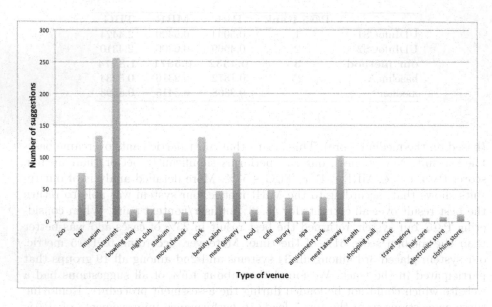

Fig. 1. Distribution of numbers of suggestion for 24 different types of venues

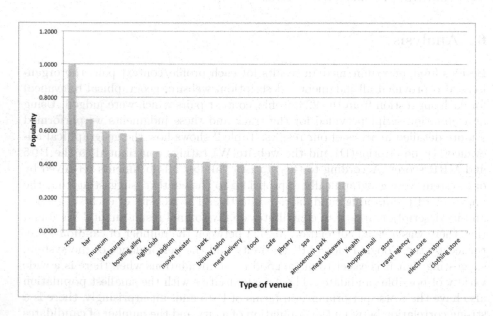

Fig. 2. Popularity of 24 different types of venues

Table 2. Geographical relevance(G), description(D) and website(W) contributions into P@5 and MRR metrics

	G	W	D	Final(WDG)
P@5	0.9363	0.5776	0.5381	0.4332
MRR	0.9675	0.7149	0.6700	0.5871

types of venues which were represented by these suggestions, amounting to a total of 24. Figure 1 presents a distribution over different types of venues. Restaurants (23%), museums (12%) and parks (11.7%) are the most common types of venues. For each venue type we calculated a popularity score, which is the fraction of "relevant" suggestions over all of the suggestions made for that type. Figure 2 demonstrates that the most popular venues for assessor were zoos, bars, museums and restaurants. This can perhaps be explained by the fact that there were few zoos recommended to users, and all of them were counted as "relevant". Restaurants, bars and museums are often suggested and are highly popular because they are very common tourist attractions and their overall popularity is perhaps not strongly affected by a visitor's interests. The popularity of venues such as travel agencies, shopping malls and electronics stores is 0, likely because these types of places are not especially attractive to tourists and are more likely to be frequented regularly by people who live in the area.

7   Conclusions and Future Work

In this paper we have described a new system, designed to take part in the TREC Contextual Suggestion track, for making context-sensitive recommendations to tourists visiting a new city. Based on analysis of results obtained from the same users who contributed the training data we have shown that the method is very effective for this problem and was able to significantly outperform the baseline system and, when compared to the 34 other competing system in the track, delivered the 3rd best result in the TREC 2013 Contextual Suggestion track. In nearly half of all contexts, our approach was able to deliver the best set of results, confirming that the choices made during the development of the system were sensible and beneficial. Our method is based on quite a simple strategy of using the descriptions of previously rated places to build user profiles, however we introduce a number of novel additions which have clearly lead to improved performance.

There are several directions for future work. Our ranking model could be easily extended by adding new features to the classifier. For example in the current ranking model we don't use information about the distance between the geolocation specified for each context and the venue, the venue rating (provided by content system) or the cuisine type of restaurants, cafs and bars. We believe that new features based on this information will allow ranking model to reflect user preferences more precisely and that subtly weighting suggestions by their distance from the user's location will lead to better acceptance of the recommendations made. It would also be interesting to make use of other content systems (such as Foursquare and TripAdvisor) to expand the list of potential candidates and the brief descriptions could perhaps be improved or tailed to the user's interests by also considering user reviews or comments from social networks. Finally, instead of assuming that the description and web site of a venue are of equal importance, we could learn specifics weights for the λ parameter in our model which could even be conditioned on the type of venue being considered.

References

1. Adomavicius, G., Tuzhilin, A.: Context-aware recommender systems. In: RecSys, pp. 335–336 (2008)
2. Ardissono, L., Goy, A., Petrone, G., Segnan, M., Torasso, P.: Intrigue: Personalized recommendation of tourist attractions for desktop and handset devices. In: Applied Artificial Intelligence, pp. 687–714. Taylor and Francis (2003)
3. Brain, D., Webb, G.I.: On the effect of data set size on bias and variance in classification learning. In: 4th Australian Knowledge Acquisition Workshop (AKAW 1999), Sydney, Australia (1999)
4. Dean-Hall, A., Clarke, C.L.A., Kamps, J., Thomas, P., Voorhees, E.: Overview of the trec 2012 contextual suggestion track. In: Text REtrieval Conference (TREC) (2012)
5. Kamps, J., Dean-Hall, A., Clarke, C., Thomas, P.: Evaluating contextual suggestion. In: 5th International Workshop on Evaluating Information Access (EVIA), Tokyo, Japan (2013)
6. Harvey, M., Ludwig, B., Elsweiler, D.: You are what you eat: Learning user tastes for rating prediction. In: Kurland, O., Lewenstein, M., Porat, E. (eds.) SPIRE 2013. LNCS, vol. 8214, pp. 153–164. Springer, Heidelberg (2013)
7. Koren, Y.: Factorization meets the neighborhood: a multifaceted collaborative filtering model. In: 14th ACM SIGKDD, pp. 426–434 (2008)
8. Ricci, F., Rokach, L., Shapira, B., Kantor, P.B. (eds.): Recommender Systems Handbook. Springer (2011)
9. Schaller, R., Harvey, M., Elsweiler, D.: Recsys for distributed events: investigating the influence of recommendations on visitor plans. In: 36th ACM SIGIR Conference, pp. 953–956. ACM (2013)
10. Voorhees, E.M.: Variations in relevance judgments and the measurement of retrieval effectiveness. In: 21st ACM SIGIR Conference (1998)
11. Witten, I.H., Frank, E.: Data Mining: Practical Machine Learning Tools and Techniques, 2nd edn. Morgan and Kaufmann (2005)
12. Xiang, B., Jiang, D., Pei, J., Sun, X., Chen, E., Li, H.: Context-aware ranking in web search. In: 33rd ACM SIGIR Conference, SIGIR 2010, pp. 451–458. ACM (2010)
13. Zhang, H., Su, J.: Naive bayesian classifiers for ranking. In: 15th ECML2004 Conference, Pisa, Italy (2004)

Optimizing Base Rankers Using Clicks
A Case Study Using BM25

Anne Schuth, Floor Sietsma, Shimon Whiteson, and Maarten de Rijke

ISLA, University of Amsterdam, The Netherlands
{anne.schuth,f.sietsma,s.a.whiteson,derijke}@uva.nl

Abstract. We study the problem of optimizing an individual base ranker using clicks. Surprisingly, while there has been considerable attention for using clicks to optimize linear combinations of base rankers, the problem of optimizing an individual base ranker using clicks has been ignored. The problem is different from the problem of optimizing linear combinations of base rankers as the scoring function of a base ranker may be highly non-linear. For the sake of concreteness, we focus on the optimization of a specific base ranker, viz. BM25. We start by showing that significant improvements in performance can be obtained when optimizing the parameters of BM25 for individual datasets. We also show that it is possible to optimize these parameters from clicks, i.e., without the use of manually annotated data, reaching or even beating manually tuned parameters.

1   Introduction

Traditional approaches to evaluating or optimizing rankers are based on editorial data, i.e., manually created explicit judgments. Recent years have witnessed a range of alternative approaches for the purpose of evaluating or optimizing rankers, which reduce or even avoid the use of explicit manual judgments. One type of approach is based on pseudo test collections, where judgments about query-document pairs are automatically generated by repurposing naturally occurring labels such as hashtags or anchor texts [1–3].

Another type of approach is based on the use of implicit signals. The use of implicit signals such as click data to evaluate or optimize retrieval systems has long been a promising alternative or complement to explicit judgments [4, 12–14, 18]. Evaluation methods that interpret clicks as absolute relevance judgments have often been found unreliable [18]. In some applications, e.g., for optimizing the click-through rate in ad placement and web search, it is possible to learn effectively from click data, using various learning to rank methods, often based on bandit algorithms. Click models can effectively leverage click data to allow more accurate evaluations with relatively little editorial data. Moreover, interleaved comparison methods have been developed that use clicks not to infer absolute judgments but to compare rankers by observing clicks on interleaved result lists [7].

The vast majority of work on click-based evaluation or optimization has focused on optimizing a linear combination of base rankers, thereby treating those rankers as black boxes [9, 10, 26]. In this paper, we try to break open those black boxes and examine

M. de Rijke et al. (Eds.): ECIR 2014, LNCS 8416, pp. 75–87, 2014.

whether online learning to rank can be leveraged to optimize the base rankers them-selves. Surprisingly, even though a lot of work has been done on improving the weights of base rankers in a combined learner, there is no previous work on online learning of the parameters of base rankers and there is a lot of potential gain from this new form of optimization. We investigate whether individual base rankers can be optimized using clicks. This question has two key dimensions. First, we aim to use clicks, an implicit signal, instead of explicit judgments. The topic of optimizing individual base rankers such as TF.IDF, BM25 or DFR has received considerable attention over the years but that work has almost exclusively used explicit judgments. Second, we work in an online setting while previous work on optimizing base rankers has almost exclusively focused on a more or less traditional, TREC-style, offline setting.

Importantly, the problem of optimizing base rankers is not the limiting case of the problem of optimizing a linear combination of base rankers where one has just one base ranker. Unlike the scoring function that represents a typical online learning to rank solution, the scoring function for a single base ranker is not necessarily linear. A clear example is provided by the well-known BM25 ranker [19], which has three parameters that are related in a non-linear manner: k_1, k_3 and b.

In this paper, we pursue the problem of optimizing a base ranker using clicks by fo-cusing on BM25. Currently, it is common practice to choose the parameters of BM25 according to manually tuned values reported in the literature, or to manually tune them for a specific setting based on domain knowledge or a sweep over a number of possible combinations using guidance from an annotated data set [5, 25]. We propose an alterna-tive by learning the parameters from click data. Our goal is not necessarily to improve performance over manually tuned parameter settings, but rather to obviate the need for manual tuning.

Specifically, the research questions we aim to answer are as follows.

RQ1. How good are the manually tuned parameter values of BM25 that are currently used? Are they optimal for all data sets on average? Are they optimal for individual data sets?

RQ2. Is it possible to learn good values of the BM25 parameters from clicks? Can we approximate or even improve the performance of BM25 achieved with manually tuned parameters?

Our contributions are (1) the insight that we can potentially achieve significant improve-ments of state-of-the-art learning to rank approaches by learning the parameters of base rankers, as opposed to treating them as black boxes which is currently the common practice; (2) a demonstration of how parameters of an individual base ranker such as BM25 can be learned from clicks using the dueling bandit gradient descent approach; and, furthermore, (3) insight into the parameter space of a base ranker such as BM25.

2   Related Work

Related work comes in two main flavors: (1) work on ranker evaluation or optimiza-tion that does not use traditional manually created judgments, and (2) specific work on optimizing BM25.

Several attempts have been made to either simulate human queries or generate relevance judgments without the need of human assessors for a range of tasks. One recurring idea is that of pseudo test collections, which consist of automatically generated sets of queries and for every query an automatically generated set of relevant documents (given some document collection). The issue of creating and using pseudo test collections goes back at least to [24]. Azzopardi et al. [2] simulate queries for known-item search and investigate term-weighting methods for query generation. Asadi et al. [1] describe a method for generating pseudo test collections for training learning to rank methods for web retrieval; they use anchor text in web documents as a source for sampling queries, and the documents that these anchors link to are regarded as relevant documents for the anchor text (query). Berendsen et al. [3] use a similar methodology for optimizing microblog rankers and build on the idea that tweets with a hashtag are relevant to a topic covered by the hashtag and hence to a suitable query derived from the hashtag.

While these methods use automatically generated labels instead of human annotations, the setting is still an offline setting; learning takes place after collecting a batch of pseudo relevant documents for a set of queries. Clicks have been used in both offline and online settings for evaluation and optimization purposes, with uses ranging from pseudo-relevance feedback [14] to learning to rank or re-rank [12, 13]. Radlinski et al. [18] found that evaluation methods that interpret clicks as absolute relevance judgments are unreliable. Using bandit algorithms, where rankers are the *arms* that can be *pulled* to observe a click as feedback, it is possible to learn effectively from click data for optimizing the click-through rate in ad placement and web search [17]. Carterette and Jones [4] find that click models can effectively leverage click data to allow more accurate evaluations with relatively little editorial data. In this paper, we use probabilistic interleave [7], an interleaved comparison method that uses clicks not to infer absolute judgments but to compare base rankers by observing clicks on interleaved result lists; we use this relative feedback not only to optimize a linear combination of base rankers, as has been done before, but also to optimize an individual ranker. Our optimization method uses this relative feedback in a dueling bandit algorithm, where *pairs of rankers* are the arms that can be pulled to observe a click as relative feedback [9, 10, 26].

Our case study into optimizing an individual base ranker using clicks focuses on BM25; a parameterized (with parameters k_1, k_3 and b) combination of term frequency (TF), inverse document frequency (IDF) and query term frequency (cf. Section 3.1). A good general introduction to this ranker can be found in [20], while detailed coverage of early experiments aimed at understanding the model's parameters can be found in [22]. Improvements to standard BM25 have previously been investigated by Svore and Burges [23], who apply BM25 to different document fields. Then a machine learning approach is used to combine the results on these different fields. However, there the parameters of BM25 are still set at a fixed value. Most similar to the work presented here is [5, 25]. There, however, the parameters of BM25 are optimized based on relevance labels, not clicks, in an offline learning, so that the parameters learned cannot be adapted while search takes place. Interestingly, over the years, different values of the key parameters in BM25 are used as manually tuned "default" values; e.g., Qin et al. [16] use $k_1 = 2.5$, $k_3 = 0$, $b = 0.8$ for the .gov collection. They use $k_1 = 1.2$, $k_3 = 7$, $b = 0.75$ for the OHSUMED collection, while Robertson and Walker [19] use $k_1 = 2.0$, $b = 0.75$.

3 Method

Today's state-of-the-art ranking models combine the scores produced by many base rankers and compute a combination of them to arrive at a high-quality ranking. In the simplest form, this combination can be a weighted sum:

$$s(q, d) = w_1 \cdot s_1(q, d) + \cdots + w_n \cdot s_n(q, d), \tag{1}$$

where w_i is the weight of each base ranker $s_i(q, d)$ that operates on the query q and document d. The base rankers may have internal parameters that influence their performance. We focus on one particular base ranker, BM25, which has three parameters that determine the weight applied to term frequency, inverse document frequency and other query or document properties in the BM25 scoring function.

Below, we first recall BM25 in full detail and then describe how we use clicks to optimize BM25's parameters.

3.1 Implementation of BM25

Several variants of BM25 are used in the literature. We use the variant that is used to compute the BM25 feature in the LETOR data set [16]. Given a query q and document d, the BM25 score is computed as a sum of scores for every term q_i in the query that occurs at least once in d:

$$BM25(q, d) = \sum_{q_i : tf(q_i, d) > 0} \frac{idf(q_i) \cdot tf(q_i, d) \cdot (k_1 + 1)}{tf(q_i, d) + k_1 \cdot (1 - b + b \cdot \frac{|d|}{avgdl})} \cdot \frac{(k_3 + 1) \cdot qtf(q_i, q)}{k_3 + qtf(q_i, q)} \tag{2}$$

The terms used in this formula are:

- $idf(q_i)$ (*inverse document frequency*): computed as $idf(q_i) := \log\left(\frac{N - df(q_i) + 0.5}{df(q_i) + 0.5}\right)$ where N is the total number of documents in the collection and $df(q_i)$ is the number of documents in which the term q_i occurs at least once;
- $tf(q_i, d)$ (*term frequency*): the number of times term q_i occurs in document d;
- $qtf(q_i, q)$ (*query term frequency*): the number of times term q_i occurs in query q;
- $\frac{|d|}{avgdl}$: the length of document d, normalized by the average length of documents in the collection;
- k_1, b and k_3: the parameters of BM25 that we want to optimize. Usually, k_1 is set to a value between 1 and 3, b is set somewhere around 0.8 and k_3 is set to 0. Note that when k_3 is set to 0 the entire right part of the product in Eq. 2 cancels out to 1.

3.2 Learning from Clicks

Most learning to rank approaches learn from explicit, manually produced relevance assessments [15]. These assessments are expensive to obtain and usually produced in an artificial setting. More importantly, it is not always feasible to obtain the assessments needed. For instance, if we want to adapt a ranker towards a specific user or a group of users, we cannot ask explicit feedback from these users as it would put an undesirable burden upon these users.

Instead, we optimize rankers using clicks. It has been shown by Radlinski et al. [18] that interpreting clicks as absolute relevance judgments is unreliable. Therefore, we use a dueling bandit approach: the candidate preselection (CPS) method. This method was shown to be state-of-the-art by Hofmann et al. [9]. It is an extension of the dueling bandit gradient descent (DBGD) method, proposed in [26]. Briefly, DBGD works as follows. The parameters that are being optimized are initialized. When a query is presented to the learning system, two rankings are generated: one with the parameters set at the current best values, another with a perturbation of these parameters. These two rankings are interleaved using probabilistic interleave [7, 8], which allows for the reuse of historical interactions. The interleaved list is presented to the user and we observe the clicks that the user produces, which are then used to determine which of the two generated rankings was best. If the ranking produced with the perturbed set of parameters wins the interleaved comparison, then the current best parameters are adapted in the direction of the perturbation. CPS is a variant of DBGD that produces several candidate perturbations and compares these on historical click data to decide on the most promising candidate. Only the ranking produced with the most promising perturbation is then actually interleaved with the ranking generated with the current best parameters and exposed to the user.

The difference between the current best ranker and the perturbed ranker is controlled by the parameter δ. The amount of adaptation of the current best ranker in case the perturbed ranker wins is controlled by a second parameter, α. Together, these parameters balance the speed and the precision with which the algorithm learns. If they are too big, the learning algorithm may oscillate, skip over optimal values and never converge to the optimum. If they are too small, the learning algorithm will not find the optimum in a reasonable amount of time.

We aim to learn the BM25 parameters k_1, b and k_3 from clicks, using the learning method described above. Because the parameters are of very different orders of magnitude, with b typically ranging between 0.45 and 0.9 and k_1 typically ranging between 2 and 25, we chose to use a separate δ and α for each parameter. This is necessary because what may be a reasonable step size for k_1 will be far too large for b. Therefore we have, for example, a separate δ_{k_1} and δ_b. This allows us to govern the size of exploration and updates in each direction.

4   Experimental Setup

We investigate whether we can optimize the parameters of a base ranker, BM25, from clicks produced by users interacting with a search engine. Below, we first describe the data we use to address this question. Then we describe how our click-streams are generated, and our evaluation setup.[1]

4.1   Data

For all our experiments we use features extracted from the .gov collection that is also included in the LETOR data set [16]. The .gov collection consists of a crawl of the .gov

---

[1] All our code is open source, available at https://bitbucket.org/ilps/lerot [21].

Table 1. Instantiations of the DCM click model used in our experiments, following [9]

	$P(click = 1\|R)$		$P(stop = 1\|R)$	
	$r = 0$	$r = 1$	$r = 0$	$r = 1$
perfect	0.0	1.0	0.0	0.0
navigational	0.05	0.95	0.2	0.9
informational	0.4	0.9	0.1	0.5
almost random	0.4	0.6	0.5	0.5

domain with about 1M documents. The six sets of queries and relevance assessments we use are based on TREC Web track tasks from 2003 and 2004. The data sets HP2003, HP2004, NP2003, and NP2004 implement navigational tasks: homepage finding and named-page finding, respectively. TD2003 and TD2004 implement an informational task: topic distillation. All six data sets contain between 50 and 150 queries and approximately 1,000 judged documents per query.

We index the original .gov collection to extract low-level features such as term frequency and inverse document frequency that are needed for BM25. While indexing, we do not perform any pre-processing (e.g., no stemming, no stop word removal). We only extract features for the documents in the LETOR data set [16]. All the data sets we use are split by query for 5-fold cross validation.

4.2   Clicks

We employ a click simulation framework analogous to that of [9]. We do so because we do not have access to a live search engine or a suitable click log. Note that, even if a click log was available, it would not be adequate since the learning algorithm is likely to produce result lists that never appear in the log.

In our click simulation framework, we randomly sample with replacement a query from the queries in the data set. The learning system is then responsible for producing a ranking of documents for this query. This ranking is then presented to a simulated user, which produces clicks on documents that can be used by the learning system to improve the ranker. In our experiments, we use the Dependent Click Model (DCM) by Guo et al. [6] to produce clicks. This click model assumes that users scan a result list from top to bottom. For each examined document, users are assumed to determine whether it potentially satisfies their information need enough to click the document. This decision is based on relevance labels manually given to each document-query combination. We model this as $P(C|R)$: the probability of clicking a document given its relevance label. After clicking, the user might continue scanning down the document list or stop, which is modeled as $P(S|R)$: the probability of stopping after a document given its relevance label. Again following [9], we instantiate $P(C|R)$ and $P(S|R)$ as in Table 1. We use four instantiations of the click model: the *perfect* click model, in which exactly every relevant document is clicked; the *navigational* click model, in which users almost only click relevant documents and usually stop when they have found a relevant document; the *informational* click model, in which non-relevant documents are also clicked quite often and users stop after finding a relevant document only half of the time; and the *almost random* click model in which there is only a very small difference in user behavior for relevant and non-relevant documents.

4.3 Learning Setup

We employ the learning approach described in Section 3.2. For CPS we use the parameters suggested by [9]: we use $\eta = 6$ candidate perturbations and we use the $\lambda = 10$ most recent queries. We initialize the weights of the BM25 model randomly. For the learning of BM25 parameters, we set $\alpha_b = 0.05$ and $\delta_b = 0.5$. We computed the average ratio between k_1 and b across the parameter values that were optimal for the different data sets, and set α_{k_1} and δ_{k_1} accordingly. This ratio was 1 to 13.3, so we set $\alpha_{k_1} = 0.665$ and $\delta_{k_1} = 6.65$. These learning parameters have been tuned on a held out development set.

Table 2. NDCG scores for various values of BM25 parameters k_1 and b, optimized for different data sets. The first and last row give the scores with parameters that have been manually tuned for the .gov and OHSUMED collections, respectively [16]. Other parameter values are chosen to produce maximal scores, printed in boldface, for the different data sets listed in the first column. For all results, $k_3 = 0$. Statistically significant improvements (losses) over the values manually tuned for .gov are indicated by $\triangle$ ($p < 0.05$) and $\blacktriangle$ ($p < 0.01$) ($\triangledown$ and $\blacktriangledown$).

	k_1	b	HP2003	HP2004	NP2003	NP2004	TD2003	TD2004	Overall
.gov	2.50	0.80	0.674	0.629	0.693	0.599	0.404	0.469	0.613
HP2003	7.40	0.80	**0.692**	0.650	0.661$\blacktriangledown$	0.591	0.423$\blacktriangle$	0.477	0.614
HP2004	7.30	0.85	0.688	**0.672$\triangle$**	0.657$\blacktriangledown$	0.575	0.423$\blacktriangle$	0.482$\triangle$	0.613
	2.50	0.85	0.671	0.613	0.682	0.579$\triangledown$	0.404	0.473	0.605$\triangledown$
	7.30	0.80	0.690	0.647	0.661$\blacktriangledown$	0.592	0.423$\blacktriangle$	0.477	0.613
NP2003	2.60	0.45	0.661	0.572$\triangledown$	**0.719**	0.635	0.374$\blacktriangledown$	0.441$\blacktriangledown$	0.607
	2.50	0.45	0.660	0.572$\triangledown$	0.718	0.635	0.374$\blacktriangledown$	0.441$\blacktriangledown$	0.607
	2.60	0.80	0.675	0.629	0.692	0.601	0.403	0.470	0.613
NP2004	4.00	0.50	0.663	0.584	0.705	**0.647$\triangle$**	0.386$\triangledown$	0.446$\blacktriangledown$	0.609
	2.50	0.50	0.663	0.573$\triangledown$	0.713	0.635	0.381$\blacktriangledown$	0.444$\blacktriangledown$	0.607
	4.00	0.80	0.680	0.645	0.683	0.605	0.414$\triangle$	0.474	0.616
TD2003	25.90	0.90	0.660	0.597	0.515$\blacktriangledown$	0.478$\blacktriangledown$	**0.456$\triangle$**	0.489$\triangle$	0.550$\blacktriangledown$
	2.50	0.90	0.676	0.607	0.672	0.560$\blacktriangledown$	0.405	0.471	0.600$\blacktriangledown$
	25.90	0.80	0.645	0.576	0.535$\blacktriangledown$	0.493$\blacktriangledown$	0.445	0.482	0.549$\blacktriangledown$
TD2004	24.00	0.90	0.664	0.604	0.520$\blacktriangledown$	0.481$\blacktriangledown$	0.449$\triangle$	**0.491$\triangle$**	0.553$\blacktriangledown$
	2.50	0.90	0.676	0.607	0.672	0.560$\blacktriangledown$	0.405	0.471	0.600$\blacktriangledown$
	24.00	0.80	0.645	0.578	0.538$\blacktriangledown$	0.496$\blacktriangledown$	0.446	0.482	0.550$\blacktriangledown$
OHSUMED	1.20	0.75	0.662$\triangledown$	0.589$\blacktriangledown$	0.703	0.591	0.398	0.461$\blacktriangledown$	0.605$\triangledown$

4.4 Evaluation and Significance Testing

As evaluation metric, we use nDCG [11] on the top 10 results, measured on the test sets, following, for instance, [23, 25]. For each learning experiment, for each data set, we run the experiment for 2,000 interactions with the click model. We repeat each experiment 25 times and average results over the 5 folds and these repetitions. We test for significant differences using the paired t-test in answering RQ1. and the independent measures t-test for RQ2..

5   Results and Analysis

We address our research questions in the following two subsections.

5.1   Measuring the Performance of BM25 with Manually Tuned Parameters

In order to answer RQ1., we compute the performance of BM25 with the parameters used in the LETOR data set [16]. The parameter values used there differ between the two document collections in the data set. The values that were chosen for the .gov collection were $k_1 = 2.5$, $b = 0.8$ and $k_3 = 0$. The values that were chosen for the OHSUMED collection were $k_1 = 1.2$, $b = 0.75$ and $k_3 = 7$. We refer to these values as the manually tuned .gov or OHSUMED parameter values, respectively. Note that the manually tuned .gov parameter values were tuned to perform well on average, over all data sets.

The results of running BM25 with the .gov manual parameters (as described in Section 4) are in the first row of Table 2. We also experiment with different values of k_1, b and k_3. We first tried a range of values for k_1 and b. For k_1 the range is from -1 to 30 with steps of 0.1 and for b the range is for -0.5 to 1 with steps of 0.05. The results are in Table 2. For each of the data sets, we include the parameter values that gave maximal nDCG scores (in bold face). For each value of k_1 and b, we show the performance on each data set and the average performance over all data sets.

The results show that when we average over all data sets, no significant improvements to the manually tuned .gov parameter values can be found. This is to be expected, and merely shows that the manual tuning was done well. However, for four out of six data sets, a significant improvement can be achieved by deviating from the manually tuned .gov parameter values *for that particular dataset*. Furthermore, taking the average optimal nDCG, weighted with the number of queries in each data set, yields an overall performance of 0.644 nDCG. Thus, it pays to optimize parameters for specific data sets.

In cases where both k_1 and b were different from the manually tuned .gov values, we also consider the results of combining k_1 with the manually tuned .gov value for b and vice-versa. E.g., when $k_1 = 4.0$ and $b = 0.5$, for the NP2004 data set the value of b has a bigger impact than the value of k_1: changing k_1 back to the manually tuned value causes a decrease of nDCG of 0.012 points, while changing b to the manually tuned value gives a decrease of 0.042 points. However, in other cases the value of k_1 seems to be more important. E.g., for the TD2003 data set we can achieve an improvement of 0.041 points by changing k_1 to 25.9, while keeping b at the manually tuned 0.8.

The bottom row in Table 2 shows the results of a BM25 ranker with the manually tuned OHSUMED parameter values. This ranker performs worse than the manually tuned .gov values averaged over all data sets, which, again, shows that it makes sense to tune these parameters, rather than just taking a standard value from the literature.

For the third parameter k_3, we performed similar experiments, varying the parameter value from 0 to 1,000. There were no significant differences in the resulting nDCG scores. The small differences that did occur favored the manually tuned value 0. The fact that k_3 hardly has any influence on the performance is to be expected, since k_3 weights the query term frequency (cf. Equation 2), the number of times a word appears in the query. For most query terms, the query term frequency is 1. Since the weight of this feature does not greatly affect the result, we omit k_3 from the rest of our analysis.

Using these results, we are ready to answer our first research question, RQ1.. The manually tuned .gov values for k_1 and b are quite good when we look at a combination of all data sets. When looking at different data sets separately, significant improvements can be reached by deviating from the values that were manually tuned for the entire collection. This shows that tuning of the parameters to a specific setting is a promising idea. Considering the last parameter k_3, the standard value was optimal for the data sets we investigated.

5.2   Learning Parameters of BM25 Using Clicks

In this section we answer RQ2.: can we learn the parameters of BM25 from clicks? We aim to learn the parameters per data set from clicks. Our primary goal is not to beat the performance of the manually tuned .gov parameters. Should optimizing a base ranker such as BM25 prove successful, i.e., reach or even beat manually tuned values, the advantage is rather that optimizing the parameters no longer requires human annotations. Furthermore, learning the parameters eliminates the need for domain-specific

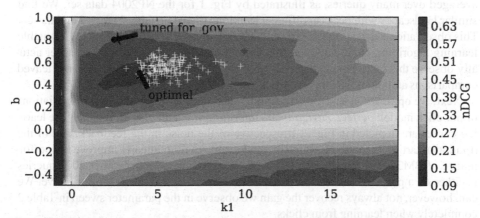

Fig. 1. Optimization landscape for two parameters of BM25, k_1 and b, for the NP2004 data set measured with nDCG. White crosses indicate where individual runs of the learning algorithm plateaued when learning from clicks produced with the *perfect* click model. For the other five data sets we experimented with we obtained a similar landscape with the peak on a different location.

84   A. Schuth et al.

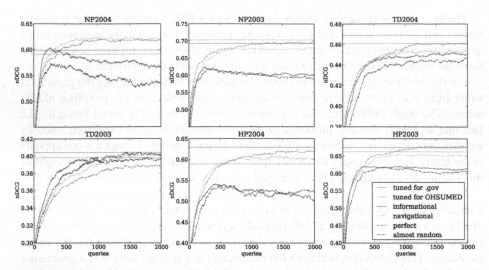

Fig. 2. Learning curves when learning the parameters of BM25 using DBGD from clicks. Measured in nDCG on a holdout data set averaged over 5-fold cross validation and 25 repetitions. The clicks used for learning are produced by the *perfect, navigational, informational* and *almost random* click model. The horizontal gray lines indicate the performance for the manually tuned .gov (solid) and OHSUMED (dotted) parameters.

knowledge, which is not always available, or sweeps over possible parameter values, which cost time and cannot be done online.

To begin, we visualize the optimization landscape for the two BM25 parameters that matter: k_1 and b. We use the data obtained from the parameter sweep described in Section 5.1. The optimization landscape is unimodal and generally smooth when averaged over many queries, as illustrated by Fig. 1 for the NP2004 data set. We find similar landscapes with peaks at different locations (listed in Table 2) for other data sets. This observation suggests that a gradient descent approach such as DBGD is a suitable learning algorithm. Note, however, that the online learning algorithm will never actually observe this landscape: it can only observe *relative* feedback from the interleaved comparisons and moreover, this feedback is observed on a *per query* basis.

Next, we optimize k_1 and b for each individual data set using the four instantiations of our click model: *perfect, navigational, informational*, and *almost random*. The learning curves are depicted in Fig. 2. Irrespective of the noise in the feedback, and of the (random) starting point, the learning method is able to dramatically improve the performance of BM25. For the *perfect* click model the final performance after 2000 queries is either on a par with the manually tuned values used by Qin et al. [16] or above. We can, however, not always recover the gain we observe in the parameter sweep in Table 2 completely when learning from clicks.

For the NP2004 data set we have plotted the final parameter values that have been learned using the *perfect* click model in Fig. 1. The final parameter values are clustered near the optimal value, indicating that the learning method is indeed capable of finding the peak of the landscape. Final parameters for individual runs using each data set are

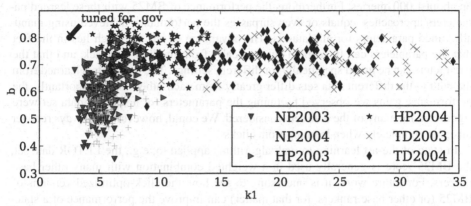

Fig. 3. Parameter settings where individual runs of the learning algorithm plateaued when learning with the *perfect* click model for 2,000 queries

depicted in Fig. 3. For each data set, the parameters converge to a different region. We also see that the manually tuned parameters are not included in any of these regions.

Performance generally degrades when clicks become less reliable. However, the performance of the navigational click model is not much lower than the performance of the perfect click model. This is a promising result, since feedback from actual users will be noisy and our learning method should be able to deal with that.

The above experiments are all initialized with random starting parameters. In cases where a good starting point is known, learning can be sped up. For example, we also initialized learning with the manually tuned .gov parameters ($k_1 = 2.5$ and $b = 0.8$) and found a plateau that was not different from the one found with random initialization. It was, however, found in fewer than 200 queries, depending on the data set.

In conclusion, we can give a positive answer to the first part of RQ2.. Learning good values for the BM25 parameters from user clicks is possible. As to the second part of RQ2., the optimized parameters learned form clicks lead to performance of BM25 that approaches, equals or even surpasses the performance achieved using manually tuned parameters for all datasets.

6   Conclusion

In this paper we investigated the effectiveness of using clicks to optimize base rankers in an online learning to rank setting. State-of-the-art learning to rank approaches use a linear combination of several base rankers to compute an optimal ranking. Rather than learning the optimal weights for this combination, we optimize the internal parameters of these base rankers. We focussed on the base ranker BM25 and aimed at learning these parameters of BM25 in an online setting using clicks.

Our results show that learning good parameters of BM25 from clicks is indeed possible. As a consequence, it is not necessary to hand tune these parameters or use human assessors to obtain labeled data. Learning with a dueling bandit gradient descent approach converges to near-optimal parameters after training with relative click feedback on about 1,000 queries. Furthermore, the performance of BM25 with these learned parameters approaches, equals or even surpasses the performance achieved using manually tuned parameters for all datasets. The advantage of our approach lies in the fact that the parameters can be learned automatically from implicit feedback, and that the parameters can be tuned specifically for different settings. The parameters learned from the data in the different data sets differ greatly from each other. More importantly, the performance gains we observed by tuning the parameters for a specific data set were significant for many of the data sets considered. We could, however, not always recover this gain completely when learning from clicks.

In state-of-the-art learning to rank algorithms applied to, e.g., the LETOR data set, the BM25 score is generally used in a weighted combination with many other base rankers. For future work, it is interesting to see how the click-optimized versions of BM25 (or other base rankers, for that matter) can improve the performance of a state-of-the-art learning to rank algorithm when BM25 is used as one query-document feature among many. There are several ways in which the process of optimizing a base ranker can be integrated with state-of-the-art online learning to rank: (1) first learn the optimal parameters of the base rankers in isolation, then use the optimized base rankers and learn the optimal weights of a linear combination of these base rankers; (2) first learn the optimal weights of a linear combination of base rankers using either optimized or default parameter values for the individual base rankers and then learn the optimal parameters of the individual base rankers based on user clicks in reply to the outcome of the ensemble learning-to-rank algorithm; (3) learn the parameters of the base rankers and the weights of the ensemble together based on user clicks in reply to the outcome of the ensemble.

Acknowledgments. This research was supported by the European Community's Seventh Framework Programme (FP7/2007-2013) under grant agreement nr 288024 (LiMoSINe project), the Netherlands Organisation for Scientific Research (NWO) under project nrs 640.004.802, 727.011.005, 612.001.116, HOR-11-10, 612.066.930, the Center for Creation, Content and Technology, the QuaMerdes project funded by the CLARIN-nl program, the TROVe project funded by the CLARIAH program, the Dutch national program COMMIT, the ESF Research Network Program ELIAS, the Elite Network Shifts project funded by the Royal Dutch Academy of Sciences, the Netherlands eScience Center under project nr 027.012.105 and the Yahoo! Faculty Research and Engagement Program.

References

[1] Asadi, N., Metzler, D., Elsayed, T., Lin, J.: Pseudo test collections for learning web search ranking functions. In: SIGIR 2011, pp. 1073–1082. ACM (2011)

[2] Azzopardi, L., de Rijke, M., Balog, K.: Building simulated queries for known-item topics: An analysis using six European languages. In: SIGIR 2007, pp. 455–462. ACM (2007)

[3] Berendsen, R., Tsagkias, M., Weerkamp, W., de Rijke, M.: Pseudo test collections for training and tuning microblog rankers. In: SIGIR 2013. ACM (2013)

[4] Carterette, B., Jones, R.: Evaluating search engines by modeling the relationship between relevance and clicks. In: NIPS 2007, pp. 217–224. MIT Press (2008)

[5] Gao, N., Deng, Z.-H., Yu, H., Jiang, J.-J.: Listopt: Learning to optimize for XML rank-
     ing. In: Huang, J.Z., Cao, L., Srivastava, J. (eds.) PAKDD 2011, Part II. LNCS, vol. 6635,
     pp. 482–492. Springer, Heidelberg (2011)
[6] Guo, F., Liu, C., Wang, Y.M.: Efficient multiple-click models in web search. In: WSDM
     2009. ACM (2009)
[7] Hofmann, K., Whiteson, S., de Rijke, M.: A probabilistic method for inferring preferences
     from clicks. In: CIKM 2011. ACM (2011)
[8] Hofmann, K., Whiteson, S., de Rijke, M.: Estimating interleaved comparison outcomes
     from historical click data. In: CIKM 2012. ACM (2012)
[9] Hofmann, K., Schuth, A., Whiteson, S., de Rijke, M.: Reusing historical interaction data for
     faster online learning to rank for IR. In: WSDM 2013. ACM (2013)
[10] Hofmann, K., Whiteson, S., de Rijke, M.: Balancing exploration and exploitation in listwise
     and pairwise online learning to rank for information retrieval. Information Retrieval 16(1),
     63–90 (2013)
[11] Järvelin, K., Kekäläinen, J.: Cumulated gain-based evaluation of IR techniques. ACM Trans-
     actions on Information Systems (TOIS) 20(4) (2002)
[12] Ji, S., Zhou, K., Liao, C., Zheng, Z., Xue, G.-R., Chapelle, O., Sun, G., Zha, H.: Global
     ranking by exploiting user clicks. In: SIGIR 2009, pp. 35–42. ACM (2009)
[13] Joachims, T.: Optimizing search engines using clickthrough data. In: KDD 2002 (2002)
[14] Jung, S., Herlocker, J.L., Webster, J.: Click data as implicit relevance feedback in web
     search. Information Processing & Management 43(3), 791–807 (2007)
[15] Liu, T.-Y.: Learning to rank for information retrieval. Foundations and Trends in Informa-
     tion Retrieval 3(3), 225–331 (2009)
[16] Qin, T., Liu, T.-Y., Xu, J., Li, H.: LETOR: A benchmark collection for research on learning
     to rank for information retrieval. Information Retrieval 13(4), 346–374 (2010)
[17] Radlinski, F., Kleinberg, R., Joachims, T.: Learning diverse rankings with multi-armed ban-
     dits. In: ICML 2008. ACM (2008a)
[18] Radlinski, F., Kurup, M., Joachims, T.: How does clickthrough data reflect retrieval quality?
     In: CIKM 2008, pp. 43–52. ACM, New York (2008b)
[19] Robertson, S., Walker, S.: Okapi at TREC-3. In: TREC-3. NIST (1995)
[20] Robertson, S., Zaragoza, H.: The probabilistic relevance framework: BM25 and beyond.
     Foundations and Trends in Information Retrieval 3(4), 333–389 (2009)
[21] Schuth, A., Hofmann, K., Whiteson, S., de Rijke, M.: Lerot: An online learning to rank
     framework. In: LivingLab 2013, pp. 23–26. ACM (2013)
[22] Sparck Jones, K., Walker, S., Robertson, S.: A probabilistic model of information retrieval:
     Development and comparative experiments. Information Processing and Management 36,
     779–808, 809–840 (2000)
[23] Svore, K.M., Burges, C.J.: A machine learning approach for improved BM25 retrieval. In:
     CIKM 2009. ACM (2009)
[24] Tague, J., Nelson, M., Wu, H.: Problems in the simulation of bibliographic retrieval systems.
     In: SIGIR 1980, pp. 236–255 (1980)
[25] Taylor, M., Zaragoza, H., Craswell, N., Robertson, S., Burges, C.: Optimisation methods
     for ranking functions with multiple parameters. In: CIKM 2006. ACM (2006)
[26] Yue, Y., Joachims, T.: Interactively optimizing information retrieval systems as a dueling
     bandits problem. In: ICML 2009 (2009)

Predicting Search Task Difficulty

Jaime Arguello

University of North Carolina at Chapel Hill, USA
jarguello@unc.edu

Abstract. Search task difficulty refers to a user's assessment about the amount of effort required to complete a search task. Our goal in this work is to learn predictive models of search task difficulty. We evaluate features derived from the user's interaction with the search engine as well as features derived from the user's level of interest in the task and level of prior knowledge in the task domain. In addition to user-interaction features used in prior work, we evaluate features generated from scroll and mouse-movement events on the SERP. In some situations, we may prefer a system that can predict search task difficulty early in the search session. To this end, we evaluate features in terms of whole-session evidence and first-round evidence, which excludes all interactions starting with the second query. Our results found that the most predictive features were different for whole-session vs. first-round prediction, that mouseover features were effective for first-round prediction, and that level of interest and prior knowledge features did not improve performance.

1 Introduction

Search engine users engage in a wide variety of search tasks. A large body of prior research focused on characterizing different types of search tasks (see Li and Belkin [10]). The motivation behind this prior work is to understand how task characteristics influence search behavior and how search systems can provide customized interactions for different task types. One important search task characteristic is *search task difficulty*. Search task difficulty refers to the user's assessment about the amount of effort required to complete the search task. In this work, we learn and evaluate predictive models of *post-task* difficulty, which refers to the user's assessment *after* completing the search task. Predicting search task difficulty has important implications for IR. First, it can help system designers determine the types of search tasks that are not well-supported by the system. Second, it can help researchers discover correlations between search task difficulty and undesirable outcomes such as search engine switching. Finally, predicting search task difficulty in real-time would enable a system to intevene and assist the user in some way.

To train and evaluate our models, we first conducted a user study to collect search-interaction data and post-task difficulty judgments from searchers. In order to collect data from a large number of search sessions and users, the study was conducted using crowdsourcing. Participants were given carefully constructed

M. de Rijke et al. (Eds.): ECIR 2014, LNCS 8416, pp. 88–99, 2014.

search tasks and asked to use our search system to find and bookmark web-pages that would be useful in completing the task. We used search tasks that we thought would cause participants to experience varying levels of difficulty. After completing each search, participants were given a post-task questionnaire that included several questions about the level of difficulty experienced while searching. Responses to these questions were averaged into single difficulty scale and this measure was used to group search sessions into *easy* and *difficult* searches. Our goal was to learn models to predict whether a search session was considered easy or difficult using behavioral measures derived from the search session. We investigate features derived from queries, clicks, bookmarks, mouse-movement and scroll events on the SERP, dwell-times, and the session duration.

Past studies also considered the task of predicting post-task difficulty using behavioral measures from the search session [12, 13]. Our work is the most similar to Liu *et al.* [13], with three main methodological differences. First, we used a larger set of search sessions for training and testing our models (600 vs. 117 [13]). Second, we used a larger number of participants (269 vs. 38 [13]), which potentially introduced more variance in search behavior. Third, we used more search tasks (20 vs. 5 [13]). Using more search tasks allowed us to avoid training and testing on search sessions from the same task. Thus, we believe that our evaluation emphasizes a model's ability to generalize to previously unseen users and tasks.

In addition to differences in methodology, we extend prior work in two directions: (1) we investigate new sources of evidence and (2) we investigate predicting task difficulty at different stages in the search session. In addition to using similar features used in prior work, we experimented with features derived from mouse movement and scrollbar events on SERPs produced by the system. To our knowledge, this is the first study to consider mouse and scroll data for predicting search task difficulty. Additionally, we experimented with features derived from the user's level of interest in the task and level of prior knowledge (domain knowledge and search experience). Our goal was not to *infer* this information about the user. Instead, as a first step, we wanted to assess the value of this information for predicting task difficulty. Thus, our level of interest and prior knowledge features were derived from responses to a pre-task questionnaire.

In certain situations, we may want the system to predict task difficulty *before* the end of the session. This may be the case, for example, if the goal to intervene and assist the user. To this end, we divided our analysis in terms of *whole-session analysis* and *first-round analysis*. Our first-round analysis excludes all search interactions starting with the second query (if any). We evaluate different types of features based on whole-session evidence and first-round evidence.

2   Related Work

A large body of prior work has focused on defining different task characteristics or dimensions (see Li and Belkin [10]). Two different, yet sometimes confused characteristics are task complexity and task difficulty. In this work, we make the same distinction made by Kim [9] and Li and Belkin [10]. Task complexity is an inherent property of the task, independent of the task doer, while task difficulty refers to a user's assessment about the amount of effort required to complete the task. In our study, we manipulated task complexity in order for our participants to experience varying levels of difficulty. Prior work found that task complexity influences task difficulty [3, 18] and we observed a similar trend in our study.

Different characterizations of task complexity have been proposed [3–5, 8, 17]. Jansen et al. [8] (and later Wu et al. [18]) defined task complexity in terms of the amount of cognitive effort and learning required to complete the task. To this end, they adopted a taxonomy of learning outcomes originally developed by Anderson and Krathwohl for characterizing educational materials [1]. We used search tasks created using this *cognitive* view of task complexity.

Several studies investigated the effects of task difficulty or complexity on search behavior [2, 8, 11, 13, 14, 18]. Results show that task difficulty and complexity affect a wide range of behavioral measures. For example, difficult tasks are associated with longer completion times, more queries, more clicks, more clicks on lower ranks, more abandoned queries, more pages bookmarked, longer landing page and SERP dwell-times, and greater use of query-operators. Given the correlation between task difficulty and different behavioral measures, prior work also focused on predicting search task difficulty [13, 12]. Liu et al. [13] combined a large set of features in a logistic regression model and were able to predict two levels of post-task difficulty with about 80% accuracy. Our work builds upon Liu et al. [13] and investigates new features derived from the whole search session and from only the first round of interactions.

Prior work also focused on predicting user actions and emotions likely to be related to task complexity and difficulty. White and Dumais [16] focused on predicting search engine switching—whether a user's next action will be to switch to a different search engine—and evaluated features from the search session, from the user's history, and from interactions from other users for the same query. Most features were found to be complementary. Field et al. [7] focused on predicting searcher frustration. Searchers were periodically asked about their level of frustration and the goal was to predict the user's response. Feild et al. combined search interaction features with physical sensor features derived from a mental state camera, a pressure sensitive mouse, and a pressure sensitive chair. Interestingly, the search interaction features were more predictive than the physical sense features.

3   User Study

In order to train models to predict task difficulty, it was necessary to run a user study to collect search interaction data and difficulty judgements. Participants were given a search task and asked to use a live search engine to find and bookmark webpages that would help them accomplish the task. The user study was run using Amazon's Mechanical Turk (MTurk).[1] Using MTurk allowed us to

---

[1] Mechanical Turk is a crowdsourcing marketplace where *requesters* can publish *human intelligence tasks* or *HITs* for *workers* to complete in exchange for compensation.

collect data from a large number of participants. Each MTurk HIT corresponded to one search session. Our HITs were implemented as external HITs, meaning that everything besides recruitment and compensation was managed by our own server. This allowed us to control the assignment of participants to search tasks and to record all user interactions with the search system. Search results were returned by our server using the Bing Web Search API. As described below, we used the same set of 20 search tasks developed by Wu *et al.* [18]. Each of the 20 search tasks was completed by 30 unique MTurk workers for a total of 600 search sessions. Tasks were assigned to participants randomly, except for two constraints: (1) participants were not allowed to "see" the same search task more than once and (2) in order to gather data from a large number of participants, each worker was not allowed to complete more than eight tasks. Each HIT was priced at $0.50 USD. To help filter malicious workers, we restricted our HITs to workers with an acceptance rate of 95% or greater and, to help ensure English language proficiency, to workers in the US.

All user interaction data was recorded at the server-side. Clicks on search results were recorded using URL re-directs. Clicking on a search result opened the landing page in an HTML frame embedded in a webpage produced by our system. In order to record landing-page dwell-times, we used Javascript and AJAX to catch focus and blur events on this page and communicate these events to our server. Similarly, we used Javascript and AJAX to record and communicate scrolls and mouse movements on the SERP.

Experimental Protocol. Upon accepting the HIT, participants were first given a set of instructions describing the goal of the HIT and the search interface (e.g., how to add/delete bookmarks and view the current set of bookmarks). After clicking a "start" button, participants were shown the search task description and were asked to carefully read the task. Following this, participants were asked to complete a pre-task questionnaire (described below). After completing the pre-task questionnaire, participants were directed to the search interface. Participants were instructed to search naturally by issuing queries and clicking on search results. Clicking a search result opened the landing page inside an HTML frame. Participants were able to bookmark a page using a button labeled "bookmark this page" located above the HTML frame. While bookmarking a page, participants were asked to provide a 2-3 sentence justification for why they bookmarked the page. Participants were not allowed to leave this field blank. At any point in the search process (either from the search interface, the landing page display, or the bookmark view page), participants were able to revisit the task description and to review the current set of bookmarks. From the book-mark view page, participants were able to delete bookmarks and to terminate the search task and proceed onto a post-task questionnaire (described below).

Pre-task Questionnaire. The pre-task questionnaire was completed immediately after reading the search task description. Participants were asked one question about their level of interest and indicated their responses on a five-point scale: How interested are you to learn more about the topic of this task?

(not at all interested, slightly interested, somewhat interested, moderately interested, and very interested). Participants were asked two questions about their level of prior knowledge and indicated their responses on a four-point scale: (1) How much do you already know about the topic of this task? (nothing, a little, some, a great deal) and (2) How many times have you searched for information about this task? (never, 1-2 times, 3-4 times, 5 times or more).

Post-task Questionnaire. The post-task questionnaire was completed after terminating the search task. Participants were asked five questions about task difficulty. The first four asked about the amount of effort expended on different search-related activities: (1) How difficult was it to *search* for information for this task? (2) How difficult was it to *understand* the information the search engine found? (3) How difficult was it to *decide* if the information the search engine found would be *useful* in completing the task? and (4) How difficult was it to determine when you had *enough information* to finish? The fifth question was designed to elicit a summative judgment about the task difficulty: (5) Overall, how difficult was the task? Responses were indicated on a five-point scale: not at all difficult, slightly difficult, somewhat difficult, moderately difficult, and very difficult. We averaged responses to these five questions to form a single difficulty measure. Participant responses indicated a strong internal consistency (Cronbach's $\alpha = .903$).

Search Tasks. We manipulated task complexity in order for participants to experience varying levels of difficulty. To accomplish this, we used the same set of 20 search tasks developed by Wu *et al.* to study the effects of task complexity on search behavior [18]. The tasks from were constructed to reflect different levels of *cognitive* complexity, which refers to the amount of learning and cognitive effort required to complete the task, and are evenly distributed across 4 topical domains (commerce, entertainment, health, and science & technology) and 5 cognitive complexity levels from Anderson and Krathwol's Taxonomy of Learning [1]:

- **Remember:** Retrieving relevant knowledge from long-term memory.
- **Understand:** Constructing meaning through summarizing and explaining.
- **Analyze:** Breaking material into constituent parts and determining how the parts relate to each other and the whole.
- **Evaluate:** Making judgments through checking and critiquing.
- **Create:** Putting elements together to form a new coherent whole.

Figure 1(a) shows the overall distribution of our difficulty scale across all 600 searches and Figure 1(b) shows the individual distributions for each cognitive complexity level. Two trends are worth noting. First, the 20 tasks were not found to be too difficult ($M = 1.749$, $SD = 0.866$). The median difficulty was 1.400. Second, more complex tasks were perceived to be more difficult, which is consistent with previous studies [3, 18]. A Kruskal-Wallis test showed a significant main effect of task complexity on difficulty ($\chi^2(4) = 36.60, p < .001$). Bonferroni-adjusted (Mann-Whitney) post-hoc tests showed significant differences between Remember (R) and all other complexity levels (U, A, E, C).

4 Predicting Task Difficulty

In this work, we cast the difficulty prediction problem as a binary classification problem. Search sessions were grouped into *easy* and *difficult* using a mean split. Search sessions with a difficulty rating equal to or lower than the mean (1.749) were considered easy and search sessions with a difficulty rating greater than the mean were considered difficult. A mean-split resulted in 358 easy and 242 difficult search sessions. We trained L2-regularized logistic regression models using the LibLinear Toolkit [6]. In prior work, logistic regression performed well for the task of predicting search task difficulty [13, 12] and related tasks such as predicting search engine switching [16] and searcher frustration [7].

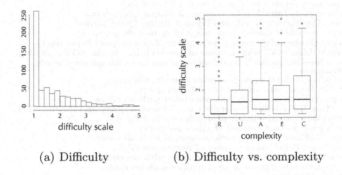

(a) Difficulty   (b) Difficulty vs. complexity

Fig. 1. Difficulty scale distribution

Features. Our models were trained to predict task difficulty as a function of a set of features. We were interested in evaluating different types of features. Thus, we organized our features into the following categories: *query features* were derived from the queries issued to the system, *click features* were derived from clicks on search results, *bookmark features* were derived from bookmarked webpages, *mouse* and *scroll features* were derived from mouse movements and scroll events on the SERP, *dwell-time features* were derived from time spent on a landing page, and *duration features* were derived from the task completion time. In addition to features derived from user-interaction data, we were interested in assessing the value of knowing the user's level of interest in the task and level of prior knowledge of the task domain (domain knowledge and search experience). We did not attempt to infer this information about the user. As a first step, we used participant responses from the pre-task questionnaire.

Before describing our features, a few concepts need clarification. A search result *click* is as instance where the participant clicked on a search result. In contrast, a result *view* is an instance where the participant actually inspected the landing page (recall that we recorded focus and blur events in our landing page display). In most browsers, users can right or mouse-wheel click a result to open the landing page in a hidden tab. In some cases, participants right or mouse-wheel clicked a result, but did not actually opened the hidden tab. A *pagination click* is one where the participant requested results beyond the top 10. A *mouseover* is an instance where the participant's mouse-pointer entered a transparent bounding-box around a search result. Finally, the *scroll position* is a number in the range [0,100] indicating the position of the scrollbar on the SERP (0 = top and 100 = bottom).

We evaluated different feature types based on whole-session and first-round evidence. The following list summarizes our features. Features associated with 'per-query' statistics were not used in our first-round analysis (which included only those interactions associated with the first query). Features included in our first-round analysis are marked with a '*'.

- **Query Features**
  - NumQueries: number of queries.
  - AvgQueryLength: average number of terms per query.
  - NumQueryTerms*: total number of query-terms.
  - UniqueQueryTerms*: total number of unique query-terms.
  - TokenTypeRatio*: NumQueryTerms / UniqueQueryTerms
  - AvgStopwords*: average percentage of stopwords per query.
  - AvgNonStopwords*: average percentage of non-stopwords per query.
  - NumAOLQueries*: total number of queries found in the AOL query-log.
  - NumQuestionQueries*: total number of queries with question words.
- **Click Features**
  - NumClicks*: total number of search results clicked.
  - AvgClicks: average number of clicks per query.
  - AvgClickRank*: average rank associated with all clicks.
  - AvgTimeToFirstClick*: average time between a query and the first click.
  - NumViews*: total number of search results viewed.
  - AvgViews: average number of views per query.
  - AvgClickRank*: average rank associated with all views.
  - NumPageClicks*: total number of pagination clicks.
  - NumAbandon*: total number of queries with no clicks
  - PercentAbandon: percentage of queries with no clicks.
- **Bookmark Features**
  - NumBook*: total number of pages bookmarked.
  - AvgBook: average number of bookmarks per query.
  - AvgBookRank*: average rank associated with all bookmarks.
  - NumQueriesWithBook: total number of queries with a bookmark
  - PercentQueriesWithBook: percentage of queries with a bookmark
  - NumQueriesWithoutBook: total number of queries without a bookmark.
  - PercentQueresWithoutBook: percentage queries with without a bookmark.
  - NumClicksWithoutBook*: total number of clicks without a bookmark.
  - PercentClicksWithoutBook: percentage of clicks without a bookmark.
  - NumViewsWithoutBook*: total number of views without a bookmark.
  - PercentViewsWithoutBook: percentage of views without a bookmark.
- **Mouse Features**
  - TotalMouseovers*: total number of mouseovers in the session.
  - AvgMouseovers: average number of mouseovers per query.
  - MaxMouseover*: max mouseover rank in the session.
  - AvgMaxMouseover: average max mouseover rank per query.
- **Scroll Features**
  - TotalScrollDistance*: total scroll distance in session.
  - AvgScrollDistance: average scroll distance per query.
  - MaxScrollPosition*: max scroll position in session.
  - AvgMaxScrollPosition: average max scroll position per query.
- **Dwell-time Features**
  - TotalDwell*: total time spent on landing pages.
  - AvgDwell*: average time spent on a landing page.
- **Duration Feature**
  - Duration*: total time to task completion.
- **Interest Feature**
  - Interest*: pre-task level of interest response.
- **Prior Knowledge Features**
  - PriorKnowledge*: pre-task level of prior knowledge response.
  - PriorSearch*: pre-task level of prior search experience response.

As described below, we trained and tested our models using cross-validation. For each individual train/test pair, all features were normalized to zero minimum and unit maximum using the training set min/max values.

5   Evaluation Methodology

We collected user-interaction and self-report data for 20 different search tasks and each search task was completed by 30 different study participants, for a total of $20 \times 30 = 600$ search sessions. Training and testing was done using cross-validation. In a production environment, search engine users are likely to engage in a large number of search tasks. For this reason, we felt it was important to not include search sessions for the same task in the training and test data. In other words, we wanted to test a model's ability to generalize to previously unseen tasks. To this end, we used 20-fold cross-validation. Each training set corresponded to the 570 search sessions associated with 19 tasks and each test set corresponded to the 30 search sessions associated with the held-out task. We can also view this as leave-one-task-out cross-validation. Regularized logistic regression uses parameter C to control the misclassification cost on the training data. Parameter C was tuned using a second-level of cross-validation. For each top-level train/test pair, we conducted a second level of 19-fold cross-validation on each training set and used the value of C with the greatest average performance. Parameter C was tuned across values of 2^x where $x = -4, -3, -2, -1, 0, 1, 2, 3, 4$.

Prediction performance was measured using average precision (AP). Logistic regression outputs a prediction confidence value (the probability that the search session is difficult). We used average precision to evaluate a model's ability to rank search sessions in descending order of difficulty. Average precision is proportional to the area under the precision-recall curve. Statistical significance was tested using an approximation of Fisher's randomization test [15]. We report the mean of AP values across our 20 training/test-set pairs. Thus, the randomization was applied to the 20 pairs of AP values for the two models being compared.

6   Results

Our goal was to evaluate different features using whole-session and first-round evidence. Before presenting our classification results, we present an analysis of each feature in isolation. Results in Table 1 show differences in feature values between easy and difficult searches, both in terms of evidence aggregated at the whole-session and first-round level. We used non-parametric Mann-Whitney U tests to compare feature values between easy and difficult searches.

In terms of whole-session evidence, most features had significant differences. Difficult searches had more interaction: longer search sessions, more queries, more clicks and bookmarks, lower-ranked clicks and bookmarks, more pagination clicks, more mouseovers and scrolls on the SERP, and lower-ranked mouseovers and scrolls on the SERP. Difficult searches also had more backtracking: more queries without a click, more clicks without a bookmark, and shorter dwell-times, which suggests more clicks where the participant quickly found the landing page not useful. Features characterizing the types of queries issued were not significantly different. These included the average query length, average number of stopwords, number of queries with a question word, and number of queries in the AOL query-log, which we used as a proxy for query popularity (easy tasks).

In terms of first-round evidence, fewer features had significant differences. There were no significant differences in the number of clicks, views, and bookmarks, and no significant differences for any of the features associated with the query. This indicates that such features become more informative after multiple queries. Interestingly, there were significant differences in the average rank associated with clicks, views, and bookmarks. Mouseover and scroll features were also significant.

Our level of interest and prior knowledge features were not based on interaction data, so their values are the same in both analyses. The observed trend is in the direction we expected (easier searches were associated with greater levels of interest and prior knowledge). However, the differences were not significant.

Figure 2 shows our classification results based on average precision (AP). The row labeled all corresponds to a model using all user-interaction features (excluding our level of interest and prior knowledge features). The rows labeled no.x correspond to models using all user-interaction features except for those in group x. The rows labeled inc.interest and inc.pk correspond to models using all user-interaction features plus our level of interest and level of prior knowledge features, respectively. Finally, the rows labeled only.x correspond to models using only those feature in group x.

As expected, using all user-interaction features (all), whole-session prediction was more effective than first-round prediction ($p < .05$). The whole-session model had access to a greater number of features and a greater number of whole-session features had significant differences between easy and difficult searches (Table 1).

In terms of whole-session evidence, in most cases, omitting a single feature type did not result in a significant drop in performance (see rows labeled no.x). Dwell time features were the only exception. Omitting dwell-time features resulted in an 8.49% drop in AP. This result suggests that our dwell-time features conveyed information not conveyed by the other features. That said, no single feature type on its own (including our dwell-time features) approached the performance of the model using all features (see rows labeled only.x). All models using a single feature type performed significantly worse. Taken together, these results suggest that given whole-session evidence, the best approach is to combine a wide range of features (including dwell-time features) in a single model.

In terms of first-round evidence, we see slightly different trends. As in the whole-session analysis, in most cases, omitting a single feature type did not result in a significant drop in performance (see rows labeled no.x). The largest drop in AP (7.77%) came from omitting bookmark features and not dwell-time features. Combined with the analysis in Table 1, this suggests that clicking and not bookmarking a page in the first round of results is highly predictive of task difficulty. In fact, using bookmark features alone (only.book) resulted in a 2.81% improvement over the model using all features.

Between mouse movement and scroll features, mouse movement features were more predictive of task difficulty. In terms of whole-session evidence, a model using only mouse features (only.mouse) performed only 5.41% worse than one

Table 1. Feature Analysis. Mean (STD) feature values for easy and difficult searches. A ▲(▼) denotes a significant increase(decrease) in the measure in difficult vs. easy searches ($p < .05$).

	Whole-Session Analysis		First-Round Analysis	
	easy	difficult	easy	difficult
Query Features				
NumQueries	1.810 (1.462)	2.373 (1.641)▲	-	-
AvgQueryLength	5.398 (2.980)	5.779 (3.702)	-	-
NumQueryTerms	9.073 (8.251)	12.448 (10.333)▲	5.415 (3.346)	5.772 (3.889)
UniqueQueryTerms	6.504 (3.666)	8.091 (5.039)▲	5.246 (2.999)	5.622 (3.549)
TokenTypeRatio	1.315 (0.628)	1.471 (0.590)▲	1.019 (0.068)	1.014 (0.044)
AvgStopwords	0.201 (0.212)	0.204 (0.196)	0.203 (0.225)	0.217 (0.225)
AvgNonStopwords	0.799 (0.212)	0.796 (0.196)	0.797 (0.225)	0.783 (0.225)
NumAOLQueries	0.286 (0.705)	0.295 (0.731)	0.165 (0.387)	0.112 (0.329)
NumQuestionQueries	0.286 (0.573)	0.336 (0.625)	0.216 (0.451)	0.224 (0.418)
Click Features				
NumClicks	3.263 (2.481)	4.618 (3.292)▲	2.289 (2.022)	2.527 (2.446)
AvgClicks	2.161 (1.739)	2.425 (2.033)	-	-
AvgClickRank	2.704 (1.737)	3.701 (3.517)▲	3.152 (2.645)	4.089 (3.819)▲
AvgTimeToFirstClick	8.613 (8.278)	8.351 (7.062)	48.425 (134.743)	63.253 (155.576)
NumViews	2.815 (2.055)	3.793 (2.623)▲	1.983 (1.644)	2.087 (1.980)
AvgViews	1.901 (1.507)	2.040 (1.703)	-	-
AvgViewRank	2.697 (1.795)	3.713 (3.499)▲	3.217 (2.756)	4.555 (3.988)▲
NumPageClicks	0.092 (0.450)	0.282 (0.937)▲	0.059 (0.381)	0.133 (0.724)▲
NumAbandon	0.294 (0.779)	0.378 (0.755)▲	0.132 (0.392)	0.149 (0.357)
PercentAbandon	0.078 (0.178)	0.106 (0.196)▲	-	-
Bookmark Features				
NumBook	2.336 (1.559)	2.722 (1.509)▲	1.681 (1.374)	1.531 (1.372)
AvgBook	1.620 (1.258)	1.548 (1.238)	-	-
AvgBookRank	2.713 (1.865)	3.900 (3.793)▲	3.425 (2.919)	4.971 (4.220)▲
NumQueriesWithBook	1.359 (0.790)	1.651 (0.905)▲	-	-
PercentQueriesWithBook	0.875 (0.229)	0.814 (0.257)▼	-	-
NumQueriesWithoutBook	0.451 (1.020)	0.722 (1.205)▲	-	-
PercentQueresWithoutBook	0.125 (0.229)	0.186 (0.257)▲	-	-
NumClicksWithoutBook	0.927 (1.521)	1.896 (2.821)▲	0.608 (1.237)	0.996 (1.721)▲
PercentClicksWithoutBook	0.184 (0.242)	0.275 (0.279)▲	-	-
NumViewsWithoutBook	0.479 (0.996)	1.071 (2.103)▲	0.303 (0.698)	0.556 (1.214)▲
PercentViewsWithoutBook	0.105 (0.193)	0.176 (0.253)▲	-	-
Mouse Features				
TotalMouseovers	23.039 (32.056)	42.602 (52.086)▲	15.602 (26.080)	22.494 (38.002)▲
AvgMouseovers	12.307 (13.160)	16.185 (15.026)▲	-	-
MaxMouseover	5.734 (5.229)	8.664 (7.845)▲	4.815 (4.889)	6.212 (6.268)▲
AvgMaxMouseovers	4.486 (3.346)	5.943 (4.432)▲	-	-
Scroll Features				
TotalScrollDistance	105.532 (161.087)	182.154 (244.690)▲	64.636 (114.955)	91.699 (147.264)▲
AvgScrollDistance	55.118 (83.464)	64.382 (74.730)▲	-	-
MaxScrollPosition	39.067 (44.027)	53.610 (45.904)▲	29.528 (40.854)	39.013 (44.387)▲
AvgMaxScrollPosition	28.626 (36.012)	34.635 (35.586)▲	-	-
Dwell-Time Features				
TotalDwell	100.577 (112.695)	91.984 (105.488)	74.736 (95.838)	67.214 (105.407)▼
AvgDwell	42.998 (50.161)	29.351 (26.185)▼	42.009 (59.925)	31.458 (47.590)▼
Duration	193.596 (145.959)	223.964 (151.590)▲	140.766 (123.834)	130.235 (128.194)▼
Interest	2.838 (1.257)	2.635 (1.114)	2.838 (1.257)	2.635 (1.114)
Prior Knowledge Features				
PriorKnowledge	1.919 (0.937)	1.834 (0.845)	1.919 (0.937)	1.834 (0.845)
PriorSearch	1.437 (0.786)	1.378 (0.703)	1.437 (0.786)	1.378 (0.703)

98   J. Arguello

Table 2. Feature Ablation Analyses. A ▼ denotes a significant drop in performance compared to all ($p < .05$).

	Whole-Session Analysis	First-Round Analysis
all	0.618	0.563
no.query	0.616 (-0.39%)	0.576 (2.28%)
no.clicks	0.617 (-0.22%)	0.551 (-2.09%)
no.book	0.616 (-0.43%)	0.519 (-7.77%)▼
no.mouse	0.616 (-0.31%)	0.568 (0.85%)
no.scroll	0.625 (1.12%)	0.562 (-0.13%)
no.dwell	0.566 (-8.49%)▼	0.558 (-0.83%)
no.duration	0.622 (0.61%)	0.561 (-0.28%)
inc.interest	0.612 (-1.08%)	0.568 (0.90%)
inc.pk	0.613 (-0.83%)	0.554 (-1.62%)
only.query	0.547 (-11.47%)▼	0.516 (-8.28%)
only.clicks	0.576 (-6.81%)▼	0.528 (-6.23%)▼
only.book	0.582 (-5.83%)▼	0.579 (2.81%)
only.mouse	0.585 (-5.41%)▼	0.519 (-7.83%)
only.scroll	0.483 (-21.88%)▼	0.490 (-12.95%)▼
only.dwell	0.526 (-14.95%)▼	0.495 (-12.00%)▼
only.duration	0.501 (-18.98%)▼	0.513 (-8.78%)
only.interest	0.467 (-24.50%)▼	0.467 (-17.06%)▼
only.pk	0.479 (-22.45%)▼	0.479 (-14.81%)▼

using all features. In terms of first-round evidence, a model using only mouse features performed only 7.83% worse.

Consistent with the analysis in Table 1, our level of interest and prior knowledge features were not highly effective for predicting search task difficulty. Including each feature type resulted in only a slight difference in performance (inc.interest and inc.pk) and neither feature set of its own approached the performance of a model using all features (only.interest and only.pk).

7 Discussion and Conclusion

We evaluated different types of features for predicting search task difficulty at different points in the session: after the whole session and after the first round of interactions. Our results suggest that following trends. First, whole-session prediction was more effective than first-round prediction. While this may not be surprising, it is an important result because a major motivation for predicting search task difficulty is to develop search assistance interventions that would have to trigger *before* the end of the session. Second, for both whole-session and first-round prediction, the best approach is to combine a wide range of features. In our results, there were no cases where a single feature type significantly outperformed the model with all features. Third, the most predictive features were different in both analyses. Dwell-time features were the most predictive for whole-session prediction and bookmark features were the most predictive for first-round prediction. With respect to bookmarks, it is worth noting that existing search engines do not typically track bookmarks. This suggests the importance of capturing explicit relevance judgements or predicting relevance judgements implicitly for the purpose of first-round prediction. Fourth, mouse-movement features were more predictive than scroll features. For first-round prediction, a model using only mouse-movement features approached the performance of the model with all features. Finally, including level of interest and prior knowledge features did not improve prediction performance.

In terms of future work, several open questions remain. Our experiment was conducted in a semi-controlled environment with simulated search tasks. Future work should consider predicting search task difficulty in a more naturalistic setting with user-initiated search tasks. In more a real-world setting, the distribution of easy vs. difficult tasks may be highly skewed and user interaction signals are likely to be noisier. Additionally, overall our tasks were not found to be too difficult. It remains to be seen whether level of interest and prior knowledge features are predictive for highly difficult search tasks.

References

1. Anderson, L.W., Krathwohl, D.R.: A taxonomy for learning, teaching, and assessing: A revision of Bloom's taxonomy of educational objectives (2001)
2. Aula, A., Khan, R.M., Guan, Z.: How does search behavior change as search becomes more difficult? In: CHI, pp. 35–44 (2010)
3. Bell, D.J., Ruthven, I.: Searchers' assessments of task complexity for web searching. In: McDonald, S., Tait, J.I. (eds.) ECIR 2004. LNCS, vol. 2997, pp. 57–71. Springer, Heidelberg (2004)
4. Byström, K., Järvelin, K.: Task complexity affects information seeking and use. Inf. Process. Manage. 31(2), 191–213 (1995)
5. Campbell, D.J.: Task complexity: A review and analysis. The Academy of Management Review 13(1), 40–52 (1988)
6. Fan, R.-E., Chang, K.-W., Hsieh, C.-J., Wang, X.-R., Lin, C.-J.: Liblinear: A library for large linear classification. JMLR 9, 1871–1874 (2008)
7. Feild, H.A., Allan, J., Jones, R.: Predicting searcher frustration. In: SIGIR, pp. 34–41 (2010)
8. Jansen, B.J., Booth, D., Smith, B.: Using the taxonomy of cognitive learning to model online searching. Inf. Process. Manage. 45(6), 643–663 (2009)
9. Kim, J.: Task difficulty as a predictor and indicator of web searching interaction. In: CHI, pp. 959–964 (2006)
10. Li, Y., Belkin, N.J.: A faceted approach to conceptualizing tasks in information seeking. Inf. Process. Manage. 44(6), 1822–1837 (2008)
11. Liu, J., Cole, M.J., Liu, C., Bierig, R., Gwizdka, J., Belkin, N.J., Zhang, J., Zhang, X.: Search behaviors in different task types. In: JCDL, pp. 69–78 (2010)
12. Liu, J., Gwizdka, J., Liu, C., Belkin, N.J.: Predicting task difficulty for different task types. ASIS&T, 16:1–16:10 (2010)
13. Liu, J., Liu, C., Cole, M., Belkin, N.J., Zhang, X.: Exploring and predicting search task difficulty. In: CIKM, pp. 1313–1322 (2012)
14. Liu, J., Liu, C., Gwizdka, J., Belkin, N.J.: Can search systems detect users' task difficulty?: some behavioral signals. In: SIGIR, pp. 845–846 (2010)
15. Smucker, M.D., Allan, J., Carterette, B.: A comparison of statistical significance tests for information retrieval evaluation. In: CIKM, pp. 623–632 (2007)
16. White, R.W., Dumais, S.T.: Characterizing and predicting search engine switching behavior. In: CIKM, pp. 87–96 (2009)
17. Wood, R.E.: Task complexity: Definition of the construct. Organizational Behavior and Human Decision Processes 37(1), 60–82 (1986)
18. Wu, W.-C., Kelly, D., Edwards, A., Arguello, J.: Grannies, tanning beds, tattoos and nascar: evaluation of search tasks with varying levels of cognitive complexity. In: IIIX, pp. 254–257 (2012)

Crawling Policies Based on Web Page
Popularity Prediction

Liudmila Ostroumova, Ivan Bogatyy, Arseniy Chelnokov,
Alexey Tikhonov, and Gleb Gusev

Yandex, Moscow, Russia
{ostroumova-la,loken17,achelnokov,altsoph,gleb57}@yandex-team.ru

Abstract. In this paper, we focus on crawling strategies for newly discovered URLs. Since it is impossible to crawl all the new pages right after
they appear, the most important (or *popular*) pages should be crawled
with a higher priority. One natural measure of page importance is the
number of user visits. However, the popularity of newly discovered URLs
cannot be known in advance, and therefore should be predicted relying on
URLs' features. In this paper, we evaluate several methods for predicting
new page popularity against previously investigated crawler performance
measurements, and propose a novel measurement setup aiming to evaluate crawler performance more realistically. In particular, we compare
short-term and long-term popularity of new ephemeral URLs by estimating the rate of popularity decay. Our experiments show that the information about popularity decay can be effectively used for optimizing
ordering policies of crawlers, but further research is required to predict
it accurately enough.

Keywords: crawling policies, new web pages, popularity prediction.

1 Introduction

Crawl ordering policy is a strategy for a crawler to choose URLs from a crawling
queue. Though a particular strategy may serve multiple purposes, it is eventually
related to one of the following two main tasks: downloading newly discovered
web pages not represented in the index, and refreshing copies of pages likely to
have important updates [15]. In this paper, we focus on the first task: crawling
newly discovered web pages (which correspond to newly discovered URLs). It
is impossible to crawl all new pages right after they appear due to very high
rate of web growth and the limited resources of even production-scale crawlers,
therefore the most important pages should be crawled with a higher priority.
There are several ways to measure the importance of a page, which all may lead
to different ordering policies and measures of crawling performance. Among other
measures of page importance, such as those based on the link structure of the
web, e.g., in-degree [11] or PageRank [1,6], the most promising methods account
for the experience of target users, whose behavior is stored, e.g., in the search
engine query logs [8,16,17]. The main idea of these approaches is to maximize

M. de Rijke et al. (Eds.): ECIR 2014, LNCS 8416, pp. 100–111, 2014.
© Springer International Publishing Switzerland 2014

the overall utility of crawled pages for the search system. In this regard, it is reasonable to count user transitions (or visits) to a web page, or its *popularity*, as the measure of importance. In this paper, we consider the number of user visits recorded in browser toolbar logs as a measure of pages' popularity. This so-called user browsing behavior-driven approach was proposed in [14]. As it was shown in [12], almost all new pages on the web are ephemeral, i.e., they are popular only during a small period of time after their creation and user interest to such pages decays rapidly over time. In this paper, we focus only on such pages with short-term interest and maximize the total number of visits these pages get after they are crawled.

The popularity of newly discovered URLs cannot be known in advance, and therefore it should be predicted relying on their features available at the time of their discovery. We analyze the problem of new page popularity prediction, in particular, we take into account the dynamics of popularity by predicting both the total popularity and the rate of popularity decay for new URLs. Crawling policies previously proposed in [14] are based on the predicted total popularity, thus they lose the temporal aspect. In fact, for two newly discovered URLs, where one will be popular today and another will gain even more popularity, but several days later, such a policy prefers to crawl the latter page first, thus losing the current traffic of the search engine. We argue that the information about the dynamics of popularity can be effectively used for optimizing ordering policies of crawlers, but it is hard to predict this dynamics.

To this end, we first predict the overall number of user visits which a newly discovered web page will gain. Unlike in [14], our prediction is a model trained by using features from different sources, including the URL itself and its domain properties. Our machine learning based framework significantly outperforms the state-of-the-art approach. In a similar way, we also predict the rate of popularity decay, which is based on the short-term popularity of pages, i.e., the number of visits during the next several days or hours after a page is discovered. Finally, we predict the temporal dynamics of pages' popularity by means of an appropriate exponential function, as it was proposed in [12].

Second, we evaluate different ordering policies based on the predicted popularity of pages. An algorithm we suggest in this paper takes into account the predicted rate of popularity decay for web pages and dynamically re-ranks the crawling queue according to the dynamics of popularity. It is worth mentioning that user browsing behavior-driven crawling policies require evaluation in a realistic experimental setup, which takes into account the dynamic nature of the task: crawl delay, appearance of new web pages, and missed popular pages that no longer have new visits. To the best of our knowledge, there was still no such evaluation conducted. We compare different crawl strategies by testing them in a realistic setup against the dynamic measure of crawl performance proposed in [12]. It turns out that ordering policies which take into account the rate of popularity decay of pages can outperform ones based on their total popularity only. This result supports our assumption that the currently popular pages are more important to crawl first in order not to lose the current traffic on the web.

To sum up, the contribution of the current study is the following:

- We address the new problem of predicting both the total popularity and the rate of popularity decay of short-lived web pages and propose an effective method for the total popularity prediction that outperforms the state-of-the-art baseline method.
- We evaluate different user browsing behavior-driven crawling policies in a realistic experimental setup and prove that strategies which take into account the dynamics of popularity outperform ones based on total popularity only, given an effective method to predict popularity decay.

The rest of the paper is organized as follows. In the next section we discuss previous research on the crawling of newly discovered web pages and page popularity prediction. In Section 3, we describe the general framework and the crawling algorithm we propose in this paper. In Section 4, we evaluate the performance of the algorithm and compare it with the baseline strategy. Finally, Section 5 concludes the paper.

2   Related Work

There are many works dealing with the prediction of popularity for different entities on the web: topics, news, users of social networks, tweets, Twitter hashtags, videos, and so on. However, only a few papers study the popularity of web pages measured in user visits. One of them [18] proposes a model which predicts future issue frequency for a given query, rate of clicks on a given document, and a given query–document pair. This model relies on the history of the past dynamics of the query issues and the clicks on the document. That is why the approach turns out to be inapplicable to the task of popularity prediction for new web pages, which do not possess any historical data yet, and, as in our case, are even not crawled yet. Another study [14] is devoted to newly discovered pages and the prediction of their future *total* traffic relying on their URL string only, which is crucial for designing a crawl ordering policy because we need to predict the importance of new pages before we even download them. Our work can be considered as a follow-up study to that paper, since we predict the popularity of new pages *in its dynamics* by combining the predicted total popularity with the rate of popularity decay. Also, our machine learning based algorithm significantly improves the state-of-the-art approach for the predicting the total popularity of pages. As the problem of URL-based page popularity prediction is relatively new, there are many studies on predicting different properties of a URL itself before downloading its content, such as web page category [10], its language [5], topic [4], or genre [2]. Some of the approaches investigated in the previous papers can be successfully used to design new features for our popularity prediction model.

The pioneering work [16] proposes to evaluate crawling performance in terms of utility of crawled pages for users of the search system for a given ranking method and a query flow stored in the log. Authors estimate the quality of

SERP averaged over all user queries and compare this quality for different crawling policies. They suggest an algorithm which can effectively re-crawl pages in order to keep their local copies up-to-date. Since the profit of re-crawling a given page is estimated by using the historical data on the previously observed profit of re-crawling this page, the ordering of newly discovered pages remains beyond the scope of the proposed approach. In contrast, our study focuses on estimating utility of the newly discovered pages, which should be predicted relying on the properties of their URLs known before we download them. The problem of ordering new URLs was analyzed in [17]. Here, as in [16], the measure of the algorithm's performance is the utility of crawled pages for a given ranking method and a query flow. In the case of newly discovered pages the expected impact must be estimated using only URL string, incoming links, domain affiliation, referring anchortext.

The crawling quality measure introduced in [16] and [17] can be interpreted as the expected number of clicks the crawled documents can gather being ranked by a fixed retrieval method in response to a query flow observed within a certain time span. Indeed, given a query workload $\mathcal{Q}$ consisting of queries and their frequencies, authors define the total impact of page p as $I(p, \mathcal{Q}) = \sum_{q \in \mathcal{Q}} f(q)I(p, q)$, where $f(q)$ is the frequency of query q and $I(p, q)$ can be interpreted as the probability that a document p will be clicked on the result page served by the ranking method in response to query q. It is also assumed that the query workload $\mathcal{Q}$ consist of real-world user queries issued during a time span close to the current moment. So far, the impact of a page p is the estimated frequency of user transitions to this page from SERP. In contrast to [16] and [17], we measure not only the current popularity of pages, but the overall contribution of these pages to search engine performance, i.e., the number of all future user visits. Therefore, our quality measure accounts for the cumulative performance we gain by crawling a particular page, not only the currently observed profit. In particular, our approach takes into account the fact that different pages become less popular with different rates.

In [12], a crawling strategy for pages with short period of user interest was suggested. Also, the problem of dividing resources between crawling of new pages and re-crawling of old pages (in order to discover links to new pages) was analyzed. However, in [12] the popularity of new pages was predicted based on the properties of their corresponding domain only (or of the page where the link to the URL was discovered, to be precise). In contrast, our prediction model is able to distinguish among different pages on one domain and pages whose URLs were found at the same hub page.

3   Algorithm

3.1   Framework

First, let us describe the general framework of the proposed algorithm. Following previous research in this area [7,16,20], we assume uniform resource cost of crawling across all pages. In other words, it takes a fixed time τ for a crawler to download one page. Therefore, every τ seconds a crawler chooses a page to crawl from its crawling queue. The crawling queue is continuously being updated as links to new pages are discovered.

The crawling queue may get updated because of several reasons. New URLs may be discovered by a crawler, or they may be mined from browser toolbar logs [3], etc. While the problem of discovering new URLs is beyond the scope of this paper, in our study new URLs are taken from browser toolbar logs.

3.2   Metric

It was proposed [12] to use the following measure of algorithm's performance. The profit $P_i(\Delta t_i)$ of crawling a page i with a delay Δt_i after its appearance is the total number of clicks this page will get on a SERP after it is crawled. Note that we consider only pages with short-term interest. Then, the quality of an algorithm at a moment t (averaged over the interval of length T) is

$$Q_T(t) = \frac{1}{T} \sum_{i:t_i+\Delta t_i \in [t-T,t]} P_i(\Delta t_i).$$

Note that the number of clicks on SERP depends on many causes, including user behavior specificity and the ranking scheme, not only the crawling algorithm. So it is not always evident how to interpret a change in clicks count when the crawler policy has been modified. For instance, should we relate it to the crawler method itself, or to specific (unseen) dependence of the ranking method on the choice among equally good crawling policies? In fact, it seems more appropriate and robust to rely on user visits mined from toolbar logs instead of clicks to measure crawler performance. So the profit $P_i(\Delta t_i)$ of crawling a page i with a delay Δt_i will further imply the number of times this page is visited after the crawling moment.

As it is described in Section 4.1, we evaluate the algorithms' performance on one-month data. Therefore, we take $T = 1$ month and the quality $Q_T(T)$ is the average profit per second in this month.

3.3   Crawling Strategy

As it was discussed in Section 3.1, every τ seconds our algorithm should choose one page from the crawling queue. The problem is how to choose a URL to crawl at each step, when the queue is dynamically changing.

We take the dynamics of popularity into account in the following way. It was shown in [12], that the profit $P_u(\Delta t)$ of crawling a page u with some delay Δt decays exponentially with time. Therefore, this profit can be approximated by $p(u)e^{-\lambda(u)\Delta t}$ with some parameters $p(u)$ and $\lambda(u)$. It is easy to see that, for $\Delta t = 0$, we have $p(u) = p(u)e^0 = P_u(0)$ that is the overall popularity of u, i.e., in our case, the total number of user visits to the page since the page was created, and $\lambda(u)$ is the rate of popularity decay.

For each page u, we predict its parameters $p(u)$ and $\lambda(u)$ as described in Section 3.4. Finally, the strategy is the following. Every τ seconds we choose the page with the maximum expected profit. Namely, for each page u, every τ seconds we compute its score, which is then used to rank all new pages in the queue, $r(u) = p(u)e^{-\lambda(u)\Delta t}$, where Δt is the time passed since the page u was discovered, and then we crawl the page with the highest score. The time when a page was discovered is used as an approximation of the time when the page was created. So, we assume that the first user visit recorded in toolbar logs is close to the time of appearance of a page.

3.4 Estimation of Popularity

In this section, we discuss the method of predicting the parameters $p(u)$ and $\lambda(u)$ for a given URL u. We learn a gradient boosted decision tree model with a list of features described further in this section.

Let us remind that $p(u)$ is the total number of toolbar visits to a page u. Since the distribution of $p(u)$ over all pages u in the dataset is heavy-tailed, there are, among others, very large values of $p(u)$. To design a crawling policy, we need to guess their order of magnitude rather than to predict them precisely. Therefore, as it is usually done for heavy-tailed distributions [19], we predict the value $a_1 = p(u)$ minimizing mean squared logarithmic error on training examples.

We also predict the popularity decay $\lambda(u)$. Let $p_t(u)$ be the number of visits during the time t (measured in hours in our experiments) after the URL u was discovered. We train a gradient boosted decision tree model which predicts the ratio of $p_t(u)$ to the number of all the visits of u, i.e., the value $a_2 = p_t(u)/p(u)$, $a_2 \in [0, 1]$ (in this case we minimize mean squared error). Then we can estimate the decay rate $\hat{\lambda}(u)$ in the following way. Due to the equation $p(u) - p_t(u) = P_u(t) \approx p(u)e^{-\lambda(u)t}$ it follows that $1 - a_2 = 1 - \frac{p_t(u)}{p(u)} \approx e^{-\lambda(u)t}$. Taking a logarithm we get $\log(1 - a_2) \approx -\lambda(u)t$, and finally we can estimate $\lambda(u)$ as $\hat{\lambda}(u) = -\frac{\log(1-a_2)}{t}$. Thus the estimated expected profit of crawling u with the delay Δt after its appearance is

$$r(u) = \hat{p}(u)e^{-\hat{\lambda}(u)\Delta t} = a_1 e^{\frac{\log(1-a_2)}{t}\Delta t}.$$

3.5 Features Used

Here we discuss the features we use for training our models that predict the parameters a_1 and a_2 for all new URLs.

For each domain in our dataset, we construct a pattern tree to organize URLs based on their syntax structures as it is described in [13]. This procedure is based on the idea that URLs with a similar syntax structure are similar to each other in terms of user visits. First, based on the syntax scheme, each URL is decomposed into "key-value" pairs. Then we start constructing procedure from the root (the root contains all URLs) and iteratively divide URLs into subgroups according to their values under a particular key. The selected key in each iteration is the one who has the most concentrated distribution of values. Please, refer to [13] for more detailed description of this procedure.

After that, we use visit and transit browsing behaviors to cut down the obtained tree (see [14]). Let $V_{in}(P)$ be the number of visits to all URLs in a pattern P and $V_{trans}(P_1, P_2)$ be the number of transitions from URLs in a pattern P_1 to URLs in a pattern P_2. By using visit-based tree-cut, we remove sibling nodes having similar visit traffic (to get lower complexity). Let U be the set of all URLs. A tree-cut plan is a set $\Gamma = \{P_1, \ldots, P_K\}$, where each P_i is a leaf node survived after the cut. The best plan minimizes the sum of model description length $L(\Gamma)$ and data description length $L(U|\Gamma)$. $L(\Gamma)$ can be approximated by

$$L(\Gamma) = \frac{K-1}{2} \log V_{in}(P_{root}).$$

For $L(U|\Gamma)$ we have

$$L(U|\Gamma) = -\sum_{u \in U} \log \left(\frac{1}{|P_k|} \frac{V_{in}(P_k)}{V_{in}(P_{root})} \right).$$

Then we use transit behavior to remove redundant nodes. Please refer to [14] for the details.

The described method matches each URL u with a URL pattern P_u that constitutes of several key-value pairs coinciding with the ones of u and several keys with blank values. For this pattern P_u, we compute the following statistics by using old URLs with the same pattern. These statistics are further used as the features of a new URL u.

1. **Transitions to the pattern**

- The number of transitions to all URLs in the pattern P: $V_{in}(P)$.
- The average number of transitions to a URL in the pattern $V_{in}(P)/|P|$, where $|P|$ is the number of URLs in P.
- The number of transitions to all URL's in the pattern P during the first t hours after their discovery: $V_{in}^t(P)$.
- The average number of transitions to a URL in the pattern P during the first t hours after its discovery: $V_{in}^t(P)/|P|$.
- The fraction of transitions to all URL's in the pattern P during the first t hours: $V_{in}^t(P)/V_{in}(P)$.

2. **Transitions from the pattern**

- The number of times URLs in the pattern act as referrers in a browser toolbar log $V_{out}(P)$.
- The average number of times a URL in the pattern acts as a referrer $V_{out}(P)/|P|$.
- The number of times URLs in the pattern act as referrers during the first t hours after their discovery $V_{out}^t(P)$.
- The average number of times a URL in the pattern acts as a referrer during the first t hours after its discovery $V_{out}^t(P)/|P|$.

– The fraction of times URLs in the pattern act as referrers during the first t hours after their discovery $V_{out}^t(P)/V_{out}(P)$.

3. Other features

– The size of the pattern $|P|$.

The first group of features directly corresponds to the popularity of pages. For example, if we want to predict the overall popularity of a page, it is reasonable to compute the average popularity $V_{in}(P)/|P|$ of "similar" URLs taken from the corresponding pattern. Similarly, the historical value $V_{in}^t(P)/V_{in}(P)$ is supposed to be correlated with $a_2 = p_t(u)/p(u)$.

The second group of features corresponds to the importance of pages defined as the number of transitions from them. Obviously, the correlation between the number of transitions from pages and their popularity in terms of visits exists, since in order to move from a page p_1 to a page p_2 a user should visit p_1 first. But, as we show in Section 4.2, the first group of features is much more important.

4 Experiments

4.1 Experimental Setup

All the experiments in this paper are based on a fully anonymized web page visits log recorded by a search engine browser toolbar, used by millions of people across different countries.

We extract all records made in the 2-month period from July 1, 2013 to August 31, 2013. From the set of all pages $\mathcal{P}$ appeared in logs during this time frame, we removed pages which have non-zero visits on June, 2013 or on September, 2013 in order to focus only on the new pages that are popular only during the considered period of time. This way, we obtained a smaller set of pages $\mathcal{P}'$. Since the algorithm of constructing the pattern trees [13], which is used by all the methods introduced in this paper, is computationally expensive, and we are not able to perform the experiments in the production scale, we restrict our dataset to the set $\mathcal{D}'$ of those domains consisting of not more than 50K pages.

As a result, our final dataset consists of 100 random domains sampled from $\mathcal{D}'$ and all pages of $\mathcal{P}'$ from these domains. URLs that were visited in July, but neither in June nor in August, were used to train the popularity prediction model. URLs visited in August, but neither in July nor in September, were used to evaluate the performance of the algorithms. Each of these two sets contains ~ 650K URLs.

4.2 Analysis of Popularity Prediction

In this section, we analyze the total popularity prediction model. In particular, we compare different orderings of pages based on their predicted popularity. As it will be shown in Section 4.3, the accurate ordering of pages according to their total popularity considerably improves the quality of a crawler.

Here the user browsing behavior-driven approach suggested in [14] is compared with the total popularity prediction model suggested in this paper. Among the four methods suggested in [14], we obtained the best results using *Average-Visit* method, i.e., the ordering of URLs according to the average number $V_{in}(P)/|P|$ of transitions to a URL in the pattern (see Section 3.5). For our method described in Section 3.4, we took $t = 24$ hours.

Table 1. The importance of features used

Feature	Importance
$V_{in}^{24}(P)/|P|$	38%
$V_{in}(P)/|P|$	29%
$V_{out}(P)/|P|$	8%
$V_{in}^{24}(P)/V_{in}(P)$	7%
$V_{in}(P)$	4%
$V_{out}(P)$	4%
$|P|$	3%
$V_{out}^{24}(P)$	2%
$V_{out}^{24}(P)/|P|$	2%
$V_{out}^{24}(P)/V_{out}(P)$	2%
$V_{in}^{24}(P)$	1%

The importance of features used is presented in Table 1. We sorted features according to the weighted contribution into our prediction model (see Section 10.13 in [9] for the description of those weights). It measures weighted improvement of the loss function over all employments of a feature during the learning process. As expected, the most important features are the average total popularity of URLs in a pattern and the average short-term popularity of URLs in a pattern.

As in [14], we rank pages according to their predicted total popularity and then, for each n, compute the percentage of traffic covered by the first

n pages. The obtained results can be seen on Figure 1. Ideal curve is obtained by sorting pages according to their actual popularity, i.e., the total number of user visits.

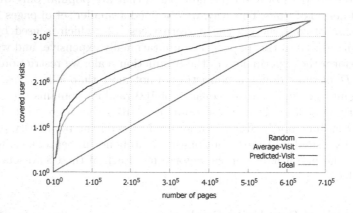

Fig. 1. Comparison of popularity-based orderings

Table 2 shows the percentage of traffic covered by top $x\%$ URLs according to different methods of total popularity prediction. The values of x considered are between 5% and 80%.

Table 2. Comparison of ordering strategies: the fraction of visits covered for different percentage of covered URLs

Algorithm	5% of URLs	10% of URLs	30% of URLs	80% of URLs
Random	5%	10%	30%	80%
Average-Visit	35%	44%	65%	85%
Predicted-Visit	42%	52%	72%	93%
Ideal	68%	74%	86%	96%

Let S_R and S_I be the areas under the curves corresponding to random and ideal orderings. Let also S_{A-V} and S_{P-V} be the areas under the curves corresponding to Average-Visit and to our prediction model. Then $\frac{S_{A-V}-S_R}{S_I-S_R} = 0.52$, while $\frac{S_{P-V}-S_R}{S_I-S_R} = 0.72$. The metric $\frac{S_O-S_R}{S_I-S_R}$ shows the advantage $S_O - S_R$ we get by an ordering O over the random ordering R compared to that advantage of the ideal ordering.

The area under the curve on Figure 1 has the following interpretation for the problem of crawling.

Proposition. *Assume that*

1. *All the N considered URLs are created at the beginning of the same time interval $[T, T + \Delta T]$ (in our case $\Delta T = 1$ month).*
2. *All the pages are visited only during this month and the visits are uniformly distributed in time.*
3. *The crawl rate of the crawler is $N/\Delta(T)$, i.e., allows to crawl exactly N pages during the considered time interval.*

Then the area S_O under the curve of a URL ordering O is the total number of times all the pages are visited after their crawling moments for the crawling policy based on the ordering O.

The measure S seems to be not natural enough in terms of crawling, since the assumptions made in proposition above are not realistic. For example, if the pages are visited only during some period of time, then it is natural to assume that their popularity decays over time during this period. Therefore, in the next section we compare different crawling strategies in a realistic experimental setup.

4.3   Crawling Strategies

Here we evaluate the performance of the crawling algorithm suggested in this paper according to the framework described in Section 3.1.

We compare the following strategies:

Average-Visit. At each step, the page with the highest total popularity is crawled. The total popularity is predicted according to Average-Visit method from [14], i.e., at each step, we crawl the page with the highest value of the average number $V_{in}(P)/|P|$ of visits to URLs in the pattern P corresponding to the considered URL (see Section 3.5).

Predicted-Visit. At each step, the page with the highest total popularity is crawled. The total popularity is predicted by our trained method.

Exponential-Predicted. In this case, we take the dynamics of popularity into account, as it was described in Section 3.3. Both the parameters a_1 and a_2 are predicted by our trained method.

Exponential-Oracle. This method is similar to the previous one, but instead of predicted a_2, we consider its actual value, i.e., $a_2 = p_t(u)/p(u)$. Comparison with this method allows to understand the influence of popularity decay on the performance of the crawler and to estimate the maximum possible profit from taking this dynamics into account.

The results obtained can be seen in Table 3. Here we compare all the algorithms with different crawl rates (number of crawls per second). For each algorithm, we measured the fraction of covered visits, i.e., $Q_T(T)$ with $T = 30$ days (defined in Section 3.2) normalized by the overall number of visits. Note that $CR = 0.1$ allows to crawl about the half of all the pages in our dataset during the considered month. If follows from Table 3 that the better prediction of the total popularity a_1 obtained by our machine learning algorithm helps to significantly improve the quality of a crawler. Also it turns out that the information about the rate of popularity decay $\lambda(u)$ (Exponential-Oracle) is important and gives even more profit. Unfortunately, we are not able to predict this parameter well enough to beat Predicted-Visit strategy.

Table 3. Comparison of crawling strategies: the fraction of covered visits for different crawl rates (CR)

Algorithm	CR=0.01 (5% URLs)	CR=0.02 (10% URLs)	CR=0.04 (20% URLs)	CR=0.1 (50% URLs)
Average-Visit	0.24	0.34	0.43	0.53
Predicted-Visit	0.32	0.42	0.51	0.60
Exponential-Predicted	0.31	0.40	0.49	0.58
Exponential-Oracle	0.36	0.44	0.54	0.64

5   Conclusion

In this paper, we addressed the problem of crawling newly discovered web pages. We suggested a machine learning based model which predicts the overall number of user visits to a page (the measure of page's importance), which uses only the features of the URL available at the time of its discovery. In addition to the total popularity of web pages, we also predicted the rate of popularity decay.

We compared different user browsing behavior-driven crawling strategies against previously investigated crawler performance measurements, and proposed a novel measurement setup aiming to evaluate crawler performance more realistically, taking into account the dynamic nature of the task: appearance of new web pages, crawl delay, and the possibility of missed visits because of it. It turns our that both the better prediction of total popularity and the information about the rate of popularity decay help to significantly improve the performance of a crawler, but it is hard to predict the rate of decay accurately enough. We leave the thorough analysis of this problem for future work.

Acknowledgements. Special thanks to Pavel Serdyukov for his careful reading and useful comments.

References

1. Abiteboul, S., Preda, M., Cobena, G.: Adaptive on-line page importance computation. In: Proc. WWW Conference (2003)
2. Abramson, M., Aha, D.: What's in a URL? Genre classification from URLs. In: Conference on Artificial Intelligence, pp. 262–263 (2012)
3. Bai, X., Cambazoglu, B.B., Junqueira, F.P.: Discovering urls through user feedback. In: Proc. CIKM Conference, pp. 77–86 (2011)
4. Baykan, E., Henzinger, M., Marian, L., Weber, I.: A comprehensive study of features and algorithms for url-based topic classification. ACM Trans. Web (2011)
5. Baykan, E., Henzinger, M., Weber, I.: Efficient discovery of authoritative resources. ACM Trans. Web (2013)
6. Cho, J., Schonfeld, U.: Rankmass crawler: a crawler with high personalized pagerank coverage guarantee. In: Proc. VLDB (2007)
7. Edwards, J., McCurley, K.S., Tomlin, J.A.: Adaptive model for optimizing performance of an incremental web crawler. In: Proc. WWW Conference (2001)
8. Fetterly, D., Craswell, N., Vinay, V.: The impact of crawl policy on web search effectiveness. In: Proc. SIGIR Conference, pp. 580–587 (2009)
9. Hastie, T., Tibshirani, R., Friedman, J.H.: The elements of statistical learning: data mining, inference, and prediction: with 200 full-color illustrations. Springer, New York (2001)
10. Kan, M.Y.: Web page classification without the web page. In: Proc. WWW Conference, pp. 262–263 (2004)
11. Kumar, R., Lang, K., Marlow, C., Tomkins, A.: Efficient discovery of authoritative resources. Data Engineering (2008)
12. Lefortier, D., Ostroumova, L., Samosvat, E., Serdyukov, P.: Timely crawling of high-quality ephemeral new content. In: Proc. CIKM Conference, pp. 745–750 (2011)
13. Lei, T., Cai, R., Yang, J.M., Ke, Y., Fan, X., Zhang, L.: A pattern tree-based approach to learning url normalization rules. In: Proc. WWW Conference, pp. 611–620 (2010)
14. Liu, M., Cai, R., Zhang, M., Zhang, L.: User browsing behavior-driven web crawling. In: Proc. CIKM Conference, pp. 87–92 (2011)
15. Olston, C., Najork, M.: Web crawling. Foundations and Trends in Information Retrieval 4(3), 175–246 (2010)
16. Pandey, S., Olston, C.: User-centric web crawling. In: Proc. WWW Conference (2005)
17. Pandey, S., Olston, C.: Crawl ordering by search impact. In: Proc. WSDM Conference (2008)
18. Radinsky, K., Svore, K., Dumais, S., Teevan, J., Bocharov, A., Horvitz, E.: Modeling and predicting behavioral dynamics on the web. In: Proc. WWW Conference, pp. 599–608 (2012)
19. Tsur, O., Rappoport, A.: What's in a hashtag?: content based prediction of the spread of ideas in microblogging communities. In: Proc. WSDM Conference, pp. 643–652 (2012)
20. Wolf, J.L., Squillante, M.S., Yu, P.S., Sethuraman, J., Ozsen, L.: Optimal crawling strategies for web search engines. In: Proc. WWW Conference (2002)

Measuring the Effectiveness of Gamesourcing Expert Oil Painting Annotations

Myriam C. Traub, Jacco van Ossenbruggen, Jiyin He, and Lynda Hardman

Centrum Wiskunde & Informatica, Science Park 123, Amsterdam, The Netherlands
firstname.lastname@cwi.nl

Abstract. Tasks that require users to have expert knowledge are difficult to crowdsource. They are mostly too complex to be carried out by non-experts and the available experts in the crowd are difficult to target. Adapting an expert task into a non-expert user task, thereby enabling the ordinary "crowd" to accomplish it, can be a useful approach. We studied whether a simplified version of an expert annotation task can be carried out by non-expert users. Users conducted a game-style annotation task of oil paintings. The obtained annotations were compared with those from experts. Our results show a significant agreement between the annotations done by experts and non-experts, that users improve over time and that the aggregation of users' annotations per painting increases their precision.

Keywords: annotations, crowdsourcing, expert tasks, wisdom of the crowd.

1 Introduction

Cultural heritage institutions place great value in the correct and detailed description of the works in their collections. They typically employ experts (e.g. art-historians) to annotate artworks, often using predefined terms from expert vocabularies, to facilitate search in their collections. Experts are scarce and expensive, so that involving non-experts has become more common. For large image archives that have been digitized but not annotated, there are often insufficient experts available, so that employing non-expert annotations would allow the archive to become searchable (see for example ARTigo[1], a tagging game based on the ESP game[2]).

In the context of a project with the Rijksmuseum Amsterdam, we take an example annotation task that is traditionally seen as too difficult for the general public, and investigate whether we can transform it into a game-style task that can be played directly, or quickly learned while playing, by non-experts. Since we need to compare the judgements of non-experts with those of experts, we picked a dataset and annotation task for which expert judgements were available.

---

[1] http://www.artigo.org/
[2] http://www.gwap.com/gwap/gamesPreview/espgame/

M. de Rijke et al. (Eds.): ECIR 2014, LNCS 8416, pp. 112–123, 2014.

We conducted two experiments to investigate the following research questions. First, we want to know how the choices of non-expert users compare to those of experts in order to estimate the suitability of the non-expert annotations as part of a professional workflow.

Second, whether users perform better later in the game, and, if so, if they do this only on repeated images or also on new images. Third, how the partial absence of the correct answer affects the user performance in order to determine whether purely non-expert input is reliable.

2 Related Work

Increasing numbers of cultural heritage institutions initiate projects based on crowdsourcing to either enrich existing resources or create new ones [1]. Two well-known projects in this field are the Steve Tagger[3] and the Your Paintings Tagger[4]. Both constitute cooperations between museum professionals and website visitors to engage visitors with museum collections and to obtain tags that describe the content of paintings to facilitate search.

A previous study, [7], suggests that expert vocabularies that are used by professional cataloguers are often too limited to describe a painting exhaustively. This gap can be closed by making use of external thesauri from domains other than art history (e.g. WordNet, a lexical, linguistic database[5]). The interface for this task, however, targets professional users.

Steve Tagger and the Your Paintings Tagger focus on enriching their artwork descriptions with information that is common knowledge (e.g. Is a flower depicted?). The SEALINCMedia project[6] focuses on finding precise information (e.g. the Latin name of a plant) about depicted objects. To achieve this, the crowd is searched for experts who are able to provide this very specific information [2] and a recommender system selects artworks that match the users' expertise.

Another example for crowdsourcing expert knowledge is "Umati". Heimerl et al. [6] transformed a vending machine into a kiosk that returns snacks for performing survey and grading tasks. The restricted access to "Umati" in the university hallway ensured that the participants possessed the necessary background knowledge to solve the presented task. While their project also aims at getting expert work done with crowdsourcing mechanisms, their approach is different from ours. Whereas they aim at attracting skilled users to accomplish the task, we give non-experts the support they need to carry out an expert task.

Since most of these approaches target website visitors or passers-by, rather than paid crowd workers on commercial platforms, they need to offer an alternative source of motivation for users. Luis von Ahn's ESP Game [9] inspired

---

several art tagging games developed by the ARTigo project[7]. These games seek to obtain artwork annotations by engaging users in gameplay.

Goldbeck et al. [4] showed that tagging behavior is significantly different for abstract compared with representational paintings. Users were allowed to enter tags freely, without being limited to the use of expert vocabularies. Since our set of images showed a similar variety in styles and periods, we also investigated whether particular features of images had an influence on the user behavior.

He et al. [5] investigated if and how the crowd is able to identify fish species on photos taken by underwater cameras. This task is usually carried out by marine biologists. In the study, users were asked to identify fish species by judging the visual similarity between an image taken from video and images showing already identified fish species.

A common challenge of tagging projects lies in transforming the large quantity of tags obtained through the crowd to high quality annotations of use in a professional environment. As Galton proved in 1907, the aggregation of the *vox populi* can lead to surprisingly exact results that are "correct to within 1 per cent of the real value" [3]. Such aggregation methods can improve the precision of user judgements [8], a feature that can potentially be used to increase the agreement between users and experts of our tagging game.

3   Experimental Setup

We investigated the categorization of paintings into subject types (e.g. landscapes, portraits, still lifes, marines), which is typically considered to be an expert task. We simplified the task by changing it into a multiple choice game with a limited, preselected set of candidates to choose from. Each included the subject type's label, a short explanation of its intended usage and a representative example image. To investigate the influence of the preselection of the candidates on the performance of the users, we carried out two experiments: a baseline condition, which always had a correct answer among the presented candidate answers, and, to simulate a more realistic setting, a condition where in 25% of the cases the correct answers had been deliberately removed.

3.1   Procedure

Users were presented with a succession of images (referred to as *query images*) of paintings that they were asked to match with a suitable subject type (see Figure 1). We supported users by showing them a pre-selection of six *candidates*. Five of these candidates represented subject types and one of them (labeled "others") could be used if the assumed correct subject type was not presented. To motivate users to annotate images correctly and to give them feedback about the "correctness"[8] of their judgements, they were awarded ten points for judgements that agree with the expert and one point for the attempt (even if incorrect).

---

[7] http://www.artigo.org/
[8] By "correct" we mean that a given judgement agrees with the expert.

Fig. 1. Interface of the art game with the large query image on the upper left. The five candidate subject types are shown below, together with the *others* candidate.

The correct answer was always presented and users got direct feedback on every judgement they made. With this experiment we wanted to find out whether (and how well) users learn under ideal conditions. We use the data of the first experiment as a baseline for comparing the results of the second experiment.

In the second experiment, the correct answer is not always presented.

3.2   Experiments Conducted

We adapted the online tagging game used for the Fish4Knowledge project [5]. On the login page of the game, we provide a detailed description of the game including screenshots, instructions and the rules of the game.

Baseline Condition - For each query image, we selected one candidate that, according to the expert ratings, represents a correct subject type and three candidates representing related, but incorrect, subject types. One candidate was chosen randomly from the remaining subject types. For cases, when there were only two related but incorrect subject types available, we showed two incorrect random ones, so the total number of candidates would remain six (including the *others* candidate). The categorization of similar subject types was done manually and is based on their similarity. An example of related subject types is *figure, full-length figure, half figure, portrait* and *allegory*.

Imperfect Condition - In this setting, the correct candidate is not presented in 25% of the cases. This is used to find out how good the learning performance of users is when the candidate selection is done by an automated technique that may fail to find a correct candidate in its top five. The selection of the candidates was the same as in the baseline experiment, for the missing correct candidate we added another incorrect candidate.

3.3   Materials

The expert dataset [10] provides annotations of subject types for the paintings of the Steve Tagger project by experts from the Rijksmuseum Amsterdam.

Table 1. Used subject types and the number of expert annotations

Subject type	Annotations
full-length figures	40
landscapes	33
half figures	13
allegories, history paintings, portraits, animal paintings, genre, kacho, figures	8
townscapes	6
flower pieces	5
marines, cityscapes, maesta, seascapes, still lifes	3

We selected 168 expert annotations for 125 paintings (Table 1). The number of annotations per painting ranged from four (for one painting) to only one (for 83 paintings). These multiple classifications are considered correct: a painting showing an everyday scene on a beach[9] can be classified as *seascapes*, *genre*, *full-length figure* and *landscapes*. This, however, makes our classification task more difficult.

Query Images - The images used as query images are a subset of the thumbnails of paintings from the Steve Tagger[10] data set. The paintings are diverse in origin, subject, degree of abstraction and style of painting. Apart from the image, we provided no further information about the painting. Within the first ten images that were presented to the user, there were no repetitions. Afterwards, images may have been presented again with a 50% chance. The repetitions gave us more insight on the performance of the users.

Candidates - A candidate consists of an image, a label (subject type) and a description. For each subject type we selected one representative image from the corresponding Wikipedia page[11]. The main criterion for the selection was that the painting should show typical characteristics. The candidates were labeled with the names of the subject types from the Art & Architecture Thesaurus[12] (AAT) which comprises in total more than 100 subject types. The representative images were intended to give users a first visual indication of which subject type might qualify and it made it easier for users to remember it. If this was not sufficient for them to judge the image, they could verify their assumption by displaying short descriptions taken from the AAT, for example:

---

[9] http://tagger.steve.museum/steve/object/280
[10] http://tagger.steve.museum/
[11] E.g.: http://en.wikipedia.org/wiki/Maesta
[12] http://www.getty.edu/research/tools/vocabularies/aat/index.html

Marines

"Creative works that depict scenes having to do with ships, shipbuilding, or harbors. For creative works depicting the ocean or other large body of water where the water itself dominates the scene, use 'seascapes'. "[13]

The descriptions of the subject types are important, as the differences between some subject types are subtle.

3.4 Participants

Participants were recruited over social networks and mailing lists. For the analysis we used 21 for the first experiment and 17 in the second one, in total 38, after removing three users who made fewer than five annotations. The majority of the participants have a technical professional background and no art-historical background. In the baseline condition, users who scored at least 400 received a small reward.

3.5 Limitations

Our image collection comprised 125 paintings, and compared with a museum's collection this is a small number. Because of the repetitions, the number of paintings that the user saw only increased gradually over time, which would have made it possible to successively introduce a larger number of images to the users. This, however, would have made it difficult to obtain the necessary ground truth.

In the available ground truth data, each painting was judged by only one expert, which prevents us from measuring agreement among experts. This measurement might have revealed inconsistencies in the data that influenced users' performance.

In realistic cases, ground truth will be available for only a small fraction of the data. To apply to such datasets, our setting needs other means of selecting the candidates. This can be realised, for example, by using the output of an imperfect machine learning algorithm, or by taking the results of another crowdsourcing platform. We think it is realistic to assume that in such settings the correct answer is not always among the results, and acknowledge that the frequency of this really happening may differ from the 25% we assumed in our second experiment.

The game did not go viral, which can mean that incentives for the users to play the game and/or the marketing could be improved.

4 Results

An overview of the results of all users of both experiments shows a large variation in number of judgements and precision (Fig. 2). Users who judged more images

---

[13] http://www.getty.edu/vow/AATFullDisplay?find=marines&logic=AND&note=
&english=N&prev_page=1&subjectid=300235692

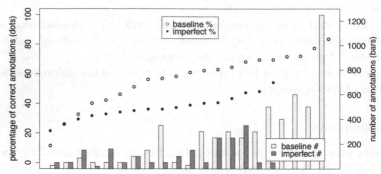

Fig. 2. Percentage of correct annotations per user (dots) and the number of annotations (bars). The users are ordered by increasing precision from left to right.

also tend to have higher precision. This might suggest that users indeed learn to better carry out the task or that well-performing users played more.

In both conditions, all users who finished at least one round of 50 images performed much better than a random selection of the candidates (with a precision of 17%), suggesting that we do not have real spammers amongst our players. On average, the precision of the users in the baseline condition (56%) is higher than in the imperfect condition (37%). This indicates that the imperfect condition is more difficult. This is in line with our expectations: in order to agree with the expert, users in the imperfect condition sometimes need to select the *other* candidate instead of a candidate subject type that might look very similar to the subject type chosen by the expert.

4.1 Agreement Per Subject Type

To understand the agreement between experts and users, we measure precision and recall per subject type. *Precision* is the number of agreedupon judgements for a subject type divided by the total judgements given by users for that subject type. *Recall* is the number of agreed-upon judgements for a subject type divided by the total judgements given by the expert for that subject type.

Both measures are visualized in confusion heatmaps (Fig.s 3 - 6). The rows represent the experts' judgements, while the columns show how the users classified the images. The shade of the cells visualizes the value of that cell as the fraction of the users' total votes for that specific subject type. Darker cells on the diagonal indicate higher agreement, while other dark cells indicate disagreement.

Some subject types score low on precision: *cityscapes* is frequently chosen by non-experts when the expert used *landscapes* or *townscapes*, while users select *history paintings* where the expert sees *figures* (Fig. 3). On the other hand, *flower pieces* and *animal paintings* score high on both precision and recall. Selecting the *others* candidate did not return points in the baseline condition, and some players reported to have noticed this and did not use this candidate afterwards. With 243 *others* judgements out of a total of 5640, it received relatively few

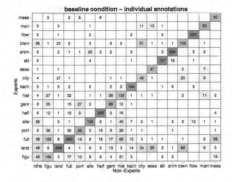

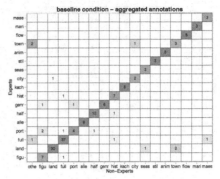

Fig. 3. Despite many deviations, the graph shows a colored diagonal representing an agreement between non-experts and experts. The task therefore seems to be difficult but still manageable for users.

Fig. 4. The "Wisdom of the Crowd" effect eliminates many deviations of the non-experts' judgements from the experts' judgements. However, there are still deviations for similar subject types such as *cityscapes* and *townscapes*.

clicks. The agreement between users and experts is substantial (Cohen's Kappa of 0.65), we see a clear diagonal of darker color.

Aggregating user judgements by using majority voting (Fig. 4), removes some deviations from the experts' judgements (Cohen's Kappa of 0.87) to almost perfect agreement. For example, all *cityscapes* judgements by users for cases where expert judged *landscapes* are overruled in the voting process and this major source of disagreement in Fig. 3 disappears. There is only one case where the expert judged *townscapes* and the majority vote of the users remained *cityscapes*. The painting description states that it shows "a dramatic bird's eye view of Broadway and Wall Street"[14] in New York. Therefore, *townscapes* cannot be the correct subject type and users were right to disagree with the expert. Most *others* judgements are largely eliminated by the majority voting. However, three paintings remain classified as *others* by the majority which indicates a very strong disagreement with the experts' judgement. One of these paintings does not show a settlement, but in an abstract way depicts a bomb store in the "interior of the mine"[15]. The other two show a carpet merchant in Cairo[16] and the "Entry of Christ into Jerusalem"[17], both being representations of large cities and therefore incorrectly categorized as *townscapes* by the expert.

In the imperfect condition, the confusion heatmaps are similar, however, the disagreement between users and experts is higher. The *others* candidate was the correct option in 25% of the cases. The users made more use of it, as shown by the higher numbers in the first column of Fig. 6. The agreement in the *allegories* column is, with 13%, even below chance. Majority voting increases the

---

[14] http://www.clevelandart.org/art/1977.43

[15] http://www.tate.org.uk/art/artworks/bomberg-bomb-store-t06998

[16] https://collections.artsmia.org/index.php?page=detail&id=10361

[17] http://tagger.steve.museum/steve/object/172

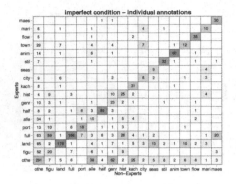

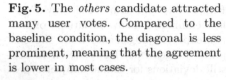

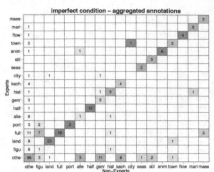

Fig. 5. The *others* candidate attracted many user votes. Compared to the baseline condition, the diagonal is less prominent, meaning that the agreement is lower in most cases.

Fig. 6. The aggregation of user votes could compensate some of the deviations from agreement, however the additional *others* candidate had a negative effect on the agreement for allegories, genre and kacho

precision, but only to 20%. The AAT defines this subject type to "express complex abstract ideas, for example works that employ symbolic, fictional figures and actions to express truths or generalizations about human conduct or experience". Therefore, it is very difficult to recognize an *allegory* as such without context information about the painting. User judgements diverging from the expert's judgements are largely removed by majority vote. The "Wisdom of the Crowd" effect, however, is not as strong as in the baseline condition. It raised the Cohen's Kappa from 0.47 to a (still) moderate agreement of 0.55.

We further analyzed the agreement of the non-experts and the experts on image level in the baseline condition. The broad range from 2% to 98% indicates very strong (dis-)agreement for some cases. In the images with the highest agreement, the relation between the depicted scenes and the subject type is intuitively comprehensible: the images with 98% agreement show flowers (*flower pieces*), monkeys (*animal painting*) and a still life (*still lifes*). An entirely different picture emerges, when we look at the images with low agreement. We presented the most striking cases to an expert from the Rijksmuseum Amsterdam to re-evaluate the experts' judgements and we identified two main reasons for disagreement: users would have needed additional information, such as the title, to classify the painting correctly; the expert annotations were incomplete or incorrect.

4.2 Performance over Time

The improvement of the users' precision over time does not necessarily mean that they have learned how to solve the problem (generalization), but that they "only" have learned the correct solution for a concrete problem (memorizing).

Memorizing - A learning effect is evident in the performance curve of the users for repeated images (Fig. 7). In the baseline condition, users had an initial

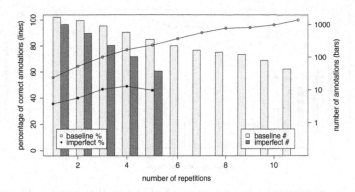

Fig. 7. Learning curves (lines) for the memorization effect of repeated images and numbers on annotations (bars) per repetition

success rate of 56% correct judgements. After seven repetitions, they judged 90% of the query images correctly. In the imperfect condition, the performance is consistently lower. The difference between the first appearance of an image (success rate of 36%) and the fifth appearance of an image (success rate of 46%) is lower than in the baseline experiment where we see an increase of 25 percent units. The lines in Fig. 7 were cut off after eleven repetitions for the baseline condition and five repetitions for the imperfect condition because the number of judgements dropped below 15. We further analyzed the results of a fixed homogeneous population of seven (baseline) and eight (imperfect) users. The outcomes were nearly identical for both conditions. These results show that users in the baseline condition improve on memorizing the correct subject type for a specific image. The differences between the two conditions indicate that users found it more difficult to learn the subject types in the imperfect condition.

Generalization - The judgement performance of users on the first appearances of images indicates whether they are able to generalize and apply the knowledge to unseen query images. If users learn to generalize, it is likely that they will improve over time at judging images that they have not seen before. Judgement precision increases throughout gameplay for both conditions (Fig. 8). While users in the baseline experiment started with a success rate of 44%, they reach 90% after about 250 images. Users in the imperfect condition started at a much lower rate of 33% and increase to 60%, after about 150 images. The declining number of images that are new to the user and the declining number of users that got so far in the game, lead to a drop in available judgements at later stages in the game. Therefore, we cut the graphs at sequence numbers 400 (baseline) and 160 (imperfect).

Our findings show that users can learn to accomplish the presented simplified expert task. This does not mean, however, that they would perform equally well if confronted with the "real" expert task. Users were given assistance by reducing the number of candidates from more than one hundred to six, they were provided

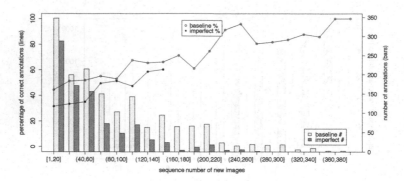

Fig. 8. Users' performance for first appearances of images that occur in different stages of the game (lines) and number of annotations

a visual key (example image) to aid memorization and a short description of the subject type. A way to increase the success rate in a realistic setting would be to train users on a "perfect" data set and after passing a predefined success threshold, introduce "imperfect" data into the game.

5   Conclusions

Our study investigates the use of crowdsourcing for a task that normally requires specific expert knowledge. Such a task could be relevant to facilitate search by improving metadata on non-textual data sets, but also in crowdsourcing relevance judgements for more complex data in a more classic IR setting.

Our main finding is that non-experts are able to learn to categorize paintings into subject types of the AAT thesaurus in our simplified set-up. We studied two conditions, one with the expert choice always present, and one in which the expert choice had been removed in 25% of the cases. Although the agreement between experts of the Rijksmuseum Amsterdam and non-experts for the first condition is higher, the agreement in the imperfect condition is still acceptably high. We found that the aggregation of votes leads to a noticeable "Wisdom of the Crowds" effect and increases the precision of the users' votes. While this removed many deviations of the users' judgements from the experts' judgements, on some images, the disagreement remained. We consulted an expert and identified two main reasons: Either the annotations by the experts were incomplete or incorrect or the correct classification required knowing context information of the paintings that was not given to the users.

The analysis of user performance over time showed that users learned to carry out the task with higher precision the longer they play. This holds for repeated images (memorization) as well as new images (generalization).

The next step is to balance the interdependencies of the three players: experts, automatic methods and gamers. We hope that reducing their weaknesses (scarce, requiring much training data, insufficient expertise) by directing the interplay of their strengths (ability to provide: high quality data, high quantity data, high quality when trained and assisted) can lead to a quickly growing collection of high quality annotations.

Acknowledgments. We would like to thank Sjaak Wouters for making his data collection available as a basis for our research, Daniel Horst for the re-evaluation of the expert judgements and all participants of the experiments for their time and effort, especially user 21 for playing 25 rounds. This publication is funded by the Dutch COMMIT/ program and EC FP7 grant 257024, the Fish4Knowledge project.

References

1. Carletti, L., Giannachi, G., McAuley, D.: Digital humanities and crowdsourcing: An exploration. In: MW 2013: Museums and the Web 2013 (2013)
2. Dijkshoorn, C., Leyssen, M.H.R., Nottamkandath, A., Oosterman, J., Traub, M.C., Aroyo, L., Bozzon, A., Fokkink, W., Houben, G.-J., Hovelmann, H., Jongma, L., van Ossenbruggen, J., Schreiber, G., Wielemaker, J.: Personalized nichesourcing: Acquisition of qualitative annotations from niche communities. In: 6th International Workshop on Personalized Access to Cultural Heritage (PATCH 2013), pp. 108–111 (2013)
3. Galton, F.: Vox populi. Nature 75(1949), 7 (1907)
4. Golbeck, J., Koepfler, J., Emmerling, B.: An experimental study of social tagging behavior and image content. Journal of the American Society for Information Science and Technology 62(9), 1750–1760 (2011)
5. He, J., van Ossenbruggen, J., de Vries, A.P.: Do you need experts in the crowd?: a case study in image annotation for marine biology. In: Proceedings of the 10th Conference on Open Research Areas in Information Retrieval, OAIR 2013, Paris, France, pp. 57–60 (2013); Le Centre De Hautes Etudes Internationales D'Informatique Documentaire
6. Heimerl, K., Gawalt, B., Chen, K., Parikh, T., Hartmann, B.: Communitysourcing: engaging local crowds to perform expert work via physical kiosks. In: Proceedings of the 2012 ACM Annual Conference on Human Factors in Computing Systems, CHI 2012, pp. 1539–1548. ACM, New York (2012)
7. Hildebrand, M., van Ossenbruggen, J., Hardman, L., Jacobs, G.: Supporting subject matter annotation using heterogeneous thesauri: A user study in web data reuse. International Journal of Human-Computer Studies 67(10), 887–902 (2009)
8. Hosseini, M., Cox, I.J., Milić-Frayling, N., Kazai, G., Vinay, V.: On aggregating labels from multiple crowd workers to infer relevance of documents. In: Baeza-Yates, R., de Vries, A.P., Zaragoza, H., Cambazoglu, B.B., Murdock, V., Lempel, R., Silvestri, F. (eds.) ECIR 2012. LNCS, vol. 7224, pp. 182–194. Springer, Heidelberg (2012)
9. von Ahn, L., Dabbish, L.: ESP: Labeling images with a computer game. In: AAAI Spring Symposium: Knowledge Collection from Volunteer Contributors, pp. 91–98. AAAI (2005)
10. Wouters, S.: Semi-automatic annotation of artworks using crowdsourcing. Master's thesis, Vrije Universiteit Amsterdam, The Netherlands (2012)

Relevance-Ranked Domain-Specific Synonym Discovery

Andrew Yates, Nazli Goharian, and Ophir Frieder

Information Retrieval Lab, Georgetown University
{andrew,nazli,ophir}@ir.cs.georgetown.edu

Abstract. Interest in domain-specific search is growing rapidly, creating a need for domain-specific synonym discovery. The best-performing methods for this task rely on query logs and are thus difficult to use in many circumstances. We propose a method for domain-specific synonym discovery that requires only a domain-specific corpus. Our method substantially outperforms previously proposed methods in realistic evaluations. Due to the difficulty of identifying pairs of synonyms from among a large number of terms, methods have traditionally been evaluated by their ability to choose a target term's synonym from a small set of candidate terms. We generalize this evaluation by evaluating methods' performance when required to choose a target term's synonym from progressively larger sets of candidate terms. We approach synonym discovery as a ranking problem and evaluate the methods' ability to rank a target term's candidate synonyms. Our results illustrate that while our proposed method substantially outperforms existing methods, synonym discovery is still a difficult task to automate and is best coupled with a human moderator.

Keywords: Synonym discovery, thesaurus construction, domain-specific search.

1 Introduction

Interest in domain-specific search has grown over the past few years. Researchers are increasingly investigating how to best search medical documents [7, 14, 16], legal documents [10, 11, 19], and patents [2, 21]. With the growing interest in domain-specific search, there is an unmet need for domain-specific synonym discovery. Domain-independent synonyms can be easily identified with resources such as thesauri, but domain-specific variants of such resources are often less common and less complete. Worse, synonyms can even be corpus-specific or specific to a subdomain within a given domain. For example, in the legal or e-discovery domain, an entity subject to e-discovery may use its own internal terms and acronyms that cannot be found in any thesaurus. In the medical domain, whether or not two terms are synonyms can depend entirely on the use case. For example, a system for detecting drug side effects might treat "left arm pain" as a synonym of "arm pain" because the arm pain is the relevant part. On the other hand, "left arm pain" would not be synonymous with "arm pain" in an electronic health record belonging to a patient who had injured her left arm.

Furthermore, domain-specific document collections (e.g., e-discovery or medical) are often significantly smaller than the collections that domain-independent synonym

M. de Rijke et al. (Eds.): ECIR 2014, LNCS 8416, pp. 124–135, 2014.
© Springer International Publishing Switzerland 2014

discovery is commonly performed on (e.g., the Web). We present a domain-specific synonym discovery method that can be used with domain-specific document collections. We evaluate our method on a focused collection consisting of 400,000 forum posts. Our results show that our method can be used to produce ranked lists that significantly reduce the effort of a human editor.

The best-performing synonym discovery methods require external information that is difficult to obtain, such as query logs [33] or documents translated into multiple languages [12, 25]. Other types of synonym discovery methods (e.g., [31, 32]) have commonly been evaluated using synonym questions from TOEFL (Test Of English as a Foreign Language), in which the participant is given a target word (e.g., "disagree") and asked to identify the word's synonym from among four choices (e.g., "coincide", "disparage", "dissent", and "deviate"). While this task presents an interesting problem to solve, this type of evaluation is not necessarily applicable to the more general task of discovering synonyms from among the many terms (n candidates) present in a large collection of documents. We address this concern by evaluating our method's and other methods' performance when used to answer domain-specific TOEFL-style questions with progressively larger numbers of incorrect choices (i.e., from 3 to 1,000 incorrect choices). While our proposed method performs substantially better than strong existing methods, neither our method nor our baselines are able to answer a majority of the questions correctly when presented with hundreds or thousands of incorrect choices. Given the difficulty of choosing a target term's synonym from among 1,000 candidates, we approach domain-specific synonym discovery as a ranking problem in which a human editor searches for potential synonyms of a term and manually evaluates the ranked list of results. To evaluate the usefulness of this approach, we use our method and several strong existing methods to rank lists of potential synonyms. Our method substantially outperforms existing methods and our results are promising, suggesting that, for the time being, domain-specific synonym discovery is best approached as a human-moderated relevance-ranking task.

Our contributions are (1) a new synonym discovery method that outperforms strong existing approaches (our baselines); (2) an evaluation of how well our method and others' methods perform on the TOEFL-style evaluations when faced with an increasing number of synonym candidates; (3) an evaluation of how well our methods and others' methods perform when used to rank a target term's synonyms; our method places 50% of a target term's synonym in the top 5% of results, whereas other approaches place 50% of a target term's synonyms in the top 40%.

2   Related Work

A variety of methods have been applied to the domain-independent synonym identification problem. Despite the limited comparisons of these methodologies, the best-performing methods are reported to use query logs or parallel corpora. We describe the existing methodologies and differentiate our approach.

Distributional Similarity. Much related work discovers synonyms by computing the similarity of the contexts that terms appear in; this is known as distributional similarity [26]. The intuition is that synonyms are used in similar ways and thus are surrounded by similar words. In [31], Terra and Clarke compare the abilities of various statistical similarity measures to detect synonyms when used along with term co-occurrence information. Terra and Clarke define a term's context as either the term windows in which the term appears or the documents in which the term appears. They use questions from TOEFL (Test Of English as a Foreign Language) to evaluate the measures' abilities to choose a target word's synonym from among four candidates. We use Terra and Clarke's method as one of our baselines (baseline 1: Terra & Clark). In [8], Chen et al. identify synonyms by considering both the conditional probability of one term's context given the other term's context and co-occurrences of the terms, but perform limited evaluation. In [27], Rybinski et al. find frequent term sets and use the term sets' support to find terms which occur in similar contexts. This approach has a similar outcome to other approaches that use distributional similarity, but the problem is formulated in terms of terms sets and support.

Distributional similarity has also been used to detect other types of relationships among words, such as hyponymy and hypernymy, as they also tend to occur in similar contexts. In [28], Sahlgren and Karlgren find terms related to a target concept (e.g., "criticize" and "suggest" for the concept "recommend") with random indexing [18], a method which represents terms as low-dimensional context vectors. We incorporate random indexing as one of our model's features and evaluate the feature's performance in our feature analysis. Brody and Lapata use distributional similarity to perform word sense disambiguation [5] using a classifier with features such as n-grams, part of speech tags, dependency relations, and Lin's similarity measure [20], which computes the similarity between two words based on the dependency relations they appear in. We incorporate Lin's similarity measure as a feature and derive features based on n-grams and part-of-speech n-grams. Strzalkowski proposes a term similarity measure based on shared contexts [30]. Carrell and Baldwin [6] use the contexts a target term appears in to identify variant spellings of a target term in medical text. Pantel et al. use distributional similarity to find terms belonging to the same set (i.e., terms which share a common hypernym) [24] by representing each term as a vector of surrounding noun phrases and computing the cosine distance between term vectors.

Lexico-syntactic Patterns. In [22], McCrae and Collier represent terms by vectors of the patterns [15] they occur in and use a classifier to judge whether term pairs are synonyms. Similarly, Hagiwara [13] uses features derived from patterns and distributional similarity to find synonyms. Hagiwara extracts dependency relations from documents (e.g., X is a direct object of Y) and use them as a term's context. Hagiwara finds that the features derived from distributional similarity are sufficient, because there is no significant change in precision or recall when adding features derived from patterns. Their analysis is logical given that lexico-syntactic patterns and distributional similarity are both concerned with the terms surrounding a target term. We use Hagiwara's method as another one of our baselines (baseline 2: Hagiwara).

Tags. Clements et al. [9] observe that in social tagging systems different user groups sometimes apply different, yet synonymous tags. They identify synonymous tags based on overlap among users/items. Other tag similarity work includes [29], which identifies similar tags that represent a "base tag". Tag-based approaches rely on the properties of tags, thus they are not applicable to domains in which tags are not used. For this reason we do not compare our method with tag-based approaches.

Web Search. Turney [32] identifies synonyms by considering the co-occurrence frequency of a term and its candidate synonym in Web search results. This method is evaluated on the same TOEFL dataset used by Terra and Clarke [31]; Terra and Clarke's method performs better. Similarly, other approaches [1, 3] rely on obtaining co-occurrence frequencies for terms from a Web search engine. We do not compare with Web search-based methods as they rely on a general corpus (the Web), whereas our task is to discover domain-specific synonyms in a domain-specific corpus.

Word Alignment. Plas [25] and Grigonytė et al. [12] observe that English synonyms may be translated to similar words in another language; they use word alignment between English and non-English versions of a document to identify synonyms within a corpus. Wei et al. [33] use word alignment between queries to identify synonyms. Similarly, word alignment can be coupled with machine translation to identify synonyms by translating text into a second language and then back into the original language (e.g., [23]). While word alignment methods have been shown to perform well, their applicability is limited due to requiring either query logs or parallel corpora. Due to this limitation, we do not use any word alignment method as a baseline; we are interested in synonym discovery methods that do not require difficult-to-obtain external data.

3   Methodology

We compare our approach against three baselines: Terra and Clarke's method [31], Hagiwara's SVM method [13], and a variant of Hagiwara's method.

3.1   Baseline 1: Terra and Clarke

In [31], Terra and Clarke evaluate how well many statistical similarity measures identify synonyms. We use the similarity measure that they found to perform best, pointwise mutual information (PMI), as one of our baselines. The maximum likelihood estimates used by PMI depend on how term co-occurrences are defined. Terra and Clarke propose two approaches: a window approach, in which two terms co-occur when they are present in the same n-term sliding window, and a document approach, in which two terms co-occur when they are present in the same document. We empirically determined that a 16-term sliding window performed best on our dataset.

With this approach the synonym of a term w_i is the term w_j that maximizes $PMI(w_i, w_j)$. Similarly, a ranked list of the synonym candidates for a term w_i can be obtained using this approach by using $PMI(w_i, w_j)$ as the ranking function.

3.2   Baseline 2: Hagiwara (SVM)

Hagiwara [13] proposes a synonym identification method based on point-wise total correlation (PTC) between two terms (or phrases treated as single terms) w_i and w_j

and a context c_k in which they both appear. Hagiwara uses syntax to define context. The RASP parser [4] is used to extract term dependency relations from documents in the corpus. A term's contexts are the (*modifier term, relation type*) tuples from the relations in which the term appears as a head word.

Hagiwara takes a supervised approach. Each pair of terms (w_i, w_j) is represented by a feature vector containing the terms' point-wise total correlations for each context as features. Features for contexts not shared by w_i and w_j have a value of 0. That is, $vector_{w_i,w_j} = \langle PTC(w_i, w_j, c_1), ..., PTC(w_i, w_j, c_n) \rangle$. We prune features using the same criteria as Hagiwara and identify synonyms by classifying each word pair as synonymous or not synonymous using SVM. We modified this approach to rank synonym candidates by ranking the results based on SVM's decision function's value.

3.3   Baseline 3: Hagiwara (Improved)

We modified Hagiwara's SVM approach to create an unsupervised approach based on similar ideas. The contexts and maximum likelihood estimates are the same as in Hagiwara's approach (described in section 3.2). Instead of creating a vector for each pair of terms (w_i, w_j), we created a vector for each term w_i and computed the similarity between these vectors. The vector for a term w_i is composed of the PMI measures between the term w_i and each context c_k. That is, $vector_{w_i} = \langle PMI(w_i, c_1), PMI(w_i, c_2), ..., PMI(w_i, c_n) \rangle$. The similarity between w_i and w_j is computed as the cosine similarity between their two vectors. Similarly, we rank synonym candidates for a term w_i by ranking vectors based on their similarity to $vector_{w_i}$.

3.4   Regression

Our approach is a logistic regression on a small set of features. We hypothesize that a supervised approach will outperform statistical synonym identification approaches since it does not rely on any single statistical measure and can instead weight different types of features. While Hagiwara's original method used supervised learning, it only used one type of contextual feature (i.e., point-wise total correlation between two terms and a context). Like Hagiwara, we construct one feature vector for each word pair. In the training set, we give each pair of synonyms a value of (+1) and each pair of words that are not synonyms a value of (-1). To obtain a ranked list of synonym candidates, the probabilities of candidates being synonyms are used as relevance scores. That is, the highest ranked candidates are those that the model gives the highest probability of being a 1.

We also experimented with SVM^{Rank} [17] and SVM, but found that a logistic regression performed similarly or better while taking significantly less time to train.

The features we used are:

1.   The number of distinct contexts both w_i and w_j appear in, normalized by the minimum number of contexts either one appears in,

$$shared\_contexts = \frac{c(w_i, w_j)}{\min(c(w_i), c(w_j))}$$

where $c(w_i)$ is the number of distinct contexts w_i appears in and $c(w_i, w_j)$ is the number of distinct contexts both w_i and w_j appear in. According to the distributional hypothesis [26], similar words should appear in the same context more often than dissimilar words do. We use Hagiwara's method as described in section 3.2 for finding contexts.

2. The number of sentences both w_i and w_j appear in, normalized by the minimum number of sentences either one appears in,

$$shared\_sentences = \frac{s(w_i, w_j)}{\min(s(w_i), s(w_j))}$$

where $s(w_i)$ is the number of windows w_i appears in and $s(w_i, w_j)$ is the number of windows both w_i and w_j appear in.

3. The cosine similarity between w_i and w_j as calculated by the Hagiwara (Improved) method, as described in section 3.3. This method weights contexts by their PMI, whereas $shared\_contexts$ weights all contexts equally.

4. The Levenshtein distance between terms w_i and w_j. Our synonym list contains phrases; that is, terms may contain multiple words (e.g., "sore_throat"). We hypothesize that this feature will be useful because synonymous phrases may share common terms (e.g., "aching_throat" and "sore_throat").

5. The probability of the target term w_i appearing in an n-gram given that the candidate term w_j appears in the n-gram. We use all n-grams of size 3 that appear in our dataset (e.g., "havelalheadache") and replace the candidate and target terms with X (e.g., "havelalX").

$$ngram\_pr = \Pr(w_i \ appears \ as \ X \mid w_j \ appears \ as \ X)$$
$$= \frac{Pr(w_i \ and \ w_j \ appear \ as \ X)}{Pr(w_j \ appears \ as \ X)}$$

6. The probability of the target term w_i appearing in a part-of-speech n-gram given that the candidate term w_j appears in the part-of-speech (POS) n-gram. As with $ngram\_pr$, we use n-grams of size 3. To construct POS n-grams, we replace the candidate and target terms with X as before and replace each term in the n-gram with its POS (e.g., "havelalX" becomes "VBPIDTIX").

$$posng\_pr = \Pr(w_i \ appears \ as \ X \mid w_j \ appears \ as \ X)$$
$$= \frac{Pr(w_i \ and \ w_j \ appear \ as \ X)}{Pr(w_j \ appears \ as \ X)}$$

7. The similarity between terms w_i and w_j as computed by Lin's information-theoretic term similarity measure (lin_sim) as described in [20]; this measure is computed using the dependency relations that terms w_i and w_j appear in.

8. The cosine distance between the vector for term w_i and the vector for term w_j as obtained using random indexing. We used the SemanticVectors (https://code.google.com/p/semanticvectors/) implementation of random indexing with the default parameters.

Features 5-7 (*ngram_pr*, *posng_pr*, and *lin_sim*) were inspired by features used in Brody and Lapata's effort on word sense disambiguation [5]; *random_indexing* was shown by Sahlgren and Karlgren to perform well at identifying related terms in [28]. We explore the utility of each feature in section 4.4.

4   Experiments

We describe our ground truth and corpus in section 4.1. In section 4.2 we evaluate the quality of our approach and various baseline methods using a more realistic variant of the TOEFL evaluation methodology commonly used in previous efforts. We approach synonym discovery problem as a ranking problem in section 4.3 and evaluate how well our approach and the baseline methods rank a target term's synonyms. Finally, we examine the impact of each feature used by our method in section 4.4.

4.1   Dataset

We focus on the medical side-effect domain in our evaluation. To evaluate our methodology and compare with existing strong approaches (i.e., our baselines), we used a corpus of medical forum posts and the MedSyn synonym list [34] as our ground truth, which contains synonyms in the medical side-effect domain. A domain-specific thesaurus is required to train the synonym discovery methods for a given domain. We removed synonyms from the list that do not occur or occur only once in our corpus because it is impossible for any of the methods to detect them. We also removed terms from the list that had no synonyms in our corpus. This left us with 1,791 synonyms, which were split into a training set (291 pairs) which was used to tune our methods, and a testing set (1,500 pairs), which was used to perform our evaluations. On average, each term in the list had 2.8 synonyms ($\sigma = 1.4$). The maximum number of synonyms per term was 11 and the minimum number was 2. Of the 1,791 synonyms that we kept, 67% of the synonyms were phrases treated as a single term (e.g., "joint_pain") and the remaining 33% were single terms (e.g., "arthralgia").

We created questions (target terms) similar to those used in TOEFL from terms in the synonym list by choosing a target term as the question (e.g., "joint pain") and choosing a synonym (e.g., "arthralgia") and non-synonymous terms (e.g., "headache", "arthritis", and "arm pain") as choices (synonym candidates). The methods' task is to identify the correct synonym from among the choices (synonym candidates) given the target term. In the general TOEFL-style evaluation (section 4.2), each question has one correct choice and n incorrect choices. In the relevance ranking evaluation (section 4.3), each question i has m_i correct choices and n incorrect choices, where m_i is the number of synonyms that question i has in the synonym list.

Our corpus was built from a crawl of 400,000 forum posts made to the Breastcancer.org discussion boards[1] and the FORCE breast cancer message boards[2]. Both Websites divide their discussion boards into topics. In keeping with our goal of identifying domain-specific synonyms, we crawled only those topics related to general discussion

---

[1] http://community.breastcancer.org/
[2] http://www.facingourrisk.org/messageboard/index.php

or to side-effects. A complete list of the pages crawled is available at *http://ir.cs.georgetown.edu/data/medposts.txt.* While this dataset is focused on the medical side-effect domain, our methods do not take advantage of any medical domain knowledge and could be applied to find synonyms within any domain. We stemmed both the synonym list and corpus with the Porter stemmer. When tokenizing our corpus and synonym list, we transformed each multi-word term in the synonym list into a single term (e.g., "joint pain" became "joint_pain"). We define synonyms as equivalent terms, including spelling variations. Synonymous phrases may be the same except for one additional word (e.g., "arm_pain" and "left_arm_pain"). We do not include separate entries in our synonym list for every morphological variant of a term, however, because the synonym list is stemmed.

4.2   General TOEFL-Style Evaluation

In related research efforts, TOEFL (Test Of English as a Foreign Language) questions have been most commonly used to measure synonym identification accuracy. Succinctly, a TOEFL question consists of a target term and four synonym candidates. The task is to identify which one of the four candidates is a synonym of the target term. To create a more realistic TOEFL-style evaluation in which methods are faced with more than four choices (synonym candidates), we created TOEFL-style questions that consisted of one target word, one correct choice, and n incorrect choices (analogous to the TOEFL evaluation when $n=3$). We let n range from 3 to 138 in multiples of 15 (3, 18, 33, 48, …, 138) and from 150 to 1050 in multiples of 100 (150, 250, …, 1050) so that we could observe the methods' performance while the questions were gradually made harder (multiples of 15) and while the questions became harder more rapidly (multiples of 100). We used five-fold cross-validation with the supervised methods. As in previous work, we measured the performance in terms of the number of questions answered correctly as the number of incorrect candidates varies (correct@n).

The results for the *general TOEFL-Style evaluation* are shown in Figure 1. We also show the expected performance of a method that randomly selects synonyms (*Random*). *Terra and Clarke*'s method quickly overtakes *Hagiwara (Improved)* as n increases. Our method, Regression, performs substantially better than *Terra & Clarke* for all values of n (about 175% better @33, @150, and @450). The performance of all methods decreases as n increases. *Hagiwara (SVM)* performs the worst among the methods (91% worse than *Regression* @150) and quickly approaches the performance of *Random*. *Hagiwara (Improved)* performs better than *Hagiwara (SVM)*, but it performs much worse than *Regression* and *Terra and Clarke* (85% worse than *Regression* @150). At $n=3$, which is equivalent to the traditional TOEFL evaluations with one correct choice and three incorrect choices, *Regression* performs 49% better (67% vs. 45%) than the next best method, *Hagiwara (improved)*. If each question's number of correct choices increases to two or three (instead of one), the methods perform similarly and *Regression* continues to substantially outperform the other methods.

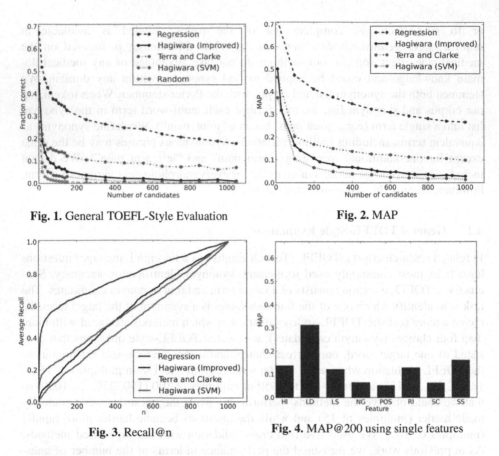

Fig. 1. General TOEFL-Style Evaluation   Fig. 2. MAP

Fig. 3. Recall@n   Fig. 4. MAP@200 using single features

While *Regression* and *Terra and Clarke* perform much better than the two *Hagiwara* methods, they do not perform well on an absolute scale. When used to find a target term's synonym from among 451 choices (450 incorrect choice and 1 correct choice), *Regression* is only correct 25% of the time; when $n=1000$, *Regression* is correct only 18% of the time. This is not accurate enough for use as a domain-specific synonym discovery method. In the next section (section 4.3), we propose a solution.

4.3 Relevance Ranking Evaluation

It is clear from the general TOEFL evaluation (section 4.2) that currently-existing methods are incapable of discovering domain-specific synonyms with acceptable accuracy. Given this observation, we propose approaching the problem of domain-specific synonym discovery as a ranking problem, in which a human editor identifies a target term's synonyms by manually reviewing a ranked list of potential synonyms. While this process does require human effort, providing a high quality ranked list significantly reduces the amount of required effort. We evaluate the methods' abilities to produce ranked lists. To do so, each method is given a target term and required to rank the term's synonym candidates. To evaluate this approach, we generated

TOEFL-style questions in which each question i has m_i correct choices and n incorrect choices, where m_i is the number of synonyms that question i has in the synonym list. That is, each m_i is fixed for each question i and n grows progressively larger. That is, there is no fixed number of correct choices as in the general TOEFL evaluation where there was only one correct choice. Instead, the number of correct choices for each question is the number of synonyms that actually exist; this is more realistic than fixing the number of correct choices in the evaluation. We started n at 10 and then allowed it to range from 100 to 1,000 in multiples of 100 (10, 100, 200, 300, ..., 1000). The quality of the ranked lists produced by each method was measured with Mean Average Precision (MAP). We used five-fold cross-validation with the supervised methods, as we did in our previous evaluations. Each method was modified to produce a ranked list of results as described in the methodology sections

The results are shown in Figure 2. In this evaluation, *Regression* outperforms *Terra and Clarke* for all values of n (57% better @10, 135% better @200, and 170% better @1000). Similarly, *Hagiwara (Improved)* outperforms *Hagiwara (SVM)* for all values of n. As in the general TOEFL evaluation, *Regression* and *Terra and Clarke* perform much better than the *Hagiwara* methods. *Regression*'s MAP remains above 0.40 for $n \leq 200$ and has a MAP of 0.27 at $n = 1000$. This suggests that *Regression* produces a ranked list that a human editor would find useful.

We measured recall@n with 1,000 candidates to explore how useful a human editor would find these ranked lists. The results are shown in Figure 3. As with MAP, *Regression* outperforms the other methods. *Regression* achieves a recall of 0.54 @50, indicating that a human editor could use *Regression*'s ranked lists to find half of a target term's synonyms by looking at only the top 50 of 1,000 results (5%). This is a sharp contrast to the other three methods, which require at least the top 400 of 1,000 results (40%) to be viewed before achieving an equivalent recall; *Regression* performs 157% better than *Terra & Clarke* @50, suggesting that our method can significantly decrease the work performed by a human editor.

4.4   Feature Analysis

We examine *Regression*'s features to determine their contribution to *Regression*'s overall performance. To do so, we analyze the features in the context of the relevance ranking evaluation. We compare the MAP@200, achieved by single features, and feature pairs. We abbreviate the name of each feature as follows: Hagiwara_improved (HI), Levenshtein_distance (LD), Lin_sim (LS), ngram_pr (NG), posng_pr (POS), random_indexing (RI), shared_contexts (SC), and shared_sentences (SS).

The performance of each single feature is shown in Figure 4. LD performs the best, which is surprising given that our corpus was stemmed. We hypothesize that LD's utility both results from synonymous terms that stem to different roots and synonymous phrases that share some terms. SS, RI, LS, and HI follow LD, but achieve MAPs approximately 50% lower than LD's. The features that use dependency relations (HI and LS) perform similar to RI, which uses term co-occurrences. NG, POS, and SC perform the worst. When pairs of features are used, the pairs containing LD perform the best. All of these pairs perform similarly, but LD-SS performs best (25% better

134   A. Yates, N. Goharian, and O. Frieder

than LD alone); it is closely followed by LD-RI. Of the feature pairs that do not contain LD, three pairs that contain SS perform the best (LS-SS, RI-SS, and SS-HI), however, they perform approximately 50% worse than the pairs containing LD. These results mirror those obtained using single features. The performance achieved by the best performing feature combinations (LD and LD-SS) cannot be achieved simply by combining baselines (e.g., HI and RI).

5   Conclusions

We proposed a new regression-based method for synonym discovery that substantially outperforms existing methods when used to rank a target term's synonym candidates. Additionally, our method performs better at a generalization of the TOEFL-style evaluation commonly used in prior work. When used to rank a target term's 100 synonym candidates, our method produces rankings with a MAP of 0.47, a 135% increase over the next-highest method. On average our method places 54% of a target term's synonyms in the top 50 of 1,000 results (5%), whereas other approaches place 50% of a target term's synonyms in the top 400 of 1,000 results (40%). While domain-specific synonym discovery is still a difficult task requiring a human editor, our method significantly decreases the number of terms an editor must review. Our method finds domain-specific synonyms, but the method itself is not domain specific. Future work could investigate the benefits of using a domain-specific method.

Acknowledgements. This work was partially supported by the US National Science Foundation through grant CNS-1204347.

References

1. Alfonseca, E., et al.: Using context-window overlapping in synonym discovery and ontology extension. In: RANLP 2005 (2005)
2. Azzopardi, L., et al.: Search system requirements of patent analysts. In: SIGIR 2010 (2010)
3. Bollegala, D.: Measuring Semantic Similarity between Words Using Web Search Engines. In: WWW 2007 (2007)
4. Briscoe, T., et al.: The second release of the RASP system. In: Proceedings of the COLING/ACL on Interactive Presentation Sessions (2006)
5. Brody, S., Lapata, M.: Good Neighbors Make Good Senses: Exploiting Distributional Similarity for Unsupervised WSD. In: COLING 2008 (2008)
6. Carrell, D., Baldwin, D.: PS1-15: A Method for Discovering Variant Spellings of Terms of Interest in Clinical Text. Clin. Med. Res. 8, 3–4 (2010)
7. Cartright, M.-A., et al.: Intentions and attention in exploratory health search. In: SIGIR 2011, p. 65 (2011)
8. Chen, L., et al.: Statistical relationship determination in automatic thesaurus construction. In: CIKM 2005 (2005)
9. Clements, M., et al.: Detecting synonyms in social tagging systems to improve content retrieval. In: SIGIR 2008 (2008)

10. Evans, D.A., et al.: E-discovery. In: CIKM 2008 (2008)
11. Ghosh, K.: Improving e-discovery using information retrieval. In: SIGIR 2012 (2012)
12. Grigonytė, G., et al.: Paraphrase alignment for synonym evidence discovery. In: COLING 2010 (2010)
13. Hagiwara, M.: A Supervised Learning Approach to Automatic Synonym Identification based on Distributional Features. In: HLT-SRWS 2008 (2008)
14. Hanbury, A.: Medical information retrieval. In: SIGIR 2012 (2012)
15. Hearst, M.A.: Automatic acquisition of hyponyms from large text corpora. In: COLING 1992, p. 539 (1992)
16. Huang, J.X., et al.: Medical search and classification tools for recommendation. In: SIGIR 2010 (2010)
17. Joachims, T.: Optimizing search engines using clickthrough data. In: KDD 2002 (2002)
18. Kanerva, P., et al.: Random indexing of text samples for latent semantic analysis. In: CogSci 2000 (2000)
19. Lewis, D.D.: Information retrieval for e-discovery. In: SIGIR 2010 (2010)
20. Lin, D.: Automatic retrieval and clustering of similar words. In: ACL/COLING 1998 (1998)
21. Lupu, M.: Patent information retrieval. In: SIGIR 2012 (2012)
22. McCrae, J., Collier, N.: Synonym set extraction from the biomedical literature by lexical pattern discovery. BMC Bioinformatics 9 (2008)
23. Nanba, H., et al.: Automatic Translation of Scholarly Terms into Patent Terms Using Synonym Extraction Techniques. In: LREC 2012 (2012)
24. Pantel, P., et al.: Web-Scale Distributional Similarity and Entity Set Expansion. In: EMNLP 2009 (2009)
25. Van Der Las, L.: Finding Synonyms Using Automatic Word Alignment and Measures of Distributional Similarity. In: COLING-ACL 2006 (2006)
26. Rubenstein, H., Goodenough, J.B.: Contextual correlates of synonymy. Commun. ACM 8(10), 627–633 (1965)
27. Rybiński, H., Kryszkiewicz, M., Protaziuk, G., Jakubowski, A., Delteil, A.: Discovering Synonyms Based on Frequent Termsets. In: Kryszkiewicz, M., Peters, J.F., Rybiński, H., Skowron, A. (eds.) RSEISP 2007. LNCS (LNAI), vol. 4585, pp. 516–525. Springer, Heidelberg (2007)
28. Sahlgren, M., Karlgren, J.: Terminology mining in social media. In: CIKM 2009 (2009)
29. Solskinnsbakk, G., Gulla, J.A.: Mining tag similarity in folksonomies. In: SMUC 2011 (2011)
30. Strzalkowski, T.: Building a lexical domain map from text corpora. In: COLING 1994 (1994)
31. Terra, E., Clarke, C.L.A.: Frequency estimates for statistical word similarity measures. In: HLT-NAACL 2003 (2003)
32. Turney, P.D.: Mining the Web for Synonyms: PMI-IR versus LSA on TOEFL. In: Flach, P.A., De Raedt, L. (eds.) ECML 2001. LNCS (LNAI), vol. 2167, pp. 491–502. Springer, Heidelberg (2001)
33. Wei, X., et al.: Context sensitive synonym discovery for web search queries. In: CIKM 2009 (2009)
34. Yates, A., Goharian, N.: ADRTrace: Detecting Expected and Unexpected Adverse Drug Reactions from User Reviews on Social Media Sites. In: Serdyukov, P., Braslavski, P., Kuznetsov, S.O., Kamps, J., Rüger, S., Agichtein, E., Segalovich, I., Yilmaz, E. (eds.) ECIR 2013. LNCS, vol. 7814, pp. 816–819. Springer, Heidelberg (2013)

Towards Generating Text Summaries
for Entity Chains

Shruti Chhabra and Srikanta Bedathur

Indraprastha Institute of Information Technology
New Delhi, India
{shrutic,bedathur}@iiitd.ac.in

Abstract. Given a large knowledge graph, discovering meaningful relationships between a given pair of entities has gained a lot of attention in the recent times. Most existing algorithms focus their attention
on identifying one or more structures –such as relationship chains or
subgraphs– between the entities. The burden of interpreting these results,
after combining with contextual information and description of relationships, lies with the user. In this paper, we present a framework that
eases this burden by generating a textual summary which incorporates
the context and description of individual (dyadic) relationships, and combines them to generate a ranked list of summaries. We develop a model
that captures key properties of a well-written text, such as coherence
and information content. We focus our attention on a special class of
relationship structures, two-length entity chains, and show that the generated ranked list of summaries have 79% precision at rank-1. Our results
demonstrate that the generated summaries are quite useful to users.

Keywords: entity chain, text summarization, relationship queries,
entity-relationship graphs.

1 Introduction

The use of large entity-relationship graphs such as DBPedia [3], Freebase [5], and
Yago [16], in various information retrieval and discovery tasks have given raise to
many challenging problems. The problem of discovering long ranging semantic
relationships between a pair or a group of entities has attracted much attention
recently. Existing solutions take as input a pair of entities and extract structures such as a chain of nodes connecting the input pair [2], or a subgraph [10]
from the underlying entity-relationship graph. While these structures are useful
in capturing the important relationships, the burden of interpreting the overall
relationship lies with the user. The attributes such as the context and relationship descriptions that make such an interpretation possible, are missing from
the extracted structure.

Efforts are being made to associate textual evidences (such as documents,
sentences, or phrases) that describe underlying relationships in a more human-
understandable form. Information extraction systems such as OpenIE [9], PATTY

M. de Rijke et al. (Eds.): ECIR 2014, LNCS 8416, pp. 136–147, 2014.

Table 1. Sample human-generated summaries for two-length entity chains

No.	Entity Chain	Human-generated Summary
1	⟨Brooke Shields, Andre Agassi, Steffi Graf⟩	Andre Agassi was married to Brooke Shields till 1999. He married Steffi Graf in 2001.
2	⟨Richard Nixon, John F. Kennedy, Lyndon Johnson⟩	John F. Kennedy defeated Richard Nixon in 1960 U.S. presidential election. Lyndon Johnson became president after Kennedy.
3	⟨Afghanistan, Opium, Europe⟩	Afghanistan is the world leading producer of Opium. The better quality Opium is smuggled to Europe.
4	⟨Charlie Sheen, Martin Sheen, West Wing⟩	Charlie Sheen is son of Martin Sheen. Martin was a part of television drama, The West Wing.
5	⟨Satwant Singh, Indira Gandhi, Operation Blue Star⟩	Indira Gandhi was killed by two of her sikh bodyguards, Satwant Singh and Beant Singh, in the aftermath of Operation Blue Star. Operation Blue Star was an Indian army's assault on the Golden Temple, ordered by Indira Gandhi.

[25], and NELL [6] routinely maintain textual evidences for relations they extract. Techniques such as support sentence retrieval [4] help to retrieve short passages or sentences from a text corpus for a given entity. Despite these, the complex task of combining these evidences available for individual relationships rests with the user.

To address these issues, we propose the idea of generating textual summaries corresponding to the extracted relationship structures. In this paper, we restrict our attention on simple chain structures consisting of two entities connected through an intermediate entity, which we call as two-length entity chains. Considering sentence-level evidence associated with edges, we develop a model that generates candidate summaries for a given two-length entity chain by combining the sentences associated with individual edges in the chain and ranks these candidate summaries.

Clearly, the summaries generated by the underlying model must satisfy the basic properties of a well-written, human generated summary. Table 1 illustrates some human-generated summaries for sample entity chains. Humans use their knowledge of writing a well formed text in presenting the facts related to given entities, which makes the summaries understandable (as presented in Table 1). The aim of our model is to generate such good quality summaries.

We conducted experiments to identify the key properties of a good quality summary. Based on the insights from literature [1][21][26] and our initial experiments [7], we argue that the following three properties are crucial to capture the intrinsic characteristics of a good summary:

- *Coherence*: Coherence refers to the "sticking together" property of the text. In other words, it is the degree to which pieces of information throughout the text are related and can be linked. For instance, entities Andre Agassi and Steffi Graf in example 1 are connected through two different relationships i.e. Andre Agassi married Steffi Graf and both are tennis players (also have played in same tournaments). However, in the presence of entity Brooke Shields, choosing marriage relationship over tennis makes the summary more coherent.
- *Succinctness*: A summary should be focused and "to the point" – i.e., it should cover only the most important aspect of text. For instance, in example summary 4 above, information such as the mention of Aaron Sorkin, the creator of the television drama West Wing is not desirable.
- *Non-Redundancy*: There should not be redundant information within the summary text. From the model summaries given in Table 1, it is clear that the sentences that form a summary do not contain any redundant information.

The model we have developed that captures these key properties is evaluated on a query set of 27 entity chains. In a preliminary user study, our summaries showed a high precision value of 79% at rank 1. The experimental analysis demonstrates the effectiveness of proposed model and also show that the three properties are helpful in generating good quality summaries.

The layout of the rest of this paper is as follows. We begin with a brief discussion of related work in Section 2. We formally describe our model for generating good summaries for a two-length entity chains in Section 3. We present our experimental framework and describe the results in Section 4, before concluding in Section 5.

2   Related Work

Researchers have been working in various dimensions of the relationship extraction problem such as *relationship queries* and *hypothesis generation* to extract connection between two or more entities. However, the problem of generating summaries for extracted connections is a relatively less explored area of research. We further discuss each dimension in following subsections.

2.1   Relationship Queries

Relationship queries refer to queries posed on entity-relationship graphs to extract connections between two or more entities. Researchers have modeled the relationships among entities in varied ways. Anyanwu and Sheth [2] defined complex relationships on RDF such as paths between entities, networks of paths, or subgraphs as *semantic associations*. Halaschek *et al.* [15] referred the problem of finding paths between entities as *ρ-path semantic associations* and proposed a system called SemDIS to discover such semantic associations in RDF data.

Kasneci *et al.* [20] further extended the problem to find relationships between a set of two or more entities as a Steiner tree computation in weighted entity-relationship graphs. Fang *et al.* [10] proposed a system called REX to find a subgraph in entity-relationship graph which connects the entity pair. They also demonstrated the necessity of including non-simple paths in the results. Srihari *et al.* [30] proposed a framework to generate ranked list of query relevant hypothesis graphs (essentially subgraphs) from concept-association graph.

2.2 Hypothesis Generation

In the mid 1980s, Swanson [32] introduced a closed-discovery framework for hypotheses generation. Given two disconnected topics, they explored MEDLINE to identify potential linkages via intermediate topics and generate interesting hypotheses. These hypotheses were identified by manual analysis. Swanson and Smalheiser [28][29][33] proposed various interesting hypotheses, which were later successfully verified by clinical studies. Srinivasan [31] proposed text mining algorithms to automatically address the problem of closed-discovery algorithms. Researchers have also studied the effectiveness of incorporating domain semantics in the discovery framework [8][17][18]. These ideas have also been imported by the IR community to find associations in web collections [19].

2.3 Summary Generation

Srihari *et al.* [30] proposed a framework to generate hypothesis graphs for two or more entities and ranked evidence trails for the graphs. The evidence trails act as summary for the graph. Jin *et al.* [19] addressed the special case of entity chains. They aim at finding most meaningful evidence trails across documents that connect topics in an entity chain. The technique was tested for topic-specific datasets, namely the 9/11 Commission report and aviation accident reports provided by NTSB, therefore the effectiveness for generic queries is unclear.

3 Entity Chain Summaries

Given a two-length entity chain, we aim to compute a ranked list of summaries by combining the textual evidences, i.e. sentences, associated with the edges in the entity chain. Consider an entity chain represented as $\langle v_1, v_2, v_3 \rangle$, thus the two edges $e_1 = (v_1, v_2)$ and $e_2 = (v_2, v_3)$ are connected by a common node v_2. Each edge e_i is associated with a set of sentences S_{e_i} (e.g., the textual evidences maintained by the information extraction system). Based on this, we can formulate two kinds of candidate summaries – first, which we call as CS_1, is generated by combining individual sentences from the sentence sets of each edge. Formally,

$$CS_1 = \{ \, l_1 \oplus l_2 \mid l_1 \in S_{e_1} \cap l_2 \in S_{e_2} \, \},$$

where $\oplus$ denotes concatenation.

The second kind of candidate summary, called CS_2, consists of rich single sentences that contain all the three entities in the entity chain. Formally, we define them as follows:

$$CS_2 = \{\ l_1 \in S_1 \wedge \phi(l_1, v_3)\ \} \cup \{\ l_2 \in S_2 \wedge \phi(l_2, v_1)\ \},$$

where $\phi(l_i, v_j)$ is a indicator function which is true when the entity v_j is mentioned in the sentence l_i.

Finally, the full candidate set of summaries CS is the union of CS_1 and CS_2 as defined above, and consists of summaries with one or two sentences. Now, the task is to rank the elements of CS.

A summary should represent a well-written piece of text, therefore, the ranking function must capture the underlying concepts and structure of a well-written document. Significant research has been made to capture the properties of a well-written text [1][21][26]. Based on these studies and results from our initial experiments [7], we identified three such properties which contribute in enhancing the quality of the summary and make it easier for the reader to comprehend the text. The properties are coherence, succinctness and non-redundancy. The coherence and succinctness properties together are important to form a good quality summary. The lower quotient of any of these may drastically degrade the quality of the summary. Moreover, it is also essential to penalize the summaries which contain redundant information. Based on this rationale, we define following ranking function to score the candidate summaries:

$$Score(m) = (1 - \alpha)(Coherence(m) * Succinctness(m)) - \alpha\ Redundancy(m),$$

where $m \in CS$. The parameter α can be adjusted based on the tolerance level for redundancy. We will now discuss how these three properties are computed in our model.

3.1   Coherence

A coherent text helps the reader to link related pieces of information and comprehend a well connected representation of the text. Various approaches coherence in automatically generated summaries have been proposed in the literature. Referring to the study by Lapata and Barzilay [23], methods based on Latent Semantic Analysis (LSA) are well correlated with human ratings of coherence in text snippets. Various researchers have used LSA to measure coherence [12][13][22]. Moreover, Foltz et al. [13] have examined and demonstrated the potential of LSA as psychological model of coherence. At the same time, LSA is well suited for our setting where the summary lengths are quite small.

LSA represents a text corpus as a matrix where each row corresponds to a unique word and each column stands for a sentence. Cells of the matrix contain the frequency of the corresponding word in the sentence. The matrix is then decomposed using Singular Value Decomposition (SVD) such that every sentence is represented as vector whose value is the sum of vectors standing for its component words.

Consider a rectangular matrix X of n terms and m sentences. SVD decomposes this $n \times m$ matrix into product of three matrices as,

$$X = T * S * P' \qquad (1)$$

where T and P are the orthogonal matrices and S is a diagonal matrix of eigenvalues. LSA uses a truncated SVD, keeping only k largest eigenvalues and associated vectors. Therefore, $X = T_k * S_k * P_k'$ where T_k and P_k represent the term vectors and sentence vectors in latent semantic space. Similarity between two sentences is directly computed using distance measures on the sentence vectors.

We define *Coherence* score as the cosine similarity between the sentence vectors. Hence, the coherence scores for summaries in set CS_1 are computed as follows,

$$Coherence(l_1 \oplus l_2) = \sum_{i=1}^{k} \frac{ls_1^i . ls_2^i}{(|ls_1^i| . |ls_2^i|)}, \qquad l_1 \oplus l_2 \in CS_1 \qquad (2)$$

where ls_1 and ls_2 are the sentence vectors of l_1 and l_2 respectively in the latent semantic space. The coherence scores of single sentence summaries in CS_2 are set to 1.

3.2 Succinctness

Succinctness captures the notion that a more unified and single purpose text results in better comprehension [14]. In our setting, we observed that while entities are the most information dense tokens of text, a text with many entities usually is contextually weaker. Therefore, we consider the number of entities present in the summary as the notion of succinctness. The *Succinctness* score for each summary in CS is computed as,

$$Succinctness(m) = \frac{1}{EntityCount(m)} \qquad (3)$$

Succinctness acts as a notion of *relevance* of single sentence summaries for the corresponding entity chain.

3.3 Non-redundancy

A precise summary should not contain any redundant information. Therefore, it is necessary to focus on the text phrases representing entity relationships in the summary. The presence of redundant relationship phrases across the sentences indicates a low quality summary. In order to handle redundancy, we take a shallow approach by considering a metric which relies on n-gram overlap between the sentences. In our model, the n-gram based similarity of sentences l_1 and l_2 is computed to account for redundancy.

$$Redundancy(l_1, l_2) = \sum_{k=1}^{n} w_k * \frac{|grams(l_1, k) \cap grams(l_2, k)|}{|grams(l_1, k) \cup grams(l_2, k)|} \qquad (4)$$

where $(l_1, l_2) \in CS_1$, $grams(s, k)$ is the set of k-grams of sentence s and w_k is the weight associated with k-gram similarity of sentences. We choose $n = 4$ and set $w_1 = 1$, $w_2 = 2$, $w_3 = 3$, and $w_4 = 4$. In case of a single sentence summary, *Redundancy* score is set to 0.

4   Experimental Results

We utilize Open Information Extraction (Open IE) [9] to identify sentence sets for edges between entity pairs. We assume the sentence sets obtained through Open IE convey correct and important relationships. LSA model is learnt on the Wikipedia corpus (published on Sep 13, 2013) containing about 4.2 million documents using Gensim [27], an open source topic modeling tool. The number of topics to be learnt is set to 400. The corpus is annotated with named entities using Stanford's CoreNLP kit [11]. The tolerance parameter for redundancy (α) in the ranking function is empirically set to 0.2.

We manually constructed a set of 27 entity chains for the purpose of evaluations. We select only those entity chains for which OpenIE provides non-empty sentence sets. The query set was designed to contain different types of entities such as person, organization, date, product, and country.

Four different summary ranking models are possible based on the following combinations of properties we proposed: (i) only coherence, (ii) coherence and succinctness, (iii) coherence and non-redundancy, and (iv) coherence, succinctness, and non-redundancy. However, we use only (ii) and (iv) in our analysis since the notion of relevance for single sentence summaries does not exist in the case of models not considering succinctness.

Top-10 output summaries are considered for user evaluation. The summaries are evaluated by three judges, who are asked to assign a grade to each summary using four levels: 1 for poor, 2 for average, 3 for good, and 4 for perfect. Following definition of these levels were given to judges prior to evaluations:

- **Perfect** when the relationships are explicitly mentioned and are topic of discussion,
- **Good** when the relationships are explicitly mentioned but they are not topic of discussion,
- **Average** when the relationships are not mentioned but can be inferred, and
- **Poor** when the relationships are neither mentioned nor can be inferred.

The performance measures – P@k and NDCG@k are calculated. While computing P@k, summaries graded as average or higher are marked as relevant. Note that in literature, ROUGE [24] family of measures is most frequently used for evaluating automatic summarization results. However, appropriate ROUGE parameterization varies with task at hand. Since the problem of generating summaries for a given entity chain has not been addressed before to the best of our knowledge, an extensive experimental study is needed before selecting the appropriate ROUGE measure. While we pursue this research direction further, in this paper we settle for an evaluation based on precision and NDCG metrics.

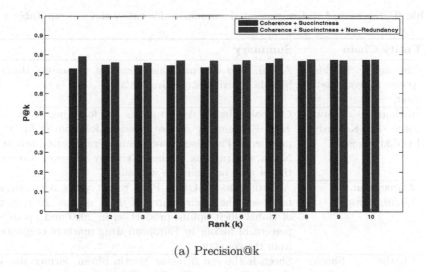

(a) Precision@k

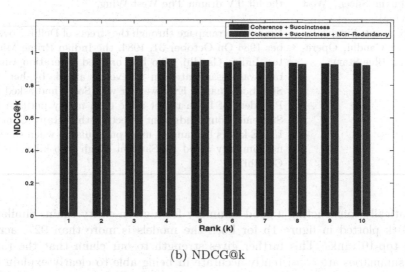

(b) NDCG@k

Fig. 1. Precision and NDCG of output summaries

Figure 1 compares the performance of the ranking model that considers only the coherence and succinctness of summaries, with the model that also includes non-redundancy defined above. In figure 1a, the average precision of summaries at different rank positions are shown. It is worth observing that the overall precision of the resulting summaries is consistently above 70% with both methods. The inclusion of non-redundancy in the model improves the precision of the result in rank-1 by more than 5%, suggesting the importance of non-redundancy property in enhancing the quality of summary. Further, the precision of the best performing model at rank-1 is 79%, which show that the model is capable of

Table 2. Top-ranked system-generated summaries for entity chains from Table 1

No.	Entity Chain	Summary
1	⟨Brooke Shields, Andre Agassi, Steffi Graf⟩	Agassi, who was previously married to actress Brooke Shields, married Steffi Graf in 2001.
2	⟨Richard Nixon, John F. Kennedy, Lyndon Johnson⟩	Campaign Issues Ask students the following: Why did John F. Kennedy choose Lyndon Johnson as his vice president? The Eisenhower years were petering out and Nixon was running against Kennedy who won because things were not going so well at all.
3	⟨Afghanistan, Opium, Europe⟩	According to UNODCCP, in recent years Afghanistan had been the main source of illicit opium: 70 percent of global illicit opium production in 2000 and up to 90 percent of heroin in European drug markets originated from the country.
4	⟨Charlie Sheen, Martin Sheen, West Wing⟩	Sheen is the son of actor Martin Sheen, former star of the hit TV drama The West Wing.
5	⟨Satwant Singh, Indira Gandhi, Operation Blue Star⟩	Hindu men rampage through the streets of Delhi, November 1984 On October 31, 1984, the Indian Prime Minister, Indira Gandhi, who had ordered Operation Bluestar, was assassinated in a revenge attack by her two Sikh bodyguards. Ever wonder why Sonia had asked the President of India to set aside on a mercy petition the Supreme Court judgment directing that Rajiv Gandhis LTTE killers be hanged, more particularly, when she had not similarly acted for Satwant Singh who killed Indira Gandhi?

automatically constructing a good summary for a given entity chain. Similarly, NDCG@k plotted in figure 1b for both the models is more than 92% across all the top-10 ranks. This further gives strength to our claim that the ranking of summaries are consistently accurate in being able to clearly explain the relationships in the given entity-chain.

Table 2 illustrates the top summary retrieved for the entity chains illustrated in Table 1 using the model that combines all the three properties we have considered. Following qualitative aspects can be inferred from the results illustrated in Table 2:

- On comparison with expected summaries presented in Table 1, the sentences in output summaries are topically similar, this illustrates the effectiveness of LSA as a coherence measure.
- Single sentence summaries are usually more coherent than two sentence summaries. Thus, it validates our assertion that single sentence summaries should be included in the candidate set.

- Summaries with lesser entities are more focused and to the point. It confirms the usefulness of incorporating succinctness property in the model.
- The summary generated for entity chain 2 weakens our assumption that the sentences from Open IE are correctly describing the relationship.

The results presented show that the proposed system effectively generates summary for an entity chain with a precision of 79% at rank 1.

4.1 Discussion

The performance of proposed framework is majorly affected by two parameters-the quality of sentence sets obtained from Open IE and evaluation dataset. In this section, we discuss the challenges associated with each of these parameters.

The quality of our results are highly dependent on the results produced by the framework generating sentence sets. As discussed above, there are various systems available such as NELL [6] and Patty [25] to extract relationship between entities. Open IE is one such system which provides list of sentences associated with the extracted relationships. But, Open IE is still restricted by the variety in type of relationships it can extract from unstructured text. Moreover, a methodology to generate sentence sets aligned more towards the task of summary generation may help in enhancing the quality of summaries. Since our work is one of the initial steps towards generating summaries for graph snippets, there is no open evaluation dataset available. As discussed above, the selection of 27 entity chains in our evaluation dataset is restricted by the output from Open IE.

5 Conclusion and Future Work

This research address the challenging problem of generating textual summaries for the two-length entity chains. The proposed system is build upon the key characteristics of a well written document. The results show that summaries generated by the system enable user to understand the underlying relationships. A high precision of 79% at rank 1 shows promise for more in-depth work.

The work presented here can be extended along various dimensions such as generating summary for generic entity chains, knowledge subgraphs. The proposed model can be augmented with other important properties of a well-written text pertaining to the problem in hand. A deeper analysis of sentences in sentence sets may help in filtering non-declarative sentences, thus enhancing quality of the summary. Further, entities in an entity pair may be connected to each other through various relationship categories such as personal, political, and professional. The ranked summaries can be diversified based on these categories so that wide range of information about the relationship can be covered.

References

1. Agrawal, R., Chakraborty, S., Gollapudi, S., Kannan, A., Kenthapadi, K.: Empowering authors to diagnose comprehension burden in textbooks. In: KDD, pp. 967–975 (2012)

2. Anyanwu, K., Sheth, A.: The ρ operator: discovering and ranking associations on the semantic web. ACM SIGMOD Record 31(4), 42–47 (2002)
3. Bizer, C., Lehmann, J., Kobilarov, G., Auer, S., Becker, C., Cyganiak, R., Hellmann, S.: Dbpedia - a crystallization point for the web of data. Web Semant. 7(3), 154–165 (2009)
4. Blanco, R., Zaragoza, H.: Finding support sentences for entities. In: SIGIR, pp. 339–346 (2010)
5. Bollacker, K., Evans, C., Paritosh, P., Sturge, T., Taylor, J.: Freebase: A collaboratively created graph database for structuring human knowledge. In: SIGMOD, pp. 1247–1250 (2008)
6. Carlson, A., Betteridge, J., Kisiel, B., Settles, B., Hruschka Jr., E.R., Mitchell, T.M.: Toward an architecture for never-ending language learning. In: AAAI Conf. on Artifical Intelligence (2010)
7. Chhabra, S., Bedathur, S.: Generating text summaries of graph snippets. In: COMAD, pp. 121–124 (2013)
8. Cohen, T., Whitfield, G., Schvaneveldt, R., Mukund, K., Rindflesch, T.: Epiphanet: An interactive tool to support biomedical discoveries. Journal of Biomedical Discovery and Collaboration 5, 21–49 (2010)
9. Etzioni, O., Fader, A., Christensen, J., Soderland, S., Mausam, M.: Open information extraction: The second generation. In: IJCAI, pp. 3–10 (2011)
10. Fang, L., Sarma, A.D., Yu, C., Bohannon, P.: Rex: explaining relationships between entity pairs. Proc. VLDB Endow. 5(3) (2011)
11. Finkel, J.R., Grenager, T., Manning, C.: Incorporating non-local information into information extraction systems by gibbs sampling. In: ACL, pp. 363–370 (2005)
12. Foltz, P.W.: Latent semantic analysis for text-based research. Behavior Research Methods, Instruments, & Computers 28(2), 197–202 (1996)
13. Foltz, P.W., Kintsch, W., Landauer, T.K.: The measurement of textual coherence with latent semantic analysis. Discourse Processes 25(2-3), 285–307 (1998)
14. Gray, W.S., Leary, B.E.: What makes a book readable. Univ. Chicago Press (1935)
15. Halaschek, C., Aleman-Meza, B., Arpinar, I.B., Sheth, A.P.: Discovering and ranking semantic associations over a large rdf metabase. In: VLDB, pp. 1317–1320 (2004)
16. Hoffart, J., Suchanek, F.M., Berberich, K., Lewis-Kelham, E., de Melo, G., Weikum, G.: Yago2: Exploring and querying world knowledge in time, space, context, and many languages. In: WWW, pp. 229–232 (2011)
17. Hristovski, D., Friedman, C., Rindflesch, T.C., Peterlin, B.: Exploiting semantic relations for literature-based discovery. In: AMIA Annual Symp., vol. 2006, pp. 349–353 (2006)
18. Hristovski, D., Kastrin, A., Peterlin, B., Rindflesch, T.C.: Combining semantic relations and dna microarray data for novel hypotheses generation. In: Proceedings of the 2009 Workshop of the BioLink Special Interest Group, International Conference on Linking Literature, Information, and Knowledge for Biology, pp. 53–61 (2010)
19. Jin, W., Srihari, R.K., Ho, H.H., Wu, X.: Improving knowledge discovery in document collections through combining text retrieval and link analysis techniques. In: ICDM, pp. 193–202 (2007)
20. Kasneci, G., Ramanath, M., Sozio, M., Suchanek, F.M., Weikum, G.: Star: Steiner-tree approximation in relationship graphs. In: ICDE, pp. 868–879 (2009)
21. Kintsch, W., Van Dijk, T.A.: Toward a model of text comprehension and production. Psychological Review 85(5), 363–394 (1978)

22. Landauer, T.K., Laham, D., Foltz, P.W.: Learning human-like knowledge by singular value decomposition: A progress report. In: NIPS, vol. 10, pp. 45–51 (1998)
23. Lapata, M., Barzilay, R.: Automatic evaluation of text coherence: Models and representations. In: IJCAI, pp. 1085–1090 (2005)
24. Lin, C.-Y.: Rouge: A package for automatic evaluation of summaries. In: Text Summarization Branches Out: Proceedings of the ACL 2004 Workshop, pp. 74–81 (2004)
25. Nakashole, N., Weikum, G., Suchanek, F.: Patty: a taxonomy of relational patterns with semantic types. In: EMNLP, pp. 1135–1145 (2012)
26. Pitler, E., Nenkova, A.: Revisiting readability: a unified framework for predicting text quality. In: EMNLP, pp. 186–195 (2008)
27. Řehůřek, R., Sojka, P.: Software framework for topic modelling with large corpora. In: LREC Workshop on New Challenges for NLP Frameworks, pp. 45–50 (2010)
28. Smalheiser, N.R., Swanson, D.R.: Indomethacin and alzheimer's disease. Neurology 46(2), 583–583 (1996)
29. Smalheiser, N.R., Swanson, D.R.: Linking estrogen to alzheimer's disease an informatics approach. Neurology 47(3), 809–810 (1996)
30. Srihari, R.K., Xu, L., Saxena, T.: Use of ranked cross document evidence trails for hypothesis generation. In: KDD, pp. 677–686 (2007)
31. Srinivasan, P.: Text mining: generating hypotheses from medline. Journal of American Society for Information Science and Technology 55(5), 396–413 (2004)
32. Swanson, D.R.: Two medical literatures that are logically but not bibliographically connected. Journal of the American Society for Information Science 38(4), 228–233 (1987)
33. Swanson, D.R., Smalheiser, N.R.: An interactive system for finding complementary literatures: a stimulus to scientific discovery. Artificial Intelligence 91(2), 183–203 (1997)

Tolerance of Effectiveness Measures to Relevance Judging Errors

Le Li[1] and Mark D. Smucker[2]

[1] David R. Cheriton School of Computer Science, Canada
[2] Department of Management Sciences, Canada
University of Waterloo, Canada

Abstract. Crowdsourcing relevance judgments for test collection construction is attractive because the practice has the possibility of being more affordable than hiring high quality assessors. A problem faced by all crowdsourced judgments – even judgments formed from the consensus of multiple workers – is that there will be differences in the judgments compared to the judgments produced by high quality assessors. For two TREC test collections, we simulated errors in sets of judgments and then measured the effect of these errors on effectiveness measures. We found that some measures appear to be more tolerant of errors than others. We also found that to achieve high rank correlation in the ranking of retrieval systems requires conservative judgments for average precision (AP) and nDCG, while precision at rank 10 requires neutral judging behavior. Conservative judging avoids mistakenly judging non-relevant documents as relevant at the cost of judging some relevant documents as non-relevant. In addition, we found that while conservative judging behavior maximizes rank correlation for AP and nDCG, to minimize the error in the measures' values requires more liberal behavior. Depending on the nature of a set of crowdsourced judgments, the judgments may be more suitable with some effectiveness measures than others, and the use of some effectiveness measures will require higher levels of judgment quality than others.

1 Introduction

Information retrieval (IR) test collection construction can require 10 to 20 thousand or more relevance judgments. The best way to obtain relevance judgments is to hire and train assessors who both originate their own search topic and then have the ability to carefully and consistently judge hundreds of potentially complex documents. There is considerable interest in utilizing crowdsourcing platforms such as Amazon Mechanical Turk to obtain relevance judgments in an affordable manner [1–3]. Crowdsourced assessors are usually *secondary* assessors, i.e. assessors who did not originate the search topics. It is well known that secondary assessors produce relevance judgments that differ from those that are or would be produced by primary assessors [4]. Whether there is a single secondary assessor, or a group of secondary assessors that are combined using sophisticated algorithms [5, 6], there will be differences.

M. de Rijke et al. (Eds.): ECIR 2014, LNCS 8416, pp. 148–159, 2014.

In this paper, we address this question: What effect do differences in judgments between primary and secondary assessors have on our ability to rank and score retrieval systems? Equivalently, what differences in judgments can various evaluation measures tolerate and still be able to match the evaluation quality produced using primary judgments?

To investigate this question, we used two sets of runs submitted to two TREC tracks. For each set of runs, we took the appropriate NIST relevance judgments (also known as qrels) and then simulated a secondary assessor to produce a set of secondary qrels that differed from the NIST, primary qrels. For each set of qrels, we produced scores for the runs using precision at 10 (P@10), mean average precision (MAP), and normalized discounted cumulated gain (nDCG). With a given effectiveness measure, e.g. MAP, we can rank the systems as per the primary and secondary qrels and then measure their rank correlation. We measured rank correlation with Yilmaz et al.'s AP Correlation (APCorr) [7]. Likewise, we can measure the accuracy of the scores produced by the secondary qrels by measuring the root mean square error (RMSE) between the two sets of scores.

To simulate the secondary assessors, we treated the NIST qrels as truth and the secondary assessor as a classifier. A classifier's performance can be understood in terms of its true positive rate (TPR) and its false positive rate (FPR). A given TPR and FPR determine both a classifier's discrimination ability and how conservative or liberal it is in its judging. For example, a conservative classifier avoids judging non-relevant documents as relevant at the cost of mistakenly judging some relevant documents as non-relevant. We used d' to measure discrimination ability and the criterion c to measure how conservative or liberal the judging behavior is [8].

We systematically varied the discrimination ability, d', and the criterion c, to produce different sets of qrels. We then evaluated the system runs submitted to the TREC 8 ad-hoc and Robust 2005 TREC tracks with these qrels and compared the results to those we obtained using the official NIST qrels. After analyzing the results, we found that:

1. In terms of rank correlation (APCorr), mean average precision (MAP) is more tolerant of errors than nDCG and P@10. In other words, MAP can obtain the same APCorr as nDCG and P@10 with assessors of a lower discrimination ability.
2. To maximize rank correlation, nDCG, MAP, and P@10 require conservative judging. Of the three measures, P@10 requires the least conservative judging and works best with the judging close to neutral. The lower the discrimination ability of the judging, the more conservative judging is required by MAP and nDCG to maximize rank correlation. MAP and nDCG appear to be sensitive to false positives.
3. Depending on the discrimination ability of the judging, it can be hard to jointly optimize APCorr and RMSE for MAP and nDCG.

The impact of these findings is that to optimize rank correlation requires attention to not only the discrimination ability of the assessors, but also to how conservative, liberal, or neutral those assessors are in their judgments. Judging schemes or consensus algorithms may need to be devised that will help produce more conservative judgments when MAP and nDCG are the targeted effectiveness measures. If P@10 is to be used as the effectiveness measure, efforts must be taken to maintain neutral judging. From a crowdsourcing point of view, it is likely that there will need to be acquired some set of high quality, primary assessor relevance judgments by which the lower quality, crowdsourced, secondary assessor relevance judgments can be calibrated to maximize rank correlation by controlling the relevance criterion used, i.e. by controlling how liberal or conservative the resulting relevance judgments are.

2   Methods and Materials

To conduct our experiments, we used the set of runs submitted to two TREC tracks. For each TREC track, we took the NIST qrels and simulated assessors of different abilities and biases as compared to the NIST qrels to produce alternative qrels. We then used these alternative qrels to evaluate the sets of runs and measure the effect that the differences in judgments had on our evaluation of the runs submitted to the tracks.

2.1   Runs Submitted to TREC Tracks and QRels

We used the runs submitted to the TREC 8 ad-hoc and Robust 2005 TREC tracks as well as the NIST qrels for each track [9, 10]. For convenience, we refer the two data sets as Robust2005 and TREC8.

Both data sets contain 50 topics. The TREC8 qrels contain 86,830 judgments of which 4,728 are relevant (5.4%). The Robust2005 qrels contain 37,798 judgments of which 6,561 are relevant (17%). TREC8 has 129 submitted runs and Robust2005 has 74 submitted runs.

2.2   Simulation of Judgments

We took the NIST qrels as truth and then simulated assessors of different abilities and biases as measured against the NIST qrels. We can describe the judging behavior of our simulated assessors in terms of their true positive rates (TPR) and false positive rates (FPR), where TPR $= TP/(TP + FN)$ and FPR $= FP/(FP + TN)$ (as shown in Table 1).

Signal detection theory allows us to separately describe the discrimination ability and the decision criterion or bias of the assessor [8]. Discrimination ability is measured as d':

$$d' = z(TPR) - z(FPR) , \tag{1}$$

and the bias of the assessor towards either liberal or conservative judging is described by the criterion c:

$$c = -\frac{1}{2}(z(TPR) + z(FPR)) , \tag{2}$$

Table 1. Confusion Matrix. "Pos." and "Neg." stand for "Positive" and "Negative" respectively.

Simulated Secondary Assessor	NIST (Primary) Assessor	
	Relevant (Pos.)	Non-Relevant (Neg.)
Relevant	TP = True Pos.	FP = False Pos.
Non-Relevant	FN = False Neg.	TN = True Neg.

where TPR and FPR are true positive rate and false positive rate of this assessor, respectively. Function z, the inverse of the normal distribution function, converts the TPR or FPR to a z score [8].

If an assessor tends to label incoming documents to be relevant to avoid missing relevant documents (but at the risk of high false positive rate), then this assessor is liberal with a negative criterion. If $c = 0$, the assessor is neutral. A conservative assessor has a positive criterion. One advantage of using this model is that the measurement of an assessor's ability to discriminate is independent of the assessor's criterion.

At a given discrimination ability d', there are many possible values for the TPR and FPR. In other words, two assessors can have the same ability to discriminate between relevant and non-relevant documents, but one may have a much higher relevance criterion than the other. The higher the relevance criterion, the more conservative the assessor. Figure 1 shows example d' curves. All of the points along a curve have the same discrimination ability. Table 2 gives the TPR and FPR for a selection of d' and c values.

Table 2. The TPR and FPR for various d' and c

	c = -1 (liberal)					c = 0 (neutral)					c = 1 (conservative)				
d'	0.5	1.0	1.5	2.0	2.5	0.5	1.0	1.5	2.0	2.5	0.5	1.0	1.5	2.0	2.5
TPR	0.89	0.93	0.96	0.98	0.99	0.60	0.69	0.77	0.84	0.89	0.23	0.31	0.40	0.50	0.60
FPR	0.77	0.69	0.60	0.50	0.40	0.40	0.31	0.23	0.16	0.11	0.11	0.07	0.04	0.02	0.01

If an assessor's d' and c are given, we can use them to calculate the TPR and FPR of this assessor using Equations 3 and 4, which are derived from Equations 1 and 2. TPR is computed as,

$$TPR = CDF(d' * 0.5 - c) , \tag{3}$$

and FPR is computed as,

$$FPR = CDF(-d' * 0.5 - c) , \tag{4}$$

where CDF is the Cumulative Density Function of the standard normal distribution $\mathcal{N}(0, 1)$.

Assuming a document's true label is given, we can generate the simulated judgment by tossing a biased coin. The probability of the assessor making an

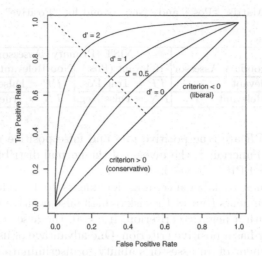

Fig. 1. Curves of equal d'. This figure is based on Figures 1.1 and 2.1 of [8].

error is calculated using the assessor's TPR and FPR rates. If the true label is relevant, the assessor makes an error with probability equal to $1 - TPR$. If the true label is non-relevant, the assessor makes an error with probability equal to FPR.

2.3   Experiment Settings

We simulated the noisy judgments of assessors by varying two variables, d' and c, as shown in Algorithm 1 and described in Sec. 2.2. What are the candidate values of d' and c for simulation? Smucker and Jethani [11] estimated the average d' and c of NIST assessors to be 2.3 and 0.37, respectively, across 10 topics in 2005 TREC Robust Track. In [12], the reported d' and c of crowdsourced assessors are 1.9 and 0.14, respectively, with the same experiment settings in [11]. If we think of the judgments from NIST assessors as the upper bound that the consensus algorithm or hired assessors could be, then the results from these two papers indicate that one assessor should have d' and c values close to NIST. Meanwhile, as shown in Fig. 1, $d' = 0$ means the assessor labels the document by tossing an unbiased coin, i.e. random guess. So, we set the range of d' as [0.5, 3] with a step size of 0.5. For the criterion c, the reported c suggests that both NIST and crowdsourced assessors are conservative with NIST assessors being more conservative than the crowdsourced workers [11, 12]. At the same time, the behaviors of liberal assessors are also worthy of investigation. So, we set the c with the range of [−3, 3] with a step size of 0.1. In total, we simulated 366 different types of assessors who make random errors based on each pair of d' and c values. While we consider c varying between −3 and 3, the likely range for c is probably at most −1 to 1. We show the range from −3 to 3 to allow trends to be better seen.

Algorithm 1. Simulate The Judgments From One Assessor

INPUT: $d', c, trueLabels$

$TPR \leftarrow CDF(d' * 0.5 - c)$ ▷ Cumulative Distribution Function of $\mathcal{N}(0,1)$

$FPR \leftarrow CDF(-d' * 0.5 - c)$

for $i = 1$: size($trueLabels$) **do**
    $judge_i \leftarrow trueLabels_i$
    $flip \leftarrow rand(0, 1)$ ▷ A random number in $[0, 1)$
    if $trueLabel_i == 0$ **then**
        if $flip \leq FPR$ **then**
            $judge_i \leftarrow 1$
        end if
    else
        if $flip > TPR$ **then**
            $judge_i \leftarrow 0$
        end if
    end if
end for
RETURN $judge$

For each simulated assessor, we repeated the simulation 100 times to generate 100 independent simulated qrels, and then averaged the performance of the simulated qrels for the assessor.

2.4 Measures

To measure the degree of correlation between the simulated and NIST assessors in IR evaluation, we evaluated the submitted runs of those two TREC tracks against qrels from simulated and NIST assessors, respectively. Three evaluation metrics were used: Precision at 10 (P@10), Mean Average Precision (MAP), and Normalized Discounted Cumulative Gain (nDCG). To be more precise, each assessor's simulated run was used as pseudo qrels to evaluate all test runs. So that each test run corresponded to one evaluation result. This evaluation result was averaged across all topics.

For example, Robust2005 has 74 test runs and we can get 74 MAPs against one qrels file. Hence, we can measure the correlation of two qrels based on the association between two lists of MAPs derived from identical test runs. The higher the correlation between the rankings produced by a set of simulated qrels and the rankings produced by the NIST qrels, the less effect the simulated errors have on the effectiveness measure. We can compare the tolerance of effectiveness measures to judging errors by measuring the correlation for each measure for a given set of simulated qrels. The higher the correlation, the more fault tolerant the effectiveness measure is.

We used the Average Precision Correlation Coefficient (APCorr) [7] to measure the correlation. APCorr is very similar to another commonly-used correlation measurement, Kendall's τ [13], but APCorr gives more weight to the errors

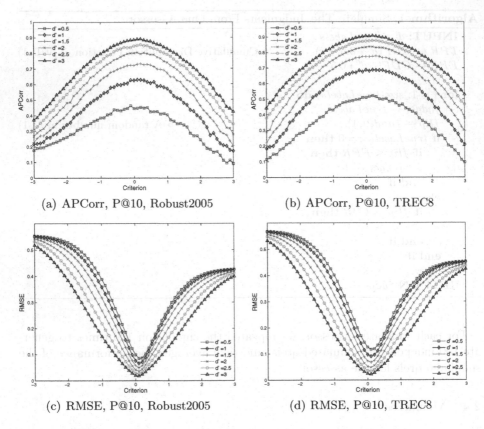

(a) APCorr, P@10, Robust2005

(b) APCorr, P@10, TREC8

(c) RMSE, P@10, Robust2005

(d) RMSE, P@10, TREC8

Fig. 2. The effects of the pseudo assessor's errors on evaluation, P@10

nearer to the top of rankings. If two ranking systems perfectly agree with each other, APCorr is 1.

Moreover, the Root Mean Square Error (RMSE) was also adopted to calculate the errors between two lists of scores. The smaller the RMSE value is, the closer two measurements are in terms of the quantity difference.

3   Results and Discussions

Results are shown in Figures 2, 3, and 4 and Tables 3 and 4. Each figure shows a different effectiveness measure on both TREC8 and Robust2005. The tables show the maximum APCorr achieved by each effectiveness measure for each of the d' values used in the experiment.

As is to be expected, the larger the d' value, the better the rank correlation (APCorr) is at a given criterion c. Recall that APCorr measures the degree to which the simulated qrels rank the retrieval systems in the same order as do the NIST qrels, and criterion values greater than 0 are conservative and those less than zero are liberal.

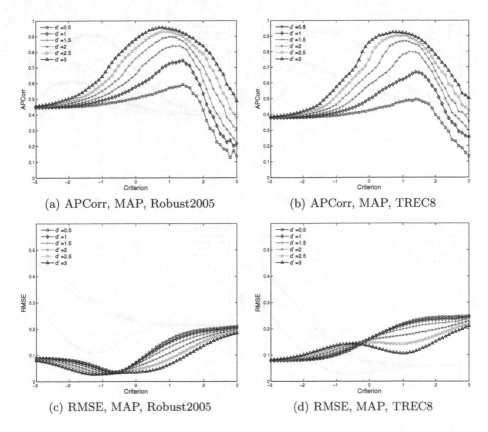

(a) APCorr, MAP, Robust2005

(b) APCorr, MAP, TREC8

(c) RMSE, MAP, Robust2005

(d) RMSE, MAP, TREC8

Fig. 3. The effects of the pseudo assessor's errors on evaluation, MAP

As can be seen in Tables 3 and 4, except for the two lowest d' values in TREC8, mean average precision (MAP) achieves the best rank correlation at a given level of discrimination ability. Indeed, MAP often has an APCorr that is near the APCorr achieved by nDCG and P@10 on the next higher d', i.e. MAP can achieve the same APCorr as the other metrics but with lower quality assessors. MAP is more fault tolerant than P@10 and nDCG on these test collections.

Evident most clearly in the Figures 2, 3, and 4, but also in Tables 3 and 4, MAP and nDCG require conservative judgments to maximize their rank correlation (APCorr). P@10 also requires conservative judgments, but the degree of conservativeness is close to neutral. As the discrimination ability decreases, MAP and nDCG require even more conservative judgments to maximize APCorr. It appears that both MAP and nDCG are sensitive to false positives. Our results reinforce those of Carterette and Soboroff [14] who also found, via a different simulation methodology, that false positives are to be avoided.

A consequence of the need for conservative judgments to maximize rank correlation is that it is hard for secondary assessors, such as crowdsourced assessors, to produce a set of qrels that can produce the same scores for MAP and nDCG

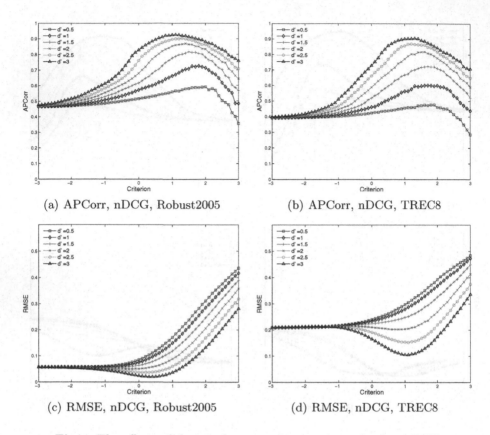

(a) APCorr, nDCG, Robust2005   (b) APCorr, nDCG, TREC8

(c) RMSE, nDCG, Robust2005   (d) RMSE, nDCG, TREC8

Fig. 4. The effects of the pseudo assessor's errors on evaluation, nDCG

as with the NIST qrels. The reason for this is that conservative judging requires missing relevant documents and judging them to be non-relevant to avoid being liberal and mistakenly judging non-relevant documents to be relevant. Both MAP and nDCG are measures over the set of known relevant documents. Thus, conservative judging results in a lower estimate of the total number of relevant documents and changes the scores of MAP and nDCG.

While at high levels of discrimination ability d', the maximum APCorr is obtained with a criterion c that also produces a near to minimum RMSE for all of the effectiveness measures. For MAP and nDCG, as d' decreases, the best criterion c for APCorr and RMSE move apart, and it becomes increasingly hard to jointly optimize for both measures.

We can also see in the figures that assessors with greater discrimination ability, d', tend to be more robust to the change of criterion c, with high values of APCorr obtained over wider ranges of c.

Meanwhile, we notice that the correlation results on TREC8 tend to be worse than that on Robust2005. Our hypothesis is that since TREC8 contains a deeper pool with more non-relevant documents than Robust2005, the number of false

Table 3. The criterion, TPR, and FPR when the APCorr is maximal for each d', Robust2005

d'	P@10				MAP				nDCG			
	c	TPR	FPR	APCorr	c	TPR	FPR	APCorr	c	TPR	FPR	APCorr
0.5	-0.1	0.637	0.440	0.462	1.4	0.125	0.049	**0.591**	2.0	0.040	0.012	0.590
1.0	0.1	0.655	0.274	0.628	1.4	0.184	0.029	**0.745**	1.8	0.097	0.011	0.725
1.5	0.2	0.709	0.171	0.731	1.3	0.291	0.020	**0.840**	1.5	0.227	0.012	0.816
2.0	0.1	0.816	0.136	0.803	1.0	0.500	0.023	**0.898**	1.4	0.345	0.008	0.868
2.5	0.2	0.853	0.074	0.854	1.0	0.599	0.012	**0.932**	1.3	0.480	0.005	0.902
3.0	0.2	0.903	0.045	0.890	0.8	0.758	0.011	**0.956**	1.1	0.655	0.005	0.926

Table 4. The criterion, TPR, and FPR when the APCorr is maximal for each d', TREC8

d'	P@10				MAP				nDCG			
	c	TPR	FPR	APCorr	c	TPR	FPR	APCorr	c	TPR	FPR	APCorr
0.5	0.2	0.520	0.326	**0.520**	1.4	0.125	0.049	0.494	1.8	0.061	0.020	0.477
1.0	0.2	0.618	0.242	**0.688**	1.4	0.184	0.029	0.668	1.7	0.115	0.014	0.601
1.5	0.3	0.674	0.147	0.777	1.2	0.326	0.026	**0.800**	1.7	0.171	0.007	0.723
2.0	0.1	0.816	0.136	0.838	1.0	0.500	0.023	**0.867**	1.6	0.274	0.005	0.818
2.5	0.2	0.853	0.074	0.877	0.8	0.674	0.020	**0.903**	1.2	0.520	0.007	0.869
3.0	0.2	0.903	0.045	0.909	0.7	0.788	0.014	**0.924**	1.3	0.579	0.003	0.905

positives is higher with TREC8 when judged with the same FPR. Another possibility is that the unique nature of the manual runs present in TREC8, which are some of the best scoring runs, make it harder to judge than Robust2005.

A somewhat surprising result occurs with MAP on TREC8 and its RMSE. As Fig. 3 shows, the highly discriminative $d' = 3.0$ qrels actually have a higher RMSE than the lower d' qrels at liberal c values less than -0.25 or so. As far as we can understand, this "inversion" of expected behavior results from the lower d' qrels having higher false positives rates that while being noisier judgments, result in MAP values that on average are closer to the NIST scores.

3.1 Limitations of Our Methods

Our existing simulation method only captures the random errors made by the assessors. Webber et al. [15] have shown that as documents are ranked lower by retrieval engines, the less likely assessors are to make false positive errors. In our simulation, the true and false positive rates do not depend on the document being judged. Likewise, we do not attempt to model crowdsourcing-specific error [16]. As such, our results cannot be used to show the discrimination ability required of assessors to obtain a desired rank correlation.

4 Related Work

Voorhees [17] conducted experiments with obtaining secondary relevance judgments using high quality NIST assessors. In these experiments, Voorhees found that even with disagreements, the rank correlation of the runs was high. Subsequent work by others has found that differing levels of assessor expertise can negatively affect the ability of secondary assessors to produce qrels that evaluate systems in the same manner as qrels produced by high-quality primary assessors [18, 19].

Most similar to our work, Carterette and Soboroff [14] hypothesized several difference models of assessor behavior that could produce judging errors compared to NIST qrels. They found that their "pessimistic" models resulted in the best rank correlation. These findings are in line with our results showing that conservative assessors are required for maximizing rank correlation. Carterette and Soboroff examined the statMAP measure, while we have looked at additional measures and discovered that P@10 does best with slightly conservative, almost neutral judging.

5 Conclusion

We simulated assessor errors by varying both their discrimination ability and their relevance criterion. We examined the effect of these errors on three effectiveness measures: P@10, MAP, and nDCG. We found that MAP is more tolerant of judging errors than P@10 and nDCG. MAP can achieve the same rank correlation with lower quality assessors. We also found that conservative assessors are preferable to achieve high correlation. In other words, it is important that assessors avoid mistakenly judging non-relevant documents as relevant. We also found that different effectiveness measures have different responses to errors in judging. For example, P@10 requires a more liberal judging behavior than does MAP and nDCG. Crowdsourced relevance judging likely will require a sample of documents judged by high quality, primary assessors to allow for the calibration of the judgments produced by crowdsourcing. Future work could involve the design of effectiveness measures specifically designed to better handle relevance judging errors.

Acknowledgments. We thank the reviewers for their helpful reviews. In particular we thank the meta-reviewer for the helpful set of references to related work. This work was supported in part by the Natural Sciences and Engineering Research Council of Canada (NSERC), in part by the facilities of SHARCNET, and in part by the University of Waterloo. Any opinions, findings and conclusions or recommendations expressed in this material are those of the authors and do not necessarily reect those of the sponsors.

References

1. Alonso, O., Mizzaro, S.: Can we get rid of TREC assessors? using Mechanical Turk for relevance assessment. In: Proceedings of the SIGIR 2009 Workshop on the Future of IR Evaluation, pp. 15–16 (July 2009)
2. McCreadie, R., Macdonald, C., Ounis, I.: Crowdsourcing blog track top news judgments at TREC. In: WSDM 2011 Workshop on Crowdsourcing for Search and Data Mining (2011)
3. Smucker, M.D., Kazai, G., Lease, M.: Overview of the TREC 2012 crowdsourcing track (2012)
4. Voorhees, E.M.: Variations in relevance judgments and the measurement of retrieval effectiveness. IPM 36, 697–716 (2000)
5. Hosseini, M., Cox, I., Milić-Frayling, N., Kazai, G., Vinay, V.: On aggregating labels from multiple crowd workers to infer relevance of documents. In: Baeza-Yates, R., de Vries, A.P., Zaragoza, H., Cambazoglu, B.B., Murdock, V., Lempel, R., Silvestri, F. (eds.) ECIR 2012. LNCS, vol. 7224, pp. 182–194. Springer, Heidelberg (2012)
6. Raykar, V.C., Yu, S., Zhao, L.H., Valadez, G.H., Florin, C., Bogoni, L., Moy, L.: Learning from crowds. The Journal of Machine Learning Research 99, 1297–1322 (2010)
7. Yilmaz, E., Aslam, J.A., Robertson, S.: A new rank correlation coefficient for information retrieval. In: SIGIR, pp. 587–594 (2008)
8. Macmillan, N.A., Creelman, C.D.: Detection theory: A user's guide. Psychology Press (2004)
9. Voorhees, E.M., Harman, D.: Overview of the Eighth Text REtrieval Conference (TREC-8). In: Proceedings of TREC, vol. 8, pp. 1–24 (1999)
10. Voorhees, E.M.: Overview of TREC 2005. In: Proceedings of TREC (2005)
11. Smucker, M.D., Jethani, C.P.: Measuring assessor accuracy: a comparison of NIST assessors and user study participants. In: SIGIR, pp. 1231–1232 (2011)
12. Smucker, M., Jethani, C.: The crowd vs. the lab: A comparison of crowd-sourced and university laboratory participant behavior. In: Proceedings of the SIGIR 2011 Workshop on Crowdsourcing for Information Retrieval (2011)
13. Kendall, M.G.: A new measure of rank correlation. Biometrika 30(1/2), 81–93 (1938)
14. Carterette, B., Soboroff, I.: The effect of assessor error on IR system evaluation. In: SIGIR, pp. 539–546 (2010)
15. Webber, W., Chandar, P., Carterette, B.: Alternative assessor disagreement and retrieval depth. In: CIKM, pp. 125–134 (2012)
16. Vuurens, J., de Vries, A.P., Eickhoff, C.: How much spam can you take? In: SIGIR 2011 Workshop on Crowdsourcing for Information Retrieval, CIR (2011)
17. Voorhees, E.: Variations in relevance judgments and the measurement of retrieval effectiveness. IPM 36(5), 697–716 (2000)
18. Bailey, P., Craswell, N., Soboroff, I., Thomas, P., de Vries, A., Yilmaz, E.: Relevance assessment: are judges exchangeable and does it matter. In: SIGIR, pp. 667–674 (2008)
19. Kinney, K., Huffman, S., Zhai, J.: How evaluator domain expertise affects search result relevance judgments. In: CIKM, pp. 591–598 (2008)

Evaluation of IR Applications
with Constrained Real Estate

Yuanhua Lv, Ariel Fuxman, and Ashok K. Chandra

Microsoft Research,
Mountain View, CA USA, 94043
{yuanhual,arielf,achandra}@microsoft.com

Abstract. Traditional IR applications assume that there is always
enough space ("real estate") available to display as many results as the
system returns. Consequently, traditional evaluation metrics were typically designed to take a length cutoff k of the result list as a parameter.
For example, one computes DCG@k, Prec@k, etc., based on the top-k
results in the ranking list. However, there are important modern ranking
applications where the result real estate is constrained to a small fixed
space, such as the search verticals aggregated in the Web search results
and the recommendation systems. For such applications, the following
tradeoff arises: given a fixed amount of real estate, shall we show a small
number of results with rich captions and details, or a larger number of
results with less informative captions? In other words, there is a tradeoff
between the length of the result list (i.e., quantity) and the informativeness of the results (i.e., quality). This tradeoff has important implications
for evaluation metrics, since it leads the length cutoff k hard to be determined a priori. In order to tackle this problem, we propose two desirable
formal constraints to capture the heuristics of regulating the quantity-
quality tradeoff, inspired by the axiomatic approach to IR. We then
present a general method to normalize the well-known Discounted Cumulative Gain (DCG) metric for balancing the quantity-quality tradeoff,
yielding a new metric, that we call Length-adjusted Discounted Cumulative Gain (LDCG). LDCG is shown to be able to automatically balance
the length and the informativeness of a ranking list without requiring an
explicit parameter k, while still preserving the good properties of DCG.

Keywords: Evaluation, Aggregated Search, Constrained Real Estate,
Quantity-Quality Tradeoff, LDCG, LNDCG.

1 Introduction

Evaluation metrics play a critical role in the field of information retrieval (IR).
Traditional IR applications assume that there is always enough space ("real
estate") available to display as many results as the system returns. To evaluate
such systems, traditional evaluation metrics were typically designed to take a
length cutoff k of the result list as a parameter. For example, one computes
DCG@k [8], Prec@k, etc., based on the top-k results in the ranking list.

M. de Rijke et al. (Eds.): ECIR 2014, LNCS 8416, pp. 160–171, 2014.

Fig. 1. Search verticals as examples of applications with constrained real estate

However, there are important modern ranking applications where the real estate available for the results is constrained to a small fixed space. For example, search engines are no longer restricted to the classical ten blue links for Web results, and they now aggregate all kinds of information (shopping, weather, news, images, entities from a knowledge repository, etc.) on the search engine result page. As an illustration, consider the snapshot in Figure 1 for the query "jon favreau director". In addition to the Web links, we can see results corresponding to images (Image Vertical), news (News Vertical), and an entity card with rich information about the famous American director Jon Favreau (Entity Vertical). Each vertical has limited real estate; for instance, the Entity Vertical results are restricted to a fixed area on the right hand side of the screen.

For each such applications, the following tradeoff arises: given the constrained amount of real estate, shall we show a small number of results with rich captions and details, or a larger number of results with less informative captions? Continuing our example of "jon favreau director", if the search engine is confident enough about the person Jon Favreau being the intended entity result, it can show a single rich entity card for him, as in Figure 1; if the search engine is not confident enough, it may instead show two more impoverished candidates: the entity for Jon Favreau and the entity for "Iron Man 3" (a film directed by Jon Favreau), as shown in Figure 2. Then, how to evaluate which one is better?

On the one hand, the evaluation of these applications does share some requirements with Web search. First, ranking order matters. In the example of query "jon favreau director", we do not want to rank "Iron Man 3" higher than the director himself. Second, we need to deal with graded relevance. In our example, the director Jon Favreau should be a "perfect" answer, and "Iron Man 3" should still be a good answer because Jon Favreau is its director. But the entity

Fig. 2. Alternative entity result for "jon favreau director"

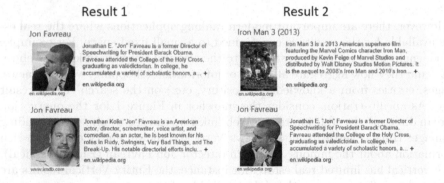

Fig. 3. Comparison of entity results for "jon favreau director"

for another person who also happens to be called Jon Favreau (i.e., Obama's speechwriter) is clearly a bad answer. This would suggest the use of DCG [8] or other existing IR metrics (e.g., [4,10]) as an evaluation metric.

On the other hand, the real estate constraints have important implications for evaluation metrics. As discussed above, we need to decide carefully if we want to show a small number of informative answers with more details (if we are confident); or give a few candidates with fewer details for each (otherwise). Both options have their pros and cons: showing a more informative result would surely delight the users if the result is correct, but it would upset them if it is incorrect; showing more results would be safer, but the more impoverished descriptions would not please users as much as the former. Therefore, to best use the constrained real estate, we need to adaptively choose an option to optimize the tradeoff between the length of the result list (quantity) and the informativeness

Table 1. All relevance judgments for query "jon favreau director" (faked)

Entity	Relevance
Jon Favreau (director)	perfect
Iron Man 3	good
Jon Favreau (speechwriter)	bad

of the results (quality) for each individual query. This poses significant challenges to traditional IR evaluation metrics. More specifically:

- No appropriate value for the length cutoff k can be easily determined a priori because the length of such a ranking list varies significantly in different alternative outputs of the ranking algorithm. In other words, a fixed cutoff parameter k could lead to poor evaluation results. Consider the examples of Figure 3. For the same query, "jon favreau director", suppose Table 1 contains all the relevance judgments. When we compute DCG@1 or NDCG@1, the result list 2 would be preferred, but when we compute DCG@2 or NDCG@2, the result list 1 would be preferred. That is, DCG@k and NDCG@k lead to inconsistent decisions depending on the value of k that we choose. Other traditional metrics share similar problems.
- The "additive" nature of DCG/NDCG and other IR metrics is not suitable in this application. To see why, compare Figure 1 and 2 again. For query "jon favreau director", the entity vertical result of Figure 1 (which shows exclusively the film director) seems more desirable than the entity vertical results of Figure 2 that show the director in the first position and the film in the second position. However, taking DCG/NDCG as an example, although both results have the same DCG@1 and NDCG@1 scores, the entity result in Figure 2 has higher DCG@2 and NDCG@2 scores than that of Figure 1.

To address these problems, we first propose two desirable formal constraints to capture the heuristics of regulating the quantity-quality tradeoff to properly evaluate IR applications with constrained result real estate. Inspired by the axiomatic approach to IR [7], we then present a general method to normalize the popular DCG metric [8] for balancing the quantity-quality tradeoff, yielding a new metric, namely Length-adjusted Discounted Cumulative Gain (LDCG). LDCG is shown to be able to automatically balance the length and the informativeness of a ranking list without requiring an explicit parameter k, while still preserving the good properties of DCG.

2 Related Work

The evaluation of IR applications with constrained real estate is a novel problem, and to the best of our knowledge, no previous work has addressed it to date. We briefly discuss some research efforts in the general IR evaluation literature that connect to our work.

Evaluation metrics play a critical role in the field of IR, and various metrics have been proposed. For example, average precision (AP) has been used extensively in TREC and other IR literature, and an extension GAP has recently proposed to extend AP to incorporate multi-graded relevance [10]; DCG and NDCG [8] have been accepted as major metrics for Web search [2]; and Expected Reciprocal Rank (ERR), an extension of the classical reciprocal rank to the graded relevance, has recently attracted much attention in Web search [4].

Since the metrics that we propose in this paper are extensions of DCG and NDCG, we now provide a brief overview of them. DCG at rank k for a ranking list is computed as [2]:

$$DCG@k = \sum_{i=1}^{k} \frac{2^{r(i)} - 1}{\log_2^{(i+1)}} \qquad (1)$$

where $r(i)$ is the relevance level of the result at rank i. However, DCG is often incomparable, and should be normalized. This can be done by sorting the relevance judgments of a query by relevance and producing the maximum possible DCG up to position k, called the ideal DCG (IDCG) at position k, as the normalization factor. The normalized DCG (NDCG) is computed as: $NDCG@k = \frac{DCG@k}{IDCG@k}$.

We can see that an explicit cutoff parameter k is required to compute DCG and NDCG. In fact, a similar cutoff parameter is also required by many other metrics, for example, average precision, precision, recall, etc. The purpose of this work is to eliminate this requirement by modeling the cutoff parameter in a "soft" way inside the metric.

One of the IR applications with constrained real estate is the search verticals in aggregated search. The evaluation of aggregated search has recently attracted attention, e.g., [1,11,12,5]. However, existing work does not address the issue of constrained result real estate and the quantity-quality tradeoff. Also the proposed metrics in our paper are more general and are not restricted to vertical search: we can potentially apply the proposed metrics to many IR applications with constrained real estate, e.g., recommendation systems.

Our work is also related to the literature on user models. An accurate user model is essential for developing a good relevance metric. In general, there are two main types of user models: position models (e.g., [6]) and cascade models (e.g., [4]). Both types of models attempt to capture the position bias of search result presentation. In contrast, our proposed length-variant user expectation is designed to approximate a threshold value for an IR metric according to the length of the result list.

The axiomatic approach has been used to develop effective IR models [7,9]. Our work adopts similar ideas to formalize the requirement of an IR evaluation scenario as formal constraints. Interestingly, a recent work [3] also uses the axiomatic approach to study IR metrics, but their work falls short when dealing with applications with constrained real estate.

3   Formal Constraints on Regulating the Quantity-Quality Tradeoff

A critical question is the following: how we can regulate the interactions between the quantity (i.e., the length of result list) and the quality (i.e., the informativeness of each result) so that we can balance the two factors? To answer this question, we first propose two desirable heuristics that a reasonable evaluation metric should implement to properly evaluate IR applications with constrained result real estate:

- First, *showing fewer results of higher relevance with more details should be preferred over showing more results, some of similar relevance and some others of lower relevance, with fewer details for each.* For example, the entity result in Figure 1 should be preferred over the entity result in Figure 2. This heuristic is used to reward the quality (informativeness) of the results. In traditional IR systems, each search result is usually displayed in a fixed space of (almost) the same size, and thus the informativeness of every result is also similar. In contrast, in the applications with constrained real estate, each result is allowed to show more or fewer details dynamically, so a relevant result that is more detailed/informative should be rewarded. To further understand this heuristic, we conducted a user study via crowdsourcing to ask the crowd to do a side-by-side comparison of a set of 174 pairs of entity search ranking lists (all with the same constrained real estate). A preliminary analysis of the user preference suggests that 82.4% people agree that one single "perfect" result is better than one "perfect" result followed by another "good" result, and that 75.8% people agree that one single "good" result is better than one "good" result followed by another "bad" result. This verifies that the proposed heuristic is desirable.
- Second, *showing more relevant results should be preferred over showing fewer results of similar relevance.* This heuristic is actually not entirely novel; it is easy to show that many current IR metrics have already implemented it. However, we still emphasize it explicitly because (1) it is widely-accepted and implemented in existing IR metrics as shown later, (2) it can be used to prevent the first heuristic from overly-penalizing the length of result list, and (3) it presumably makes sense in our applications since the former covers more diverse information with no relevance degradation.

Next, in order to analytically diagnose the limitations of current IR metrics, we propose two formal constraints to capture the above two heuristics of regulating the quantity-quality tradeoff so that it is possible to apply them to any IR metrics analytically.

We first define some key notations. Let L_1 and L_2 be two ranking lists. Assume $|L_1| = |L_2|$, where $|L_1|$ and $|L_2|$ are the sizes of the space taken to show L_1 and L_2 respectively, i.e., the real estate. And assume $L_1 = < d_1 >$, and $L_2 = < d_2, d_3 >$, where d_1, d_2, and d_3 are three results with relevance degrees $R(d_1)$, $R(d_2)$, and $R(d_3)$, respectively. We denote $E(\cdot)$ as a reasonable evaluation metric. Then the two constraints are defined as follows:

- **C1:** If $R(d_1) = R(d_2) > R(d_3)$, then $E(L_1) > E(L_2)$.
- **C2:** If $R(d_1) = R(d_2) = R(d_3) > 0$ [1], then $E(L_1) < E(L_2)$.

The proposed two constraints are useful to ensure the quantity-quality tradeoff. When either one is violated, the metric would likely not perform fairly for the

---

[1] We assume the relevance degree to be 0 for "bad" result.

Table 2. Constraint analysis results for different IR metrics

	DCG	NDCG	GAP	ERR	LDCG	LNDCG
C1	No	No	No	No	Yes	Yes
C2	Cond	Cond	Cond	Cond	Yes	Yes

evaluation of IR applications with constrained real estate $|L|$, and there should be room to improve the metric through improving its ability of satisfying the corresponding constraint. We analyze some typical traditional IR metrics that support graded relevance, including DCG/NDCG [8], ERR [4], and GAP [10].

- Under the condition in C1: Due to the additive nature of DCG, ERR, and GAP, it is easy to show that, DCG@$k(L1)$ <= DCG@$k(L2)$, NDCG@$k(L1)$ <= NDCG@$k(L2)$, GAP@$k(L1)$ <= GAP@$k(L2)$, and ERR@$k(L1)$ <= ERR@$k(L2)$. This shows that none of these existing evaluation metrics can satisfy C1, no matter what value the length cutoff parameter k is.
- Under the condition in C2: Similarly due to their additive nature, when $k > 1$, we can also see that DCG@$k(L1)$ < DCG@$k(L2)$, NDCG@$k(L1)$ < NDCG@$k(L2)$, GAP@$k(L1)$ < GAP@$k(L2)$, and ERR@$k(L1)$ < ERR@$k(L2)$, which is consistent with C2. However, when $k = 1$, it is not surprising that DCG@$k(L1)$ = DCG@$k(L2)$, NDCG@$k(L1)$ = NDCG@$k(L2)$, GAP@$k(L1)$ = GAP@$k(L2)$, and ERR@$k(L1)$ = ERR@$k(L2)$, which is inconsistent with C2. These analysis results show that the existing evaluation metrics can only satisfy C2 conditionally when an appropriate length cutoff parameter k is chosen a prior which itself is nontrivial.

Due to space limitations, we cannot show all the analysis details in this paper, but the results are presented in Table 2.

4   A Quantity-Quality Balanced Metric

The analysis above shows that traditional IR metrics do not satisfy the proposed quantity-quality regulation constraints. In order to properly evaluate applications with constrained real estate, we need to add the quantity-quality tradeoff into the evaluation metric. However, we do not want the addition of this property to change other desirable properties (e.g., ranking order and multi-graded relevance awareness) possessed by these widely-accepted metrics.

We propose a general approach to achieve this goal by adopting the current additive metrics to measure the quality of a ranking list, while developing a length (quantity) normalization component to normalize the quality score. In this paper, we choose the popular DCG [8,2] as the quality measure, and propose a novel method for length-normalization based on an intuition of length-variant user expectation.

4.1 Length-Variant User Expectation

User experience relies not only on the quality of the results themselves, but also on the a priori expectation that the users have on those results. For example, when user expectation is low, a ranking list could be satisfactory even if its quality is not so good; in contrast, when user expectation is high, a result list could still be unsatisfactory even if its quality is already very good.

In this section, we explore this notion, and approximate it as a threshold score for the quality of a ranking list: failing to achieve this threshold renders the results unsatisfactory to the user. This can also be regarded as some sort of a "lower-bound" on the user expectation.

Intuitively, user expectation depends on the length of a ranking list: users would generally have higher expectations when they are shown a longer list of results. That is, user expectation increases with the length of the result list. On the other hand, user expectation would become less sensitive to the length of the result list as the list becomes longer.

Based on this intuition, we approximate a length-variant user expectation for a ranking list. Specifically, we propose a model that relies on a single assumption: a necessary condition for a user to be satisfied with the results is that *there must be at least one relevant result appearing in the list, and that this relevant result can be ranked at any position according to some given position priors* (which may reflect the fact that the user would expect the relevant result to appear at a high position.)

4.2 Length Normalized DCG and NDCG

As we choose DCG as our quality measure, what would be the user expected DCG (i.e., the threshold DCG score), for a ranking list with N results? Based on the assumption articulated in the previous section, the threshold score can be met when there is at least one relevant result in the ranking list. When the relevant result occurs at position i, DCG can be calculated as:

$$DCG(i) = \frac{2^{rel} - 1}{\log_2^{(i+1)}} \tag{2}$$

where rel represents the relevance level of the relevant result. We can simply assume $rel = 1$, since it does not influence the relative score as we will show.

Let $p(i)$ be a function representing the "prior" belief of the user regarding what position holds the relevant result. For example, $p(1) = 0.5$ means that the user expects the relevant result to occur at the first position with a probability of 0.5. For the purposes of estimating the user's expected DCG, without loss of generality, we take a prior proportional to DCG's discounting factor when $i \leq M$, and take a prior of 0 when $i > M$, where M is the maximum number of results allowed to show in the result real estate. That is,

$$p(i) \propto \begin{cases} \frac{1}{\log_2^{(i+1)}} & \text{if } i \leq M \\ 0 & \text{otherwise} \end{cases} \tag{3}$$

For a ranking list of length N, the user's expected DCG can be defined as the expected value of $DCG(i)$ under $p(i)$:

$$E[DCG] = \sum_{i=1}^{N} p(i) \cdot DCG(i) = Z(M, rel) \cdot \sum_{i=1}^{N} \left(\frac{1}{\log_2^{(i+1)}} \right)^2 \qquad (4)$$

where $Z(M, rel)$ is a normalization factor, which is a constant. More precisely,

$$Z(M, rel) = \frac{2^{rel} - 1}{\sum_{i=1}^{M} \frac{1}{\log_2^{(i+1)}}} = \frac{1}{\sum_{i=1}^{M} \frac{1}{\log_2^{(i+1)}}} \qquad (5)$$

It is easy to see that $\frac{\partial E[DCG]}{\partial N} > 0$ and $\frac{\partial^2 E[DCG]}{\partial^2 N} < 0$, which shows that the user's expected DCG is monotonically increasing with N, but the increasing speed decreases with N. These characteristics are consistent with our intuitions that a user would expect more from a longer result list, and that if the result list is already long, the user's expectation would not be so sensitive to N.

In order to evaluate the user's satisfaction on a ranking list, we compare the standard DCG score against the user's expected DCG score. Specifically, we normalize the standard DCG by dividing it using the user's expected DCG, leading to a Length-adjusted DCG (LDCG). Formally,

$$LDCG = \frac{DCG}{E[DCG]} = \frac{DCG}{Z(M, rel) \cdot \sum_{i=1}^{N} \left(\frac{1}{\log_2^{(i+1)}} \right)^2} \propto \frac{DCG}{\sum_{i=1}^{N} \left(\frac{1}{\log_2^{(i+1)}} \right)^2} \qquad (6)$$

which essentially normalizes DCG based on the square sum of the document discounting factors, but does not change other existing properties of DCG. One may notice that if we use a uniform probability for $p(i)$ in Formula 3, we will essentially use the sum of the document discounting factors as the normalization component; however, this will lead LDCG to violate the constraint C2 due to the over penalization of the list length. Nonetheless, using DCG's discounting factor or other decreasing functions as $p(i)$ is a more natural choice because $p(i)$ should represent the "prior" belief of the user regarding what position holds the relevant result.

Similar to NDCG, we can further rescale the score range of LDCG to $[0, 1]$ using the ideal LDCG (ILDCG) score: $ILDCG = \max \left\{ \frac{DCG}{E[DCG]} \right\}$. It is not hard to see that, if we use integer relevance labels (e.g., 0, 1, $\cdots$) and as long as the ranking list has no more than 25 results (which is arguably always the case for the constrained real estate), ILDCG would be obtained when the ranking list covers and only covers all of the highest relevant answers in the judgment set. For example, if we use relevance labels "perfect", "good", and "bad", each is associated with some integer value, the ILDCG will be obtained when all and only perfect results (or all good answers when there is no perfect answer) are included in the ranking list. Therefore, ILDCG can be calculated as:

$$ILDCG = \frac{\sum_{i=1}^{R} \frac{2^m - 1}{\log_2^{(i+1)}}}{Z(M, rel) \cdot \sum_{i=1}^{R} \left(\frac{1}{\log_2^{(i+1)}} \right)^2} \qquad (7)$$

where R indicates the smaller one between M (the maximum number of results allowed) and the number of relevance judgments with the highest relevance degree m in the judgment set of the current query. Finally, with this ILDCG, the Length-adjusted NDCG is derived:

$$LNDCG = \frac{LDCG}{ILDCG} = \frac{DCG}{\sum_{i=1}^{N}\left(\frac{1}{\log_2^{(i+1)}}\right)^2} \cdot \frac{\sum_{i=1}^{R}\left(\frac{1}{\log_2^{(i+1)}}\right)^2}{\sum_{i=1}^{R}\frac{2^{rel_m}-1}{\log_2^{(i+1)}}} \quad (8)$$

where $Z(M, rel)$ can been mathematically eliminated in the calculation of LNDCG. It shows that *we do not really need to choose values for M and rel*.

From Formulas 6 and 8, we can see that, unlike DCG and NDCG, LDCG and LNDCG do not require any explicit length cutoff parameter, because they already model the length of the ranking list into the metric. Furthermore, it is not hard to verify that LDCG and LNDCG satisfy both formal constraints introduced in Section 3 unconditionally, as reported in Table 2.

5 Analysis

We use our running example query "jon favreau director" to compare the scores of LDCG/LNDCG and DCG/NDCG@k for several scenarios of interest. All the relevance judgments have been presented in Table 1, and we use the following settings of relevance labels: perfect= 2, good= 1, and bad= 0. We assume the maximum number of results allowed to show is $M = 3$ [2]. The comparison results have been shown in Table 3.

Our first observation is that the DCG/NDCG scores at different cutoff values vary significantly in the different cases, confirming our statement that it is hard to determine an appropriate parameter k in a constrained real estate application. Take Scenarios 6 and 7 as examples: when we compute DCG@2 or NDCG@2, Scenario 7 will be preferred, but when we compute DCG@3 or NDCG@3, Scenario 6 will be preferred, leading to inconsistent evaluation decisions. In contrast, the proposed LDCG/LNDCG, which can essentially be regarded as an aggregation of DCG/NDCG scores for different k values, produces a single score without relying on any explicit parameter k, suggesting that LDCG/LNDCG would be more stable and consistent for IR applications with constrained result real estate.

Second, LDCG/LNDCG is more effective than DCG/NDCG for length penalization. Comparing Scenarios 1 and 2, we can see that LDCG/LNDCG prefers cases where only perfect answers are shown in a clean way by means of length penalization. Another interesting comparison is between Scenario 4 and Scenario 5: Scenario 5 would lead to worse user experience than Scenario 4, because the former is exactly the latter with an additional bad result. DCG/NDCG does not

---

[2] We emphasize again that LDCG and LNDCG do not require any length cutoff parameter, as shown in Section 4.2. Yet we set M to some value just to help readers understand the LDCG calculation in Table 3 using the full-fledged Formula 6.

Table 3. Comparison of LDCG with DCG@k and LNDCG with NDCG@k for several scenarios of interest

	Ranking Lists	DCG			LDCG	NDCG			LNDCG
		@1	@2	@3		@1	@2	@3	
1	Jon Favreau (director)	3	3	3	6.40	1	0.82	0.82	1
2	Jon Favreau (director) Iron Man 3	3	3.63	3.63	5.54	1	1	1	0.87
3	Jon Favreau (director) Jon Favreau (speechwriter)	3	3	3	4.58	1	0.82	0.82	0.72
4	Iron Man 3 Jon Favreau (director)	1	2.89	2.89	4.41	0.33	0.79	0.79	0.69
5	Iron Man 3 Jon Favreau (director) Jon Favreau (speechwriter)	1	2.89	2.89	3.74	0.33	0.79	0.79	0.59
6	Iron Man 3 Jon Favreau (speechwriter) Jon Favreau (director)	1	1	2.5	3.23	0.33	0.27	0.69	0.51
7	Jon Favreau (speechwriter) Jon Favreau (director) Iron Man 3	0	1.89	2.39	3.09	0	0.52	0.66	0.48
8	Jon Favreau (speechwriter) Jon Favreau (director)	0	1.89	1.89	2.88	0	0.52	0.52	0.45
9	Jon Favreau (speechwriter) Iron Man 3 Jon Favreau (director)	0	0.63	2.13	2.76	0	0.17	0.59	0.43
10	Iron Man 3	1	1	1	2.13	0.33	0.27	0.27	0.33
11	Iron Man 3 Jon Favreau (speechwriter)	1	1	1	1.52	0.33	0.27	0.27	0.24
12	Jon Favreau (speechwriter) Iron Man 3	0	0.63	0.63	0.96	0	0.17	0.17	0.15

deal with such cases well by giving the same scores to both scenarios at all k values, but LDCG/LNDCG does a clearly better job to score Scenario 4 higher than Scenario 5.

6   Conclusions

We proposed new IR metrics, namely Length-adjusted DCG and NDCG (LDCG and LNDCG), to evaluate IR applications with constrained result real estate. LDCG/LNDCG extend DCG/NDCG to be able to automatically achieve length normalization without requiring a length cutoff parameter k that is typically required by traditional IR metrics, while still preserving the good properties of DCG/NDCG. Our preliminary analysis shows that LDCG and LNDCG work better than existing metrics for evaluating IR applications with constrained real estate. In the future, we will conduct a thorough empirical evaluation of the proposed LDCG and LNDCG based on user studies and large-scale query log analysis.

Acknowledgments. We thank Jay He, Jiayuan Huang, Dhyanesh Narayanan, and Bo Zhao for their helpful discussions. We also thank the anonymous reviewers for their useful comments.

References

1. Arguello, J., Diaz, F., Callan, J., Carterette, B.: A methodology for evaluating aggregated search results. In: Clough, P., Foley, C., Gurrin, C., Jones, G.J.F., Kraaij, W., Lee, H., Mudoch, V. (eds.) ECIR 2011. LNCS, vol. 6611, pp. 141–152. Springer, Heidelberg (2011)
2. Burges, C., Shaked, T., Renshaw, E., Lazier, A., Deeds, M., Hamilton, N., Hullender, G.: Learning to rank using gradient descent. In: Proceedings of the 22nd International Conference on Machine Learning, ICML 2005, pp. 89–96 (2005)
3. Busin, L., Mizzaro, S.: Axiometrics: An axiomatic approach to information retrieval effectiveness metrics. In: Proceedings of the 4th International Conference on Theory of Information Retrieval: Advances in Information Retrieval Theory, ICTIR 2013 (2013)
4. Chapelle, O., Metlzer, D., Zhang, Y., Grinspan, P.: Expected reciprocal rank for graded relevance. In: Proceedings of the 18th ACM Conference on Information and Knowledge Management, CIKM 2009, pp. 621–630 (2009)
5. Chuklin, A., Schuth, A., Hofmann, K., Serdyukov, P., de Rijke, M.: Evaluating aggregated search using interleaving. In: Proceedings of the 22nd ACM International Conference on Conference on Information and Knowledge Management, CIKM 2013, pp. 669–678 (2013)
6. Craswell, N., Zoeter, O., Taylor, M., Ramsey, B.: An experimental comparison of click position-bias models. In: Proceedings of the 2008 International Conference on Web Search and Data Mining, WSDM 2008, pp. 87–94 (2008)
7. Fang, H., Zhai, C.X.: An exploration of axiomatic approaches to information retrieval. In: Proceedings of the 28th Annual International ACM SIGIR Conference on Research and Development in Information Retrieval, SIGIR 2005, pp. 480–487 (2005)
8. Järvelin, K., Kekäläinen, J.: Cumulated gain-based evaluation of ir techniques. ACM Trans. Inf. Syst. 20(4), 422–446 (2002)
9. Lv, Y., Zhai, C.: Lower-bounding term frequency normalization. In: Proceedings of the 20th ACM International Conference on Information and Knowledge Management, CIKM 2011, pp. 7–16 (2011)
10. Robertson, S.E., Kanoulas, E., Yilmaz, E.: Extending average precision to graded relevance judgments. In: Proceedings of the 33rd International ACM SIGIR Conference on Research and Development in Information Retrieval, SIGIR 2010, pp. 603–610 (2010)
11. Zhou, K., Cummins, R., Lalmas, M., Jose, J.: Evaluating large-scale distributed vertical search. In: Proceedings of the 9th Workshop on Large-Scale and Distributed Informational Retrieval, LSDS-IR 2011, pp. 9–14 (2011)
12. Zhou, K., Cummins, R., Lalmas, M., Jose, J.M.: Evaluating aggregated search pages. In: Proceedings of the 35th International ACM SIGIR Conference on Research and Development in Information Retrieval, SIGIR 2012, pp. 115–124 (2012)

Dissimilarity Based Query Selection for Efficient Preference Based IR Evaluation

Gabriella Kazai and Homer Sung

Microsoft
Cambridge, UK and Bellevue, US

Abstract. The evaluation of Information Retrieval (IR) systems has recently been exploring the use of preference judgments over two lists of search results, presented side-by-side to judges. Such preference judgments have been shown to capture a richer set of relevance criteria than traditional methods of collecting relevance labels per single document. However, preference judgments over lists are expensive to obtain and are less reusable as any change to either side necessitates a new judgment. In this paper, we propose a way to measure the dissimilarity between two sides in side-by-side evaluation experiments and show how this measure can be used to prioritize queries to be judged in an offline setting. Our proposed measure, referred to as Weighted Ranking Difference (WRD), takes into account both the ranking differences and the similarity of the documents across the two sides, where a document may, for example, be a URL or a query suggestion. We empirically evaluate our measure on a large-scale, real-world dataset of crowdsourced preference judgments over ranked lists of auto-completion suggestions. We show that the WRD score is indicative of the probability of tie preference judgments and can, on average, save 25% of the judging resources.

Keywords: Side-by-side evaluation, preference judgments, weighted ranked difference measure, query prioritization, judging cost reduction.

1 Introduction

When collecting relevance judgments for the evaluation of Information Retrieval (IR) systems, the established method is to assess the relevance of individual documents to a given query, independently of other retrieved documents [15]. Recently, a number of studies explored an alternative method for the comparative evaluation of IR systems, which is based on preference judgments over two search result lists, shown side-by-side to relevance judges [14,10,2,5,9]. This method allows to capture the interaction among search results, for example their diversity, as part of the relevance criteria [10,2,9,12]. Radlinski et al. [10,11] demonstrated that even subtle differences in retrieval results could be measured with preference based comparisons. Thomas and Hawking [14] reported high levels of accuracy in users preferring the top ranked results, when shown Google's first and second page of results side by side. Preference judgments have also been shown to be more reliable than independent absolute labels in terms of improved inter-assessor agreement levels [5].

M. de Rijke et al. (Eds.): ECIR 2014, LNCS 8416, pp. 172–183, 2014.

However, there are a number of significant problems with collecting preference judgments over lists of documents, e.g., as search results: 1) they require considerably more judging resources due to the complexity of the task, and 2) they result in considerably less reusable judgments due to the fact that any change to either of the two lists can have a potential impact on the preference direction, thus necessitating a new judgment. So, while, on the one hand, preference judgments over search result lists are needed to capture a richer set of relevance dimensions, on the other hand, they are a less efficient use of judging resources. Approaches to economize their use are thus highly desirable.

Towards this end, in this paper, we propose a way to measure the dissimilarity between two sides in side-by-side evaluation experiments and show how this measure can be used to guide the judging process, selecting the queries to be judged by human assessors. Our intuition and hypothesis is that queries with lower dissimilarity scores are less likely to be able to discriminate between the two sides that are being compared, i.e., are more likely to result in a tie or 'no preference' judgment. Given such a dissimilarity measure, we can then prioritize queries with higher dissimilarity and thus with higher potential to impact the overall evaluation. Our proposed measure, referred to as Weighted Ranking Difference (WRD), takes into account both the ranking differences and the similarity of the documents across the two sides, where a document may, for example, be a URL or a query suggestion. We empirically evaluate our measure on a large-scale, real-world dataset of nearly 600k crowdsourced preference judgments over 58k pairs of auto-completion suggestion lists. We show that WRD's dissimilarity score is indicative of the probability of tie preference judgments and can, on average, save 25% of the judging resources by only judging queries with high WRD scores.

The problem of comparing two ranked lists and quantifying their similarities or correlations has been extensively studied in IR. The most commonly used rank correlation measures include the well-known Kendall's τ correlation coefficient and a top-weighted variant, proposed by Yilmaz et al. [17], called τ_{AP}. These correlation statistics are often adopted to measure the similarity between rankings of IR systems, for example, derived based on different sets of relevance judgments (qrels) [15]. Other extensions of Kendall's correlation include methods to associate different weights with different errors in the rankings [13,6]. An alternative top-weighted measure, called M, was proposed by Bar-Ilan et al. [3], which measures the similarity of two ranked lists as the normalized sum of the differences in the reciprocal ranks of each item in the two lists. Items that only appear in one of the lists are assumed to occur at the end of the other list. The rank-biased overlap (RBO) measure, proposed by Webber et al. [16], uses a convergent series of weights to achieve top-weighting. In this paper, we adopt and extend the M measure [3] to incorporate the similarity of the items themselves. For example, consider two lists of search suggestions that are being compared as possible suggestions for a given query that a user is entering into a search engine, where one list contains the suggestion "restaurant" while the other has "restaurants". Rather than treating these as two different items (or allowing them to be the same item), our goal is to measure their similarity and incorporate this score into the overall measure of similarity (or dissimilarity) over the two lists. We then use the obtained dissimilarity scores to prioritize the set of side-by-side comparisons for which to obtain preference judgments from human judges with the aim to reduce judging costs.

This problem is related to the issue of constructing traditional test collections efficiently, which has been investigated in the literature, e.g., [1,4,18,7,8]. Approaches include various document or query selection methods for reducing the number of judgments needed to accurately estimate the average system performance. For example, Guiver et al. [7] showed that it is possible to reproduce the results of an exhaustive evaluation by using a reduced, but representative, set of queries. Hosseini et al. [8] developed a query selection approach to enhance the re-usability of a test collection when previously un-judged documents are retrieved by systems. Common to all these methods is the availability of an initial set of judgments, where the objective is then to select a subset of queries or documents to obtain additional judgments for. Unlike these works, in this paper, we focus on a completely offline setting, where no judgments are available and aim to derive a prioritized ordering of queries to be judged a-priori, reducing the number of judgments needed independently of the labels themselves. We also study the properties of our metric and its relationship with the preference judgments and the overall evaluation metric. Our main contributions are the proposed dissimilarity metric for side-by-side experiments and its empirical evaluation on a large-scale dataset.

The next section introduces our dissimilarity metric and examines some of its properties. In section 3, we study the metric's behavior in a large-scale, real-world preference judgment experiment and empirically evaluate its ability to reduce judging costs.

2   Measuring Dissimilarity

In this section we propose a new metric to measure the dissimilarity between two sides in a side-by-side experiment. Our goal is to measure the difference between two lists of documents, both in terms of differences in ranking order and differences between the documents themselves. The documents may simply be query suggestions or URLs of search results, or more complex documents such as the full captions of search results. We start by describing the desired properties of a new metric, then introduce two related metrics and our final metric that builds on the former two and provide illustrative examples of their behaviors.

2.1   Desiderata

The purpose of our new metric is to measure the dissimilarity of two ranked lists of documents, taking into account both the ranking similarities and the similarities between the items themselves. While these two aspects could be measured independently, our goal is a single metric that captures both aspects in a single score, allowing to consider document similarity as an inherent property.

Furthermore, as with many ranked lists of documents, our metric should consider differences at the top of the rankings with higher weight. For example, if two lists differed at only one rank position, the reported dissimilarity should be higher when the difference is found closer to the top. Given two ranked lists of completely different documents, our measure should report total dissimilarity, as the score of 1. When the two sides are identical, the metric should report 0.

Before we detail our proposed metric, we first introduce two metrics that measure only one of the two aspects of dissimilarity mentioned above: Ranking Difference (RD), which only considers ranking differences, and Minimal Dissimilarity (mD), a set based measure, which only considers the pairwise dissimilarity of the documents between the two sides. We will then build on these two measures to define our Weighted Ranking Difference (WRD) metric.

2.2 Ranking Difference (RD)

We define our RD dissimilarity measure for two ranked lists of documents as a function of the distance between the rank positions that the same documents occur at on the two sides. More specifically, let L and R be two ranked lists of documents, those on the left side, $L = l_1, l_2, \ldots, l_m$, and those on the right, $R = r_1, r_2, \ldots, r_n$. A document in L, R may be a string, e.g., a query suggestion, a URL, etc. RD is then calculated as the difference between the reciprocal of the ranks at which documents that are common to both sides occur at, plus the reciprocal of the ranks that contain documents that only occur on one side, normalized by the maximum dissimilarity. The maximum dissimilarity is simply when no documents are common between the two sides, and is thus equal to the sum of all reciprocal ranks over both sides:

$$ RD_{L,R} = \frac{\sum_{(i,j)\in C} |\frac{1}{f(i)} - \frac{1}{f(j)}| + \sum_{i \notin C} \frac{1}{f(i)} + \sum_{j \notin C} \frac{1}{f(j)}}{2\sum_{k=max(i,j)} \frac{1}{f(k)}}, \tag{1} $$

where C is the set of rank-pairs of common documents between L and R, i.e., when $l_i = r_j$, and $f(p)$ is a function of rank position p, such as $f(p) = p$ or $f(p) = log(1 + p)$. With $f(p) = p$, differences lower down the ranking are weighted less (i.e., more top heavy) than when using $f(p) = log(1 + p)$.

The RD measure above is a modified version of the M measure [3], where the items that only appear in one list are not appended to the end of the other list. Unlike M, RD measures dissimilarity and results in a score of 0 when the two sides are identical, and a score of 1 when no common documents occur across the two sides. Note that this measure relies on a binary notion of 'matching documents' between the two sides, i.e., documents are either common to the two sides or not, regardless of the matching function used, e.g., exact or fuzzy matching. For example the suggestions "restaurant" and "restaurants" or the URLs "m.facebook.com" and "www.facebook.com" may be treated either as the same items or as pairs of different items, but, either way, their level of similarity is lost. An alternative method, that addresses this issue, is described next.

2.3   Minimal Dissimilarity (mD)

We define our mD metric as the average of the minimal pairwise document dissimilarity scores. Given two lists of documents, L and R, we first calculate dissimilarity scores for each document pairs across the two sides (l_i, r_j). Note that we assume the two sides are of equal length, and treat empty rank positions as empty string documents. Then, using a greedy algorithm, we select the pairs of documents that are the most similar to each other, S, and calculate mD as the average of the dissimilarity scores in S:

$$mD_{L,R} = \frac{\sum_{(i,j) \in S} d_{ij}}{|S|}, \tag{2}$$

where d_{ij} is the dissimilarity score of documents l_i and r_j, which may be calculated using any document-document similarity measures or edit distance algorithms. In this paper, since our data set contains query suggestions, which are short strings, we adopt the well-known Levenshtein distance for measuring the difference between two documents: the minimum number of single-character edits (insertion, deletion, substitution) required to change one string into the other. The greedy algorithm works by selecting from all pairs of (l_i, r_j) documents, the one with the minimum dissimilarity score. It then removes all other document pairings where either l_i or r_j are a member of the pairing, before selecting the next minimally dissimilar document pair. The mD measure results in a score of 0 when the two sides are identical, and leads to a score of 1 when the documents across the two sides are completely dissimilar.

The advantage of the mD metric is that it does capture the degree of dissimilarity between the documents, e.g., "restaurant" and "restaurants". However, mD is a set based measure, which ignores ranking differences. Next, we propose a metric that combines both document and ranking dissimilarities.

2.4   Weighted Ranking Difference (WRD)

We define our WRD metric as the dissimilarity between two ranked lists of documents, L and R. As with mD, we first calculate the pairwise document dissimilarity scores and derive the set of most similar document pairs over the two sides: S. Then, we calculate WRD as the sum of the weighted differences between the reciprocal of the associated ranks and the weighted sum of the reciprocal ranks:

$$WRD_{L,R} = \frac{\sum_{(i,j) \in S}\left((1 - d_{ij})\left|\frac{1}{f(i)} - \frac{1}{f(j)}\right| + d_{ij}\left(\frac{1}{f(i)} + \frac{1}{f(j)}\right)\right)}{2\sum_{k=max(i,j)} \frac{1}{f(k)}}; \tag{3}$$

where S is the rank-pairs of the most similar documents $(l_i \in L, r_j \in R)$, d_{ij} is the document dissimilarity weight associated with the documents l_i and r_j, and $f(p)$ is a function of the rank position p, e.g., $f(p) = p$ or $f(p) = log(1 + p)$.

2.5   Illustrative Examples

Table 1 lists a number of example cases of different document rankings of length four and the resulting scores for three baseline metrics, Kendall's τ, the τ_{AP}

(APcorr) [17], and the M measure[1] [3], and our three metrics, with two different $f(p)$ functions: $f(p) = p$ and $f(p) = log(1 + p)$. All dissimilarity metrics correctly report 0 when the two sides are identical and 1 when the two sides contain completely different documents. The Kendall scores are 1 for identical, 0 for random, and -1 for reverse rankings. For the rest of the examples, the dissimilarity scores vary considerably. In rows 2-5, different pairs of documents from the left are swapped on the right side. We see that mD, which is insensitive to ranking changes, reports 0 for all these cases. Both RD and WRD, however, are affected and, as desired, report higher dissimilarity for the case when the swap occurs higher to the top of the ranking. Rows 6-10 demonstrate the advantages of WRD over mD and RD. In each case, we simply add an additional character to one of the documents on the right. Thus, we introduce a small difference between the two documents ranked at the same positions on the two sides. RD consistently overestimates the dissimilarity between the two sides, when each pair of non-exact matching strings are treated as different items. As before, mD recognizes the similarities between the two documents, e.g., 'a' and 'as', but reports the same levels of dissimilarity regardless of where in the rankings the documents appear. WRD is the only metric that reflects both the similarity of the documents and is sensitive to the rank position, reporting higher dissimilarity when the similar documents are closer to the top. Figure 1c plots the metric scores for rows 1-11 (ID on x axis).

Table 1. Metric scores for illustrative examples: lists of query suggestions on left (L) and right (R) in side-by-side experiments

ID	$L : l_1, l_2, l_3, l_4$	$R : r_1, r_2, r_3, r_4$	τ	τ_{AP}	M	RD	RDLog	mD	WRD	WRDLog
1	a,b,c,d	a,b,c,d	1	1	0	0	0	0	0	0
2	a,b,c,d	**b,a**,c,d	0.67	0.78	0.39	0.24	0.14	0	0.24	14
3	a,b,c,d	a,b,**d,c**	0.67	0.33	0.06	0.04	0.03	0	0.04	0.03
4	a,b,c,d	**d**,b,c,**a**	-0.67	-0.67	0.58	0.36	0.22	0	0.36	0.22
5	a,b,c,d	**d,c, b, a**	-1	-1	0.71	0.44	0.27	0	0.44	0.27
6	a,b,c,d	**as**, b, c, d	-0.40	-0.42	0.62	0.48	0.39	0.13	0.24	0.20
7	a,b,c,d	a,**bs**, c, d	0	-0.25	0.23	0.24	0.25	0.13	0.12	0.12
8	a,b,c,d	a, b, **cs**, d	0.40	0	0.10	0.16	0.20	0.13	0.08	0.10
9	a,b,c,d	a, b, c, **ds**	0.80	0.50	0.04	0.12	0.17	0.13	0.06	0.08
10	a,b,c,d	**as, bs, cs, ds**	0	0	1	1	1	0.50	0.50	0.50
11	a,b,c,d	e, f, g, h	0	0	1	1	1	1	1	1

To get a better feel for the behavior of our WRD metrics, Figure 1a shows the obtained WRDLog scores for a parameter sweep over d_{ij} when the pairs of documents in S are at the same ranks on both sides. We can see that 1) lower ranks contribute less (decreasing height of columns towards rank 10), and 2) more similar document pairs contribute less to the overall dissimilarity score (range of scores in each column). Figure 1b shows the distribution of the WRDLog scores

---

[1] We use $1 - M$ here for the sake of comparability.

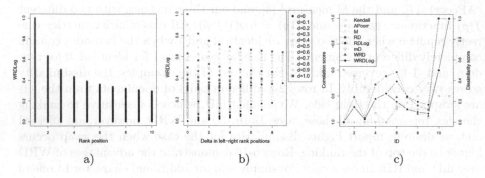

a)   b)   c)

Fig. 1. a) WRDLog scores for sweep of d_{ij} in $[0, 1]$ when matched documents are at same ranks on both sides, b) WRDLog scores for sweeps of left-right rank differences for matched documents, for a series of fixed d_{ij} weights, and c) rank correlation and dissimilarity metric scores for examples in rows 1-11 (ID on x axis) of Table 1

for sweeps of left vs right rank differences between the paired documents in S, for a series of fixed d_{ij} weights. We observe that 1a) for more similar document pairs, i.e., lower values of d_{ij}, the larger the difference between the left and right ranks, the higher the contribution to WRDLog, and that 1b) the closer one of the rank positions to the top, the larger its contribution to WRDLog, i.e., column of values for the same d_{ij} at a given rank-delta. On the other hand, 2a) when matched documents are more dissimilar, i.e., higher values of d_{ij}, then the closer they rank on both sides, the higher their dissimilarity contribution, where again 2b) top ranks weigh more.

3   Empirical Evaluation

In this section, we apply our metrics in a large-scale, real-world side-by-side evaluation scenario with two aims: 1) to examine the relationship between the metrics and the collected preference labels, and 2) to experiment with using our dissimilarity metric to reduce the number of judgments and thus judging costs, by prioritizing the queries to be judged based on the obtained dissimilarity scores.

3.1   Experiment Data

We use preference labels collected in a series of side-by-side experiments over ranked lists of auto-completion query suggestions from two IR systems. Each experiment involves the following steps. First a set of queries is sampled from the Bing query logs for which the auto-completion search suggestions are collected from the two systems, using only parts of the queries, i.e., prefixes (representing the part of the query a user may have typed into a search engine up to a given point in time). Preference labels over the pairwise lists of auto-completions (four suggestions per side), shown side by side, are then crowdsourced from crowd

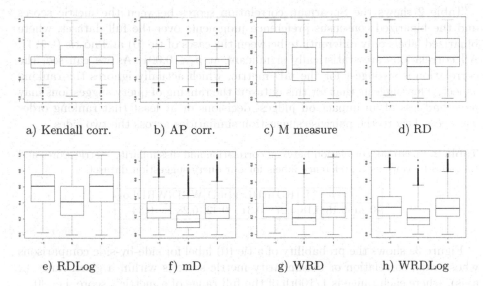

a) Kendall corr.	b) AP corr.	c) M measure	d) RD

e) RDLog	f) mD	g) WRD	h) WRDLog

Fig. 2. Distribution of correlation/dissimilarity metric scores (y axis) per consensus preference label class (x axis) for our query suggestion dataset

workers on Clickworker[2], a crowdsourcing platform similar to Amazon's Mechanical Turk. The side on which one system's auto-completion query suggestions are shown is randomized, so that each system has the same probability of appearing on both sides. The preference judgments are collected on a 3 point categorical scale: left/right side preferred or no preference. We first convert the obtained preference judgments to reflect preferences for one system over the other and then map them to a 3 point numerical scale: $(-1, 0, 1)$, where 0 means tie, and $-1,1$ is preference for one system or the other. Finally, we calculate consensus labels by taking the single most frequent label for each query.

Our full dataset contains a total of 58,519 queries (side-by-side comparisons) for which a total of 596,450 preference judgments were collected from 659 workers, i.e., an average of 10.19 labels per query and nearly 905 judgments per worker. 24.4% of the consensus labels are tie (0) judgments.

3.2   Relationship to Preference Labels

Figure 2 shows the distribution of the different metric scores and the consensus preference labels. All metrics demonstrate some level of relationship with the preference judgments: tie preference judgments tend to be associated with more similar (less dissimilar) metric scores compared with judgments when one side is preferred over the other. This confirms our initial intuition that similar pairs of lists are more likely to result in a tie judgment.

---

[2] www.clickworker.com

Table 2 shows the Spearman correlation scores between the metric scores and the binarized consensus preference judgments over the full dataset, where binarized labels only differentiate between the cases of tie (0) and no-tie $(-1, 1)$. All correlations are statistically significant at $p < 0.001$. As we can see, the correlation is strongest for the mD metric, which actually ignores the ranking aspect. This suggests that for this dataset, the ranking of query suggestions may have had less of an impact on judges' decisions or at least that ranking order was secondary to the pairwise suggestion similarities across the two sides.

Table 2. Spearman correlation between correlation and dissimilarity metric scores and tie vs no-tie consensus preference labels for our query suggestion dataset

τ	τ_{AP}	M	RD	RDLog	mD	WRD	WRDLog
-0.20	-0.18	0.18	0.27	0.31	0.42	0.28	0.34

Figure 3a shows the probability of a tie (0) label for side-by-side comparisons whose rank correlation or dissimilarity metric score is within a given range (x axis), where each range is 1/100th of the full range of a metric's score, i.e., $[0, 1]$ for dissimilarity metrics and $[-1, 1]$ for rank correlation statistics. For the RD, mD and WRD metrics, we see that the probability of observing ties in a subset tends to decrease with increasing dissimilarity scores. The trend is the cleanest and strongest for the mD and WRDLog metrics. The Kendall correlation based metrics and the M measure result in a lot more noise, where the metric scores do not reflect the probability of observing tie labels. Figure 3b shows the probability of tie labels, calculated over gradually reduced subsets of the full dataset, where at each step, we remove 1/100th of the full dataset with the lowest dissimilarity or rank correlation scores. Thus, as we gradually remove more similar cases, if the dissimilarity metric is predictive of ties, then we expect the probability of tie labels in the remaining dataset to decrease. This is indeed observed for the RD, mD and WRD metrics, again with mD and WRDLog obtaining best results. These findings suggests that the RD, mD and WRD metrics may be used to predict if a side-by-side experiment is likely to result in a tie judgment, which in turn can alleviate the need for a judgment.

3.3   Relationship to WinLoss

Given a set of preference judgments, we can calculate which IR system won overall in a side-by-side experiment by simply counting the number of wins, losses and ties for that system. More specifically, the Win-Loss measure reports the percentage of times when one system is preferred over the other, minus when the other system is preferred, and the total number of comparisons:

$$WinLoss = \frac{win - loss}{win + loss + tie}\%   \tag{4}$$

Given this definition, the relationship between the dissimilarity metric scores and a judgment's contribution to the final WinLoss evaluation score is the inverse of the probability of tie labels, i.e., the plot in Figure 3a flipped on its x axis.

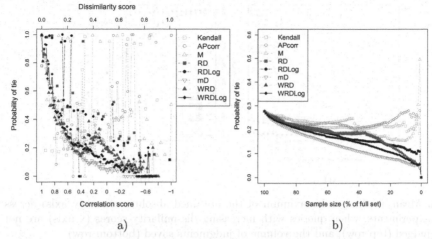

Fig. 3. Probability of tie preference judgments for a) subsets with dissimilarity metric or rank correlation score in a given range, e.g., [0.2, 0.4] and b) gradually reduced sized subsets of the highest dissimilarity side-by-side cases

3.4   Dissimilarity Based Query Sampling for Judgment

We saw that our dissimilarity metrics are indicative of the judges' binarized preference judgments in our auto-completion side-by-side experiments. This suggests that a sampling strategy that prioritizes queries with higher dissimilarity scores could lead to reduced judging costs. In an offline setting, this requires to establish a threshold value of dissimilarity a-priori, defining the cutoff beyond which queries with lower scores do not need to be judged, ideally without hurting the WinLoss estimate for the full dataset. In this section, we run a set of simulation experiments to evaluate the impact of such a strategy.

We take a sample of n queries from the full dataset and calculate WinLoss over this set; this is the score we would get if we judged all n queries. We then subsample the n queries based on our dissimilarity metric scores, removing queries with a dissimilarity value below a given threshold and calculate WinLoss again for the remaining queries (thus we are effectively 'not collecting judgments' for the set of queries with a dissimilarity score below the threshold). We gradually increase the threshold, thus collecting judgments for less and less queries and calculate the resulting error as the difference between the original WinLoss score and the WinLoss score obtained over the reduced query set. We repeat the sampling process 1,000 times, each time, randomly selecting 5,000 queries from the full 58k set and repeating the dissimilarity based sub-sampling process.

Figure 4 plots the mean RMSE values across the 1,000 experiments. We see that the error increases faster for mD, WRD and WRDLog compared to the other metrics, but at the same time, these metrics also reduce the sample size much faster. Another consideration here is that the rank correlation metrics are actually more likely to remove queries that would have been judged as non-tie, but removing similar amounts of 1 and -1 preference labels seemingly reduces

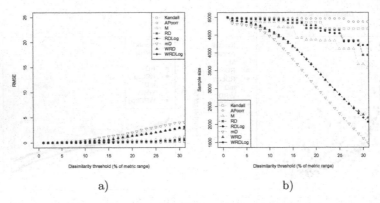

a)   b)

Fig. 4. Mean, median and maximum of the obtained absolute errors (y axis) across 1,000 experiments, when queries with increasing dissimilarity scores (x axis) are not being judged (top row), and the volume of judgments saved (bottom row)

the error (a property of WinLoss). With these limitations in mind, we find that a saving of 25% of the judging costs (same as the proportion of ties in our dataset) leads to a WinLoss error of one point when using the WRDLog dissimilarity score to select which queries to judge.

4   Conclusions

In this paper, we proposed a new metric, WRD, to measure the dissimilarity of two ranked lists of documents in side-by-side experiments that can reflect both document dissimilarities and ranking differences. Through a set of simulation experiments over a large preference judgment dataset, we empirically showed the benefits of WRD, its correlation with the preference judgments and its possible use to prioritize queries to be judged, leading to a significant amount of saving in judging resources that can benefit any large-scale side-by-side evaluation efforts.

So far, however, we only considered an offline evaluation scenario where no judgments are observed. Our future work aims to extend the use of WRD in an online framework to update the priority order of queries with each observed new judgment, combining the offline dissimilarity score and the observed impact on the final evaluation score. We will also explore other datasets, where the ranking order of documents is expected to have a greater impact on the preference judgments, for example, side-by-side experiments on search engine result pages. Extensions of other metrics, such as RBO [16] to incorporate item similarity or alternative WinLoss metrics is left for future research.

References

1. Aslam, J.A., Pavlu, V., Yilmaz, E.: A statistical method for system evaluation using incomplete judgments. In: Proc. of the 29th ACM SIGIR Conference, SIGIR 2006, pp. 541–548. ACM, New York (2006)

2. Bailey, P., Craswell, N., White, R.W., Chen, L., Satyanarayana, A., Tahaghoghi, S.M.: Evaluating search systems using result page context. In: Proc. of the Third Symposium on Information Interaction in Context, IIiX 2010, pp. 105–114. ACM, New York (2010)

3. Bar-Ilan, J., Mat-Hassan, M., Levene, M.: Methods for comparing rankings of search engine results. Comput. Netw. 50(10), 1448–1463 (2006)

4. Carterette, B., Allan, J., Sitaraman, R.: Minimal test collections for retrieval evaluation. In: Proc. of the 29th ACM SIGIR Conference, SIGIR 2006, pp. 268–275. ACM, New York (2006)

5. Chandar, P., Carterette, B.: Using preference judgments for novel document retrieval. In: Proc. of the 35th ACM SIGIR Conference, SIGIR 2012, pp. 861–870. ACM, New York (2012)

6. Fagin, R., Kumar, R., Sivakumar, D.: Comparing top k lists. SIAM J. Discret. Math. 17(1), 134–160 (2004)

7. Guiver, J., Mizzaro, S., Robertson, S.: A few good topics: Experiments in topic set reduction for retrieval evaluation. ACM Trans. Inf. Syst. 27(4), 21:1–21:26 (2009)

8. Hosseini, M., Cox, I.J., Milic-Frayling, N., Vinay, V., Sweeting, T.: Selecting a subset of queries for acquisition of further relevance judgements. In: Amati, G., Crestani, F. (eds.) ICTIR 2011. LNCS, vol. 6931, pp. 113–124. Springer, Heidelberg (2011)

9. Kim, J., Kazai, G., Zitouni, I.: Relevance dimensions in preference-based ir evaluation. In: Proc. of the 36th ACM SIGIR Conference, SIGIR 2013, pp. 913–916. ACM, New York (2013)

10. Radlinski, F., Bennett, P.N., Carterette, B., Joachims, T.: Redundancy, diversity and interdependent document relevance. SIGIR Forum 43(2), 46–52 (2009)

11. Radlinski, F., Craswell, N.: Comparing the sensitivity of information retrieval metrics. In: Crestani, F., Marchand-Maillet, S., Chen, H.-H., Efthimiadis, E.N., Savoy, J. (eds.) SIGIR 2010, pp. 667–674. ACM (2010)

12. Sanderson, M., Paramita, M.L., Clough, P., Kanoulas, E.: Do user preferences and evaluation measures line up? In: Proc. of the 33rd ACM SIGIR Conference, SIGIR 2010, pp. 555–562. ACM, New York (2010)

13. Shieh, G.: A weighted kendall's tau statistic. Statistics and Probability Letters 39, 17–24 (1998)

14. Thomas, P., Hawking, D.: Evaluation by comparing result sets in context. In: Proc. of the 15th ACM International Conference on Information and Knowledge Management, CIKM 2006, pp. 94–101. ACM, New York (2006)

15. Voorhees, E.M., Harman, D.K. (eds.): TREC: Experimentation and Evaluation in Information Retrieval. MIT Press, Cambridge (2005)

16. Webber, W., Moffat, A., Zobel, J.: A similarity measure for indefinite rankings. ACM Trans. Inf. Syst. 20, 1–20 (2010)

17. Yilmaz, E., Aslam, J.A., Robertson, S.: A new rank correlation coefficient for information retrieval. In: Proc. of the 31st ACM SIGIR Conference, SIGIR 2008, pp. 587–594. ACM, New York (2008)

18. Zhu, J., Wang, J., Vinay, V., Cox, I.J.: Topic (query) selection for IR evaluation. In: Proc. of the 32nd ACM SIGIR Conference, SIGIR 2009, pp. 802–803. ACM, New York (2009)

Blending Vertical and Web Results

A Case Study Using Video Intent

Damien Lefortier[1,2], Pavel Serdyukov[1], Fedor Romanenko[1], and Maarten de Rijke[2]

[1] Yandex, Moscow
{damien,pavser,fedor}@yandex-team.ru
[2] ISLA, University of Amsterdam
derijke@uva.nl

Abstract. Modern search engines aggregate results from specialized verticals into the Web search results. We study a setting where vertical and Web results are *blended* into a single result list, a setting that has not been studied before. We focus on video intent and present a detailed observational study of Yandex's two video content sources (i.e., the specialized vertical and a subset of the general web index) thus providing insights into their complementary character. By investigating how to blend results from these sources, we contrast traditional federated search and fusion-based approaches with newly proposed approaches that significantly outperform the baseline methods.

1 Introduction

Modern search engines integrate results from specialized search services, or verticals, into their Web search results. This task is called aggregated search [11] and is usually decomposed into *vertical selection* [1] and *result merging* [3, 13]. Aggregated search aims at diversifying the search engine result page (SERP) in order to provide results for different possible information needs, or intents, in different types of content (e.g., images, videos). Previous work often assumes a so-called block search result presentation, where vertical results are grouped together in a block and placed amongst the generic web results; see Fig. 1 (left). Instead, we assume a search result presentation in which homogeneous vertical and Web results are *blended* into a single result list, as in Fig. 1 (right), *before* being placed amongst other results, a setting that has not been studied before. In this setting, vertical and Web results blended together should be homogeneous because incoherent result presentation can result in user dissatisfaction [5].

We focus on video intent, i.e., the user intent that is satisfied by videos (in the same way that image intent is satisfied by images). Video intent is broad and diverse with queries such as "*games of thrones watch video*" or "*the great gatsby.*" To answer such queries commercial search engines have two content sources: (1) A specialized video vertical that typically only contains videos from video hosting services, such as Youtube; we refer to this vertical as the *Video* vertical. (2) The subset of pages from the general web index with at least one video embedded in them; one can easily identify such pages and we refer to the set of such pages as the *WebVideo* source. These two sources correspond to different products and have different underlying ranking systems.

M. de Rijke et al. (Eds.): ECIR 2014, LNCS 8416, pp. 184–196, 2014.

Is one of these two sources sufficient? To what extent do the Video and Web-Video sources complement each other as sources for answering the video intent in terms of user satisfaction? We investigate these questions through a case study of Yandex's two video content sources. We examine the hypothesis that the two sources satisfy different video information needs and complement each other in this respect, i.e., that both sources are required in order to more effectively satisfy information needs with a video intent. One striking example of such a need (query) is *"games of thrones watch video,"* for which a user might want to watch the latest episode of the series

ranking **A**	ranking **B**
Web_1 ———	Web_1 ———
Web_2^* ———	$Video_1$ ··········
$Video_1$ ··········	Web_2^* ———
$Video_2$ ··········	$Video_2$ ··········
$Video_3$ ··········	Web_4^* ———
Web_3 ———	$Video_3$ ··········
Web_4^* ———	Web_3 ———
Web_5 ———	Web_5 ———

Fig. 1. Traditional block (left) vs. blended (right) search result presentations, where video vertical documents are shown as dotted lines, while pages from the general web index that have at least one video embedded in them are marked with *

(such videos are usually ranked sufficiently high only in Video) or to browse through all previous episodes to watch some of them (such hub pages are usually included only in WebVideo): these information needs are actually sub-intents of the video intent. These information needs are both satisfied by watching a video online, but the videos are typically found in different types of source (Video and Webvideo) so that both types of source are required to address queries with a video intent.

Below, we start with an observational comparison of the two types of video content source that we consider. We find that they complement each other and that both are required to best answer video information needs. For many queries, the two sources return highly relevant, but different results and, in such cases, presenting results from both sources is better for diversity than using a single source. In some cases, one of the two sources does not provide highly relevant results even for queries with a strong, dominant video intent. Therefore, by itself query intent is not enough to properly decide when to present results from a particular source, and the relevance of the results w.r.t. sub-intents of the underlying video intent must be taken into account. I.e., we also need to consider the ability of a particular source to answer some classes of queries.

These observations motivate us to address the following algorithmic problem: which results from the two sources should be presented on the SERP? Specifically, we consider the following *blending problem*: given the result lists from the two sources, produce the best possible result list in terms of overall relevance, where documents from each source can be interleaved (blended) together, subject to the constraint that no change is allowed in the relative order of results from the same source (see below).

The main challenge with our blending problem lies in the fact that document scores are not comparable across the two sources, because they have completely different underlying ranking systems. Our blending problem is similar to an aggregated or federated search problem, but it differs in the following essential ways. First, we blend documents and not blocks, as has been done so far in aggregated search [3, 13]. Second, compared to a federated search setting we exploit additional knowledge about each source (e.g., document scores returned by each source plus relevance assessments for training our models). Third, we do not want to change the relative order of results from the same source; we assume that the base rankers underlying the sources are the best experts about their own documents, so we do not overrule them [3, 16]; we refer to this constraint as the *no re-ordering constraint*. According to our experiments, our methods do not improve by relaxing this constraint; doing so only slows down the computation.

We consider a number of methods from federated search and data fusion. Their performance is found to be far below what could be achieved given the quality of the sources being blended. We therefore consider two types of alternative method. One predicts the quality of the top results from each source for a given query in order to blend results, but without directly using the document scores. The second is based on learning to rank (LTR); we propose three approaches: point-wise, pair-wise and list-wise. Our most effective blending approach belongs to the first type; it yields a significant +13.20% improvement in ERR@5 scores compared to using the best vertical only.

The main contributions of this paper are the following. (1) We present a detailed observational study of two video content sources thus providing insights into their complementary character. (2) We propose and evaluate different methods for blending results from these two sources into a single result list.

2   Related Work

There are several tasks where documents are retrieved from multiple systems and merged into a single SERP. We review them and explain how our blending problem differs.

Aggregated Search. Prior work on aggregated search and result merging [3, 13] assumes a block search result presentation, so that, in our case, results from the Video vertical would be grouped together in a block and placed amongst the generic web results; see Fig. 1 (left). We, instead, assume a search result presentation in which vertical and Web results are blended into a single result list (right), a setting that has not been studied before. Aggregated search methods are not usable to solve our problem, because we blend documents and not blocks, as is customary in aggregated search [3, 13].

Federated Search. In federated search [16], of which aggregated search is a possible application, documents can come from several search engines. Popular result merging methods, such as CORI [4], use a combination of local collection scores and document scores to compute a global score for each document that is then used to rank results globally. We use variations of CORI as baselines (see §3.1), but not the original CORI method because it relies on term usage statistics from each collection. Such statistics are meaningful only when collections have documents of the same nature, which is not the case here as WebVideo indexes web pages, while Video indexes video metadata [18], which implies the use of a specialized ranking system designed for the video retrieval task. Other popular methods (e.g., SAFE [17]) rely on a centralized index and are not applicable to our task because such an index is unavailable to us.

Data Fusion. In many data fusion methods, results in the lists being fused come from a *single* collection, but are retrieved using different strategies. These methods, either unsupervised (e.g., CombSUM [6]) or supervised (e.g., λ-Merge [15], where result lists are retrieved using the same ranker unlike in our setting), are based on a voting principle and therefore some intersection between results is assumed. In our case, the intersection between the two video content sources is really small (see §6.1). Hence, fusion methods are not readily applicable to our blending problem. Still, we use some merging algorithms from data fusion as some of our baselines (see §3.1).

Learning to Rank. Learning to rank (LTR) approaches are not straightforward to use here because of the different feature spaces (one per source), which must be merged somehow. This was extensively discussed in [3], where different ways to construct feature vectors were proposed and compared in the context of block-ranking (ordering blocks of Web and vertical results in response to a query). In [3], the variant performing the best is one that makes a copy of each feature for each vertical, thus allowing the LTR model to learn a vertical-specific relationship between features and relevance. In this paper, we follow Arguello et al. [3] and also use *one copy* of each source feature for every source, which yields better performance according to our experiments as well. Then, to solve our problem of blending results, we propose a pair-wise approach inspired by [3, 15], but adapted to our problem as well as a point-wise approach similar in aim to the method described in [19], where it was outperformed by *Round-Robin* [21], and a list-wise approach that is the most fitted to our problem and allows us to use features from [12], where query performance was predicted.

In sum, we contribute a new problem (our blending problem) for which existing federated search and fusion methods are ill-fitted, plus algorithms to address the problem.

3 Blending Methods

In this section, we describe different methods to address our blending problem. Recall that we do not change the relative order of documents that come from the same source. For example: W_1, W_2, V_1, V_2 is an allowed blending, while V_1, W_3, V_2, W_2 is disallowed as W_2 and W_3 were re-ordered, where V (W) means Video (WebVideo) and S_i is the ith result from S. Given this constraint, there are, for a blended list of N results, 2^N possible blendings of the two sources.

3.1 Baselines

We use the following baselines: *Round-Robin* [21], *Raw Score Merging* [10], and several variations of CORI merging [4]. The CORI result merging method computes a *global score* for each document from each source using a linear combination of the collection score C and the original document score D returned by each source:

$$\text{global score}(D) = \frac{D' + 0.4 \cdot D' \cdot C'}{1.4}$$

where C' (respectively D') is C (respectively D) MinMax-normalized to $[0, 1]$. The global score of each document is then used to rank results globally. We cannot use the CORI collection selection algorithm [4] to compute C (see §2). Instead we propose two variations, where D is the original document score and C is defined as follows:

CORI-Size. We define C to be the number of documents retrieved from each source (this was already proposed in [14]).

CORI-ML. We define C to be the predicted quality of each source as defined in §3.2, which makes this baseline our contribution as well.

3.2 Approaches Based on Source Quality

These methods consist of two steps: (1) we predict the quality of the top results from each source for a given query in order to (2) blend results from each source using this predicted quality (as in *CORI-ML*). Such methods are common in federated search (see §2) and the main problem lies in the uncertainty of how this predicted quality best translates into a weight for estimating the global score of each document. We therefore propose methods that do not use the original document scores returned by each source directly to overcome such issues (as opposed to baselines described in §3.1). Note that our methods are not suitable for a federated search setting as they are not readily applicable to more than two sources. Let $q_S(\overline{v}_q, \theta)$ be the *source quality function* of a source S with parameters θ, and where $\overline{v}_q$ is the vector of source features (see §4.1) for a query q. Using q_{Video} and q_{WebVideo}, we define the following blending methods:

Source-Binary. For each query, we show results from one source only, i.e., from Video if $q_{\text{Video}} > q_{\text{WebVideo}}$, and otherwise from WebVideo.

Source-KMeans. Let $x = q_{\text{Video}} - q_{\text{WebVideo}}$ be the predicted difference in quality of the two sources for a given query. For each possible difference x, the best uniform blending (i.e., the same blending for all queries) is found out of all possible blendings satisfying the *no re-ordering constraint*. E.g., for $x = 0.01$, it is V_1, W_1, V_2, W_2, while for $x = 0.6$, it is V_1, V_2, V_3, V_4. To find this blending, for a difference x, we take the k queries in the training data whose predicted difference in quality of the two sources is closest to x. For each query, we compute the score of each possible blending according to the retrieval metric used to train q_S. We then take the uniform blending with the highest average performance among these k queries. Here, k is a parameter that is determined experimentally on the training data.

We also investigated methods based on clustering the predicted values of quality (e.g., 0–0.20, 0.20–0.45, 0.45–1.0) and on defining a uniform blending depending on the class of each source. However, such methods did not yield better results than the *Source-KMeans* method and are omitted for brevity. We train q_S to predict ERR@N (where N is defined for the retrieval metric being used) by minimizing the mean squared error of the difference between the predicted and actual value of ERR for labeled result lists in the training data using Gradient Boosted Regression Trees (GBRT) [7]. Using ERR here yields better results than NDCG, both for ERR and NDCG.

3.3 Learning to Rank Approaches

Point-Wise. The first of the three LTR approaches that we describe is a *point-wise* approach. In other words, we predict the *global score* of each document from each source that is then used to rank results globally using one independent model per source. Using one unified model does not improve the performance according to preliminary experiments on the training data, and is less scalable as the number of sources increases. Let $f_S(\overline{v}_q, \overline{w}_{qd}, \theta)$ be the *scoring function* of documents from a source S with parameters

Algorithm 1. Point-wise LTR blending algorithm

input : Result list from S, f_S, θ_S for $S \in \{$Video, WebVideo$\}$, $\overline{v}_q$, $\overline{w}_{qd}$, N
output: Blended result list

BlendedResultList $\longleftarrow$ $\{\}$; MaxScore $\longleftarrow$ 0
for *tuple* $\in$ *NAryCartesianProduct($\{0,1\}$)* **do**
    Blending $\longleftarrow$ $\{\}$ i $\longleftarrow$ 0 j $\longleftarrow$ 0
    for $x \in$ *tuple* **do**
        if $x == 0$ **then**
            Blending[i + j] = ResultList$_{\text{WebVideo}}$[i] `//` i^{th} `result from WebVideo`
            i $\longleftarrow$ i + 1
        else
            Blending[i + j] = ResultList$_{\text{Video}}$[j]   `//` j^{th} `result from Video`
            j $\longleftarrow$ j + 1
    Score $\longleftarrow$ Metric@N(Blending, f_{Video}, θ_{Video}, f_{WebVideo}, θ_{WebVideo}, $\overline{v}_q$, $\overline{w}_{qd}$)
    if *Score* > *MaxScore* **then**
        BlendedResultList $\longleftarrow$ Blending
        MaxScore $\longleftarrow$ Score
return BlendedResultList

θ, where $\overline{v}_q$ and $\overline{w}_{qd}$ are, respectively, the vector of source features (see §4.2) and the vector of document features (see §4.3) for a query q and document d. This function predicts the global score of each document from source S. Using f_{Video} and f_{WebVideo}, we can easily define a blending algorithm, where the blended result list is the one with the highest value of the retrieval metric being used according to predicted scores; see Algorithm 1. Due to the *no re-ordering constraint*, we cannot simply rank documents by their predicted score. We train f_S to predict the score of each document obtained through judgements (see §5.2) by minimizing the mean squared error using GBRT [7].

Pair-Wise. For this method, we use a *pair-wise* approach to LTR. In other words, we predict the *preference* between any pair of documents from Video *and* WebVideo in a unified model. Let $f(\overline{v}_q, \overline{w}_{qd}, \theta)$ be a *document preference function* with parameters θ, where $\overline{v}_q$ and $\overline{w}_{qd}$ are, respectively, the vector of source features (see §4.2) and the vector of document features (see §4.3) for query q and document d. This function predicts the *preference* between any pair of documents d_x and d_y coming from either Video or WebVideo. Here, $f(\overline{v}_q, \overline{w}_{qd_x}, \theta) > f(\overline{v}_q, \overline{w}_{qd_y}, \theta)$ means that d_x should be preferred over d_y, i.e., is predicted to be more relevant. Using f, we can easily define a blending algorithm, where the blended result list is obtained by ranking documents according to the predicted preferences and by satisfying the *no re-ordering constraint*; see Algorithm 2. We train f by directly optimizing for the retrieval metric being used to assess the blended result list (e.g., ERR or NDCG); for this purpose, we use YetiRank [8]. We use one copy of each source feature for each source, which yields better performance than using one single feature according to our experiments (as in [3]).

List-Wise. In this method, we use a *list-wise* approach to LTR. In other words, we predict the *preference* between any pair of blendings satisfying the *no re-ordering*

Algorithm 2. Pair-wise LTR blending algorithm

---

input : Result list from S for $S \in \{$Video, WebVideo$\}$, f, θ ,$\bar{v}_q$, $\bar{w}_{qd}$, N
output: Blended result list
BlendedResultList $\longleftarrow \{\}$ i $\longleftarrow 0$ j $\longleftarrow 0$
for _ $\longleftarrow$ 0 *to* $N - 1$ **do**
$\quad$ $d_w \longleftarrow$ ResultList$_{\text{WebVideo}}$[i] $\quad$// i^{th} result from WebVideo
$\quad$ $d_v \longleftarrow$ ResultList$_{\text{Video}}$[j] $\quad$// j^{th} result from Video
$\quad$ **if** $f(\bar{v}_q, \bar{w}_{qd_w}, \theta) > f(\bar{v}_q, \bar{w}_{qd_v}, \theta)$ **then**
$\quad\quad\mid$ BlendedResultList[i + j] $= d_w$
$\quad\quad\mid$ i $\longleftarrow$ i + 1
$\quad$ **else**
$\quad\quad\mid$ BlendedResultList[i + j] $= d_v$
$\quad\quad\mid$ j $\longleftarrow$ j + 1

return BlendedResultList

---

constraint. This approach is the best fit for our problem and is applicable due to this constraint, which drastically limits the number of possible blendings. Let $f(\bar{v}_q, \bar{u}_b, \theta)$ be a *blending preference function* with parameters θ, where $\bar{v}_q$ and $\bar{u}_b$ are, respectively, the vector of source features (see §4.2) and the vector of blended result list features (see §4.4) for a query q and blending b. This function predicts the *preference* between any pair of blendings b_x and b_y from all possible blendings satisfying the constraint mentioned at beginning of §3. Here, $f(\bar{v}_q, \bar{u}_{b_x}, \theta) > f(\bar{v}_q, \bar{u}_{b_y}, \theta)$ means that b_x should be preferred over b_y, i.e., is predicted to be more relevant. Using f, we can easily define a blending algorithm, where the blended result list is obtained by using the most preferred blending as predicted by f. This algorithm is similar to Algorithm 1, but, in this case, the score of each blending is computed by f directly using blended result list features and without document features. We train f using GBRT [7], where the target score of each blending in the training data is computed using the retrieval metric used.

4 Features

We present the features used in the blending methods proposed in §3.

4.1 Query-Source Features (Source Quality-Based)

These are the features for a given query-source pair used by a Yandex system to decide which vertical results to present and how to integrate them into the Web results. More than one hundred features are currently used by this system, including, in particular, click-through, vertical-intent and hit count features (similar features were used in [1, 2]). For space reasons, we do not describe them here and instead, in §6.2, we discuss only the strongest features. We also use the following features, where the ones from [15] are marked with *: ListMax$_S$, ListMin$_S$, ListMean$_S$*, ListStd$_S$*, ListSkew$_S$*, which are respectively, the maximum, minimum, mean, standard deviation and skew of original documents scores the result list taken from each source S.

4.2 Query-Source Features (LTR)

In our LTR approaches, we use all the features described in 4.1, as well as the following features: PredictedERR@N_S: the predicted value of ERR@N for each source S as computed by q_S (§3.2), where $N \in \{1, 3, 5, 10\}$.

4.3 Query-Source-Document Features

These are the features for a given query-source-document triplet (q, S, d), with features from [15] marked with *: (1) Score*: the original document score returned by S. (2) Rank*: the position of d in the results from S. (3) NormScore$^*_{[0,1]}$: the score of d returned by S after the top 10 results were shifted and scaled in the interval $[0, 1]$ using MinMax normalization. (4) IsTopN*: a binary feature to indicate if d is within the top N results, where $N \in \{1, 3, 5, 10\}$. (5) IsVideo: a binary feature to indicate if S is Video.

4.4 Blended Result List Features

These are the features for a given *blended* result list b: (1) List*: the same features as defined in §4.1, but over b. (2) NumFrom{Video, WebVideo}: the number of results from each source in b. (3) CosineSimilarity(S, b): the cosine similarity [12] between each source S and b. We do not use the KL similarity from [12], because it is not well defined in our case. (4) CombinedPredictedERR@N: the combined predicted value of ERR for the top N results ($\overline{\text{ERR@}N_S}$) from each source S from [12], where $N \in \{1, 3, 5, 10\}$, as: $\text{sim}_{cosine}(\text{Video}, b) \cdot \overline{\text{ERR@}N_{\text{Video}}} + \text{sim}_{cosine}(\text{WebVideo}, b) \cdot \overline{\text{ERR@}N_{\text{WebVideo}}}$.

5 Experimental Setup

5.1 Research Questions

In §6.1, we investigate the following observational question: (RQ1) To what extent do the Video and WebVideo complement each other as sources for answering the video intent in terms of user satisfaction? When investigating which results from the two sources should be presented on the SERP in §6.2, we investigate the following research questions: (RQ2) How effective are source quality-based approaches when compared against common federated search methods to solve this problem? (RQ3) How effective are LTR approaches when compared against common federated search methods to solve this problem?

5.2 Data Set

We randomly sampled 1,000 queries from the live stream of search queries to Yandex that had Video results aggregated into the SERP. For each query, the top 10 results of the Video and WebVideo sources were independently assessed for relevance by one judge using a 5-grade system (*Perfect, Excellent, Good, Fair, Bad*) and the same assessor instructions (video intent) for both types of content.

5.3   Oracle Baselines and Comparisons

We use the following *oracle* blending methods, which utilize relevance judgments of each document and satisfy the *no re-ordering constraint*: (1) *Best uniform blending:* We use the best uniform blending for all queries. (2) *Best blending:* We show the best blending for each query. This method is the upper-bound of any blending method. (3) *Best source:* We show results only from the best source for each query. Here, the *best* (blending, source) refers to the one giving the best results in terms of the retrieval metric being used. We compare all methods mentioned so far, plus those listed in §3, and the Video vertical (i.e., only return results from this vertical) against the WebVideo source (the best source on average). Comparisons are based on average performance across all queries using 10-fold cross-validation.

5.4   Metrics and Significance Testing

We use the well-known ERR and NDCG measures as metrics. Statistical significance is tested using a one-tailed paired t-test (at the $p < 0.05$ level). Regarding the k parameter of the *Source-KMeans* blending method, we tuned it manually using the training data and use $k = 10$ in all our experiments.

6   Results

6.1   Observational

We start with our observational research question (RQ1). Let us first give an overview of the results returned by each source. Results from Video include relevant documents (at least *Fair*) from 481 hosts with 71% of the documents coming from the top 5 hosts, while results from WebVideo include relevant documents (at least *Fair*) from 1,850 hosts with only 536 of them from the top host. The intersection between the top 10 results of each source is really small: 68% of queries have no intersection, 17% have one common document, 9% have two common documents, and 6% have three or four common documents (never more). In other words, the two sources seem quite different, which is unsurprising considering the specialized nature of Video in comparison with WebVideo.

Let us now compare the performance of each source (Video and WebVideo) with the oracle blending methods from §5.3. The results are shown in Table 1. Video gives

Table 1. Relative effectiveness of the Video vertical and oracle blending methods when compared against WebVideo. A ▲ (▼) denotes significantly better (worse) performance compared to *(Oracle) Best source*. Bold face indicates best performance per metric.

	ERR@5	ERR@10	NDCG@5	NDCG@10
Video	−5.16%▼	−0.57%▼	−2.68%▼	−1.34%▼
WebVideo	−	−	−	−
(Oracle) Best uniform blending	+4.05%▼	+9.63%▼	+3.81%▼	+7.82%▼
(Oracle) Best blending	**+26.06%**	**+26.65%**	**+33.27%**	**+37.29%**
(Oracle) Best source	+22.25%▼	+22.58%▼	+28.61%▼	+30.77%▼

−5.16% in ERR@5 and −2.68% in NDCG@5 compared to WebVideo, which means that the specialized video vertical is worse, in general, at answering the video intent. When blending results from both sources, the oracle method, as the upper bound, gives +26.06% in ERR@5 and +33.27% in NDCG@5 (RQ1). This large improvement shows that, indeed, the two sources complement each other and that both are required in order to best answer the video intent. The best uniform blending mix (V_1, W_1, V_2, V_3, W_2 for ERR@5) gives only +4.05% in ERR@5 and +3.81% in NDCG@5, so a blending algorithm must be query dependent to get closer to the upper bound. Using the best source for each query gives as much as +22.25% in ERR@5 and +28.61% in NDCG@5.

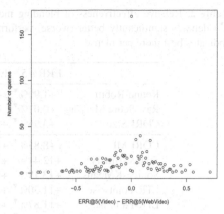

Fig. 2. Number of queries vs. difference in ERR@5 between Video and WebVideo

For a significant number of queries, the difference between the relevance of the top results from each source is quite small, as shown in Fig. 2 (notice 170 queries at 0). Moreover, as results from the sources are different, presenting results from both sources will lead to an increase in *diversity*. In such cases, e.g., for "*games of thrones watch video*," each source contains results for different sub-intents of the main video intent and a single source would therefore likely lead to user dissatisfaction. Hence, presenting results from both sources while keeping the quality of the blending at the same level is better when each has highly relevant results (RQ1); below, we examine how diverse the blended result lists are for the blending methods from §3.

We also find that, in some cases, one of the two sources does not provide highly relevant results even for queries with a strong dominant video intent as shown in Fig. 2 and Table 3. E.g., for "*watch anime video online*," where Video does not return highly relevant results as opposed to WebVideo due to the query being really ambiguous. Therefore, by itself the intent of the query is not enough to properly decide when to present results from a particular video source, and the relevance of the results w.r.t. sub-intents of the video intent underlying the query must be taken into account.

6.2 Algorithmic

Next, we turn to our algorithmic research questions (RQ2, RQ3). The results obtained are shown in Table 2. Of the federated search and fusion-based approaches from §3.1, *CORI-ML* performs best and gives +8.85% in ERR@5 and +10.76% in NDCG@5. Next, our source quality-based approaches from §3.2 perform well as *Source-KMeans* (the best one) gives a +13.19% in ERR@5 and +15.59% in NDCG@5, thus outperforming *CORI-ML* by a large margin (RQ2). The fact that *Source-KMeans* is not significantly better than *Source-Binary* @5 can be explained by the fact that even queries with small predicted differences in quality of the two sources can have really different best blendings, especially close to 0. Our LTR approaches perform well, but do not outperform

194   D. Lefortier et al.

Table 2. Relative effectiveness of blending methods when compared against WebVideo. A ▲ (▼) denotes significantly better (worse) performance compared to *Source-KMeans*. Bold face indicates best score per metric.

	ERR@5	ERR@10	NDCG@5	NDCG@10
Round-Robin	+1.90%▼	+7.20%▼	+2.43%▼	+4.44%▼
Raw Score Merging	+0.02%▼	+0.27%▼	+0.21%▼	+1.75%▼
CORI-Size	+3.96%▼	+10.71%▼	+4.69%▼	+9.71%▼
CORI-ML	+8.85%▼	+13.73%▼	+10.76%▼	+15.20%▼
Source-Binary	+12.44%	+13.49%▼	**+15.73%**	+17.01%▼
Source-KMeans	**+13.19%**	**+14.88%**	+15.59%	**+18.07%**
LTR-Pointwise	+11.30%▼	+14.01%▼	+14.11%▼	+17.14%
LTR-Pairwise	+11.87%▼	+14.82%▼	+13.80%▼	+17.38%
LTR-Listwise	+11.88%▼	–	+15.20%	–

Source-KMeans. Our best LTR approaches are *LTR-Listwise* and *LTR-Pairwise*, which yield +11.88% and +11.87% in ERR@5 and +15.20% and +14.82% in NDCG@5, respectively (RQ3). Results at @10 are consistent with these findings, although *LTR-Listwise* was not used @10 due to the combinatorial explosion. Even though there is no significant difference between the three LTR methods, *LTR-Pointwise* scales much better as the number of sources increases due to the fact that two independent models are used, and its training process is much faster.

Next, we examine content-type diversity of the top 5 results pro-

Table 3. Content-type diversity of the top 5 results produced by each of the blending methods (using ERR@5)

	The number of results @5 from video hosting sites					
	0	1	2	3	4	5
Video	0%	0%	0%	0%	0%	100%
WebVideo	44%	21%	23%	7%	4%	1%
(Oracle) Best blending	11%	11%	10%	10%	12%	46%
Round-Robin	0%	0%	26%	47%	21%	6%
Raw Score Merging	44%	21%	23%	8%	3%	1%
CORI-Size	3%	19%	24%	26%	20%	8%
CORI-ML	14%	20%	21%	19%	17%	9%
Source-Binary	29%	12%	10%	3%	2%	44%
Source-KMeans	23%	12%	12%	7%	13%	33%
LTR-Pointwise	28%	12%	15%	10%	10%	25%
LTR-Pairwise	26%	13%	14%	10%	8%	29%
LTR-Listwise	19%	14%	11%	11%	15%	30%

duced by each of the blending methods, i.e., the percentage of result lists with 0 to 5 results from video hosting sites (using ERR@5). The results are shown in Table 3. The content-type diversity of all our source quality-based and LTR approaches is similar. Compared with the oracle method as the upper bound, we observe that our methods have a similar top 5 content-type diversity, but return slightly more result lists with 0 results from video hosting sites and fewer with 5 results. Investigating ways to directly take diversity into account in our blending methods is future work.

Finally, we turn to the contribution of individual features. The top 10 features according to their weighted contribution to q_{Video} from §3.2 (one of the two models used in *Source-KMeans*) using ERR@5 are shown in Table 4; see [9, §10.13] for the description of those weights. We observe that List* features perform really well, which is consistent with previous studies [20]. Other well-performing features are: the number of documents returned

Table 4. Top 10 features according to their contribution to q_{Video} from §3.2 (ERR@5)

Rank	Feature	Score	Rank	Feature	Score
1	ListMax	6.70	6	ListSkew	3.56
2	ListMean	6.00	7	WebDCG	3.04
3	ListMin	4.54	8	IsFilm	2.55
4	ListStd	4.41	9	VideoSites	2.51
5	NumDocs(Video)	4.40	10	VideoTop3MaxRel	2.32

by the Video vertical, WebDCG (i.e., DCG of original document scores from Web-Video), IsFilm (i.e., whether the query is about a film), VideoSites (i.e., whether the sites in the top results are known to be relevant for video queries) and VideoTop3MaxRel (i.e., the maximum of the top 3 original document scores from Video). Results for $q_{WebVideo}$ (the second model used in *Source-KMeans* and *Source-Binary*) are consistent with these findings.

7 Conclusion

We have introduced and studied a setting where vertical and Web results are *blended* into a single result list. We have focused on video intent and presented a detailed observational study of a Yandex's two video content sources. The two sources that we consider complement each other and both are required in order to best answer video information needs. An oracle-based upper bound gives +26.06% improvement in ERR@5 and +33.27% in NDCG@5 compared to the case where only WebVideo (the best performing source) is used. For a large number of queries, presenting results from both sources is better for diversity than using a single source.

These observations motivated us to investigate the following algorithmic problem: which results from the two sources should be presented on the SERP? We proposed source quality-based methods and learning to rank methods to address this problem. Our source quality-based approaches perform well (+13.20% in ERR@5 for the best one) and outperform the best baseline approach, a variation of CORI, by a large margin (+8.85% in ERR@5). Our LTR approaches perform well (+11.88% in ERR@5 for the best one), outperforming the best baseline approach but not the source quality-based approaches. Compared with the oracle method as our upper bound, we observe that our methods have a similar top 5 content-type diversity.

Acknowledgments. This research was partially supported by the European Community's Seventh Framework Programme (FP7/2007-2013) under grant agreement nr 288024 (LiMoSINe project), the Netherlands Organisation for Scientific Research (NWO) under project nrs 640.004.802, 727.011.005, 612.001.116, HOR-11-10, the Center for Creation, Content and Technology (CCCT), the QuaMerdes project funded by the CLARIN-nl program, the TROVe project funded by the CLARIAH program, the Dutch national program COMMIT, the ESF Research Network Program ELIAS, the Elite Network Shifts project funded by the Royal Dutch Academy of Sciences (KNAW), the Netherlands eScience Center under project number 027.012.105 and the Yahoo! Faculty Research and Engagement Program.

References

[1] Arguello, J., Callan, J., Diaz, F.: Classification-based resource selection. In: CIKM 2009, pp. 1277–1286. ACM (2009a)

[2] Arguello, J., Diaz, F., Callan, J., Crespo, J.-F.: Sources of evidence for vertical selection. In: SIGIR 2009, pp. 315–322. ACM (2009b)

[3] Arguello, J., Diaz, F., Callan, J.: Learning to aggregate vertical results into web search results. In: CIKM 2011, pp. 201–210. ACM (2011)

[4] Callan, J.P., Lu, Z., Croft, W.B.: Searching distributed collections with inference networks. In: SIGIR 1995, pp. 21–28. ACM (1995)

[5] Dumais, S., Cutrell, E., Chen, H.: Optimizing search by showing results in context. In: Proceedings of the SIGCHI Conference on Human Factors in Computing Systems, pp. 277–284. ACM (2001)

[6] Fox, E.A., Shaw, J.A.: Combination of multiple searches. In: TREC 1993, pp. 243–243. NIST (1994)

[7] Friedman, J.H.: Greedy function approximation: a gradient boosting machine. In: Annals of Statistics, pp. 1189–1232 (2001)

[8] Gulin, A., Kuralenok, I., Pavlov, D.: Winning the transfer learning track of Yahoo!'s learning to rank challenge with YetiRank. J. Machine Learning Research 14, 63–76 (2011)

[9] Hastie, T., Tibshirani, R., Friedman, J.J.H.: The elements of statistical learning, vol. 1. Springer, New York (2001)

[10] Kwok, K., Grunfeld, L., Lewis, D.: TREC-3 ad-hoc, routing retrieval and thresholding experiments using PIRCS. In: TREC 1993, pp. 247–247. NIST (1995)

[11] Lalmas, M.: Aggregated search. In: Advanced Topics in Information Retrieval, pp. 109–123. Springer (2011)

[12] Markovits, G., Shtok, A., Kurland, O., Carmel, D.: Predicting query performance for fusion-based retrieval. In: CIKM 2012, pp. 813–822. ACM (2012)

[13] Ponnuswami, A., Pattabiraman, K., Wu, Q., Gilad-Bachrach, R., Kanungo, T.: On composition of a federated web search result page. In: WSDM 2011, pp. 715–724. ACM (2011)

[14] Rasolofo, Y., Abbaci, F., Savoy, J.: Approaches to collection selection and results merging for distributed information retrieval. In: CIKM 2001, pp. 191–198. ACM (2001)

[15] Sheldon, D., Shokouhi, M., Szummer, M., Craswell, N.: LambdaMerge: merging the results of query reformulations. In: WSDM 2011, pp. 795–804. ACM (2011)

[16] Shokouhi, M., Si, L.: Federated search. Foundations and Trends in Information Retrieval 5(1), 1–102 (2011)

[17] Shokouhi, M., Zobel, J.: Robust result merging using sample-based score estimates. ACM Transactions on Information Systems (TOIS) 27(3), 14 (2009)

[18] Snoek, C.G., Worring, M.: Concept-based video retrieval. Foundations and Trends in Information Retrieval 2(4), 215–322 (2008)

[19] Tjin-Kam-Jet, K.-T.T., Hiemstra, D.: Learning to merge search results for efficient distributed information retrieval. In: DIR 2010 (2010)

[20] Vinay, V., Cox, I.J., Milic-Frayling, N., Wood, K.: On ranking the effectiveness of searches. In: SIGIR 2006, pp. 398–404. ACM (2006)

[21] Voorhees, E.M., Gupta, N.K., Johnson-Laird, B.: Learning collection fusion strategies. In: SIGIR 1995, pp. 172–179. ACM (1995)

Personalizing Aggregated Search

Stanislav Makeev, Andrey Plakhov, and Pavel Serdyukov

Yandex, Moscow, Russia
{stasd07,finder,pavser}@yandex-team.ru

Abstract. Aggregated search nowadays has become a widespread technique of generating Search Engine Result Page (SERP). The main task of aggregated search is incorporating results from a number of specialized search collections (often referenced as verticals) into a ranked list of Web-search results. To proceed with the blending algorithm one can use a variety of different sources of information starting from some textual features up to query-log data. In this paper we study the usefulness of personalized features for improving the quality of aggregated search ranking. The study is carried out by training a number of machine-learned blending algorithms, which differ by the sets of used features. Thus we not only measure the value of personalized approach in the aggregated search context, but also find out which classes of personalized features outperform the others.

1   Introduction

Modern search engines often provide access to specialized search services or verticals which allow a user to obtain results of a certain media type (e.g., video, images) or those dedicated to a specific domain (e.g., news, weather). However, the results of these verticals may be integrated into the core document ranking for some queries. This method has been widely used in recent years by leading commercial search engines and is usually referred to as aggregated (or federated) search.

Among all the benefits of using this approach, we would like to highlight two items. The first one is providing the user with the opportunity to get relevant results of a certain type directly on the search engine result page (SERP). The second one is composing a result page covering a large variety of user intents, which is important while dealing with the queries, which do not explicitly specify the user's information need. Both advantages can be observed in Figure 1, which shows top of the search results (including Images and Videos verticals results) produced by the Yandex search engine in response to the query [metallica].

However, to the best of our knowledge, none of the previous works has been devoted to estimating the personal relevance of verticals results for specific users. In this study we propose to retrieve this kind of information from the user search history and use it for blending results from different verticals. The main contributions of our paper are the following:

M. de Rijke et al. (Eds.): ECIR 2014, LNCS 8416, pp. 197–209, 2014.
© Springer International Publishing Switzerland 2014

Fig. 1. SERP with Images and Video verticals results

- A novel framework for personalized aggregated search;
- Machine-learned personalized verticals ranking function which significantly improves baseline verticals ranking;
- Three classes of personalized features which are supposed to improve verticals ranking and comprehensive analysis of their influence;
- Evaluation of the impact of the personalized approach for the verticals ranking quality for different classes of queries and users.

The rest of the paper has the following structure. In Section 2 we discuss previous studies related to our research. Section 3 describes our evaluation framework for personalized aggregated search. In Section 3.1 we describe the features we used to construct the baseline ranker and in Section 3.2 we propose three classes of personalized features aimed to improve the verticals ranking algorithm. All the evaluated re-ranking functions are presented in Section 3.3. We describe our experimental methodology and evaluation datasets in Section 3.4. The results of these experiments with some discussion are reported in Section 4. Conclusions and inspirations for the further work can be found in Section 5.

2 Related Work

In this section we would like to discuss prior works devoted to both aggregated search techniques and to personalizing ranking systems. All of these works were focused separately on either web search personalization or aggregated search, but none of the previous studies investigated the value of personalization in the context of aggregated search. Some common approaches to the verticals results ranking problem are reported in Section 2.1 and Section 2.2 gives a brief review of the papers devoted to the search results personalization.

2.1 Vertical Selection

One of the most important problems regarding aggregated search is the problem of finding verticals relevant to the user request and placing their results properly on the SERP. Traditionally this problem is addressed by training a machine-learned model based on the features which are supposed to detect the vertical relevance to the query (for instance, [1] or [2]). In this section we would like to observe the most important groups of such features proposed to the date. This classification is rather arbitrary for some features, because they actually can be assigned to more than one group.

Query data.[1, 2, 11, 12]. Features of this type provide some information about vertical relevance using only the text of the query. One of the most common approaches to this problem is to measure the query likelihood given the vertical language models. These models are usually built on the vertical documents or the queries, for which the vertical was the final destination according to user clicks. This type of features also includes some simple features based on the query text such as query length, regular expressions-based features etc.

Vertical data. [1, 2, 11, 12]. Usually the vertical provides scores estimating its relevance for the query based on the properties of its indexed documents collection. E.g., for text collections there might be features like text relevance or text similarity between the query and top retrieved documents. For some types of verticals, utilizing features like the hit count might also be more or less useful.

Click-through data. These features are built upon the history of the user search behaviour including clicks, skips, etc. There is a wide range of ways to make use of such data in general, for instance, to use it for calibrating existing classifiers [7, 8]. Ponnuswami *et al.* [11] use click-through rates as features for training a model on pairwise judgements also obtained from clicks and skips.

Moreover, we can single out the *Web data* (see [11]) features as a separate class. These are features obtained from the organic web search results like text-relevance, click-through features of the web documents, etc., which can be useful for determining the correct placements of vertical results in respect to web search results. But, to the best of our knowledge, none of the previous studies utilized any information about specific users to build a verticals ranking system.

2.2 Personalization

Search results personalization always implies extending search context with the user-specific information. For instance, Bennet *et al.* [4] used geographical data to enrich the search context. Collins-Thompson *et al.* [6] personalized results ranking by utilizing information about the user reading level. Kharitonov *et al.*

[10] proposed to take advantage of the information about the user's gender to better disambiguate the intents corresponding to the query. The work which seems to have the most importance to the presented one is [13], in which the authors propose personalized versions of the web search diversification algorithms. However, the results of this paper concern only web documents ranking and do not touch the problem of ranking heterogeneous results from different verticals.

3   Our Approach

The personalization of verticals ranking was performed following the same methodology as in [4], [6] or [10]. The original ranking algorithm presented 10 web results and a number of vertical results injected in between. Note that a vertical result could sometimes be presented as a block of top-ranked documents of a certain type as in Figure 1. To avoid this ambiguity hereinafter by "vertical result" we mean a block of no less than one vertical document.

The original result pages satisfied the following constraints. First, there could only be no more than one vertical result inserted for each vertical. Second, the vertical results could only be embedded into four slots — above the first web result, between the third and the fourth results, between the sixth and the seventh results and after the tenth web result. Similar constraints were described in a number of previous works, for instance, in [1]. Figure 1 shows the top of the SERP satisfying these constraints. In our experiments we tried to address a more general problem allowing our re-ranking functions to blend the presented web and vertical results in any order violating the above-described constraints, if necessary. We considered only the queries for which at least one vertical result was presented (see Section 3.4 for details of experimental settings), so we actually re-ranked from 11 up to 14 heterogeneous results. In our experiments we worked with the following set of verticals: Images, Video, Music, News, Dictionaries, Events and Weather, but our approach could be easily applied to any other set of verticals.

To proceed with the re-ranking functions we composed the multidimensional feature vector $\phi(q, u, r_i)$ for each result r_i corresponding to query q issued by user u. Note, that result r_i could be either web result or vertical result for any vertical. If the result was relevant for the user, we labeled this vector with 1, otherwise it was labeled with 0. After that we used the point-wise approach to train the ranking model and re-ranked all the results according to the model score. In the following sections we provide the description of the feature vectors construction for different experiment settings, but, first, we introduce the notation used throughout this paper.

Let us denote each observed vertical as $V_j, j = 1, \ldots, N$. For each result r_i there is indicator function $I(r_i)$ that outputs j if r_i is the result of vertical V_j and 0 if it is the web result. Let $\chi(r_i)$ be the formal function that outputs r_i, if $I(r_i) = 0$ and $V_{I(r_i)}$ otherwise. By $C(q, u, r)$ we denote the number of clicks made by user u on specific result r (vertical or web one) for query q. $C_{30}(q, u, r)$ and $C_{100}(q, u, r)$ are the counts of clicks with the dwell time more than 30 and 100 seconds respectively. By $C_{l,30}(q, u, r)$ we mean the number of clicks which were the last clicks on the results for the respective query q and had the dwell time more than 30 seconds. By means of this notation we can define $C(\bullet, u, r)$ as $\sum_i C(q_i, u, r)$, where q_i are all the queries issued by user u during the observed period of time. The aggregated values like $C(q, \bullet, r)$ or $C(q, \bullet, \bullet)$ can be defined in the same way. Moreover, we need to define $C(q, u, V_j)$ as the sum of the clicks of user u on all the results of the vertical V_j presented for query q during the observed period of time. These results could actually be different at different moments, so this value should be treated as "clicks on vertical". $C(q, u, \bullet_V)$ is defined as the sum of $C(q, u, V_j)$ for all verticals.

The number of times any result r was *seen* by the user u following the query q will be referred to as $S(q, u, r)$. We consider the result as *seen* in one of the following cases: 1) if it was placed on the first position, or 2) if it was clicked, or 3) if a document positioned lower was clicked (see, for instance, In accordance with the definitions above it is easy to proceed with definitions of values like $S(q, u, v)$, $S(q, u, \bullet)$, etc. Similar notation could be found in [9].

3.1 Baseline Features

Here we describe the construction of user-independent baseline feature vector $\phi_B(q, r)$. The first element of $\phi_B(q, r)$ is $I(r)$, so that a learning method is always informed of the type of the result (i.e. whether it is a web, or, for instance, images or news result). The features unavailable for a certain result type were set to zero and the first element of $\phi_B(q, r)$ hence indicated such situations. To build a competitive baseline, we implemented the following features, representative of the ones mentioned in Section 2.

Query data. Following [11], a boolean variable indicating whether the query is considered to be navigational was included into the baseline feature set. For each vertical V_j we also built the unigram vertical language model L_j. Each model was built based on the queries for which the result from V_j was clicked with the dwell time more than 30 seconds, which is a rather accurate and widely used (e.g., in [4, 5, 9, 10]) indicator of result relevance. So, if r was the vertical result and $I(r) = j$ we added query likelihood given by L_j to the feature vector $\phi_B(q, r)$. If r was the web result we appended zero to $\phi_B(q, r)$ (and the machine learning algorithm was informed about the result type due to the first coordinate of $\phi_B(q, r)$). We preferred query texts to document texts to build our models because some of our verticals operated with nontextual content and we wanted to treat the verticals consistently. Another reason is that the models constructed in such a manner have very similar semantics to the keyword features and thus provided our re-ranking functions with the signal of this type too. We also added the query length as a feature.

Vertical data and *Web data.* The first feature of this type is the result position in the original ranking. The result relevance score estimated by the original ranking algorithm only for web documents was also utilized as a feature. Note that our baseline set of features includes the features needed to produce the non-personalized version of the vertical relevance score, so, for the sake of proper comparison of personalized and non-personalized approaches, we measure the relevance of verticals explicitly and do not obtain it elsewhere.

Search-log data. The next features we used were the following five click-based ones:

$$F^c = \frac{C(q, \bullet, \chi(r))}{S(q, \bullet, \chi(r))}, F^c_{30} = \frac{C_{30}(q, \bullet, \chi(r))}{S(q, \bullet, \chi(r))}, F^c_{100} = \frac{C_{100}(q, \bullet, \chi(r))}{S(q, \bullet, \chi(r))}$$

$$F^c_{l,30} = \frac{C_{l,30}(q, \bullet, \chi(r))}{S(q, \bullet, \chi(r))}, F^c_{\%} = \frac{C(q, \bullet, \chi(r))}{C(q, \bullet, \bullet)}$$

Actually similar features were already used by the search engine original ranking algorithm, but we explicitly added them to $\phi_B(q, r)$ to emphasize the effect of the personalized approach. If $\chi(r)$ is V_j, these features give us information about the clicks on the vertical results. So they could also be considered as **Vertical data** features.

3.2   Personalization Features

We propose three classes of vertical-related personalization features with some intuitive reasons for their potential utility.

Aggregated search need. This set of features is supposed to describe whether the user is interested in aggregated search results in general and prefers them to core web results. Vertical results usually have presentation which differs from the common web results, and it can affect the user experience. Theses features were supposed to reflect the user's attitude to such changes. In order to express this intuition, we proceeded by utilizing the historic click information. Particularly the set of features is as follows:

$$F^u = \frac{C(\bullet, u, \bullet_V)}{S(\bullet, u, \bullet_V)}, F^u_{30} = \frac{C_{30}(\bullet, u, \bullet_V)}{S(\bullet, u, \bullet_V)}, F^u_{100} = \frac{C_{100}(\bullet, u, \bullet_V)}{S(\bullet, u, \bullet_V)}$$

$$F^u_{l,30} = \frac{C_{l,30}(\bullet, u, \bullet_V)}{S(\bullet, u, \bullet_V)}, F^u_{\%} = \frac{C(\bullet, u, \bullet_V)}{C(\bullet, u, \bullet)}$$

We denote the vector of these five features as $\phi_a(u)$.

Certain vertical preference. This set of properties describes the user's demand for obtaining results of a certain type throughout all the queries. We believe that it highly correlates with the user's interests and adding features of this type to the user's personalization profile can help to disambiguate some ambiguous queries for a concrete user.

The first feature of this class expresses the difference between the user's unigram language model (built on the queries issued by the user during the observed period of time) and the language model for the result's vertical, described in Section 3.1. The difference is calculated as Kullback-Leibler divergence $\sum_{w \in W} P_{V_j}(w) * log \frac{P_{V_j}(w)}{P_u(w)}$, where $V_j = \chi(r_i)$. If $I(r_i) = 0$ this feature is set to zero.

Another way to express the motivation above is to utilize click information. We propose the next set of features for this purpose:

$$F^{uv} = \frac{C(\bullet, u, V_j)}{S(\bullet, u, V_j)}, F^{uv}_{30} = \frac{C_{30}(\bullet, u, V_j)}{S(\bullet, u, V_j)}, F^{uv}_{100} = \frac{C_{100}(\bullet, u, V_j)}{S(\bullet, u, V_j)}$$

$$F^{uv}_{l,30} = \frac{C_{l,30}(\bullet, u, V_j)}{S(\bullet, u, V_j)}, F^{uv}_{\%} = \frac{C(\bullet, u, V_j)}{C(\bullet, u, \bullet)}$$

Here $j = I(r_i)$. The feature vector of these six features is denoted as $\phi_c(u, r_i)$.

Vertical navigationality. On the other hand, the user's needs may not match her overall preferences for some particular queries. For example, the results from News or Weather vertical may be more relevant than the results from Image vertical for the user living in Amsterdam and issuing query [Amsterdam] despite her usual preference for Images.

The click-based features reflecting this intuition are evaluated as follows:

$$F^{quv} = \frac{C(q, u, V_j)}{S(q, u, V_j)}, F^{quv}_{30} = \frac{C_{30}(q, u, V_j)}{S(q, u, V_j)}, F^{quv}_{100} = \frac{C_{100}(q, u, V_j)}{S(q, u, V_j)}$$

$$F^{quv}_{l,30} = \frac{C_{l,30}(q, u, V_j)}{S(q, u, V_j)}, F^{quv}_{\%} = \frac{C(q, u, V_j)}{C(q, u, \bullet)}$$

Again, here $j = I(r_i)$ and the feature vector of these five features is denoted as $\phi_n(q, u, r_i)$.

We added the absolute values of the respective clicks and shows to each of these feature vectors (for instance, $S(\bullet, u, \bullet_V)$ and $C(\bullet, u, \bullet)$ to $\phi_a(u)$ and so on). By this our models were informed of the user's activity level in respect to the verticals results which could also have useful signal for the learning algorithm. Another reason for adding these features is that the remaining features become more reliable with the larger values of activity features. So providing the learning algorithm with such information is also helpful for the whole learning process. Note, that these feature vectors make sense only for verticals results, so if $I(r_i) = 0$, then all the elements of these three feature vectors were equal to zero.

3.3 Re-ranking Functions

We trained a number of re-ranking functions differing in the sets of features they use. To train the models we used a proprietary version of Gradient Boosted Decision Trees-based algorithm configured for minimizing MSE (similar proprietary GBDT-based algorithms were used, for instance, in [3, 11]). Although the choice of the algorithm is not central to this work we selected a Decision Trees-based algorithm to utilize the intuition that we wanted our ranking functions to be vertical-sensitive and use some features only if they were available for the certain result type (for instance, personalization features were not available for web results). The similar results are expected to be provided by any other reasonable Decision Trees-based algorithm. We used the same learning algorithm parameters (shrinkage rate, size of trees) as the ones used for training Yandex production ranking function.

The baseline ranking function R_B was trained according to the scheme described at the beginning of Section 3 using features vector $\phi_B(q,r)$ (see Section 3.1). This feature vector includes the position of the result in the original ranking which represents a production ranking of the Yandex search engine. On the other hand, it includes the representative set of the features mentioned in Section 2.1. So this set of features provides us with a very competitive baseline.

To evaluate the potential of personalization for improving aggregated search and to estimate the strength of different personalization feature classes we also trained 4 more ranking functions. R_{acn} function was trained on the concatenation of the feature vectors ϕ_B, ϕ_a, ϕ_c, ϕ_n. R_{ac} was trained on the concatenation of the feature vectors ϕ_B, ϕ_a, ϕ_c, R_{an} — on vectors ϕ_B, ϕ_a, ϕ_n and R_{cn} — on vectors ϕ_B, ϕ_c, ϕ_n.

3.4   Dataset and Experiment Protocol

To perform our experiments we collected user sessions from Yandex search logs. For each query these logs contained the query itself, the top results returned by the search engine in response to the query and the click information about the results. All of the users of this search engine were assigned with a special anonymous UID (User Identifier) cookie, which was also stored in the logs and allowed us to distinguish actions performed by different users. Our dataset consisted of eight weeks of user sessions logged during May and June, 2012. We considered only the search results, which contained at least one vertical result.

Since we were going to evaluate personalized features it was impossible to use the assessors' judgements so we had to retrieve information about results relevance from the search logs. Following the recent publications on web search personalization (see Section 2.2 for references), we considered the result to be relevant for the particular query if it was clicked with the dwell time more than 30 seconds or the click on the result was the last user action in the whole session. Otherwise we considered the presented result as irrelevant.

To obtain not biased results, in our experiments we followed the same protocol as the one described in [5]. To build the training and the test sets we used the sessions from the seventh and the eighth weeks of our observations respectively. We constructed the datasets in such a manner in order not to test the models on the sessions from the same period as the sessions used for learning which could have biased the results. For both of the datasets we considered only the queries for which at least one result had a positive judgement and the vertical result was *seen* in accordance with the definition of *seen* from Section 3. To build the search logs features (both personalized and non-personalized) for the training set we used the sessions from weeks 1-6, and for the test set — from weeks 2-7, so the amount of the information for training and testing was equal.

Our final user pool consisted of the users who saw a result from any vertical at least 5 times during the both periods of feature collection (weeks 1-6 and 2-7). Note, that we did not filter the users by their activity during the test period, which could have biased the obtained results as well. Such filtering gave us about 30 million distinct users. Both the training and the test sets consisted of approximately 100 million queries, and about 70% of these queries were issued by the users from our user pool. As the whole user pool was excessive for the evaluation process, we randomly selected 10% of the collected profiles for the experiments. Due to the random sampling this subset reflected all characteristics of the whole user profiles pool. Thus, our final user pool consisted of about 3 million users and the training and the test sets contained search results for about 7 million queries.

All the ranking functions were trained and then evaluated in the same manner using the user-specific version of five-fold cross-validation, which was set up as follows. The final users profiles pool was split into 5 folds. During each cross-validation iteration four user profiles folds were used for training and the remaining fold was retained for testing. To train the model we took the sessions from the training set (week 7), which were committed by the users from the four learning folds. After that the model was tested on the sessions from the test dataset (week 8) made by the users from validation fold. This procedure was then repeated five times, so that each fold was used exactly once as the validation data. This form of cross-validation ensures us that the obtained results are not biased towards the users we used for learning.

4   Results

The re-ranking quality was measured by the mean of average precision of the re-ranked documents for the queries in each test fold (Mean Average Precision, MAP) and then averaged over folds. We do not report the exact values of this averaged MAP due to proprietary nature of this information. Instead, we only show the relative improvements of the personalized algorithms over the baseline ranking, in accordance with the previous studies, e.g. [4–6, 10].

We also measure the performance of our models over a number of query stream subsets with different potential to benefit from personalization. Recent works (for instance, [4], [5] or [9]) show that the changes in query click entropy highly correlate with the profit which could be obtained by personalizing the documents ranking for this query. We adapted common click entropy to the needs of the aggregated search, i.e. we did not use the click probability for each specific result. Instead we used the aggregated probabilities of a click on all the web results and the probability of a click for each vertical. More formally: $P_{V_j} = \frac{C(q,\bullet,V_j)}{C(q,\bullet,\bullet)}, P_{web} = 1 - \sum_j P_{V_j}, CE_A(q) = \sum_j (-P_{V_j} * log_2(P_{V_j})) - P_{web} * log_2(P_{web})$.

4.1   Overall Performance

The overall results of all our models are presented in Table 1. The column header indicates the applied model. Symbol ▲ denotes the 99% (p-value < 0.01) statistical significance level of improvements over the baseline according to the Wilcoxon paired signed-rank test for each of the five test folds. Symbol ▼ means that the

Table 1. Re-ranking results for test dataset

	R_{acn}, %	R_{cn},%	R_{an},%	R_{ac},%
All	3.00 ▲	13.42	2.92 ▲ ▼	2.31 ▲ ▼	2.94 ▲ ▼
$CE < 0.5$	1.90 ▲	6.69	1.86 ▲	1.35 ▲ ▼	1.86 ▲
$0.5 \leq CE < 1$	3.01 ▲	14.43	2.93 ▲ ▼	2.43 ▲ ▼	2.93 ▲ ▼
$1 \leq CE$	3.30 ▲	14.49	3.18 ▲ ▽	2.25 ▲ ▼	3.19 ▲ ▼

corresponding model performs significantly worse than R_{acn} model on each fold
with p-value < 0.01. ▽ means the same with p-value < 0.05.

The right part of R_{acn} column shows the improvements for the queries, where
MAP of the rankings produced by the baseline R_b model and personalized R_{acn}
model differs. Such queries comprise approximately 29% of the query stream of
each testing fold. MAP grows for 18% of the stream so we improved the ranking
for 62% of the queries affected by our re-ranking.

From this table we can see that all four personalized models significantly improve ranking quality. The degree of improvement varies through different query
classes, but it grows noticeably with the growth of click entropy which agrees
with the previous studies. As for personalizing features strength, omitting the
Certain vertical preference features worsens the model performance the most.
Excluding the other two feature types leads to less quality decline, but nevertheless significantly worsens the model performance for the whole query stream
and for most of the subsets.

The following sections provide more detailed analysis of the performance of
R_{acn} model depending on the query, the user or presented verticals.

4.2   Query-Level Analysis

First, we would like to consider the dependence of the impact of personalized
approach on the changes in click entropy in more detail (however, overall performance of the personalized models for the queries with known click entropy
always improves on average). Figure 2 shows MAP improvements as a function of adapted click entropy (see the beginning of Section 4 for the definition
of adapted click entropy). This graph shows that in general growth of entropy
leads to increased effect of personalization for verticals ranking. We can see positive effect even for the queries with low entropy and despite the decline in MAP
improvements to the left of 1, the average growth of MAP in the range between
0.5 and 1 still surpasses this value counted for the [0, 0.5]-interval as shown in
Table 1.

As it was mentioned above, our re-ranking affected 29% of the query stream
of each testing fold, which consisted of approximately 1.2 million queries. On the
other hand, each test fold consisted of approximately 680,000 unique queries. The
share of unique queries, affected by our re-ranking, is 32%, 61% of which was
re-ranked positively. Note, that the same query issued by different users might
have been re-ranked in different ways or not re-ranked at all. We considered a

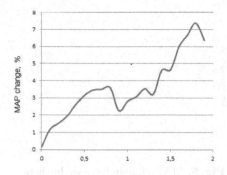

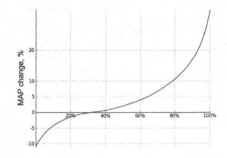

Fig. 2. MAP change as a function of entropy

Fig. 3. Distribution of MAP change for unique queries (ordered by MAP change)

unique query affected if the rankings of R_b and R_{acb} models differed for any of its occurrences in the dataset. Figure 3 shows the average re-ranking performance of the personalized model for the share of the unique queries (issued at least 5 times) between the 5th and 95th percentiles of the sample of such queries sorted by MAP growth. The occurrences of the query were counted during the eighth week of observations.

4.3 User-Level Analysis

One more valuable aspect of the analysis of personalized models is their influence on distinct users. As the 3 million user pool was split into 5 non-overlapping folds, each user-fold consisted of approximately 600,000 users. However, the corresponding part of the test dataset contained the queries issued by only about 450,000 users, since not all the observed users were active during the eighth week of observations. The sessions of 54% of these users were affected by our re-ranking and for 64% of them the personalized re-ranking had a positive effect. Figure 4 shows the distribution of average MAP growth for the share of the users between the 5th and 95th percentiles of the sample of the users who issued at least 5 queries during the eighth week.

To find out the classes of the users which differ by the effect of personalization we split the users into buckets depending on the number of times each user saw verticals results during the feature collection period. Thus, if a user saw verticals results k times, she was assigned to the bucket numbered $\lfloor (k-5)/5 \rfloor$. We selected top of such buckets by the number of assigned users and present average MAP changes inside a bucket as a function of the bucket number in Figure 5. Although the average MAP inside each bucket grows, we can see that the value of this growth highly depends on the number of a bucket and therefore on the number of times the user had seen any vertical. It is important to highlight here, that this number is calculated during the feature collection period, so it can be used in the learning process, for instance, by training different models for users with different levels of aggregated search-related activity.

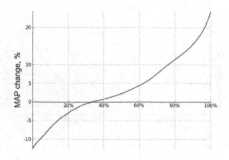   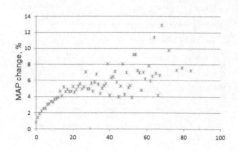

Fig. 4. Distribution of MAP change for users (ordered by MAP change)   **Fig. 5.** MAP change for user buckets

4.4   SERP Analysis

As the next direction of our analysis we studied how the effect of personalization depends on the verticals presented on the SERP. First, we measured how the personalized re-ranking changed the order of vertical results if the SERP contained at least two of them. For this purpose we calculated MAP considering only verticals results and obtained 1.24% growth (p-value < 0.01 on each fold). We also studied the dependence of MAP growth on the number of presented verticals on the SERP and got the following results: for 1 vertical result presented (75% queries) MAP grows by 2.72%, for 2 presented verticals (22% queries) MAP grows by 3.80%, for 3 presented vertical results the growth is 4.31% (2.5% queries) and for 4 results — 3.43% (approximately 0.5% queries). All the changes are significant with p-value < 0.01 for each 5 folds. We also studied which verticals benefit more or less from personalization approach and found out that for Video and Weather verticals profit is the most (5.35% and 8.2%), but for Dictionaries and Events verticals we failed to achieve significant improvements.

5   Conclusion and Future Work

In this paper we studied the novel problem of personalized aggregated search and proposed the framework with three classes of personalized features. It allowed us to build the ranking function outperforming very competitive baseline by 3% in Mean Average Precision score on the whole aggregated search query stream. The proposed re-ranking function changed the results for 29% queries, improving the ranking for 61% of them. We also found out that *Certain vertical preference* features class is the most important class for building competitive personalized ranker, however the other two classes are significantly important too.

Natural future directions of related research may include extending presented feature classes by new features or proposing new classes aiming to improve personalized aggregated search ranking. Especially valuable could be features improving the ranking for the users who have not seen vertical results many times as we showed that personalized ranking for such users improves baseline less. This research can also be extended by learning the decay effect for personalizing features since the user's interests can vary highly in the course of time.

References

1. Arguello, J., Diaz, F., Callan, J.: Learning to aggregate vertical results into web search result. In: Proceedings of CIKM 2011(2011)
2. Arguello, J., Diaz, F., Callan, J., Crespo, J.F.: Sources of evidence for vertical selection. In: Proceedings of SIGIR 2009 (2009)
3. Arguello, J., Diaz, F., Paiement, J.F.: Vertical selection in the presence of unlabeled verticals. In: Proceedings of SIGIR 2010 (2010)
4. Bennett, P.N., Radlinski, F., White, R.W., Yilmaz, E.: Inferring and using location metadata to personalize web search. In: Proceedings of SIGIR 2011 (2011)
5. Bennett, P.N., White, R.W., Chu, W., Dumais, S.T., Bailey, P., Borisyuk, F., Cui, X.: Modeling the impact of short- and long-term behavior on search personalization. In: Proceedings of SIGIR 2012 (2012)
6. Collins-Thompson, K., Bennett, P.N., White, R.W., de la Chica, S., Sontag, D.: Personalizing web search results by reading level. In: Proceedings of CIKM 2011 (2011)
7. Diaz, F.: Integration of news content into web results. In: Proceedings of WSDM 2009 (2009)
8. Diaz, F., Arguello, J.: Adaptation of offline vertical selection predictions in the presence of user feedback. In: Proceedings of SIGIR 2009 (2009)
9. Dou, Z., Song, R., Wen, J.R.: A large-scale evaluation and analysis of personalized search strategies. In: Proceedings of WWW 2007 (2007)
10. Kharitonov, E., Serdyukov, P.: Gender-aware re-ranking. In: Proceedings of SIGIR 2012 (2012)
11. Ponnuswami, A., Pattabiraman, K., Wu, Q., Gilad-Bachrach, R., Kanungo, T.: On composition of a federated web search result page: using online users to provide pairwise preference for heterogeneous verticals. In: Proceedings of WSDM 2011 (2011)
12. Styskin, A., Romanenko, F., Vorobyev, F., Serdyukov, P.: Recency ranking by diversification of result set. In: Proceedings of CIKM 2011 (2011)
13. Vallet, D., Castells, P.: Personalized diversification of search results. In: Proceedings of SIGIR 2012 (2012)

HetPathMine: A Novel Transductive Classification Algorithm on Heterogeneous Information Networks

Chen Luo[1], Renchu Guan[1], Zhe Wang[1,*], and Chenghua Lin[2]

[1] College of Computer Science and Technology, Jilin University,
Changchun 130012, China
rackingroll@163.com, {wz2000,guanrenchu}@jlu.edu.cn
[2] School of Natural and Computing Sciences, University of Aberdeen, UK
chenghua.lin@abdn.ac.uk

Abstract. Transductive classification (TC) using a small labeled data to help classifying all the unlabeled data in information networks. It is an important data mining task on information networks. Various classification methods have been proposed for this task. However, most of these methods are proposed for homogeneous networks but not for heterogeneous ones, which include multi-typed objects and relations and may contain more useful semantic information. In this paper, we firstly use the concept of meta path to represent the different relation paths in heterogeneous networks and propose a novel meta path selection model. Then we extend the transductive classification problem to heterogeneous information networks and propose a novel algorithm, named HetPathMine. The experimental results show that: (1) HetPathMine can get higher accuracy than the existing transductive classification methods and (2) the weight obtained by HetPathMine for each meta path is consistent with human intuition or real-world situations.

1 Introduction

Information network is an efficient way to represent relational data in data mining tasks [1]. For example, a co-author relationship dataset can be represented as a co-author network and a web-page connection dataset could construct a WWW network. The network research has been attracted many attentions in recent years [2–5]. Among these researches, transductive classification [6] (TC) is one of the popular method to extract knowledge with the help of a small amount of labeled data [7]. Most of these algorithms are proposed for the homogeneous information networks [6, 7], however, the real world is full of heterogeneous networks which include various types of objects and relations. In this paper, we move transductive classification problem to heterogeneous networks [1, 8], which have more meaningful information than homogeneous ones.

---

* Corresponding author.

M. de Rijke et al. (Eds.): ECIR 2014, LNCS 8416, pp. 210–221, 2014.

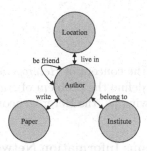

Fig. 1. Network Schema for a Sample Heterogeneous Network

Up till now, many transductive classification methods have been proposed on information networks. For example, learning with local and global consistency algorithm (LLGC) [6] utilizes the network structure to propagate the labels to the unlabeled data. For networked data, it is one of the most popular transductive classification method. The weighted-vote relational neighbor classifier (wvRN), proposed in [9], is another widespread method, which determines the label by considering all the neighbors label. All these algorithms are proposed for homogeneous networks. In 2010, Ming *et.al* proposed GNetMine[10], which is a classification method on heterogeneous networks. GNetMine assumes that all the objects in the network have the same classification criterion. However, most of the heterogeneous networks with different types of objects may have different classification criteria. For example, a network with four types objects: people, institute, location, and paper. The network schema is showed in Figure 1. Author can be classified into research area, but institute and location cannot be classified with this criterion.

In this paper, we extend the transductive classification into heterogeneous networks by utilizing the relation paths existed in the network. The relationships in heterogeneous network include not only the relations but also the relation path. Considering network in Figure 1, two authors can be connected by not only the "author-author" (friendship) relationship but also the "author-institute-author" or "author-paper-paper" relation path. Meta-path [11], proposed as a topology measure for heterogeneous information networks, depicted the different relation paths in heterogeneous networks. We use this concept to model the different relationship in heterogeneous information networks. In addition, a novel meta path selection model is proposed to calculate the different weight of each relation paths. Finally, by using the different weight of each meta path, we propose a transductive classification framework, named HetPathMine, on the heterogeneous information network.

The rest of this paper is organized as follow: some concepts used in this paper is introduced in section 2, we present the HetPathMine algorithm in section 3. Experimental result is showed in section 4. Finally, we conclude our work in section 5.

2   Background

2.1   Problem Definition

In this section, we introduce the concept of *Heterogeneous Information Network* and *class*. Then we formally define the problem of *transductive classification in heterogeneous network*. Following [8], the Heterogeneous Information Network is defined as follows:

Definition 1 (Heterogeneous Information Network [8]). *Suppose we have m types of data objects, denoted by $X_1 = \{x_{11}, ..., x_{1n_1}\}, ..., X_m = \{x_{m1}, ..., x_{mn_m}\}$, a heterogeneous information network is in the form of a graph $G = \langle V, E, W \rangle$, where $V = \bigcup_{i=1}^{m} X_i$ and $m \geq 2$, E is the set of links between any two data objects of V ,and W is the set of weight values on the links. G reduces to a homogeneous network, when $m = 1$.*

A *class* on heterogeneous information network is defined as follow:

Definition 2 (Class). *Given a heterogeneous information network $G = \langle V, E, W \rangle$, a subset of V: V' is a class in G, when $V' \subseteq V \in X_i$, and X_i is the target type for classification.*

It is pointed out that, our definition of *class* is different from the definition in [10]. In this paper, we only consider a *class* as a set of objects within the same type. It is meaningless to take objects of different type into same *class*. In addition, different type has very different semantic meaning and may have different classification criteria.

After defining the concept of class in this paper, following [10], we define the transductive classification in heterogeneous information network as follow:

Definition 3 (Transductive classification in heterogeneous information network). *Given a heterogeneous information network $G = \langle V, E, W \rangle$, suppose V' is a subset of V and $V' \subseteq V \in X_t$, where X_t is the target type for classification, and each data object O in V' is labeled with a value γ indicating which class O should be in. The classification task is to predict the labels for all the unlabeled objects $V - V'$.*

2.2   Meta Path

In [11], the author uses the concept meta-path to denote the different relations and relation paths in heterogeneous information networks. Follow [11], the meta path is defined as follow:

Definition 4 (Meta Path [11]). *Given a heterogeneous information network $G = \langle V, E, W \rangle$, A meta path P is in the form of $A_1 \xrightarrow{R_1} A_2 \xrightarrow{R_2} ...A_{l-1} \xrightarrow{R_{l-1}} A_t$, which defines a composite relation $P = R_1 \circ R_2 \circ ... \circ R_{l-1}$ between two objects, and $\circ$ is the composition operator on relations.*

Note that, in this paper, we only consider the meta path that $A_1 = A_t$. After defining the concept of meta path, we need to know how to measure it. Here, we introduce a topology measure, PathSim, proposed in [6], for measuring the meta-path: Given a meta path, denoted as P, the PathSim between two objects $s, t \in X_t$ can be calculated as follow:

$$S_P^{PathSim}(s,t) = \frac{2 * S_P^{PathCount}(s,t)}{S_P^{PathCount}(s,:) + S_{P-1}^{PathCount}(:,t)}$$

In the above, $S_P^{PathCount}$ is a Path Count Measure [6] and it can be calculated as the number of path instances between s and t. P^{-1} denotes the inverse meta path [6] of P, $S_P^{PathCount}(s,:)$ denotes the path count value following P starting with s, and $S_{P-1}^{PathCount}(:,t)$ denotes the path count value following P^{-1} ending with t.

3 The HetPathMine Framework

Transductive classification, as a relational learning [12] method, utilizes the structure of the network data to predict labels. There are two assumptions in transductive classification. First, it assumes that the network resulting from a social process often possess a high amount of influence [13]. Such that linked nodes may have a high possibility to have the same label. Second, transductive classification, as a semi-supervised learning algorithm, some pre-labeled information is obtained before the learning process. And the classification result should be consisted with the pre-labeled information [10].

In this section, we propose a novel transductive classification method on heterogeneous networks based on the meta path, named HetPathMine. The HetPathMine algorithm will consist with the two assumptions introduced before. We firstly introduce the Meta-path selection model. Secondly, we extend the transductive classification to the heterogeneous networks. Finally, the detailed steps of HetPathMine is introduced.

3.1 Meta Path Selection Model

Each meta-path can represent a relationship in the heterogeneous networks [14]. For example, in Figure 1, the co-author relationship can be represented by the meta path: "author-paper-author", and co-institute relationship can be represented by the meta path: "author-institute-author". If we want to classify the authors into different research area, the co-author relationship will play a very important role, while the co-institute relationship will bias the classifying process. As a result, we need to know different weight for each meta path. In [14], the author proposed an algorithm which can find the different weight of each meta-path. This process is a called meta-path selection process [14]. However, the model, proposed in [14], is too complicated and has high-computational. In [15], the author proposed a relation extraction model for multi-relational network. Similar to [15], in this paper, we propose a novel meta path selection model.

Given a set of meta paths, denoted as $P = [p_1, p_2, ..., p_d]$, where d is the number of meta paths used in our algorithm. Then, the weight for each meta path is denoted as $B = [\beta_1, \beta_2, ..., \beta_d]$. We use a linear regression algorithm [16] to calculate B. The cost function $L(B)$ is given bellow:

$$L(B) = \left\| \sum_{i=1}^{n} \sum_{j=1}^{n} (\mathbb{R}(x_{t,i}, x_{t,j}) - \sum_{d=1}^{m} \beta_d s_{p_k}^{PathSim}(x_{t,i}, x_{t,j})) \right\| + \mu \left\| \sum_{k=1}^{d} \beta_k \right\| \quad (1)$$

where n is the number of labeled objects. $x_{t,i}, x_{t,j} \in X_t$, and X_t is the target type for classification. $S_{p_k}^{PathSim}(x_{t,i}, x_{t,j})$ is the PathSim measure for meta path p_k. $\mathbb{R}$ is a relation matrix and can be calculated as follow:

$$\mathbb{R}(x_{t,i}, x_{t,j}) = \begin{cases} 1 & x_{t,i}, x_{t,j} \text{ are labeled as the same label} \\ 0 & \text{otherwise} \end{cases} \quad (2)$$

The first term of (1) ensure that the weight for each meta path can follow the pre-labeled information. The second term is a *smoothness constraint*. The weight for each term is captured by μ, a positive parameter. Then $B = [\beta_1, \beta_2, ..., \beta_d]$ can be calculated by solving the following problem:

$$B^* = \arg max_{B=[\beta_1, \beta_2, ..., \beta_d]} L(\beta) \quad (3)$$

Many useful methods can be used to solve such problem [16], in this research we use gradient descent method [16] to solve this problem. After having all the weights for each meta path, we can use them to model the transductive classification on heterogeneous information network.

3.2 Meta Path-Based Transductive Classification Model

In this section, we extend the transductive classification model to the heterogeneous network by considering different meta paths. As introduced before, there are two assumptions in transductive classification tasks: (1) objects have a tight relationship tend to have the same label. (2) The classification result should consist with the pre-labeled information. In [6], the author proposed a transductive classification model on homogeneous networks. We extend the model to heterogeneous network as follow:

$$\varrho(F) = \frac{1}{2} \sum_{k=1}^{d} (\sum_{i=0}^{n} \sum_{j=0}^{n} \beta_k W_{ij}^k \left\| \frac{F_i}{\sqrt{D_{ii}}} - \frac{F_j}{\sqrt{D_{jj}^k}} \right\|^2) + \lambda \sum_{i=0}^{n} \sum_{i=0}^{n} \|F_i - Y_i\|^2 \quad (4)$$

In the above, n is the number of objects within the target type X_i, W^k is the similarity matrix for the target type reduced by the $k-th$ meta path. F is a $n*p$ matrix, where p is the number of class, and $F(i, j)$ denotes the probability of $i-th$ object belonging to the $j-th$ class. Y is also a $n*p$ matrix which denotes the pre-labeled information in the network, D^k is a diagonal matrix with its $(i, i) - th$ element equal to the sum of the $i - th$ row of W^k. $B = [\beta_1, \beta_2, ..., \beta_d]$ is calculated by Eq. (3). The first term of this cost function is the *smoothness constrain*, which follows the first assumption. The second term is the *fitting constrain*, which follows the second assumption introduced before. These two constraints are captured by the parameter λ.

Then, we can get result as follow:

$$\frac{\partial \varrho}{\partial F} = F^* - F^*(\sum_{k=1}^{m} \beta_k S^k) + \mu(F^* - Y) \tag{5}$$

where $S = D^{k^{-1/2}} W^k D^{k^{-1/2}}$.

As the process in [6], F^* can be directly obtained without iterations as follow:

$$F^* = \beta(I - aS_{com})^{-1}Y \tag{6}$$

where $a = \frac{1}{1+\mu}$, $a = \frac{\mu}{1+\mu}$ and $S_{com} = \sum_{k=1}^{m} \beta_k S^k$.

Noticing that, when the network is a homogeneous information network, then $S_{com} = S$, the formula becomes $F^* = \beta(I - aS)^{-1}Y$, which is the same as the formula in homogeneous networks. In other words, our model is a generalized form for the homogeneous one.

After getting F, we can obtain labels of all the objects $x_{t,i} \in X_t$ as follow:

$$Lable(x_{t,i}) = \max\{F(i, 1), F(i, 2), ..., F(i, n)\} \tag{7}$$

3.3 The Detailed Steps of HetPathMine

After having the calculation method for each relevant variable, the complete framework of HetPathMine is then summarized as the following steps:

Step1. Given a heterogeneous information network $G = \langle V, E, W \rangle$,target type X_t for classification, a set of meta paths $P = [p_1, p_2, ..., p_d]$ and labeled information Y.

Step2. Use Eq. (2) to calculate the relation matrix $\mathbb{R}$.

Step3. Use Eq. (3) and the labeled information Y to obtain the weight of each meta-path: $B = [\beta_1, \beta_2, ..., \beta_d]$.

Step4. Use Eq. (6) to obtain the classification result matrix F^*.

Step5. Use Eq. (7) to obtain the final label of all the objects $x_{t,i} \in X_t$.

It is pointed out that, the output of HetPathMine not only contains the classification result but also the different weight $B = [\beta_1, \beta_2, ..., \beta_d]$ for the selected meta paths, B can be used in many data mining tasks [17, 18].

3.4 Complexity Analysis

In this section, we consider the time complexity of the HetPathMine. All the topological features are calculated at the beginning of our framework, the time complexity is $O(k_p n^2)$. Here k_p is the number of meta paths selected for Het-PathMine, and $n = |X_t|$ is the number of target objects. For the Meta Path selection model, the time complexity is $O(t_1 \sum_m |T_m|)$. Here t_1 is the number of iterations, and m is the dimension of the training dataset. For the Meta Path-Based transductive classification model, F can be calculated directly, then time complexity is $O(1)$. Then the overall time complexity of our framework is $O(k_p n^2) + O(t_1 \sum_m |T_m|) + O(1)$.

4   Experimental Results

In this section, we represent an empirical study of the effectiveness of HetPath-Mine compared with several baseline algorithms and state-of-the-art algorithms based on DBLP dataset.

4.1   Datasets

We use the DBLP dataset [1] for performance test. DBLP, computer science bibliography database, which has been used in many research papers [1, 10, 19], is a typical heterogeneous information network. There are four types of objects in this network: paper, author, term, and conference. Links between author and paper defined by the relation of "write" and "written by", denoted as "$write^{-1}$". Relation between Term and Paper is "mention" and "mentioned by", denoted as "$mention^{-1}$". Relation between Paper and Conference is "publish" and "published by", denoted as "$publish^{-1}$". The network schema is showed in Figure 2.

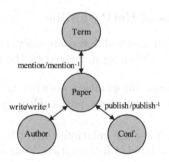

Fig. 2. Network Schema for a Sample Heterogeneous Network

We extract a sub network of DBLP, "four-area dataset [10]", which contains 18 major conferences in four areas: Data Mining, Database, Information Retrieval and Machine Learning. Each area contains several top conferences. They are KDD, ICDM, SDM and PAKDD in Data Mining, SIGMOD, VLDB, ICDE, PODS and EDBT in Database, SIGIR, ECIR, WSDM and WWW, CIKM in Information Retrieval and NIPS, ICML, AAAI and IJCAI in Machine Learning. Four sub-datasets are used in our experiment:

---

[1] http://www.informatik.uni-trier.de/~ley/db/

DataSet-1 top 100 authors in the DBLP within the 18 major conferences, and the corresponding papers published by these authors after 2007.

DataSet-2 top 500 authors in the DBLP within the 18 major conferences, and the corresponding papers published by these authors after 2007.

DataSet-3 top authors in the DBLP within the 18 major conferences, and the corresponding papers published by them after 2007.

DataSet-4 top authors in the DBLP within the 18 major conferences, and the corresponding papers.

Top authors means that the number of papers published by these authors are larger than the other authors. In these datasets, the term is extracted from the paper titles. The labeled information or the ground truth is obtained from the "four-area dataset [10]". The summary of these four data sets is showed in Table 1.

Table 1. Summary of the Four Sub-DataSet

DataSet	Authors	Papers	Terms	Conferences
DataSet-1	100	3212	3984	18
DataSet-2	500	8194	7561	18
DataSet-3	1479	10298	8831	18
DataSet-4	3109	12873	10239	18

We choose 3 kinds of meta-paths in the experiment: "$author \overset{write}{\rightarrow} paper \overset{write^{-1}}{\rightarrow} author$", "$author \overset{write}{\rightarrow} paper \overset{publish^{-1}}{\rightarrow} conference \overset{publish}{\rightarrow} paper \overset{write^{-1}}{\rightarrow} author$", "$author \overset{write}{\rightarrow} paper \overset{mention}{\rightarrow} term \overset{mention^{-1}}{\rightarrow} paper \overset{write^{-1}}{\rightarrow} author$". The semantic meaning of these three meta paths is showed in table 2. Note that, "$A - P - A$" is short for "$author \overset{write}{\rightarrow} paper \overset{write^{-1}}{\rightarrow} author$", "$A - P - C - P - A$" is short for "$author \overset{write}{\rightarrow} paper \overset{publish^{-1}}{\rightarrow} conference \overset{publish}{\rightarrow} paper \overset{write^{-1}}{\rightarrow} author$", "$A - P - T - P - A$" is short for "$author \overset{write}{\rightarrow} paper \overset{mention}{\rightarrow} term \overset{mention^{-1}}{\rightarrow} paper \overset{write^{-1}}{\rightarrow} author$".

Table 2. Summary of the Three Meta Paths

ID	Meta Path	Description
1	A-P-A	Tow author are co-author relationship
2	A-P-C-P-A	Two author published their paper in the same conference
3	A-P-T-P-A	Two author write papers is mentioned the same term

4.2 Methods for Comparison and Evaluation

Methods for Comparison. Two baselines are used in this paper: The first baseline is the Learning with Local and Global Consistency (LLGC) algorithm proposed in [6], which utilizes link structure to propagate labels to the rest of the network. The second baseline is Weighted-vote Relational Neighbor classifier (wvRN) proposed in [9, 20]. These two algorithms are widely used in network classification. As LLGC and wvRN is designed for homogeneous networks, we use all the three homogeneous networks reduced by the three corresponding relation paths (A-P-A, A-P-C-P-A, A-P-T-P-A) to run each algortihm.

One state-of-the-art method, GNetMine [10], is also used in our experiment. GNetMine proposed by Ming et.al. This method handles the classification problem in heterogeneous information network. For the input, we use three types of link (A-P, P-C, P-T) and three four of object (A, P, C, T). The same as HetPathMine, the pre-labeled information is only distributed for the target type objects.

Evaluation Method. We use accuracy [14] to evaluate the effectiveness of classification. The accuracy [14] measure, which is calculated as the percentage of target objects going to the correct class, is defined as follow:

$$Accuracy = \frac{\sum_{i=1}^{k} a_i}{n}$$

where a_i is the number of data objects classified to its corresponding true class, k is the number of class and n is the number of data objects in the dataset.

4.3 Result and Analysis

The result is showed in Figure 3. The results obtained by LLGC and wvRN are sharply different by using different meta paths. When using meta path: "A-P-A" or "A-P-T-P-A", the result is very bad. Compared with these two homogeneous algorithm (LLGC and wvRN), HetPathMine can obtain highly accuracy results in different situations. From this point of view, HetPathMine is an efficient method for classification.

On the other hand, HetPathMine can obtain more accuracy results in different dataset compared with the state-of-the-art algorithm, GNetMine. We can see from the result, in each dataset, HetPathMine keeps a high accuracy and has a nearly 5% improvement in accuracy compared with GNetMine.

Overall, HetPathMine performs the best on all datasets with different seed fractions. It is more stable than LLGC and wvRN and more effectiveness than GNetMine. The result showed in Figure 3 also demonstrates that by mining heterogeneous network, one can get more meaningful result than the homogeneous ones.

4.4 Case Study on Meta-path Selection

In this section, we study the learned weights for each meta path by HetPathMine. The rank for the three selected meta paths used in our test is showed in Table.3.

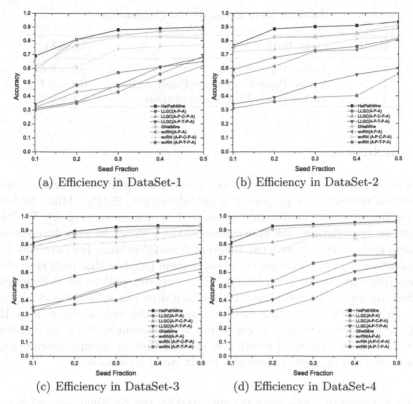

Fig. 3. Performances under different seed fractions (0.1, 0.2, 0.3, 0.4 and 0.5) in each class are tested. Seed fraction determines the percentage of labeled information. If the seed fraction is 0.1, and the number of object is 1000, then there are 100 objects labeled in the dataset. For each given seed fraction, we average the results over 10 different selections. The classification result is showed in Figure. In our experiment, we set $\mu = 0.01$ in the meta path selection model, and set $\lambda = 2$ for the transductive classification model.

For the ease of illustrate, the weights for each meta path is mapped to the range $[0, 1]$. The meta path *"author − paper − conference − paper − author"* always has a highest weight compared with the other two meta paths. This is consistent with the human intuition: different conference have different interesting topic. Such as ICML is mainly for machine learning and ECIR is mainly for information retrieval. Authors intend to submit their papers to the conference which has their interesting topics. The meta path *"author − paper − author"* also has a high weight in HetPathMine too. This means that two authors are very likely to have similar research interest if they have co-author relationship. The meta path *"author − paper − term − paper − author"* has the lowest weight in HetPathMine. This is consistent with real world scenarios: it is not rare that two papers from different areas can have the same term. For example, the words "algorithm" and "method" appear in many papers from different areas.

Table 3. Meta Paths Weight Comparison

Rank	Meta Path	Weight
1	Author-Paper-Author	$0.5 \sim 0.6$
2	Author-Paper-Conference-Paper-Author	$0.3 \sim 0.4$
3	Author-Paper-Term-Paper-Author	$0.1 \sim 0.2$

5 Conclusion

In this paper, we study the transductive classification problem in heterogeneous information network and propose a novel algorithm, HetPathMine. Different from the transductive classification method of homogeneous network, HetPathMine utilize the relation path information of the network by introduce the concept meta path. On the other hand, different from GNetMine, HetPathMine have distinct advantages in managing the classification problem that each type of object has different classification criteria. The experimental result demonstrates the effectiveness of HetPathMine: (1) HetPathMine can achieve good accuracy in comparison with the existing methods, and (2) the weight obtained by HetPathMine for each meta path is consistent with human intuition or real-world situations. The experimental result also showed that by mining heterogeneous information network, more meaningful result could be obtained.

Acknowledgement. This work is supported by the National Natural Science Foundation of China (NSFC) under Grant No. 61373051; the National Science and Technology Pillar Program (Grant No. 2013BAH07F05), Jilin Province Science and Technology Development Program (Grant No. 20111020); Project of Science and Technology Innovation Platform of Computing and Software Science (985 engineering), and the Key Laboratory for Symbolic Computation and Knowledge Engineering, Ministry of Education, China.

References

1. Sun, Y., Han, J.: Mining heterogeneous information networks: principles and methodologies. Synthesis Lectures on Data Mining and Knowledge Discovery 3(2), 1–159 (2012)
2. Gao, J., Liang, F.E.: On community outliers and their efficient detection in information networks. In: KDD 2010, pp. 813–822. ACM (2010)
3. Taskar, B., Abbeel, P., Koller, D.: Discriminative probabilistic models for relational data. In: UAI 2002, pp. 485–492. Morgan Kaufmann Publishers Inc. (2002)
4. Castells, M.: The rise of the network society: The information age: Economy, society, and culture, vol. 1 (2011), Wiley.com
5. Even, S.: Graph algorithms. Cambridge University Press (2011)
6. Zhou, D., Bousquet, O.E.: Learning with local and global consistency. In: Advances in Neural Information Processing Systems, vol. 16(16), pp. 321–328 (2004)

7. Wu, M., Schölkopf, B.: Transductive classification via local learning regularization. In: AISTATS 2007, pp. 628–635 (2007)
8. Sun, Y., Han, J.E.: Rankclus: integrating clustering with ranking for heterogeneous information network analysis. In: EDBT 2009, pp. 565–576. ACM (2009)
9. Macskassy, S.A., Provost, F.: A simple relational classifier. Technical report, DTIC Document (2003)
10. Ji, M., Sun, Y., Danilevsky, M., Han, J., Gao, J.: Graph regularized transductive classification on heterogeneous information networks. In: Balcázar, J.L., Bonchi, F., Gionis, A., Sebag, M. (eds.) ECML PKDD 2010, Part I. LNCS, vol. 6321, pp. 570–586. Springer, Heidelberg (2010)
11. Sun, Y., Han, J., Yan, X., Yu, P.S., Wu, T.: Pathsim: Meta path-based top-k similarity search in heterogeneous information networks. In: VLDB 2011 (2011)
12. Getoor, L., Taskar, B.: Introduction to statistical relational learning. The MIT Press (2007)
13. La Fond, T., Neville, J.: Randomization tests for distinguishing social influence and homophily effects. In: WWW 2010, pp. 601–610. ACM (2010)
14. Sun, Y., Norick, B., Han, J., Yan, X., Yu, P.S., Yu, X.: Integrating meta-path selection with user-guided object clustering in heterogeneous information networks. In: KDD 2012, pp. 1348–1356. ACM (2012)
15. Cai, D., Shao, Z., He, X., Yan, X., Han, J.: Mining hidden community in heterogeneous social networks. In: LinkKDD 2005, pp. 58–65. ACM (2005)
16. Montgomery, D.C., Peck, E.A., Vining, G.G.: Introduction to linear regression analysis, vol. 821. Wiley (2012)
17. Mintz, M.E.: Distant supervision for relation extraction without labeled data. In: ACL 2009, pp. 1003–1011. Association for Computational Linguistics (2009)
18. Nguyen, T.V.T., Moschitti, A., Riccardi, G.: Convolution kernels on constituent, dependency and sequential structures for relation extraction. In: EMNLP 2009, pp. 1378–1387. Association for Computational Linguistics (2009)
19. Sun, Y.E.: Co-author relationship prediction in heterogeneous bibliographic networks. In: ASONAM 2011, pp. 121–128. IEEE (2011)
20. Macskassy, S.A., Provost, F.: Classification in networked data: A toolkit and a univariate case study. The Journal of Machine Learning Research 8, 935–983 (2007)

Leveraging Dynamic Query Subtopics for Time-Aware Search Result Diversification*

Tu Ngoc Nguyen and Nattiya Kanhabua

L3S Research Center / Leibniz Universität Hannover
Appelstrasse 9a, Hannover 30167, Germany
{tunguyen,kanhabua}@L3S.de

Abstract. Search result diversification is a common technique for tackling the problem of ambiguous and multi-faceted queries by maximizing query aspects or subtopics in a result list. In some special cases, subtopics associated to such queries can be temporally ambiguous, for instance, the query US Open is more likely to be targeting the tennis open in September, and the golf tournament in June. More precisely, users' search intent can be identified by the popularity of a subtopic with respect to the time where the query is issued. In this paper, we study search result diversification for time-sensitive queries, where the temporal dynamics of query subtopics are explicitly determined and modeled into result diversification. Unlike aforementioned work that, in general, considered only static subtopics, we leverage dynamic subtopics by analyzing two data sources (i.e., query logs and a document collection). By using these data sources, it provides the insights from different perspectives of how query subtopics change over time. Moreover, we propose novel time-aware diversification methods that leverage the identified dynamic subtopics. A key idea is to re-rank search results based on the freshness and popularity of subtopics. To this end, our experimental results show that the proposed methods can significantly improve the diversity and relevance effectiveness for time-sensitive queries in comparison with state-of-the-art methods.

1 Introduction

A significant fraction of web search queries are ambiguous, or contain multiple aspects or subtopics [7]. For example, the query apple can refer to a kind of fruit or a company selling computer products. Moreover, the underlying aspects of the query apple inc can be a new Apple product, software updates or it latest press releases. While it is difficult to identify user's search intent for multi-faceted queries, it is common to present results with a high coverage of relevant aspects. This problem has been well studied in aforementioned work on search result diversification [1,5,6,10,15,16]. However, previous work only consider a set of static subtopics without taking into account the temporal dynamics of query subtopics.

---

* This work was partially funded by the European Commission FP7 under grant agreement No.600826 for the ForgetIT project (2013-2016).

M. de Rijke et al. (Eds.): ECIR 2014, LNCS 8416, pp. 222–234, 2014.

In this paper, we study the search result diversification of *temporally ambiguous, multi-faceted queries*, where the relevance of query subtopics is highly time-dependent. For example, when issuing the query kentucky derby in April, relevant aspects are likely to be about "festival" or "food" referring to the Kentucky Derby Festival, which occurs two weeks before the stakes race. However, at the end of May, other facets like "result" and "winner" should be more relevant to than pre-event aspect. Identifying dynamic subtopics for temporally ambiguous, multi-faceted queries is essential for time-aware search result diversification. In order that, we explicitly extract dynamic subtopics and leverage them into diversifying retrieved results. To the best of our knowledge, none of the aforementioned works considers the temporal changes in query subtopics before.

Our contributions in this paper are as follows. We study the temporal dynamics of subtopics for queries, which are *temporally ambiguous* or *multi-faceted*. We analyze the temporal variability of query subtopics by applying subtopic mining techniques at different time periods. In addition, our analysis results reveal that the popularity of query aspects changes over time, which is possibly the influence of a real-world event. The analysis study is based on two data sources, namely, query logs and a temporal document collection, where time information is available. To this end, we propose different time-aware search result diversification methods, which leverage dynamic subtopics and show the performance improvement over the existing non time-aware methods.

2   Dynamic Subtopic Mining

In this section, we present our methodology in modeling and mining temporal subtopics from two different datasets. The mined subtopics are input for our time-aware diversification approach. Figure 1 depicts the pipeline process of this paper.

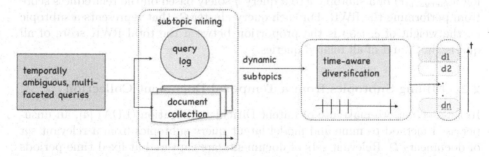

Fig. 1. Pipeline for dynamic subtopic mining and time-aware diversification

2.1   Mining Subtopics from Query Logs

In our work, we followed a state-of-the-art finding related queries technique proposed in [9]. We applied Markov random walk with restart (RWR) on the weighted bipartite graph composed of two sets of nodes, namely, queries and URLs. The bipartite graph is constructed using the history information with regards to different time points. Our model for dynamic subtopic mining assigns each subtopic a *temporal weight* that reflects the probability of the relevance of a subtopic at the particular time.

Models. In this work, we use real-world Web search logs composed of a set of queries Q, a set of URLs D_{url} and click-through information S. Each query $q \in Q$ is made up of query terms, the hitting time q_t (when a user issues the query) and a set of subtopic C. A clicked URL $u_q \in D_{url}$ refers to a web document returned as an answer for a given query q, which has been clicked by the user. The click-through information composes of a query q, a clicked URL u_q, the position on result page, and its timestamps. A weighted directed bipartite graph $G = (Q \cup D_{url}, E)$, where edges E represent the click-through information from a query $q \in Q$ to a URL $u \in D$. Edges are weighted using click frequency - inverse query frequency (CF-IQF) model. CF-IQF compensates for common clicks on less frequent but distinguished URLs over common clicks on frequent URLs. A subtopic $c \in C$ is represented as a bag of words and associates to a temporal weight $w(c)$, where w is a weighting function.

Clustering Subtopic Candidates. Random walk with restart on the click-through graph provides us a set of related queries. However, these related queries can be duplicated or near-duplicated in their semantics. To achieve finer-grained query subtopics at hitting time q_t, we cluster the acquired queries in a similar approach proposed in [17]. The steps are as follows: (1) Construct a query similarity matrix (using lexical, click and semantic similarity), (2) Cluster related queries (using Affinity Propagation technique), and (3) Extract dynamic query subtopics. Due to the limited space in this paper, readers can refer to [17] for detailed description of the steps.

Temporal Subtopic Weight. We calculate the subtopic weight from query log $w_{query\_log}(c)$ of a subtopic c to a query q solely based on the relatedness score from performing the RWR. For each query cluster $\mathbb{C}_i$ that represents a subtopic c_i, the weight of c, $w(c)$ is the proportion between the total RWR score of all queries in $\mathbb{C}_i$ and of all related queries.

2.2 Mining Subtopics from a Temporal Document Collection

In this section, we make use of Latent Dirichlet Allocation (LDA) [4], an unsupervised method to mine and model latent query subtopics from a relevant set of documents D. Relevant sets of documents are captured at fixed time periods in order to measure the variance of the mined latent subtopics over time. Here, a subtopic $c \in C$ is modeled as multinomial distribution of words, a document $d \in D$ composes of a mixture of topics.

Estimating Number of Subtopics. Deciding the optimum number of subtopics is an important task for assessing the overall query subtopic dynamics. The number of subtopics is expected to change when mining it at different time points.

In this work, we follow the approach that proposed by Arun et al. [2] to identify the number of latent subtopics that are naturally present in each partition.
The non-optimum number of subtopics produces the high divergence between the
salient distributions derived from two matrix factors (compose of topics-words
and documents-topics). In our case, we set the number of topics in a pre-defined
range from γ to δ, the chosen number of topic is the one with the minimum KL-
divergence value.

Temporal Subtopic Weight. We estimate the weight of a mined subtopic at
every hitting time. The weight $w_{docs}(c)$ of a subtopic c reflects the probability
that a given query q implies the subtopic c. The temporal distribution that
specifies the probability that a given query belongs to a subtopic c, $Pr(c|q)$ derives
from the popularity of the subtopic in the studied time slice of the document
collection. It is calculated as the proportion between the total probabilities of
all documents belongs to a subtopic $Pr(c|d)$ and the number of documents in
the time slice. $Pr(c|d)$ is calculated from the Dirichlet prior topic distribution of
LDA.

3   Time-Aware Diversification

Most of the existing diversification approaches mentioned in this paper deploy a
greedy approximation approach. We examine three state-of-the-art diversification models (i.e., IA-Select [1], xQuaD [16] and topic-richness [10]). We aim to
maximize the utility of the models by fostering recent documents in the ranking,
with the assumption that the recency level of a subtopic is linearly proportional
to its temporal popularity.

temp-IA-Select. The objective function of IA-Select can be expressed using a
probabilistic model as:

$$f_s(d) = \sum_c Pr(q|d)Pr(d|c)Pr(c|q) \prod_{d\prime \in S} (1 - Pr(q|d\prime)Pr(d\prime|c)) \tag{1}$$

where S is the selected set of diversified documents from the original result set.
Our assumption is our temporal mined subtopics are fresh subtopics and the
subtopics tend to favor recent documents. We propose an exponential distribution on the probability of documents $Pr(d)$ with regards to a subtopic c. The
document-subtopic probability $Pr(d|c)$ at time t_d, defined as $Pr_{t_d}(d|c)$ is calcu-
lated in Equation 2.

$$Pr_{t_d}(d|c) = Pr(c|d)Pr(d|t_d) = Pr(c|d) \cdot \lambda \cdot e^{-\lambda \cdot t_d} \tag{2}$$

We apply $Pr_{t_d}(d|c)$ into the probabilistic objective function of IA-Select to achieve
our time-aware objective function (temp-IA-Select), described in Equation 3.
With this approach, a document d which is published closer to the hitting time
t_q, in essence, has a shorter age t_d will be weighted higher than the one with the
same $Pr(c|d)$. Note that for this setting, we do not account for time to calcu-
late the document-query probability, $Pr(d|q)$, that remains unchanged over time.

Table 1. *Temporally ambiguous*, multi-faceted AOL queries

harry potter	apple	ncaa	clinton	selection	civil war
mlb	final four	easter	election	march madness	obama
oscar	kentucky derby	nasdaq	nfl	cannes	hurricane
mccain	olympics	opening day	kentucky	tiger	presidential election
triple crown	iraq	preakness	us open	euro 2008	spain

Our intuition is to leverage only exponential distribution of a document d towards certain subtopic d in favoring recent documents in the task of diversifying search results (according to the mined subtopics C).

$$f_s(d) = \sum_c Pr(c|q)Pr(q|d)Pr(c|d) \cdot \lambda \cdot e^{-\lambda \cdot t_d} \prod_{d\prime \in S} (1 - Pr(q|d\prime)Pr(c|d\prime) \cdot \lambda \cdot e^{-\lambda \cdot t_{d\prime}})$$

(3)

temp-xQuaD. Analogously, we modified the probabilistic model of xQuaD. Different from IA-Select, xQuaD introduces the parameter α, to control the trade-off between relevance and diversity. The objective function of **temp-xQuaD** is given in Equation 4.

$$f_s(d) = (1 - \alpha)Pr(d|q) + \alpha \sum_c Pr(c|q)Pr(c|d) \cdot \lambda \cdot e^{-\lambda \cdot t_d}$$
$$\prod_{d\prime \in S} (1 - Pr(c|d\prime) \cdot \lambda \cdot e^{-\lambda \cdot t_{d\prime}})$$

(4)

temp-topic-richness. Differently, the subtopics in topic-richness are modeled as a set of different data sources. The objective function of topic richness model is the generalization of IA-Select and xQuaD framework. Hence, we inject the temporal factor into the model analogously to what we did with **temp-IA-Select** and **temp-xQuaD**.

4   Experiments

In this section, we first investigate the quality of the subtopic mining from multiple sources. We then evaluate the performance of our time-aware diversification models on top of the mined subtopics on different metrics.

4.1   Evaluating Subtopics Mined from Query Log

The dataset used is the query log of AOL search engine (from March to May 2006), which has more than 30 million queries and 20 million click information. We time partition the query log into 12 accumulated parts (each approx. one week length), according to 12 simulated fixed hitting times. We manually selected 30 *temporally ambiguous*, multi-faceted queries, as shown in Table 1 for further studies.

Preliminary Analysis in the AOL Query Log. Due to the short time span (three months) of AOL, we decide to present a small study on the dynamic query aspects at a specific period. Table 2 shows the subtopics of the query ncaa at three

Table 2. Top-10 query subtopics (ordered by temporal weights) of *ncaa* in AOL

March 14th	March 31th	April 07th
0.0132 · march madness schedule	0.0100 · oakland raiders	0.0122 · ncaa women's basketball tournament
0.0117 · ncaa basketball tournament	0.0090 · ncaaw	0.0053 · ncaa basketball tournament
0.0068 · nfl draft	0.0042 · tito francona	0.0049 · cbs sports line
0.0048 · selection sunday	0.0031 · ncaa brackets	0.0033 · ncaaw
0.0037 · oakland raiders	0.0029 · ncaa division ii	0.0031 · ncaa final four
0.0032 · 2006 ncaa tournament bracket	0.0024 · andy goram	0.0029 · ncaa wrestling
0.0026 · brad hopkins released nfl	0.0024 · lakers	0.0028 · march madness bracket
0.0023 · roger clemens	0.0024 · ncaa women's basketball tournament bracket	0.0019 · ncaa basketball results
0.0021 · ncaa division ii	0.0021 · ncaa basketball brackets	0.0009 · andy goram
0.0014 · college basketball	0.0021 · nit brackets	0.0009 · ncaa division ii

Table 3. Statistics on subtopics quality

	Mean	Variance	Kurtosis
Coherence	0.23	0.11	3.402
Distinctness	0.831	0.072	1.031
Plausibility	0.158	0.193	4.227
Completeness	0.654	0.316	-1.029

different hitting times: *March 14th, March 31th* and *April 7th*. We speculate that the change in temporal aspects of a query tends to bind to the appearance of a happening event. We observe that the subtopics retrieved on *March 14th* reflect the *pre-event* aspects of the event march madness[1], e.g., *2006 ncaa tournament bracket, march madness schedule.* When mining subtopics on March 31st, a new subtopic emerges (*ncaa women's basketball tournament bracket*) refers about a new sub-event that occurs later on March 18th. We capture subtopics with post and late-event aspects when mining them on April 7th, i.e., *ncaa basketball results* for *post-event, ncaa final four* for *late-event.*

Evaluation Metrics. As we work on a specific set of queries, there is no standard test collection to evaluate the quality of the subtopic mining. We hence applied a novel assessment proposed by Radlinski *et al.* [14], where they defined four distinct metrics (i.e., coherence, distinctness, plausibility and completeness). Coherence indicates the level of explicitness is a subtopic, distinctness measures the distance between subtopics in terms of the relevant documents (URLs). Plausibility indicates how many percent of users who issue query q are satisfied with subtopic c. while completeness shows how complete is the set of subtopics C. In the query log context, we assume the relevance of a URL as being clicked on.

Experimental Results. We report the detailed results for coherence, distinctness, plausibility and completeness in Table 3. We empirically report the results of 10 (out of 30) queries that are most active in the AOL time span (by analyzing the time series) the queries based on statistical measurements i.e., mean, variance and kurtosis. All the reported results are in the range from 0 to 1, with 1 as the best result. The high distinctness result indicates the high diversity of our mined subtopics in the query log. The result for completeness is 0.654, meaning

---

[1] Note that march madness is the synonym of the ncaa basketball tournament that is held every March and final four refers to the later stage of the tournament.

our mined subtopics give a good coverage to the possible aspects of a query. The rather low results for coherence and plausibility (β reported at 0.5) are expected because the experiment is conducted based only on the query log. We focus on mining facets of the query and the distance between facets and its original query can be long. The similarity metric for the measurement is based only on Jaccard Similarity between sets of documents, without looking into the document contents (that are not available), also contributes to the low result.

4.2 Evaluating Subtopics Mined from a Temporal Document Collection

We used the TREC Blogs08 as a temporal document collection to conduct the experiment. The collection is crawled from January 2008 to February 2009, with more than 28 millions web documents in total (more than 70 thousand per day). For our task, the accurate and trustworthy timestamps of documents are necessary. However, timestamps associated to web documents in general are less reliable due to its decentralized nature and there is no standard for time and date [11]. We determined the timestamp of a blog document based on the three sources (ranked by reliability order), namely, document content, document URL and the crawled date. For each query, we construct a different LDA model on its partitioned results (there are 14 according to the 14 months² of Blogs08) to mine the latent subtopics. The training data in time slice t_i is the top 2000 relevant English documents D_{t_i} returned by Okapi BM25 retrieval model, such that for a document $d \in D$, $pubDate(d) \in t_i$. The optimum number of subtopics is determined on-the-fly in the range from (lower bound) $\gamma = 5$ to (upper bound) $\delta = 20$. The subtopic models are trained using Mallet³, an open source topic modeling tool, with its default parameters.

Preliminary Analysis in the Blogs08 Collection. We adopt the metrics proposed by [12] to measure the similarity between two latent LDA topics in order to assess the dynamic or variance between subtopics mined from LDA from two consecutive time slices t and t+1. The idea is to measure the similarity between two latent topics modeled by LDA, ϕ_i^t at time slice t and ϕ_j^{t+1} at the consecutive time slice $t + 1$.

Figure 2a demonstrates the dynamics of latent subtopics mined by LDA over time for the queries ncaa, windows, apple, mlb and final four. The dynamics of these queries are visualized by measuring the distance between two sets of subtopics mined from two consecutive time slices. We first observe the different level of subtopic dynamics of the four queries over time. The subtopic dynamics of query windows or apple are rather lower compared with ncaa or mlb. Kulkarni *et al.* [13] has initially analyzed the temporal correlations between the developments of query volume (or document content) and the query subtopics. In Figure 2b, we present the time series of these queries constituted using the query

---

² A finer-grained granularity can be achieved by mining from other source (e.g. query log), here we use monthly to acquire sufficient training data for our temporal queries.

³ http://mallet.cs.umass.edu/

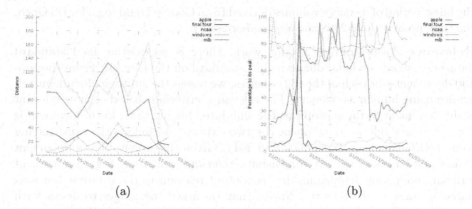

(a)   (b)

Fig. 2. Temporal dynamics of LDA subtopics measured by (a) JS Divergence and (b) normalized query volumes from Google Trend

volumes retrieved from Google Trend[4] at the same time period with Blogs08. The query volumes are normalized by the peaked volume. We further observe that queries with low dynamic level tend to have stable high query volumes over time, i.e., windows or apple while queries of high dynamic level (i.e., ncaa or mlb) tend to have sudden peaks in query volume. The query final four has a similar query volume development with ncaa, however, has a lower level. It is because final four only indicates a specific stage of the NCAA basketball tournament. The coverage of its aspects are then narrower than the query ncaa. A quantitative analysis of the correlation is left for future work.

Evaluation Metric and Result. Perplexity is used to evaluate the quality of our trained LDA models (the ability of a model to generalize documents). The better the generalization performance of a model is, the lower is the perplexity score of its output (towards zero). We use holdout validation with 90% of the data for training and 10% for testing. We obtained the average perplexity of 0.00715 (variance is 0.000437).

4.3   Diversification Models

Due to the time gap between the AOL query log (March to May 2006) and Blogs08 collections (crawled from January 2008 to February 2009), we exclude the subtopics mined from AOL query log. The latent LDA subtopics mined from Blogs08 are the sole source of subtopics in this experiment. We take the top-10 words from the word probabilities of a LDA topic as an explicit representation of the subtopic. We evaluate the effectiveness of diversification models at diversifying the search results produced by Okapi BM25 retrieval model. Only English documents with content-length of more than 300 characters are accepted in the final top-100 results. For each query, we choose a studying time-point (as a simulated hitting time) based on

---

[4] http://www.google.com/trends/

the burst period of its query volumes derived from Google Trend (e.g., for US Open, the time-point is *June 2008* and *September 2008*).

Relevance Assessments. In this work, there is no existing gold standard dataset. Instead, we build our own gold standard on the Blogs08. From the top-100 documents for each of the (30) queries, we assess the subtopic-document relevance using human assessment. The relevance criteria is based on how relevant is the document to the subtopic at the simulated hitting time. Each document is given a binary relevance judgment (by two experts), as follows the same setting from TREC Diversity Track 2009 and 2011. Given this orientation, a document is assessed based on the two dimensions, *relevance* and *time*. E.g., a document written about some happening that is content relevant to the subtopic but outdated is considered irrelevant. Notice that we asked the judges to assess with regards to different hitting times (simulated by monthly granularity)[5].

Evaluation Metrics. To evaluate the performance of our time-aware models, we use three different metrics (i.e., α-nDCG, Precision-IA and ERR-IA) that account for both the diversity and relevance of the results. In our evaluation, all metrics are computed following the standard practice in the TREC 2009 and TREC 2011 Web track [7,8]. In particular, α-nDCG is computed at $\alpha = 0.5$, in order to give equal weights to both relevance and diversity. We made a slight difference that in TREC 2009 Web track where they consider all query aspects equally important. We set the subtopics weight based on our dynamic subtopic measurement.

State-of-the-Art Model Performance. We measure the performance of the four state-of-the-art models: MMR, xQuaD, IA-Select and the topic richness model. The results are shown in Figures 3. For xQuaD, IA-Select and topic-richness, we use the mined temporal subtopics and their temporal weights as input (we skip their static methods (e.g., via Open Directory Project) since it is irrelevant in our case). We denote this change to the models with (*) symbol. We observe that measuring with α-nDCG@k, xQuaD*, IA-Select* and topic-richness model outperform MMR, while MMR shows certain increase over the baseline where there is no diversity re-ranking. We observe the same fashion when measuring with Precision-IA@k and ERR-IA@k. The results are expected since MMR does not account for subtopics when diversifying top-k result, it just tries to maximize the content gap between the top-k documents.

Time-Aware Parameter Optimization. The recency rate parameter λ is tuned to optimize the diversification models. We test λ in a wide range from 0.01 to 0.40. The parameter value with highest performance in terms of α-nDCG@k, ERR-IA@k and Precision-IA@k is chosen as best parameter value for the latter experiments. We choose k to be 10 in this set of experiments, as 10 is the common cutoff level in relevant diversity tasks [7,8]. We obtained λ equals to 0.04 as the optimal value of the experiments.

---

[5] The judgment is available at: www.l3s.de/~tunguyen/ecir2014_dataset.zip

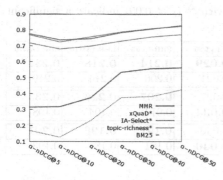 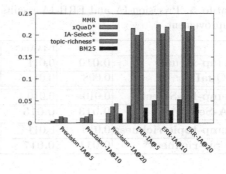

Fig. 3. Ranking results of baseline models, * models are with dynamic subtopic mining

Table 4. α-nDCG results with $^\triangle$ ($p < 0.05$), $^{\triangle\triangle}$ ($p < 0.01$) indicate a significant improvement

	$\alpha - nDCG@5$	$\alpha - nDCG@10$	$\alpha - nDCG@20$	$\alpha - nDCG@30$	$\alpha - nDCG@40$	$\alpha - nDCG@50$
temp-xQuaD	**$0.783^\triangle$**	**$0.737^\triangle$**	**$0.758^\triangle$**	**$0.805^{\triangle\triangle}$**	**$0.820^\triangle$**	**$0.847^\triangle$**
xQuaD*	0.699	0.687	0.706	0.751	0.772	0.789
temp-IA-Select	**0.781**	**$0.739^{\triangle\triangle}$**	**$0.755^{\triangle\triangle}$**	**$0.798^{\triangle\triangle}$**	**$0.822^{\triangle\triangle}$**	**$0.836^\triangle$**
IA-Select*	0.738	0.698	0.718	0.760	0.790	0.807
temp-topic-richness	**0.697**	**0.662**	**$0.686^\triangle$**	**$0.731^\triangle$**	**$0.753^\triangle$**	**$0.769^\triangle$**
topic-richness*	0.654	0.638	0.660	0.702	0.727	0.741

Diversification Performance. In these experiments, we aim to evaluate our time-aware models to answer our stated research question whether taking time into account that favors recency can improve the performance of the state-of-the-art diversification models. Tables 4 and 5 represent the results of the state-of-the-art and our time-aware models for α-nDCG and the two metrics Precision-IA and ERR-IA at different cutoffs respectively. The results for α-nDCG show that temp-XQuaD significantly ($p < 0.05$) outperforms the state-of-the-art xQuaD all cut-offs (with $p < 0.01$ at k = 30). temp-xQuaD also achieves better results for Precision-IA and ERR-IA, however the results are not significant. One intuitive reason is that, different from α-nDCG that is influenced by the diversity of the top-k document result, Precision-IA and ERR-IA is more sensitive on document ranking, while we only test on the top-100 documents. The margin value can become significant when testing with top-1000 documents for the two metrics. Similar to temp-xQuaD, temp-IA-Select surpass IA-Select in overall, significantly outperforms the state-of-the-art IA-Select when measuring by α-nDCG at the cutoff k = 10, 20, 30, 40 and 50. temp-IA-Select also gives better yet not significant performance when measured by ERR-IA. However, temp-IA-Select does not surpass the original IA-Select for Precision-IA. The results of Precision-IA at cutoff k = 5, 10 and 20 show a slight decrease in performance of temp-IA-Select. We also report the results for temp-topic-richness and topic-richness in a similar fashion. Overall, our time-aware models exceed their originated state-of-the-art diversification models in most of the experimental settings. temp-xQuaD

Table 5. Precision-IA and ERR-IA results with $^\triangle$ ($p < 0.05$) indicate a significant improvement

	P-IA@5	P-IA@10	P-IA@20	ERR-IA@5	ERR-IA@10	ERR-IA@20
temp-xQuaD	**0.010**	0.011	**0.029**	**0.214**	**0.218**	**0.232**$^\triangle$
xQuaD*	0.008	0.011	0.021	0.206	0.214	0.219
temp-IA-Select	**0.010**	0.010	0.027	**0.207**	**0.216**	**0.235**
IA-Select*	**0.013**	**0.013**	**0.034**	0.014	0.194	0.198
temp-topic-richness	0.010	0.011	0.030	**0.191**	**0.196**	**0.201**
topic-richness*	**0.011**	**0.017**	**0.040**	0.181	0.188	0.193

is the most consistent algorithm that outperforms xQuaD and gives better results among the six tested algorithms. On the other hand, even though surpassing the based model, temp-topic-richness gives a lower performance compared to the other two time-aware diversification models. However, the model is meant for taking subtopics from multiple sources, its performance could be enhanced if we account for other sources of subtopics (i.e., query log).

5   Related Work

Studying the temporal dynamics of subtopics has been addressed in some recent works [19,20]. Whiting *et al.* [19] considered event-driven topics as a prominent source of high temporal variable subtopics (search intent). They proposed an approach (in the absence query log) to present query intents by sections in the Wikipedia article. They further linked the temporal variance of intents (reflected by query volumes) with the change activity of the article sections. The proposed approach has certain limitations where the temporal dynamics and complexity in content structure of a Wikipedia article (where the subtopics are mined) is left un-tapped. Zou *et al.* [20], in another aspect, studied the effects of such subtopic temporal dynamics for the task of diversity evaluation. They conducted a small study on the Wikipedia disambiguation pages to analyze the changes in a subtopic popularity (the number of page views) over time. They concluded that such temporal dynamics impact the traditional diversity metrics for ambiguous queries, where the subtopic popularity is considered static over time. On the other hand, Berberich *et al.* [3] aimed to diversify search results over time, for those queries that are temporally ambiguous (i.e., the relevant time is un-known). Their proposed model, therefore, ignores the underlying intents of such queries and solely focuses on diversifying the relevant time periods of such queries. Styskin *et al.* [18] proposed a machine learning approach to identify recency-sensitive queries. Their large-scale experiments on real (recency-sensitive) queries show that promoting recent results (to the extent proportional to the query's recency level) to the result sets increases users' satisfaction.

6   Conclusions

In this paper, we studied the problem of diversifying search results for temporally ambiguous, multi-faceted queries. For such queries, the popularity and relevance of their corresponding subtopics are highly time-dependent, that is, the temporal dynamics of query subtopics can be observed. We determined dynamic subtopics by analyzing two data sources (i.e., query log and a document collection), which provides interesting insights for the identified temporal subtopics. Moreover, we proposed three time-aware diversifying methods that take into account the recency aspect of subtopics for re-ranking. The experimental results show that leveraging temporal subtopics as well as recency can improve the diversification performance (diversity and relevance) and outperform the baselines significantly, for temporally ambiguous, multi-faceted queries.

References

1. Agrawal, R., Gollapudi, S., Halverson, A., Ieong, S.: Diversifying search results. In: Proceedings of WSDM 2009 (2009)
2. Arun, R., Suresh, V., Veni Madhavan, C.E., Narasimha Murthy, M.N.: On finding the natural number of topics with latent dirichlet allocation: some observations. In: Zaki, M.J., Yu, J.X., Ravindran, B., Pudi, V. (eds.) PAKDD 2010, Part I. LNCS, vol. 6118, pp. 391–402. Springer, Heidelberg (2010)
3. Berberich, K., Bedathur, S.: Temporal diversification of search results. In: SIGIR 2013 Workshop on Time-aware Information Access (TAIA 2013) (2013)
4. Blei, D.M., Ng, A.Y., Jordan, M.I.: Latent dirichlet allocation. J. Mach. Learn. Res. 3, 993–1022 (2003)
5. Carbonell, J., Goldstein, J.: The use of MMR, diversity-based reranking for reordering documents and producing summaries. In: Proceedings of SIGIR 1998 (1998)
6. Carterette, B., Chandar, P.: Probabilistic models of ranking novel documents for faceted topic retrieval. In: Proceedings of CIKM 2009 (2009)
7. Clarke, C.L.A., Craswell, N., Soboroff, I.: Overview of the TREC 2009 web track. In: TREC (2009)
8. Clarke, C.L.A., Craswell, N., Soboroff, I., Voorhees, E.M.: Overview of the TREC 2011 web track. In: TREC (2011)
9. Craswell, N., Szummer, M.: Random walks on the click graph. In: Proceedings of SIGIR 2007 (2007)
10. Dou, Z., Hu, S., Chen, K., Song, R., Wen, J.-R.: Multi-dimensional search result diversification. In: Proceedings of WSDM 2011 (2011)
11. Kanhabua, N., Nørvåg, K.: Improving temporal language models for determining time of non-timestamped documents. In: Christensen-Dalsgaard, B., Castelli, D., Ammitzbøll Jurik, B., Lippincott, J. (eds.) ECDL 2008. LNCS, vol. 5173, pp. 358–370. Springer, Heidelberg (2008)
12. Kim, D., Oh, A.: Topic chains for understanding a news corpus. In: Gelbukh, A. (ed.) CICLing 2011, Part II. LNCS, vol. 6609, pp. 163–176. Springer, Heidelberg (2011)
13. Kulkarni, A., Teevan, J., Svore, K.M., Dumais, S.T.: Understanding temporal query dynamics. In: Proceedings of WSDM 2011 (2011)

14. Radlinski, F., Szummer, M., Craswell, N.: Metrics for assessing sets of subtopics. In: Proceedings of SIGIR 2010 (2010)
15. Rafiei, D., Bharat, K., Shukla, A.: Diversifying web search results. In: Proceedings of WWW 2010 (2010)
16. Santos, R.L., Macdonald, C., Ounis, I.: Exploiting query reformulations for web search result diversification. In: Proceedings of WWW 2010 (2010)
17. Song, W., Zhang, Y., Gao, H., Liu, T., Li, S.: HITSCIR system in NTCIR-9 subtopic mining task (2011)
18. Styskin, A., Romanenko, F., Vorobyev, F., Serdyukov, P.: Recency ranking by diversification of result set. In: Proceedings of CIKM 2011 (2011)
19. Whiting, S., Zhou, K., Jose, J., Lalmas, M.: Temporal variance of intents in multi-faceted event-driven information needs. In: Proceedings of SIGIR 2013 (2013)
20. Zhou, K., Whiting, S., Jose, J.M., Lalmas, M.: The impact of temporal intent variability on diversity evaluation. In: Serdyukov, P., Braslavski, P., Kuznetsov, S.O., Kamps, J., Rüger, S., Agichtein, E., Segalovich, I., Yilmaz, E. (eds.) ECIR 2013. LNCS, vol. 7814, pp. 820–823. Springer, Heidelberg (2013)

A Study of Query Term Deletion Using Large-Scale E-commerce Search Logs

Bishan Yang[1], Nish Parikh[2], Gyanit Singh[2], and Neel Sundaresan[2]

[1] Computer Science Department, Cornell University, USA
[2] eBay Research Labs, USA

Abstract. Query term deletion is one of the commonly used strategies for query rewriting. In this paper, we study the problem of query term deletion using large-scale e-commerce search logs. Specifically, we focus on queries that do not lead to user clicks and aim to predict a reduced and better query that can lead to clicks by term deletion. Accurate prediction of term deletion can potentially help users recover from poor search results and improve shopping experience. To achieve this, we use various term-dependent and query-dependent measures as features and build a classifier to predict which term is the most likely to be deleted from a given query. Our approach is data-driven. We investigate the large-scale query history and the document collection, verify the usefulness of previously proposed features, and also propose to incorporate the query category information into the term deletion predictors. We observe that training within-category classifiers can result in much better performance than training a unified classifier. We validate our approach using a large collection of query sessions logs from a leading e-commerce site and demonstrate that our approach provides promising performance in query term deletion prediction.

1 Introduction

Popular e-commerce site like Amazon and eBay see a large number of queries from their users wanting to buy products listed on these sites. It is not uncommon that certain percentage of the search request lead to no results. Such phenomenon happens more often in search engines that have limited collection of data with less redundant information than in general web search engines like Google [16]. For example, the query "small carry on bag for air plane" results in no search results while the query "carry on bag" leads to many relevant items in a real-world commercial search engine. In general, queries that contain extraneous terms can easily confuse the search engine and result in retrieval failure. This phenomenon can be seen in other sites like Netflix, LinkedIn, and Youtube as well. E-commerce search engines need to solve the important problem of recovery from no results. Typically they do by deleting terms from the original query. Picking the right set of terms to drop is an important problem.

In this paper, we investigate the problem of query term deletion in e-commerce queries. The goal is to delete extraneous terms from queries to improve the

M. de Rijke et al. (Eds.): ECIR 2014, LNCS 8416, pp. 235–246, 2014.

retrieved results. When there is no result for a query, term deletion can be performed until some results are retrieved. As deleting a term may result in information loss, we aim to select the term that is the least important regarding to the essential search needs of the users. To learn a good estimator of which query term is the less important, we make use of a large collection of query logs from a real-world commercial search engines, and collect pairs of user-generated query transitions where the first query does not result in user clicks but the second one does and the second query is reformulated by removing one term from the first one. We train a term deletion predictor by taking into account the information in the query history and the corpus as well as the current query context, and evaluate our approach on the collected query transition data.

Query term deletion has been explored in the past [9,12] in the information retrieval community. Various measures such as deletion frequency in history and inverse term frequency were proposed as indicators or features to predict how likely a term is deleted. Inspired by the previous work, we represent a term given its containing query using three types of features: linguistic features of the term; the deletion history of the term; the relevance of the term with respect to the query context. The selection of features mainly comes from the investigation of the large-scale query history and the corpus statistics. We then train a classifier using these features to assign higher scores to terms that are more likely to be deleted. We find that it is important to compute features conditioned on the query category and train within-category classifiers. To obtain query categories, we apply a multi-class query classifier [15]. Taking into account the category information allows us to predict term deletion according to the search goal as the query category captures the high-level search intent of the user.

To evaluate our approach, we construct a gold-standard dataset from large-scale search logs in a leading e-commerce site. We use data from query reformulations via term deletion by the actual users and the term deletion results in better click activities (from no clicks to at least one click). Experimental results show that our approach provides promising performance for predicting term deletion compared to various existing baselines. It indicates that our approach has the potential to be very useful in improving queries that lead to low recall in search results or no results in the e-commerce sites.

2   Related Work

Query term deletion has been explored in web search to improve the document retrieval performance. Rule-based approaches select words to delete from a given query using historical deletion pattens [9] or syntactic, paraphrase-based and frequency-based attributes [19]. The performance improvement provided by these approaches is limited because they ignore the query context from which the term is deleted. [17] proposed a strategy that deletes terms with the highest Mutual Information (MI) value. [11] formulated the query term deletion problem as ranking of sub-queries, and used Mutual information (MI) to select top-ranked sub-queries and then chosen the final sub-query based on the information in the snippets of the text from the top-ranked documents retrieved by the sub-queries. Although the MI value is expected to capture the context information of the query, it is limited in that it is only based on term pairs but not the query as a whole. [12] trained a classifier to rank all sub-queries using MI, IDF and other query-quality-related features computed from documents from an initial retrieval. It is expensive to realize as it requires additional retrieval for feature computing.

Query term deletion also closely relates to query-term weighting. [13] proposed a regression-based framework for weighing terms in long queries. It utilizes term-level linguistic features, corpus-based features such as TF, DF and context features extracted from the surrounding words. [18] applied regression to estimate the necessity of a term. The necessity value measures the percentage of relevant documents retrieved by a term. Term- and query-dependent features such as synonymy and topic centrality were used for supervised training. Such approaches are less direct in predicting query term deletion and harder to realize since finding the optimal weight for each query term is difficult.

Other relevant work includes identifying key concepts in queries. [1] used linguistic and statistical properties to identify core terms in the description queries. [3] extracted key concepts (noun phrases) from verbose queries using a supervised learning approach with features computed from query and corpus statistics. Such approaches have limitations in that key concepts may not be expressed as noun phrases in queries (especially for queries that do not have correct grammatical structures) and the training data requires expensive efforts to obtain.

The problem of improving zero-recall queries in commercial search engines has been previously studied. [16,14] discussed the problem of query rewriting for e-commerce queries that lead to zero recall, and showed that query rewriting techniques that help improve web document retrieval performance do not work well in the e-commerce domain. [8] showed that long and rare queries on product inventory introduces unique challenges for query suggestion methods. Our approach addresses zero-recall queries in e-commerce search engines. With effective term deletion prediction, our approach can be potentially very useful in helping users recover from zero-recall queries.

There has been a large body of work demonstrating the usefulness of search session data for various query applications. [9] demonstrated the effectiveness of using search session logs to predict word deletion. Jones et al.[10] employed user session information to produce query suggestions for ad selection. [5,4,7] mined query refinements based on the session co-occurrence information.

3   Dataset and Terminology

Our dataset consists of a large sample from 19-month search logs from a leading e-commerce site from years 2011-12. We organize search sessions by users and order users' activity by time. In the following we introduce the terminologies used through out the paper.

Definition 1. *A **query** q consists of a sequence of terms $t_1, t_2, ..., t_{|q|}$.*

Definition 2. *A **query trail** consists of a sequence of user-issued queries and their associated clicks within a search session.*

Definition 3. *A **successful query transition** consists of two consecutive queries q_1 and q_2 at the end of a query trail where q_1 and q_2 are both keyword-based queries, and they satisfy the following constraints: (1) q_1 does not lead to any click activity. (2) q_2 leads to one or more than one clicks. (3) q_2 is user-typed. (4) q_2 does not appear in the query trail before.*

Definition 4. *A **term deletion instance** is a successful query transition ($q_1 \rightarrow q_2$) where $|q_1| - |q_2| = 1$ and $q_1 \subset q_2$ (the set of terms in q_1 contains the set of terms in q_2).*

Definition 5. *A **Null-to-non-Null instance** is a term deletion instance ($q_1 \rightarrow q_2$) where q_1 leads to no search results.*

We sampled a collection of 88 million term deletion instances from the data, and split the data into a training set and a test set according to time [1]: instances from the last month are for testing and all the instances from previous months are for training. 30% of the considered data set contains Null-to-non-Null instances. We will report results on Null-to-non-Null instances and all term deletion instances but focus our discussion on the Null-to-non-Null instances.

Given an unsuccessful user query q (leading to no clicks), our goal is to predict a reduced form of the query by deleting one term from q. We denote the reduced query as q'. Ideally q' leads to more relevant results and can result in one or more user clicks.

For each term deletion instance ($q_1 \rightarrow q_2$) in our training set, we construct $|q_1|$ training instances: (t_i, q_1, y_i), $i \in 1, 2, ..., |q_1|$, where t_i is a term in query q_1, and y_i is a binary indicator $y_i = \delta(t_i = q_2 - q_1)$, that is, $y_i = 1$ if t_i is the deleted term chosen by the user, otherwise $y_i = 0$. For each instance ($q_1' \rightarrow q_2'$) in our test set, we predict a binary label for each term in q_1' given the query q_1'. We evaluate the prediction results by looking at q_2', the reduced query formulated by the actual user. We will distinguish queries of different lengths in our approach since the longer the query is the harder it is for search engines to understand [2], and thus the term deletion problem is harder [2].

4   Approach

In this section, we first briefly describe the supervised learning model we used for query term deletion problem. We then describe the features we used for estimating the term deletion probability, and lastly the training and prediction procedures.

---

[1] We divided the training/testing datasets based on time because it aligns with the real application scenario.

[2] Queries with length greater or equal to four only make up one fifth of our data.

Given training instances $\{(t_i, q, y_i)\}_{i=1}^{N}$, we train a logistic regression model to predict the probability of each term t_i being deleted given the term itself and its query context q,

$$p(y_i = 1 | t_i, q) = \frac{1}{1 + \exp(-\beta f(t_i, q))} \tag{1}$$

where $f(t_i, q)$ denotes a vector of features

$$f(t_i, q) = (f_1(t_1, q), f_2(t_2, q), ..., f_k(t_k, q)),$$

and β is a parameter vector estimated by the model.

We hope to construct term-level features that are query-dependent. To incorporate the query-dependent information, we classify each query into a predefined category and compute features depending on the query category. The query category provides important context information for query term deletion prediction and also allows the same term to have different level of importance given its appearing context. Take the queries *gap wool blazer jacket 4* and *vintage spark gap transmitter* as examples, the word *gap* conveys very different meanings and has different levels of importance in the two queries.

Most e-commerce services provide hierarchical categories to assist users better explore their product listing more effectively and efficiently. For example, eBay provides different categories, e.g. books, toys, computers, etc. These categories provide valuable information for matching users' search intent to the products they are interested in. In this work, we consider 38 meta-categories from the e-commerce site. During training, we can observe the query category by looking at the click-through records. During testing, we apply a multi-class query category classifier built on the historical queries that lead to user clicks to infer the possible category of the test query [15].

During feature construction, we assume the query category information is available for all queries. Given a query $q = t_1 t_2 ... t_n$ and its category c, we construct features for each term t_i that capture three types of information: (1) the linguistic and lexical properties of t_i, (2) the deletion history of t_i in category c, (3) the surrounding context of t_i in query q.

4.1 Linguistic and Lexical Features

We consider the lexical form of the term t_i as well as its linguistic properties as features, including binary indicators of being a conjunction, an adjective and numeric. We also make use of a dictionary of brand names [3] and create an indicator feature to indicate whether the considered term is a part of a brand name. In addition, we build a term importance predictor based on the product titles and descriptions in the inventory and use the importance value as a feature. Specifically, the importance score of a term is measured as the probability of the

---

[3] It is constructed using product titles in the product inventory, and covers around 45% of the training data

term occurring in the titles conditioned on its probability occurring in the descriptions. The higher the value is the more important the term is for describing the product. This predictor alone can be used for predicting term deletion (the term with the least importance score gets deleted). We will use it as a baseline to our approach in Section 5.

4.2 History-Based Features

We draw on previous work on rule-based term deletion [9] and consider rules that are effective in identifying the deleted terms as features. For a given term t_i, these include (1) **Deletion history**, the deletion frequency of t_i in historical queries under the query category c. It is defined as $p(delete(t_i)|contains(t_i), c) = td/tq$, where td is the number of times word t_i being deleted and tq is the number of times word t_i being seen; (2) **Rareness**: the rareness of t_i in historical queries under the query category c. It is defined as $\log(N/df)$, where N is the total number of queries in category c and df is the number of queries in c where term t_i appears. Intuitively, the more rare the word is, the more discriminative it is for expressing the search intent; (3) **isRightmost**: a binary feature that describes whether the term is the rightmost word in the query. This feature is created based on the observation that users tend to delete the rightmost word in the query [4].

4.3 Context Features

The textual context of the considered term t_i in the given query q provides strong clues for determining whether the term is likely to be deleted. For example, the term **card** is more likely to be deleted in query *nintendo club game card case* than in query *16gb p vita memory card*. Another example is the term **new** in queries *oem new sony dualshock* and *reebok pnk new delhi*, it is reasonable to delete "new" from the first query but not the second query. To incorporate the surrounding context, we consider the following features: (1) **Collocations**, the lexical forms of the neighboring words t_{i-1} and t_{i+1}; (2) **Point-wise Mutual Information**, a binary feature derived from the PMI score between two terms in the query. For each term pair (t_a, t_b) in query q, we compute $pmi(t_a, t_b) = \log \frac{p(t_a, t_b)}{p(t_a)p(t_b)}$ using the frequencies of term t_a and t_b in historical queries under category c. Then we find the pair of terms with the highest pmi score

$$(t_p, t_q) = \arg \max_{t_a, t_b \in q} pmi(t_a, t_b).$$

If $t_i = t_p$ or $t_i = t_q$, then the pmi feature equals 1, otherwise equals 0. The intuition is that if t_i is highly associated with another term t_j in query q, then it

---

[4] Predicting the rightmost word will give us around 40% accuracy on queries of length greater than four, which is much better than the random baseline (less than 20%). The explanation maybe that users tend to build up queries from left to right and put the most important words first than less important words at the end.

may be unreasonable to delete t_i and keep t_j. (3) **Mutual information with query categories**, a binary feature that captures whether t_i is the most relevant to category c among the other terms in q. For each term in q, we compute its mutual information score with the query category c, and identify the term with the highest mutual information score:

$$t = \arg\max_{t' \in q} MI(t', c),$$

where $MI(t', c) = \sum_{t'} \sum_{c} p(t', c) \log \frac{p(t',c)}{p(t')p(c)}$. If t_i equals t, then the feature has value 1, otherwise it has value 0.

4.4   Within-Category Training and Prediction

We find that it is important to training different term deletion predictors for different query categories. Training a unified predictor by combining training data from different categories results in much poorer performance than training within-category predictors. This is because query category provides important information for the query context and allows the prediction of term deletion to be more query-dependent.

During prediction, we apply the in-category logistic regression predictor to every term in the test query q' and predict the term with the highest deletion probability to be deleted: $t = \arg\max_i p(y_i' = 1 | t_i', q')$. For evaluation, we compare this term to the term that is deleted by the actual user in our gold-standard data. If they are the same then the prediction is correct, otherwise it is not correct.

5   Experiments

As described in Section 3, we construct our evaluation dataset from the query logs and split it into a training set and a test set. We further split the training set and the test set according to the length of the considered queries (the length of query q_1 in the term deletion instance $(q_1 \rightarrow q_2)$). Table 1 shows the data statistics of the datasets.

Table 1. Data statistics

Data	Training	Test
Length=2	11,833,859	567,411
Length=3	34,133,105	1,705,659
Length $\geq$ 4	38,078,382	2,208,426
Length=2(Null-to-non-Null)	3,252,655	154,528
Length=3(Null-to-non-Null)	9,236,936	447,807
Length $\geq$ 4(Null-to-non-Null)	12,801,525	717,887

5.1 Baselines and Measures

We consider the following methods as baselines:

1. **Random:** Delete a random word.
2. **Leftmost:** Delete the leftmost word. The assumption is that most users tend to put optional adjectives on the left and key noun phrases on the right.
3. **Rightmost:** Delete the rightmost word. The assumption is that most users tend to write important words first and then less important words to the right.
4. **Deletion Frequency (DF):** Delete the word which was deleted the most frequently in the past.
5. **Conditional Deletion Frequency (CDF):** Delete the word which was deleted the most frequently with respect to the times it appears in queries.
6. **Term Importance Baseline (TIMP):** Delete the least important word based on the term importance score computed by the method described in Section 4.1.

The first four baselines were previously proposed methods used for query term deletion [9]. They are all rule-based approaches using historical deletion pattens. **TIMP** is a learning-based method for query term deletion.

   We trained our term deletion predictors using L1-regularized logistic regression with default parameters using the LIBLINEAR [6] package. We use accuracy as the performance measure. For each deletion instance $(q_1 \rightarrow q_2)$ in the test data, if the predicted word $t \in q_1$ is the same as the actual deleted word $q_1 - q_2$, then we consider the prediction is correct for the test instance. The accuracy is computed as

$$Accuracy = \frac{\#\text{correctly predicted instances}}{\#\text{test instances}}$$

5.2 Results

Table 2 shows the performance of query term deletion prediction on the subset of test data that contains the Null-to-non-Null instances. The trend of the results for the whole dataset is the same. The results are reported for queries of different lengths, and each accuracy score is averaged on all the meta-categories. In general, the performance drops as the length of queries increases. This is expected, as long queries are usually rare and harder to understand. We can see that our method significantly outperforms all the baselines. The *Rightmost* baseline performs surprisingly well, especially better than *DF* that uses the history information. This implies that the rightmost position can be a better indicator for term deletion than the absolute term deletion frequency. *CDF* performing better than *DF* indicates that we should take into account the frequency of the word in the queries when predicting its deletion probability. *TIMP* makes use of corpus statistics to infer term importance, but did not yield much improvement comparing to the history-based baselines. The superior performance of our approach is due to the integration of the history- and corpus-based indicators and the incorporation of the context information of the considered query.

Table 2. Accuracy (%) results of different methods on Null-to-non-Null instances

Method	len=2	len=3	len $\geq$ 4
Random	50.29	33.28	21.97
LeftMost	31.2	28.74	22.15
RightMost	68.8	42.48	31.14
DF	63.22	36.82	24.9
CDF	68.34	54.16	44.49
TIMP	65.25	52.36	48.1
Our approach	**81.87**	**69.95**	**62.38**
Our approach (top 2)	–	–	**84.35**

Table 3. Accuracy (%) results on Null-to-non-Null instances in different categories

Category	len=2	len=3	len$\geq$ 4	len$\geq$ 4 (Top 2)
Clothing, Shoes & Accessories	81.79	57.9	51.19	77.57
Toys & Hobbies	84.08	60.79	57.61	83.01
Computers/Tablets & Networking	81.77	59.74	52.2	76.37
Musical Instruments & Gear	**86.03**	63.41	58.05	82.79
Books	84.95	**66.12**	**62.95**	**84.1**

We also evaluate our method using a more relaxed accuracy measure (denoted as *top 2* in the table) which considers a prediction to be correct if the top 2 predicted terms (the two terms with the highest deletion probabilities) contain the gold standard deleted term. Using such measure gives us promising prediction results – the accuracy can go up to 84.35%. In practice, this implies that given a zero-recall query, with 84.35% accuracy, the system can formulate a better query for the user by removing either the top predicted term or the second best predicted term provided by our model. These two choices can both be represented to the users in the form of query suggestions.

Table 3 shows the performance of our model in a subset of 38 meta categories. We can see that the performance results vary across different categories. Some categories demonstrate much better prediction performance than the others, e.g. the "Books" category has the best performance for queries of length $\geq$ 3. Note that this is not due to the available size of the training data, e.g. the performance in the "computer" categories is consistently lower than the performance in the Toy category while their training size is similar. We also observe that by training within-category predictors, the averaged accuracy on Null-to-non-Null instances is much higher than the averaged accuracy obtained by training a unified predictor disregarding the category information. This further confirms that query term deletion prediction is very query-category-dependent, and taking into account the query category information is important.

In the following experiments, we will focus on term deletion prediction on long queries (using the subset of training and test data with queries of length $\geq$ 4) and demonstrate the effectiveness of our approach on long queries.

Null-to-non-Null Instances vs. All Instances. First, we study the performance of our method on both Null-to-non-Null instances and all term deletion instances. Figure 1 shows the results. We can see that in general the performance on Null-to-non-Null instances is better than the performance on all the instances. This implies that it is easier to infer the deleted term in zero-result queries than queries that yield some search results. In other words, the deletion behavior of users who try to recover from a zero-result query is more predictable. The averaged accuracy across different categories is 62.38% for Null-to-non-Null instances, much better than the accuracy 57.72% on all the term deletion instances. For the Top 2 measure, the average accuracy is 84.35% versus 81.74%.

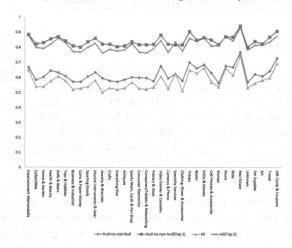

Fig. 1. Performance on Null-to-non-Null instances vs. all instances on long queries

Table 4. Feature weights estimated by logistic regression (on the Clothing, Shoes & Accessories category)

Features	Weights
BRAND	1.978346456
Adjective	1.321605972
Numerical	1.311431391
MI	1.041425089
PMI	0.459417038
Rareness	0.054978431
isRightmost	-1.834808818
Deletion history	-3.610057573

Feature Analysis. We also examine the effect of various features by looking at their weights estimated by logistic regression. Table 4 lists some features and their weights output by logistic regression trained in the Clothing category [5]. In

---

[5] We only show the analysis on the Clothing category due to space limitation. The learned feature weights vary across different categories since we train a separate model for each category.

general, the larger the weight is, the more indicative the feature is for predicting the term to be kept, in contrast, the smaller the weight is, the more indicative the feature is for predicting the term to be deleted. For example, *Brand* is a strong indicator for keeping the term while *Deletion history* and *isRightmost* both contribute to predicting the term to be deleted. Also, as expected, context features MI and PMI play a role in predicting the term to be kept. The general trend of the importance of the history-based and context-based features is similar across different categories. However, the weights of the linguistic and lexical features can vary a lot in different categories due to the usage of domain-dependent language. For example, *Adjective* has positive weight in the "Clothing" category but has negative weight in other categories such as "Computers".

6 Conclusion

In this paper, we study the problem of query term deletion in e-commerce queries. Especially we focus on real-world queries that were transformed by users via term deletion and results in better click activities. We formulate the problem of query term deletion as selecting the least important term from the query. We use various term- and query-dependent measures as features and build a classifier to predict the probability of each term being deleted given a specific query. To better model the conditional probability, we compute the features statistics conditioned on the query category and train within-category term deletion predictors. Experiments on a large-scale e-commerce query data demonstrates that our data-driven approach provides promising performance for predicting term deletion and can potentially be useful for helping users recover from zero-recall queries.

Identifying extraneous terms in queries is a challenging task. The analysis of a large-scale query term deletion dataset in the real world provides valuable insights into the understanding of the problem. For future work, we would like to combine query term deletion with other forms of query rewriting such as query substitution and query expansion for achieving better search experience for the users. We also want to investigate the incorporation of more linguistically-informed features for better understanding of the relevance of the query terms to the search intent.

References

1. Allan, J., Callan, J., Croft, W., Ballesteros, L., Broglio, J., Xu, J., Shu, H.: Inquery at trec-5. In: Center for Intelligent Information Retrieval, Dept. of Computer Science, University of Massachusetts, Amherst, Mass (1996)
2. Bailey, P., White, R., Liu, H., Kumaran, G.: Mining historic query trails to label long and rare search engine queries. ACM Transactions on the Web (TWEB) 4(4), 15 (2010)
3. Bendersky, M., Croft, W.B.: Discovering key concepts in verbose queries. In: Proceedings of SIGIR, pp. 491–498 (2008)

4. Chien, S., Immorlica, N.: Semantic similarity between search engine queries using temporal correlation. In: Proceedings of the 14th International Conference on World Wide Web, pp. 2–11. ACM (2005)
5. Cucerzan, S., Brill, E.: Extracting semantically related queries by exploiting user session information. Technical report, Technical report, Microsoft Research (2005)
6. Fan, R.-E., Chang, K.-W., Hsieh, C.-J., Wang, X.-R., Lin, C.-J.: Liblinear: A library for large linear classification. The Journal of Machine Learning Research 9, 1871–1874 (2008)
7. Fonseca, B., Golgher, P., Pôssas, B., Ribeiro-Neto, B., Ziviani, N.: Concept-based interactive query expansion. In: Proceedings of the 14th ACM International Conference on Information and Knowledge Management, pp. 696–703. ACM (2005)
8. Hasan, M.A., Parikh, N., Singh, G., Sundaresan, N.: Query suggestion for e-commerce sites. In: Proceedings of the Fourth ACM International Conference on Web Search and Data Mining, WSDM 2011, pp. 765–774 (2011)
9. Jones, R., Fain, D.C.: Query word deletion prediction. In: Proceedings of SIGIR, pp. 435–436 (2003)
10. Jones, R., Rey, B., Madani, O., Greiner, W.: Generating query substitutions. In: Proceedings of the 15th International Conference on World Wide Web, WWW 2006, New York, NY, USA, pp. 387–396 (2006)
11. Kumaran, G., Allan, J.: A case for shorter queries, and helping users create them. In: HLT-NAACL, pp. 220–227 (2007)
12. Kumaran, G., Carvalho, V.R.: Reducing long queries using query quality predictors. In: Proceedings of SIGIR, pp. 564–571 (2009)
13. Lease, M., Allan, J., Croft, W.B.: Regression rank: Learning to meet the opportunity of descriptive queries. In: Boughanem, M., Berrut, C., Mothe, J., Soule-Dupuy, C. (eds.) ECIR 2009. LNCS, vol. 5478, pp. 90–101. Springer, Heidelberg (2009)
14. Parikh, N., Sundaresan, N.: Inferring semantic query relations from collective user behavior. In: Proceedings of the 17th ACM Conference on Information and Knowledge Management, pp. 349–358. ACM (2008)
15. Shen, D., Ruvini, J.D., Somaiya, M., Sundaresan, N.: Item categorization in the e-commerce domain. In: Proceedings of the 20th ACM International Conference on Information and Knowledge Management, pp. 1921–1924. ACM (2011)
16. Singh, G., Parikh, N., Sundaresan, N.: Rewriting null e-commerce queries to recommend products. In: Proceedings of the 21st International Conference Companion on World Wide Web, WWW 2012 Companion, pp. 73–82 (2012)
17. Wu, H., Fang, H.: An exploration of query term deletion. In: Proceedings of the ECIR 2011 Workshop on Information Retrieval Over Query Sessions (2011)
18. Zhao, L., Callan, J.: Term necessity prediction. In: Proceedings of the 19th ACM International Conference on Information and Knowledge Management, CIKM 2010, pp. 259–268 (2010)
19. Zukerman, I., Raskutti, B., Wen, Y.: Query expansion and query reduction in document retrieval. In: Proceedings of the 15th IEEE International Conference on Tools with Artificial Intelligence, pp. 552–559. IEEE (2003)

Detecting Missing Content Queries in an SMS-Based HIV/AIDS FAQ Retrieval System

Edwin Thuma[1,2], Simon Rogers[1], and Iadh Ounis[1]

[1] School of Computing Science, University of Glasgow, Glasgow, UK
{simon.rogers,iadh.ounis}@glasgow.ac.uk
[2] Department of Computer Science, University of Botswana, Gaborone, Botswana
thumae@dcs.gla.ac.uk, thumae@mopipi.ub.bw

Abstract. Automated Frequently Asked Question (FAQ) answering systems use pre-stored sets of question-answer pairs as an information source to answer natural language questions posed by the users. The main problem with this kind of information source is that there is no guarantee that there will be a relevant question-answer pair for all user queries. In this paper, we propose to deploy a binary classifier in an existing SMS-Based HIV/AIDS FAQ retrieval system to detect user queries that do not have the relevant question-answer pair in the FAQ document collection. Before deploying such a classifier, we first evaluate different feature sets for training in order to determine the sets of features that can build a model that yields the best classification accuracy. We carry out our evaluation using seven different feature sets generated from a query log before and after retrieval by the FAQ retrieval system. Our results suggest that, combining different feature sets markedly improves the classification accuracy.

Keywords: Frequently Asked Question, Missing Content Queries, Text Classification.

1 Introduction

Mobile phones have emerged as the platform of choice for providing services such as banking [18], payment of utility bills [26] and learning (M-Learning) [1] in the developing world. This is because of their low cost and high penetration in the market [1,7]. In order to take advantage of this high mobile phone penetration in Botswana, we have developed an SMS-Based HIV/AIDS FAQ retrieval system that can be queried by users to provide answers on HIV/AIDS related queries. The system uses, as its information source the full HIV/AIDS FAQ question-answer booklet provided by the Ministry of Health (MOH) in Botswana for its IPOLETSE[1] call centre. This FAQ question-answer booklet is made up of 205 question-answer pairs organised into eleven chapters of varying sizes. For example, there is a chapter on "Nutrition, Vitamins and HIV/AIDS" and a chapter on "Men and HIV/AIDS". Below is an example of a question-answer pair entry that can be found in Chapter Eight, "Introduction to ARV Therapy":

---

[1] http://www.hiv.gov.bw/content/ipoletse

M. de Rijke et al. (Eds.): ECIR 2014, LNCS 8416, pp. 247–259, 2014.

Question : What is the importance of taking ARV therapy if there is no cure for AIDS?

Answer : Although ARV therapy is not a cure for AIDS, it enables you to live a longer and more productive life if you take it the right way. ARV therapy is just like treatment for chronic illnesses such as diabetes or high blood pressure.

For the remainder of this paper, we will refer to a question-answer pair as the FAQ document and the set of all 205 FAQ documents as the FAQ document collection. The users' SMS messages will be referred to as queries.

Because of the constrained nature of mobile phone displays, our system does not return a ranked list of FAQ documents for each SMS query. Instead our system adopts an iterative interaction strategy proposed in [24]. However, in order to reduce the number of iterations between the user and the system, we differ with this earlier strategy by returning the whole FAQ document to the user at each iteration and not the question part only. For example, for each SMS query sent by the user, the system ranks the FAQ documents in the FAQ document collection. The top ranked FAQ document is returned to the user. If the user is satisfied that this FAQ document matches the SMS query, the user respond with "YES" or remain idle and the interaction terminates. If the user is not satisfied, they reply with "NO", and the system then displays the next highest ranked FAQ document and the process is repeated until the user respond with "YES".

One key problem in this domain is that there is no guarantee that there will be a relevant FAQ document for all user queries [22,23]. This may result in users iterating with our system for longer, wasting time and their SMS credit. It is for this reason that we propose to deploy a classifier that can detect those queries for which there are no relevant FAQ documents in the collection. In their earlier work, Yom-Tov et al. [25] defined Missing Content Queries (MCQs) as those queries for which there are no relevant documents in the collection and non-MCQs as those that have relevant documents. In this work, we will use this definition by Yom-Tov et al. Detecting $MCQs$ can be useful for both the user and the information supplier in an SMS Based HIV/AIDS FAQ retrieval system. The user can be informed if the FAQ document collection does not contain the relevant FAQ document for the user query rather than returning non-relevant FAQ documents [16]. On the other-hand, the information supplier can note the kind of information need that is of interest to the user but not addressed by the current information source (FAQ document collection) [25]. Armed with this knowledge, the information supplier can update the FAQ document collection by adding or modifying FAQ documents entries.

Previous work on $MCQs$ detection by Yom-Tov et al. [25] , Hogan et al. [13] and Leveling [17] relied on classification to detect queries that do not have relevant FAQ documents in the FAQ document collection. In their work, Yom-Tov et al. and Hogan et al. trained their classifiers using features generated by query difficulty estimators. Leveling [17] on the other-hand generated the features for training the classifier during the retrieval phase on the training data and examined the top five documents (e.g scores for the top five documents).

In this work, we will follow the work in [13,16,17,25] by tackling the detection of $MCQs$ in an HIV/AIDS FAQ retrieval systems as a binary classification problem. This paper attempts to contribute to the current state-of-the-art in missing content detection for an SMS-Based FAQ retrieval system by first analysing and evaluating different feature sets in order to determine the best combination of features that can be used to build a model that would yield the highest classification accuracy. We will carry out a thorough evaluation using two different datasets. The first dataset is a collection of HIV/AIDS documents and a query log of HIV/AIDS-related $MCQs$ and $non - MCQs$ collected in Botswana over a period of 3 months. In Section 3.2, we describe how we collected this query log. The other dataset is a collection of FAQ documents from the Forum for Information Retrieval Evaluation (FIRE2012)[2] English monolingual SMS-Based FAQ retrieval training data and the associated query log (SMS queries). The FIRE2012 dataset has 7251 FAQ documents in total. We use two different datasets in order to be able to make a general conclusion. Seven different feature sets will be created for these datasets. The first feature set (baseline) will be made up of the actual query strings of $non - MCQs$ and $MCQs$ as features for the training and testing instances. The second feature set will be made up of features deployed by Leveling [17] while the third feature set will be made up of the features deployed by Hogan et al. [13] and some additional query difficulty predictors. We then create four additional feature sets by combining the previous three feature sets in order to determine whether combining these feature sets would yield a better classification accuracy. Three different classifiers in WEKA [9], namely Naive Bayes [15], RandomForest [2] and S-SVC (Support Vector Classification) [4] will be trained and tested on this feature sets to evaluate their effectiveness in classifying $non - MCQs$ and $MCQs$. In this paper, we also investigate whether increasing the size of the training set would yield a better classification accuracy.

The rest of this paper is organised as follows: We survey related work in Section 2, followed by a description of our methodology in Section 3. In Section 4, we describe our experimental setting. We then present experimental results and evaluation in Section 5, followed by concluding remarks in Section 6.

2   Related Work

Earlier work on the detection of missing content queries ($MCQs$) was first introduced by Yom-Tov et al. [25] on their investigation of the applications of query difficulty estimation. In their experiment, they artificially created 166 $MCQs$ by deleting the relevant documents for 166 queries from the TREC-8 collection that had a 200 description-part queries and 200 title-part queries. They then trained a tree-based estimator to classify $MCQs$ and $non - MCQs$ using the complete set of 400 queries. In their experiment, they used a query difficulty estimator trained by analysing the overlap between the results of the full query and the results of its sub-queries to pre-filter easy queries before identifying $MCQs$

---

[2] http://www.isical.ac.in/~fire/faq-retrieval/2012/data.html

with a tree-based classifier. Their results suggest that identifying $MCQs$ can be improved by combining the MCQ classifier (tree-based classifier) with a query difficulty estimator. They also reported that when the MCQ classifier is used alone to detect $MCQs$, it groups together easy queries and $MCQs$ thus yielding worse results. They suggested that this can be alleviated by pre-filtering easy queries using a query difficulty estimator.

Hogan et al. [13] combined 3 different lists of $MCQs$ generated through three different approaches and then applied a simple majority voting approach to identify $MCQs$. The first list of candidate $MCQs$ was generated using an approach proposed by Ferguson et al. [8] for determining the number of relevant documents to use for query expansion. In this approach, a score for each query was produced based on the inverse document frequency (IDF) component of the BM25 score for each query without taking into consideration the term frequency and the document length. First, the maximum score possible for any document was calculated as the sum of the IDF scores for all the query terms. Following this approach, documents without all the query terms will have a score less than the maximum score. A threshold was then used to determine if a query should be added to the list of candidate $MCQs$. They added queries that had all their document scores below 70 % of the maximum score to this list.

The second list of candidate $MCQs$ was generated by training a k-nearest-neighbour classifier to identify $MCQs$ and $non - MCQs$. The features used to train this classifier included query performance estimators (Average Inverse Collection Term Frequency (AvICTF) [12], Simplified Clarity Score (SCS)) [11], the derivates of the similarity score between collection and query (SumSCQ, AvSCQ, MaxSCQ) [27], result set size and the un-normalised BM25 document scores for the top five documents. Their classifier achieved 78 % accuracy on the FAQ SMS training data using a leave-one-out validation. The third list of candidate $MCQs$ was generated by simply counting the number of term overlaps for each incoming query and the highest ranked documents (For example, if the query consists of more than one term and had only one term in common with the document, that query was marked as MCQ). Hogan et al. used the held-out training data to evaluate their approach and they concluded that combining the three lists of candidate $MCQs$ through a simple majority voting yielded better results.

Leveling [17] viewed the detection of missing content queries as a classification problem. In his approach, he trained an IB1 classifier as implemented in TiMBL [6] using numeric features generated during the retrieval phase on the training data (FIRE2011 SMS-Based FAQ retrieval monolingual English data) to distinguish between $MCQs$ and $non - MCQs$. The features used for training were comprised of the result set size for each query, the raw BM25 document scores for the top five documents (5 features), the percentage difference of the BM25 document scores between the consecutive top 5 documents (4 features), the normalised BM25 document scores for the top five retrieved documents (5 features) and the term overlap scores for the SMS query and the top 5 retrieved documents (5 features). Their approach essentially yielded a binary classifier that can determine whether a query is MCQ or $non - MCQ$. This approach is much simpler compared to the approach proposed by Hogan et al. [13] because it relies on a single classifier instead of relying on several classifiers. Leveling evaluated this approach using a leave-one-out validation approach which is supported by TiMBL and reported a classification accuracy of 86.3 % for $MCQs$ with the best performing system. Such a high classification accuracy for $MCQs$ resulted in a very low classification accuracy of 56.0 % for $non - MCQs$.

Misclassification of a large number of $non - MCQs$ can be costly in an HIV/AIDS FAQ retrieval system. In this work, our goal is to minimise this misclassification accuracy. In order to achieve this, we differ with previous work, by first analysing and evaluating different feature sets in order to determine the best combination of features that would yield the best classification accuracy.

3   Methodology

We begin Section 3.1 by outlining our research questions, followed by Section 3.2 where we describe how we collected and identified non-$MCQs$ and $MCQs$. We then describe how we created the training and testing instances in Section 3.3.

3.1   Research Questions

- **R1:** Which types of features produce the highest classification accuracy when classifying $MCQs$ and $non - MCQs$?
- **R2:** Does combining different types of feature sets, produce a better classification accuracy when classifying the $MCQs$ and the $non-MCQs$, compared to classifying using any individual feature set?
- **R3:** Does increasing the size of the training set for the $MCQs$ and the $non - MCQs$ yield a higher classification accuracy?
- **R4:** Do we get comparable classification accuracy when these feature sets are generated using a different dataset?

3.2   Collecting and Identifying Missing Content and Non-missing Content Queries

A study was conducted in Botswana to collect SMS queries on the general topic of HIV/AIDS. In this study, 85 participants were recruited to provide SMS queries. Having provided the SMS queries, they then used a web-based interface to find the relevant FAQ documents from the FAQ document collection using the SMS queries. This provided us with SMS queries linked to the appropriate FAQ documents in the collection. The SMS queries that could be matched to an FAQ documents were labelled as $non - MCQ$. Those that could not be matched to an FAQ document were labelled as $MCQs$. In total, 957 SMS queries were collected of which 750 $(non - MCQs)$ could be matched to an FAQ document in the collection. The remaining 207 $(MCQs)$ did not match anything in the collection. In this work, we investigate how to detect these $MCQs$ in an Automated

HIV/AIDS FAQ retrieval system. In order to investigate the robustness of our approach, we also used a second dataset of 707 SMS queries (540 $non - MCQs$ and 167 $MCQs$) that we randomly selected from the FIRE2012 English Monolingual SMS query dataset. This dataset had 4476 SMS queries. We selected only a fraction of these SMS queries to use in our experimental evaluation because we had to manually correct them for spelling errors.

The main difference between the two datasets (HIV/AIDS and FIRE2012 datasets) is that the FIRE2012 dataset covers several topics (Railways, telecommunication, health, career counselling and general knowledge e.t.c) while the HIV/AIDS dataset only has one topic, HIV/AIDS. Moreover, the $MCQs$ for the HIV/AIDS dataset are on-topic (related to HIV/AIDS only) while the $MCQs$ for the FIRE2012 dataset has both on-topic and off-topic $MCQs$. Both the HIV/AIDS and the FIRE2012 SMS queries were manually corrected for spelling errors so that such a confounding variable does not influence the outcome of our experiments. In the next section, we describe how we created the training and testing instances using this query log to answer the above research questions.

3.3 Creating Training and Testing Instances for Missing Content and Non-missing Content Queries

In this work, we used the HIV/AIDS and the FIRE2012 SMS query log that we categorised as described in Section 3.2. Seven different feature sets were created for our experimental investigation and evaluation. In order for us to be able to answer research question $R1$, we created three different feature sets, $fSet1$, $fSet2$, $fSet3$ as described below:

$fSet1$: Instances in this feature set were represented by a vector of attributes representing word count information from the text contained in the query strings. Below is an example of an instance, first represented as a query string and then as a vector of attributes representing word count information of this query string.

Query String: what does aids stand for?

Word Count: 23 1,159 1,212 1,488 1,591 1.

In our example above, the attributes in this vector are separated by commas and each attribute is made up of two parts, the attribute number, and the word count information. For example the attribute *"23 1"* denotes that the term *what* is attribute number 23 in the string vector and this term only appears once.

$fSet2$: For this feature set, the training and testing instances were created using the approach proposed by Leveling [17]. In particular, numeric attributes generated during the retrieval phase of the FAQ documents by the aforementioned $non - MCQs$ and $MCQs$ were used in this feature set. For each query, we performed retrieval on the FAQ Retrieval Platform described in Section 4.1 to extract attributes for identifying $non - MCQs$ and $MCQs$. In total, we created 957 instances for the feature set ($fSet2$) with the HIV/AIDS SMS queries and 707 instances for the feature set ($fSet2$) with the FIRE2012 SMS queries. These instances were assigned their corresponding class label ($non - MCQs$) and ($MCQs$).

fSet3: For this feature set, the training and testing instances were created using eight different query difficulty estimation predictors. Seven of these predictors were pre-retrieval predictors and these were : Average Pointwise Mutual Information (AvPMI) [10], Simplified Clarity Score (SCS) [11], Average Inverse Collection Term Frequency (AvICTF) [12], Average Inverse Document Frequency (AvIDF) [10] and the derivatives of the similarity score between the collection and the query (SumSCQ, AvSCQ, MaxSCQ) [27]. One post-retrieval predictor was used, the Clarity Score (CS) [5]. For each query, the FAQ Retrieval Platform described in Section 4.1 was used to generate the score for each query difficulty estimation predictor.

Combined Feature Sets: We created four additional feature sets by combining the above feature sets ($fSet1$, $fSet2$ and $fSet3$) in order to answer research question, **R2**. The feature sets were simply combined by merging the corresponding instances. These four additional feature sets were : *fSet1+fSet2, fSet1+fSet3, fSet2+fSet3 and fSet1+fSet2+fSet3*

To enable us to answer research question **R3**, we randomly split the feature set ($fSet1 + fSet2 + fSet3$) 10 times into training and testing sets. For each training/testing split, we created two training sets, one containing 50 % of the data (instances) and the other containing 75 % of the data. The training set with 75 % of the data was the superset of the training set with 50 % of the data. The remaining 25 % of the data was made the testing set. In total, we had 20 different training sets, 10 containing 50 % of the data and the other 10 containing 75 % of the data. The training data with 50 % of the data shared the same testing set with its superset containing 75 % of the data.

4   Experimental Setting

We begin Section 4.1 by describing the HIV/AIDS FAQ Retrieval Platform used for generating the features for the training and testing instances, followed by a description on how we train and classify $non-MCQs$ and $MCQs$ in Section 4.2.

4.1   FAQ Retrieval Platform

For our experimental evaluation, we used the Terrier-3.5[3] [19] Information Retrieval (IR) platform with BM25 [21]. All the HIV/AIDS and the FIRE2012 FAQ documents used in this study were first pre-processed before indexing and this involved tokenising the text and stemming each token using the full Porter [20] stemming algorithm. To filter out terms that appear in a lot of FAQ documents, we did not use a stopword list during the indexing and the retrieval process. Instead, we ignored the terms that had low Inverse Document Frequency (IDF) when scoring the documents. Indeed, all the terms with term frequency higher than the number of the FAQ documents (205) were considered to be low IDF terms. Earlier work in [17] has shown that stopword removal using a stopword

---

[3] http://terrier.org/

list from various IR platforms like Terrier-3.5 can affect retrieval performance in SMS-Based FAQ retrieval. The normalisation parameter for BM25 [21] was set to its default value of $b = 0.75$.

4.2 Training and Classifying Missing Content and Non-missing Content Queries

Three different classifiers in WEKA, namely Naive Bayes, RandomForest and C-SVC were deployed in our experimental evaluation. Evidence from previous works suggest that RandomForest and Support Vector Classifiers achieve excellent performance compared to naive bayes across a wide variety of binary classification problems and evaluation metrics [3]. We used three classifiers on the labelled feature sets created in Section 3.3 to train and classify $non-MCQs$ and $MCQs$. For each feature set, we created 10 random splits of training and testing sets. For each training/testing split, each training set was made up of 75 % of the data while the remaining 25 % of the data was for testing. All the attributes in these training and testing sets were scaled between -1 and 1. Different Kernels were used for C-SVC. A linear kernel was used for the feature sets with a large number of attributes (String) and a Radial Basis Function (RBF) kernel was used on the feature sets with few attributes.

The regularization parameter C and the kernel parameter γ for the RBF kernel were chosen through a grid-search strategy [14]. This involved performing a 10-fold cross validation on the labelled data with various pairs of (C, γ) and selecting the pair that gave the best classification accuracy. The same grid-search strategy was deployed to select the parameters for RandomForest. The C and the γ parameter for the RBF kernel were set to 1.0 and 0.9 respectively while the C parameter for the linear kernel was set to 0.7. For RandomForest, we set the number of trees to 10 for each feature set while the number of random features for creating the trees varied and were 5 and 10 for $fSet3$ and $fSet2$, respectively and 30 when using $fSet1$. In this experimental evaluation, we will define the $non-MCQs$ as the positive class and the $MCQs$ as the negative class. Table 1 shows a confusion matrix for the outcome of this two-class problem.

Table 1. Confusion matrix for a 2-class problem

		Predicted Class	
		$non-MCQs$ (+ve)	$MCQs$ (-ve)
Actual Class	$non-MCQs$ (+ve)	True Positive (TP)	False Negative (FN)
	$MCQs$ (-ve)	False Positive (FP)	True Negative (TN)

5   Experimental Results and Evaluation

Table 2 summarises the overall classification accuracy for all the feature sets. The sensitivity measures the proportion of the true positive instances (recall for TP) correctly classified as $non-MCQs$. The specificity measures the proportion of the true negatives instances (recall for TN) correctly classified as $MCQs$. It

Table 2. The overall (for the 10 random splits) classification accuracy of all the feature sets

Dataset	Feature Set	Classifier	Sensitivity	Specificity	Accuracy(%)	ROC area	Kappa
HIV/AIDS	fSet1	NB	0.935	0.546	85.06∗	0.84∗	0.522∗
		RF	0.983	0.406	85.79∗	0.857∗	0.481∗
		C-SVC	0.959	0.454	84.95∗	0.833∗	0.4812∗
FIRE2012	fSet1	NB	0.957	0.431	83.31◦	0.807◦	0.457◦
		RF	0.974	0.341	82.41	0.782	0.3935
		C-SVC	0.956	0.449	83.59◦	0.832◦	0.4709◦
HIV/AIDS	fSet2	NB	0.953	0.058	75.97	0.604	0.016
		RF	0.935	0.121	75.86	0.639	0.072
		C-SVC	0.999	0.005	78.37	0.502	0.0055
FIRE2012	fSet2	NB	0.836	0.593	77.86	0.767	0.4107
		RF	0.937	0.443	82.09	0.793	0.4333
		C-SVC	0.941	0.437	82.23	0.811	0.43382
HIV/AIDS	fSet3	NB	0.891	0.473	80.04⋆	0.796⋆	0.3821⋆
		RF	0.937	0.348	80.98⋆	0.777⋆	0.337⋆
		C-SVC	0.969	0.251	81.40⋆	0.748⋆	0.2867⋆
FIRE2012	fSet3	NB	0.95	0.311	79.97	0.748	0.3199
		RF	0.958	0.389	82.37	0.737	0.4146
		C-SVC	0.978	0.380	82.63	0.759	0.4619
HIV/AIDS	fSet1 + fSet2	NB	0.923	0.304	78.89⊛	0.694⊛	0.2672⊛
		RF	0.989	0.271	83.39⊛	0.813⊛	0.3465⊛
		C-SVC	0.952	0.449	84.33⊛	0.841⊛	0.4647⊛
FIRE2012	fSet1 + fSet2	NB	0.876	0.581	80.62◁	0.771◁	0.4596◁
		RF	0.981	0.329	82.74	0.836	0.3939
		C-SVC	0.961	0.479	84.72◁	0.857◁	0.5097◁
HIV/AIDS	fSet1 + fSet3	NB	0.900	0.507	81.50⊛	0.828⊛	0.4274⊛
		RF	0.975	0.425	85.89⊛	0.903⊛	0.4837⊛
		C-SVC	0.953	0.493	85.37⊛	0.866⊛	0.5083⊛
FIRE2012	fSet1 + fSet3	NB	0.954	0.347	81.05	0.717	0.3643
		RF	0.989	0.383	84.58◁	0.83◁	0.4655◁
		C-SVC	0.948	0.521	84.72◁	0.735◁	0.5256◁
HIV/AIDS	fSet2 + fSet3	NB	0.903	0.435	80.15	0.774	0.2672
		RF	0.944	0.324	80.98	0.774	0.3230
		C-SVC	0.959	0.271	80.77	0.776	0.2854
FIRE2012	fSet2 + fSet3	NB	0.893	0.563	81.52•	0.807•	0.4705•
		RF	0.948	0.479	83.78•	0.812•	0.4869•
		C-SVC	0.954	0.593	86.88•	0.862•	0.6001•
HIV/AIDS	fSet1 + fSet2 + fSet3	NB	0.919	0.464	82.03⊛	0.804⊛	0.4191⊛
		RF	0.979	0.314	83.49⊛	0.887⊛	0.3754⊛
		C-SVC	0.948	0.502	85.16⊛	0.871⊛	0.5072⊛
FIRE2012	fSet1 + fSet2 + fSet3	NB	0.913	0.545	82.60◁	0.794◁	0.4871◁
		RF	0.972	0.413	84.02◁	0.864◁	0.4653◁
		C-SVC	0.965	0.515	85.86	0.866	0.5504

can be seen from Table 2 that the different feature sets yield fairly reasonable recall rates for the TP instances. In particular, the recall rates for TP (sensitivity) ranges from 0.891 to 0.999 for the HIV/AIDS dataset and from 0.836 to 0.989 for the FIRE2012 dataset. To put these values into perspective, these translate to between 668 and 749 correctly classified instances from a total of 750 instances for the HIV/AIDS dataset. In contrast, our classifiers did not perform well for the TN instances. Fairly low recall rates (specificity) for the TN instances were observed. Depending on the feature set and the classifier used, the specificity ranged from 0.005 to 0.546 for the HIV/AIDS dataset and from 0.311 to 0.593 for the FIRE2012 dataset. These values translate to between 1 and 113 correctly classified TN instances from a total of 207 TN instances for the HIV/AIDS and between 52 and 99 from a total of 167 for the FIRE2012 dataset.

Our empirical evaluation suggests that all the feature sets performed well for the $non - MCQs$. For the $MCQs$, the best performing feature set only yielded roughly 50% classification accuracy. When we compare our results with previous works, we observe that our classifiers perform fairly poorly in the detection of $MCQs$. One plausible explanation for this dissimilarly is that there were more $non - MCQs$ (majority class) used as training instances than there were $MCQs$ (minority class). In previous works, the majority class was the $MCQs$. It is interesting to note that the majority class in our study and in previous works yielded a higher classification accuracy compared to the minority class.

To answer research question $\boldsymbol{R1}$, we used an unpaired t-test to analyse the classification accuracy between the following 10 random splits, $(fSet1$ and $fSet2)$, $(fSet1$ and $fSet3)$, and $(fSet2$ and $fSet3)$. $fSet1$ provides a significantly higher classification accuracy (unpaired t-test, $p < 0.05$) compared to the other feature sets as denoted by $*$ for the HIV/AIDS dataset and $\diamond$ for the FIRE2012 dataset in Table 2. Also observed were significantly higher (unpaired t-test, $p < 0.05$) Kappa statistic and ROC area (AUC) for $fSet1$. The kappa statistic measures the agreement of prediction with the true class and a value of 1 signifies total agreement and a value of 0 signifies total disagreement. The ROC area on the other-hand signifies the overall ability of the classifier to identify $MCQs$ and $non - MCQs$. The best classifier has an area of 1.0 and a classifier with an area of 0.5 or lower is considered ineffective.

A comparison between $fSet2$ and $fSet3$ was also made using unpaired t-test and it was observed that $fSet3$ gives a better classification accuracy for the HIV/AIDS dataset as denoted by $\star$ in Table 2. No significant difference in classification accuracy was observed between $fSet2$ and $fSet3$ for the FIRE2012 dataset. This disparity between the HIV/AIDS and the FIRE2012 dataset when we compare the classification accuracy between $fSet2$ and $fSet3$ suggests that the retrieval scores and word overlap information (used in $fSet2$) are not good discriminators for the on-topic $MCQs$ (research question $\boldsymbol{R4}$). Although $fSet2$ did not perform well for the on-topic $MCQs$ (TN instances, HIV/AIDS dataset), it performed well for the off-topic $MCQs$ (TN instances FIRE2012 dataset) as depicted by higher specificity values.

There was a significantly higher classification accuracy observed when $fSet1$ was combined with the other feature sets (research question $\boldsymbol{R2}$). This is denoted by $\circledast$ and $\triangleleft$ in Table 2 for the HIV/AIDS and FIRE2012 dataset, respectively (unpaired $t - test < 0.5$ for $((fSet1 + fSet2)$ and $fSet2)$, $((fSet1 + fSet3)$ and $fSet3)$ and $((fSet1 + fSet2 + fSet3)$ and $(fSet2 + fSet3))$). Similar findings were observed when $fSet3$ was combined with $fSet2$ as denoted by $\bullet$, (unpaired $t - test < 0.5$ for $((fSet2 + fSet3)$ and $fSet3))$.

A paired t-test was used to analyse whether increasing the size of the training instances increases the classification accuracy ($\boldsymbol{R3}$). The results, as shown in Table 3, indicate a significant improvement in classification accuracy as denoted by $*$ (paired t-test, $p < 0.05$) when the training set is increased by a quarter (25%) of the data (from 50% of the data to 75% of the data).

Table 3. The overall classification accuracy for $(fSet1 + fSet2 + fSet3)$. One training set contains 50% of the data (instance) and the other contains 75% of the data. All the values depicted, range from 0 to 1 except the accuracy which is expressed as a percentage.

Dataset	Feature Set	Training Set Size	Classifier	Sensitivity	Specificity	Accuracy(%)	ROC area	Kappa
HIV/AIDS	$fSet1 + fSet2 + fSet3$	50%	NB	0.924	0.454	82.24	0.816	0.4192
			RF	0.976	0.271	82.34	0.86	0.3213
			C-SVC	0.96	0.478	85.58	0.889	0.5075
FIRE2012	$fSet1 + fSet2 + fSet3$	50%	NB	0.898	0.473	79.77	0.757	0.3984
			RF	0.972	0.305	82.47	0.837	0.3509
			C-SVC	0.943	0.462	82.88	0.832	0.4598
HIV/AIDS	$fSet1 + fSet2 + fSet3$	75%	NB	0.923	0.493	82.98∗	0.822∗	0.4526∗
			RF	0.983	0.266	82.75∗	0.884∗	0.3281∗
			C-SVC	0.956	0.56	87.04∗	0.886∗	0.5747∗
FIRE2012	$fSet1 + fSet2 + fSet3$	75%	NB	0.92	0.539	83.02∗	0.798∗	0.494∗
			RF	0.972	0.443	84.72∗	0.882∗	0.4952∗
			C-SVC	0.967	0.497	85.57∗	0.85∗	0.537∗

6 Conclusions

This study set out to determine the best classifier to deploy in our already existing SMS-Based HIV/AIDS FAQ retrieval system for detecting $MCQs$ and $non - MCQs$. Several research questions were addressed to achieve the above goal. Our result suggest that the most important feature set (**R1**) for building such a classifier is $fSet1$, which is a set of attributes representing word count information from the text contained in the query strings. It also emerged from this study that the classification accuracy of a classifier built using other feature sets ($fSet2$ and $fSet3$) can be improved further by combining these feature sets with $fSet1$ (**R2**), in particular $fSet3$ (feature sets generated by query difficulty predictors). In the future, we will investigate better ways on how to combine these feature sets, in order to improve the classification accuracy.

In addition, we also investigated whether increasing the training set size would yield a better classification accuracy. A significant increase in accuracy, ROC area and Kappa statistic was observed when the training set was increased by a quarter (25%) of the data (from 50% of the data to 75% of the data). The other finding to emerge from this study is that some feature sets work best for some datasets and perform poorly on other datasets (**R4**). As our results suggest in Table 2, $fSet2$ does not perform well when the $MCQs$ are on-topic ($MCQs$ related to the FAQ document collection) as in the case of the HIV/AIDS dataset. This feature set does however perform well when the $MCQs$ are off-topic ($MCQs$ not related to the FAQ document collection) as in the case of the FIRE2012 dataset. Other feature sets ($fSet1$ and $fSet3$) do however perform well across these different collections.

References

1. Bornman, E.: The Mobile Phone in Africa: Has It Become a Highway to the Information Society or Not? Contemp. Edu. Tech. 3(4) (2012)
2. Breiman, L.: Random forests. Machine Learning 45(1) (2001)

3. Caruana, R., Niculescu-Mizil, A.: An Empirical Comparison of Supervised Learning Algorithms. In: Proc. of ICML (2006)
4. Chang, C.-C., Lin, C.-J.: LIBSVM: A library for Support Vector Machine. ACM Trans. Intell. Syst. Technol. 2(3) (2011)
5. Cronen-Townsend, S., Zhou, Y., Croft, W.B.: Predicting Query Performance. In: Proc. of SIGIR (2002)
6. Daelemans, W., Zavrel, J., Sloot, K.V.D., Bosch, A.V.D.: TiMBL: Tilburg Memory-Based Learner - version 4.3 - Reference Guide (2002)
7. Donner, J.: Research Approaches to Mobile Use in the Developing World: A Review of the Literature. The Info. Soc. 24(3) (2008)
8. Ferguson, P., O'Hare, N., Lanagan, J., Smeaton, A.F., McCarthy, K., Phelan, O., Smyth, B.: CALRITY at the TREC 2011 Microblog Track. In: Proc. of TREC (2011)
9. Hall, M., Frank, E., Holmes, G., Pfahringer, B., Reutemann, P., Witten, I.H.: The WEKA Data Mining Software: an Update. SIGKDD Explor. Newsl. 11(1) (2009)
10. Hauff, C., Murdock, V., Baeza-Yates, R.: Improved Query Difficulty Prediction for the Web. In: Proc. of CIKM (2008)
11. He, B., Ounis, I.: Inferring Query Performance using Pre-Retrieval Predictors. In: Proc. of SPIRE (2004)
12. He, B., Ounis, I.: Query Performance Prediction. Info. Syst. 31(7) (2006)
13. Hogan, D., Leveling, J., Wang, H., Ferguson, P., Gurrin, C.: DCU@FIRE 2011: SMS-based FAQ Retrieval. In: Proc. of FIRE (2011)
14. Hsu, C.-W., Chang, C.-C., Lin, C.-J.: A Practical Guide to Support Vector Classification (2010)
15. John, G.H., Langley, P.: Estimating Continuous Distributions in Bayesian Classifiers. In: Proc. of UAI (1995)
16. Lane, I., Kawahara, T., Matsui, T., Nakamura, S.: Out-of-Domain Utterance Detection Using Classification Confidences of Multiple Topics. IEEE Transact. on Aud. Speech, and Lang. Process. 15(1) (2007)
17. Leveling, J.: On the Effect of Stopword Removal for SMS-Based FAQ Retrieval. In: Bouma, G., Ittoo, A., Métais, E., Wortmann, H. (eds.) NLDB 2012. LNCS, vol. 7337, pp. 128–139. Springer, Heidelberg (2012)
18. Medhi, I., Ratan, A., Toyama, K.: Mobile-Banking Adoption and Usage by Low-Literate, Low-Income Users in the Developing World. In: Proc. of IDGD (2009)
19. Ounis, I., Amati, G., Plachouras, V., He, B., Macdonald, C., Lioma, C.: Terrier: A High Performance and Scalable Information Retrieval Platform. In: Proc. of OSIR at SIGIR (2006)
20. Porter, M.F.: An Algorithm for Suffix Stripping. Elec. Lib. Info. Syst. 14(3) (1980)
21. Robertson, S., Zaragoza, H.: The Probabilistic Relevance Framework: BM25 and Beyond. Found. Trends Info. Retr. 3(4) (2009)
22. Sneiders, E.: Automated FAQ Answering: Continued Experience with Shallow Language Understanding. Question Answering Systems. In: Proc. of AAAI Fall Symp. (1999)
23. Sneiders, E.: Automated FAQ Answering with Question-Specific Knowledge Representation for Web Self-Service. In: Proc. of HSI (2009)
24. Thuma, E., Rogers, S., Ounis, I.: Evaluating Bad Query Abandonment in an Iterative SMS-Based FAQ Retrieval System. In: Proc. of OAIR (2008)

25. Yom-Tov, E., Fine, S., Carmel, D., Darlow, A.: Learning to Estimate Query Difficulty: Including Applications to Missing Content Detection and Distributed Information Retrieval. In: Proc. of SIGIR (2005)
26. Zhang, M., Dodgson, M.Y.: High-tech Entrepreneurship in Asia: Innovation, Industry and Institutional Dynamics in Mobile Payments. Edward Elgar Publishing, Inc. (2007)
27. Zhao, Y., Scholer, F., Tsegay, Y.: Effective Pre-Retrieval Query Performance Prediction Using Similarity and Variability Evidence. In: Macdonald, C., Ounis, I., Plachouras, V., Ruthven, I., White, R.W. (eds.) ECIR 2008. LNCS, vol. 4956, pp. 52–64. Springer, Heidelberg (2008)

Cross-Language Pseudo-Relevance Feedback Techniques for Informal Text

Chia-Jung Lee and W. Bruce Croft

Center for Intelligent Information Retrieval
School of Computer Science, University of Massachusetts, Amherst, USA
{cjlee,croft}@cs.umass.edu

Abstract. Previous work has shown that pseudo relevance feedback (PRF) can be effective for cross-lingual information retrieval (CLIR). This research was primarily based on corpora such as news articles that are written using relatively formal language. In this paper, we revisit the problem of CLIR with a focus on the problems that arise with informal text, such as blogs and forums. To address the problem of the two major sources of "noisy" text, namely translation and the informal nature of the documents, we propose to select between inter- and intra-language PRF, based on the properties of the language of the query and corpora being searched. Experimental results show that this approach can significantly outperform state-of-the-art results reported for monolingual and cross-lingual environments. Further analysis indicates that inter-language PRF is particularly helpful for queries with poor translation quality. Intra-language PRF is more useful for high-quality translated queries as it reduces the impact of any potential translation errors in documents.

Keywords: Informal text, discussion forum, cross-language information retrieval, pseudo-relevance feedback.

1 Introduction

The task of cross-lingual information retrieval (CLIR) attempts to bridge the mismatch between source and target languages using approaches such as query and document translation. Previous techniques [1][31] addressed CLIR using different benchmark collections[1] that are written with relatively formal language (e.g., news articles). There are, however, many applications that involve relatively informal language. Social sites like discussion forums often contain slang or abbreviations that will result in frequent translation errors. Effective CLIR for this type of data may require modifications to the existing techniques.

Informal text poses several problems for machine translation (MT). It is likely that both document and query translations contain a significant proportion of mistranslated and untranslated terms. An example query from our data set is: *"What are people saying about real-name tweeting registration that was imposed*

---

[1] CLEF and NTCIR are two examples.

M. de Rijke et al. (Eds.): ECIR 2014, LNCS 8416, pp. 260–272, 2014.
© Springer International Publishing Switzerland 2014

on March 16th 2012 in China" in source language English. The term *tweeting* was translated into 啁啾 (chirping) in the target language, Chinese. The cause of this mistranslation lies in the parallel collection used to train the MT engine, where the only instances of *tweeting* referred to birds *chirping*. Similarly, correct document translation is complicated by the creativity of social site users. For example, 微博 (Weibo) is often written as 围脖 (surrounding neck) in forum documents. This is an amusing pun because the pronunciation of both terms is identical in Chinese. These types of noise, in addition to the noise produced by translation errors, can easily result in significant topic drift in translated queries and collections.

In this paper, we explore techniques to improve CLIR for a collection of informal text. We propose a new technique based on feature-driven cross-lingual pseudo relevance feedback that expands translated queries with intra-language or inter-language feedback terms. Given source queries Q_s and target corpus C_t as input, both query and document translation are performed to generate the translated $Q_{T(s)}$ and $C_{T(t)}$. We propose to use inter- and intra-language PRF to reduce noise produced by poor query and document translation, respectively. Intra-language PRF extracts terms from top ranked documents in C_t retrieved by $Q_{T(s)}$. This type of feedback helps to mitigate any drop in performance due to poor document translations. For poorly translated queries, we consider the opportunity of recovering semantics from PRF performed on source language. Accordingly, inter-language PRF first retrieves documents in $C_{T(t)}$ using Q_s, based on which it then locates the aligned documents in C_t and extracts feedback terms. For example, despite *tweeting* being translated to *chirping*, retrieval on the source corpus is able to discover documents discussing *real-name registration*. We may recover the lost term, *tweeting*, by locating their parallel documents in the target language and directly extracting terms from these documents.

We evaluate existing techniques and our selective PRF model using a recently created collection of web forum posts. Test queries were manually created and associated relevant posts were judged using a pooling technique across multiple retrieval techniques. We explore the language pair English and Chinese[2]. Our aim is to select between intra- and inter-language PRF for each query. Experimental results show that this selective PRF model can significantly improve retrieval performance over several monolingual and cross-lingual baselines.

The remainder of this paper is laid out as follows. Section 2 discusses related work on CLIR and PRF. Section 3 describes the proposed approach. We show the evaluation results in Section 4 and Section 5 concludes the paper.

2   Related Work

Figure 1 provides an overview of CLIR approaches that have been studied. For clarity, we denote the languages used in original query and document collection respectively as L_1 and L_2. Thus the translated query and document would be

---

[2] The crawled forum data is Chinese, the queries are posed in English, and documents were judged in Chinese.

generated in languages L_2 and L_1. Query translation [5][27] is one of the most common methods of bridging the language gap. This approach first translates the original topics into L_2, performs monolingual retrieval on the language-compatible index, and produces a single ranked list of documents. Alternatively, document translation [3][15] translates the original document collection into topic language L_1 and the original queries are used for retrieval. Hybrid approaches [6] consider both query and document translation. As a retrieval system generates a final single ranked list, approaches that conduct multiple retrieval runs need to merge lists using fusion techniques [9][18][25]. Gey et al [11] proposed bypassing the merge process by building a single index that contains documents in all languages and concatenating a source query with all language translations as a new query. Chen et al [5], however, showed that the combined approach is empirically less effective than the query translation approach. Studies have also been conducted on different translation techniques [23][33].

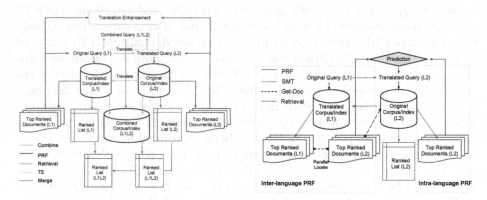

Fig. 1. An overview of related work on CLIR

Fig. 2. An overview of the proposed approach

The practice of PRF has been effectively applied in various monolingual and cross-lingual IR environments. For monolingual retrieval, a number of approaches [16][29][32] have been proposed to improve the performance and robustness of PRF. Metzler et al. [21] proposed a feature-based approach, latent concept expansion, to model term dependencies. For cross-lingual settings, PRF can be applied in different retrieval stages of pre-translation, post-translation [25] and the combination of both [1][19][26]. Chen and Gey [6] further demonstrated that rank fusion of returned lists obtained from L_1 and L_2 is able to improve retrieval performance in most circumstances. In addition to expanding queries, as shown at the top of Figure 1, He and Wu [13] used PRF to enhance query translation by adjusting translation probabilities and resolving out-of-vocabulary terms.

Recent research has shown that monolingual retrieval performance can be improved by the use of another assisting language. Chinnakotla et al [7] used a second language to improve the performance of PRF based on a framework "MultiPRF". Similarly, Na and Ng [22] proposed translating documents into an auxiliary language which then served as a semantically enhanced representation for supplementing the original bag of words.

3   Our Approach

Query expansion using PRF provides a method for automatic local analysis [30]. A typical instantiation is to retrieve an initial set of documents with the original query, assume the top retrieved documents are relevant, and expand the original query with the significant terms extracted from those pseudo-relevant documents. The newly constructed query is used to retrieve the final result.

Although PRF has been shown to be helpful for CLIR [1], it is not clear how effective PRF can be for a collection of informal text and its associated queries. Queries of different characteristics may benefit differently from different feedback techniques. In particular, when query translation quality is poor, we hypothesize that inter-language PRF is more useful, considering the *tweeting* example in Section 1. On the other hand, intra-language PRF can be more helpful when query translation quality is good as it reduces the impact of any potential translation errors in documents. Document translation errors can happen often, particularly in a collection of informal text. For example, for the query *How should I maintain a car?*, we identify some documents losing relevance because 汽车的脚, meaning "feet" of a car, was used to refer to tires of a car in Chinese. Yet it has nothing to do with car tires after it is translated into *motor vehicles of feet*. Translation of named entities can also cause problems in translated documents. For example, 富士康 (*Foxconn*) is translated into *fuji kang* in our system.

Noting that queries can benefit differently from different types of PRF, we propose to select between inter- and intra-language PRF for each individual query, as opposed to previous work that mostly focused on applying the same technique to all query instances. Figure 2 provides a high level overview of our approach. In the following, we denote a user-issued query as Q_{L_1} in source language L_1 and the aim is to retrieve a set of documents from corpus C_{L_2} in target language L_2. Query and document translations are performed to generate Q_{L_2} and C_{L_1} resulting in a bilingual dataset.

3.1   Intra-language and Inter-language PRF

Intra-language PRF is an implementation of PRF using only the target language L_2. The translated queries Q_{L_2} are issued against the target language index I_{L_2} of the collection C_{L_2}. The set of pseudo relevance feedback terms are then extracted from the top ranked documents. Metzler and Croft [21] further proposed Latent Concept Expansion (LCE) that models term dependencies during expansion and have showed significant improvements in retrieval effectiveness. The terminology, *intra-language* PRF (intra-PRF), is used to indicate that other language, L_1, is not used in the process, as shown in the right part of Figure 2.

Inter-language PRF (inter-PRF) modifies the PRF framework to use the translated collection. We construct inter-PRF by first retrieving documents from the translated corpus C_{L_1} using source queries Q_{L_1}, as shown in the left part of Figure 2. One difference from intra-PRF is that the retrieved documents are now written in source language L_1, meaning that the feedback terms cannot be directly compared against target language index I_{L_2}. We therefore locate the target language documents aligned with the retrieved set and extract significant terms directly from those documents in C_{L_2}. These terms are then used to expand the translated query, Q_{L_2}, for the second retrieval pass.

3.2 Selecting between Intra-PRF and Inter-PRF

Another contribution of this paper is to compare the relative utility of these different feedback techniques and properly integrate them together. We hypothesize that inter-PRF is more helpful for queries that are malformed due to poor query translation. Meanwhile, intra-PRF is better for correctly translated queries and can compensate for document translation errors. To combine relevance signals from inter- and intra-PRF, we formulate the problem as a classification task, such that either intra-PRF or inter-PRF is selected as feedback to the input query using a per-query basis.

To integrate the two sets of PRF terms, we propose to estimate the weights $\{\phi_i\}$ that assess the importance of each component, constrained by $\sum_i \phi_i = 1$ and $\phi_i \geq 0$, into the retrieval model as shown in Equation 1:

$$
\begin{aligned}
sc(Q, D) &= \theta \cdot g_q(Q, D) + \bar{\theta} \cdot g_p(PRF, D) \\
&= \theta \cdot g_q(Q, D) + \bar{\theta} \cdot \sum_{\substack{i \in \{intra, \\ inter\}}} \phi_i \cdot g_p(PRF_i, D)
\end{aligned}
\tag{1}
$$

In Equation 1, $g_q(\cdot)$ and $g_p(\cdot)$ are retrieval functions taking query or feedback terms as input, based on which the functions search the index and produce relevance scores. Parameter $\theta \in [0, 1]$ controls the belief in relevance estimators and $\bar{\theta} = 1 - \theta$. For brevity, Q denotes a query in target language L_2, and PRF_{intra} and PRF_{inter} respectively represent the set of corresponding PRF terms.

We further need to estimate the weights $\{\phi_i\}$ guiding the model to incorporate PRF terms either from the intra- or inter-language method. Given a retrieval performance metric $m(.)$, we can represent the effectiveness of a query with PRF by $m(Q, PRF_{intra})$ or $m(Q, PRF_{inter})$. We then define the relative effectiveness of using intra- and inter-PRF as $\hat{m} = m(q, PRF_{intra}) - m(q, PRF_{inter})$. Intuitively, a query is more suited to intra-PRF if $\hat{m}$ is positive while inter-PRF is more suitable if $\hat{m}$ is negative. A classification decision is then naturally guided by the sign function $sign(\hat{m})$ where labels $+1$ and -1 respectively stand for the choice of intra- and inter-PRF. Supposing a classifier c is established based on $sign(\hat{m})$, we use a binary belief estimator $\mathbb{1}_c$ that takes the predicted results from c as input and outputs a binary value 0 when $c = -1$ or 1 when $c = +1$. Replacing $\{\phi_i\}$ in Equation 1 with output of indicator function $\mathbb{1}_c$ we can obtain a rigidly-classified retrieval model.

In addition, we can estimate $\{\phi_i\}$ probabilistically such that the estimation would alleviate the penalty of binary mis-classification. Specifically, we transform the Support Vector Machine (SVM) prediction results into posterior probabilities [10][24], and incorporate these probabilities into the retrieval model indicating the importance of each PRF set. Alternatively, the hyper-parameter learning framework proposed [2] could be used for learning the cross-lingual weights. We leave this as future work.

The retrieval functions $g_q(\cdot,\cdot)$ and $g_p(\cdot,\cdot)$ can be modeled using a wide variety of methods. Metzler and Croft [20] proposed an effective family of retrieval models using Markov Random Field (MRF) for term dependency modeling. This approach has consistently outperformed bag-of-words retrieval models. We use the Sequential Dependency Model (SDM) [20], an instantiation of the MRF model, to compute $g_q(Q, D)$ for a query Q and a document D. For PRF terms, we adopt unigram query likelihood to compute $g_p(\cdot,\cdot)$ as dependency between feedback terms is less likely to improve effectiveness.

3.3   Features

The feature set, consisting a total number of 42 features, is used to select between intra- and inter-PRF for each query.

We compute the number of nouns, verbs, and named entities in Q_{L_1} and Q_{L_2}. The degree of conformity between these statistical distributions can be good indicators about how well the syntactic structure and semantics have been retained after the translation process.

He and Ounis [12] proposed using pre-retrieval predictors for inferring query performance. For a query predicted to have high performance, using PRF from the same side of language could be more effective than that of another side. We adopt the simplified query clarity score (SCS) and query scope (QS) from [12] to characterize both Q_{L_1} and Q_{L_2}. More details can be found in [12].

We compute the collection frequency and inverse document frequency of each query term $t_i \in Q_{L_1}$ and $t_i \in Q_{L_2}$. We then generate the features by taking the minimum, maximum, and average frequencies among the results of each language respectively. We additionally include the standard deviation and the max min ratio of *idf* as features as in [12]. Features based on the query length of Q_{L_1} and Q_{L_2} are included to provide an estimate of the length variation after translation.

Intuitively, for a query, if the retrieved document set for each language is similar, the translation quality for both the query and the document set is likely to be high. Accordingly, we compute the intersection of the retrieved document ranked lists produced by $Q_{L_1} \to C_{L_1}$ and $Q_{L_2} \to C_{L_2}$ at ranks 10, 100, 500, and 1000. More sophisticated post-retrieval predictors such as [8][14] can be used.

We consider the degree of collection coherence between query terms and feedback terms. Denoting a feedback term for a query Q as z_i and a query term as t_j, we compute Pointwise Mutual Information (PMI) between the pairwise combinations of z_i and t_j for query instance Q. PMI provides a semantic similarity measure that sorts lists of important neighbor words from a large corpus.

4   Experiments

4.1   Experimental Setup and Dataset

Document Set: All of our experiments are conducted over a collection of Chinese (L_2) language forum posts. All posts were translated into English (L_1) using state-of-the-art machine translation techniques, as part of the Broad Operational Language Technology project (BOLT)[3]. The original data set of 287,783 threads was collected by the Linguistic Data Consortium (LDC)[4]. Threads may contain multiple posts. By splitting the threads into posts, we obtain 2,416,869 posts (documents). The dataset spans a wide range of discussions and documents with informal language are prevalent across the entire collection.

Query and Relevance Judgments: We manually constructed 50 natural language English queries Q_{en} and aim to retrieve documents from the Chinese forum collection. With forum posts being the search target, we created queries that request suggestions or opinions on current events/topics that are common in social media such as discussion forums. We generated relevance judgments based on a three level assessment $\{0, 1, 2\}$, representing non-relevant, partially relevant and relevant. The judgments were done by two bilingual assessors and a document is assigned the maximum relevance level of the two. Relevance judgments were collected using a pooling method based on multiple well-known retrieval approaches. A total number of 2,495 judged documents were collected (49.9 posts/query), among which 1,072 were judged relevant (21.44 posts/query). The ratio of the number of relevant documents to the total number of judged is relatively high partly because quoting and reusing are popular in discussion forums.

Statistical Machine Translation of Queries: Several machine translation tools were used to provide the noisy channels that we are investigating in this paper. Our first translation engine was the open source tool Moses and Giza++ [5] that implements the statistical (or data-driven) approach using sentence-aligned English and Chinese Sougo news articles corpus. Our alternative translations for queries come from the BOLT project and the Google Translate tool. Our three translated query sets are denoted as Q_{mos}, Q_{bolt} and Q_{gt}. In total, we explore 4 types of instances of the 50 query topics, including 1 original query (English) and 3 query translation instances (Chinese).

Other Open Source Tools: Indexes are created for both original C_{L_2} and translated C_{L_1} corpora using Indri[6]. We use a support vector machine (SVM) [4] to classify when a query should use inter- or intra-PRF. Predictions are conducted using 10-fold cross validation. We extract linguistic features using

---

[3] http://www.darpa.mil/Our_Work/I2O/Programs/
   Broad_Operational_Language_Translation_(BOLT).aspx
[4] http://www.ldc.upenn.edu/
[5] http://www.statmt.org/moses/index.php?n=Main.HomePage
[6] http://www.lemurproject.org/indri/

the tools built by the Stanford Natural Language Processing Group[7] and Harbin Institute of Technology [8] for Chinese NER.

Retrieval Setup: We evaluate the results using the top 1000 retrieved documents, and report mean average precision (MAP), precision@ 10 (P@10) and normalized discounted cumulative gain@k (n@10). We use Dirichlet smoothing with $\mu = 2500$ and fix θ in Equation 1 as 0.8. PRF parameters are set to use top 10 documents and top 20 feedback terms. Note that these parameters can be tuned to improve performance. In this paper, we only apply the selective cross-lingual PRF approach to the Chinese side (i.e., the original corpus). This is because relevance feedback can be less useful to the English side as the document translation quality is often poor due to the informal text.

4.2   Retrieval Experiments

In this section, we report retrieval performance of the proposed methods and baselines. We compare our approach to two strong monolingual baseline approaches including SDM [20] and LCE [21]. We also consider two strong cross-lingual retrieval methods that fuse two monolingual ranked lists produced using LCE. As shown in Equation 2, we implement CombSUM $C_{sum}(Q, D)$ and CombMNZ $C_{mnz}(Q, D)$ [17][28] with four kinds of revised score functions $S_R(\cdot, \cdot)$. These include raw similarity score, normalized raw score, logarithm of normalized raw score and rank-based score [18]. $C_{mnz}(Q, D)$ revises $C_{sum}(Q, D)$ by considering the binary existence of document D_{L_i} in top k retrieved documents, where k is a parameter set to 500 in our experiments. w is set to 0.5 assuming no prior knowledge on language side estimation.

$$C_{sum}(Q, D) = wS_R(Q_{L_1}, D) + \bar{w}S_R(Q_{L_2}, D)$$
$$C_{mnz}(Q, D) = \left(\sum_{L_i} I(D_{L_i} \in \{TopK\}) \right) \cdot C_{sum}(Q, D) \tag{2}$$

Recall that we have four groups of queries Q_{mos}, Q_{bolt}, Q_{gt} and Q_{en}. In addition to the proposed approach, we generate an oracle run by selecting intra- or inter-PRF according to the true label $\hat{m}$ of the query. The oracle run defines an upper bound for the approach under these experimental settings. Note that, SDM+PRF_{intra} essentially implements LCE as discussed in Section 3.1. For English queries Q_{en}, we report the performance only on SDM and SDM+PRF_{intra}. In this case, we are performing monolingual retrieval against English (L_1).

Table 1 shows the retrieval results of the proposed approaches, where the best performance for each query set is underlined [9]. For all translated query sets, the classification-based retrieval models SDM+PRF$_{svm}$ (rigid) and SDM+PRF$_{psvm}$ (probabilistic) consistently show significant improvements over the strong baselines SDM and/or SDM+PRF$_{intra}$, and approach oracle performance.

---

[7] http://nlp.stanford.edu/
[8] http://ir.hit.edu.cn/ltp/
[9] Oracle runs are not considered.

Table 1. Retrieval performance of the proposed and baseline methods. All SDM+PRF$_i$ is abbreviated as PRF$_i$. † and ⋆ denote significant difference over SDM and SDM+PRF$_{intra}$ with $p < 0.05$.

Query	Metric	SDM	PRF$_{intra}$	PRF$_{inter}$	PRF$_{svm}$	PRF$_{psvm}$	PRF$_{ora}$	C_{mnz}	C_{sum}
	MAP	.2155	$.2330^\dagger$	$.2709_\star$	$.2721^\dagger_\star$	$.2753^\dagger_\star$	$.2786^\dagger_\star$	.2617	.2616
Q_{mos}	P@10	.2860	$.3060^\dagger$	$.3560_\star$	$.3580^\dagger_\star$	$.3610^\dagger_\star$	$.3640^\dagger_\star$	.3440	.3340
	n@10	.2780	$.2977^\dagger$	$.3408_\star$	$.3412^\dagger_\star$	$.3450^\dagger_\star$	$.3439^\dagger_\star$	.3235	.3235
	MAP	.2827	.2975	$.3081^\dagger$	$.3150^\dagger_\star$	$.3100^\dagger$	$.3194^\dagger_\star$	.3136	.3111
Q_{bolt}	P@10	.3200	.3380	$.3760_\star$	$.3760_\star$	$.3800^\dagger_\star$	$.3840^\dagger_\star$	.3720	.3680
	n@10	.3219	.3377	$.3627_\star$	$.3778_\star$	$.3792^\dagger_\star$	$.3789^\dagger_\star$	.3641	.3662
	MAP	.3419	.3542	.3540	$.3602^\dagger$	$.3612^\dagger_\star$	$.3658^\dagger_\star$	.3542	.3419
Q_{gt}	P@10	.4100	$.4380^\dagger$	$.4320^\dagger$	$.4410^\dagger$	$.4500^\dagger_\star$	$.4480^\dagger_\star$	.4360	.4360
	n@10	.3960	$.4274^\dagger$	$.4220^\dagger$	$.4303^\dagger$	$.4303^\dagger$	$.4362^\dagger_\star$	.4095	.4080
	MAP	.2284	$.2525^\dagger$						
Q_{en}	P@10	.3200	.3340						
	n@10	.3005	$.3154^\dagger$						

The benefit of each technique investigated varies between different query sets of different translation quality. We first quantify query translation quality by manually assigning a tag of being good or poor. A query is labelled *poor* when the assessor could not even guess the original intent of the translated query. A query is labelled *good* when the original intent is partially or fully preserved. Figure 3 shows that Google Translate Q_{gt} outperforms Q_{bolt}, and Q_{bolt} achieves better quality than Q_{mos}. Note that better performance can be tuned for Moses and Giza++ by considering different parameters. In Table 1, for Q_{mos}, the most poorly translated query set, the advantages of SDM+PRF_{inter} are clearly apparent. This validates our hypothesis that translations with poor quality would benefit more from inter-PRF. For high quality translations such as Q_{gt}, intra-PRF is more effective than inter-PRF. For Q_{bolt}, the performance ordering resembles that of Q_{mos} with smaller gap between inter- and intra-PRF.

The two rightmost columns of Table 1 report the retrieval performance for the best result among the four scoring methods using CombSUM and CombMNZ. While CombMNZ and CombSUM are strong benchmarks, the results show that SDM+PRF_{psvm} significantly outperforms the fusion approaches in most cases. Consistent with [18], CombMNZ slightly outperforms CombSUM and normalized functions can be more effective than unnormalized or rank-based similarities.

4.3 Per-Query Retrieval Variation

One of the main purposes of this paper is to demonstrate how queries are influenced differently by different PRF techniques. Figure 4 shows the relative performance gain of applying intra- or inter-PRF against a baseline method SDM (y-axis) using a per-query basis (x-axis). That is, for each query in Q_{bolt}, the y-axis displays either Δ_{inter} or Δ_{intra} as defined in Equation 3, where $m(\cdot)$ computes the metric MAP. Figure 4 is sorted according to non-decreasing Δ_{inter}.

$$\Delta_i = m(SDM + PRF_i) - m(SDM), i \in \{inter, intra\} \qquad (3)$$

We observe, from Figure 4, that there is a small fraction of queries that receive negative feedback from inter- or intra-PRF (i.e., $\Delta_{inter} < 0$ or $\Delta_{inter} < 0$). This can be because either the query is too hard or the translation is too poor. On the other hand, queries can be improved using the two techniques at the same time, providing an explanation to why SDM+PRF_{psvm} may outperform SDM+PRF_{oracle} in some cases. More importantly, the trends of Δ_{inter} and Δ_{intra} often intersect each other across the query set, indicating proper integration of inter- and intra-PRF can result in overall optimized performance as previously shown in Table 1.

	Good	Poor
Q_{mos}	30 (60%)	20 (40%)
Q_{bolt}	38 (76%)	12 (24%)
Q_{gt}	48 (96%)	2 (4%)
Avg	38.7 (77%)	11.3 (23%)

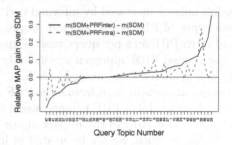

Fig. 3. Translation quality (number of *poor* or *good* query instances) for different query sets

Fig. 4. Relative MAP gain of cross-lingual PRF techniques over SDM for each Q_{bolt} query

4.4 Combining Inter- and Intra-PRF

We now explore the effects of combining inter- and intra-PRF using linear weighting. Figure 5 shows the retrieval performance based on our original weighted retrieval model (Equation 1). We vary the weight ϕ_{intra} for intra-PRF from 0 to 1 by steps of 0.05, and $\phi_{inter} = 1 - \phi_{intra}$. Note that the weight configurations are fixed across all queries, as opposed to the prediction framework where the PRF class for each query would be rigidly or probabilistically selected. For all three types of translated query sets, we observe a general trend that over-weighting ϕ_{intra} results in decreased retrieval performance. For Q_{mos}, we find the performance negatively correlates with the increase of ϕ_{intra}, indicating that inter-PRF outperforms intra-PRF consistently for queries of poor translation quality. For Q_{gt}, the increase of ϕ_{intra} initially improves retrieval effectiveness. The peak performance is reached around $\phi_{intra} \cong 0.3$ and further increase of ϕ_{intra} results in performance drop. The performance trend for Q_{bolt} is similar to Q_{mos}.

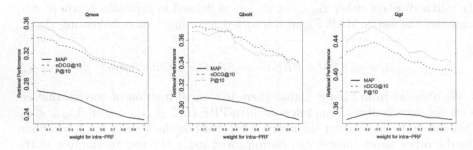

Fig. 5. Retrieval performance for weighted combination of inter- and intra-PRF

5   Conclusions

We investigate the task of CLIR on a large collection of forum posts. The translation noise is increased by informal text used in discussion forums. We consider two types of PRF for CLIR and propose a solution that selects between inter- and intra-PRF on a per-query basis. Experimental results show that the selective cross-lingual PRF approach significantly improves the performance over strong monolingual and cross-lingual baselines. We find that queries with poor translation quality benefit most from inter-PRF as document translation is more reliable in such cases. Intra-PRF is more useful for queries with good query translation accuracy as it reduces the impact of any translation error in documents.

In future work, we are interested in investigating other types of queries and other techniques for CLIR. We also consider the potential of integrating thread information in the discussion forum for better smoothing.

Acknowledgments. This work was supported in part by the Center for Intelligent Information Retrieval and in part by IBM subcontract #4913003298 under DARPA prime contract #HR001-12-C-0015. Any opinions, findings and conclusions or recommendations expressed in this material are those of the authors and do not necessarily reflect those of the sponsor.

References

1. Ballesteros, L., Croft, W.B.: Resolving ambiguity for cross-language retrieval. In: Proc. of SIGIR, SIGIR 1998, pp. 64–71 (1998)
2. Bendersky, M., Croft, W.B.: Modeling higher-order term dependencies in information retrieval using query hypergraphs. In: Proc. of SIGIR, pp. 941–950
3. Braschler, M., Schäuble, P.: Experiments with the eurospider retrieval system for clef 2000. In: Peters, C. (ed.) CLEF 2000. LNCS, vol. 2069, pp. 140–148. Springer, Heidelberg (2001)
4. Chang, C.-C., Lin, C.-J.: LIBSVM: A library for support vector machines. ACM Transactions on Intelligent Systems and Technology 27, 27:1–27:27 (2011)
5. Chen, A.: Multilingual information retrieval using english and chinese queries. In: Peters, C., Braschler, M., Gonzalo, J., Kluck, M. (eds.) CLEF 2001. LNCS, vol. 2406, pp. 44–58. Springer, Heidelberg (2002)

6. Chen, A., Gey, F.C.: Multilingual information retrieval using machine translation, relevance feedback and decompounding. Inf. Retr. 7(1-2), 149–182 (2004)
7. Chinnakotla, M.K., Raman, K., Bhattacharyya, P.: Multilingual prf: english lends a helping hand. In: Proc. of SIGIR, pp. 659–666
8. Collins-Thompson, K., Bennett, P.N.: Predicting query performance via classification. In: Gurrin, C., He, Y., Kazai, G., Kruschwitz, U., Little, S., Roelleke, T., Rüger, S., van Rijsbergen, K. (eds.) ECIR 2010. LNCS, vol. 5993, pp. 140–152. Springer, Heidelberg (2010)
9. Cormack, G.V., Clarke, C.L.A., Buettcher, S.: Reciprocal rank fusion outperforms condorcet and individual rank learning methods. In: Proc. of SIGIR, SIGIR 2009, pp. 758–759 (2009)
10. Cortes, C., Vapnik, V.: Support-vector networks. Machine Learning 20(3), 273–297 (1995)
11. Gey, F.C., Jiang, H., Petras, V., Chen, A.: Cross-language retrieval for the clef collections - comparing multiple methods of retrieval. In: Peters, C. (ed.) CLEF 2000. LNCS, vol. 2069, pp. 116–128. Springer, Heidelberg (2001)
12. He, B., Ounis, I.: Inferring query performance using pre-retrieval predictors. In: Apostolico, A., Melucci, M. (eds.) SPIRE 2004. LNCS, vol. 3246, pp. 43–54. Springer, Heidelberg (2004)
13. He, D., Wu, D.: Translation enhancement: a new relevance feedback method for cross-language information retrieval. In: Proc. of CIKM, pp. 729–738 (2008)
14. Kurland, O., Shtok, A., Carmel, D., Hummel, S.: A unified framework for post-retrieval query-performance prediction. In: Amati, G., Crestani, F. (eds.) ICTIR 2011. LNCS, vol. 6931, pp. 15–26. Springer, Heidelberg (2011)
15. Lam-Adesina, A.M., Jones, G.J.F.: Exeter at clef 2003: Experiments with machine translation for monolingual, bilingual and multilingual retrieval. In: CLEF, pp. 271–285 (2003)
16. Lavrenko, V., Croft, W.B.: Relevance based language models. In: Proc. of SIGIR, SIGIR 2001, pp. 120–127 (2001)
17. Lee, J.H.: Combining multiple evidence from different properties of weighting schemes. In: Proc. of SIGIR, pp. 180–188 (1995)
18. Lee, J.H.: Analyses of multiple evidence combination, pp. 267–276. ACM Press (1997)
19. Levow, G.-A.: Issues in pre- and post-translation document expansion: untranslatable cognates and missegmented words. In: Proc. of AsianIR 2003, pp. 77–83 (2003)
20. Metzler, D., Croft, W.B.: A markov random field model for term dependencies. In: Proc. of SIGIR, SIGIR 2005, pp. 472–479 (2005)
21. Metzler, D., Croft, W.B.: Latent concept expansion using markov random fields. In: Proc. of SIGIR, SIGIR 2007, pp. 311–318 (2007)
22. Na, S.-H., Ng, H.T.: Enriching document representation via translation for improved monolingual information retrieval. In: Proc. of SIGIR, pp. 853–862 (2011)
23. Oard, D.W.: A comparative study of query and document translation for cross-language information retrieval. In: Proc. of the Third Conference of the Association for Machine Translation in the Americas (1998)
24. Platt, J.C.: Probabilities for SV Machines, pp. 61–74 (2000)
25. Qu, Y., Eilerman, A.N., Jin, H., Evans, D.A.: The effects of pseudo-relevance feedback on mt-based. In: CLIR, RIAO 2000, Content-Based Multi-Media Information Access, CSAIS, pp. 46–60 (2000)

26. Rogati, M., Yang, Y.: Cross-lingual pseudo-relevance feedback using a comparable corpus. In: Peters, C., Braschler, M., Gonzalo, J., Kluck, M. (eds.) CLEF 2001. LNCS, vol. 2406, pp. 151–157. Springer, Heidelberg (2002)
27. Savoy, J.: Report on CLEF-2001 experiments: Effective combined query-translation approach. In: Peters, C., Braschler, M., Gonzalo, J., Kluck, M. (eds.) CLEF 2001. LNCS, vol. 2406, pp. 27–43. Springer, Heidelberg (2002)
28. Shaw, J.A., Fox, E.A., Shaw, J.A., Fox, E.A.: Combination of multiple searches. In: TREC-2, pp. 243–252 (1994)
29. Voorhees, E.M.: Query expansion using lexical-semantic relations. In: Proc. of SIGIR, SIGIR 1994, pp. 61–69 (1994)
30. Xu, J., Croft, W.B.: Query expansion using local and global document analysis. In: Proc. of SIGIR, pp. 4–11 (1996)
31. Nie, J.Y., Simard, M., Isabelle, P., Dur, R., De Montréal, U.: Cross-language information retrieval based on parallel texts and automatic mining of parallel texts from the web. In: Proc. of SIGIR, pp. 74–81 (1999)
32. Zhai, C., Lafferty, J.: Model-based feedback in the language modeling approach to information retrieval. In: Proc. of CIKM 2001, pp. 403–410 (2001)
33. Zhu, J., Wang, H.: The effect of translation quality in mt-based cross-language information retrieval. In: Proc. of ACL, ACL-44, pp. 593–600 (2006)

Hierarchical Multi-label Conditional Random Fields for Aspect-Oriented Opinion Mining

Diego Marcheggiani[1], Oscar Täckström[2,*], Andrea Esuli[1], and Fabrizio Sebastiani[1]

[1] Istituto di Scienza e Tecnologie dell'Informazione
Consiglio Nazionale delle Ricerche
56124 Pisa, Italy
firstname.lastname@isti.cnr.it
[2] Swedish Institute of Computer Science
164 29 Kista, Sweden
oscar@sics.se

Abstract. A common feature of many online review sites is the use of an overall rating that summarizes the opinions expressed in a review. Unfortunately, these document-level ratings do not provide any information about the opinions contained in the review that concern a specific aspect (e.g., cleanliness) of the product being reviewed (e.g., a hotel). In this paper we study the finer-grained problem of *aspect-oriented* opinion mining *at the sentence level*, which consists of predicting, for all sentences in the review, whether the sentence expresses a positive, neutral, or negative opinion (or no opinion at all) about a specific aspect of the product. For this task we propose a set of increasingly powerful models based on conditional random fields (CRFs), including a hierarchical multi-label CRFs scheme that jointly models the overall opinion expressed in the review and the set of aspect-specific opinions expressed in each of its sentences. We evaluate the proposed models against a dataset of hotel reviews (which we here make publicly available) in which the set of aspects and the opinions expressed concerning them are manually annotated at the sentence level. We find that both hierarchical and multi-label factors lead to improved predictions of aspect-oriented opinions.

1 Introduction

Sharing textual reviews of products and services is a popular social activity on the Web. Some websites (e.g., Amazon, TripAdvisor[1]) act as hubs that gather reviews on competing products, thus allowing consumers to compare them. While an overall rating (e.g., a number of "stars") is commonly attached to each such review, only a few of these websites (e.g., TripAdvisor) allow reviewers to include *aspect-specific* ratings, such as distinct ratings for the Value and Service provided by a hotel.

The overall and the aspect-specific ratings may help the user to perform a first screening of the product, but they are of little use if the user wants to actually read the *comments* about specific aspects of the product. For example, a low rating for the Rooms aspect of a hotel may be due to the small size of the room or to the quality of the furniture; different issues may be of different importance to different persons. In this case the user may have to read a lot of text in order to retrieve the relevant information.

---

* Currently employed by Google Research. Contact: oscart@google.com
[1] http://www.amazon.com/, http://www.tripadvisor.com/

M. de Rijke et al. (Eds.): ECIR 2014, LNCS 8416, pp. 273–285, 2014.

Overall rating: ★★★★★	Aspect-specific opinions	
Title: Good vlue [sic], terrible service	Value: Positive	Service: Negative
OK the value is good and the hotel is reasonably priced, but the service is terrible.	Value: Positive	Service: Negative
I was waiting 10 min at the erception [sic] desk for the guy to figure out whether there was a clean room available or not.	Checkin: Negative	Service: Negative
That place is a mess.	Service: Negative	
Rooms are clean and nice, but bear in mind you just pay for lodging, service does not seem to be included.	Cleanliness: Positive	Service: Negative

Fig. 1. An example hotel review annotated with aspect-specific opinions at the sentence level

Opinion mining research [9] has frequently considered the problem of predicting the overall rating of a review [14] or the ratings of its individual aspects [5]. While these are interesting research challenges, their practical utility is somewhat limited, since this information is often already made explicit by the reviewers in the form of an ordinal score. Our goal is instead to build an automatic system that, given a sentence in a review and one of the predefined aspects of interest, (a) predicts if an opinion concerning that aspect is expressed in the sentence, and (b) if so, predicts the polarity of the opinion (i.e., positive, neutral/mixed, or negative). This is a *multi-label* problem: a sentence may be relevant for (i.e., contain opinions concerning) zero, one, or several aspects at the same time, and the opinions contained in the same sentence and pertaining to different aspects may have different polarities. For example, *the room was spacious but the location was horrible* expresses a positive opinion for the Rooms aspect and a negative opinion for the Location aspect, while the remaining aspects are not touched upon.

The contribution of this study is twofold. First, inspired by the "coarse-to-fine" opinion model of [11] we develop an increasingly powerful set of multi-label conditional random field (CRF) schemes [6] that jointly model the overall, document-level opinion expressed by a review together with the aspect-specific opinions expressed at the sentence level. Our models are thus able to also predict the document-level ratings. However, as already pointed out, these ratings are of smaller practical interest, because they are often explicitly provided by the reviewers, whereas the aspect-level predictions are often not available and the sentence-level annotations (i.e., the indication of which sentences justify the aspect-level ratings) are never available. The use of a conditional model for this task is in contrast with previous work in this area, which has focused on generative models, mostly based on Latent Dirichlet Allocation, with strong independence assumptions [7,12,17,18]. This problem has also been tackled via supervised learning methods in [8]; like ours, this work relies on CRFs to model the structure of the reviews, but is unable to cater for sentences that are relevant to more than one aspect at the same time, which is a strong limitation. Two works that are close in spirit to ours are [7,18], and they may be considered the "generative counterparts" of our approach.

Second, we present (and make publicly available) a new dataset of hotel reviews that we have annotated with aspect-specific opinions at the sentence level. A previous dataset annotated by opinion at the sentence level exists [16], but the dataset introduced here also adds the aspect dimension and has a multi-label nature. Only very recently, and after we created our dataset, a dataset similar to ours has been presented [7], in which elementary discourse units (EDUs), which can be sub-sentence entities, are annotated using a *single-label* model. The dataset of [7] is composed of 65 reviews, with a total of 1541 EDUs. Our dataset annotates 442 reviews, with a total of 5799 sentences.

The evaluation of generative models is often based on unannotated datasets [12,18], and thus only on a qualitative analysis of the generated output. We believe that our dataset will be a valuable resource to fuel further research in the area by enabling a quantitative evaluation, and thus a rigorous comparison of different models.

1.1 Problem Definition

Before describing our approach, let us define the task just introduced more formally. Let $\mathbb{A}$ be a discrete set of aspect labels and let $\mathbb{Y}$ be a discrete set of opinion labels. Given a review $\mathbf{x} \in \mathbf{X}$ composed of T consecutive segments, we seek to infer the values of the following variables: first, the overall opinion $y_o \in \mathbb{Y}$ expressed in $\mathbf{x}$; second, the opinion $y_t^a \in \mathbb{Y} \cup \{\text{No-op}\}$ expressed concerning aspect a in segment t, for each segment $t \in \{1, ..., T\}$ and each aspect $a \in \mathbb{A}$ (where No-op stands for "no opinion"). This is a *multi-label* problem, since each segment t can be assigned up to $|\mathbb{A}|$ different opinions.

To model these variables we assume a feature vector $\mathbf{x}_t$ representing review segment t and a feature vector $\mathbf{x}_o$ representing the full review. For our experiments, reported in Section 3, we use a dataset of hotel reviews; we take segments to correspond to sentences, and we take $\mathbb{Y} = \{\text{Positive, Negative, Neutral}\}$ and $\mathbb{A} = \{\text{Rooms, Cleanliness, Value, Service, Location, Check-in, Business, Food, Building, Other}\}$. However, we want to stress that the proposed models are flexible enough to incorporate arbitrary sets of aspects and opinion labels, and to use a different type of segmentation.

2   Models, Inference and Learning

Previous work on aspect-oriented opinion mining has focused on generative probabilistic models [7,17,18]. Thanks to their generative nature, these models can be learnt without any explicit supervision. However, at the same time they make strong independence assumptions on the variables to be inferred, which is known to limit their performance in the supervised scenario considered in this study. Instead, we turn to CRFs — a general and flexible class of structured conditional probabilistic models. Specifically, we propose a hierarchical multi-label CRF model that jointly models the overall opinion of a review together with aspect-specific opinions at the segment level. This model is inspired by the *fine-to-coarse* opinion model [11], which was recently extended to a partially supervised setting [16].[2] However, while previous work only takes opinion into

---

[2] While we only consider the supervised scenario in this study, our model is readily extensible to the partially supervised setting by treating a subset of the fine-grained variables as latent.

account, we jointly model both sentence-level opinion and aspect, as well as overall review opinion. Below, we introduce a sequence of increasingly powerful CRF models (that we implemented using Factorie [10].) for aspect-specific opinion mining, leading up to the full hierarchical sequential multi-label model.

2.1   CRF Models of Aspect-Oriented Opinion

A CRF models the conditional distribution $p(\mathbf{y}|\mathbf{x})$ over a collection of output variables $\mathbf{y} \in \mathbf{Y}$, given an input $\mathbf{x} \in \mathbf{X}$, as a globally normalized log-linear distribution [6]:

$$p(\mathbf{y}|\mathbf{x}) = \frac{1}{Z(\mathbf{x})} \prod_{\Psi_c \in \mathbf{F}} \Psi_c(\mathbf{y}_c, \mathbf{x}_c) \propto \prod_{\Psi_c \in \mathbf{F}} \Psi_c(\mathbf{y}_c, \mathbf{x}_c), \tag{1}$$

where $\mathbf{F}$ is the set of factors and $Z(\mathbf{x}) = \sum_{\mathbf{y} \in \mathbf{Y}} p(\mathbf{y}|\mathbf{x})$ is a normalization constant. In this study, $\mathbf{y} = \{y_o\} \cup \{y_t^a : t \in [1, T], a \in \mathbb{A}\}$. Each factor $\Psi_c(\mathbf{y}_c, \mathbf{x}_c) = \exp(\mathbf{w} \cdot \mathbf{f}(\mathbf{y}_c, \mathbf{x}_c))$ scores a set of variables $\mathbf{y}_c \subset \mathbf{y}$ by means of the parameter vector $\mathbf{w}$ and the feature vector $\mathbf{f}(\mathbf{y}_c, \mathbf{x}_c)$. The models described in what follows differ in terms of their factorization of Equation (1) and in the features employed.

Linear-Chain Baseline Model. As a baseline model, we take a simple first-order linear-chain CRF (LC) in which a separate linear chain over opinions at the segment level is defined for each aspect. This model is able to take into account sequential dependencies between segment opinions [11,13] specific to the same aspect, whereas opinions related to different aspects are assumed to be independent. Formally, the LC model factors as

$$p(\mathbf{y}|\mathbf{x}) \propto \prod_{a \in \mathbb{A}} \prod_{t=1}^{T} \Psi_s(y_t^a, \mathbf{x}_t) \prod_{t=1}^{T-1} \Psi_\frown(y_t^a, y_{t+1}^a), \tag{2}$$

where $\Psi_s(y_t^a, \mathbf{x}_t)$ models the aspect-specific opinion of the segment at position t and $\Psi_\frown(y_t^a, y_{t+1}^a)$ models the transition between the aspect-specific opinion variables at position t and $t + 1$ in the linear chain corresponding to aspect a.

Multi-label Models. The assumption of the LC model that the aspect-specific opinions expressed in each segment are independent of each other may be overly strong for two reasons. First, only a limited number of aspects are generally addressed in each segment. Second, when several aspects are mentioned, it is likely that there are dependencies between them based on discourse structure considerations. To address these shortcomings, we propose to model the dependencies between aspect-specific opinion variables within each segment, by adopting the multi-label pairwise CRF formulation of [4].

We first consider the *Independent Multi-Label* (IML) model, in which there are factors between the opinion variables within a segment, while each segment is independent from each other. In terms of Equation (1), the IML model factors as

$$p(\mathbf{y}|\mathbf{x}) \propto \prod_{t=1}^{T} \prod_{a \in \mathbb{A}} \Psi_s(y_t^a, \mathbf{x}_t) \prod_{b \in \mathbb{A} \backslash \{a\}} \Psi_m(y_t^a, y_t^b), \tag{3}$$

where $\Psi_m(y_t^a, y_t^b)$ is the pairwise multi-label factor, which models the interdependence of the opinion variables corresponding to aspects a and b at position t. Note that this factor ignores the input, considering only the interaction of the opinion variables.

To allow for sequential dependencies between segments, the IML model can naturally be combined with the LC model. This yields the *Chain Multi-Label* (CML) model:

$$p(\mathbf{y}|\mathbf{x}) \propto \prod_{t=1}^{T} \prod_{a \in \mathbb{A}} \Psi_s(y_t^a, \mathbf{x}_t) \prod_{b \in \mathbb{A} \backslash \{a\}} \Psi_m(y_t^a, y_t^b) \prod_{t=1}^{T-1} \Psi_\frown(y_t^a, y_{t+1}^a). \tag{4}$$

Hierarchical (Multi-label) Models. Thus far, we have only modeled the aspect-specific opinions expressed at the segment level. However, many online review sites ask users to provide an overall opinion in the form of a numerical rating as part of their review. As shown by [11,16], jointly modeling the overall opinion and the segment-level opinions in a hierarchical fashion can be beneficial to prediction at both levels.

The LC, IML and CML models can be adapted to include the overall rating variable in a hierarchical model structure analogous to that of the "coarse-to-fine" opinion model of [11]. This is accomplished by adding the following two factors to the three models above: the overall opinion factor $\Psi_o(y_o, \mathbf{x}_o)$, which models the overall opinion with respect to the input; and the pairwise factor $\Psi_h(y_t^a, y_o)$, which connects the two levels of the hierarchy by modeling the interaction of the aspect-specific opinion variable at position t and the overall opinion variable.

By combining the shared product of factors $\Phi(y_o, y_t^a, \mathbf{x}) = \Psi_s(y_t^a, \mathbf{x}_t) \cdot \Psi_o(y_o, \mathbf{x}_o) \cdot \Psi_h(y_t^a, y_o)$ with the LC, IML and CML models, we get the *Linear-Chain Overall* (LCO) model:

$$p(\mathbf{y}|\mathbf{x}) \propto \prod_{t=1}^{T} \prod_{a \in \mathbb{A}} \Phi(y_o, y_t^a, \mathbf{x}) \prod_{t=1}^{T-1} \Psi_\frown(y_t^u, y_{t+1}^u), \tag{5}$$

the *Independent Multi-Label Overall* (IMLO) model:

$$p(\mathbf{y}|\mathbf{x}) \propto \prod_{t=1}^{T} \prod_{a \in \mathbb{A}} \Phi(y_o, y_t^a, \mathbf{x}) \prod_{b \in \mathbb{A} \backslash \{a\}} \Psi_m(y_t^a, y_t^b), \tag{6}$$

and the *Chain Multi-Label Overall* (CMLO) model:

$$p(\mathbf{y}|\mathbf{x}) \propto \prod_{t=1}^{T} \prod_{a \in \mathbb{A}} \Phi(y_o, y_t^a, \mathbf{x}) \prod_{b \in \mathbb{A} \backslash \{a\}} \Psi_m(y_t^a, y_t^b) \prod_{t=1}^{T-1} \Psi_\frown(y_t^a, y_{t+1}^a). \tag{7}$$

2.2 Model Features

The joint problem of aspect-oriented opinion prediction requires model features that help to discriminate opinions and aspects, as well as opinions specific to a particular aspect. In the experiments of Section 3 we use both word and word bigram identity features, as well as a set of polarity lexicon features based on the General Inquirer (GI) [15], MPQA [20], and SentiWordNet (SWN) [2] lexicons. The numerical polarity values of these lexicons are mapped into the set {Positive, Negative, Neutral}. The mapped lexicon values are used to generalize word bigram features by substituting the matching words of the bigram with the correspondent polarity. For example, the bigram *nice hotel* is generalized to the bigram *SWN:positive hotel*, from looking up *nice* in SentiWordNet. These features are used both with segment-level and review-level factors; see Table 1.

Table 1. The collection of model factors and their corresponding features, see Section 2.1 for details on notation. Feature vectors: x_t: {words, bigrams, SWN/MPQA/GI bigrams, χ^2 lexicon matches} in the t:th segment in review x; x_o: {words, bigrams, SWN/MPQA/GI bigrams} in x.

Factor	Description	Features
$\Psi_s(y_t^a, x_t)$	Segment aspect-opinion	$x_t \otimes y_t^a \otimes a$
$\Psi_o(y_o, x_o)$	Overall opinion	$x_o \otimes y_o$
$\Psi_\frown(y_t^a, y_{t+1}^a)$	Segment aspect-opinion transition	$y_t^a \otimes y_{t+1}^a \otimes a$
$\Psi_m(y_t^a, y_t^b)$	Multi-label segment aspect-opinion	$y_t^a \otimes y_t^b \otimes a \otimes b$
$\Psi_h(y_t^a, y_o)$	Hierarchical overall / segment aspect-opinion	$y_t^a \otimes y_o \otimes a$

In addition to these features we use an aspect-specific lexicon obtained via the algorithm proposed in [18]; this is an algorithm that iteratively builds a set of aspect-specific words by adding to it words that co-occur with the words already present in it, and where co-occurrence is detected via the χ^2 measure. We use the output of this algorithm to create what we call the χ^2 *lexicon*, in which each word is associated with the (normalized) frequency with which the word is used to describe a certain aspect.

2.3 Inference and Learning

While the maximum a posteriori (MAP) assignment $y^* \in Y$ and factor marginals can be inferred exactly in the LC and LCO models by means of variants of the Viterbi and forward-backward algorithms [11], exact inference is not tractable in the remaining models due to a combinatorial explosion and to the presence of loops in the graph structure. Instead, we revert to approximate inference via Gibbs sampling (see, e.g., [3]).

All models are trained to approximately minimize the Hamming loss over the training set $D = \{(x^{(i)}, y^{(i)})\}_{i=1}^N$ using the SampleRank algorithm [19], which is a natural fit to sampling-based inference.[3] Briefly put, with SampleRank the model parameters w

---

[3] While inference and learning algorithms are likely to impact results, this decision brings about no substantial loss of generality, since the focus of this study is on comparing model structures.

are updated locally after each draw from the Gibbs sampler by taking an atomic gradient step with respect to the local Hamming loss incurred by the sampled variable setting. This procedure is repeated for a number of epochs until the ℓ_2-norm of the sum of the atomic gradients from the epoch is below a threshold ϵ; in each epoch every variable in the training set is sampled in turn. For the experiments in Section 3, the SampleRank learning rate was fixed to $\alpha = 1$ and the gradient threshold to $\epsilon = 10^{-5}$.

After fitting the model parameters to the training data, at test time we perform 100 Gibbs sampling epochs to find an approximate MAP assignment $\mathbf{y}^*$ for input $\mathbf{x}$ with respect to the distribution $p(\mathbf{y}|\mathbf{x})$.

3   Experiments

In this section we study the proposed models empirically. After discussing our evaluation strategy, we describe and discuss the creation of a new dataset of hotel reviews, which has been manually annotated with aspect-specific opinion at the sentence level. Finally, we compare the proposed models quantitatively by their performance on this dataset.

Evaluation Measures. When evaluating system output and comparing human annotations below, we view the task as composed of the following two subtasks:

Aspect identification: for each segment and for each aspect, predict if there is any opinion expressed towards the aspect in the segment. Since each of these aspect-specific tasks is a binary problem, for this subtask we adopt the standard F_1 evaluation measure.

Opinion prediction: for each segment and each applicable (true positive) aspect for the segment, predict the opinion expressed towards the aspect in the segment. Since opinions are placed on an ordinal scale, as an evaluation measure we adopt *macro-averaged mean absolute error* (MAE^M) [1], a measure for evaluating ordinal classification that is also robust to the presence of imbalance in the dataset.

Let $\mathbf{T}$ be the correct label assignments and let $\widehat{\mathbf{T}}$ be the corresponding model predictions. Let $\mathbf{T}_j - \{y_i : y_i \in \mathbf{T}, y_i = j\}$ and let n be the number of unique labels in $\mathbf{T}$. The macro-averaged mean absolute error is defined as

$$\text{MAE}^M(\mathbf{T}, \widehat{\mathbf{T}}) = \frac{1}{n} \sum_{j=1}^{n} \frac{1}{|\mathbf{T}_j|} \sum_{y_i \in \mathbf{T}_j} |y_i - \hat{y}_i| \tag{8}$$

This is suitable for evaluating the overall review-level opinion predictions. However, when evaluating the aspect-specific opinions at the segment level, we instead report $\text{MAE}^M(\mathbf{T}_{\widehat{\mathbf{I}}_a}, \widehat{\mathbf{T}}_{\widehat{\mathbf{I}}_a})$, where $\widehat{\mathbf{I}}_a$ is the sequence of indices of segment opinion labels that were predicted as true positive for aspect a and $\mathbf{T}_{\widehat{\mathbf{I}}_a}$ is the set of true positive opinion labels for aspect a.

Inter-annotator Agreement Measures. We also use F_1 and MAE^M to assess inter-annotator agreement, by computing the average of these measures over all pairs of annotators. While F_1 and the micro-averaged version of MAE are both symmetric, the use of macro-averaging makes MAE^M asymmetric, i.e., switching the predicted labels with the gold standard labels may change the outcome. This is problematic when

280   D. Marcheggiani et al.

Table 2. Number of opinion expressions at the sentence level, broken down by aspect and opinion. Out of 5799 annotated sentences, 4810 sentences contain at least one opinion-laden expression.

	Other	Service	Rooms	Clean.	Food	Location	Check-in	Value	Building	Business	NotRelated	Total
Pos	893	513	484	180	287	435	93	188	185	23	63	*3344*
Neg	353	248	287	66	127	51	56	87	62	3	40	*1377*
Neu	167	40	111	5	82	38	12	35	22	4	350	*866*
Total	1413	801	882	251	496	524	161	310	269	30	453	*5134*

used to measure inter-annotator agreement, since no annotator can be given precedence over the others (unless they have different levels of expertise). We thus symmetrize the measure by treating each annotator in turn as the gold standard and by averaging the corresponding results. This yields the *symmetrized macro-averaged mean absolute error*:

$$\text{sMAE}^M(\mathbf{T}, \widehat{\mathbf{T}}) = \frac{1}{2}\left(\text{MAE}^M(\mathbf{T}, \widehat{\mathbf{T}}) + \text{MAE}^M(\widehat{\mathbf{T}}, \mathbf{T})\right) \tag{9}$$

3.1 Annotated Dataset

We have produced a new dataset of manually annotated hotel reviews[4]. Three equally experienced annotators provided sentence-level annotations of a subset of 442 randomly selected reviews from the publicly available TripAdvisor dataset [18]. Each review comes with an overall rating on a discrete ordinal scale from 1 to 5 "stars".

The annotations are related to 9 aspects often present in hotel reviews. In addition to the 7 aspects explicitly present (at the review level) in the TripAdvisor dataset (Rooms, Cleanliness, Value, Service, Location, Check-in, and Business), we decided to add 2 other aspects (Food and Building), since many comments in the reviews refer to them. Furthermore, the "catch-all" aspects Other and NotRelated were added, for a total of 11 aspects. Other captures those opinion-related aspects that cannot be assigned to any of the first 9 aspects, but which are still about the hotel under review. The NotRelated aspect captures those opinion-related aspects that are not relevant to the hotel under review. In what follows, segments marked as NotRelated are treated as non-opinionated.

The annotation distinguishes between Positive, Negative and Neutral/Mixed opinions. The Neutral/Mixed label is assigned to opinions that are about an aspect without expressing a polarized opinion, and to opinions of contrasting polarities, such as *The room was average size* (neutral) and *Pricey but worth it!* (mixed). The annotations also distinguish between explicit and implicit opinion expressions, i.e., between expressions that refer directly to an aspect and expressions that refer indirectly to an aspect by referring to some other property/entity related to the aspect. For example, *Fine rooms* is an explicitly expressed positive opinion concerning the Rooms aspect, while *We had great views over the East River* is an implicitly expressed positive opinion concerning the Location aspect.

---

[4] At `http://nemis.isti.cnr.it/~marcheggiani/datasets/` the interested reader may find both the dataset and a more detailed explanation of it.

Table 3. Inter-annotator agreement results. Top 3 rows: segment-level aspect agreement, expressed in terms of F_1 (higher is better). Bottom 3 rows: segment-level opinion agreement (restricted to the true positive aspects for each segment), expressed in terms of sMAE^M (lower is better).

	Other	Service	Rooms	Clean.	Food	Location	Check-in	Value	Building	Business	Avg
Overall	.607	.719	.793	.733	.794	.795	.464	.575	.553	.631	.675
Implicit	.167	.123	.263	.111	.306	.286	.061	.131	.095	.333	.188
Explicit	.479	.684	.706	.739	.741	.710	.481	.560	.521	.624	.625
Overall	.308	.219	.191	.114	.234	.259	.003	.202	.150	.029	.171
Implicit	.167	.000	.000	.000	.074	.061	.000	.000	.000	.000	.030
Explicit	.262	.167	.147	.064	.190	.119	.000	.179	.092	.000	.122

Out of the 442 reviews, 73 reviews were independently annotated by all three annotators so as to facilitate the measurement of inter-annotator agreement, while the remaining 369 reviews were subdivided equally among the annotators. These 369 reviews were then partitioned into a training set (258 reviews, 70% of the total) and a test set (111 reviews, 30% of the total). The data were split by selecting reviews for each subset in an interleaved fashion, so that each subset constitutes a minimally biased sample both with respect to the full dataset and with respect to annotator experience.

Table 2 shows, for each aspect and for each opinion type, the number of segments annotated with a given aspect and a given opinion type (across the unique reviews and averaged across the shared reviews). Both opinions and aspects show a markedly imbalanced distribution. As expected, the imbalance with respect to opinion is towards the Positive label. In terms of aspects, the Rooms, Service and Other aspects dominate.

Inter-annotator Agreement. We use the 73 shared reviews (943 sentences) to measure the agreement between the 3 annotators with respect to both aspects and opinions, using F_1 and symmetrized MAE^M. For each aspect we separately measure the agreement on implicit and explicit opinionated mentions, and the agreement on mentions of both types.

From the agreement results in Table 3 (top) we see a large disagreement with respect to implicit opinions. However, the agreement overall (disregarding the explicit/implicit distinction) is higher than the agreement on explicit opinions in isolation. This suggest that, while it is difficult for annotators to separate implicit from explicit opinions, separating opinionated mentions from non-opinionated mentions is easier. In what follows we thus ignore the distinction between implicit and explicit opinions.

Table 3 (bottom 3 rows) shows the agreement on the true positive opinion annotations, that is, the agreement on the opinions with respect to those aspects on which the two annotators agree. Closer inspection of the data shows that, as could be expected, the disagreement mainly affects the pairs Neutral–Positive and Neutral–Negative.

3.2 Results

All models were trained on the training set described in Section 3.1, for a total of 258 reviews. Below we describe two separate evaluations. First, we compare the different

Table 4. Aspect-oriented opinion prediction results for different CRF models averaged across five experiments with 5 different random seeds. Top 6 rows: segment-level aspect prediction results in terms of F_1 (higher is better). Bottom 6 rows: segment-level opinion prediction results (restricted to the true positive aspects for each segment) in terms of MAE^M (lower is better).

	Other	Service	Rooms	Clean.	Food	Location	Check-in	Value	Building	Business	Avg
LC	.499	.606	.662	.700	.579	.623	.329	**.395**	.298	.000	.469
IML	**.542**	.597	**.664**	**.732**	**.605**	.668	**.371**	.373	**.363**	.000	.491
CML	.489	**.645**	.655	.708	.605	**.673**	.327	.408	.358	**.076**	.494
LCO	.515	.586	.661	.697	.582	.611	.301	.384	**.368**	.173	**.488**
IMLO	.513	.621	**.685**	.702	.593	.614	**.370**	.363	.348	.040	.485
CMLO	**.531**	**.629**	.663	**.706**	**.602**	**.618**	.271	**.393**	.350	.081	.485
LC	.526	.721	.572	1.000	.566	.932	**.644**	**.616**	.693	.000	.627
IML	.520	**.659**	**.494**	**.956**	**.377**	.939	.670	.700	.668	.000	.598
CML	**.492**	.681	.613	.978	.482	**.906**	.735	.691	**.377**	.000	.595
LCO	.482	.626	**.398**	1.000	.633	**.903**	.690	.490	.233	.000	.546
IMLO	**.473**	**.615**	.398	1.000	**.457**	.970	**.343**	**.469**	.269	.000	**500**
CMLO	.499	.626	.428	1.000	.711	.906	.536	.552	**.232**	.000	.549

models by their accuracy on the test set (111 reviews). Since training is non-deterministic due to the use of sampling-based inference, we report the average over five trials with different random seeds. Second, we compare the best-performing model to the human annotators on the set of 73 reviews independently annotated by all three annotators.

Comparison among Systems. As shown in Table 4, the multi-label and hierarchical models outperform the LC baseline in both aspect identification and opinion prediction. In particular, the multi-label models (IML, CML) significantly outperform the baseline on both subtasks, which shows the importance of modeling the interdependence of different aspects and their opinions within a segment. On the other hand, combining both multi-label and transition factors in the hierarchical model (CMLO) leads to worse predictions compared to only including the multi-label factors (IMLO) or the transition factors. We hypothesize that this is due to inference errors, where the more complex graph structure causes the Gibbs sampler to converge more slowly. Furthermore, while the hierarchical models provide a significant improvement compared to their non-hierarchical counterparts in terms of opinion prediction, modeling both the overall and segment-level opinions is not helpful for aspect identification. This is not too surprising, given that the overall opinion contains no information about aspect-specific opinions.[5]

---

[5] In addition to the reported experiments, we performed initial experiments with models that also included variables for overall opinions with respect to specific aspects. However, including these variables hurts performance at the segment level. We hypothesize that this is because reviewers often rate multiple aspects while only discussing a subset of them in the review text.

Table 5. Comparison between the best-performing model (IMLO) and the human annotators with IMLO results averaged over five runs (F_1 for the top two rows, sMAE^M for the bottom two rows)

	Other	Service	Rooms	Clean.	Food	Location	Check-in	Value	Building	Business	*Avg*
Human	.607	.719	.793	.795	.553	.575	.794	.464	.733	.631	*.675*
IMLO	.479	.585	.606	.614	.536	.673	.407	.429	.208	.190	*.473*
Human	.308	.219	.191	.259	.150	.202	.234	.003	.114	.029	*.171*
IMLO	.676	.498	.445	.142	.451	.704	.212	.387	.025	.415	*.396*

The overall review-level opinion prediction results (not shown in Table 4) are in line with the segment-level results. The IMLO model (.504) outperforms the LCO baseline (.518), as measured with MAE^M. However, as with the segment-level predictions, including both multi-label and transition factors in the hierarchical model (CMLO) hurts overall opinion prediction (.544).

Comparison among Humans and System. We now turn to a comparison between the best-performing model (IMLO) and the human annotators, treating the model as a fourth annotator when computing inter-annotator agreement. This allows us to assess how far our model is from human-level performance. Table 5 clearly shows that much work remains to be done for both subtasks. The aspects Building and Business are difficult to detect for the automatic system, while a human identifies them with ease. We believe that the reasons for the poor performance may be different for the two aspects. For the Business aspect, the reason is likely the scarcity of training annotations, whereas for the Building aspect the reason may be lexical promiscuity (that is, a hotel building may be described by a multitude of features, such as interior, furniture, architecture, etc.).

Interestingly, the system identifies the Value aspect at close to human level, but performs dramatically worse on its opinion prediction. We suggest that this is because assessing the value of something coined in absolute terms (for example, that a $30 room is cheap) requires world knowledge (or feature engineering).

4   Conclusions

We have considered the problem of aspect-oriented opinion mining at the sentence level. Specifically, we have devised a sequence of increasingly powerful CRF models, culminating in a hierarchical multi-label model that jointly models both the overall opinion expressed in a review and the set of aspect-specific opinions expressed in each sentence of the review. Moreover, we have produced a manually annotated dataset of hotel reviews in which the set of relevant aspects and the opinions expressed concerning these aspects are annotated for each sentence; we make this dataset publicly available with the hope to spur further research in this area. We have evaluated the proposed models on this dataset; the empirical results show that the hierarchical multi-label model outperforms a strong comparable baseline.

References

1. Baccianella, S., Esuli, A., Sebastiani, F.: Evaluation measures for ordinal regression. In: Proceedings of the 9th IEEE International Conference on Intelligent Systems Design and Applications (ISDA 2009), Pisa, IT, pp. 283–287 (2009)
2. Baccianella, S., Esuli, A., Sebastiani, F.: SentiWordNet 3.0: An enhanced lexical resource for sentiment analysis and opinion mining. In: Proceedings of the 7th Conference on Language Resources and Evaluation (LREC 2010), Valletta, MT (2010)
3. Bishop, C.M.: Pattern recognition and machine learning. Springer, Heidelberg (2006)
4. Ghamrawi, N., McCallum, A.: Collective multi-label classification. In: Proceedings of the 14th ACM International Conference on Information and Knowledge Management (CIKM 2005), Bremen, DE, pp. 195–200 (2005)
5. Hu, M., Liu, B.: Mining and summarizing customer reviews. In: Proceedings of the 10th ACM SIGKDD International Conference on Knowledge Discovery and Data Mining (KDD 2004), Seattle, US, pp. 168–177 (2004)
6. Lafferty, J., McCallum, A., Pereira, F.: Conditional random fields: Probabilistic models for segmenting and labeling sequence data. In: Proceedings of the 18th International Conference on Machine Learning (ICML 2001), Williamstown, US, pp. 282–289 (2001)
7. Lazaridou, A., Titov, I., Sporleder, C.: A Bayesian model for joint unsupervised induction of sentiment, aspect and discourse representations. In: Proceedings of the 51st Annual Meeting of the Association for Computational Linguistics (ACL 2013), Sofia, BL, pp. 1630–1639 (2013)
8. Li, F., Han, C., Huang, M., Zhu, X., Xia, Y.J., Zhang, S., Yu, H.: Structure-aware review mining and summarization. In: Proceedings of the 23rd International Conference on Computational Linguistics (COLING 2010), Bejing, CN, pp. 653–661 (2010)
9. Liu, B.: Sentiment analysis and opinion mining. Morgan & Claypool Publishers, San Rafael (2012)
10. McCallum, A., Schultz, K., Singh, S.: Factorie: Probabilistic programming via imperatively defined factor graphs. In: Proceedings of the 23rd Annual Conference on Neural Information Processing Systems (NIPS 2009), Vancouver, CA, pp. 1249–1257 (2009)
11. McDonald, R., Hannan, K., Neylon, T., Wells, M., Reynar, J.: Structured models for fine-to-coarse sentiment analysis. In: Proceedings of the 45th Annual Meeting of the Association for Computational Linguistics (ACL 2007), Prague, CZ, pp. 432–439 (2007)
12. Moghaddam, S., Ester, M.: ILDA: Interdependent LDA model for learning latent aspects and their ratings from online product reviews. In: Proceedings of the 34th ACM SIGIR International Conference on Research and Development in Information Retrieval (SIGIR 2011), Bejing, CN, pp. 665–674 (2011)
13. Pang, B., Lee, L.: A sentimental education: Sentiment analysis using subjectivity summarization based on minimum cuts. In: Proceedings of the 42nd Annual Meeting of the Association for Computational Linguistics (ACL 2004), Barcelona, ES, pp. 271–278 (2004)
14. Pang, B., Lee, L., Vaithyanathan, S.: Thumbs up? Sentiment classification using machine learning techniques. In: Proceedings of the 7th Conference on Empirical Methods in Natural Language Processing (EMNLP 2002), Philadelphia, US, pp. 79–86 (2002)
15. Stone, P.J., Dunphy, D.C., Smith, M.S.: The General Inquirer: A Computer Approach to Content Analysis. The MIT Press, Cambridge (1966)
16. Täckström, O., McDonald, R.: Discovering fine-grained sentiment with latent variable structured prediction models. In: Clough, P., Foley, C., Gurrin, C., Jones, G.J.F., Kraaij, W., Lee, H., Mudoch, V. (eds.) ECIR 2011. LNCS, vol. 6611, pp. 368–374. Springer, Heidelberg (2011)

17. Titov, I., McDonald, R.T.: A joint model of text and aspect ratings for sentiment summarization. In: Proceedings of the 46th Annual Meeting of the Association for Computational Linguistics (ACL 2008), Columbus, US, pp. 308–316 (2008)
18. Wang, H., Lu, Y., Zhai, C.: Latent aspect rating analysis on review text data: A rating regression approach. In: Proceedings of the 16th ACM SIGKDD International Conference on Knowledge Discovery and Data Mining (KDD 2010), Washington, US, pp. 783–792 (2010)
19. Wick, M., Rohanimanesh, K., Bellare, K., Culotta, A., McCallum, A.: SampleRank: Training factor graphs with atomic gradients. In: Proceedings of the 28th International Conference on Machine Learning (ICML 2011), Bellevue, US, pp. 777–784 (2011)
20. Wilson, T., Wiebe, J., Hoffmann, P.: Recognizing contextual polarity in phrase-level sentiment analysis. In: Proceedings of the Conference on Human Language Technology and Empirical Methods in Natural Language Processing (HLT/EMNLP 2005), Vancouver, CA, pp. 347–354 (2005)

Generating Pseudo-ground Truth
for Predicting New Concepts in Social Streams

David Graus, Manos Tsagkias, Lars Buitinck, and Maarten de Rijke

ISLA, University of Amsterdam, Amsterdam, The Netherlands
{d.p.graus,e.tsagkias,l.j.buitinck,derijke}@uva.nl

Abstract. The manual curation of knowledge bases is a bottleneck in fast paced domains where new concepts constantly emerge. Identification of nascent concepts is important for improving early entity linking, content interpretation, and recommendation of new content in real-time applications. We present an unsupervised method for generating pseudo-ground truth for training a named entity recognizer to specifically identify entities that will become concepts in a knowledge base in the setting of social streams. We show that our method is able to deal with missing labels, justifying the use of pseudo-ground truth generation in this task. Finally, we show how our method significantly outperforms a lexical-matching baseline, by leveraging strategies for sampling pseudo-ground truth based on entity confidence scores and textual quality of input documents.

1 Introduction

We increasingly harvest the power of knowledge bases to interpret the content generated around us. This is achieved via semantic linking, a process that identifies mentions of real-life entities or concepts in text, and links them to concepts in a knowledge base (KB) [20]. Core to its success is the extensive coverage of today's KBs; they span the majority of popular and well established concepts. For most content domains this coverage is enough; however it does not provide a solid basis for domains that refer to "long-tail" entities or where new entities are constantly born: e.g. in news and social streams. Here, entities emerge (and sometimes disappear) before editors of a KB reach consensus on whether an entity should be included. Identifying newly emerging entities that will become concepts in a knowledge base is important in knowledge base population and complex filtering tasks, where users are not just interested in any entity, but also in attributes like impact or importance.

The target users we have in mind are media analysts in an online reputation management setting who track entities that can impact the reputation of their customer in social streams, e.g., Twitter. Our problem is related to named entity detection, classification and disambiguation, with the additional constraint that an entity should have "impact" or be important. Although impact or importance are hard to model because they depend on the context of a task, we argue that entities that are included in a knowledge base are more important than those that are not, and use this signal for modeling the importance of an entity.

Named entity recognition is a natural approach for identifying newly emerging entities, that are not in the KB. However, current models fall short as they do not account

M. de Rijke et al. (Eds.): ECIR 2014, LNCS 8416, pp. 286–298, 2014.

for the importance, or impact of the entity. In this paper, we present an unsupervised method for generating pseudo-ground truth for training a named entity recognizer to predict new concepts. Our method is applicable to any trainable model for named entity recognition. In addition, our method is not restricted to a particular class of entities, but can be trained to predict any type of concept that is in the KB.

The challenge here is two-fold: (a) how to model the attribute of importance, and (b) the system needs to adapt to its input (social streams) and the updates in the knowledge base; content that is eligible for addition in Wikipedia today, may no longer be in the future, which renders static training annotations unusable.

Our approach answers both challenges. For the first challenge, we carefully craft the training set of a named entity recognizer to steer it towards identifying new concepts. That is, we leverage prior knowledge of important concepts, to identify new concepts that are likely to share the same attributes. Just as a named entity recognizer trained solely on English person-type entities will recognize only such entities, a named entity recognizer trained on entities with referent KB concepts is likely to recognize only this type of entity. For the second challenge, we provide an unsupervised method for generating pseudo-ground truth from the input stream. This way, we are not dependent on human annotations that are necessarily limited and domain and language specific, and newly added knowledge will be automatically included.

We focus on social streams because of the fast paced evolution of content and its unedited nature, which make it a challenging setting for predicting which entities will feature in a knowledge base. The main research question we seek to answer is: *What is the utility of our sampling methods for generating pseudo-ground truth for a named entity recognizer?* We measure utility within the task of predicting new concepts from social streams as the prediction effectiveness of a named entity recognizer trained using our method. We also study the impact of prior knowledge, in our second research question: *What is the impact of the size of prior knowledge on predicting new concepts?* Our main contribution is a method that uses entity linking for generating training material for a named entity recognizer.

2   Approach

We view the task of identifying new concepts in social streams as a combination of an *entity linking* (EL) problem and a *named-entity recognition and classification* (NERC) problem. We visualize our method in Fig. 1. Starting from a document in a document stream, we extract sentences, and use an EL system to identify referent concepts in each sentence. If any is identified, the sentence is pooled as a candidate training example for NERC (we refer to this type of sentence as a *linkable sentence*), otherwise it is routed to NERC for identifying new concepts (*unlinkable sentences*): an underlying assumption behind our method is that the first place to look for new entities is the set of unlinkable sentences. Most of our attention in this paper is devoted to training NERC. Two ideas are important here. First, we extend the distributional hypothesis [11] (i.e., words that occur in the same contexts tend to have similar meaning) from words to entities and concepts; we hypothesize that new entities that should be included in the knowledge base occur in similar contexts as current knowledge base concepts. Second, we apply

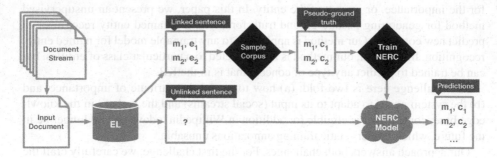

Fig. 1. Our approach for *generating pseudo-ground truth* for training a NERC method, and for *predicting new concepts*

EL on the input stream and transform its output into pseudo-ground truth for NERC; this results in an unsupervised way of generating pseudo-ground truth, with the flexibility of choosing any type of entity or concept described in the KB.

3   Unsupervised Generation of Pseudo-ground Truth

We start with the output of an entity linking method[1] [19, 23], on a sentence of a document in the stream. This output is our source for generating training material. The output consists of tuples of *entity mentions* and *referent entities* (m, e pairs in Fig. 1).

Since we are allowed to use generic corpora from any domain, e.g., news, microblog posts, we may expect to have noise in our pseudo-ground truth. We apply various sampling methods to select sentences that make up a high quality training corpus. These sampling methods are described in Section 4.

After sampling, we convert the remaining sentences in a format suitable for input for NERC. This format consists of the *entity span*; the sequence of tokens that refers to an entity, i.e. the entity mention, and *entity class* for each linked entity (m, c pairs in Fig. 1). To denote the entity span, we apply the BIO-tagging scheme [24], where each token is tagged with whether it is the **B**eginning of an entity, **I**nside an entity, or **O**utside one, so that a document like *Kendrick Lamar and A$AP Rocky. That's when I started listening again. Thanks to Brendan.* becomes: *$Kendrick_B$ $Lamar_I$ and_O $A\$AP_B$ $Rocky_I$. $That's_O$ $when_O$ I_O $started_O$ $listening_O$ $again_O$. $Thanks_O$ to_O $Brendan_B$.* The final step is to assign a class label to an entity. For this, we need to move from a concept to a concept class. As not all knowledge bases associate a concept class to their concepts, we use DBPedia for looking up the concept and extracting the concept's DBPedia ontology class, if any; see Section 5 for details. Our example then becomes: *$Kendrick_{B\text{-}PER}$ $Lamar_{I\text{-}PER}$ and_O $A\$AP_{B\text{-}PER}$ $Rocky_{I\text{-}PER}$. $That's_O$ $when_O$ I_O $started_O$ $listening_O$ $again_O$. $Thanks_O$ to_O $Brendan_{B\text{-}PER}$.* Now we can proceed and train NERC with our generated pseudo-ground truth. We do so using a two-stage approach [4] where the recognition stage is implemented using the fast structured perceptron algorithm [7].[2]

---

[1] http://semanticize.uva.nl

[2] https://github.com/larsmans/seqlearn

Table 1. Nine features used for sampling documents from which we train a NERC system

Feature	Description	Feature	Description
n_mentions	Number of usernames (@)	avg_token_len	Average token length
n_hashtags	Number of hashtags (#)	tweet_len	Length of tweet
n_urls	Number of URLs	density	Density as in [13]
ratio_upper	Percentage of uppercased chars	personal	Contains personal
ratio_nonalpha	Percentage of non-alphanumeric chars		pronouns (I, me, we, etc.)

4   Sampling Pseudo-ground Truth

In this section, we present two methods for sampling pseudo-ground truth: (a) sampling based on the EL system's confidence score for a detected entity, and (b) sampling based on the textual quality of an input document.

Sampling based on entity linker's confidence score. Typically, entity linkers return a confidence score with each entity mention (n-gram) they are able to link to a knowledge base concept. These confidence scores can be used to rank possible concepts for an n-gram, but also for pruning out entity mention-concept pairs about which the linker is not confident. Although the scale of the confidence score is dependent on the model behind the entity linker, the scores can be normalized over the candidate concepts for an entity, e.g., using linear or z-score normalization. We use the *SENSEPROB* [23] metric as confidence score. This score is calculated by combining the probability of an n-gram being used as an anchor text on Wikipedia, with the *commonness* [21] score.

Sampling based on textual quality of an input document. Taking the textual quality of content into account has proved helpful in a range of tasks. Based on [18, 26], we consider nine features indicative of textual quality; see Table 1. While not exhaustive, our feature set is primarily aimed at social streams as our target document stream (see §5) and suffices for providing evidence on whether this type of sampling is helpful for our purposes. Based on these features, we compute a final score for each document d as $score(d) = \frac{1}{|F|} \sum_{f \in F} \frac{f(d)}{\max_f}$, where F is our set of feature functions (Table 1) and $\max_f$ is the maximum value of f we have seen so far in the stream of documents. Since all features are normalized in $[0, 1]$, $score(d)$ has this same range. As a sanity check, we rank documents from the MSM2013 [1] dataset using our quality sampling method and list the top-3 and bottom-3 scoring documents in Table 2. Top scoring documents are longer and denser in information than low scoring documents. We assume that these documents are better examples for training a NERC system.

In the next section, we follow a linear search approach to sampling training examples as input for NERC. First, we find an optimal threshold for confidence scores, and fix it. For sampling based on textual quality, we turn to the MSM2013 dataset to determine sampling thresholds. We calculate the scores for each tweet, and scale them to fall between [0,1]. We then plot the distribution of scores, and bin this distribution in three parts: tweets that fall within a single standard deviation of the mean are considered *normal*, tweets to the left of this bin are considered *noisy*, whilst the remaining tweets

Table 2. Ranking of documents in the MSM2013 dataset based on our quality sampling method. Top ranking documents appear longer and denser in information than low ranking documents.

Top-3 quality documents
"Watching the History channel, Hitler's Family. Hitler hid his true family heritage, while others had to measure up to Aryan purity."
"When you sense yourself becoming negative, stop and consider what it would mean to apply that negative energy in the opposite direction."
"So. After school tomorrow, french revision class. Tuesday, Drama rehearsal and then at 8, cricket training. Wednesday, Drama. Thursday ... (c)"

Bottom-3 quality documents
*Toni Braxton ˜ He Wasnt Man Enough for Me _HASHTAG_ _HASHTAG_? _URL_ RT _Mention_*
"tell me what u think The GetMore Girls, Part One _URL_."
this girl better not go off on me rt

to the right of the distribution are considered *nice*. We repeat this process for our tweet corpus, using the bin thresholds gleaned from the MSM2013 set.

5 Experimental Setup

In addressing the new concept prediction in document streams problem, we concentrate on developing an unsupervised method for generating pseudo-ground truth for NERC and predicting new concepts. In particular, we want to know the effectiveness of our unsupervised pseudo-ground truth (UPGT) method over a random baseline and a lexical matching baseline, and the impact on effectiveness of our two sampling methods. To answer these questions, we conduct both optimization and prediction experiments.

Dataset. As a document stream, we use tweets from the TREC 2011 Microblog dataset [16], a collection of 4,832,838 unique English Twitter posts. This choice is motivated by the unedited and noisy nature of tweets, which can be challenging for prediction. Our knowledge base (KB) is a subset of Wikipedia from January 4, 2012, restricted to concepts that correspond to the NERC classes person (PER), location (LOC), or organization (ORG). We use DBPedia to perform selection, mapping the DBPedia classes *Organisation*, *Company*, and *Non-ProfitOrganisation* to ORG, *Place*, *PopulatedPlace*, *City*, and *Country* to LOC, and *Person* to PER.[3] Our final KB consists of 1,530,501 concepts.

Experiment I: Sampling pseudo-ground truth. To study the utility of our sampling methods, we turn to the impact of setting a threshold on the entity linker's confidence score (Experiment Ia.) and the effectiveness of our textual quality sampling (Experiment Ib.). In Experiment Ia., we do a sweep over thresholds from 0.1 up to 0.9, using

---

[3] http://mappings.dbpedia.org/server/ontology/classes/

the same threshold for both the generation of pseudo-ground truth and evaluating the prediction effectiveness of new concepts. Lower thresholds allow low confidence entities in the pseudo-ground truth, and likely generates more data at the expense of noisy output. We emphasize that we are not interested in the correlation between noise and confidence score, but rather in the performance of finding new concepts given the EL system's configuration. In Experiment Ib., we compare our methods performance with differently sampled pseudo-ground truths, containing nice, normal, or noisy tweets.

Experiment II: Prediction experiments. To answer our second research question, and study the impact of prior knowledge on detecting new concepts, we compare the performance of our method (UPGT) to two baselines: a *random baseline* (RB) that extracts all n-grams from test tweets and considers them new concepts, and a *lexical-matching baseline* (NB) that follows our approach, but generates pseudo-ground truth by applying lexical matching of KB entity titles, instead of an EL system, and refrains from sampling based on textual quality. For this experiment, we use the optimal sampling parameters for generating pseudo-ground truth from our previous experiment, i.e., include linked entities with a confidence score higher than 0.7, and use only normal tweets. As we will show in Section 6, the threshold of 0.7 balances performance with high recall of entities in the pseudo-ground truth.

Evaluation. We evaluate the quality of the generated pseudo-ground truth on the effectiveness of a NERC system trained to predict new concepts. As measuring the addition of new concepts to the knowledge base is non-trivial, we consider a retrospective scenario: Given our KB, we random sample concepts to yield a smaller KB (KB_s). This KB_s simulates the available knowledge at the present point in time, whilst KB represents the future state. By measuring how many concepts we are able to detect in our corpus that feature in KB, but not KB_s, we can measure new concept prediction. We create KB_s by taking random samples of 20–90% the size of KB (measured in concepts), in steps of 10%. We repeat each sampling step ten times to avoid bias.

We generate test sets and pseudo-ground per KB_s. We link the corpus of tweets using KB_s, and yield two sets of tweets: (a) tweets that contain new concepts, and (b) tweets with linked concepts, analog to the *unlinked* and *linked sentences* in Fig. 1. The size of these two sets depend on the size of KB_s and makes the comparison of results difficult across different KB_s. We cater for this bias by randomly sampling 10,000 tweets from both the test set and the pseudo-ground truth and repeating our experiments ten times.[4] Ground truth is then assembled by linking the corpus of tweets using KB. This ground truth consists of 82,305 tweets, with 12,488 unique concepts.

We evaluate the effectiveness of our method in two ways: (a) the ability of NERC to generalize from our pseudo-ground truth, and (b) the accuracy of our predictions. For the first, we compare the predicted concept *mentions* to those in our ground truth, akin to traditional NERC evaluation. For the second we take the set of correct predictions (true positives), and link each mention to the referent concept in the ground truth. This allows us to measure what we're actually interested in: the fraction of newly discovered concepts. For both types of evaluation, we report on average precision and recall over

---

[4] Using the smallest KB_s (20%) results in about 15,000 tweets in the pseudo-ground truth.

100 runs per KB_s. Statistical significance is tested using a two-tailed paired t-test and is marked as ▲ for significant differences for $\alpha = .01$.

6 Results

Experiment I: Sampling pseudo-ground truth. Our first experiment aims to answer RQ1: *What is the utility of our sampling methods for generating pseudo-ground truth for a named entity recognizer?* We fix the KB_s at 50%. We start by looking at the ability of NERC to generalize from our pseudo-ground truth, measured on two aspects: (i) effectiveness for identifying mentions of new concepts, and (ii) predicting new concepts.

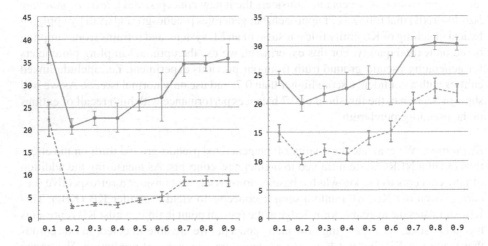

Fig. 2. Experiment Ia. Impact of confidence score on UPGT. Effectiveness of identifying mentions of new concepts (left). New concept prediction effectiveness (right). Threshold set on the confidence score in the x-axis. Precision (solid line) and recall (dotted) are shown in the y-axis.

In Experiment Ia., we look at our confidence-based sampling method. For identifying mentions of new concepts, we find that effectiveness peaks at 0.1 confidence threshold with a precision of 38.84%, dips at 0.2 and slowly picks up to 35.75% as threshold increases (Fig. 2, left). For new concept prediction, effectiveness positively correlates with the threshold. Effectiveness peaks at the 0.8 confidence threshold, statistically significantly different from 0.7 but not from 0.9 (Fig. 2, right).

Interestingly, besides precision, recall also shows a positive correlation with thresholds. This suggests that in new concept prediction, missing training labels are likely to have less impact on performance than generating incorrect, noisy labels. This is an interesting finding as it sets the concept prediction task apart from traditional NERC, where low recall due to incomplete labeling is a well-understood challenge.

Next, we turn to the characteristics of the pseudo-ground truth that results for each of these thresholds, and provide an analysis of their potential impact on effectiveness. We find that more data through a larger pseudo-ground truth allows NERC to better generalize and predict a larger number of new concepts.

This claim is supported by the number of predicted concepts per threshold in Table 3. We find a similar trend as in the precision and recall graph above: the number of predicted concepts peaks for the threshold at 0.1 (6,653 concept mentions), and drops between 0.2 and 0.4, and picks up again from 0.5 reaching another local maximum at 0.8. The increasing number of predicted concept mentions with stricter thresholds indicates that the NERC model is more successful in learning patterns for separating concepts from noisy labels. This may be due to the entity linker linking only those entities it is most confident about, providing a clearer training signal for NERC.

Table 3. Number of predicted concept mentions per threshold on the confidence score

Threshold	0.1	0.2	0.3	0.4	0.5	0.6	0.7	0.8	0.9
Predictions	6,653	1,500	1,618	1,512	1,738	2,025	2,662	2,713	2,614
Ground truth	11,429	11,533	11,291	11,078	10,955	10,935	10,799	10,881	10,855

For the rest of our experiments we use a threshold of 0.7 on confidence score because it is deemed optimal in terms of trade-off between performance and quantity of concepts in pseudo-ground truth.

In Experiment Ib., we study three different textual quality-based sampling strategies: we consider only normal tweets (i), nice tweets (ii), or both normal and nice tweets (iii); see Section 4. For reference, we also report on the performance achieved when no textual quality sampling is used. We keep KB$_S$ fixed at 50%, and use 0.7 for confidence threshold.

Table 4. Experiment Ib. Precision and recall for three sampling strategies based on textual quality of documents: nice, normal, normal+nice. We also report on effectiveness of a system that uses no sampling for reference. Boldface indicates best performance. Statistical significance is tested against the previous sampling method, e.g., nice to normal.

	Mention		Concept	
Sampling	Precision	Recall	Precision	Recall
No sampling	34.26±2.65	8.21±0.83	29.63±1.67	20.25±1.56
Normal+nice	45.50±4.71▲	12.97±2.03▲	36.22±2.07▲	24.86±1.69▲
Normal	66.09±3.86▲	30.94±3.46▲	44.62±1.51▲	**32.20±1.67▲**
Nice	**70.36±3.07▲**	**30.98±3.25**	**45.99±1.34▲**	29.69±1.79

Textual quality-based sampling turns out to be twice as effective as no sampling on both identifying mentions of new concepts and new concept prediction. Among our sampling strategies, nice proves to be the most effective with a precision of 70.36% for concept mention identification.

In new concept prediction, the performance of nice and normal strategies hovers around the same levels. In terms of recall, nice and normal methods are on par, outperforming both other strategies. The success of nice and normal sampling methods can be attributed to the fact that a more coherent and homogeneous training corpus allows the NERC model to more easily learn patterns.

Experiment II: Impact of prior knowledge Next, we seek to answer RQ2: *What is the impact of the size of prior knowledge on predicting new concepts?* We use the optimal combination of our sampling methods from the previous experiments, i.e., a confidence threshold of 0.7, and the normal textual quality sampling. We again look at the effectiveness of our methods in identifying new concept mentions and predicting new concepts.

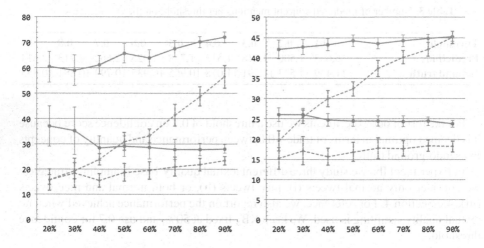

Fig. 3. Our method (UPGT, blue line) versus the lexical baseline (red) for both identifying mentions of new concepts (left) and new concept detection (right). Knowledge base size is on the *x*-axis, and precision (solid lines) and recall (dotted lines) are marked on the *y*-axis.

Fig. 3 shows the effectiveness of our methods as a function of the size of the knowledge base. For identifying mentions of new concepts, our method (UPGT, blue line) constantly and statistically significantly outperforms both lexical (red line) and random baselines (not shown). In terms of recall, the lexical baseline is on par with UPGT for KB sizes up to 30%. The random baseline shows very low precision for both identifying mentions of new concepts, and new concept prediction over all KB_s, but almost perfect recall—which is expected given that it assigns all possible n-grams as new concept mentions and new concepts (0.69% precision and 65% recall for entity mention identification, 1.82% precision and 94.95% recall for new concept prediction).

Next, we take a closer look at the results. The lexical baseline's recall increases slightly when more prior knowledge is added to the knowledge base. This is expected behavior because, as we saw in our previous experiment, the pseudo-ground gets more diverse labels, and helps the NERC to generalize. Looking at the number of unique concepts in the pseudo-ground truth, we find that the number increases with the size of KB_s. UPGT assigns labels to 2,500 unique concepts at 20% KB_s, which tops at 11,000 unique concepts at 90%. These numbers are lower for the lexical baseline (1,800 at

20% KB_s, and 7,000 at 90% KB_s). However, for both methods, the number of concepts in the ground truth stays around the same. The gradual improvement in precision and recall of UPGT for increasing KB_s can be attributed to a broader coverage for labeling (observed through looking at the prior entities), and the main distinction between the lexical baseline and UPGT: more strict labeling through leveraging the entity linker's confidence score.

Finally, to better understand the performance of our method, we have looked at the set of correct predictions (new concepts), and false positives, or incorrectly identified new concepts. Our analysis revealed examples of out of knowledge base entities, that were not included in the initial KB, highlighting the challenging setting of evaluating the task.

On the whole, however, our method is able to deal with missing labels and incomplete data, as observed through its consistent and stable precision, justifying our assumption that data is incomplete by design.

7   Related Work

The problem we concentrate on, and the method we propose for approaching it relate to literature in (a) entity linking in document streams, (b) training methods on automatically annotated data, and (c) predicting new concepts.

Document streams. Lexical matching-based entity linking approaches have shown to be successful in the challenging genre of social streams [19], and have shown to be suitable for adaptation to new genres and languages [23]. They provide a strong baseline, and as an added advantage are independent of intricate NLP pipelines, linguistic features, etc. Cassidy et al. [6] expand on this approach by considering "coherence" between entities to aid disambiguation. Guo et al. [12] propose a weakly supervised method for detecting so-called NIL entities, but this cannot handle or recognize out-of-KB entities. In addition, the noisy character of social streams degrades the effectiveness of NER methods [9, 10], and current approaches largely tailor the NLP pipeline to Twitter and heavily rely on large amounts of labeled data [3, 25]. We generate large amounts of training data for these types of system to improve NER effectiveness on social streams.

Automatically generated pseudo-ground truth. Several attempts have been made to either simulate or generate human annotations. Kozareva [15] uses regular expressions for generating training data for NERC. Zhou et al. [28] generate training data by considering Wikipedia links as positive examples, and consider each other entity that may be referred to by the same anchor as negative examples Nothman et al. [22] leverage the anchor text of links inbetween two same articles in different Wikipedia translations for training a multilingual NERC system. Wu et al. [27] investigate generating training material for NERC from one domain and testing on another. Becker et al. [2] study the effects of sampling automatically generated data for training NER. Our setting differs from settings considered before, and our approach to automatically generating ground truth by using entity linking is new too.

Predicting new concepts. Viewed abstractly, our task is similar to named entity normalization in user generated content [14], and named entity disambiguation in streams [8], but the conditions are different because of the lack of "context" and discussion structure (e.g., comments on an article). Our task is also different because of our focus on knowledge bases and the emergence of unknown entities. Bunescu [5] study out-of-Wikipedia entity detection by setting a threshold on their candidate ranker. Lin et al. [17] leverage n-gram statistics from Google Books for predicting new concepts. Our method generates training data solely from the input stream and a knowledge base and does not depend on third sources which may have different evolution rate.

8   Conclusions

We tackled the problem of predicting new concepts in social streams. We presented an unsupervised method for generating pseudo-ground truth to train NERC for detecting entities that are likely to become concepts in a knowledge base. Our method uses the output of an entity linker to generate training material for NERC. We introduced two sampling methods, based on the entity linker confidence, and the textual quality of an input document. We found that sampling by textual quality improves performance of NERC and consequently our method's performance in new concept prediction. As setting a higher threshold on the entity linker's confidence score for generating pseudo-ground truth results in fewer labels but better performance, we show that the NERC is better able to separate noise from entities that are worth including in a knowledge base. The entity linker's confidence score is an effective signal for this separation. Our sampling methods significantly improve detection of knowledge base worthy entities.

In the case of a small amount of prior knowledge , i.e., size of the available KB, our method is able to cope with missing labels and incomplete data, as observed through its consistent and stable precision, justifying our proposed method that assumes incomplete data by design. This finding furthermore suggests the scenario of an increasing rate of new concept prediction, as more data is fed back to the KB. Additionally, we found that a larger number of entities in the KB allows for setting a desirable stricter threshold on the confidence scores, and leads to improvements in both precision and recall. This finding suggests an adaptive threshold that takes prior knowledge into account could prove effective.

Our proposed method can be applied with any trainable NERC model and entity linker that is able to return a confidence score for a linked entity. In addition, our method is suitable for domain and/or language adaptation as it does not rely on language specific features or sources.

Acknowledgments. This research was partially supported by the European Community's Seventh Framework Programme (FP7/2007-2013) under grant agreement nr 288024 (LiMoSINe project), the Netherlands Organisation for Scientific Research (NWO) under project nrs 640.004.802, 727.011.005, 612.001.116, HOR-11-10, the Center for Creation, Content and Technology (CCCT), the QuaMerdes project funded by the CLARIN-nl program, the TROVe project funded by the CLARIAH program, the Dutch national program COMMIT, the ESF Research Network Program ELIAS, the

Elite Network Shifts project funded by the Royal Dutch Academy of Sciences (KNAW), the Netherlands eScience Center under project number 027.012.105 and the Yahoo! Faculty Research and Engagement Program.

References

[1] Basave, A.E.C., Varga, A., Rowe, M., Stankovic, M., Dadzie, A.-S.: Making sense of microposts (#msm2013) concept extraction challenge. In: MSM 2013, pp. 1–15 (2013)

[2] Becker, M., Hachey, B., Alex, B., Grover, C.: Optimising selective sampling for bootstrapping named entity recognition. In: ICML-LMV 2005, pp. 5–11 (2005)

[3] Bontcheva, K., Derczynski, L., Funk, A., Greenwood, M.A., Maynard, D., Aswani, N.: TwitIE: An open-source information extraction pipeline for microblog text. In: RANLP 2013, ACL (2013)

[4] Buitinck, L., Marx, M.: Two-stage named-entity recognition using averaged perceptrons. In: Bouma, G., Ittoo, A., Métais, E., Wortmann, H. (eds.) NLDB 2012. LNCS, vol. 7337, pp. 171–176. Springer, Heidelberg (2012)

[5] Bunescu, R.: Using encyclopedic knowledge for named entity disambiguation. In: EACL 2006, pp. 9–16 (2006)

[6] Cassidy, T., Ji, H., Ratinov, L.-A., Zubiaga, A., Huang, H.: Analysis and enhancement of wikification for microblogs with context expansion. In: COLING 2012, pp. 441–456 (2012)

[7] Collins, M.: Discriminative training methods for hidden markov models: Theory and experiments with perceptron algorithms. In: ACL 2002, pp. 1–8 (2002)

[8] Davis, A., Veloso, A., da Silva, A.S., Meira Jr., W., Laender, A.H.F.: Named entity disambiguation in streaming data. In: ACL 2012, pp. 815–824 (2012)

[9] Derczynski, L., Maynard, D., Aswani, N., Bontcheva, K.: Microblog-genre noise and impact on semantic annotation accuracy. In: HT 2013, pp. 21–30. ACM (2013)

[10] Finkel, J.R., Manning, C.D., Ng, A.Y.: Solving the problem of cascading errors: approximate bayesian inference for linguistic annotation pipelines. In: EMNLP 2006, pp. 618–626 (2006)

[11] Firth, J.R.: A Synopsis of Linguistic Theory, 1930-1955. In: Studies in Linguistic Analysis, pp. 1–32 (1957)

[12] Guo, S., Chang, M.-W., Kiciman, E.: To link or not to link? a study on end-to-end tweet entity linking. In: NAACL 2013, pp. 1020–1030 (2013)

[13] Hu, M., Sun, A., Lim, E.-P.: Comments-oriented document summarization: understanding documents with readers' feedback. In: SIGIR 2008, pp. 291–298. ACM (2008)

[14] Jijkoun, V., Khalid, M., Marx, M., de Rijke, M.: Named entity normalization in user generated content. In: AND 2008 (2008)

[15] Kozareva, Z.: Bootstrapping named entity recognition with automatically generated gazetteer lists. In: EACL-SRW 2006, pp. 15–21. ACL (2006)

[16] Lin, J., Macdonald, C., Ounis, I., Soboroff, I.: Overview of the TREC 2011 Microblog track. In: TREC 2011 (2012a)

[17] Lin, T., Mausam, Etzioni, O.: No noun phrase left behind: detecting and typing unlinkable entities. In: EMNLP-CoNLL 2012, pp. 893–903 (2012b)

[18] Massoudi, K., Tsagkias, M., de Rijke, M., Weerkamp, W.: Incorporating query expansion and quality indicators in searching microblog posts. In: Clough, P., Foley, C., Gurrin, C., Jones, G.J.F., Kraaij, W., Lee, H., Mudoch, V. (eds.) ECIR 2011. LNCS, vol. 6611, pp. 362–367. Springer, Heidelberg (2011)

[19] Meij, E., Weerkamp, W., de Rijke, M.: Adding semantics to microblog posts. In: WSDM 2012, pp. 563–572 (2012)

[20] Mihalcea, R., Csomai, A.: Wikify!: linking documents to encyclopedic knowledge. In: CIKM 2007, pp. 233–242 (2007)
[21] Milne, D., Witten, I.H.: Learning to link with wikipedia. In: CIKM 2008, pp. 509–518 (2008)
[22] Nothman, J., Ringland, N., Radford, W., Murphy, T., Curran, J.R.: Learning multilingual named entity recognition from wikipedia. Artificial Intelligence 194, 151–175 (2013)
[23] Odijk, D., Meij, E., de Rijke, M.: Feeding the second screen: Semantic linking based on subtitles. In: OAIR 2013 (2013)
[24] Ramshaw, L., Marcus, M.: Text chunking using transformation-based learning. In: Third Workshop on Very Large Corpora. ACL (1995)
[25] Ritter, A., Mausam, O.E., Clark, S.: Open domain event extraction from twitter. In: KDD 2012 (2012)
[26] Weerkamp, W., de Rijke, M.: Credibility-inspired ranking for blog post retrieval. Information Retrieval 15(3-4), 243–277 (2012)
[27] Wu, D., Lee, W.S., Ye, N., Chieu, H.L.: Domain adaptive bootstrapping for named entity recognition. In: EMNLP 2009, pp. 1523–1532. ACL (2009)
[28] Zhou, Y., Nie, L., Rouhani-Kalleh, O., Vasile, F., Gaffney, S.: Resolving surface forms to wikipedia topics. In: COLING 2010, pp. 1335–1343 (2010)

Thumbnail Summarization Techniques
for Web Archives

Ahmed AlSum and Michael L. Nelson

Computer Science Department, Old Dominion University, Norfolk VA, USA
{aalsum,mln}@cs.odu.edu

Abstract. Thumbnails of archived web pages as they appear in common browsers such as Firefox or Chrome can be useful to convey the nature of a web page and how it has changed over time. However, creating thumbnails for all archived web pages is not feasible for large collections, both in terms of time to create the thumbnails and space to store them. Furthermore, at least for the purposes of initial exploration and collection understanding, people will likely only need a few dozen thumbnails and not thousands. In this paper, we develop different algorithms to optimize the thumbnail creation procedure for web archives based on information retrieval techniques. We study different features based on HTML text that correlate with changes in rendered thumbnails so we can know in advance which archived pages to use for thumbnails. We find that SimHash correlates with changes in the thumbnails ($\rho = 0.59, p < 0.005$). We propose different algorithms for thumbnail creation suitable for different applications, reducing the number of thumbnails to be generated to 9% – 27% of the total size.

1   Introduction

A thumbnail is a small image that represents a web page as it is rendered in a web browser such as Firefox or Chrome. Representing web pages with thumbnails has been used in various research such as: the visualization of the web search results [1–3], the visualization of recommended and similar pages such as SimilarWeb.com, and helping in revisitation and remembrance of the web page [4, 5]. Using thumbnails in web archives and temporal data has been studied [6, 7], but the creation cost of the thumbnails in the web archive has not yet been studied.

There are currently a few web archives that create thumbnails for mementos (archived versions of web pages). For example, the UK Web Archive and Archive.is provide a partial list of thumbnails per URI. Archive-It enables the partners to generate thumbnails for quality assurance purposes.

Figure 1 shows two examples of a TimeMap (a list of all the available mementos for a URI) for VisitWales.com. The HTML bubble interface in the Internet Archive[1] is in the back with the red border while in the front is the thumbnail view of the same URI in the UK Web Archive[2]. The thumbnail view gives the user an idea about the

---

[1] http://web.archive.org/web/*/http://visitwales.com/
[2] http://www.webarchive.org.uk/ukwa/
target/56197146/source/subject

M. de Rijke et al. (Eds.): ECIR 2014, LNCS 8416, pp. 299–310, 2014.
© Springer International Publishing Switzerland 2014

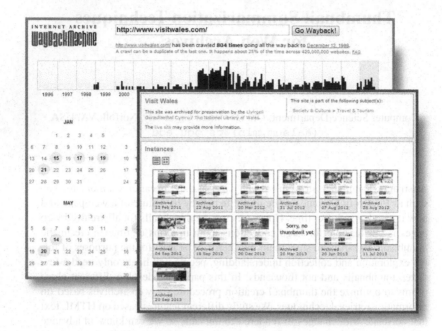

Fig. 1. Different representations for VisitWales.com TimeMap

website layout and its evolution through time. Also, it helps the user in searching and revisitation for specific mementos of this URI. For example, VisitWales.com did not change from Jul 2012 to Sep 2012; the website layout was the same from Feb 2011 to Jun 2013 while it had a slight change for the rest of the timeline. This visual representation of the website through the time helps the user to select the desired mementos in the TimeMap. The thumbnail with the text "Sorry, no thumbnail yet" could have been generated for several reasons, but a likely explanation is that the HTML was not successfully processed by the thumbnail generator.

Creating the thumbnails in web archive has a number of challenges. First, scalability in time, the computation of the thumbnail requires the rendering of the web page including retrieving all the images, style sheets, and running any javascript. Second, scalability in space, the thumbnail (as metadata information about the web page) requires storage and our experiment shows that the average size for thumbnail with 64x64 pixels is 3KB and with 600x600 pixels is 133KB while the average size of HTML text is 39KB. Based on the Internet Archive estimated size of 240 billions mementos [8], they would need an additional 335 TB to store a 64x64 thumbnail for each memento, it will reach 14.5 PB to store a 600x600 thumbnail for each memento. Finally, page quality, the construction of the archived web page may not be successful on the time of building the thumbnail due to the missing embedded resources.

Using the thumbnails also has some limitations. Some TimeMaps have too many mementos for browsing. For example, the Internet Archive has over 17,000+ mementos for cnn.com over a 14 year span. So even if the Internet Archive has thumbnails for all the mementos, some form of sampling or partitioning will be necessary because the cognitive load of processing all the mementos will be beyond what the user can handle [9]. In a personal digital library, Graham et al. [10] suggested 25 images per view panel because displaying all images did not help in conveying an overview of the album.

In this paper, we use the HTML text of the mementos and the crawler log information to predict the visual change in the URI through the time in order to select the significant set of mementos that can summarize the TimeMap. We explore various features that could be extracted from the HTML pages and select the most relevant to the visual effect, and then we propose different techniques for online and offline thumbnail creation.

The rest of this paper is organized as follows. Section 2 describes the related work. Section 3 describes the dataset that we use in our experiment. Section 4 explores the different features that extracted from HTML text or crawler log. Section 5 discusses the selection algorithms that is used to estimate the representative thumbnails for TimeMap. Section 6 concludes with a summary and future work for this study.

2   Related Work

Few web archives provide a thumbnail view for its content. The UK Web Archive uses thumbnails in timeline and 3D wall visual browsing [11]. Reed archives[3], National Taiwan University Web Archiving System (NTUWAS) [12], and Archive.is provide either a partial or full list of thumbnails for each memento. Archive-It generates on-demand thumbnails as part of the quality assurance process but it is only accessible for their partners. Web Archive Service at California Digital Library[4] uses thumbnails to represent the collections. The usage of thumbnails in the web archives has been studied in few applications. Soman et al. [13] developed ArcSpread which took a query from the user, extracted related information from the Stanford WebBase [14] using a Hadoop cluster, and displayed the results in spread sheet style. ArcSpread used both images and web page thumbnails to express the matched web pages. Jatowt et al. [15] proposed *"Past Web Browser"* to present TimeMap page snapshots as a slideshow of memento thumbnail. It has been extended to *Page History Explorer* [16] for visualizing and comparing the history of web pages. Zoetrope [7] provided a timeline visualization to show the duration and frequency of web pages on specific website. Padia et al. [6] implemented new visualization techniques for Archive-It collections entitled *"image plot with histogram"* in which they represented each page with its thumbnail.

Thumbnails help with the presentation of temporal data. Tsang et al. [17] proposed the concept of *"Temporal Thumbnail"* where they included the time dimension into the space of 3D model. They showed that the Temporal Thumbnails were effective for quickly analyzing large amounts of viewing data. Stoev and Straßer [18] studied the visualization of historical data in existence of time dimension. They studied two models; navigation with 4D, and navigation with one fixed dimension.

Thumbnails have applications on the web. SimilarWeb.com and StumbleUpon.com have recommendation pages function that uses the page thumbnails to represent the recommended pages to help the user to determine the page

---

[3] http://www.reedarchives.com/
[4] http://webarchives.cdlib.org/

relevancy. Janssen [19] used fixed thumbnail per document (called an Icon Document) to visualize the search results from UpLib digital library system[5]. Janssen studied the computation of the icon size and the decoration of the icon with labels (e.g., creation date).

The thumbnail can enhance the users' ability to find the information. Lam and Baudisch [20] proposed a *"Summary Thumbnail"* that generated customized thumbnail for web pages to increase the readability of the text on embedded/small screens devices. They discovered that the preservation of the page layout allowed the users to better detect useful information. Woodruff et al. [1] showed that using a mixture of textual and image representations of the web page increased the user prediction of the page effectiveness.

Aula et al. [21] compared thumbnail and textual summaries by surveying the users between different types of page previews. The study showed that the combination of textual information with the thumbnails increased the user estimation of the page content and usefulness. Also, they found that the recognition improved when the thumbnail size was 200x250 pixels. Kaasten et al. [4] and Teevan et al. [5] discovered that thumbnails of size 208x208 pixels or above are effective to remember an exact page.

The selection of the representative image from a set of images depends on comparison techniques between the images itself. These techniques are different from our proposed techniques as we depend on the HTML text to select the representative thumbnail image. AutoAlbum [22] clustered personal photographs into album using time-based and content-based clustering. Coelho and Ribeiro [23] used image filter to get an abstracted version of the images to reduce the dataset to 10% of the original size. Chu et al. [24] studied the near duplicate detection technique for images to cluster a set of images and select a representative image from them. Kherfi and Ziou [25] used the image clustering to organize a large set of images and to provide the user with an overview of the collection's content. They used probabilistic models based on predefined keywords or visual features. Graham et al. [10] studied a set of images that had an attached timestamp (e.g., images from digital camera). They provided various clustering and summarization techniques such as summarization by time and clustering based on the time and location. They developed a calendar browser with specific 25 photos per panel. They selected the images based on the number of images each month.

3 Dataset

We built a collection of TimeMaps using the homepage URIs for the companies listed in the 2013 Fortune 500 list[6]. For each URI, we retrieved the TimeMap from the Internet Archive Wayback Machine. Of those 500 URIs, we have 488 TimeMaps since 12 are not archived due to robots.txt exclusion. The total number of mementos was 499,540. The mean number of mementos per TimeMap is 1023 and the median is 685. For each memento, we downloaded the HTML text from IA and we used PhantomJS[7] to capture a screenshot for each memento.

---

[5] http://uplib.parc.com/
[6] http://money.cnn.com/magazines/
fortune/fortune500/2013/full_list/
[7] http://phantomjs.org/

4   Exploration of Features

In this section, we explore various features that can be used to predict the change in the visual representation of the web page. The features could be obtained from the crawler log (e.g., CDX[8] file for Heritrix crawler) or from the HTML text for the memento.

4.1   Web Page Similarity Features

There have been several methods proposed for calculating similarity between web pages [26–29]. In this section, we explore various features and techniques to estimate the resulting difference in thumbnail images based on the HTML source of the mementos.

SimHash Similarity. SimHash [28] is a fingerprint technique to calculate the near-duplicates between web pages. We used 64-bit SimHash fingerprints with $k = 4$. We calculate the SimHash for different parts of the web page. First, we calculate SimHash for the full HTML text. Then, we use boilerpipe library [30] to extract different parts of the web page such as: the main content from the web page, all the text (even the template text), templates including the text, and the template excluding the text (just HTML structure). For each subsequent mementos in the TimeMap, we computed the Hamming Distance between their SimHash fingerprints.

Levenshtein Distance between HTML DOM Tree. Each web page can be expressed in a DOM tree. We can compare between two web pages by calculating the difference between their DOM trees. In this feature, we transform each HTML page into a DOM tree (using jsoup[9]), then we calculate the Levenshtein distance between both trees by calculating the number of operations (insert, delete, and replace) to turn one tree into the other [31].

Embedded Resources. The change in the embedded resources that construct the web page affects the visual appearance of the rendered web page. We extract the embedded resources for each memento and calculate the difference between each two pages. We calculate the total number of new resources that have been added and the resources that have been removed. Then, we divide the total number based on the embedded resource type (e.g., image, style sheet, javascript), so we have the specific number of addition and removal for each category. For example, the difference between M_{t1} and M_{t2} could be: addition of 5 resources (2 javascript files and 3 images) and removal of 2 resources (1 javascript and 1 image).

Memento Datetime. Memento Datetime is the datetime that the memento was crawled (or observed) by the web archive. This information is already available in the CDX file. The similarity is calculated based on the difference in seconds, a low number indicates high similarity.

---

[8] http://archive.org/web/researcher/cdx_file_format.php
[9] http://jsoup.org/

4.2 Experiment

The success criteria for each feature is how well it can predict the visual difference between two mementos. First, we generate a thumbnail for each memento in each TimeMap. Then, we calculate the difference by comparing the number of different pixels between each two thumbnails using SciPy[10]. In order to compare two thumbnails, we resize them into different dimensions: 64x64, 128x128, 256x256, and 600x600. We calculated the Manhattan distance and Zero distance between each pair. Finally, we calculate the correlation between the similarity of the web pages (based on features in section 4.1) and the difference between the thumbnails.

4.3 Results

Figure 2 shows histogram of the correlation (using Pearson's ρ) (on x-axis) between some features and the image difference calculated by Manhattan distance using thumbnail size 600x600 and the number of TimeMap that achieved this level (on y-axis). Generally, all the features showed a positive relationship that range from weak to strong. The best correlation has been found between SimHash fingerprints for the original HTML text showed a positive correlation, 70% of the TimeMap has a correlation $\rho \geq 0.5$ (on average $\rho = 0.59, p < 0.005$). The Levenshtein distance between the HTML DOM tree has a high correlation on average ($\rho = 0.57, p < 0.005$). We use the SimHash similarity because the computation is faster than the HTML DOM tree distance.

Our results show interesting features. First, the calculation of the visual difference between the images is not affected by the thumbnail size. The results are consistent between the different thumbnails sizes. Second, we repeat the SimHash calculation on different parts of the web page, the SimHash similarity between the full HTML text shows the highest correlation over the rest of web page parts. Third, however the datetime and embedded resources show a weak positive correlation, however it can be used with other strong features to get a better prediction.

5 Selection Algorithms

In this section, we discuss three algorithms to select a list of representative thumbnails for individual TimeMaps.

5.1 Threshold Grouping

In this algorithm, we use feature f for TimeMap $TM = \{M(t_1), M(t_2), ..M(t_n)\}$. The initial step is to divide the TM into groups G, each group has only two subsequent mementos $M(t_i)$ and $M(t_{i-1})$. For each G, we compute the $diff(f)$ between $M(t_i)$ and $M(t_{i-1})$. If $diff(f, M(t_i), M(t_{i-1})) < \alpha$, we will eliminate one of the mementos M in the group. Then, the new list is sorted by time and we repeat the grouping again. We will continue the process until we reach that for every pair of mementos $diff(f, M(t_i), M(t_{i-1})) > \alpha$. Figure 3 shows the first step of the threshold grouping.

---

[10] http://docs.scipy.org/doc/scipy/reference/index.html

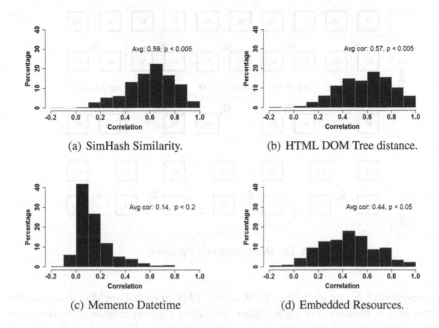

(a) SimHash Similarity.   (b) HTML DOM Tree distance.

(c) Memento Datetime   (d) Embedded Resources.

Fig. 2. Histogram for the correlation between Thumbnail difference and various features

The value of α can be configured and it may depend on other factors. Figure 4 shows the relationship between the change in the SimHash threshold (on x-axis) and the reduction of the TimeMap and the loss of image difference (on y-axis). The image loss is calculated as the Manhattan distance between the selected thumbnail and the eliminated thumbnail. We defined the optimum point as the smallest TimeMap size with the least image difference loss. Our empirical study shows that for SimHash similarity, we find that $\alpha = 0.05$ is the optimum value where it decreases the TimeMap to 27% of its original size with the loss of 27% of the image differences.

We notice here that with SimHash threshold at 0, which means removes all the duplicate snapshots only, we still have loss in image differences. It happens because even if the HTML text may be the same, the rendered image may not be visually identical. The reason for that on the live web may be different advertisements, different results from javascript (e.g., picking a random image each time), etc. The web archive environment adds more causes such as: the embedded resources may not be archived, change of embedded resources (e.g., update in style sheet without changing the URI), or the web archive server can not render the page at that time (cf. "Sorry, no thumbnail yet" in figure 1).

5.2 K Clustering

In this algorithm, the web archive will select K thumbnails for each TimeMap. K could be absolute, relative, or an expression based on other parameters (e.g., rate of change, crawling frequency). The algorithm will depend on the a set of features F. To get the

306   A. AlSum and M.L. Nelson

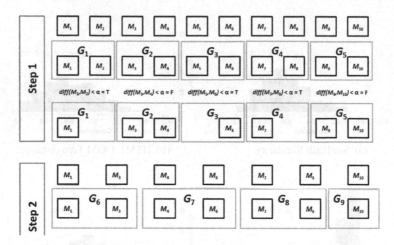

Fig. 3. Threshold Grouping algorithm

most representative K, we apply the "*K-Medoids*" [32] on each TimeMap to cluster the mementos into K clusters. For each cluster, we select random thumbnail to represent the cluster. We apply the algorithm on two sets of features, $F_1 = \{SimHash\}$ and $F_2 = \{SimHash, Memento - Datetime\}$, we repeat the algorithm on $K \in \{5 : 200\}$. Figure 5 shows the average sum square error and the average image loss in both cases. The red dots describes the average sum square error for each cluster, the red line is the power regression for the sum square error. The blue dots describes the average image loss, and the blue line is its power regression. The black line is a linear cost function based on the time taken to generate the K thumbnails. Notice that the increasing number of clusters decreases the error value but also it increases the cost of creating the thumbnails. Also, the sum square error and image loss error trends are the same which align with our results at section 4.3. Using two features (figure 5(b)) gives lower error rate in both sum square error and image loss which means a better representation of the TimeMap. Based on our empirical study, we find the optimal values for $K = 25..40$, which decreases the size of the average TimeMap by 9% to 12% of its original size.

5.3   Time Normalization

The drawback of the previous algorithms is that it did not take the time dimension into the consideration. In this algorithm, we will apply normalization per time for the TimeMap. We will divide the TimeMap into fixed time-slots T and select random k thumbnails from each slot. The advantage of this algorithm is that it is easy to implement because it depends only on the CDX files independent from the web pages itself. Figure 6 shows the reduction of each TimeMap after selecting $k = 1$ and $T = 1\ month$. It reduced the TimeMap size on average to 23% of its original size.

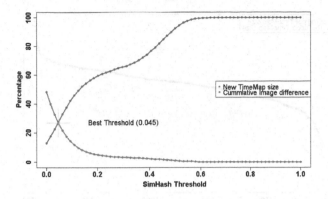

Fig. 4. Optimum SimHash threshold point for Threshold Grouping algorithm

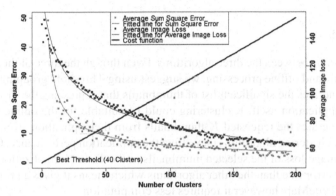

(a) Best K using SimHash feature only.

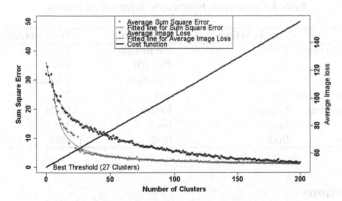

(b) Best K using SimHash and Memento-Datetime features.

Fig. 5. Best K value for K Clustering algorithm

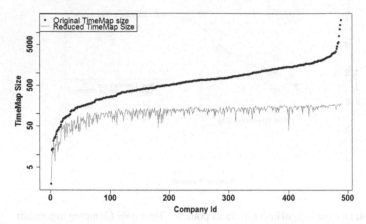

Fig. 6. Reduction of the TimeMap size after applying k thumbnail per time-slot algorithm

5.4   Analysis

Table 1 compares between the three algorithms. Even though the three of them could be used for online and offline processing, we suggest using Threshold grouping for offline because it generates the significant list of thumbnails that represents the TimeMap disregards the application itself. K clustering could be suitable for the online processing because it generates the requested K thumbnails from the application, even the extensive nature of clustering algorithms may affect the application performance. Comparing the average image loss of the selected thumbnails in the Time Normalization technique shows higher error rate than the other algorithms which means it gives a poor representation of the TimeMap, however it requires less computation.

Table 1. Comparison between the selection algorithms

	Threshold Grouping	K clustering	Time Normalization
TimeMap Reduction	27%	9% to 12%	23%
Image Loss	28	78 - 101	109
# Features	1 feature	1 or more	1 feature
Preprocessing required	Yes	Yes	No
Efficient processing	Medium	Extensive	Light
Incremental	Yes	No	Yes
Online/offline	Both	Both	Both

6   Conclusions

We studied effective methods to generate thumbnails for the web archive corpus. We explored various similarity features, we found that SimHash and Levenshtein distance between HTML DOM trees had $\rho = 0.59$ and $\rho = 0.57$ correlation with the visual difference between two mementos. We suggest using SimHash as it is efficiently computed from the HTML text. We proposed three algorithms to select K representative thumbnails from the TimeMap. The algorithms decreased the TimeMap size from 9% to 27% of its original size with the minimum loss in thumbnails differences. The techniques could be used with online and offline processing.

Our next step in this research will be to integrate an exploratory TimeMap service in ArcLink that will apply the algorithms defined above to selectively generate thumbnails for TimeMaps to facilitate collection understanding.

References

1. Woodruff, A., Faulring, A., Rosenholtz, R., Morrsion, J., Pirolli, P.: Using thumbnails to search the Web. In: Proceedings of the SIGCHI Conference on Human Factors in Computing Systems, CHI 2001, pp. 198–205 (2001)
2. Kules, B., Wilson, M.L., Shneiderman, B.: From Keyword Search to Exploration: How Result Visualization Aids Discovery on the Web. Technical report, HCIL-2008-06 (2008)
3. Treharne, K., Powers, D.M.W.: Search Engine Result Visualisation: Challenges and Opportunities. In: Proceedings of 13th International Conference on Information Visualisation, pp. 633–638 (2009)
4. Kaasten, S., Greenberg, S., Edwards, C.: How People Recognise Previously Seen Web Pages from Titles, URLs and Thumbnails. In: People and Computers XVI - Memorable Yet Invisible SE, pp. 247–265. Springer, Heidelberg (2002)
5. Teevan, J., Cutrell, E., Fisher, D., Drucker, S.M., Ramos, G., André, P., Hu, C.: Visual Snippets: Summarizing Web Pages for Search and Revisitation. In: Proceedings of the SIGCHI Conference on Human Factors in Computing Systems, CHI 2009, pp. 2023–2032. ACM (2009)
6. Padia, K., AlNoamany, Y., Weigle, M.C.: Visualizing digital collections at Archive-It. In: Proceedings of the 12th ACM/IEEE-CS Joint Conference on Digital Libraries, JCDL 2012, pp. 15–18 (2012)
7. Adar, E., Dontcheva, M., Fogarty, J., Weld, D.S.: Zoetrope: interacting with the ephemeral web. In: Proceedings of the 21st Annual ACM Symposium on User Interface Software and Technology, UIST 2008, pp. 239–248 (2008)
8. AlSum, A., Nelson, M.L.: ArcLink: Optimization Techniques to Build and Retrieve the Temporal Web Graph. Technical report, arXiv: 1305.5959 (2013)
9. Mayer, R.E., Moreno, R.: Nine ways to reduce cognitive load in multimedia learning. Educational Psychologist 38(1), 43–52 (2003)
10. Graham, A., Garcia-Molina, H., Paepcke, A., Winograd, T.: Time as essence for photo browsing through personal digital libraries. In: Proceedings of the Second ACM/IEEE-CS Joint Conference on Digital Librariesm, JCDL 2002, pp. 326–335 (2002)
11. Hockx-Yu, H.: The Past Issue of the Web. In: Proceedings of 3rd International Conference on Web Science, WebSci 2011, pp. 1–8 (2011)
12. Chen, K., Chen, Y., Ting, P.: Developing National Taiwan University Web Archiving System. In: Proceedings of 8th International Web Archiving Workshop, IWAW 2008 (2008)
13. Soman, S., Chhajta, A., Bonomo, A., Paepcke, A.: ArcSpread for Analyzing Web Archives. Technical report. Stanford InfoLab (2012)
14. Cho, J., Garcia-Molina, H., Haveliwala, T., Lam, W., Paepcke, A., Raghavan, S., Wesley, G.: Stanford WebBase Components and Applications. ACM Transactions on Internet Technology 6(2) (2006)

15. Jatowt, A., Kawai, Y., Nakamura, S., Kidawara, Y., Tanaka, K.: Journey to the past: proposal of a framework for past web browser. In: Proceedings of the 17th Conference on Hypertext and Hypermedia, HYPERTEXT 2006, pp. 135–144. ACM (2006)
16. Jatowt, A., Kawai, Y., Tanaka, K.: Page History Explorer: Visualizing and Comparing Page Histories. IEICE Transactions on Information and Systems E94-D(3), 564–577 (2011)
17. Tsang, M., Morris, N., Balakrishnan, R.: Temporal Thumbnails: rapid visualization of time-based viewing data. In: Proceedings of the Working Conference on Advanced Visual Interfaces, AVI 2004, pp. 175–178 (2004)
18. Stoev, S.L., Straßer, W.: A case study on interactive exploration and guidance aids for visualizing historical data. In: Proceedings of the Conference on Visualization, VIS 2001, pp. 485–488 (2001)
19. Janssen, W.C.: Document Icons and Page Thumbnails: Issues in Construction of Document Thumbnails for Page-Image Digital Libraries. In: Heery, R., Lyon, L. (eds.) ECDL 2004. LNCS, vol. 3232, pp. 111–121. Springer, Heidelberg (2004)
20. Lam, H., Baudisch, P.: Summary thumbnails: Readable Overviews for Small Screen Web Browsers. In: Proceedings of the SIGCHI Conference on Human Factors in Computing Systems, CHI 2005, pp. 681–690 (2005)
21. Aula, A., Khan, R.M., Guan, Z., Fontes, P., Hong, P.: A comparison of visual and textual page previews in judging the helpfulness of web pages. In: Proceedings of the 19th International Conference on World Wide Web, WWW 2010, pp. 51–59. ACM Press (2010)
22. Platt, J.C.: AutoAlbum: clustering digital photographs using probabilistic model merging. In: Proceedings of IEEE Workshop on Content-based Access of Image and Video Libraries, pp. 96–100 (2000)
23. Coelho, F., Ribeiro, C.: Image abstraction in crossmedia retrieval for text illustration. In: Baeza-Yates, R., de Vries, A.P., Zaragoza, H., Cambazoglu, B.B., Murdock, V., Lempel, R., Silvestri, F. (eds.) ECIR 2012. LNCS, vol. 7224, pp. 329–339. Springer, Heidelberg (2012)
24. Chu, W.T., Lin, C.H.: Automatic selection of representative photo and smart thumbnailing using near-duplicate detection. In: Proceeding of the 16th ACM International Conference on Multimedia, MM 2008, pp. 829–832 (October 2008)
25. Kherfi, M.L., Ziou, D.: Image Collection Organization and Its Application to Indexing, Browsing, Summarization, and Semantic Retrieval. IEEE Transactions on Multimedia 9(4), 893–900 (2007)
26. Henzinger, M.: Finding near-duplicate web pages. In: Proceedings of the 29th Annual International ACM SIGIR Conference on Research and Development in Information Retrieval, SIGIR 2006, pp. 284–291 (2006)
27. Broder, A., Glassman, S.: Syntactic clustering of the web. Computer Networks and ISDN Systems 29(8-13) (1997)
28. Charikar, M.S.: Similarity estimation techniques from rounding algorithms. In: Proceedings of the Thiry-Fourth Annual ACM Symposium on Theory of Computing, STOC 2002, pp. 380–388 (2002)
29. Manku, G.S., Jain, A., Das Sarma, A.: Detecting near-duplicates for web crawling. In: Proceedings of the 16th International Conference on World Wide Web, WWW 2007, pp. 141–149 (2007)
30. Kohlschütter, C., Fankhauser, P., Nejdl, W.: Boilerplate detection using shallow text features. In: Proceedings of the Third ACM International Conference on Web Search and Data Mining, WSDM 2010, pp. 441–450 (2010)
31. Pawlik, M., Augsten, N.: RTED: a robust algorithm for the tree edit distance. In: Proceedings of the VLDB Endowment, vol. 5(4), pp. 334–345 (December 2011)
32. Park, H.S., Jun, C.H.: A simple and fast algorithm for K-medoids clustering. Expert Systems with Applications 36(2, pt. 2), 3336–3341 (2009)

CiteSeer^x: A Scholarly Big Dataset

Cornelia Caragea[1,4], Jian Wu[2,5], Alina Ciobanu[3,6], Kyle Williams[2,5],
Juan Fernández-Ramírez[1,7], Hung-Hsuan Chen[1,5], Zhaohui Wu[1,5],
and Lee Giles[1,2,5]

[1] Computer Science and Engineering
[2] Information Sciences and Technology
[3] Computer Science
[4] University of North Texas, Denton, TX, USA
[5] Pennsylvania State University, University Park, PA, USA
[6] University of Bucharest, Bucharest, Romania
[7] University of the Andes, Bogota, Colombia
ccaragea@unt.edu, {jxw394,giles}@ist.psu.edu,
alina.ciobanu@my.fmi.unibuc.ro, {kwilliams,hhchen,zzw109}@psu.edu,
jp.fernandez29@uniandes.edu.co

Abstract. The CiteSeer^x digital library stores and indexes research articles in Computer Science and related fields. Although its main purpose is to make it easier for researchers to search for scientific information, CiteSeer^x has been proven as a powerful resource in many data mining, machine learning and information retrieval applications that use rich metadata, e.g., titles, abstracts, authors, venues, references lists, etc. The metadata extraction in CiteSeer^x is done using automated techniques. Although fairly accurate, these techniques still result in noisy metadata. Since the performance of models trained on these data highly depends on the quality of the data, we propose an approach to CiteSeer^x metadata cleaning that incorporates information from an external data source. The result is a subset of CiteSeer^x, which is substantially cleaner than the entire set. Our goal is to make the new dataset available to the research community to facilitate future work in Information Retrieval.

Keywords: CiteSeer^x, Scholarly Big Data, Record Linkage.

1 Introduction

As science advances, scientists around the world continue to produce large numbers of research articles, which provide the technological basis for worldwide dissemination of scientific discoveries. Online digital libraries such as DBLP, CiteSeer^x, Microsoft Academic Search, ArnetMiner, arXiv, ACM Digital Library, Google Scholar, and PubMed that store research articles or their metadata, have become a medium for answering questions such as: how research ideas emerge, evolve, or disappear as a topic; what is a good measure of quality of published works; what are the most promising areas of research; how authors connect and influence each other; who are the experts in a field; and what works are similar.

M. de Rijke et al. (Eds.): ECIR 2014, LNCS 8416, pp. 311–322, 2014.

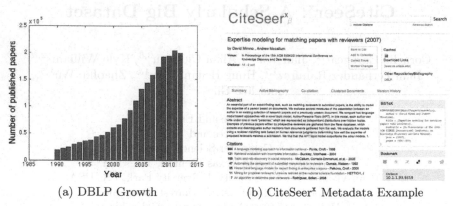

(a) DBLP Growth   (b) CiteSeer^x Metadata Example

Fig. 1. (a) The growth in the number of research papers published between 1990 and 2011, extracted from DBLP; (b) An example of CiteSeer^x metadata.

Unfortunately, our ability to manually process and filter the huge amount of information available in digital libraries lags far behind the number of research articles available today. Figure 1(a) shows the growth in the number of research articles published between 1990 and 2011, extracted from the DBLP dataset.

Recent developments in data mining, machine learning, and information retrieval made it possible to transform the way we analyze research articles on a web-wide scale. CiteSeer^x [1] has been proven as a powerful resource in many applications such as text classification [2,3,4], collective classification [5], document and citation recommendation [6,7,8,9,10], author name disambiguation [11], expert search [12], collaborator recommendation [13], paper and slides alignment [14], author influence [15], joint modeling of documents' content and interests of authors [16], and entity resolution [17].

To extract metadata from each article, the CiteSeer^x project uses automated techniques [18,19]. An example of metadata for an entry in CiteSeer^x (i.e., a research article) is shown in Figure 1(b). The metadata contains information such as the title of an article, the authors, venue where the article was published, the year of publication, article's abstract, and the references list. Among the automated techniques for metadata extraction used in CiteSeer^x, Han et al. [18] employed a Support Vector Machine based classification method to extract information from the header of research papers. Councill et al. [19] used Conditional Random Fields to segment reference strings from plain text, and used heuristic rules to identify reference strings and citation contexts. Although fairly accurate, these automated techniques still make mistakes during the metadata extraction process. Thus, the metadata inherently contains noise, e.g., errors in the extraction of the year of publication.

In contrast, DBLP provides *manually* curated metadata. However, DBLP metadata is not as rich as that provided by CiteSeer^x. For example, DBLP does not provide an article's abstract or references strings, which are crucial in applications such as citation recommendation and topic evolution, that use either the citation graph and/or textual information.

In this paper, we present a record linkage based approach to building a *scholarly big dataset* that uses information from DBLP to automatically remove noise in CiteSeer^x. Inspired by the work on record linkage by Bhattacharya and Getoor [20], we study the usage of both the similarity between paper titles and the similarity between the authors' lists of "similar" papers and their number of pages.

The contributions of the research described here are two-fold. First, we present and study an approach to building a scholarly dataset derived from CiteSeer^x that is substantially cleaner than the entire set. Second, the new dataset will be made available to the research community and can benefit many research projects that make use of rich metadata such as a citation graph and textual information available from the papers' abstracts. The dataset will be maintained and updated regularly to compile the dynamic changes in CiteSeer^x.

The rest of the paper is organized as follows. We present related work in Section 2. In Section 3, we describe our approach to CiteSeer^x data cleaning. The evaluation and the characteristics of the new dataset are presented in Section 4. We conclude the paper with a summary and discussion of limitations of our approach, and directions for future work in Section 5.

2   Related Work

Record linkage refers to the problem of identifying duplicate records that describe the same entity across multiple data sources [21] and has applications to data cleaning, i.e., discovering and removing errors from data to improve data quality [22], and information integration, i.e., integrating multiple heterogeneous data sources to consolidate information [23].

Many approaches to record linkage and its variants, e.g., duplicate detection or deduplication, co-reference or entity resolution, identity uncertainty, and object identification, have been studied in the literature. Several research directions considered are: the classification of record pairs as "match" or "not match" [21,24,25,26]; design of approximate string matching algorithms [27], adaptive algorithms that learn string similarity measures [24,25], iterative algorithms that use attribute as well as linked objects similarities [17,20]; object identification and co-reference resolution [23,28,29]; near-duplicates detection by identifying f-bit fingerprints that differ from a given fingerprint in at most k-bit positions [30]; and algorithms for reference matching in the bibliometrics domain [28].

Our work builds upon previous works on record linkage and uses information from DBLP to improve the quality of CiteSeer^x metadata records. An approach to building a citation network dataset using DBLP was proposed by Arnet-Miner[1]. However, our approach results in a scholarly dataset that contains richer metadata compared with ArnetMiner's DBLP citation dataset, e.g., it contains the *citation contexts* for a paper's references list.

To our knowledge, this is the first attempt to generate a *scholarly big dataset* that consists of cleaner CiteSeer^x metadata, that will be made available to the

---

[1] http://arnetminer.org/DBLP_Citation

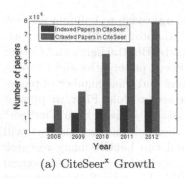

Total number of indexed documents:	2,356,568
Unique documents (after clustering):	1,957,448
Authors in collection:	7,511,558
Unique authors (identified by name):	2,447,515
Disambiguated authors:	304,833
Total number of citations:	52,185,298
Unique records (citations+documents):	15,375,541

(a) CiteSeer^x Growth   (b) CiteSeer^x Statistics

Fig. 2. (a) The growth in the number of crawled documents as well as in the number of documents (or research papers) indexed by CiteSeer^x between 2008 and 2012; (b) Statistics of the entire CiteSeer^x dataset

research community, and will aim at facilitating future work in Information Retrieval. Through the merger of CiteSeer^x and DBLP, errors in CiteSeer^x metadata fields will be corrected. Only clean CiteSeer^x records with a match in DBLP will be retained in the dataset, whereas the rest will be filtered out.

3   Approach to Data Cleaning

We first present the characteristics of the entire collection of articles indexed in CiteSeer^x, and then describe our approach to data cleaning and its evaluation.

3.1   Characteristics of the Entire CiteSeer^x Dataset

The CiteSeer^x dataset is rapidly growing in size. Figure 2(a) shows the increase in both the number of crawled documents as well as the number of documents (or research papers) indexed by CiteSeer^x during the last five years. As can be seen from the figure, the number of crawled documents has increased from less than two million to almost eight million, whereas the number of indexed documents has increased from less than one million to more than two million. Note that because CiteSeer^x only crawls open-access documents (e.g., those available from authors' webpages), many of the crawled documents are manuscripts.

As of May 2013, the total number of documents indexed in CiteSeer^x is $\approx$2.35M. After clustering to remove multiple versions of a paper, there are $\approx$1.9M unique papers. The number of authors in collection, i.e., authors (with repetition) from all documents is $\approx$7.5M, with the number of unique authors being $\approx$2.4M (i.e., authors without repetition), and the number of disambiguated authors being $\approx$300K (e.g., "Andrew McCallum", "Andrew K. McCallum" and "A. McCallum" are disambiguated to the same author, whereas "Wei Hong" is disambiguated to multiple authors by their affiliations). The number of citations in collection, i.e., all papers with repetition that occur in the references list of all documents is $\approx$52M, and the number of unique records, i.e., unique papers and citations, is $\approx$15M. The exact numbers are summarized in Figure 2(b).

Algorithm 1. CiteSeer^x Data Cleaning

Input:
$\mathcal{C} \leftarrow$ CiteSeer^x metadata entries; $\mathcal{D} \leftarrow$ DBLP metadata entries;
$\theta \leftarrow$ a predefined threshold; **boolean** checkAuthors, checkPageCount.
Output: A subset of CiteSeer^x cleaned using DBLP, $\mathcal{C}^{\mathcal{D}}$.
Index the fields in $\mathcal{D}$ into an inverted index;
$\mathcal{C}^{\mathcal{D}} = \phi$;
for all entries $e \in \mathcal{C}$ **do**
   $t_e \leftarrow$ getTitle(e); // Query t_e = title of entry e
   $\mathcal{D}_e \leftarrow$ retrieve(t_e); // $\mathcal{D}_e$ contains the hits from DBLP for the query title t_e
   for all $d \in \mathcal{D}_e$ **do**
      $t_d \leftarrow$ getTitle(d); // t_d = title of d
      $s \leftarrow sim(t_d, t_e)$;
      if $s \geq \theta$ **then**
         if (checkAuthors = false) && (checkPageCount = false) **then**
            $m \leftarrow$ match(d, e); // write the fields from d to e
         end if
         $a_e \leftarrow$ getAuthors(e); $a_d \leftarrow$ getAuthors(d);
         if (checkAuthors = true) && (checkPageCount = false) && ($a_e \subseteq a_d$) **then**
            $m \leftarrow$ match(d, e); // write the fields from d to e
         end if
         $p_e \leftarrow$ PageCount(e); $p_d \leftarrow$ PageCount(d);
         if (checkAuthors = false) && (checkPageCount = true) && ($p_e \approx p_d$) **then**
            $m \leftarrow$ match(d, e); // write the fields from d to e
         end if
         $\mathcal{C}^{\mathcal{D}} = \mathcal{C}^{\mathcal{D}} \cup m$;
      end if
   end for
end for
return $\mathcal{C}^{\mathcal{D}}$ // The subset of CiteSeer^x cleaned using DBLP

3.2 Building a Cleaner CiteSeer^x Dataset

Our approach to building a cleaner CiteSeer^x dataset is to merge metadata information from CiteSeer^x and DBLP. The procedure for merging these two sources of information is shown in Algorithm 1. The input of the algorithm is given by two sets $\mathcal{C}$ and $\mathcal{D}$ of CiteSeer^x and DBLP metadata entries, respectively, and a threshold θ. The output is $\mathcal{C}^{\mathcal{D}}$, a merged dataset of CiteSeer^x and DBLP entries, obtained by performing record linkage.

The algorithm starts by indexing the entries in $\mathcal{D}$ into an inverted index. Specifically, titles and authors from DBLP are indexed using Solr 4.3.0, an indexing platform that is built using Apache Lucene[2]. Next, the algorithm iterates through all the entries e in $\mathcal{C}$ and treats each CiteSeer^x title as a query that is used to search against the DBLP indexed entries. For each query title t_e, a candidate set $\mathcal{D}_e$ of DBLP entries is retrieved in the following way: we extract the n-grams from the query title and retrieve all entries from DBLP that have at least one n-gram in common with the query (an n-gram is defined as a sequence of n contiguous tokens in a title). More precisely, for the bi-gram "expertise modeling" in the query "Expertise modeling for matching papers with reviewers.", we retrieve all DBLP entries that contain the bi-gram in their titles. In experiments, we used various

---

[2] http://lucene.apache.org

values of n for the length of an n-gram ($n = 1, 2, 3, 4$, and $|t_e|$, where $|t_e|$ represents the length, in the number of tokens, of t_e).

In our record linkage algorithm, we studied the effect of using only title similarity on the matching performance as well as using the similarity between other attributes of entries such as papers' author lists *or* their page count, in addition to title similarity. After the candidate set $\mathcal{D}_e$ is retrieved, the algorithm computes the similarity between the query title t_e and the titles t_d of entries in $\mathcal{D}_e$. If the similarity is greater than θ and if only title information is required, the algorithm performs the match between t_e and t_d and adds the match to $\mathcal{C}^{\mathcal{D}}$. Otherwise, if additional information is required besides title similarity, the algorithm checks one of the following conditions: (1) if the list of authors a_e of e is included in the list of authors a_d of d, or (2) if the page count p_e of c is approximately the same as the page count p_d of d. If the condition is satisfied, a match between e and d is performed. More precisely, the fields available in DBLP overwrite those in CiteSeer^x (the fields in CiteSeer^x with no equivalent in DBLP are kept the same). If more matches are found for a CiteSeer^x entry, which satisfy the condition, the DBLP entry with the highest similarity score with the query title is returned as a match. If no DBLP match is found, the CiteSeer^x record is not added to $\mathcal{C}^{\mathcal{D}}$. The algorithm terminates with the set $\mathcal{C}^{\mathcal{D}}$ of CiteSeer^x records that are merged with their equivalent entries from DBLP.

In our implementation, for condition (1), we considered only the authors' last names to avoid ambiguities caused by various uses of authors' first names in the two data sources (e.g., "A. McCallum" vs. "Andrew McCallum" in CiteSeer^x and DBLP, respectively). Author disambiguation applied to both CiteSeer^x and DBLP will be considered in future, for further versions of $\mathcal{C}^{\mathcal{D}}$.

For condition (2), we extracted the page count from DBLP from the field: $< \text{pages} > \text{page\_start} - \text{page\_end} < /\text{pages} >$. For CiteSeer^x entries, we used PDFBox[3] to determine the page count directly from the pdf file of each document (note that the pdf is available for each document in CiteSeer^x).

4   Evaluation of the Proposed Approach

To evaluate our approach, we randomly sampled 1000 documents from CiteSeer^x and manually identified their matching records in DBLP. We describe the manual labeling process first and then present the experimental design and results.

4.1   Manual Labeling of the Random CiteSeer^x Sample

For each of the records in the random sample, we searched DBLP for a "true match". Specifically, we used the *actual* paper title found from the pdf of the paper to query the Solr index and retrieve a candidate set of DBLP records (we again used Solr 4.3.0 to index the DBLP fields). In determining the true DBLP match for a CiteSeer^x document, we also used information about authors, year, and venue, obtained from the profile page of the document (through

---

[3] http://pdfbox.apache.org

the CiteSeer^x web service), as well as the number of pages, obtained either by checking the pdf of the document or by using PDFBox. If a matching decision was difficult to make (e.g., due to a one-to-many relation, one CiteSeer^x record and multiple DBLP search results), we downloaded the corresponding pdfs and made a side-by-side comparison. The general criteria for decision making are:

1. Matched papers must have the same title and author lists;
2. The order of importance for other metadata is venue > year > page count;
3. If only title and authors are available, the original papers are downloaded and compared side-by-side.

After completing the manual labeling of all documents in the sample, we found 236 documents had true matches in DBLP. Since this sample is randomly selected in the DOI space, it reflects the average properties of the entire CiteSeer^x sample.

4.2   Experimental Design

Our experiments are designed around the following questions. *How does the matching performance vary when we vary the threshold θ on the similarity between a query title and the DBLP entries relevant to the query and what is the effect of the similarity measure on the matching performance?* High thresholds impose high overlaps of words between two titles, whereas low thresholds allow for low overlaps, which can handle CiteSeer^x titles that are wrongly extracted (e.g., when an author name is included in the title, or when only the first line is extracted for a title that spans multiple lines). In experiments, we varied the threshold θ from 0.5 to 0.9, in steps of 0.1. We also experimented with two similarity measures: Jaccard and cosine, and found Jaccard to perform better than cosine (see the Results section). In subsequent experiments, we used Jaccard.

The next question is: *Is the proposed approach to data cleaning computationally efficient?* The most expensive part of Algorithm 1 is the Jaccard similarity calculation (Jaccard similarity is given by the number of common words between two titles divided by the total number of unique words). The larger the size of the retrieved document set is (i.e., the size of $\mathcal{D}_e$), the more Jaccard similarity computations are needed. We experimented with several techniques for query construction to control the size of $\mathcal{D}_e$. We compared the matching performance for all tested query construction techniques to determine what is the *most time efficient* and *highly accurate* among these techniques. Specifically, the query construction techniques to retrieve the candidate set $\mathcal{D}_e$ of DBLP entries for a query t_e are n-gram queries, with $1 \leq n \leq |t_e|$, defined as follows:

- n-gram query: $\mathcal{D}_e = \{d \in \text{DBLP} \mid d \text{ has } at\ least \text{ one } n\text{-gram in common with } t_e\}$, $1 \leq n \leq |t_e|$, where $|t_e|$ is the length of t_e, in the number of tokens.

The use of $|t_e|$-gram (called AND) query results in a small size of the set $\mathcal{D}_e$, whereas the use of the unigram (called OR) query results in a large size of the set. The n-gram query ($2 \leq n \leq |t_e| - 1$) presents a tradeoff between AND and OR. For each query type above, we also experimented with the setting where we

Table 1. The matching performance, using only titles, for various values of θ, for Jaccard (j) and cosine (c) similarities, and AND, 3-gram, and OR query types

θ	AND				3-gram				OR			
	Prec.	Recall	F1-score	time	Prec.	Recall	F1-score	time	Prec.	Recall	F1-score	time
0.5j	0.694	0.691	0.692	55	0.627	0.826	0.713	106	0.519	0.826	0.637	3722
0.6j	0.744	0.691	0.716	42	0.713	0.809	0.758	65	0.692	0.809	0.746	4100
0.7j	0.756	0.682	0.717	44	0.746	0.797	0.770	67	0.740	0.797	0.767	4088
0.8j	0.768	0.674	0.718	43	0.765	0.746	0.755	64	0.762	0.746	0.754	4064
0.9j	0.772	0.661	0.712	43	0.769	0.678	0.721	65	0.769	0.678	0.721	3616
0.5c	0.652	0.699	0.675	42	0.484	0.847	0.616	63	0.274	0.843	0.414	4158
0.6c	0.679	0.699	0.689	43	0.561	0.839	0.672	66	0.410	0.835	0.55	4102
0.7c	0.688	0.691	0.689	44	0.648	0.818	0.723	67	0.580	0.818	0.678	4201
0.8c	0.733	0.686	0.709	43	0.726	0.809	0.766	66	0.715	0.809	0.759	4116
0.9c	0.763	0.669	0.713	43	0.763	0.725	0.743	67	0.763	0.725	0.743	4178

removed the stop words from titles, which significantly reduces the size of $\mathcal{D}_e$. In experiments, we found that the matching performance with and without stop words are similar or the same, whereas the time spent to compute the Jaccard similarities is significantly reduced, by a factor of 3, when stop words are removed (the size of $\mathcal{D}_e$ is much smaller). Hence, we report the results without stop words.

Our last question is: *What is the effect of using the similarity between other attributes such as author lists or page count, in addition to title similarity?* We experimented with the following settings for match identification: title only, title + authors, and title + page count.

To evaluate the performance of Algorithm 1 for matching CiteSeer^x and DBLP entries, we report precision, recall and F1-score on the manually annotated sample of 1000 randomly selected CiteSeer^x entries. Precision gives the fraction of matches correctly identified by the algorithm among all matches identified by the algorithm, whereas recall gives the fraction of matches correctly identified by the algorithm among all actual matches. F1-score gives the harmonic mean of precision and recall. With the best setting that we obtained on the manually annotated sample, we then ran experiments on the entire CiteSeer^x. We report the characteristics of the newly constructed dataset, i.e., the resulting set from the merger of CiteSeer^x and DBLP.

4.3 Results

The Effect of Varying the Threshold θ, the Similarity Measure, as Well as the Query Type on the Matching Performance. Table 1 shows the comparison of CiteSeer^x-DBLP matching performance for various values of the threshold θ on the similarity between a query and a DBLP title, ranging from 0.5 to 0.9 in steps of 0.1, using two similarity measures, Jaccard (j) and cosine (c), on the manually labeled sample of 1000 CiteSeer^x entries. The results are shown for three query types, AND, 3-gram, and OR, with stop words removed, using only title similarity (i.e., no author or page count information is considered).

The time (in seconds) taken by each experiment to finish is also given in the table. As can be seen from the table, as we increase θ, and thus, impose a higher word overlap between a query and a DBLP title, the recall decreases, whereas the precision increases, in most cases, for all query types and for both Jaccard and cosine similarities. Hence, lower θ can handle better wrongly extracted titles (as shown by the higher recall), but too low θ allows matching entries that have some word overlap between their titles, which in fact are not true matches (as shown by the lower precision). Moreover, it can be seen from the table that, in most cases, Jaccard similarity yields better results compared with cosine similarity.

From the table, we can also see that the matching performance for the 3-gram query is generally better compared with both AND and OR queries. Although the time (in seconds) spent by 3-gram queries is slightly worse than that of AND queries, it is significantly shorter than that of OR queries. The average number of DBLP hits for a CiteSeer^{x} query, i.e., the size of $\mathcal{D}_e$, is $81,625$ for OR, and it drops substantially to 212 for 3-gram, and to 131 for AND. Hence, much less computations are needed for the 3-gram and AND queries compared with OR. For the 3-gram query, we performed experiments when the stop words were not removed and found that the matching performance remains the same with that of removing stop words. However, the time (in seconds) spent to finish an experiment increased by a factor of 5. The value $n = 3$ for the n-gram query was empirically selected based on comparisons with 2-gram and 4-gram queries. These results, using Jaccard similarity, are shown in Table 2. As can be seen from the table, for $\theta = 0.7$, the 3-gram query results in the highest F1-score of 0.77, with a slight increase in time compared with 4-gram query.

Although the size of the candidate set $\mathcal{D}_e$ is significantly reduced through an AND query, its use seems to be quite limiting since it requires that every word in the title extracted by CiteSeer^{x} must be matched for a document to be retrieved and added to $\mathcal{D}_e$. While the AND query will not affect retrieval for incomplete titles, however, if the extractor mistakenly appended an author name to a title, no DBLP hits are found, and hence, the DBLP cannot be used to fix the error in metadata. The AND query increases the precision, but results in a decrease in recall. The OR query overcomes the limitations of the AND query, however, the size of retrieved DBLP documents for which Jaccard similarity needs to be computed increases significantly. The 3-gram queries provide a good tradeoff between the size of the retrieved documents and the matching performance.

The Effect of Using Author and Page Count on the Matching Performance. Table 3 shows, using a 3-gram query and Jaccard similarity, the comparison of the matching performance when only title information is used (Title Only) with the matching performance when additional information is used, i.e., papers' author lists (Title+Authors) or page count (Title+Pages). As can be seen from the table, the recall is fairly high using Title Only compared with Title+Authors and Title+Pages, but the precision drops. Title+Pages generally achieves the highest precision, e.g., 0.904 for $\theta = 0.9$. Thus, as more information is used, the precision increases, however, at the expense of decreasing recall. A potential explanation for low recall could be noisy extraction of authors' names

Table 2. The matching performance, using only titles, for 2 − 4-gram queries

θ	2-gram				3-gram				4-gram			
	Prec.	Recall	F1-score	time	Prec.	Recall	F1-score	time	Prec.	Recall	F1-score	time
0.5j	0.560	0.826	0.668	156	0.627	0.826	0.713	106	0.662	0.805	0.726	53
0.6j	0.695	0.809	0.748	149	0.713	0.809	0.758	65	0.722	0.792	0.755	50
0.7j	0.743	0.797	0.769	152	0.746	0.797	0.770	67	0.747	0.779	0.763	51
0.8j	0.762	0.746	0.754	140	0.765	0.746	0.755	64	0.766	0.737	0.751	51
0.9j	0.769	0.678	0.721	143	0.769	0.678	0.721	65	0.769	0.677	0.720	52

Table 3. The matching performance, using title only, title+authors, and title+page count information

θ	Title Only				Title+Authors				Title+Pages			
	Prec.	Recall	F1-score	time	Prec.	Recall	F1-score	time	Prec.	Recall	F1-score	time
0.5j	0.627	0.826	0.713	106	0.819	0.631	0.713	66	0.802	0.551	0.653	64
0.6j	0.713	0.809	0.758	65	0.835	0.623	0.714	65	0.869	0.534	0.661	64
0.7j	0.746	0.797	0.770	67	0.847	0.610	0.709	67	0.875	0.534	0.663	66
0.8j	0.765	0.746	0.755	64	0.873	0.581	0.697	66	0.888	0.504	0.643	66
0.9j	0.769	0.678	0.721	65	0.868	0.530	0.658	66	0.904	0.479	0.626	66

in CiteSeer^x or the mismatch between the page count. The page count is an additional evidence for record matching, which is independent of the metadata extraction quality. If two records have the same page count in addition to *similar* titles, they are likely to refer to the same document. However, during the manual inspection, we found that many matched papers did not have the same page count (e.g., often an extra page is added as a cover page from the institution or research lab). The CiteSeer^x version was generally a manuscript. In experiments, we allowed ±1 from the page count. Further investigation of this will be considered in future.

If the page count or authors' list for a paper cannot be extracted in one of the data sources, the CiteSeer^x entry is skipped. If there is a one-to-many relationship between CiteSeer^x and DBLP, the algorithm matches the CiteSeer^x entry with the one from DBLP with the highest Jaccard similarity score.

5   Summary, Discussion, and Future Directions

We presented an approach to CiteSeer^x metadata cleaning that uses information from DBLP to clean metadata in CiteSeer^x. In addition to using title similarity to perform record linkage between CiteSeer^x and DBLP, we studied the use of additional information such as papers' author lists or page count. Results of experiments on a random sample of 1000 CiteSeer^x entries show that the proposed approach is a promising solution to CiteSeer^x metadata cleaning.

One of the major limitations of our approach is that it assumes the titles in CiteSeer^x are extracted correctly. However, there are many titles that are wrongly extracted. For example, we found: (i) titles that contain tokens that do not occur at all in the actual title[4]; (ii) titles that contain only one stop word, "and" or "the"[5]; (iii) titles that are incomplete[6]; and (iv) titles that contain other tokens besides the title tokens, such as author, venue, or year[7]. The algorithm will fail to find a matching record in DBLP for (i) and (ii) since the retrieved candidate DBLP set is not relevant to the actual title or the Jaccard similarity does not exceed the predefined threshold θ. For (iii) and (iv), it is possible that the algorithm will find a match in DBLP if θ is too low. Author or page count could help improve precision, however at the expense of decreasing recall (potentially due to noise in authors' name extraction or difference in page count). In future, we plan to use information for other external data sources such as IEEE and Microsoft Academic Search to improve data quality in CiteSeer^x. Additional information, e.g., venue names will be investigated in future as well.

5.1 Dataset Characteristics, Sharing and Maintenance

We generated a *scholarly big dataset* of cleaner CiteSeer^x metadata records. The dataset is made available to the research community, along with our Java implementation at http://www.cse.unt.edu/~ccaragea/citeseerx. We ran our implementation on the entire CiteSeer^x with the setting that had the highest performance on the random sample of 1000 documents (i.e., title only, 3-gram query, $\theta = 0.7$, Jaccard). The total number of matches found between CiteSeer^x and DBLP is $630, 351$ and the total time taken to finish is $184, 749$ seconds. The entries in the newly constructed dataset are xml files that contain articles' metadata. Regular updates will be done to integrate new articles crawled and indexed by CiteSeer^x.

In addition to the newly constructed dataset, we will provide the xml files for the entire CiteSeer^x data repository and let the researchers decide what setting they prefer to use for the dataset generation, using our Java implementation.

References

1. Giles, C.L., Bollacker, K., Lawrence, S.: Citeseer: An automatic citation indexing system. In: Digital Libraries 1998, pp. 89–98 (1998)
2. Lu, Q., Getoor, L.: Link-based classification. In: ICML (2003)
3. Peng, F., Schuurmans, D.: Combining naive bayes and n-gram language models for text classification. In: Sebastiani, F. (ed.) ECIR 2003. LNCS, vol. 2633, pp. 335–350. Springer, Heidelberg (2003)
4. Caragea, C., Silvescu, A., Kataria, S., Caragea, D., Mitra, P.: Classifying scientific publications using abstract features. In: SARA (2011)
5. Sen, P., Namata, G.M., Bilgic, M., Getoor, L., Gallagher, B., Eliassi-Rad, T.: Collective classification in network data. AI Magazine 29(3), 93–106 (2008)

---

[4] http://citeseerx.ist.psu.edu/viewdoc/summary?doi=10.1.1.239.1803
[5] http://citeseerx.ist.psu.edu/viewdoc/summary?doi=10.1.1.63.1066
[6] http://citeseerx.ist.psu.edu/viewdoc/summary?doi=10.1.1.169.7994
[7] http://citeseerx.ist.psu.edu/viewdoc/summary?doi=10.1.1.267.8099

6. Zhou, D., Zhu, S., Yu, K., Song, X., Tseng, B.L., Zha, H., Giles, C.L.: Learning multiple graphs for document recommendations. In: Proc. of WWW 2008 (2008)
7. Caragea, C., Silvescu, A., Mitra, P., Giles, C.L.: Can't see the forest for the trees? a citation recommendation system. In: Proceedings of JCDL 2013 (2013)
8. Huang, W., Kataria, S., Caragea, C., Mitra, P., Giles, C.L., Rokach, L.: Recommending citations: translating papers into references. In: CIKM (2012)
9. Küçüktunç, O., Saule, E., Kaya, K., Çatalyürek, Ü.V.: Diversified recommendation on graphs: pitfalls, measures, and algorithms. In: WWW (2013)
10. Nallapati, R.M., Ahmed, A., Xing, E.P., Cohen, W.W.: Joint latent topic models for text and citations. In: Proceedings of KDD 2008 (2008)
11. Treeratpituk, P., Giles, C.L.: Disambiguating authors in academic publications using random forests. In: Proc. of JCDL, JCDL 2009 (2009)
12. Gollapalli, S.D., Mitra, P., Giles, C.L.: Similar researcher search in academic environments. In: JCDL (2012)
13. Chen, H.H., Gou, L., Zhang, X., Giles, C.L.: Collabseer: a search engine for collaboration discovery. In: Proceedings of JCDL 2011 (2011)
14. Kan, M.Y.: Slideseer: a digital library of aligned document and presentation pairs. In: Proceedings of JCDL 2007 (2007)
15. Kataria, S., Mitra, P., Caragea, C., Giles, C.L.: Context sensitive topic models for author influence in document networks. In: Proceedings of IJCAI 2011 (2011)
16. Rosen-Zvi, M., Griffiths, T., Steyvers, M., Smyth, P.: The author-topic model for authors and documents. In: Proceedings of UAI 2004 (2004)
17. Bhattacharya, I., Getoor, L.: A latent dirichlet model for unsupervised entity resolution. In: SDM (2006)
18. Han, H., Giles, C.L., Manavoglu, E., Zha, H., Zhang, Z., Fox, E.A.: Automatic document metadata extraction using support vector machines. In: JCDL (2003)
19. Councill, I.G., Giles, C.L., Yen Kan, M.: Parscit: An open-source crf reference string parsing package. In: Intl. Language Resources and Evaluation (2008)
20. Bhattacharya, I., Getoor, L.: Iterative record linkage for cleaning and integration. In: DMKD (2004)
21. Fellegi, I.P., Sunter, A.B.: A theory for record linkage. Journal of the American Statistical Association 64, 1183–1210 (1969)
22. Rahm, E., Do, H.H.: Data cleaning: Problems and current approaches. IEEE Data Engineering Bulletin 23 (2000)
23. Tejada, S., Knoblock, C.A., Minton, S.: Learning object identification rules for information integration. Journal Information Systems (2001)
24. Bilenko, M., Mooney, R.J.: Adaptive duplicate detection using learnable string similarity measures. In: Proceedings of KDD 2003 (2003)
25. Cohen, W.W., Richman, J.: Learning to match and cluster large high-dimensional data sets for data integration. In: Proceedings of KDD 2002 (2002)
26. Winkler, W.E.: Methods for record linkage and bayesian networks. Technical report, Statistical Research Div., U.S. Bureau of the Census (2002)
27. Cohen, W.W., Ravikumar, P., Fienberg, S.E.: A comparison of string distance metrics for name-matching tasks. In: IJCAI, pp. 73–78 (2003)
28. Pasula, H., Marthi, B., Milch, B., Russell, S., Shpitser, I.: Identity uncertainty and citation matching. In: NIPS. MIT Press (2003)
29. McCallum, A., Wellner, B.: Toward conditional models of identity uncertainty with application to proper noun coreference. In: IIWeb (2003)
30. Manku, G.S., Jain, A., Das Sarma, A.: Detecting near-duplicates for web crawling. In: Proc. of WWW 2007 (2007)

"User Reviews in the Search Index? That'll Never Work!"

Marijn Koolen

Institute for Logic, Language and Computation,
University of Amsterdam, The Netherlands

Abstract. Online book search services allow users to tag and review books but do not include such data in the search index, which only contains titles, author names and professional subject descriptors. Such professional metadata is a limited description of the book, whereas tags and reviews can describe the content in more detail and cover many other aspects such as quality, writing style and engagement. In this paper we investigate the impact of including such user-generated content in the search index of a large collection of book records from Amazon and LibraryThing. We find that professional metadata is often too limited to provide good recall and precision and that both user reviews and tags can substantially improve performance. We perform a detailed analysis of different types of metadata and their impact on a number of topic categories and find that user-generated content is effective for a range of information needs. These findings are of direct relevance to large online book sellers and social cataloguing sites.

Keywords: Book Search, Metadata, Social Media, User-Generated Content.

1 Introduction

Book search behaviour on the web has become more complex with the growing amount of book information generated through social media. Readers can use this user-generated (UGC) content to help them select interesting, engaging, well-written and fun books to read. Yet most book search services, such as on GoodReads, LibraryThing (LT), Amazon and online bookshops, as well as libraries, only allow search via book titles, author names and professional metadata in the form of a small set of descriptive terms from a controlled vocabulary to describe the content of the book. But users often want to read other readers' opinions and summaries before deciding which book they want to read. For the more subjective and non-topical aspects, such retrieval systems are insufficient and force users to browse through many reviews to find the right information or turn to online networks of book readers to ask suggestions. In this paper, we investigate the impact of including UGC to the retrieval index and compare it directly with the professional metadata. Our main research question is:

- What is the impact of professional metadata and user-generated content on book search?

M. de Rijke et al. (Eds.): ECIR 2014, LNCS 8416, pp. 323–334, 2014.
© Springer International Publishing Switzerland 2014

324   M. Koolen

Professional metadata is often based on controlled vocabularies, with trained indexers assigning a small number of descriptive terms to capture the main topics of a book, with all books described in roughly the same amount of detail. With vocabulary control, descriptive terms give access to all and only books relevant to those terms. With UGC this is very different. Popular books are tagged and reviewed more often than obscure books, there is no vocabulary control and reviews can be long or short, useful or useless, honest or misleading or anywhere in between. It is unclear how UGC affects retrieval if added to the index. This leads to the following more specific research questions:

- How do professional metadata and user-generated content compare in terms of vocabulary size and frequency distribution?
- How do they compare in terms of retrieval effectiveness?

We conduct our analysis in the context of the INEX Social Book Search (SBS) Track [11][1]. The track uses a collection of book descriptions from Amazon, with subject headings and user reviews, enriched with user tags from LT. The topics and relevance judgements come from the LT discussion forums. Together, these provide a natural scenario of book search on the web. Most library searches involve keyword searching [7] so we argue our methodology and topic set are appropriate for this evaluation.

This paper is organised as follows: In Section 2 we discuss related work. Next, in Section 3 we detail the SBS Track and quantitatively compare the different types of book information. We explain our experimental setup in Section 4, discuss the evaluation results in Section 5 and provide further analysis in Section 6. Finally, we draw conclusions in Section 7.

2   Related Work

The Cranfield tests for IR evaluation [4] showed that indexing natural language terms from documents was at least as effective as formal indexing schemes with controlled languages. However, controlled vocabularies still hold the potential to improve completeness and accuracy of search results by providing consistent and rigorous index terms and ways to deal with synonymy and homonymy [12, 17]. Lu et al. [13] compared LT tags and Library of Congress Subject Headings (LCSH). They find that social tags can improve accessibility to library collections. A similar finding was reported by Sterken [16]. Peter J. Rolla [15] found that tags increase subject access but introduce noise because of personal and idiosyncratic tags. Yi and Chan [21] explored the possibility of mapping user tags to LCSH. With word matching they can link two-thirds of all tags to LC subject headings. In subsequent work [20], they use semantic similarity between tags and headings to automatically apply headings to tagged resources. This urges us to compare subject headings and user tags in an actual retrieval setting.

Searchers often find it difficult to use controlled vocabularies effectively [7]. On top of that, searchers and indexers might use different terms because they

---

[1] https://inex.mmci.uni-saarland.de/tracks/books/

have different perspectives. Buckland [2] describes the differences between vocabularies of authors, cataloguers, searchers, queries as well as the vocabulary of
syndetic structure. With all these vocabularies used in a single process, there are
many possibilities for mismatches. This is especially pertinent to book search if
UGC is indexed, which adds the vocabularies from multiple other users. One of
the interesting aspects of UGC in this respect is that it has a smaller gap with the
vocabulary of searchers [14]. Golovchinsky et al. [6] constructed queries based
on annotations of articles and showed they lead to better perform than feedback
based on the user's relevance judgements.

Golder and Huberman [5] finds that tags have different organising functions.
Describing what (or who) it is about, what it is and who owns it, refining categories, qualities or characteristics, self referencing and task organising. [1] conducted a user study to find out why LT members tag. For 42%, one of the top
3 reasons to help others find a book, for 73% it is to help their own collection
management. Retrieval seems a strong motivation for assigning tags.

Some recent work already looked at social metadata for book search. Kazai
and Milic-Frayling [9] incorporated social approval votes for book IR using external resources that refer to books in the corpus—such as lists from libraries
and publishers and lists of bestsellers and award winning books. They find that
social approval votes can improve a BM25F baseline that indexes both full-text
and MARC[2] records. Koolen et al. [10] compared the effectiveness of professional
metadata and UGC for book search, but used very minimal professional metadata,
which does not reflect the amount in actual library catalogues. They focused on
building test collections for book search based on topics and book suggestions
from discussion forums. In this paper, we analyse the difference between the two
types of metadata in more detail, using professional metadata from two national
libraries and identifying specific strengths and weaknesses of each metadata type.

3 Social Book Search

There is a large social component to the way readers discover and select books to
read. Chandler [3] analysed which channels are used by readers on GoodReads
to select their next books to read and found that suggestions from offline and
online friends are important sources for discovering books. Koolen et al. [10]
showed that social book search and suggestion represents a different task from
traditional search and recommendation tasks used in IR evaluation. On the LT
discussion forums, members discuss many aspects of books and regularly start
topics to ask for book recommendations. Other members join the topic thread
with their suggestions [11]. These topic threads are a form of social book search.
Instead of browsing or searching a book catalogue, readers rely on the book
knowledge of their peers to discover interesting and fun books to read. The
received suggestions are often a subset of all possibly relevant books. Someone
asking for good historical fiction set in Tudor England will receive suggestions
reflecting what the forum members consider to be good historical fiction, even

---

[2] http://www.loc.gov/marc/

though the LT catalogue contains many more historical fiction books set in Tudor England. The catalogue will not help the reader to identify the best, most fun or most engaging work, which is a reason for the reader to turn to the forums. The information need is more complex than what the catalogue can answer. Such needs contain aspects that are typically covered in reviews but not in professional metadata. However, users cannot search on the content of reviews and tags directly, but can only navigate through individual tags and browse through the reviews of a selected book. It is not clear what information from reviews and tags they use, but it is assumed that searchers use this data to determine whether they want to read a book or not. In terms of library catalogue objectives [18], user reviews can help searchers *choosing* which of the relevant items to read but by excluding reviews from the search index they cannot help *finding or locating* relevant items.

3.1   Comparing Data

How do professional metadata and user-generated content compare in terms of vocabulary size and frequency distribution? The SBS Track uses the INEX Amazon/LibraryThing (A/LT) collection, which contains book records from Amazon, including user reviews, and enriched with UGCfrom LT for a set of 2.8 million books. Each book is identified by ISBN and the record is marked up in XML [11]. Additional information for a large portion of those books is provided by records from the British Library (BL, 1.15 million records) and the Library of Congress (LoC, 1.25 million records). These records contain more subject headings per book than the Amazon records.

Traditional catalogues are based on controlled vocabularies to ensure consistency and rigorousness and to resolve ambiguity. A lot of care has been given to designing systems that give precise and complete results [17]. Social catalogues lack this control, which makes them easier to use for both cataloguers and searchers—they are not restricted in how they express themselves—but often leads to inconsistent use of terminology, ambiguity and inaccurate and incomplete descriptions [19]. Social cataloguing allows a much larger crowd to contribute than traditional methods, with the potential to catalogue a larger corpus of books in more detail. Both methods have advantages and disadvantages, but it is not clear how these affect retrieval performance.

Descriptors from controlled vocabularies are usually not repeated, so there is little information in the term frequency distribution. It is possible in principle to use multiple occurrences of a controlled subject term to reflect the degree to which the described object is about or related to the concept of the subject term. Why is frequency or degree not used in controlled vocabulary access points? It would require a way to measure the degree of relevance of a subject term, but thereby also allow relevance ranking. The indexer would have to analyse each book in much more detail, assigning terms for less prominent topics in the book and to determine the degree of relevance of each subject term. Currently, only subject headings for the main topics of a book are added. In natural language descriptions such as reviews, the term frequency distribution contains more information to measure

Table 1. Total and mean (median) number of types and tokens per metadata field

Field	total # types	# tokens	per doc. # types	# tokens
Book title	196,977	14,265,785	5 (4)	5 (5)
Author	224,554	7,652,715	3 (2)	3 (2)
Am. subject	87,827	5,225,797	2 (2)	2 (2)
BL/LoC LCSH	96,460	17,307,446	4 (3)	6 (3)
LT tags (set)	216,515	46,865,010	13 (6)	17 (7)
LT tags (bag)	216,515	251,868,997	13 (6)	91 (8)
Am. review	2,601,520	1,184,800,633	170 (0)	426 (0)

term relevance, but with the open nature of the web, there is no guarantee that a review contains a good description of the book. We quantitatively compare the subject headings from Amazon and from BL and LoC (BL/LoC) with the user reviews and tags as well as the book titles and author names. In Table 1 we see the mean (median) number of word types and tokens per metadata field. The numbers shown are based on single words after stopword removal and Krovetz stemming (see the next section on the experimental setup) and averaged over all 2.8 million records. For the LT tags, the bag of tags takes into account the number of users who assigned a tag to a book and the set of tags counts each tag only once per book. The second columns shows the number of term types in the index, i.e. the full vocabulary per metadata type. There are 196,977 distinct terms in the 2.8M book titles. The Amazon and BL/LoC headings have a smaller vocabulary than the other metadata types. The tag vocabulary is similar to the book title and author vocabulary, while the reviews, not surprisingly, have a far larger vocabulary. The third column shows the number of tokens. The BL/LoC subject headings have more than three times the number of tokens of the Amazon headings. Amazon records do not reflect the amount of detail of traditional library catalogues. The tags and especially the reviews have many more tokens, since multiple users can add tags and reviews to a book and reviews are often longer than subject headings. Columns four and five show the mean (median) number of types and tokens per book description. The book titles have a mean (median) of 5 (4) distinct terms and 5 (5) total terms. The author names are shorter. The Amazon subject headings are very short. The BL/LoC headings have a mean of 6 term tokens but 4 term types. In other words, there is some term repetition in subject headings. The tags and reviews show a more skewed distribution. A small number of books have a large number of tags and a lot of review text. This skew has an impact on the ranking produced by standard retrieval models.

4   Experimental Setup

For the professional metadata we use the library catalogue records of the BL and the LoC. For some books we have records from both libraries. In most of these cases, the subject headings are identical. In the cases where they are not identical, we use both headings, which increases the number of headings and thereby the richness of the subject descriptors. For indexing we use Indri[3], Krovetz stemming and stopword removal. For retrieval we use the standard language model with Dirichlet smoothing ($\mu = 2500$) without any Indri-specific belief operators. We created indexes for four types of descriptions:

Title contains only the book titles.
Am. subject contains the subject headings from the Amazon records.
BL/LoC contains the merged subject headings of the BL and LoC records.
Review contains all Amazon reviews per book.
Tag contains all LT user tags, where the number of users who assigned a tag t to book b is taken as the term frequency $tf(t, d)$.

As Table 1 showed, there is a large difference in term frequency distributions. Therefore, we create two types of indexes:

TF$_1$ based on the term types, that is, for each term the term frequency in the index is 1.
TF$_a$ based on all term tokens, that is, for each term the term frequency in the index is based on the term frequency in the document.

This allows us to investigate the impact of term repetition in descriptions on retrieval effectiveness. The distribution of user reviews and tags is skewed: popular books tend to have many more reviews and tags than obscure books, with a larger set of terms and more term repetition. When matching query terms against document representations based on UGC, popular books have a higher probability of being retrieved and ranked highly than less popular books. To study this impact, we employ a popularity prior probability, based on the number of reviews per book, which we assume is an (imperfect) indicator of popularity. The review prior $P_r(d)$ is computed as the number of reviews $r(d)$ for book d divided by the total number of reviews r_C in the entire collection C. This is combined with the language model score as follows:

$$P(d|q) = P_r(d) \cdot P(q|d) \tag{1}$$

With this popularity prior we can investigate how much of the retrieval effectiveness of reviews and tags comes from differences in popularity.

5   Evaluation

The 2012 SBS task uses a set of 94 topics taken from the LT discussion forums. Members often link the book title they suggest to the catalogue record of the book on LT. The records of those books contain ISBNs associated with these books, which are used to map suggestions on the forum to books in the document

---

[3] Url: http://www.lemurproject.org/indri/

Table 2. Evaluation results for runs on the different indexes using the INEX 2012 SBS topics

Run	nDCG@10				R@1000			
	TF_1	$TF_1\ P_r$	TF_a	$TF_a\ P_r$	TF_1	$TF_1\ P_r$	TF_a	$TF_a\ P_r$
Title	0.0278	0.0722	0.0281	0.0726	0.2175	0.2270	0.2197	0.2277
Author	0.0213	0.0363	0.0120	0.0348	0.1043	0.1033	0.1043	0.1033
Am. subject	0.0027	0.0153	0.0075	0.0214	0.0354	0.0442	0.0376	0.0410
BL/LoC	0.0156	0.0576	0.0203	0.0583	0.1814	0.1936	0.1681	0.1975
Review	0.0184	0.0195	0.0951	0.1242	0.2465	0.1731	0.4579	0.4209
Tag	0.0124	0.0532	0.0835	0.0907	0.2655	0.2622	0.3628	0.3281

collection of the SBS Track. These suggestions are used as relevance judgements. The relevance judgements have a graded scale, with $rv = 1$ for all suggestions with two exceptions. Suggested books that the topic creator already had in her personal catalogue before starting the topic are not relevant ($rv = 0$) and suggestions that the topic creator adds to her personal catalogue after it was suggested on the forum are the most relevant ($rv = 4$) [11].

The evaluation results for the metadata types are shown in Table 2. The official evaluation measure for the task is nDCG@10. We also show recall at rank 1000 (R@1000). Book titles are generally more effective for retrieval than author names and subject headings. Performance on the BL/LoC subject headings is much better than on the Amazon subject headings. Without term frequency differentiation, UGC is not as effective as book titles and author names for precision, but with a larger number of distinct terms it gives better recall. Not surprisingly, using the term frequency information (columns 4, 5, 8 and 9) has little impact on metadata types that have little term repetition, i.e., book titles, author names and subject headings. By far the most effective are the full user reviews and tags, with the reviews resulting in the highest scores for precision and recall. Reviews and tags contain good descriptive terms for retrieving relevant books. The review prior P_{rev} leads to improved precision on all runs except on the Review TF_1 index. This shows that the good performance of reviews is not just based on the inherent popularity information, but it does play a role. However, it hurts R@1000 on the Review and Tag indexes, which is probably due to the most popular books matching any query and getting boosted to the top by their high review prior. In other words, the information needs on the forum have a specific topical focus which is easily lost by the popularity prior.

The effectiveness of UGCis partly derived from the inherent popularity signal and partly from the term distribution. Relevant search terms occur more frequently in reviews and tags than other terms. Even with the presence of misleading, off-topic and badly written reviews, the user reviews still provide a better book representation for retrieval than subject headings, book titles and author names.

What happens if we use multiple types of metadata? Indexes typically contain both book titles, author names and subject headings. Table 3 shows results

Table 3. Evaluation results for combined metadata using the INEX 2012 SBS topics

Fields	nDCG@10	R@1000
Title+Author+BL/LoC	0.0366	0.3457
Title+Author+BL/LoC+Review	0.1344	0.5294
Title+Author+BL/LoC+Tag	0.1020	0.4391
Title+Author+BL/LoC+Review+Tag	0.1640	0.5639

for runs on indexes with combinations of metadata, with full term frequency information and no review prior. The combination of title and author and subject headings leads to improvements over the individual types in both precision and recall, but is well below the performance of the Review and Tag indexes. Adding the reviews and tags leads to improvements over all indexes of the individual types. Both reviews and tags improve the book representation. Combining all metadata gives the best performance, reflecting the importance of poly-representation [8] and suggests the reviews and tags are complementary to each other.

To summarise, professional metadata, including subject headings, provide limited information on the content of books and no information on the quality, popularity and interestingness of books. User reviews and tags can improve retrieval performance when they are included in the search index. Even without any filtering of personal or idiosyncratic tags and unhelpful or misleading reviews, UGC adds more descriptive text to improve recall and better relevance cues for precision through term repetitions.

6   Analysis

The differences in performances between the subject headings on the one hand and the user reviews and tags on the other hand beg for further analysis. A lot of professional effort is spent on creating, maintaining, updating and assigning subject headings and library studies have often stated that UGC is too messy and unreliable to be a suitable alternative for searching, but our findings point in the opposite direction. In this section we break down the results and try to identify in which cases UGC performs better than subject headings and vice versa. We analyse the topic titles and categorise them on the types of aspects they cover, namely formal metadata (name of author, book or series) and content (genre, subject). The formal metadata topics we divide further into sets of topics that ask for the best *edition* of a specific work, best *starting point* for an author or series or for *similar* works to a named work or author. For instance, topic 29129 asks for the best *edition* of 'Canterbury Tales', topic 30984 asks for a good *starting point* for the works of 'Eudora Welty' and topic 10392 asks for books *similar* to 'David Copperfield'. The content-related topics we divide into sets that ask for books on a certain genre (topic 115958 asks for *western horror* recommendations), books about a certain subject, which we further split into named entity subjects (topic 50302 asks for books about *Cedar Creek* which is

Table 4. Evaluation results of the BL/LOC, Review and Tag indexes per topic category

	#tpcs	nDCG@10			Recall$_{set}$		
		BL/LoC	Review	Tag	BL/LoC	Review	Tag
Metadata	25	0.0524	0.1753	0.0860	0.1795	0.7344	0.5860
Edition	3	0.0000	0.0534	0.0000	0.2731	0.5880	0.5185
Similar	8	0.0476	0.0552	0.0847	0.0849	0.6729	0.3638
Episode	14	0.0664	0.2701	0.1051	0.2136	0.8008	0.7274
Metadata & Subject	2	0.0000	0.1034	0.0665	0.4445	1.0000	0.5000
Content	69	0.0108	0.0625	0.0823	0.4174	0.6652	0.6667
Genre	11	0.0000	0.0292	0.1025	0.2972	0.6811	0.7521
Genre & Subject	15	0.0064	0.0795	0.0663	0.3874	0.6961	0.6906
Subject Named	12	0.0080	0.0970	0.0779	0.3201	0.6467	0.6247
Subject Unnamed	3	0.0000	0.0096	0.0200	0.6563	0.8937	0.9542
Subject	43	0.0151	0.0651	0.0962	0.4586	0.6503	0.6364
Named	34	0.0191	0.0720	01216	0.5333	0.6749	0.7018
Unnamed	9	0.0000	0.0391	0.0000	0.1763	0.5574	0.3896
All	94	0.0203	0.0951	0.0835	0.3560	0.6902	0.6422

the name of both a place and an event) and non-entity subjects (topic 26348 asks for books on *portraiture*). Topics can also contain a combination of formal metadata and content (genre and/or subject).

These categories present a broad and challenging but natural set of search tasks. Finding similar books is a form of item-item recommendation. Finding editions of a work is a form of known-item search and the subject-related topics resemble classical topic-relevance search tasks, but with an element of recommendation, as not all books on a subject are suggested.

Performance per category is shown in Table 4. For recall, subject headings score better on content-related topics than on formal metadata-related topics, while for precision they score better on formal metadata-related topics and very badly on subject related-topics. This is surprising, given that their purpose is to support subject access while formal access is supported by title information. The Review scores high on Metadata related topics, suggesting that reviews typically contain metadata terms such book titles and author names. Tags are particularly effective for genre-related topics and subject-topics. When the topic combines both subject and genre, reviews are at least as effective. Both reviews and tags contain relevant genre and subject terms. For precision, tags score well on named subject terms, but fail to score for general subjects. In sum, UGC is not only effective for traditional subject search tasks, but also for known-item search, genre search and complex combinations of these tasks. Adding UGC to the document representations for indexing increases retrieval effectiveness for a broad range of tasks.

On the last row of Table 4, we see set-based recall over all 94 topics. On the BL/LoC index recall is 0.36, on the review index 0.69 and on the tag index 0.64. That is, on average, only 36% of the suggested books can be found through subject

Table 5. Set-based performance with single subject headings optimised for precision, recall and F_1

Optimised for	Precision	Recall	F_1
Precision	0.8878	0.1966	0.2563
Recall	0.1897	0.5307	0.1436
F_1	0.6402	0.2473	0.2818

headings, while the majority of suggestions is found through reviews and tags. This has consequences for potential use of precision devices like low smoothing (missing query terms is heavily punished, resulting in coordination level ranking). The low recall on the BL/LoC index offers little potential for improving precision and might have to be increased first through devices like pseudo-relevance feedback.

Is there a relation between the per-topic recall of the three indexes, i.e. do they perform well on the same topics? For that we look at the per-topic set-based recall distribution of the BL/LoC, Review and Tag indexes and compute the Kendall's τ ranking correlation between runs on the BL/LoC, review and tag indexes. Between the BL/LoC and Review indexes the correlation is low ($\tau = 0.35$), meaning that the topics on which the BL/LoC index scores highest are different from the topics on which the Review index scores highest. Between the BL/LoC and Tag index this correlation is stronger (0.60) indicating that the subject headings are more similar to tags than to reviews. The correlation between reviews and tags is 0.61. For recall, tags are somewhere in between subject headings and reviews.

6.1   Selecting Subject Headings

Subject headings are not meant to be used (only) for keyword searches. In library catalogues, users can select specific headings from a list or from the headings assigned to a specific book. The heading will show the user a list of catalogue items to which the heading is assigned. A good subject heading may be different from a keyword query. This prompts the question whether the user could do better by selecting an appropriate heading instead typing her own query.

In Table 5 we show the set-based precision, recall and F_1 score when the user would choose the optimal subject heading for precision (row 1), recall (row 2) or F_1 (row 3). High precision can be achieved but at the cost of recall. In many cases, the subject heading is very specific and leads to a single book. If the user aims for high recall, the subject headings is more general and contains many books not relevant to her information need. This mismatch between available subject headings and users' information needs is often ignored in analyses of keyword search success in library catalogues.

7   Conclusions

In this paper we investigated the impact of indexing professional metadata and user-generated content on book search.

We first compared professional metadata and UGC in terms of vocabulary size and frequency distribution. Professional book descriptions have a small number of subject headings representing the content of the book. The number of descriptive terms is small and has little term frequency information to signal what the most relevant terms are. The amount of descriptive text is relatively uniform across all books. With UGC this is more skewed, with popular books described with many more terms than obscure books. Descriptive terms are often repeated, thereby providing relevance cues. This popularity effect is a larger for reviews than for tags—assigning tags takes less effort than writing a review—so reviews give access to fewer books than tags, but they do provide a richer set of descriptive terms.

Next, we compared professional metadata and UGC in terms of retrieval effectiveness. Book titles, author names and subject headings lead to low precision, as they provide few cues to what the best, most relevant books are on a topic. Both reviews and tags lead to better performance compared to the professional metadata across many different topic types. Combining them further improves performance. Reviews and tags provide a large number of descriptive terms, leading to higher recall and the relevant terms are repeated more frequently leading to higher precision. On top of that, the skewed distribution of UGC functions like an inherent popularity prior, favouring books that more users want to read.

What is the impact of professional metadata and user-generated content on book search? Professional metadata is too limited to allow effective retrieval of books. The lack of term repetition leads to weak relevance cues and the small number of descriptive terms does not cover the content and non-content aspects of books in sufficient detail. User reviews and tags are more effective for retrieval and to some extent complementary to each other and to professional metadata. Moreover, reviews can contain important information that is lacking from professional metadata. Subject headings also allow users to browse through sets of books on the same topic in a controlled way, but for keyword searches UGC is much more effective.

These findings are of direct relevance to large online book shops and cataloguing sites such as Amazon, GoodReads, LT as well as the numerous online national and international library catalogues. Although including UGC unmodified to the search index opens the search process up to exploitation by spammers and commercial interests, companies with book search services should be aware of the value of tags and reviews for retrieval and can perhaps find safe ways of including (parts of) the user data in the index. Although we have only looked at books, we expect these findings generalise to others product categories where metadata, reviewing and searching are similar in nature, like music and films.

References

[1] Bartley, P.: Book tagging on librarything: How, why, and what are in the tags? Proceedings of the American Society for Information Science and Technology 46(1), 1–22 (2009)

[2] Buckland, M.: Vocabulary as a Central Concept in Library and Information Science. In: Proceedings of CoLIS3 (1999)

[3] Chandler, O.: How Consumers Discover Books Online (2012),
http://cdn.oreillystatic.com/en/assets/1/event/73/How

[4] Cleverdon, C.W.: The Cranfield tests on index language devices. Aslib 19, 173–192 (1967)

[5] Golder, S.A., Huberman, B.A.: Usage patterns of collaborative tagging systems. Journal of Information Science 32(2), 198–208 (2006)

[6] Golovchinsky, G., Price, M.N., Schilit, B.N.: From reading to retrieval: free-form ink annotations as queries. In: SIGIR 1999, pp. 19–25 (1999)

[7] Gross, T., Taylor, A.G.: What Have We Got to Lose? The Effect of Controlled Vocabulary on Keyword Searching Results. College & Research Libraries 66(3) (2005)

[8] Ingwersen, P.: Polyrepresentation of information needs and semantic entities: Elements of a cognitive theory for information retrieval interaction. In: Croft, W.B., van Rijsbergen, C.J. (eds.) SIGIR, pp. 101–110. ACM/Springer (1994)

[9] Kazai, G., Milic-Frayling, N.: Effects of social approval votes on search performance. In: Third International Conference on Information Technology: New Generations, pp. 1554–1559 (2009)

[10] Koolen, M., Kamps, J., Kazai, G.: Social Book Search: The Impact of Professional and User-Generated Content on Book Suggestions. In: CIKM 2012. ACM (2012)

[11] Koolen, M., Kazai, G., Kamps, J., Doucet, A., Landoni, M.: Overview of the INEX 2012 Social Book Search Track. In: Geva, S., Kamps, J., Schenkel, R. (eds.) INEX 2011. LNCS, vol. 7424, pp. 1–29. Springer, Heidelberg (2012)

[12] Lancaster, F.W.: Vocabulary control for information retrieval, 2nd edn. Information Resources Press, Arlington (1986)

[13] Lu, C., Jung-ran, P., Hu, X.: User tags versus expert-assigned subject terms: A comparison of librarything tags and library of congress subject headings. Journal of Information Science 36(6), 763–779 (2010)

[14] Mathes, A.: Folksonomies - cooperative classification and communication through shared metadata (December 2004)

[15] Rolla, P.J.: User Tags versus Subject Headings. Library Resources & Technical Services 53(3), 174–184 (2009)

[16] Sterken, V.: Folksonomy as a thing for a library: An analysis of user generated metadata in Library Thing. Cahiers de la Documentation–Bladen Voor Documentatie 1, 9 (2009)

[17] Svenonius, E.: Unanswered questions in the design of controlled vocabularies. Journal of the American Society for Information Science 37(5), 331–340 (1986)

[18] Svenonius, E.: The Intellectual Foundation of Information Organization. MIT Press (2000) ISBN 0-262-19433-3

[19] Voss, J.: Tagging, folksonomy & co - renaissance of manual indexing? CoRR, abs/cs/0701072 (2007)

[20] Yi, K.: A semantic similarity approach to predicting library of congress subject headings for social tags. JASIST 61(8), 1658–1672 (2010)

[21] Yi, K., Chan, L.M.: Linking folksonomy to Library of Congress subject headings: an exploratory study. Journal of Documentation 65(6), 872–900 (2009)

A Scalable Gibbs Sampler
for Probabilistic Entity Linking

Neil Houlsby[1],* and Massimiliano Ciaramita[2]

[1] University of Cambridge
nmth2@cam.ac.uk
[2] Google Research, Zürich
massi@google.com

Abstract. Entity linking involves labeling phrases in text with their
referent entities, such as Wikipedia or Freebase entries. This task is challenging due to the large number of possible entities, in the millions, and
heavy-tailed mention ambiguity. We formulate the problem in terms of
probabilistic inference within a topic model, where each topic is associated with a Wikipedia article. To deal with the large number of topics
we propose a novel efficient Gibbs sampling scheme which can also incorporate side information, such as the Wikipedia graph. This conceptually
simple probabilistic approach achieves state-of-the-art performance in
entity-linking on the Aida-CoNLL dataset.

1 Introduction

Much recent work has focused on the 'entity-linking' task which involves annotating phrases, also known as *mentions*, with unambiguous identifiers, referring
to *topics*, *concepts* or *entities*, drawn from large repositories such as Wikipedia
or Freebase. Mapping text to unambiguous references provides a first scalable
handle on long-standing problems such as language *polysemy* and *synonymy*, and
more generally on the task of *semantic grounding* for language understanding.

Most current approaches use heuristic scoring rules or machine-learned models to rank candidate entities. In contrast, we cast the entity-linking problem as
inference in a probabilistic model. This probabilistic interpretation has a number
of advantages: (i) The model provides a principled interpretation of the objective function used to rank candidate entities. (ii) One gets automatic confidence
estimates in the predictions returned by the algorithm. (iii) Additional information can be incorporated into the algorithm in a principled manner by extending
the underlying model rather than hand tuning the scoring rule. (iv) In practice, probabilistic inference is often found to be less sensitive to the auxiliary
parameters of the algorithm. Finally, our method has the advantage of being
conceptually simple compared to many state-of-the-art entity-linking systems,
but still achieves comparable, or better, performance.

The model underlying the linking algorithm presented here is based upon Latent Dirichlet Allocation (LDA) [1]. In a traditional LDA model, the topics have

---

* Work carried out during an internship at Google.

M. de Rijke et al. (Eds.): ECIR 2014, LNCS 8416, pp. 335–346, 2014.
© Springer International Publishing Switzerland 2014

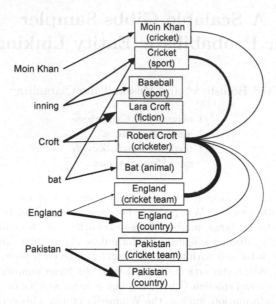

Fig. 1. Example of document-Wikipedia graph

no inherent interpretation; they are simply collections of related words. Here we construct an LDA model in which each topic is associated with a Wikipedia article. Using this 'Wikipedia-interpretable' LDA model we can use the topic-word assignments discovered during inference directly for entity linking. The topics are constructed using Wikipedia, and the corresponding parameters remain fixed. This model has one topic per Wikipedia article, resulting in over 4 million topics. Furthermore, the vocabulary size, including mention unigrams and phrases, is also in the order of millions.To ensure efficient inference we propose a novel Gibbs sampling scheme that exploits sparsity in the Wikipedia-LDA model. To better identify document-level consistent topic assignments, we introduce a 'sampler-memory' heuristic and propose a simple method to incorporate information from the Wikipedia in-link graph in the sampler. Our model achieves the best performance in entity-linking to date on the Aida-CoNLL dataset [2].

2   Background and Related Work

Much recent work has focused on associating textual mentions with Wikipedia topics [2–9]. The task is known as *topic annotation, entity linking* or *entity disambiguation*. Most of the proposed solutions exploit sources of information compiled from Wikipedia: the link graph, used to infer similarity measures between topics, anchor text, to estimate how likely a string is to refer to a given topic, and finally, to a lesser extent so far, local textual content.

Figure 1 illustrates the main intuitions behind most annotators' designs. The figure depicts a few words and names from a news article about cricket. Connections between strings and Wikipedia topics are represented by arrows whose line weight represents the likelihood of that string mentioning the connected topic. In this example, a priori, it is more likely that "Croft" refers to the fictional character rather than the cricket player. However, a similarity graph induced from Wikipedia[1] would reveal that the cricket player topic is actually densely connected to several of the candidate topics on the page, those related to cricket (again line weight represents the connection strength). Virtually all topic annotators propose different ways of exploiting these ingredients.

Extensions to LDA for modeling both words and *observed* entities have been proposed [10, 11]. However, these methods treat entities as *strings*, not linked to a knowledge base. [5, 12, 13] propose LDA-inspired models for documents consisting of words and mentions being generated from distributions identified with Wikipedia articles. Only Kataria *et al.* investigate use of the Wikipedia category graph as well [12]. These works focus on both training the model and inference using Gibbs sampling, but do not exploit model sparsity in the sampler to achieve fast inference. Sen limits the topic space to $17k$ Wikipedia articles [13]. Kataria *et al.* propose a heuristic topic-pruning procedure for the sampler, but they still consider only a restricted space of $60k$ entities. Han and Sun propose a more complex hierarchical model and perform inference using incremental Gibbs sampling rather than with pre-constructed topics [5]. Porteous *et al.* speed up LDA Gibbs sampling by bounding on the normalizing constant of the sampling distribution [14]. They report up to 8 times speedup on a few thousand topics. Our approach exploits sparsity in the sampling distribution more directly and can handle millions of topics. Hansen *et al.* perform inference with, fewer, fixed topics [15]. We focus upon fast inference in this regime. Our algorithm exploits model sparsity without the need for pruning of topics. A preliminary investigation of a full distributed framework that includes re-estimation of the topics for the Wikipedia-LDA model is presented in [16].

3   Entity Linking with LDA

We follow the task formulation and evaluation framework of [2]. Given an input text where entity mentions have been identified by a pre-processor, e.g. a named entity tagger, the goal of a system is to disambiguate (link) the entity mentions with respect to a Wikipedia page. Thus, given a snippet of text such as "[Moin Khan] returns to lead [Pakistan]" where the NER tagger has identified entity mentions "Moin Kahn" and "Pakistan", the goal is to assign the cricketer id to the former, and the national cricket team id to the latter.

We are given a collection of D documents to be annotated, $\mathbf{w}_d$ for $d = 1, \ldots, D$. Each document is represented by a bag of L_d words, taken from a vocabulary of size V. The entity-linking task requires annotating only the mentions, and not the other words in the document (content words). Our model does not distinguish these, and will annotate both. As well as single words, mentions can be N-gram phrases as in the example "Moin Kahn" above. We assume the segmentation has already been performed using an NER tagger. Because the

---

[1] The similarity measure is typically symmetric.

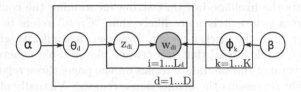

Fig. 2. Graphical model for LDA

model treats mentions and content words equally, we use the term 'word' to refer to either type, and it includes phrases.

The underlying modeling framework is based upon LDA, a Bayesian model, commonly used for text collections [1]. We review the generative process of LDA below, the corresponding graphical model is given in Figure 2.

1. For each topic k, sample a distribution over the words $\phi_k \sim \text{Dir}(\beta)$.
2. For each document d sample a distribution over the topics $\theta_d \sim \text{Dir}(\alpha)$.
3. For each content word i in the document:
   (a) Sample a topic assignment: $z_i \sim \text{Multi}(\theta_d)$.
   (b) Sample the word from topic z_i: $w_j \sim \text{Multi}(\phi_{z_i})$.

The key modeling extension that allows LDA to be used for entity-linking is to associate each topic k directly with a single Wikipedia article. Thus the topic assignments z_i can be used directly to annotate entity mentions. Topic identifiability is achieved via the model construction; the model is built directly from Wikipedia such that each topic corresponds to an article (details in Section 5.1). After construction the parameters are not updated, only inference is performed.

Inference in LDA involves computing the topic assignments for each word in the document $\mathbf{z}_d = \{z_1, \ldots, z_{L_d}\}$. Each z_i indicates which topic (entity) is assigned to the word w_i. For example, if $w_i = $ "Bush", then z_i could label this word with the topic "George Bush Sn.", "George Bush Jn.", or "bush (the shrub)" etc. The model must decide on the assignment based upon the context in which w_i is observed. LDA models are parametrized by their topic distributions. Each topic k is a multinomial distribution over words with parameter vector ϕ_k. This distribution puts high mass on words associated with the entity represented by topic k. In our model each topic corresponds to a Wikipedia entity, therefore the number of topic-word distributions, K, is large ($\approx$ 4M).

To characterize uncertainty in the choice of parameters most LDA models work with distributions over topics. Therefore, instead of storing topic multinomials ϕ_k (as in EDA [15]) we use Dirichlet distributions over the multinomial topics. That is, $\phi_k \sim \text{Dir}(\lambda_k)$, where λ_k are V-dimensional Dirichlet parameter vectors. The set of all vectors $\lambda_1, \ldots, \lambda_K$ represents the model. These Dirichlet distributions capture both the average behavior and the uncertainty in each topic. Intuitively, each element λ_{kv} governs the prevalence of vocabulary word v in topic k. For example, for the topic "Apple Inc." λ_{kv} will be large for words such as "Apple" and "Cupertino". The parameters need not sum to one, $||\lambda_k||_1 \neq 1$, but the greater the values, the lower the variance of the distribution, that is, the more it concentrates around its mean topic.

Most topics will only have a small subset of words from the large vocabulary associated with them, that is, topic distributions are *sparse*. However, the model would not be robust if we were to rule out all possibility of assigning a particular topic to a new word – this would correspond to setting $\lambda_{kv} = 0$. Thus, each parameter takes at least a small minimum value β. Due to the sparsity, most λ_{kv} will take value β. To save memory we represent the model using 'centered' parameters, $\hat{\lambda}_{kv} = \lambda_{kv} - \beta$, most of which take value zero, and need not be stored explicitly. Formally, α, β are scalar hyper-parameters for the symmetric Dirichlet priors; they may be interpreted as topic and word 'pseudo-counts' respectively.

4   Efficient Inference with a Sparse Gibbs Sampler

The English Wikipedia contains around 4M articles (topics). The vocabulary size is around 11M. To cope with this vast parameter space we build a highly sparse model, where each topic only explicitly contains parameters for a small subset of words. Remember, during inference *any* topic could be associated with a word due to the residual probability mass from the hyper-parameter β.

The goal of probabilistic entity disambiguation is to infer the distribution over topics for each word, that is, compute $p(\mathbf{z}|\mathbf{w}, \hat{\lambda}_1, \ldots, \hat{\lambda}_K)$. This distribution is intractable, therefore, one must perform approximate Bayesian inference. Two popular approaches are to use Gibbs sampling [17] or variational Bayes [1]. We use Gibbs sampling, firstly, because it allows us to exploit model sparsity, and secondly, it provides a simple framework into which we may incorporate side information in a scalable manner. During inference, we wish to compute the topic assignments. To do this, Gibbs sampling involves sampling each assignment in turn conditioned on the other current assignments and the model, $z_i \sim p(z_i|\mathbf{z}^{\backslash i}, w_i, \hat{\lambda}_1, \ldots \hat{\lambda}_K) = \int p(z_i|w_i, \theta_d, \hat{\lambda}_1, \ldots \hat{\lambda}_K)p(\theta_d|z^{\backslash i})d\theta_d$. Here, we integrate (collapse) out θ_d, rather than sample this variable in turn. Collapsed inference is found to yield faster mixing in practice for LDA [17, 18].

We adopt the sampling distribution that results from performing variational inference over all of the variables and parameters of the model. Although we only consider inference of the assignments with fixed topics here, this sampler can be incorporated into a scalable full variational Bayesian learning framework [16], a hybrid variational Bayes – Gibbs sampling approach originally proposed in [19]. Following [1, 16, 19], the sampling distribution for z_i is:

$$p(z_i = k|\mathbf{z}^{\backslash i}, w_i \lambda_1, \ldots, \lambda_K) \propto (\alpha + N_k^{\backslash i}) \exp\{\Psi(\beta + \hat{\lambda}_{kw_i}) - \Psi(V\beta + \sum_v \hat{\lambda}_{kv})\}, \quad (1)$$

where $N_k^{\backslash i} = \sum_{j \neq i} \mathbb{I}[z_j = k]$ counts the number of times topic k has been assigned in the document, not including the current word w_i. $\Psi()$ denotes the Digamma function. The sampling distribution is dependent upon both the current word w_i and the current topic counts $N_k^{\backslash i}$, therefore, naïvely one must re-compute its normalizing constant for every Gibbs sample. The distribution has K terms, and so this would be very expensive in this model. We therefore propose using the following rearrangement of Eqn. (1) that exploits the model and topic-count sparsity to avoid performing $\mathcal{O}(K)$ operations per sample:

$$p(z_i = k|\mathbf{z}^{\backslash i}, w_i, \hat{\lambda}_1, \ldots, \hat{\lambda}_K) \propto \underbrace{\frac{\alpha \exp\{\Psi(\beta)\}}{\kappa'_k}}_{\mu_k^{(d)}} + \underbrace{\frac{\alpha \kappa_{kw_i}}{\kappa'_k}}_{\mu_k^{(v)}} + \underbrace{\frac{N_k^{\backslash i} \exp\{\Psi(\beta)\}}{\kappa'_k}}_{\mu_k^{(c)}} + \underbrace{\frac{N_k^{\backslash i} \kappa_{kw_i}}{\kappa'_k}}_{\mu_k^{(c,v)}},$$

(2)

where $\kappa_{kw} = \exp\{\Psi(\beta + \hat{\lambda}_{kw})\} - \exp\{\Psi(\beta)\}$ and $\kappa'_k = \exp\{\Psi(V\beta + \sum_v \hat{\lambda}_{kv})\}$ are transformed versions of the parameters. Clearly $\hat{\lambda}_{kv} = 0$ implies $\kappa_{kv} = 0$. κ'_k is dense. The distribution is now decomposed into four additive components: $\mu_k^{(d)}, \mu_k^{(v)}, \mu_k^{(c)}, \mu_k^{(c,v)}$, whose normalizing constants can be computed independently. $\mu_k^{(d)}$ is dense, but it can be pre-computed once before sampling. For each word we have a term $\mu_k^{(v)}$ which only has mass on the topics for which $\kappa_{kv} \neq 0$; this can be pre-computed for each unique word v in the document, again just once before sampling. $\mu_k^{(c)}$ only has mass on the topics currently observed in the document, i.e. those for which $N_k^{\backslash i} \neq 0$. This term must be updated at every sampling iteration, but this can be done incrementally. $\mu_k^{(c,v)}$ is non-zero only for topics which have non-zero parameters and counts. It is the only term that must be fully recomputed at every iteration. To compute the normalizing constant of the Eqn. (2), the normalizer of each component is computed when the component is constructed, and so all $\mathcal{O}(K)$ sums are performed in the initialization.

Algorithm 1 summarizes the sampling procedure. The algorithm is passed the document $\mathbf{w}_d$, initial topic assignment vector $\mathbf{z}_d^{(0)}$, and transformed parameters κ'_k, κ_{kv}. Firstly, the components of the sampling distribution in (2) that are independent of the topic counts ($\mu^{(d)}, \mu^{(v)}$) and their normalizing constants ($\mathcal{Z}^{(d)}, \mathcal{Z}^{(v)}$) are pre-computed (lines 2-3). This is the only stage at which the full dense K–dimensional vector $\mu^{(d)}$ needs to be computed. Note that one only computes $\mu_k^{(v)}$ for the words in the current document, not for the entire vocabulary. In lines 4-5, two counts are initialized from $\mathbf{z}^{(0)}$. N_{ki} contains the number of times topic k is assigned to word w_i, and N_k counts total number of occurrences of each topic in the current assignment. Both counts will be sparse as most topics are not sampled in a particular document. While sampling, the first operation is to subtract the current topic from N_k in line 8. Now that the topic count has changed, the two components of Eqn. (2) that are dependent on this count ($\mu_k^{(c)}, \mu_k^{(c,v)}$) are computed. $\mu_k^{(c)}$ can be updated incrementally, but $\mu_k^{(c,v)}$ must be re-computed as it is word-dependent. The four components and their normalizing constants are summed in lines 13-14, and a new topic assignment to w_i is sampled in line 15. N_{ki} is incremented in line 17 if burn-in is complete (due to the heuristic initialization we find $B = 0$ works well). If the topic has changed since the previous sweep then N_k is updated accordingly (line 20).

The key to efficient sampling from the multinomial in line 15 is to visit μ_k in order $\{k \in \mu_k^{(c,v)}, k \in \mu_k^{(c)}, k \in \mu_k^{(v)}, k \in \mu_k^{(d)}\}$. A random schedule would require on average $K/2$ evaluations of μ_k. However, if the distribution is skewed, with most of the mass on the topics contained in the sparse components, then much fewer evaluations are required if these topics are visited first. The degree of skewness is governed by the initialization of the parameters, and the priors α, β. In our experiments (see Section 6) we found that we visited on average 4-5 topics per iteration. Note that we perform no approximation or pruning, we still sample from the exact distribution $\text{Multi}(\mu/\mathcal{Z})$. After completion of the Gibbs sweeps, the distribution of the topic assignments to each word is computed empirically from the sample counts in line 24.

Algorithm 1. Efficient Gibbs Sampling

1: **input:** $(\mathbf{w}_d, \mathbf{z}_d^{(0)}, \{\kappa_{kv}\}, \{\kappa_k'\})$
2: $\mu_k^{(d)} \leftarrow \alpha e^{\Psi(\beta)}/\kappa_k', \ \mathcal{Z}^{(d)} \leftarrow \sum_k \mu_k^{(d)}$ ▷ Pre-compute dense component of Eqn. (2).
3: $\mu_k^{(v)} \leftarrow \alpha\kappa_{kv}/\kappa_k', \ \mathcal{Z}^{(v)} \leftarrow \sum_k \mu_k^{(v)} \ \forall v \in \mathbf{w}_d$
4: $N_{ki} \leftarrow \mathbb{I}_{z_i^{(0)}=k}$ ▷ Initial counts.
5: $N_k \leftarrow \sum_{i=1}^{L_d} N_{ki}$
6: **for** $s \in 1, \ldots, S$ **do** ▷ Perform S Gibbs sweeps.
7: $\quad$ **for** $i \in 1, \ldots, L_d$ **do** ▷ Loop over words in document.
8: $\quad\quad N_k^{\backslash i} \leftarrow N_k - \mathbb{I}_{z_i=k}$ ▷ Remove topic z_i from counts.
9: $\quad\quad \mu_k^{(c)} \leftarrow N_k^{\backslash i} e^{\Psi(\beta)}/\kappa_k'$ ▷ Compute sparse components of Eqn. (2).
10: $\quad\quad \mu_k^{(c,v)} \leftarrow N_k^{\backslash i} \kappa_{kw_i}/\kappa_k'$
11: $\quad\quad \mathcal{Z}^{(c)} \leftarrow \sum_k \mu_k^{(c)}$ ▷ Compute corresponding normalizing constants.
12: $\quad\quad \mathcal{Z}^{(c,v)} \leftarrow \sum_k \mu_k^{(c,v)}$
13: $\quad\quad \mu_k \leftarrow \mu_k^{(d)} + \mu_k^{(v)} + \mu_k^{(c)} + \mu_k^{(c,v)}$
14: $\quad\quad \mathcal{Z} \leftarrow \mathcal{Z}^{(d)} + \mathcal{Z}^{(v)} + \mathcal{Z}^{(c)} + \mathcal{Z}^{(c,v)}$
15: $\quad\quad z_i^{(s)} \sim \text{Multi}(\{\mu_k/\mathcal{Z}\}_{k=1}^K)$ ▷ Sample topic.
16: $\quad\quad$ **if** $s > B$ **then** ▷ Discard burn in.
17: $\quad\quad\quad N_{z_i^{(s)}i} \leftarrow N_{z_i^{(s)}i} + 1$ ▷ Update counts.
18: $\quad\quad$ **end if**
19: $\quad\quad$ **if** $z_i^{(s)} \neq z_i^{(s-1)}$ **then**
20: $\quad\quad\quad$ update N_k for $k \in \{z_i^{(s)}, z_i^{(s-1)}\}$ ▷ Update incrementally.
21: $\quad\quad$ **end if**
22: $\quad$ **end for**
23: **end for**
24: $p(z_i = k|w_i) \leftarrow \frac{1}{S-B} N_{ki}$
25: **return:** $p(z_i = k|w_i)$ ▷ Return empirical distribution over topics.

4.1   Incorporating Memory and the Wikipedia Graph

When working with very large topic spaces, the sampler will take a long time to explore the full topic space and an impractical number of samples will be required to achieve convergence. To address this issue we augment the sampler with a 'sampler memory' heuristic and information from the Wikipedia graph.

After a good initialization (see Section 5.2), to help the sampler stay on track we include the current sample in the topic counts when evaluating (2). Allowing the sampler to 'remember' the current assignment assists it in remaining in regions of good solutions. With memory the current effective topic-count is given by $N_k^{\backslash i} \leftarrow N_k \mathrm{coh}(z_k|w_i)$. An even better solution might be to include here an appropriate temporal decaying function, but we found this simple implementation yields strong empirical performance already.

We also exploit the Wikipedia-interpretability of the topics to readily include the graph into our sampler to further improve performance. Intuitively, we would like to weight the probability of a topic by a measure of its consistency with the other topics in the document. This is in line with the Gibbs sampling approach where, by construction, all other topic assignments are known. For this purpose we use the following coherence score [4] for the word at location i:

$$\mathrm{coh}(z_k|i) = \frac{1}{|\{\mathbf{z}_d\}| - 1} \sum_{k' \in \{\mathbf{z}_d\} \backslash i} \mathrm{sim}(z_k, z_{k'}). \qquad (3)$$

where $\{\mathbf{z}_d\}$ is the set of topics in the assignment $\mathbf{z}_d$, and $\mathrm{sim}(z_k, z_{k'})$ is the 'Google similarity' [20] between two Wikipedia pages. We include the coherence score by augmenting $N_k^{\backslash i}$ in Eqn. 2 with this weighting function, i.e. line 8 in Algorithm 1 becomes $N_k^{\backslash i} \leftarrow (N_k - \mathbb{I}_{z_i=k})\mathrm{coh}(z_k|w_i)$.

Notice that the contributions of the graph-coherence and memory components are incorporated into the computation of the normalizing constant. Incorporating the graph and memory directly into the sampler provides cheap and scalable extensions which yield improved performance. However, it would be desirable to include such features more formally in the model, for example, by including the graph via hierarchical formulations, or appropriate document-specific priors α in stead of the memory. We leave this to future research.

5   Model and Algorithmic Details

5.1   Construction of the Model

We construct models from the English Wikipedia. An article is an admissible topic if it is not a disambiguation, redirect, category or list page. This step selects approximately 4M topics. Initial candidate word strings for a topic are generated from its title, the titles of all Wikipedia pages that redirect to it, and the anchor text of all its incoming links (within Wikipedia). All strings are lower-cased, single-character mentions are ignored. This amounts to roughly 11M words and 13M parameters. Remember, 'words' also includes mention phrases. This initialization is highly sparse - for most word-topic pairs, $\hat{\lambda}_{kv}$ is set to zero. The parameters $\hat{\lambda}_{kv}$ are initialized using the empirical distributions from Wikipedia counts, that is, we set $\hat{\lambda}_{kv} = P(k|v) - \beta = \frac{\mathrm{count}(v,k)}{\mathrm{count}(v)} - \beta$. Counts are collected from titles (including redirects) and anchors. We found that initializing the parameters using $P(v|k)$, rather than $P(k|v)$ yields poor performance because the normalization by count(k) in this case penalizes popular entities too heavily.

5.2 Sampler Initialization

A naive initialization of the Gibbs sampler could use the topic with the greatest parameter value for a word $z_i^{(0)} = \arg\max_k \lambda_{kv}$, or even random assignments. We find that these are not good solutions because the distribution of topics for a word is typically long-tailed. If the true topic is not the most likely one, its parameter value could be several orders of magnitude smaller than the primary topic. Topics have extremely fine granularity and even with sparse priors it is unlikely that the sampler will converge to the the right patterns of topic mixtures in reasonable time. We improve the initialization with a simpler, but fast, heuristic disambiguation algorithm, TagMe [4]. We re-implement TagMe and run it to initialize the sampler, thus providing a good set of initial assignments.

6 Experiments

We evaluate performance on the CoNLL-Aida dataset, a large public dataset for evaluation of entity linking systems [2]. The data is divided in three partitions: train (946 documents), test-a (216 documents, used for development) and test-b (231 documents, used for blind evaluation). We report *micro-accuracy*: the fraction of mentions whose predicted topic is the same as the gold-standard annotation. There are 4,788 mentions in test-a and 4,483 in test-b. We also report *macro-accuracy*, where document-level accuracy is averaged over the documents.

6.1 Algorithms

The baseline algorithm (Base) predicts for mention w the topic k maximizing $P(k|w)$, that is, it uses only empirical mention statistics collected from Wikipedia.This baseline is quite high due to the skewed distribution of topics – which makes the problem challenging. TagMe* is our implementation of TagMe, that we used to initialize the sampler. We also report the performance of two state-of-the-art systems: the best of the Aida systems on test-a and test-b, extensively benchmarked in [2] (Aida13)[2], and finally the system described in [9] (S&Y13) which reports the best micro precision on the CoNLL test-b set to date. The latter reference reports superior performance to a number of modern systems, including those in [21, 2, 3]. We also evaluate the contributions of the components to our algorithm. WLDA-base uses just the sparse sampler proposed in Section 4. WLDA-mem includes the sampler memory, and WLDA-full incorporates both the memory and the graph.

---

[2] We report figures for the latest best model ("r-prior sim-k r-coh") from the Aida web site, http://www.mpi-inf.mpg.de/yago-naga/aida/. We are grateful to Johannes Hoffart for providing us with the development set results of the Aida system.

Table 1. Accuracy on the CoNLL-Aida corpus. In each row, the best performing algorithm, and those whose performance is statistically indistinguishable from the best, are highlighted in bold. Error bars indicate ±1 standard deviation. An empty cell indicates that no results are reported.

	Base	TagMe*	Aida13	S&Y13	WLDA-base	WLDA-mem	WLDA-full
				test-a			
Micro	70.76	76.89	**79.29**	-	75.21 ± 0.57	78.99 ± 0.50	**79.65 ± 0.52**
Macro	69.58	74.57	**77.00**	-	74.51 ± 0.55	**76.10 ± 0.72**	**76.61 ± 0.72**
				test-b			
Micro	69.82	78.64	82.54	**84.22**	78.75 ± 0.54	**84.88 ± 0.47**	**84.89 ± 0.43**
Macro	72.74	78.21	81.66	-	79.18 ± 0.71	**83.47 ± 0.61**	**83.51 ± 0.62**

6.2   Hyper-Parameters

We set hyper-parameters, α, β and S using a greedy search that optimizes the sum of the micro and macro scores on both the train and test-a partitions. Setting α, β is a trade-off between sparsity and exploration. Smaller values result in sparser sampling distributions but larger α allows the model to visit topics not currently sampled and larger β lets the model sample topics with parameter values $\hat{\lambda}_{kv}$ equal to zero. We found that comparable performance can be achieved using a wide range of values: $\alpha \in [10^{-5}, 10^{-1}]$, $\beta \in [10^{-7}, 10^{-3})$. Regarding the sweeps, performance starts to plateau at $S = 50$. The robustness of the model's performance to these wide ranges of hyper-parameter settings advocates the use of this type of probabilistic approach. As for TagMe's hyper-parameters, in our experiments ϵ and τ values around 0.25 and 0.01 respectively worked best.

6.3   Results and Discussion

Table 1 summarizes the evaluation results. Confidence intervals are estimated using bootstrap re-sampling, and statistical significance is assessed using a unpaired t-test at the 5% significance level. Overall, WLDA-full, produces state-of-the-art results on both development (test-a) and blind evaluation (test-b). Table 1 shows that Base and Tagme*, used for model construction and sampler initialization respectively, are significantly outperformed by the full system. TagMe includes the information contained in Base and performs better, particularly on test-a. The gap between TagMe and WLDA-full is greatest on test-b. This is probably because the parameters are tuned on test-a, and are kept fixed for test-b and the proposed probabilistic method is more robust to the parameter values. The inclusion of memory produces a large performance gains and inclusion of the graph adds some further improvements, particularly on test-a.

In all cases we perform as well as, or better than, the current best systems. This result is particularly remarkable due to the simplicity of our approach. The S&Y13 system addresses the broader task of entity linking and named entity recognition. They train a supervised model from Freebase, using extensively engineered feature vectors. The Aida systems incorporate a significant amount of knowledge from the YAGO ontology, that is, they also know the *type* of the entity being disambiguated. Our algorithm is conceptually simple and requires no training or additional resources beyond Wikipedia, nor hand crafting of features or scoring rules. Our approach is based upon Bayesian inference with a model created from simple statistics taken from Wikipedia. It is therefore remarkable that we are performing favorably against the best systems to date an this provides strong motivation to extend this probabilistic approach further.

Inspection of errors on the development partitions reveals scenarios in which further improvements can be made. In some documents, a mention can appear multiple times with different gold annotations. E.g. in one article, 'Washington' appears multiple times, sometimes annotated as the city, and sometimes as USA (country); in another, 'Wigan' is annotated both as the UK town and its rugby club. Due to the 'bag-of-words' assumption, LDA is not able to discriminate such cases and naturally tends to commit to one assignment for all occurrences of a string in a document. Local context could help disambiguate these cases. Within our sampling framework it would be straightforward to incorporate contextual information e.g. via up-weighting of topics using a distance function.

7 Conclusion and Future Work

Topic models provide a principled, flexible framework for analyzing latent structure in text. These are desirable properties for a whole new area of work that is beginning to systematically explore semantic grounding with respect to web-scale knowledge bases such as Wikipedia and Freebase. We have proposed a Gibbs sampling scheme for inference in a static Wikipedia-identifiable LDA model to perform entity linking. This sampler exploits model sparsity to remain efficient when confronted with millions of topics. Further, the sampler is able to incorporate side information from the Wikipedia in-link graph in a straightforward manner. To achieve good performance it is important to construct a good model and initialize the sampler sensibly. We provide algorithms to address both of these issues and report state-of-the-art performance in entity-linking.

We are currently exploring two directions for future work. In the first, we seek to further refine the parameters of the model λ_{kv} from data. This requires training an LDA model on huge datasets, for which we must exploit parallel architectures [16]. In the second, we wish to simultaneously infer the segmentation of the document into words/mentions and the topic assignments through use of techniques such as blocked Gibbs sampling.

Acknowledgments. We would like to thank Michelangelo Diligenti, Yasemin Altun, Amr Ahmed, Marc'Aurelio Ranzato, Alex Smola, Johannes Hoffart, Thomas Hofmann and Kuzman Ganchev for valuable feedback and discussions.

References

1. Blei, D.M., Ng, A.Y., Jordan, M.I.: Latent Dirichlet Allocation. JMLR 3, 993–1022 (2003)
2. Hoffart, J., Yosef, M.A., Bordino, I., Fürstenau, H., Pinkal, M., Spaniol, M., Taneva, B., Thater, S., Weikum, G.: Robust disambiguation of named entities in text. In: EMNLP, pp. 782–792. ACL (2011)
3. Kulkarni, S., Singh, A., Ramakrishnan, G., Chakrabarti, S.: Collective annotation of wikipedia entities in web text. In: SIGKDD, pp. 457–466. ACM (2009)
4. Ferragina, P., Scaiella, U.: TagMe: On-the-fly annotation of short text fragments (by wikipedia entities). In: CIKM, pp. 1625–1628. ACM (2010)
5. Han, X., Sun, L.: An entity-topic model for entity linking. In: EMNLP-CoNLL, pp. 105–115. ACL (2012)
6. Mihalcea, R., Csomai, A.: Wikify!: Linking documents to encyclopedic knowledge. In: CIKM, pp. 233–242. ACM (2007)
7. Milne, D., Witten, I.H.: Learning to link with Wikipedia. In: CIKM, pp. 509–518. ACM (2008)
8. Ratinov, L.A., Roth, D., Downey, D., Anderson, M.: Local and global algorithms for disambiguation to wikipedia. In: ACL, vol. 11, pp. 1375–1384 (2011)
9. Sil, A., Yates, A.: Re-ranking for joint named-entity recognition and linking. In: CIKM (2013)
10. Newman, D., Chemudugunta, C., Smyth, P.: Statistical entity-topic models. In: SIGKDD, pp. 680–686. ACM (2006)
11. Kim, H., Sun, Y., Hockenmaier, J., Han, J.: Etm: Entity topic models for mining documents associated with entities. In: 2012 IEEE 12th International Conference on Data Mining (ICDM), pp. 349–358. IEEE (2012)
12. Kataria, S.S., Kumar, K.S., Rastogi, R.R., Sen, P., Sengamedu, S.H.: Entity disambiguation with hierarchical topic models. In: SIGKDD, pp. 1037–1045. ACM (2011)
13. Sen, P.: Collective context-aware topic models for entity disambiguation. In: Proceedings of the 21st International Conference on World Wide Web, pp. 729–738. ACM (2012)
14. Porteous, I., Newman, D., Ihler, A., Asuncion, A., Smyth, P., Welling, M.: Fast collapsed gibbs sampling for latent dirichlet allocation. In: SIGKDD, pp. 569–577. ACM (2008)
15. Hansen, J.A., Ringger, E.K., Seppi, K.D.: Probabilistic explicit topic modeling using wikipedia. In: Gurevych, I., Biemann, C., Zesch, T. (eds.) GSCL. LNCS, vol. 8105, pp. 69–82. Springer, Heidelberg (2013)
16. Houlsby, N., Ciaramita, M.: Scalable probabilistic entity-topic modeling. arXiv preprint arXiv:1309.0337 (2013)
17. Griffiths, T.L., Steyvers, M.: Finding scientific topics. PNAS 101(suppl. 1), 5228–5235 (2004)
18. Teh, Y.W., Newman, D., Welling, M.: A collapsed variational bayesian inference algorithm for latent dirichlet allocation. In: NIPS, vol. 19, p. 1353 (2007)
19. Mimno, D., Hoffman, M., Blei, D.: Sparse stochastic inference for latent dirichlet allocation. In: Langford, J., Pineau, J. (eds.) ICML, pp. 1599–1606. Omni Press, New York (2012)
20. Milne, D., Witten, I.: An effective, low-cost measure of semantic relatedness obtained from Wikipedia links. In: AAAI Workshop on Wikipedia and Artificial Intelligence (2008)
21. Cucerzan, S.: Large-scale named entity disambiguation based on wikipedia data. In: EMNLP-CoNLL, vol. 7, pp. 708–716 (2007)

Effective Kernelized Online Learning
in Language Processing Tasks

Simone Filice[1], Giuseppe Castellucci[2], Danilo Croce[3], and Roberto Basili[3]

[1] DICII
[2] DIE
[3] DII
University of Roma, Tor Vergata
00133 Roma, Italy
{filice,castellucci}@ing.uniroma2.it,
{croce,basili}@info.uniroma2.it

Abstract. Kernel-based methods for NLP tasks have been shown to enable robust and effective learning, although their inherent complexity is manifest also in Online Learning (OL) scenarios, where time and memory usage grows along with the arrival of new examples. A state-of-the-art *budgeted* OL algorithm is here extended to efficiently integrate complex kernels by constraining the overall complexity. Principles of Fairness and Weight Adjustment are applied to mitigate imbalance in data and improve the model stability. Results in Sentiment Analysis in Twitter and Question Classification show that performances very close to the state-of-the-art achieved by batch algorithms can be obtained.

1   Introduction

Given the growing interactivity needed by Web applications, Information Retrieval challenges such as Question Answering (QA) [17], or Sentiment Analysis (SA) [26] over Web or microblog sources [34] are increasingly interesting. In these tasks, as well as in real-time marketing, semantic web-search or exploratory data analysis, the application of Natural Language Processing (NLP) techniques is crucial. Natural Language Learning (NLL) systems deal with the acquisition of models in order to turn texts into meaningful structures. In NLL, traditional and effective paradigms, such as Support Vector Machines or Maximum Entropy models [13], are effectively applied in a *batch* fashion: they require all examples to be available while inducing the model. However, daily interactions are more natural in the Web and suggest to explore dynamic techniques such as Online Learning (OL) [20,6]. These algorithms, that evolve from the Rosenblatt's Perceptron [27], induce models that can be updated when negative feedback is available, in order to correctly account for new instances and improve accuracy, without complete re-training. In this way, the resulting OL model *follows* its target problem and adapts dynamically. This makes online schemas very appealing in Web scenarios, although performance drops can be observed with respect to the corresponding batch learning algorithms.

M. de Rijke et al. (Eds.): ECIR 2014, LNCS 8416, pp. 347–358, 2014.

The robustness in NLL systems depends also on the suitability of the adopted linguistic features whereas manual encoding is usually carried out by experts. Kernel methods [32] enable an implicit acquisition of such information. They have been largely employed in NLP, such as in [5,22,36,35,8], in order to provide statistical models able to separate the problem representation from the learning algorithm. However, kernel methods are not suitable for large datasets, due to time and space complexity. In large margin learning algorithms, classification complexity may prevent kernel adoption in real world applications, as the required time for a single classification depends on the number of *Support Vectors* (SVs) [30]. This led to an unbounded complexity that is in contrast with the online paradigm, as it should provide a (potentially) never-ending learning. An effective way to cope with this problem introduces budgeted-version of such algorithms [4,24,33,10], binding the maximum number of SVs.

In this paper, we adapt and improve the Budgeted Passive Aggressive algorithm [33] in order to provide an effective and efficient way to design NLL systems. We first investigate the idea of *Fairness* in order to balance the importance of different involved classes. Then, we introduce the notion of *Weight Adjustment*, consisting in a consolidation of SVs. It improves the usability of off-the-shelf kernel functions within OL algorithms. From one hand, it allows a fast system design, by avoiding a manual feature engineering. On other hand, we combine the robustness of state-of-the-art kernels with the efficiency of Budgeted OL algorithms. The complexity of the resulting system is thus bounded, allowing to control the computational cost at the best achievable quality. We evaluate this method in two NLP tasks: Sentiment Analysis in Twitter and Question Classification. In both settings, results close to the state-of-the-art are achievable and they are comparable with one of the most efficient SVM implementation. Results are straightforward considering that no manually coded resource (e.g. WordNet or a Polarity Lexicon) has been used. We mainly exploited distributional analysis of unlabeled corpora.

In the rest of the paper, Section 2 describes the Budgeted Online Learning algorithm. In Section 3 employed kernel functions for robust NLL are discussed. In Section 4 experimental evaluation is discussed.

2   Efficient Kernel-Based Passive Aggressive Algorithm

The Passive Aggressive (PA) learning algorithm [6] is one of the most popular online approaches and it is generally referred as the state-of-art online method. When an example is misclassified, the model is updated with the hypothesis most similar to the current one, among the set of classification hypotheses that correctly classify the example.

Let $(\mathbf{x}_t, y_t)$ be the t-th example where $\mathbf{x}_t \in \mathbb{R}^d$ is a feature vector in a d-dimensional space and $y_t \in \{+1, -1\}$ is the corresponding label. Let $\mathbf{w}_t \in \mathbb{R}^d$ be the current classification hypothesis. The PA classification function is $f(\mathbf{x}) = \mathbf{w}^T \mathbf{x}$. The learning procedure starts setting $\mathbf{w}_1 = (0, \ldots, 0)$, and after receiving $\mathbf{x}_t$, the new classification function $\mathbf{w}_{t+1}$ is the one that minimizes the following objective function[1] $Q(\mathbf{w})$:

---

[1] We are referring to the PA-I version in [6].

$$Q(\mathbf{w}) = \frac{1}{2} \|\mathbf{w} - \mathbf{w}_t\|^2 + C \cdot l(\mathbf{w}; (\mathbf{x}_t, y_t)) \tag{1}$$

where the first term $\|\mathbf{w} - \mathbf{w}_t\|$ is a measure of how much the new hypothesis differs from the old one, while the second term $l(\mathbf{w}, (\mathbf{x}_t, y_t))$ is a proper loss function[2] assigning a penalty cost to an incorrect classification. C is the aggressiveness parameter that balances the two competing terms in Equation 1. Minimizing $Q(\mathbf{w})$ corresponds to solving a constrained optimization problem, whose closed form solution is the following:

$$\mathbf{w}_{t+1} = \mathbf{w}_t + \alpha_t \mathbf{x}_t, \quad \alpha_t = y_t \cdot \min\left\{C, \frac{H(\mathbf{w}_t; (\mathbf{x}_t, y_t))}{\|\mathbf{x}_t\|^2}\right\} \tag{2}$$

After a wrong prediction of an example $\mathbf{x}_t$, a new classification function $\mathbf{w}_{t+1}$ is computed. It is the result of a linear combination between the old $\mathbf{w}_t$ and the feature vector $\mathbf{x}_t$. The linear PA is extremely attractive as the classification and the updating steps have a computational complexity of $\mathcal{O}(d)$, i.e. a single dot product in the d-dimensional space. However, these algorithms cannot directly learn non-linear classification functions.

The kernelized version of the PA algorithm overcomes this limitation, and enables to use structured data, like syntactic trees using tree kernel functions [5]. Generic data representations x can be exploited using an implicit mapping $\phi(x)$ into a Reproducing Kernel Hilbert Space $\mathcal{H}$ operated by a proper kernel function $k(\cdot, \cdot)$. When a misclassification occurs, the model is updated to embed the "problematic" example. At step t, the classification function is $f_t(x) = \sum_{i \in SV_t} \alpha_i k(x_i, x)$, while SV_t is the set of the support vector indices. The updating step corresponds to adding a new support vector:

$$f_{t+1}(x) = f_t(x) + \alpha_t k(x_t, \cdot), \quad \alpha_t = y_t \cdot \min\left\{C, \frac{H(f_t; (x_t, y_t))}{\|x_t\|_{\mathcal{H}}^2}\right\}$$

where $\|\cdot\|_{\mathcal{H}}$ is the norm.

Unfortunately, the kernelized PA algorithm (as any other OL schemas) cannot be used against large datasets. The growth of SVs has a significant drawback in terms of computational complexity and memory usage. All the SVs and their weights must be stored, and each new prediction requires the computation of the kernel function between the example and all current SVs. Several works dealt with such issue, as [4,33,10]. In particular, [33] proposed a budgeted version of the PA algorithm that strictly binds the number of support vectors to a predefined budget B. In their proposal, the updating step of the PA algorithm changes when B is reached: the optimal solution of the PA problem f^* cannot be employed as it would require $B + 1$ support vectors; thus the best estimation of f^* in the space spanned by B support vectors must be chosen. Before adding a new support vector, an old r one must be removed, obtaining a new function f_{t+1}^r. Thus, a new constraint to the minimization problem $Q(\mathbf{w})$ is considered, resulting in:

---

[2] In this work we will consider the hinge loss $H(\mathbf{w}; (\mathbf{x}_t, y_t)) = max(0, 1 - y_t \mathbf{w}^T \mathbf{x}_t)$.

$$f_{t+1}^r = \operatorname*{argmin}_{f^r} \tfrac{1}{2} \| f^r - f_t \|_{\mathcal{H}}^2 + C \cdot \xi$$

$$s.t. \quad 1 - y_t \mathbf{w}^T \mathbf{x}_t \leq \xi, \; \xi \geq 0$$

$$f^r = f_t - \underbrace{\alpha_r k(x_r, \cdot)}_{SV\ elimination} + \underbrace{\sum_{i \in V} \beta_i k(x_i, \cdot)}_{weights\ modification} \tag{3}$$

where r is the index of the support vector to be removed, t is the index of the new support vector to be added and $V \subseteq SV_t \cup \{t\} - \{r\}$ is the set of the indices of SVs whose weights can be modified. Let f_{t+1}^r be the solution of Equation 3 for a given r; using a brute force approach, the support vector to be removed r^* and the corresponding new classification function f_{t+1} are selected as the ones that minimize Equation 1: $r^* = \operatorname*{argmin}_{r \in SV_t \cup \{t\}} Q(f_{t+1}^r)$. In [33] f_{t+1}^r has been solved in closed-form and the overall updating complexity is shown[3] to be $\mathcal{O}(B|V|^2)$. The choice of V (i.e. the set of SVs whose weights can be modified) has a deep impact on the computational complexity of this step. Three different Budgeted Passive Aggressive (BPA) policies are proposed:

- BPA-Simple (BPA-S): $V = \{t\}$
- BPA-Projecting (BPA-P): $V = SV_t \cup \{t\} - \{r\}$
- BPA-Nearest-Neighbor (BPA-NN): $V = \{t\} \cup NN(r)$ where $NN(r)$ is the index of the nearest neighbor of x_r.

The third one is shown to be a good trade-off between the accuracy and computational complexity. The main idea is to preserve the information provided by the support vector r to be removed, by projecting it into the space spanned by the support vectors in V. The space spanned by r is similar to the one spanned by its nearest neighbor, so the BPA-NN is supposed to provide good performances.

2.1   Improving Robustness through Fairness and Weight Adjustment

The original formulation of the PA algorithm makes no distinction between positive and negative examples. In many real-world classification domains, examples are not equally distributed among the classes. Thus, very imbalanced datasets can penalize less frequent classes. A common solution is data sampling [31] in order to obtain a more balanced distribution. However, these methods discard some informative examples. A different approach directly modifies the learning algorithm and emphasizes the contribution of the less frequent examples. In [21] the original SVM formulation is slightly modified splitting the empirical risk term into a positive part and a negative one. Each is weighted by parameters C_+ and C_- that substitute the original regularization parameter C: in this way it is possible to adjust the cost of false positives vs. false negatives. Accordingly, we reformulated the original objective function:

$$\operatorname*{argmin}_{\mathbf{w}} \tfrac{1}{2} \| \mathbf{w} - \mathbf{w}_t \|^2 + C(y_t) \cdot \xi$$

$$s.t. \; 1 - y_t \mathbf{w}^T \mathbf{x}_t \leq \xi, \; \xi \geq 0, \text{where } C(y_t) = \begin{cases} C_+ \text{ if } y_t = +1 \\ C_- \text{ if } y_t = -1 \end{cases}$$

---

[3] It applies when kernel computations are cached.

The parameters C_+ and C_- can be chosen so that the potential total cost of the false positives equals the potential total cost of the false negatives, i.e. the ratio C_+/C_- is equal to the ratio between the overall number of negative and positive training examples. It allows to introduce a sort of **fairness** in the learning phase. During the experimental evaluation we will refer to the experiment settings as $fair$ BPA (F-BPA) if the aggressiveness parameters are chosen as the above ratio, and as $unfair$ BPA if $C_+ = C_-$.

Another drawback of OL algorithms is that generally they do not perform as good as batch learning ones, due to their inherent simplicity. For instance, the updating step of the PA algorithm performs a local optimization that considers a single example at a time. Batch learning algorithms, e.g. SVM, perform a global optimization over the whole training set. Changing the model at each misclassification has two consequences in OL: from one side, it enables the algorithm to track a shifting concept; on the other side, it can produce instability in the learning process. Outliers or mislabeled examples can affect the current model and different orders of the same training examples can produce quite different solutions. The PA approach mitigates this instability modifying the current classification hypothesis as less as possible, but it does not completely solve this issue. A common solution is performing multiple iterations on the training data. A new iteration (i.e. epoch) involves the computation of new kernel operations that have not been evaluated during the first epoch, thus losing the OL advantages. In fact, let SV_t be the set of the indices of the current SVs and let x_t be the current training example such that $t \notin SV_t$ (i.e. x_t is not a support vector); then, during the second epoch, for each support vector x_i such that $i > t$, a new kernel computation $k(x_i, x_t)$ must be computed. In order to overcome these drawbacks, we propose an effective variation to the standard multiple iteration method. Instead of revisiting the whole training set, only the support vector set is exploited again. We will refer to this approach as **weight adjustment** to emphasize the fact that the support vector set does not change, and only the weights α_i of the support vectors can be adjusted changing their contribution to the classification function. From a computational perspective, caching the kernel operation of the first epoch, any additional computation must be performed. We will indicate as A-BPA a standard Budgeted Passive Aggressive Algorithms which performs a weight adjustment, and AF-BPA is the corresponding fair version.

3   Combining Semantic Kernels

Combination of classifiers is commonly used to improve the performances joining the strengths of many methods. Another strategy is to combine similarity functions among different representations in a kernel combination. It is still a valid kernel and can thus be integrated in learning algorithms [30]. We decided to apply three kernel functions, each emphasizing a specific aspect.

Bag of Word Kernel. (*BOWK*) A first kernel function exploits pure lexical information, expressed as the word overlap between texts. It is very common in Information Retrieval, since [29], where documents are represented as vectors whose dimensions correspond to different terms. A boolean weighting is applied: each dimension represents an indicator of the presence or not of a word in the text. The kernel function is the cosine similarity between vector pairs.

Lexical Semantic Kernel (LSK). A kernel function generalizes the lexical information, without exploiting any manually coded resource. Lexical information is obtained by a co-occurrence Word Space built accordingly to the methodology described in [28]. First, a word-by-context matrix M is computed through a large scale corpus analysis. Then, *Latent Semantic Analysis* [18] technique is applied as follows. The matrix M is decomposed through Singular Value Decomposition (SVD). The original statistical information about M is captured by the new k-dimensional space, which preserves the global structure while removing low-variant dimensions, i.e. distribution noise. Thus, every word is projected in the reduced space and a sentence is represented by applying a *linear combination*. The resulting kernel is the cosine similarity between vector pairs, as in [7].

Smoothed Partial Tree Kernel. (SPTK) Tree kernels exploit syntactic similarity through the idea of convolutions among substructures. Any tree kernel evaluates the number of common substructures between two trees T_1 and T_2 without explicitly considering the whole fragment space [5]. Its general equation is $TK(T_1, T_2) = \sum_{n_1 \in N_{T_1}} \sum_{n_2 \in N_{T_2}} \Delta(n_1, n_2)$ where N_{T_1} and N_{T_2} are the sets of the T_1's and T_2's nodes respectively, and $\Delta(n_1, n_2)$ is equal to the number of common fragments rooted in the n_1 and n_2 nodes. In the SPTK formulation [8] the function Δ emphasizes lexical nodes. SPTK main characteristic is its ability to measure the similarity between syntactic tree structures, which are partially similar and whose nodes can differ but are semantically related. One of the most important outcomes is that SPTK allows "embedding" external lexical information in the kernel function only through a similarity function σ among lexical nodes, namely words. Such lexical information can be automatically acquired though a distributional analysis of texts. The $\sigma(n_1, n_2)$ function between lexical item is measured within a Word Space. As in [8] the Grammatical Relation Centered Tree[4] (GRCT) is used to represent the syntactic information. Grammatical Relations are central nodes from which dependencies are drawn and all the other features of the central node, i.e. lexical surface form and its POS-Tag, are added as additional children.

A kernel combination αBOWK + βLSK + γSPTK combines the lexical properties of BOWK (generalized by LSK) and the syntactic information captured by the SPTK.

4   Experimental Evaluations

Experimental evaluations over the Twitter Sentiment Analysis [34] and Question Classification [36] tasks are described. We aim at verifying that: (i) the combination of the proposed online learning schema with *Fairness* and *Weight*

---

[4] Notice that in [8] other trees are presented. In this work we selected the GRCT as it provides the most explicit syntactic information.

Adjustment techniques achieves results comparable with batch ones; (ii) the number of kernel computation can be reduced with reasonable budgets.

4.1 Sentiment Analysis in Twitter

Web 2.0 and Social Network technologies allow users to generate contents on blogs, forums and new forms of communication (such as micro-blogging) writing their opinion about facts, things, events. Twitter[5] represents an intriguing source of information as it is used to share opinions and sentiments about brands, products, or situations [14]. Tweet analysis represents a challenging task for Data Mining and Natural Language Processing systems: tweets are short, informal and characterized by their own particular language.

When applied to tweets, traditional approaches to Sentiment Analysis [26] show significant performance drops as they use to focus on larger and well written texts, e.g. product reviews. Some recent works tried to model the sentiment in tweets [25,16,9,1]. Specific approaches are used, such as hand coded resources and artificial feature modeling, in order to achieve good accuracy levels. To assure a fast deployment of robust systems, our approach exploits kernel formulations to enable learning algorithms to extract the useful information. We applied a Multi-Kernel approach considering the BOWK and LSK above described. The Word Space used within the LSK is acquired through the analysis of a generic corpus made of 3 million of tweets. We built a matrix M, whose rows are vectors of pairs ⟨lemma, POS⟩ of the downloaded tweets. Columns of M are the contexts of such words in a short window ($[-3, +3]$) to capture paradigmatic relations. The most frequent 10,000 items are selected along with their left and right 20k contexts. Point-wise-mutual information is used to score M entries. Finally, the SVD reduction is applied, with a dimensionality cut of $k = 250$. We used only lexical information, as the low quality of parse trees of tweets [11] would compromise the Tree Kernel contribution.

A pre-processing stage is applied to reduce data sparseness thus improving the generalization capability of learning algorithms. In particular, a normalization step is applied on the text: fully capitalized words are converted in lowercase; reply marks are replaced with the pseudo-token USER, hyperlinks by LINK, *hashtags* by HASHTAG and emoticons by special tokens[6]; any character repeated more than three times are normalized (e.g. "*nooo!!!!!*" is converted into "*noo!!*"). Then, an almost standard NLP syntactic processing chain [3] is applied. Evaluation is carried out on the SemEval-2013 Task 2 corpus [34]. In particular, we focus on the *Message Polarity Classification*: it deals with the classification of a tweet with respect to three classes *positive*, *negative* and *neutral*. Training dataset is composed by $10,205$ annotated examples, while test dataset is made of $3,813$ examples[7]. Classifier parameters are tuned with a Repeated Random Sub-sampling

---

[5] http://www.twitter.com
[6] We normalized 113 well-known emoticons in 13 classes.
[7] Training dataset is made of $3,786$ positive, $1,592$ negative and $4,827$ neutral examples; test dataset is made of $1,572$ positive, 601 negative and $1,640$ neutral.

Validation, consisting in a 10-fold validation strategy on a subset of the training data split according to a 70%-30% proportion. As results may depend on the order of training data, 10 different models are acquired on 10 shuffled training data.

Table 1. Sentiment Analysis in Twitter Results. *Saving* is the percentage of kernel computations avoided with respect to a SVM classifier implemented in SvmLight [15]. SVM requires 141 million of kernel computations, achieving 0.654 F1. Notice that BPA/AF-BPA *Saving* is the same, as Adjustment does not add any new kernel computation.

Budg.	BPA				AF-BPA				Saving BPA	Saving AF-BPA
	BOW_{lin}	LSK_{lin}	BOW_{lin} + LSK_{lin}	BOW_{lin} + LSK_{rbf}	BOW_{lin}	LSK_{lin}	BOW_{lin} + LSK_{lin}	BOW_{lin} + LSK_{rbf}	BOW_{lin} + LSK_{rbf}	
100	.396±.04	.451±.04	.444±.08	.430±.06	.394±.03	.346±.07	.456±.06	**.471±.03**	98%	98%
250	.408±.03	.469±.03	.462±.06	.444±.06	.418±.03	.454±.06	**.505±.02**	.438±.06	95%	95%
500	.443±.02	.480±.02	.513±.03	**.516±.04**	.412±.04	.459±.06	.503±.05	**.516±.04**	89%	89%
750	.448±.03	.495±.03	.500±.04	.503±.06	.426±.04	.451±.06	**.542±.04**	.527±.03	84%	84%
1000	.446±.02	.488±.02	.517±.03	.525±.03	.475±.02	.511±.04	.542±.03	**.552±.02**	80%	80%
1500	.448±.04	.496±.04	.516±.04	.506±.05	.471±.02	.517±.04	**.555±.02**	.591±.03	71%	70%
2000	.463±.03	.494±.03	.530±.04	.520±.05	.490±.03	.564±.02	**.585±.02**	.572±.02	62%	62%
2500	.478±.03	.501±.03	.525±.04	.532±.04	.498±.03	.585±.01	**.599±.01**	.595±.03	55%	54%
3000	.484±.02	.496±.02	.548±.05	.544±.05	.541±.02	.582±.01	.600±.02	**.605±.03**	49%	47%
3500	.496±.01	.521±.01	.546±.05	.568±.04	.548±.01	.574±.01	.609±.01	**.619±.01**	43%	41%
4000	.496±.03	.513±.03	.559±.05	.572±.04	.556±.02	.573±.01	.621±.01	**.623±.01**	38%	35%
4500	.504±.01	.538±.01	.564±.05	.574±.05	.569±.02	.573±.02	.618±.02	**.628±.01**	35%	30%
5000	.505±.02	.535±.02	.572±.04	.579±.04	.576±.01	.574±.01	**.610±.01**	.604±.02	32%	24%
6000	.505±.02	.543±.02	.581±.04	.593±.04	.545±.01	.573±.01	.626±.01	**.633±.01**	28%	18%
7000	.510±.02	.548±.02	.580±.04	.595±.04	.576±.01	.572±.01	.634±.01	**.640±.01**	28%	14%
8000	.516±.02	.550±.02	.580±.04	.597±.04	.575±.01	.574±.01	.635±.01	**.643±.01**	28%	14%
9000	.516±.02	.550±.02	.580±.04	.597±.04	.578±.01	.574±.01	.636±.01	**.643±.01**	28%	14%
10000	.516±.02	.550±.02	.580±.04	.597±.04	.577±.01	.575±.01	.636±.01	**.642±.01**	28%	14%
-	.516±.02	.550±.02	.580±.04	.597±.04	.577±.01	.575±.01	.635±.01	**.643±.01**	28%	14%

In Table 1 results of different kernels and different budget values are reported. We investigated the contribution of a linear kernel on top of BOWK (BOW_{lin}), and latent representation (LSK_{lin}) as well as their linear combination ($BOW_{lin} + LSK_{lin}$). Moreover, a combination considering a gaussian kernel over the latent semantic representation is investigated ($BOW_{lin} + LSK_{rbf}$). Results are reported in terms of mean F1-Measure between the positive and negative classes, consistently with results reported in [34]: the same metric used in SemEval challenge. 10 different models have been trained by shuffling the training set and the result mean is reported. We report also the standard deviation of the F1 to verify the stability of the proposed solution with respect to training order. The standard BPA and the AF-BPA learning are here considered. All kernels benefits from the adoption of AF-BPA in terms of pure performance, i.e. the mean F1-Measure improves of 5 F1 points. For almost all budget values the standard deviation is lower, so reducing the dependence on training order. Moreover, AF-BPA overtakes best results of BPA at low budget levels, i.e. 3000. Due to lack of space, A-BPA and F-BPA results are not reported; however, their performance is between the BPA and the AF-BPA, demonstrating that both Fairness and Weight Adjustment individually have a valuable contribution.

As a comparison, we trained a SVM classifier: it achieves 0.654 F1 after about 141 millions of kernel computations[8]. The AF-BPA achieves 0.643 F1, with a performance drop of only 1.6%. It is straightforward considering that all advantages of online learning are preserved. At $B = 4000$ computational cost is reduced, saving about 35% kernel computations with respect to SVM, achieving 0,623 F1.

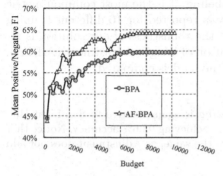

Fig. 1. Mean Positive/Negative F1

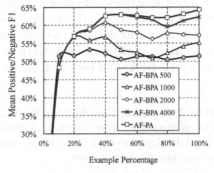

Fig. 2. Learning curve

Performances of AF-BPA and BPA are shown in Figure 1 where the improvement of the proposed method is noticeable since low budgets. We also report the generalization capability in poor training conditions, summarized by the learning curve in Figure 2. Performance at $B = 4000$ is close to the unbudgeted version at different training set size. Results are promising even with smaller budgets, e.g. $B = 1000$. The lack of performance drop (when the budget is "full") is important here, as it suggests that the model would be stable through time, adapting itself to shifting concepts. Finally, the AF-BPA and the used kernels are competitive with respect to state-of-the art systems that participated in SemEval-2013 Task [34]. The AF-BPA would have ranked 4th over 36 systems, with respect to systems trained with the same conditions, i.e. not using external annotated material. It is straightforward, considering that better systems in the challenge used hand-made lexicon [34]. Limiting the number the support vectors to $B = 4000$ the AF-BPA would have ranked $7^t h$.

4.2 Question Classification

Question Classification (QC) is usually applied in Question Answering systems to map the question into one of k classes of answers, thus constraining the search. In these experiments, we used the UIUC dataset [19]. It is composed by a training set of 5,452 and a test set of 500 questions[9], organized in 6 classes (like ENTITY or

---

[8] We compare SVM and OL by the number of kernel computations, as the computational cost of the learning algorithm itself is negligible if compared to novel kernel computations. More details on specific kernels in [7,8].

[9] http://cogcomp.cs.illinois.edu/Data/QA/QC/

HUMAN). It has been already shown the contribution of (structured) kernel based learning within batch algorithms for this task, since [36,23]. Here, we want to study an online setting with a kernel combination αBOWK + βLSK + γSPTK to robustly combine lexical and syntactic information[10]. Lexical similarity within LSK and SPTK is derived through the distributional analysis on the UkWaC [2] corpus. The same classifier and Word Space settings of the Sentiment Analysis scenario is here applied. In Table 2 results are reported in terms of Accuracy, i.e. the percentage of test examples obtaining a correct labeling; it is the most common metric adopted in QC. Again, the mean Accuracy is reported for 10 different training shuffles. Standard deviation as indicator of the robustness of our solution is reported. Here, traditional BPA, the contribution of Weight Adjustment (A-BPA), Fairness (F-BPA) and both (AF-BPA) are the models compared.

Table 2. Question Classification Results. *Saving* is again relative to SvmLight [15]. SVM requires 40 million of kernel computations, achieving 93,6% accuracy. Notice that BPA/A-BPA and F-BPA/AF-BPA *Savings* are the same, as Adjustment does not add any new kernel computation.

Budget	Standard			withFairness		
	BPA	A-BPA	Saving	F-BPA	AF-BPA	Saving
100	**79,2%±3,9%**	73,3%±4,9%	92%	75,5%±5,6%	64,1%±5,7%	92%
250	**82,1%±2,4%**	74,4%±8,6%	81%	80,3%±3,0%	79,2%±5,9%	81%
500	86,9%±1,9%	82,5%±4,0%	66%	**87,5%±1,7%**	83,4%±4,4%	66%
750	87,3%±2,3%	84,8%±3,8%	53%	**88,7%±1,7%**	87,9%±1,3%	53%
1000	86,2%±3,8%	86,9%±2,4%	44%	88,3%±1,5%	**88,5%±1,1%**	42%
1500	87,4%±2,9%	87,5%±1,4%	32%	89,6%±0,5%	**90,8%±0,6%**	30%
2000	88,1%±2,4%	88,1%±1,7%	29%	90,1%±0,6%	**91,1%±0,5%**	24%
3000	87,9%±2,1%	88,2%±1,7%	28%	90,1%±0,5%	**90,8%±0,4%**	21%
4250	87,9%±2,1%	88,2%±1,7%	28%	90,1%±0,5%	**90,8%±0,5%**	21%
-	87,9%±2,1%	88,2%±1,7%	28%	90,1%±0,5%	**90,9%±0,4%**	21%

Results confirm that Fairness improves the discrimination between the 6 classes by an absolute 3%, even if a significant imbalance is faced[11]. Accuracy also benefits from the adoption of Weight Adjustment, as shown by the improvements of A-BPA and AF-BPA. Adjustment is also useful to reduce standard deviation, especially with respect to standard BPA. Moreover, AF-BPA reaches better results at low budget levels, i.e. 750. As a comparison, we trained a SVM classifier, which achieves a 93,6% after about 40 millions of kernel computations. The AF-BPA shows only a 3% accuracy drop with a 30% reduction of kernel computations, and a 6% with 53% reduction, i.e. only $B = 750$. When Adjustment and Fairness are applied within extremely low budgets, performance drops. Fairness tends to assign higher weights to less represented classes: when only few examples can be stored, the classifier tends to prefer such classes, increasing the overall error probability. Weight Adjustment instead leads to data over-fitting acting on few Support Vectors and to worse results.

---

[10] In our experiments, α, β and γ are set to 1.

[11] Examples in **Entity** class are 1250 while **Abbreviation** are 86.

5   Conclusion

In this paper, an extension of the Budgeted Passive Aggressive Algorithm is proposed to enable robust and efficient Natural Language Learning processes based on semantic kernels. The proposed principles of Fairness and Weight Adjustment obtain results that are close to the state-of-the-art achieved by batch versions of the same algorithm. This confirms that the proposed OL learner can be applied to real world linguistic tasks. Opinions and sentiments in fact change over time and are strongly language-driven resulting as highly volatile and dynamic. Batch learning is unsuited in these cases as full re-training would be infeasible most of the times. The ability to continuously learn binding the overall complexity, coupled with the Weight Adjustment stage, allows to preserve accuracy by saving in computational costs: in Question Classification we observed only a 6% drop with a 53% saving with respect to SVM. Future work will address concept shifts against ad-hoc datasets. Moreover, extension to other Kernels, or Kernel Combinations, is foreseen [12].

References

1. Agarwal, A., Xie, B., Vovsha, I., Rambow, O., Passonneau, R.: Sentiment analysis of twitter data. In: Proceedings of LASM, pp. 30–38 (2011)
2. Baroni, M., Bernardini, S., Ferraresi, A., Zanchetta, E.: The wacky wide web: a collection of very large linguistically processed web-crawled corpora. Language Resources and Evaluation 43(3), 209–226 (2009)
3. Basili, R., Zanzotto, F.M.: Parsing engineering and empirical robustness. Nat. Lang. Eng. 8(3), 97–120 (2002)
4. Cesa-Bianchi, N., Gentile, C.: Tracking the best hyperplane with a simple budget perceptron. In: Lugosi, G., Simon, H.U. (eds.) COLT 2006. LNCS (LNAI), vol. 4005, pp. 483–498. Springer, Heidelberg (2006)
5. Collins, M., Duffy, N.: Convolution kernels for natural language. In: Proceedings of Neural Information Processing Systems (NIPS 2001), pp. 625–632 (2001)
6. Crammer, K., Dekel, O., Keshet, J., Shalev-Shwartz, S., Singer, Y.: Online passive-aggressive algorithms. Journal of Machine Learning Research 7, 551–585 (2006)
7. Cristianini, N., Shawe-Taylor, J., Lodhi, H.: Latent semantic kernels. J. Intell. Inf. Syst. 18(2-3), 127–152 (2002)
8. Croce, D., Moschitti, A., Basili, R.: Structured lexical similarity via convolution kernels on dependency trees. In: Proceedings of EMNLP, Scotland, UK (2011)
9. Davidov, D., Tsur, O., Rappoport, A.: Enhanced sentiment learning using twitter hashtags and smileys. In: COLING, pp. 241–249 (2010)
10. Dekel, O., Shalev-Shwartz, S., Singer, Y.: The forgetron: A kernel-based perceptron on a budget. SIAM J. Comput. 37(5), 1342–1372 (2008)
11. Foster, J., Çetinoglu, Ö., Wagner, J., Roux, J.L., Hogan, S., Nivre, J., Hogan, D., van Genabith, J.: #hardtoparse: Pos tagging and parsing the twitterverse. In: Analyzing Microtext (2011)
12. Gönen, M., Alpaydin, E.: Multiple kernel learning algorithms. Journal of Machine Learning Research 12, 2211–2268 (2011)
13. Jaakkola, T., Meila, M., Jebara, T.: Maximum entropy discrimination. In: Solla, S.A., Leen, T.K., Müller, K.R. (eds.) NIPS, pp. 470–476. The MIT Press (1999)

14. Jansen, B.J., Zhang, M., Sobel, K., Chowdury, A.: Twitter power: Tweets as electronic word of mouth. J. Am. Soc. Inf. Sci. Technol. 60(11), 2169–2188 (2009)
15. Joachims, T.: Learning to Classify Text Using Support Vector Machines. Kluwer Academic Publishers (2002)
16. Kouloumpis, E., Wilson, T., Moore, J.: Twitter sentiment analysis: The good the bad and the omg? In: ICWSM (2011)
17. Kwok, C.C., Etzioni, O., Weld, D.S.: Scaling question answering to the web. In: World Wide Web, pp. 150–161 (2001)
18. Landauer, T., Dumais, S.: A solution to plato's problem: The latent semantic analysis theory of acquisition, induction and representation of knowledge. Psychological Review 104 (1997)
19. Li, X., Roth, D.: Learning question classifiers: the role of semantic information. Natural Language Engineering 12(3), 229–249 (2006)
20. Littlestone, N.: Learning quickly when irrelevant attributes abound: A new linear-threshold algorithm. In: Machine Learning, pp. 285–318 (1988)
21. Morik, K., Brockhausen, P., Joachims, T.: Combining statistical learning with a knowledge-based approach - a case study in intensive care monitoring. In: ICML, pp. 268–277. Morgan Kaufmann Publishers Inc., San Francisco (1999)
22. Moschitti, A., Pighin, D., Basili, R.: Tree kernels for semantic role labeling. Computational Linguistics 34 (2008)
23. Moschitti, A., Quarteroni, S., Basili, R., Manandhar, S.: Exploiting syntactic and shallow semantic kernels for question/answer classification. In: Proceedings of ACL 2007 (2007)
24. Orabona, F., Keshet, J., Caputo, B.: The projectron: a bounded kernel-based perceptron. In: Proceedings of ICML 2008, pp. 720–727. ACM, USA (2008)
25. Pak, A., Paroubek, P.: Twitter as a corpus for sentiment analysis and opinion mining. In: LREC (2010)
26. Pang, B., Lee, L.: Opinion mining and sentiment analysis. Found. Trends Inf. Retr. 2(1-2), 1–135 (2008)
27. Rosenblatt, F.: The perceptron: A probabilistic model for information storage and organization in the brain. Psychological Review 65(6), 386–408 (1958)
28. Sahlgren, M.: The Word-Space Model. Ph.D. thesis, Stockholm University (2006)
29. Salton, G., Wong, A., Yang, C.: A vector space model for automatic indexing. Communications of the ACM 18 (1975)
30. Shawe-Taylor, J., Cristianini, N.: Kernel Methods for Pattern Analysis. Cambridge University Press, New York (2004)
31. Van Hulse, J., Khoshgoftaar, T.M., Napolitano, A.: Experimental perspectives on learning from imbalanced data. In: Proceedings of the ICML. ACM, USA (2007)
32. Vapnik, V.N.: Statistical Learning Theory. Wiley-Interscience (1998)
33. Wang, Z., Vucetic, S.: Online passive-aggressive algorithms on a budget. Journal of Machine Learning Research - Proceedings Track 9, 908–915 (2010)
34. Wilson, T., Kozareva, Z., Nakov, P., Ritter, A., Rosenthal, S., Stoyonov, V.: Semeval-2013 task 2: Sentiment analysis in twitter. In: Proceedings of the 7th International Workshop on Semantic Evaluation (2013)
35. Zanzotto, F.M., Pennacchiotti, M., Moschitti, A.: A machine learning approach to textual entailment recognition. Natural Language Engineering 15-04 (2009)
36. Zhang, D., Lee, W.S.: Question classification using support vector machines. In: Proceedings of SIGIR 2003, pp. 26–32. ACM, New York (2003)

On Inverted Index Compression
for Search Engine Efficiency

Matteo Catena[1], Craig Macdonald[2], and Iadh Ounis[2]

[1] GSSI - Gran Sasso Science Institute, INFN
Viale F. Crispi 7, 67100 L'Aquila, Italy
matteo.catena@gssi.infn.it
[2] School of Computing Science, University of Glasgow,
Glasgow G12 8QQ, UK
{craig.macdonald,iadh.ounis}@glasgow.ac.uk

Abstract. Efficient access to the inverted index data structure is a key aspect for a search engine to achieve fast response times to users' queries. While the performance of an information retrieval (IR) system can be enhanced through the compression of its posting lists, there is little recent work in the literature that thoroughly compares and analyses the performance of modern integer compression schemes across different types of posting information (document ids, frequencies, positions). In this paper, we experiment with different modern integer compression algorithms, integrating these into a modern IR system. Through comprehensive experiments conducted on two large, widely used document corpora and large query sets, our results show the benefit of compression for different types of posting information to the space- and time-efficiency of the search engine. Overall, we find that the simple Frame of Reference compression scheme results in the best query response times for all types of posting information. Moreover, we observe that the frequency and position posting information in Web corpora that have large volumes of anchor text are more challenging to compress, yet compression is beneficial in reducing average query response times.

1 Introduction

The ubiquitous *inverted index* data structure remains a key component of modern search engines [1]. Indeed, for each unique indexed term, the inverted index contains a *posting list*, where each *posting* contains the occurrences information (e.g. frequencies, and positions) for documents that contain the term. To rank the documents in response to a query, the posting lists for the terms of the query must be traversed, which can be costly, especially for long posting lists.

Different orderings of the postings in the posting lists change both the algorithm for ranked retrieval and the underlying representation of postings in the inverted index, such as how they are compressed. For instance, the postings for each term can be sorted in order of impact, allowing ranked retrieval to be short-circuited once enough documents have been retrieved [2]. However, search engines reportedly [1, 3, 4] use the traditional static *docid* ordering, where each posting list is ordered by ascending document id, which permits a reduced inverted index size and efficient retrieval [5]. The appropriate compression of this classical docid ordering of posting lists is the focus of our work.

M. de Rijke et al. (Eds.): ECIR 2014, LNCS 8416, pp. 359–371, 2014.

Various schemes have been proposed to ensure the time- and space-efficient compression of the inverted index posting lists. Indeed, as ranking is data-intensive due to the traversing of posting lists, maximising decompression speed while minimising the size of the posting lists is necessary to ensure quick query response times. As docid ordered posting lists contain integer numbers, such compression schemes (or *codecs*) can be operated on a stream of postings – known as *oblivious* – (for example, Elias-Gamma [6]), or on groups of postings – known as *list-adaptive* [7] – (e.g. PForDelta [8]). There are a few comparisons of such integer compression schemes on inverted indexes. Yan et al. [9] analysed the compression ratio and decompression speed of many schemes while dealing with the compression of doc ids and term frequencies extracted from the GOV2 corpus using the AOL query log dataset. A similar work has been done by Lemire et al. [10], studying the codecs behaviour for the ClueWeb09 corpus, but focusing just on document ids. While dealing with smaller datasets, Delbru et al. [11] provided a comparison of codecs integrating these into an actual search engine with positional information. However, missing from the literature is a study that both deploys compression codecs into a realistic information retrieval (IR) system and measures performances on different, large and well-known document corpora and query sets. Moreover, the benefit of different compression algorithms for each of the different types of *payload* data within a posting – ids, term and field frequencies, positions – has not been comprehensively studied.

For the above reasons, this paper provides a thorough comparison and analysis of the response time and index size benefits of modern posting list compression codecs, deployed within an actual modern search engine platform – namely Terrier [12][1] – and examining different posting payload information. Indeed, the contributions of this paper are three-fold: Firstly, we separately analyse the compression benefits of document ids, term frequencies, field frequencies and positions, in terms of average query response time and index size; Secondly, we analyse how different term distributions caused by anchor text affect the achievable compression; Finally, we use these thorough experiments on two standard corpora, namely GOV2 and ClueWeb09 to derive best practices for index compression. The results of our study demonstrate that the contribution of compression in time- and space-efficiency varies greatly depending on the type of posting payload information, while leading us to suggest the use of the simpler Frame of reference (FOR) [24] list-adaptive algorithms for the best benefit to average query response times.

This paper is structured as follows: Section 2 provides a background on efficient ranked retrieval; Section 3 summarises the compression codecs we analyse later within an IR system; Section 4 describes our experimental setup. Sections 5 through 8 show our experimental results; Conclusions and final remarks follow in Section 9.

2   Background

While the effectiveness of a search engine at satisfying the users' information needs is critical, efficiency is also a key aspect for IR, not least because techniques that typically enhance effectiveness can degrade the efficiency of the search engine, i.e. its ability to respond to user queries in a timely fashion [13]. Indeed, efficiency is also important to a search engine, as users are not willing to wait long for queries to be answered [14].

---

[1] http://terrier.org

The architecture of a modern IR system can be roughly divided into three layers [18]: the index with its corresponding *(de)compression* layer; the *matching* layer, which identifies documents to rank from the inverted index; and the uppermost *re-ranking* layer, which applies features and learning to rank techniques to obtain the final ranking of documents.

Techniques to enhance the efficiency of an IR system have been proposed at each layer. For instance, at the matching layer, dynamic pruning techniques such as WAND [15] enhance efficiency by omitting the scoring of documents that cannot reach the final retrieved set. In the top-most re-ranking layer, Cambazoglu et al. [16] showed how learning to rank models could be simplified to enhance their efficiency. Similarly, Wang et al. showed how various features could be avoided [13] or their efficiency cost reduced [17] while minimising any negative impact on effectiveness.

On the other hand, in this paper, we focus on the lowest architecture layer, and in particular, on the compression of the posting lists within the inverted index. Each posting contains a *payload* set of integers, representing the occurrence information of a term within a given document. Indeed, a typical posting contains the docid of the matching document, the frequency of occurrence in the document (often within different *fields*, such as title, URL, body, and anchor text [18]), and – to facilitate phrasal and proximity matching – *positional* information (as a set of ascending integers for *each* posting).

The compression of the posting lists within the inverted index ensures that as much as possible of the inverted index can be kept in the higher levels of the computer memory hierarchy – indeed, many search engines reportedly keep the entire inverted index in main memory [1], as in our experiments. Hence, a compression scheme should not only be time-efficient (i.e. inexpensive to decompress), but also space-efficient (i.e. high compression), to minimise the necessary required computing resources while answering queries. While a range of existing compression schemes or *codecs* have been proposed in the literature (see review in the next section), a comprehensive study addressing their actual impact upon search engine retrieval efficiency has not previously been addressed in the literature. Indeed, while several compression schemes were compared in [9] for document ids and term frequencies extrapolated from the GOV2 corpus against the AOL query log dataset, their study was primarily focused on the decompression speed in terms of integers per second, instead of the resulting impact on query response time. Moreover, they did not consider other posting payload information, such as field frequencies and positions, nor did they investigate how codecs behave on different corpora. On the other hand, while Lemire et al. [10] did investigate compression codecs on both the GOV2 and ClueWeb09 corpora, they were focused only on docid compression, and did not analyse the impact on search engine efficiency in terms of query response time. Finally, while Delbru et al. [11] integrated different compression codecs into an actual IR system with document ids, term frequencies and positional information, their experiments with synthetically generated queries do not necessarily reflect a realistic search engine workload. Hence, in this work, we review a wide-range of posting list compression codecs from the recent literature (Section 3) and later thoroughly empirically compare their time- and space-efficiency for a realistic query workload, across different posting payload information (Sections 5-8).

Table 1. The presented codecs divided by features

Codec	Bitwise	Byte aligned	Word aligned	Oblivious	List-adaptive
Unary/Gamma	✓			✓	
Golomb/Rice	✓				✓
Variable-byte		✓		✓	
Simple family			✓		✓
FOR/PFOR			✓		✓

3   Compression Techniques

Inverted index compression has been common for some time. For example, one common practice while storing a posting list is to use *delta-gaps* (or *d-gaps*) where possible [5], i.e. to record the differences between monotonically increasing components (such as ids or positions) instead of their actual values. Indeed, by using notations different from binary word-aligned, smaller numbers lead to smaller representations.

Many different compression algorithms or *codecs* have been designed to encode integers. We can distinguish codecs using attributes such as the nature of their output, in terms of bitwise, byte-aligned or word-aligned. Moreover, we differentiate between codecs that compress every value on its own, and those which compress values in groups, called *oblivious* and *list-adaptive* methods respectively [7]. In the following, we provide a short description of the oblivious (Section 3.1) and list-adaptive (Section 3.2) codecs that we analyse in this paper. Their attributes are summarised in Table 1.

3.1   Oblivious Codecs

When a set of integers is given to an oblivious codec for compression, it encodes each value on its own, without considering its value relative to the rest of the set. A desirable side effect is that every single value can be decompressed separately, or by the decompression of just the preceding values if d-gaps are used. On the other hand, such codecs ignore global information about the set, which can help to attain a better compression ratio. The oblivious compression algorithms we deploy in our experiments are briefly described below.

Unary and Gamma: Unary and Gamma codecs are two bitwise, oblivious codecs. Unary represents an integer x as $x - 1$ one bits and a zero bit (ex.: 4 is 1110). While this can lead to extremely large representations, it is still advantageous to code small values. Gamma, described in [6], represents x as the Unary representation of $1 + \lfloor \log_2 x \rfloor$ followed by the binary representation of $x - 2^{\lfloor \log_2 x \rfloor}$ (ex.: 9 is 1110 001).

Variable Byte: Variable byte codec [19] is a byte-aligned, oblivious codec. It uses the 7 lower bits of any byte to store a partial binary representation of the integer x. It then marks the highest bit as 0 if another byte is needed to complete the representation, or as 1 if the representation is complete. For example, 201 is 10000001 01001001. While this codec may lead to larger representations, it is usually faster than Gamma in term of decompression speed [20].

3.2   List-Adaptive Codecs

A list-adaptive codec compresses integers in blocks, exploiting aspects such as the prox-imity of values in the compressed set. This information can be used to improve com-pression ratio and/or decompression speed. However, this also means that an entire block must be decompressed even when just a single posting is required from it (e.g. for partial scoring approaches such as WAND [15]). Moreover, it is possible to obtain a larger output than the input when there are not enough integers to compress, because extra space is required in the output to store information needed at decompression time. When there are too few integers to be compressed, this header information can be larger in size than the actual payload being compressed. Below, we provide short descriptions of the list-adaptive codecs that we investigate in this work.

Golomb and Rice: Golomb codec is a bitwise and list-adaptive compression scheme [21]. Here, an integer x is divided by b. Then, Unary codec is used to store the quotient q while the remainder r is stored in binary form. b is chosen depending on the integers being compressed. Usually, $b = 0.69 * avg$ where avg is the average value of the numbers being treated [5]. In the Rice codec [22], b is a power of two, which means that bitwise operators can be exploited, permitting more efficient implementations at the cost of a small increase in the size of the compressed data. Nevertheless, Golomb and Rice coding are well-known for their decompression inefficiency [7, 10, 23], and hence, we omit experiments using these codecs from our work.

Simple Family: This family of codecs, firstly described in [23], stores as many integers as possible in a single word. It is possible using the first 4 bits of a word to describe the organisation of the remaining 28 bits. For example, in a word we can store $\{509, 510, 511\}$ as three 9-bits values, with the highest 4 bits of the word reflecting this configuration, at a cost of one wasted bit.

Frame Of Reference (FOR): Proposed by Goldstein et al. [24], FOR compresses in-tegers in blocks of fixed size (e.g. 128 elements). It computes the gap between the maximum M and the minimum m value in the block, then stores m in binary notation. The other elements are saved as the difference between them and m using b bits each, where $b = \lceil \log_2(M + 1 - m) \rceil$.

Patched Frame Of Reference (PFOR): FOR may lead to a poor compression in pres-ence of outliers: single large values that force an increase of the bit width b on all the other elements in the block. To mitigate this issue, Patched frame of reference (PFOR) has been proposed [8]. This approach chooses b to be reasonable for most of the ele-ments in the block, treating these as in FOR. The elements in a range larger than 2^b are treated as *exceptions*. In the original approach, those are stored at the end of the output, not compressed. The unused b bits are used to store the position of the next exception in the block. If b bits are not enough to store the position of the next exception, an ad-ditional one is generated. This means that one of the values is treated as an exception even if it could have been encoded using b bits.

More recent implementations of PFOR treat the exceptions differently. NewPFD [9] stores the exception positions in a dedicated area of its output, but divides them in two parts: b bits are normally stored, as for the normal values, while the remaining $32-b$ bits are compressed using a Simple family codec. OptPFD [9] works similarly, but chooses b to optimise the compression ratio and the decompression speed. FastPFOR [10], instead, reserves 32 different areas of its output to store exceptions. Each area contains those exceptions that can be encoded using the same number of bits. Outliers in the same exception area are then compressed using a FOR codec, to improve both the compression ratio and the decompression speed.

As discussed above, a thorough empirical study of compression codecs for realistic search engine workloads remains missing from the literature. Hence, in this paper, we experiment to address the following research question: How does the compression of (a) document ids, (b) term frequencies, (c) field frequencies and (d) term position information within inverted index posting lists affect the response times of a search engine under a realistic query load.

To address this research question, we integrate the above discussed compression codecs within a single modern IR system, and measure both the compression achieved, as well as its efficiency in retrieving a realistic set of queries, on two standard widely used research corpora. In the next section, we describe the experimental setup used to achieve this comparison.

4 Experimental Setup

We now discuss our experimental setup in terms of the IR system, corpora and queries (Section 4.1) as well as the setup of the compression codecs (Section 4.2).

4.1 IR System, Documents and Queries

We experiment using the compression codecs in Table 1 (except Golomb/Rice) by analysing their compression and decompression of inverted indices for two standard TREC corpora, namely the GOV2 and ClueWeb09 (cat. B) corpora. The GOV2 corpus of about ~25M documents represents a search engine for the .gov domain, while the ClueWeb09 corpus of ~50M Web documents is intended to represent the first tier of a general Web search engine. Each corpus is indexed using the Terrier IR platform [12], saving positions of terms in the documents and their frequencies for four fields in each document, namely title, the URL, the body and the anchor text of incoming hyperlinks. All terms have the Porter stemmer applied, and stopwords have been removed. The statistics of the resulting indices are shown in Table 2. In line with common practices [9], both corpora are reordered such that docids are assigned lexicographically by URL. Finally, d-gaps are used to represent document ids and term positions in postings for all compression codecs.

For retrieval, we use BM25 to retrieve 1000 documents using an exhaustive DAAT retrieval strategy[2]. Indeed, such a setting represents the first stage of a typical retrieval infrastructure, where additional features (e.g. proximity) would later be computed for application by a learned model [18]. For experiments on the GOV2 corpus, we use the

---

[2] We avoid applying dynamic pruning techniques such as WAND, to prevent skipping becoming a confounding variable within our experiments.

Table 2. Statistics about the GOV2 and ClueWeb09 corpora, with the average standard deviations ($\bar{\sigma}$) for values computed for blocks of 1024 postings

	GOV2		ClueWeb09	
	#	$\bar{\sigma}$	#	$\bar{\sigma}$
Number of documents	25,205,178	-	50,220,423	-
Number of postings	4,691,479,151	-	12,725,738,385	-
Number of positions	16,581,352,732	100.10	37,221,252,036	234.64
Number of Tokens	16,794,511,752	7.73	52,031,310,320	26.10
for title field	131,519,548	0.08	326,022,392	0.19
for URL field	426,062,666	0.05	356,811,926	0.10
for body field	15,612,059,739	5.66	26,046,648,801	4.38
for anchor field	624,869,799	2.46	11,024,667,766	23.73

first 1000 queries from the TREC 2006 Terabyte track efficiency task - these queries were sampled from a search engine query log with clicks into the .gov domain. For ClueWeb09, we use the first 1000 queries from the MSN 2006 query log. The average lengths of the queries in the two sets are 3.4 and 2.4, respectively.

Experiments are conducted on a dedicated, otherwise idle, 8-core (@2.13GHz) Intel Xeon processor, with 32GB RAM. The posting lists related to the query terms are loaded into main memory at the beginning of each experiment. We measure the size of the compressed inverted index, as well as the direct query response times experienced by the IR system while retrieving the query sets. In order to warm-up the execution environment, each query set is run ten times for each experiment, and the response times only measured for the last run. The mean response times for each corpus exhibited by Terrier with its baseline compression are 1.34 seconds for GOV2, and 1.55 for ClueWeb09. While these are higher than would be permissible for an interactive retrieval system, they reflect a fair comparison without resorting to distributed retrieval strategies, and without the network latencies that would then be inherent in measuring response times. Moreover, we note that while each of our postings contains four additional term frequencies for each field compared to those reported in [13], our response times are markedly faster, despite Wang et al. using the Terrier compression library.

4.2 Compression Codecs

Our baseline compression is represented by the default configuration of Terrier, which uses Gamma codec for the compression of ids and positions and Unary codec for the compression of term frequencies and field frequencies. The remaining codecs are integrated within the Terrier platform, but based on different open source implementations. As a variable byte codec, we use the Apache Hadoop[3] implementation, VInt. For classical PFOR [9, 25], we use the implementation in the Kamikaze project[4]. Finally, the JavaFastPFOR package[5] provides implementations of several codecs, namely FOR, NewPFD, OptPFD, FastPFOR and Simple 16 [26] (a member of the Simple family).

---

[3] http://hadoop.apache.org/
[4] http://sna-projects.com/kamikaze/
[5] https://github.com/lemire/JavaFastPFOR

Table 3. Average query response time μ (in seconds/query) and inverted index size β (in GB) for GOV2 and ClueWeb09. Ids and term frequencies are compressed by the shown codec; field frequencies are always compressed using Unary; positions are always compressed using Gamma. The best improvement in each column is emphasised.

Name	Document Ids								Term frequencies							
	GOV2				ClueWeb09				GOV2				ClueWeb09			
	μ	%	β	%	μ	%	β	%	μ	%	β	%	μ	%	β	%
Baseline	1.34	-	32.38	-	1.55	-	80.91	-	1.34	-	32.38	-	1.55	-	80.91	-
VInt	1.25	-6.7	35.02	8.2	1.56	0.6	84.94	5.0	1.23	-8.2	37.23	15.0	1.69	9.0	95.83	18.4
Simple16	1.24	-7.5	32.57	**0.6**	1.44	-7.1	80.78	**-0.2**	1.23	-8.2	31.65	**-2.3**	1.51	-2.6	81.50	**0.7**
FOR	1.23	-8.2	33.31	2.9	1.40	**-9.7**	81.98	1.3	1.21	**-9.7**	32.21	-0.5	1.50	-3.2	84.23	4.1
PForDelta	1.23	-8.2	33.11	2.3	1.43	-7.7	81.87	1.2	1.24	-7.5	32.29	-0.3	1.53	-1.3	83.59	3.3
NewPFD	1.22	**-9.0**	32.98	1.9	1.42	-8.4	81.30	0.5	1.22	-9.0	32.03	-1.1	1.49	**-3.9**	82.63	2.1
OptPFD	1.22	**-9.0**	32.84	1.4	1.42	-8.4	80.91	0.0	1.22	-9.0	31.98	-1.2	1.50	-3.2	82.46	1.9
FastPFOR	1.23	-8.2	33.23	2.6	1.42	-8.4	81.87	1.2	1.22	-9.0	32.08	-0.9	1.51	-2.6	83.27	2.9

To allow meaningful comparisons of compression ratios, we compress blocks of integers (document ids, term frequencies, field frequencies, positions) separately, even for bitwise oblivious codecs. In our experiments, blocks are *typically* 1024 postings, following [11]. Hence, while iterating over a posting list, our system reads a block of postings, decompresses it and consumes the postings before moving to the next block. However, for small postings lists (and the remainder of each posting list) with less than 1024 elements, block sizes can be smaller, but no less than 128 elements [9]. For blocks smaller than 128, (P)FOR codecs have to *fall back* to other codecs: we use variable byte as per [10].

In the following experiments, we consider four types of the posting list payload (ids, term frequencies, fields, positions) separately. This means that in every experiment run, only one of these payload types is encoded using the analysed compression codec, while the others remain compressed as in the baseline configuration used by Terrier, but still decoded at querying time. Experimental results for document ids, term frequencies, field frequencies and positions follow in Sections 5-8, respectively. Each section evaluates the codecs in terms of the overall inverted index size, as well as in terms of benefit to the average query response times compared to the baseline system.

5   Document ids Compression

In this section, we address the compression of the document ids, while compressing the other components of each posting as in the baseline (Unary for frequencies, Gamma for positions). The left hand side of Table 3 reports the index size (β, in GB) and mean query response times (μ, in seconds) of various compression codecs for both GOV2 and ClueWeb09. Percentage differences from the baseline compression codec (in this case, Gamma for document ids) are also shown.

On analysing Table 3, firstly, for the GOV2 corpus, we observe that each studied codec results in an inverted index larger than the baseline Gamma codec. A similar observation is obtained for ClueWeb09, where only Simple16 reduces the size of the index. Indeed, the byte-aligned VInt codec results in a 5-8% larger index. Overall, the increases in the index size are generally smaller for ClueWeb09 than GOV2. On the other hand, query response times are markedly reduced by all compression codecs (with a single exception of VInt on ClueWeb09). Indeed, using OptPFD and NewPFD, the query response times for GOV2 are reduced by 9%, while the simpler FOR reduces the query response times by 9.7% for ClueWeb09, with other PFOR-based implementations not far behind. However, this is at the cost of an increased index size ranging from $\sim$1.3% (ClueWeb09) to $\sim$3% (GOV2) with respect to the baseline.

Overall, in addressing the first part of our research question concerning document id compression, we find that the (P)FOR-based codecs exhibit the fastest query response times behaviour. Indeed, the efficiency of the PFOR-based codecs are as expected from [10]. However, no response time benefit for FastPFOR is observed, despite its larger index size compared to other PFOR-based codecs.

6 Term Frequencies Compression

The right hand side of Table 3 shows our results for the compression of term frequencies. Document ids, fields and positions remain as in the baseline. On analysing Table 3, we note that differently from the ids compression, it is possible to slightly reduce the inverted index size with respect to the baseline for the GOV2 corpus. In this case, the best compression (β) is obtainable using the Simple16 codec. For the FOR and PFOR codecs, index sizes are decreased (from -0.3% to -1.2% compared to the baseline), with FOR generally performing worse than any other PFOR implementations, with the single exception of PForDelta. However, in contrast, for the ClueWeb09 corpus, all compression codecs actually increase the index size compared to using Unary for the compression of term frequencies. This may be explained by the different distribution of term frequency values in the two corpora. Indeed, since ClueWeb09 is a sample of the general Web, many of its documents have a higher inlink distribution than in GOV2, with a corresponding effect on the distribution of anchor text. This leads to a greater variability of frequency values (see the $\bar{\sigma}$ columns in Table 2) and hence to a more difficult compression.

Next, in terms of average query response time (μ), for GOV2, FOR exhibits the best query response time, although other PFOR codecs are very similar, except the original PForDelta, which is slower than Simple16 and VInt. For the ClueWeb09 corpus, the benefit of using (P)FOR or Simple16 codecs to reduce the average query response time can still be observed. However, this benefit is less than that observed for GOV2. For instance, on GOV2, FOR exhibits a -9.7% reduction in average response time, but only -3.9% for ClueWeb09.

In summary, in addressing the second part of our research question concerning term frequency compression, we have observed that similar to docid compression, more advanced codecs can improve query response times (up to 9.7% faster), however, they may not result in decreased index size. Indeed, our results show that the compression of term frequencies for corpora that encapsulate higher linkage patterns (in our case, ClueWeb09) is challenging. This has not been observed by previous literature, such as [9].

Table 4. Field frequencies and positions are compressed by the codec shown in the table; document ids are always compressed using Gamma; term frequencies are always compressed using Unary. Notation as in Table 3.

| Name | Field frequencies | | | | | | | | Positions | | | | | | | |
| | GOV2 | | | | ClueWeb09 | | | | GOV2 | | | | ClueWeb09 | | | |
	μ	%	β	%	μ	%	β	%	μ	%	β	%	μ	%	β	%
Baseline	1.34	-	32.38	-	1.55	-	80.91	-	1.34	-	32.38	-	1.55	-	80.91	-
VInt	1.29	-3.7	45.69	41.1	2.55	64.5	117.85	45.7	1.32	-1.5	38.51	18.9	1.76	13.5	95.83	18.4
Simple16	1.23	-8.2	32.21	**-0.5**	1.51	-2.6	82.07	**1.4**	1.09	-18.7	35.54	9.8	1.47	-5.2	81.5	**0.7**
FOR	1.19	**-11.2**	32.82	1.4	1.47	**-5.2**	84.98	5.0	1.01	**-24.6**	33.23	2.6	1.36	**-12.3**	84.23	4.1
PForDelta	1.24	-7.5	32.74	1.1	1.53	-1.3	84.17	4.0	1.06	-20.9	34.21	5.7	1.48	-4.5	83.59	3.3
NewPFD	1.19	**-11.2**	32.73	1.1	1.48	-4.5	83.82	3.6	1.01	**-24.6**	33.35	3.0	1.37	-11.6	82.63	2.1
OptPFD	1.20	-10.4	32.71	1.0	1.50	-3.2	83.76	3.5	1.04	-22.4	32.2	-0.6	1.39	-10.3	82.46	1.9
FastPFOR	1.21	-9.7	32.75	1.1	1.49	-3.9	84.40	4.3	1.04	-22.4	31.89	**-1.5**	1.39	-10.3	83.27	2.9

7   Fields Compression

Next, we discuss the compression of field frequencies. Recall that our indices contain separate frequency counts for four different fields, namely title, URL, body and anchor text. We firstly highlight the layout of the compressed postings. In particular, two variants are plausible: (i) compression of the stream of all frequencies, regardless of field; and (ii) compression of each field separately. However, our initial experiments showed a ~3% improvement in the achieved compression by the separate compression of the frequencies of each field, with a corresponding ~4% improvement in response time. These results are intuitive as the distribution of frequencies across different fields are very different, as illustrated by the field statistics in Table 2 - indeed, terms occurring in URLs and titles typically have a frequency of 1, while terms occurring in anchor text and body follow Zipfian distributions, and hence exhibit markedly higher variances. For this reason, the following experiments use the same codec for each field, but separate the actual compression of each field.

The results for the compression of field frequencies are shown in the left hand side of Table 4. For the GOV2 corpus, the compression of field frequencies using modern integer compression codecs shows promising benefits in term of query response time compared to the baseline. However, this is generally at the price of a slight increase in the inverted index size. Indeed, while Simple16 can somewhat reduce the index size and VInt increases it enormously, the FOR and PFOR codecs increase the inverted index size by around ~1% of the baseline. This tradeoff in index size achieves a marked decrease in response time, which ranges from -7% to -11% for the (P)FOR codecs.

For ClueWeb09, the response time benefits and compression size benefits are less marked. The FOR and PFOR codecs are between ~3% and ~5% faster than the baseline, with the only exception of PForDelta, which is below the ~2% improvement attained by Simple16. Moreover, using a basic codec such as VInt for large corpora such as ClueWeb09 is not a feasible option: it increases the average response time by more than 50% ($1.55 \rightarrow 2.55$ seconds in Table 4). Finally, similar to our observation for term

frequencies, the compression of field frequencies on the ClueWeb09 corpus is challenging. Overall, we find that for the compression of field frequencies, the FOR codec gives
the best benefit to query response time, but not the best compression.

8   Positions Compression

The right hand side of Table 4 shows the results for the compression of term positions.
The compression of positions shows the most promising results. In particular, a $\sim-25\%$
decrease in average query response time can be attained for the GOV2 corpus along
with a $\sim-12\%$ decrease for the ClueWeb09 corpus, using FOR. However, we note contrasting observations between GOV2 and ClueWeb09. In particular, most of the codecs
attain better query response time improvements on GOV2 than on ClueWeb09. This
may be explained in the higher amount of anchor text in the ClueWeb09 corpus, which
is also represented in the position information of each posting. Nevertheless, there are
considerable response time benefits compared to the baseline for both corpora, which
is not reported in the previous literature.

We postulate that the compressibility of positions is due to the tendency of terms
to occur in clusters within a document, i.e. if a term occupies a certain position in the
text, there is a higher probability that it appears again soon after [7]. This suits well
the list-adaptive codecs, while it is less exploitable by oblivious codecs. In fact, VInt
demonstrates the smallest decrease in response time, and usually increases the index
size compared to the baseline. Overall, since positions represent a considerable portion
of data within a posting list (see Table 2 for the number of positions w.r.t. the number
of postings in both the corpora), this explains why improvements in their compression
provide marked response time benefits.

Finally, with respect to the index sizes, the compression of positions using codecs
other than Gamma leads to increased space usage. However, some reductions in size are
possible for the GOV2 corpus using FastPFOR or OptPFD. To summarise, in addressing
the final part of our research question, we find that FOR or NewPFD represent good
codecs for low query response times.

9   Conclusions

In this paper, we experimented upon two large corpora using a realistic query load to
determine the query response time benefits of a range of modern integer compression
codecs directly integrated within an IR system. Our thorough experiments addressed
different types of posting payload information: document ids, term frequencies, field
frequencies and term positions. Our findings allow a realistic estimate of the actual
contribution, in term of query response time, when using compression for the different types of payload within a posting list. For instance, we highlighted the importance
of the appropriate compression of term positions in benefiting query response times.
Moreover, we investigated the role of anchor text in achieving high index compression
and reduced query response times. In general, the list-adaptive codecs demonstrated a
good tradeoff in inverted index size and query response time. As our best practice recommendation, we found that the simpler Frame of Reference (FOR) [8] codec results in

fast average response times for all types of posting list payload, slightly outperforming the more complex but state-of-the-art PFOR implementations. Finally, we note that the gains obtainable from the compression of the various information within a posting list are cumulative. Indeed, when using the best aforementioned codecs for each payload type, we attain an improvement of $\sim 37\%$ and $\sim 27\%$ in query response time, for GOV2 and ClueWeb09 respectively, at a cost of an inverted index size increase of $\sim 6\%$ and $\sim 12\%$ w.r.t. the default configuration of Terrier.

References

1. Dean, J.: Challenges in building large-scale information retrieval systems: invited talk. In: Proc. WSDM 2009 (2009)
2. Anh, V.N., Moffat, A.: Pruned query evaluation using pre-computed impact scores. In: Proc. SIGIR 2006 (2006)
3. Moffat, A., Webber, W., Zobel, J., Baeza-Yates, R.: A pipelined architecture for distributed text query evaluation. Inf. Retr. 10 (2007)
4. Broccolo, D., Macdonald, C., Orlando, S., Ounis, I., Perego, R., Tonellotto, N.: Load-sensitive selective pruning for distributed search. In: Proc. CIKM 2013 (2013)
5. Witten, I.H., Bell, T.C., Moffat, A.: Managing Gigabytes: Compressing and Indexing Documents and Images, 1st edn. (1994)
6. Elias, P.: Universal codeword sets and representations of the integers. Trans. Info. Theory 21(2) (1975)
7. Yan, H., Ding, S., Suel, T.: Compressing term positions in web indexes. In: Proc. SIGIR 2009 (2009)
8. Zukowski, M., Heman, S., Nes, N., Boncz, P.: Super-scalar RAM-CPU cache compression. In: Proc. ICDE 2006 (2006)
9. Yan, H., Ding, S., Suel, T.: Inverted index compression and query processing with optimized document ordering. In: Proc. WWW 2009 (2009)
10. Lemire, D., Boytsov, L.: Decoding billions of integers per second through vectorization. Software: Practice and Experience (2013)
11. Delbru, R., Campinas, S., Samp, K., Tummarello, G.: Adaptive frame of reference for compressing inverted lists. Technical Report 2010-12-16, DERI (2010)
12. Ounis, I., Amati, G., Plachouras, V., He, B., Macdonald, C., Lioma, C.: Terrier: A High Performance and Scalable IR Platform. In: Proc. OSIR 2006 (2006)
13. Wang, L., Lin, J., Metzler, D.: Learning to efficiently rank. In: Proc. SIGIR 2010 (2010)
14. Shurman, E., Brutlag, J.: Performance related changes and their user impacts. In: Velocity: Web Performance and Operations Conference (2009)
15. Broder, A.Z., Carmel, D., Herscovici, M., Soffer, A., Zien, J.: Efficient query evaluation using a two-level retrieval process. In: Proc. CIKM 2003 (2003)
16. Cambazoglu, B.B., Zaragoza, H., Chapelle, O., Chen, J., Liao, C., Zheng, Z., Degenhardt, J.: Early exit optimizations for additive machine learned ranking systems. In: Proc. WSDM 2010 (2010)
17. Wang, L., Lin, J., Metzler, D.: A cascade ranking model for efficient ranked retrieval. In: Proc. SIGIR 2011 (2011)
18. Macdonald, C., Santos, R.L., Ounis, I., He, B.: About learning models with multiple query dependent features. Trans. Info. Sys. 13(3) (2013)
19. Williams, H.E., Zobel, J.: Compressing integers for fast file access. The Computer Journal 42 (1999)

20. Scholer, F., Williams, H.E., Yiannis, J., Zobel, J.: Compression of inverted indexes for fast query evaluation. In: Proc. SIGIR 2002 (2002)
21. Golomb, S.: Run-length encodings. Trans. Infor. Theory 12(3) (1966)
22. Rice, R., Plaunt, J.: Adaptive variable-length coding for efficient compression of spacecraft television data. Trans. Communication Technology 19(6) (1971)
23. Anh, V.N., Moffat, A.: Inverted index compression using word-aligned binary codes. Inf. Retr. 8(1) (2005)
24. Goldstein, J., Ramakrishnan, R., Shaft, U.: Compressing relations and indexes. In: Proc. ICDE 1998 (1998)
25. Zhang, J., Long, X., Suel, T.: Performance of compressed inverted list caching in search engines. In: Proc. WWW 2008 (2008)
26. Zhang, J., Suel, T.: Efficient search in large textual collections with redundancy. In: Proc. WWW 2007 (2007)

Exploring the Space of IR Functions

Parantapa Goswami[1], Simon Moura[1], Eric Gaussier[1], Massih-Reza Amini[1], and Francis Maes[2]

[1] Université Grenoble Alps, CNRS - LIG/AMA
Grenoble, France
`firstname.lastname@imag.fr`
[2] D-Labs, Paris, France
`francis@d-labs.fr`

Abstract. In this paper we propose an approach to discover functions for IR ranking from a space of simple closed-form mathematical functions. In general, all IR ranking models are based on two basic variables, namely, term frequency and document frequency. Here a grammar for generating all possible functions is defined which consists of the two above said variables and basic mathematical operations - addition, subtraction, multiplication, division, logarithm, exponential and square root. The large set of functions generated by this grammar is filtered by checking mathematical feasibility and satisfiability to heuristic constraints on IR scoring functions proposed by the community. Obtained candidate functions are tested on various standard IR collections and several simple but highly efficient scoring functions are identified. We show that these newly discovered functions are outperforming other state-of-the-art IR scoring models through extensive experimentation on several IR collections. We also compare the performance of functions satisfying IR constraints to those which do not, and show that the former set of functions clearly outperforms the latter one[1].

Keywords: IR Theory, Function Generation, Automatic Discovery.

1 Introduction

Developing new term-document scoring functions that outperform already existing traditional scoring schemes is one of the most interesting and popular research area in theoretical information retrieval (IR). Many state-of-the-art IR scoring schemes have been developed since the dawn of IR research, such as the vector space model [19], the language model [17], BM25 [18], and, more recently, the HMM [15], DFR [1] and information-based models [2]. All these scoring schemes were developed along what one could call a "theoretical line", in which theoretical principles guide the development of the scoring function, and then the developed function is assessed on different standard IR test collections. There

---

[1] The program code and the list of generated scoring functions can be found on
`http://ama.liglab.fr/resourcestools/code/ir-functions-generation/`

M. de Rijke et al. (Eds.): ECIR 2014, LNCS 8416, pp. 372–384, 2014.

is however a chance, or more precisely a risk, that some high performing scoring schemes will not come into light through such an approach, as they are not so intuitive and/or are not so easily explainable theoretically. Quoting [9]: *There is no guarantee that existing ranking functions are the best/optimal ones available. It seems likely that more powerful functions are yet to be discovered.*

These considerations have led researchers to explore the space of IR functions in a more systematic way, even though such attempts have always been limited by the complexity of the search space (of infinite dimension and containing potentially all real functions) with regard to the current computational power. The first attempts to this exploration were based on genetic programming and genetic algorithms, which were seen as a way to automatically learn IR functions by exploring parts of the solution space stochastically [12,16,6]. [9] applied genetic programming to discover optimal personalized ranking functions for each individual query or a consensus ranking function for a group of queries. The approach shows the power of automated function discovery using machine intelligence tools. But, being a non-deterministic method, the solutions generated by these genetic programming approaches are often difficult to analyze. Moreover, there is another issue associated with genetic programming, namely "code bloat" – the fact that almost all genetic programming algorithms have a tendency to produce larger and larger functions along the iterations. Thus there is a high risk of missing simple functions of high quality. To this end, [5] provides metrics to measure the distance between rank lists generated by different solutions, and thus explaining their position is the solution space. More recently, researchers have focused on particular function forms as linear combinations or well-defined kernel functions, the parameters of which are learned from some training data. This approach has been highly successful in IR, through the various "learning to rank" methods proposed so far: pointwise approaches, e.g. [4], pairwise approaches, e.g. [3,11,13], or list wise approaches, e.g. [20].

Even though all the methods mentioned above enlarge the space of scoring functions, they are still limited in two aspects: first, they usually assume that the IR scoring function takes a particular form (e.g. linear or polynomial), and they require some training set in order to learn the parameters of the function given a particular collection. Two questions, directly addressed in the current study, thus remain open: (a) is it possible to explore the space of IR scoring functions in a more systematic (i.e. exhaustive) way? (b) Is it possible to find a function that behaves well on all (or most) collections, and thus dispenses from re-training the function each time a new collection is considered? To answer those questions, we introduce an automatic discovery approach based on the systematic exploration of a search space of simple closed-form mathematical functions. This approach is inspired from the work of [14] on multi-armed bandit problems and is here coupled with the use of heuristic IR constraints [10] to prune the search space and so, limiting the computational requirements. Such a possibility was mentioned in [6] but has not been tried to the best of our knowledge.

The remainder of the paper is organized as follows: Section 2 introduces the function generation process we have followed, whereas Sections 3 and 4 present the experiments and results obtained. Finally, Section 5 concludes the paper.

2   Function Generation

In this section we present the proposed exploration strategy for function generation and their validation that we deploy to find a set of candidate scoring functions. These functions assign positive scores to a document $\mathbf{d}$ and a query term $\mathbf{w}$ and are involved in the retrieval status value of a query-document pair $(\mathbf{d}, \mathbf{q})$. Two variables are at the basis of classical IR scoring functions: term frequency $(t_{\mathbf{w}}^{\mathbf{d}})$ and document frequency $(\mathcal{N}_{\mathbf{w}})$, to score a document $\mathbf{d}$ with respect to a query term $\mathbf{w}$. However it is well known that normalized versions of these variables yield better results. For example language models use relative term counts [17] and the BM25 model uses the Okapi normalization [18]. Any such normalization scheme can be used with our approach. For this work, we selected a common scheme, which is the one used in DFR and in information based models [2]. Thus, we consider the following variables (for notations see Table 1):

- Normalized term frequency $x_{\mathbf{w}}^{\mathbf{d}} = t_{\mathbf{w}}^{\mathbf{d}} \log \left(1 + c \frac{l_{avg}}{l_{\mathbf{d}}}\right)$. Here $c \in \mathbb{R}$ is a multiplying factor. This variable incorporates both $t_{\mathbf{w}}^{\mathbf{d}}$ and $l_{\mathbf{d}}$. For simplicity, unless otherwise stated it is written as x from now on;
- Normalized document frequency $y_{\mathbf{w}} = \frac{\mathcal{N}_{\mathbf{w}}}{\mathcal{N}}$. For simplicity, unless otherwise stated it is written as y from now on;
- A constant real valued parameter $k \in \mathbb{R}$.

Table 1. Notations

Notation	Description
$t_{\mathbf{w}}^{\mathbf{d}}$	# of occurrences of term $\mathbf{w}$ in document $\mathbf{d}$, term frequency
$t_{\mathbf{w}}^{\mathbf{q}}$	# of occurrences of term $\mathbf{w}$ in query $\mathbf{q}$
$x_{\mathbf{w}}^{\mathbf{d}}$	normalized version of term frequency
$\mathcal{N}_{\mathbf{w}}$	# of documents in the collection containing $\mathbf{w}$, document frequency
$y_{\mathbf{w}}$	normalized version of document frequency
$\mathcal{N}$	# of documents in a given collection
$l_{\mathbf{d}}$	Length of document $\mathbf{d}$ in # of terms
l_{avg}	Average length of documents in a given collection

We hence define a function as the combination of the basic quantities x, y and k, and unary (logarithm, exponentiation with respect to e, square root, unary negation) and binary (addition, subtraction, multiplication, division and exponentiation with respect to any real number) operations. The grammar we use to generate syntactically correct functions is given in Figure 1. The $-(.)$ signifies the unary negation operation (e.g. $-y$). Thus, a function g may be a binary expression $\mathbb{B}(g, g)$, or a unary expression $\mathbb{U}(g)$, or a symbol $\mathbb{S}$.

$$g ::= \mathbb{B}(g, g) \mid \mathbb{U}(g) \mid \mathbb{S}$$
$$\mathbb{B} ::= + \mid - \mid \times \mid \div \mid pow$$
$$\mathbb{U} ::= log \mid exp \mid sqrt \mid -(.)$$
$$\mathbb{S} ::= x \mid y \mid k$$

Fig. 1. Grammar $\mathcal{G}$ to generate scoring functions

After a first combination of these quantities and operations we look at the validity of the generated functions. This validity verification is mainly three fold.

- *Domain of definition*, where we verify that all the operations used in a function are well defined. Bad operations include logarithm or square root of a negative number and division by zero.
- *Positiveness*, where we check that a generated function is positive valued. In fact, under all normal circumstances raw term and document frequency values are strictly positive. Hence $x \in \mathbb{R}$ and $x > 0$. Moreover, $\mathcal{N}_{\mathbf{w}} \leq \mathcal{N}$, which gives $y \in \mathbb{R}$ and $0 \leq y \leq 1$
- *IR constraints*, where we look if the generated function satisfies or not the heuristic IR constraints proposed by the community [10,2]. That is for the generated function g, if we have:

$$\frac{\partial g(x,y)}{\partial t_{\mathbf{w}}^d} > 0, \; \frac{\partial^2 g(x,y)}{(\partial t_{\mathbf{w}}^d)^2} < 0, \; \frac{\partial g(x,y)}{\partial \mathcal{N}_{\mathbf{w}}} < 0, \; \frac{\partial g(x,y)}{\partial l_{\mathbf{d}}} < 0$$

From the definitions of x and y, it can be shown $\frac{\partial x}{\partial t_{\mathbf{w}}^d} > 0$, $\frac{\partial x}{\partial l_{\mathbf{d}}} > 0$ and $\frac{\partial y}{\partial \mathcal{N}_{\mathbf{w}}} > 0$. Hence it is sufficient that g satisfies the following constraints:

$$\frac{\partial g(x,y)}{\partial x} > 0, \; \frac{\partial^2 g(x,y)}{\partial x^2} < 0, \; \frac{\partial g(x,y)}{\partial y} < 0$$

We now define the *length* of a function as the number of symbols or operators present in that function. As for example the function $sqrt(x/y)$ has a length 4, where $sqrt()$ and the *division* are two operators and x, y are two symbols present. Similarly $sqrt(x) * exp(-y)$ has a length 6.

A function generated by grammar $\mathcal{G}$ is said to be a candidate function if it survives all the validity verification steps described earlier. Algorithm 1 specifies the iterative length-limited strategy used here to generate the set of all candidate scoring functions till length $length_{max}$ (denoted by $\mathcal{C}_V$) and it works as follows. Suppose $\mathcal{S}_{\mathcal{G}}$ is the space of all possible functions generated by grammar $\mathcal{G}$ and in a particular iteration $\mathcal{S} \subset \mathcal{S}_{\mathcal{G}}$ is the set of already generated functions with a length less than or equal to $length_{curr}$ where $length_{curr} < length_{max}$ and $\mathcal{S} = \{g_1, g_2, \ldots, g_{|\mathcal{S}|}\}$. Next iteration expands the set $\mathcal{S}$ by creating new functions. A new function $g_{|\mathcal{S}|+1}$ is created by appending another operation or symbol to any function $g_i \in \mathcal{S}$. As for example, starting from an initial empty set $\mathcal{S} = \{\}$, the function $sqrt(x/y)$ is generated by the following steps:

$$(g_1 = x) \rightarrow (g_2 = y) \rightarrow (g_3 = g_1/g_2) \rightarrow (g_4 = sqrt(g_3))$$

Once a new function $g_{|\mathcal{S}|+1}$ is generated its validity is checked. If it passes all three steps, it is included in the set of generated candidate scoring functions $\mathcal{C}_V$, otherwise it is rejected. For the purpose of our experimental study, the algorithm also stores the functions until length $length_{max}$ which do not satisfy heuristic IR constraints but otherwise are valid (denoted by $\mathcal{C}_N$).

3   Experimental Setup

We conducted a number of experiments aimed at validating those functions which respect the IR constraints and also comparing these functions with respect to classical IR models.

Algorithm 1 has been implemented using *python*(www.python.org). The symbolic mathematics library of python, *SymPy*(sympy.org) is used to symbolically verify *domain of definition, positiveness* and *heuristic IR constraints*. Here the maximum considered length is 8. Table 2 shows the number of candidate functions available at each length till length 8 and also the corresponding generation time. The point to be noted is that functions with length less than 4 do not pass the three steps of validity testing.

Algorithm 1. Generating candidate scoring functions

Input   : maximum length $length_{max}$, the grammar $\mathcal{G}$
Output   :
  − set of candidate functions $\mathcal{C}_V$ till $length_{max}$
  − set of functions which do not satisfy *heuristic IR constraints* but pass other two validity tests, $\mathcal{C}_N$
Initialization: $\mathcal{C}_V \leftarrow \{\,\}, \mathcal{C}_N \leftarrow \{\,\}, \mathcal{S} \leftarrow \{\,\}$

for $length_{curr} \in \{1, 2, \dots, length_{max}\}$ **do**
  repeat
      A new function $g_{|\mathcal{S}|+1}$ is created by any of the following rules:
        − Append a symbol (variable or constant): $g_{|\mathcal{S}|+1} = x, y$ or k
        − Append a new unary operation: $g_{|\mathcal{S}|+1} = \mathbb{U}(g_i), i \in [1, |\mathcal{S}|]$
        − Append a new binary operation: $g_{|\mathcal{S}|+1} = \mathbb{B}(g_i, g_j), i, j \in [1, |\mathcal{S}|]$
      $\mathcal{S} \leftarrow \mathcal{S} \cup \{g_{|\mathcal{S}|+1}\}$
      if $g_{|\mathcal{S}|+1}(x, y)$ *satisfies* domain of definition *test* **AND** $g_{|\mathcal{S}|+1}(x, y)$ *satisfies* positiveness *test* **then**
          if $g_{|\mathcal{S}|+1}(x, y)$ *satisfies* heuristic IR constraints **then**
            | $\mathcal{C}_V \leftarrow \mathcal{C}_V \cup \{g_{|\mathcal{S}|+1}\}$
          else
            | $\mathcal{C}_N \leftarrow \mathcal{C}_N \cup \{g_{|\mathcal{S}|+1}\}$
          end
      end
  until *all the functions till $length_{curr}$ in $\mathcal{S_G}$ are generated;*
end

Table 2. Number of candidate functions and generation times for different lengths

length	total number of generated functions by $\mathcal{G}$	number of candidate functions $|\mathcal{C}_V|$	generation time
4	42	2	≈1 sec
5	328	10	≈1 min
6	2378	100	≈5 min
7	16447	638	≈30 min
8	49989	4657	≈1 day

Table 3. Statistics of various collections used in our experiments, sorted by size

Collection	$\mathcal{N}$	l_{avg}	Index size	#queries
GOV2	25,177,217	646	19.6 GB	100
WT10G	1,692,096	398	1.3 GB	100
TREC-3	741,856	261	427.7 MB	50
TREC-4	567,529	323	379.0 MB	50
TREC-5	524,929	339	378.0 MB	50
TREC-6,7,8	528,155	296	373.0 MB	50
CLEF-3	169,477	301	126.2 MB	60

The candidate functions are tested using CLEF (www.clef-campaign.org) and a large number of TREC (trec.nist.gov) collections. Basic statistics of the collections used are provided in Table 3. We appended TREC-9 and TREC-10 Web tracks to experiment with WT10G, and TREC-2004 and TREC-2005 Terabyte tracks for experimenting with GOV2.

Experiments are performed on Terrier IR platform v3.5 (terrier.org) as all standard modules are integrated. We implemented our models inside this framework and used other necessary standard modules by Terrier, mainly the indexing and the evaluation components. The preprocessing steps in creating an index include stemming using Porter stemmer and removing stop-words using the stopword list provided by Terrier. For comparison purpose three standard IR models are used, namely Okapi BM25 (denoted by BM), Dirichlet language model (denoted by LM) and log-logistic model of information model family (denoted by LG). These models are used with default parameter values, as well as optimized values. For the former, values of the parameters are the values provided as default in Terrier. That is $b = 0.75, k_1 = 1.2, k_3 = 8.0$ for BM, $\mu = 2500$ for LM and $c = 1.0$ for LG. For optimized version, the parameters are optimized from a set of values using 5 fold cross validation meaning that the query set is sequentially partitioned into 5 subsets. Of the 5 subsets, a single subset is retained for testing the model, and the remaining four subsets are used as training the parameters of each model (b and k_1 for BM, μ for LM and c for LG). The cross-validation process is then repeated 5 times, with each of the 5 subsets used exactly once for testing. Each time a querywise average precision and precision at 10th document is calculated for each set. After 5 folds, average precision of all the queries are obtained and Mean Average Precision (MAP) is calculated. Similarly average of precision at 10th document for each query is obtained and average of the quantity (P@10) is reported.

4 Results

From Algorithm 1, two sets of functions are produced, namely the set of all valid candidate functions ($\mathcal{C}_V$) and the set of functions which do not satisfy heuristic IR constraints, but are valid otherwise ($\mathcal{C}_N$). We first begin our investigation over the comparison of functions within $\mathcal{C}_V$ and $\mathcal{C}_N$, and then compare functions in $\mathcal{C}_V$ with respect to the classical IR functions.

Table 4. Average MAP and P@10 of the set of valid of $\mathcal{C}_V$ and non-valid $\mathcal{C}_N$ functions

Datasets	MAP		P@10	
	$\mathcal{C}_N$	$\mathcal{C}_V$	$\mathcal{C}_N$	$\mathcal{C}_V$
CLEF-3	0.1615	**0.3067**	0.1308	**0.2376**
TREC-3	0.0449	**0.1506**	0.0929	**0.3027**
TREC-5	0.0189	**0.0762**	0.0364	**0.1407**
TREC-6	0.1038	**0.1625**	0.1437	**0.2715**
TREC-7	0.0627	**0.1234**	0.1393	**0.2688**
TREC-8	0.0826	**0.1638**	0.1517	**0.2951**

4.1 Constraint Validation

From each of the sets $\mathcal{C}_V$ and $\mathcal{C}_N$, 10 subsets are created, each subset containing 100 randomly selected sample functions chosen from each initial set ($\mathcal{C}_V$ or $\mathcal{C}_N$) without replacement. These samples are tested on CLEF-3 and TREC-3,5,6,7,8. For each function MAP and P@10 are noted and they are averaged over all 100 functions within a single sample set. Finally, average performance over 10 sample sets are reported. Table 4 shows the average MAP and P@10 of 10 sample sets drawn from both $\mathcal{C}_V$ and $\mathcal{C}_N$ over these collections. As it can be seen the average MAP measure of functions in $\mathcal{C}_V$ are 6% to 14% higher than the MAP measures of functions in $\mathcal{C}_N$, and the difference is even more striking with the P@10 measures. These results empirically validate the IR constraints, and are in line with other empirical studies which aimed to test the validity of these constraints [8,7].

4.2 Function Validation

As shown in Table 2 there are a total of 5407 valid functions from length 4 to length 8. Testing all these functions and getting the best performing functions on all the collections is time consuming. Hence we chose a simple strategy which consists in selecting the 500 best performing functions, among 5407, on CLEF-3 with respect to the MAP measure and testing these functions on the remaining datasets. We first limit our analyzes over the TREC-3,5,6,7,8 collections. Each function is ranked based on MAP (and P@10) within a collection. An average rank of each function is estimated by taking the average of all the ranks of that function on the testing collections. Note that functions with lower average ranks are better performing. Average ranks of standard models, BM, LM and LG with respect to these 500 functions over the test collections are also considered. For this phase of experiment, we take the default values $k = 1.0, c = 1.0$ for the generated functions[2] and the default hyperparameter values for BM, LM and LG, as mentioned in the previous section. Table 5 shows the top 7 functions along with standard IR models with respect to the average rank over 5 test collections, TREC-3,5,6,7,8. Here we replaced k with 1 and presented the simplified functions. We denote these functions by using an exponent P or M for whether they are the best performing functions with respect to MAP or P@10, and d to indicate

---

[2] Note that the parameter c is used in the definition of x.

that they are used with their default values. We note that the 7 best ranked functions over the 5 TREC collections are better ranked than all the classical IR models. And that the first ranked function $(x,y) \mapsto e^{\sqrt{\log\left(\frac{x+y}{y}\right)}}$ is 2 to 6 times better ranked (with respect to P@10 and MAP) than the best standard IR model. In each case we statistically compare the performance of standard models with first ranked function in terms of MAP using a paired two sided t-test at 0.05 level. A $\uparrow$ indicates that the corresponding model is statistically significantly worse than the first ranked function. Whereas a $\downarrow$ indicates the opposite.

In tables 6 and 7 we show the rank and respectively the MAP and the P@10 measures of each function on TREC-3,5,6,7,8 collection. Here we see that the ranks of the 7 best functions over different collections stay approximately on

Table 5. Best functions with respect to their average ranks on TREC-3,5,6,7,8

	on MAP			on P@10	
functions	denoted by	rank_{avg}	functions	denoted by	rank_{avg}
$e^{\sqrt{\log\left(\frac{x+y}{y}\right)}}$	f_1^{M-d}	4.8	$e^{\sqrt{\log\left(\frac{x+y}{y}\right)}}$	f_1^{P-d}	19.8
$\sqrt{\frac{\log(1+x)}{\sqrt{y}}}$	f_2^{M-d}	14.8	$\log\left(-x+\frac{x+y}{y}\right)$	f_2^{P-d}	24.6
$\sqrt{\frac{\sqrt{xy}}{y}}$	f_3^{M-d}	15.6	$\sqrt{x+\sqrt{\frac{x}{y}}}$	f_3^{P-d}	25.2
$\sqrt{y+\sqrt{\frac{x}{y}}}$	f_4^{M-d}	17.8	$\sqrt{\sqrt{x}+\sqrt{\frac{x}{y}}}$	f_4^{P-d}	27.6
$\sqrt{\sqrt{\frac{x}{y}}.e^{-y}}$	f_5^{M-d}	18.8	$\sqrt{\frac{\log(1+x)}{\sqrt{y}}}$	f_5^{P-d}	27.8
$\sqrt{\sqrt{x}+\sqrt{\frac{x}{y}}}$	f_6^{M-d}	19.0	$\log\left(\frac{x+y}{y}\right)$	f_6^{P-d}	30.0
$\log\left(-x+\frac{x+y}{y}\right)$	f_7^{M-d}	20.0	$\log\left(\frac{x}{y}+\sqrt{e}\right)$	f_7^{P-d}	34.2
LGD	LG_d	26.0	LGD	LG_d	36.8
BM25	BM_d	108.8	BM25	BM_d	43.6
LM_{Dir}	LM_d	129.6	LM_{Dir}	LM_d	208.4

Table 6. MAP based ranks of functions with default parameter values (first value in the parenthesis) and their corresponding MAP (second value in the parenthesis)

TREC-3	TREC-5	TREC-6	TREC-7	TREC-8
BM_d (1; .252)	f_1^{M-d} (1; .140)	f_1^{M-d} (1; .249)	f_4^{M-d} (5; .194)	f_1^{M-d} (1; .256)
f_5^{M-d} (2; .252)	f_2^{M-d} (3; .139)	f_6^{M-d} (2; .248)	f_3^{M-d} (6; .194)	f_7^{M-d} (6; .255)
f_3^{M-d} (3; .250)	f_5^{M-d} (4; .138)	f_4^{M-d} (4; .247)	f_5^{M-d} (8; .194)	LG_d (11; .255)
f_2^{M-d} (4; .249)	BM_d (5; .138)	f_3^{M-d} (5; .247)	f_6^{M-d} (9; 194)	f_6^{M-d} (30; .252)
f_1^{M-d} (5; .249)	LM_d (10; .137)	f_2^{M-d} (7; .246)	f_2^{M-d} (13; .193)	f_{d4}^{M} (46; .250)
f_4^{M-d} (7; .249)	f_3^{M-d} (13; .137)	LG_d (16; .245)	f_1^{M-d} (16; .192)	f_2^{M-d} (47; .249)
LM_d (8; .249)	f_7^{M-d} (14; .137)	f_5^{M-d} (19; .244)	f_7^{M-d} (41; .189)	f_3^{M-d} (51; .249)
f_7^{M-d} (9; .248)	f_4^{M-d} (27; .136)	f_7^{M-d} (30; .244)	LG_d (49; .188)	f_5^{M-d} (61; .248)
f_6^{M-d} (12; .246)	LG_d (40; .135)	BM_d (252; .232)$\uparrow$	LM_d (64; .186)	BM_d (181; .241)$\uparrow$
LG_d (14; .245)	f_{d6}^{M} (42; .134)	LM_d (381; .222)$\uparrow$	BM_d (105; .183)	LM_d (185; 0.240)$\uparrow$

the same range, while the ranks of IR models may vary a lot. For example, considering the MAP, BM with its default values is ranked first on TREC-3 while it is ranked 252nd on TREC-6 (Table 6). We find the same result by looking at the ranks of different models with respect to their P@10 measure on different TREC collections (Table 7). As the language model which is ranked first on TREC-3 and TREC-5 ; it has the lowest rank over the three other TREC collections.

Table 7. P@10 based ranks of functions with default parameter values (first value in the parenthesis) and their corresponding P@10 (second value in the parenthesis)

TREC-3	TREC-5	TREC-6	TREC-7	TREC-8
LM_d (1; .532)	LM_d (1; .276)	f_5^{P-d} (1; .418)	f_1^{P-d} (6; .432)	f_6^{P-d} (4; .474)
f_3^{P-d} (6; .516)	f_3^{P-d} (6; .248)	BM_d (2; .414)	f_5^{P-d} (11; .430)	f_2^{P-d} (5; .474)
BM_d (9; .514)	f_1^{P-d} (47; .236)	f_4^{P-d} (7; .402)	f_6^{P-d} (16; .428)	LG_d (7; .474)
f_4^{P-d} (12; .506)	f_4^{P-d} (49; .234)	f_1^{P-d} (12; .400)	f_7^{P-d} (17; .428)	f_7^{P-d} (8; .472)
f_2^{P-d} (13; .504)	f_2^{P-d} (53; .234)	f_3^{P-d} (15; .398)	f_2^{P-d} (18; .428)	BM_d (14; .472)
f_1^{P-d} (15; .504)	BM_d (63; .232)	f_6^{P-d} (29; .396)	LG_d (26; .428)	f_1^{P-d} (19; .468)
f_5^{P-d} (25; .496)	f_5^{P-d} (73; .228)	f_7^{P-d} (33; .396)	f_4^{P-d} (27; .426)	f_5^{P-d} (29; .460)
f_6^{P-d} (27; .496)	f_6^{P-d} (74; .228)	f_2^{P-d} (34; .396)	f_3^{P-d} (58; .422)	f_3^{P-d} (41; .458)
f_7^{P-d} (28; .496)	LG_d (75; .228)	LG_d (47; .396)	BM_d (130; .418)	f_4^{P-d} (43; .458)
LG_d (29; .496)	f_7^{P-d} (85; .226)	LM_d (483; .346)	LM_d (285; .392)	LM_d (272; .432)

Table 8. Best optimized functions based on average rank on TREC-3,5,6,7,8

on MAP			on P@10		
functions	denoted by	$rank_{avg}$	functions	denoted by	$rank_{avg}$
$\sqrt{\frac{\log(1+x)}{\sqrt{y}}}$	f_1^{M-o}	3.0	BM25	BM_o	9.0
$e^{\sqrt{\log\left(\frac{x+y}{y}\right)}}$	f_2^{M-o}	8.4	$\log\left(-x + \frac{x+y}{y}\right)$	f_1^{P-o}	9.2
$\sqrt{y + \sqrt{\frac{x}{y}}}$	f_3^{M-o}	8.8	$\log\left(\frac{x}{y} + \sqrt{e}\right)$	f_2^{P-o}	10.0
$\sqrt{\frac{\sqrt{xy}}{y}}$	f_4^{M-o}	9.0	$\log\left(\frac{x+y}{y}\right)$	f_3^{P-o}	10.4
$\sqrt{\sqrt{\frac{x}{y}}.e^{-y}}$	f_5^{M-o}	10.0	$\sqrt{\sqrt{x} + \sqrt{\frac{x}{y}}}$	f_4^{P-o}	11.0
$\sqrt{1 + \sqrt{\frac{x}{y}}}$	f_6^{M-o}	10.4	LM_{Dir}	LM_o	12.2
$\sqrt{\sqrt{x} + \sqrt{\frac{x}{y}}}$	f_7^{M-o}	11.0	$\sqrt{x + \sqrt{\frac{x}{y}}}$	f_5^{P-o}	12.8
LM_{Dir}	LM_o	16.4	LGD	LG_o	13.0
BM25	BM_o	16.6	$\log\left(\frac{x+2y}{y}\right)$	f_6^{P-o}	13.4
LGD	LG_o	17.6	$\sqrt{1 + \sqrt{\frac{x}{y}}}$	f_7^{P-o}	13.6

We now consider the top 25 functions among the best performing functions based on average ranks they have, and optimize their hyper-parameter c using 5 fold cross validation and again considering $k = 1$. In order to maintain our comparisons, we also consider the optimized versions of the standard models, BM, LM and LG using again 5 fold cross validation sets. Table 8 shows the top optimized functions along with optimized standard IR models with respect to the average rank over TREC-3,5,6,7,8. Here the exponent or the index o indicates that the function or the standard IR model are used with their optimized values.

In tables 9 and 10 we show detailed ranks of the optimized versions of different functions with respect to their MAP and P@10 measures on different TREC-3,5,6,7,8 collections. Although, the difference between the average ranks of the optimized IR models and the top 7 generated functions have decreased, but still, the simple generated functions are better ranked than the standard IR models. We also performed experiments on WT10G and GOV2 datasets using the same 25 selected functions found previously with the default and optimized values of their parameters. On GOV2, function 2 behaves well with a difference of 5% in MAP with the second best model when optimizing the parameters. However on WT10G the language model seems to be the best model when using optimizing or using its

Table 9. MAP based ranks of functions with optimized parameter values (first value in the parenthesis) and their corresponding MAP (second value in the parenthesis)

TREC-3	TREC-5	TREC-6	TREC-7	TREC-8
BM_o (1; .273)↓	f_1^{M-o} (1; .141)	f_1^{M-o} (1; .255)	f_5^{M-o} (4; .195)	f_1^{M-o} (1; .262)
LM_o (2; .269)↓	LM_o (2; .141)	f_7^{M-o} (2; .250)	f_1^{M-o} (6; .194)	f_2^{M-o} (2; .261)
f_5^{M-o} (3; .260)	f_4^{M-o} (3; .139)	f_6^{M-o} (4; .249)	f_6^{M-o} (7; .193)	BM_o (10; .259)
f_4^{M-o} (4; .258)	f_3^{M-o} (4; .138)	f_3^{M-o} (7; .248)	f_3^{M-o} (9; .193)	LG_o (15; .258)
f_3^{M-o} (5; .257)	f_5^{M-o} (5; .138)	f_9^{M-o} (9; .248)	f_4^{M-o} (10; .193)	f_3^{M-o} (18; .257)
f_1^{M-o} (6; .256)	f_2^{M-o} (6; .138)	f_2^{M-o} (12; .248)	f_4^{M-o} (12; .192)	f_4^{M-o} (19; .257)
f_7^{M-o} (7; .253)	f_6^{M-o} (8; .137)	f_5^{M-o} (15; .247)	f_7^{M-o} (15; .191)	f_7^{M-o} (22; .256)
f_6^{M-o} (8; .252)	f_7^{M-o} (9; .135)	LG_o (20; 245)	BM_o (16; .191)	f_5^{M-o} (24; .256)
f_2^{M-o} (10; .251)	LG_o (20; .133)	LM_o (27; .243)↑	LG_o (20; .190)	f_6^{M} (25; .255)
LG_o (13; .246)↑	BM_o (28; .126)↑	BM_o (28; .232)↑	LM_o (25; .189)	LM_o (26; .254)

Table 10. P@10 based ranks of functions with optimized parameter values (first value in the parenthesis) and their corresponding P@10 (second value in the parenthesis)

TREC-3	TREC-5	TREC-6	TREC-7	TREC-8
BM_o (1; .562)	LM_o (1; .274)	f_4^{P-o} (1; .410)	f_1^{P-o} (1; .444)	BM_o (1; .464)
f_5^{P-o} (2; .560)	f_5^{P-o} (2; .258)	f_1^{P-o} (2; .406)	f_3^{P-o} (3; .442)	f_2^{P-o} (3; .458)
LM_o (3; .558)	BM_o (4; .250)	f_3^{P-o} (6; .404)	LG_o (4; .442)	f_7^{P-o} (4; .458)
f_4^{P-o} (7; .538)	f_4^{P-o} (9; .240)	f_2^{P-o} (8; .404)	f_2^{P-o} (5; .440)	f_3^{P-o} (5; .456)
f_1^{P-o} (17; .490)	f_2^{P-o} (13; .238)	f_6^{P-o} (9; .404)	f_7^{P-o} (6; .440)	f_6^{P-o} (7; .456)
f_3^{P-o} (19; .490)	f_7^{P-o} (14; .238)	LG_o (11; .404)	f_6^{P-o} (7; .438)	LG_o (9; .456)
LG_o (20; .484)	f_1^{P-o} (15; .238)	f_5^{P-o} (15; .400)	BM_o (13; .430)	f_1^{P-o} (11; .452)
f_2^{P-o} (21; .482)	f_3^{P-o} (19; .236)	f_7^{P-o} (18; .400)	LM_o (17; .424)	LM_o (13; .452)
f_6^{P-o} (22; .476)	LG_o (21; .236)	BM_o (26; .394)	f_5^{P-o} (18; .416)	f_4^{P-o} (19; .444)
f_7^{P-o} (25; .472)	f_6^{P-o} (22; .234)	LM_o (27; .390)	f_4^{P-o} (19; .412)	f_5^{P-o} (27; .438)

Table 11. MAP and P@10 measures of different functions and IR models with their default values on WT10G and GOV2 datasets

WT10G		GOV2	
MAP	P@10	MAP	P@10
LM_d (.204)	f_5^{P-d} (.300)	f_1^{M-d} (.291)	LM_d (.555)
f_7^{M-d} (.196)	f_1^{P-d} (.299)	f_7^{M-d} (.289)	f_7^{P-d} (.544)
LG_d (.194)	LM_d (.293)	LG_d (.288)	f_1^{P-d} (.543)
f_1^{M-d} (.194)	f_4^{P-d} (.292)	LM_d (.280)	f_2^{P-d} (.542)
f_4^{M-d} (.187)	BM_d (.291)	f_2^{M-d} (.274)	f_6^{P-d} (.541)
f_3^{M-d} (.187)	f_6^{P-d} (.287)	BM_d (.274)	LG_d (.541)
f_5^{M-d} (.187)	LG_d (.287)	f_6^{M-d} (.265)	BM_d (.538)
f_6^{M-d} (.186)	f_2^{P-d} (.284)	f_3^{M-d} (.262)	f_5^{P-d} (.535)
f_2^{M-d} (.186)	f_7^{P-d} (.284)	f_4^{M-d} (.261)	f_4^{P-d} (.525)
BM_d (.184)	f_3^{P-d} (.256)	f_5^{M-d} (.260)	f_3^{P-d} (.471)

Table 12. MAP and P@10 measures of different functions and IR models with their optimized values on WT10G and GOV2 datasets

WT10G		GOV2	
MAP	P@10	MAP	P@10
LM_o (.473)↓	LM_o (.372)	f_2^{M-o} (.302)	f_2^{P-o} (.557)
f_6^{M-o} (.466)	f_2^{P-o} (.368)	f_1^{M-o} (.295)	f_6^{P-o} (.553)
f_3^{M-o} (.463)	BM_o (.362)	LM_o (.294)	f_1^{P-o} (.551)
f_4^{M-o} (.462)	f_4^{P-o} (.359)	LG_o (.288)	LM_o (.551)
f_5^{M-o} (.454)	f_3^{P-o} (.354)	BM_o (.284)	f_3^{P-o} (.550)
f_7^{M-o} (.453)	f_6^{P-o} (.352)	f_7^{M-o} (.274)	LG_o (.541)
f_2^{M-o} (.449)	f_7^{P-o} (.350)	f_3^{M-o} (.271)	f_7^{P-o} (.539)
f_1^{M-o} (.438)	LG_o (.349)	f_6^{M-o} (.271)	BM_o (.531)
BM_o (.421)	f_1^{P-o} (.347)	f_4^{M-o} (.270)	f_4^{P-o} (.505)
LG_o (.414)	f_5^{P-o} (.337)	f_5^{M-o} (.269)	f_5^{P-o} (.469)

default parameter values. Though it seems that the simple strategy of selecting the first 500 functions over CLEF-3 has its limits on larger collections, but the selected functions are still competitive on MAP and P@10 with respect to standard models over these large datasets.

5 Conclusion

In this paper we have addressed the problem of exploring the space of simple IR functions with the goal to discover some promising IR scoring functions. To do so, we proposed a systematic iterative approach to explore the search space till some given length and to identify the set of candidate scoring functions which are mathematically valid and satisfy heuristic IR constraints. We tested the functions obtained on a variety of standard IR test collections.

Our results show that, if one wants to make use of an efficient IR scoring function without tuning parameters on a collection, then one should use:

$$\text{ES-LG}(x, y) = e^{\sqrt{\log\left(\frac{x+y}{y}\right)}}$$

where the name ES-LG derives from the fact that this function consists in the exponential of the square root of the log-logistic function. ES-LG is consistently above (for the MAP) all other ones; furthermore, the difference with standard IR functions is significant in several cases. This result is all the more interesting that the "complexity" of this function (measured by its length) is smaller than the one of BM_d and LM_d, which can, in theory, be generated by our method using a different term frequency normalization but nevertheless require additional computing resources as they involve more operators. In the situation where it is possible to optimize the value of some parameters (i.e. when relevance judgments are readily available), our results are more contrasted: the difference in ranks between the best functions and the standard ones is not as important as before and if in several cases the difference is significant in favor of the discovered functions, it is, in other cases, in favor of standard functions. All in all, there is no real difference in this case. As ES-LG is ranked second in this setting, we recommend its use in all cases if one is interested in the MAP. It can of course be used as an additional feature in learning to rank approaches.

Acknowledgment. This work was supported in part by the ANR project Class-Y, the Mastodons project Garguantua, and the LabEx PERSYVAL-Lab ANR-11-LABX-0025.

References

1. Amati, G., Van Rijsbergen, C.J.: Probabilistic models of information retrieval based on measuring the divergence from randomness. ACM Trans. Inf. Syst. 20(4), 357–389 (2002)
2. Clinchant, S., Gaussier, E.: Information-based models for ad hoc ir. In: Proceedings of the 33rd ACM SIGIR Conference (2010)
3. Cohen, W.W., Schapire, R.E., Singer, Y.: Learning to order things. Journal of Artificial Intelligence Research 10(1), 243–270 (1999)
4. Crammer, K., Singer, Y.: Pranking with ranking. In: Advances in Neural Information Processing Systems (NIPS 14), pp. 641–647. MIT Press (2001)
5. Cummins, R., O'Riordan, C.: Evolved term-weighting schemes in information retrieval: an analysis of the solution space. Artif. Intell. Rev. 26(1-2), 35–47 (2006)
6. Cummins, R., O'Riordan, C.: Evolving local and global weighting schemes in information retrieval. Inf. Retr. 9(3), 311–330 (2006)
7. Cummins, R., O'Riordan, C.: Analysing Ranking Functions in Information Retrieval Using Constraints. In: Information Extraction from the Internet, CreateSpace Independent Publishing Platform (August 2009)
8. Cummins, R., O'Riordan, C.: Measuring constraint violations in information retrieval. In: Proceedings of the 32nd SIGIR, pp. 722–723 (2009)
9. Fan, W., Gordon, M.D., Pathak, P.: A generic ranking function discovery framework by genetic programming for information retrieval. Inf. Process. Manage. 40(4), 587–602 (2004)

10. Fang, H., Tao, T., Zhai, C.: A formal study of information retrieval heuristics. In: Proceedings of the 27th ACM SIGIR Conference (2004)
11. Freund, Y., Iyer, R., Schapire, R.E., Singer, Y.: An efficient boosting algorithm for combining preferences. Journal of Machine Learning Research (2003)
12. Gordon, M.: Probabilistic and genetic algorithms in document retrieval. Commun. ACM 31(10), 1208–1218 (1988)
13. Joachims, T.: Optimizing search engines using clickthrough data. In: Proceedings of the 8th ACM SIGKDD, pp. 133–142 (2002)
14. Maes, F., Wehenkel, L., Ernst, D.: Automatic discovery of ranking formulas for playing with multi-armed bandits. In: Sanner, S., Hutter, M. (eds.) EWRL 2011. LNCS, vol. 7188, pp. 5–17. Springer, Heidelberg (2012)
15. Metzler, D., Croft, W.B.: A markov random field model for term dependencies. In: SIGIR, pp. 472–479 (2005)
16. Pathak, P., Gordon, M.D., Fan, W.: Effective information retrieval using genetic algorithms based matching functions adaptation. In: HICSS (2000)
17. Ponte, J.M., Croft, W.B.: A language modeling approach to information retrieval. In: Proceedings of the 21st ACM SIGIR Conference (1998)
18. Robertson, S.E., Zaragoza, H.: The probabilistic relevance framework: BM25 and beyond. Foundations and Trends in Information Retrieval 3(4), 333–389 (2009)
19. Salton, G., McGill, J.: Introduction to Modern Information Retrieval. McGraw-Hill, New York (1983)
20. Valizadegan, H., Jin, R., Zhang, R., Mao, J.: Learning to rank by optimizing ndcg measure. In: Advances in Neural Information Processing Systems (NIPS 22), pp. 1883–1891 (2009)

Metric Spaces
for Temporal Information Retrieval

Matteo Brucato[1] and Danilo Montesi[2]

[1] School of Computer Science, University of Massachusetts, Amherst, MA, USA
matteo@cs.umass.edu
[2] Department of Computer Science and Engineering, University of Bologna, Italy
montesi@cs.unibo.it

Abstract. Documents and queries are rich in temporal features, both
at the meta-level and at the content-level. We exploit this information
to define *temporal scope similarities* between documents and queries in
metric spaces. Our experiments show that the proposed metrics can
be very effective for modeling the relevance for different search tasks,
and provide insights into an inherent asymmetry in temporal query
semantics. Moreover, we propose a simple ranking model that combines the temporal scope similarity with traditional keyword similarities.
We experimentally show that it is not worse than traditional keyword-
based rankings for non-temporal queries, and that it improves the overall
effectiveness for time-based queries.

1 Introduction

The amount of available digital information is constantly increasing—a phenomenon that many refer to as *big data*. As more and more information becomes available year after year, its variety and richness in terms of *temporal
aspects* become more manifest. A recent research area aimed at incorporating
temporal aspects in modern information retrieval systems is *temporal information retrieval* (TIR) [1]. By its literature, it is clear how time comes into play in
many different facets and forms. For instance, what kind of temporal features
should we consider? How can we define the "temporal needs" of users, and the
"temporal intent" of queries? Some of these issues have been explored by the
research community [2,3,4], but further works are still needed as we are far from
having widely-accepted solutions.

In this work, we explore ways to improve traditional ranking models by considering what we call the *temporal scopes* of documents and queries. These indicate
which periods of time the documents are about and which periods of time users
are interested in when issuing time-based queries. For example, news stories very
often refer to periods of time close to the publication dates, chapters from history
books may refer to any past period, and tweets from the Twitter social network
might conversely have very narrow time scopes.

There are quite a number of search tasks in which the temporal scope is of
great importance. For instance, imagine an expert user (e.g. a librarian, a his-
torian, or a philologist) who is searching a digital library or a digital historical

M. de Rijke et al. (Eds.): ECIR 2014, LNCS 8416, pp. 385–397, 2014.
© Springer International Publishing Switzerland 2014

archive to find information about a specific period of time. To be able to only filter documents by their creation date would be too limiting. So would be treating temporal expressions such as "last year" as simple search terms, since they indicate specific time periods and their meaning changes with time.

In this paper, we propose a new model for temporal information retrieval based on metric spaces. We study temporal aspects of documents and queries, and we use temporal expressions [5] extracted from their texts to model their temporal scope. References to temporal information are represented in the temporal domain as time intervals, and temporal similarities between documents and queries are defined according to them. We evaluate the effectiveness of our model through a series of experiments, whose results are twofold. First, they confirm that exploiting temporal information can enhance the effectiveness of traditional keyword-only models for temporal queries. Second, they provide insights into an inherent asymmetry in temporal query semantics, confirming our intuition which led us to the definition of generalized metrics for modeling the temporal relevance for different search tasks.

2   Related Work

The time dimension has been extensively studied in temporal databases [6]. More recently, its importance has also been acknowledged in information retrieval [1]. While earlier works mostly concentrated on exploiting temporal meta-information (such as the creation date of documents) [7], there is a more recent interest in considering temporal expressions extracted from the text to improve the effectiveness of ranking algorithms [8]. Recent advances in natural language processing (NLP) have made it possible to effectively identify and interpret these expressions in a variety of texts (see [9] for an overview of the problem). Nowadays, there are ready-to-use tools for extracting [10] and normalizing (i.e. interpreting) [11] them easily and reliably.

There are several aspects of temporal information retrieval that current research has been focusing on. The *temporal intent* of textual queries has been discussed, for instance, in [2]. The *implicit time* of textual search queries has been studied in [12,3], among others. First attempts to linearly combine non-temporal scores with temporal ones have been presented in [7,13]. Moreover, several workshops on time-related aspects of information retrieval have been recently organized, such as [14] and [15], thus showing the attention given to this topic by current research.

More importantly, there exists previous work aimed at showing how to improve the effectiveness of search engines on temporal queries. For instance, in [4] and [7] the authors presented different language models to address different temporal information needs. Our work differs from all existing works because it introduces the first non-probabilistic ranking model for the temporal scope of documents and queries in temporal information retrieval based on (generalized) metric spaces.

3   Motivation

Time is an important aspect in every collection of documents. It is a ubiquitous dimension that can be interrelated with all sorts of information. For instance, it can be attributed to events ("when" they take place) and facts ("when" they are true or false). It is so intrinsic that it is often taken for granted and, thus, disregarded.

At the **meta-level**, a document has a *creation date*, a *publication date*, a *revision date*, and so on. But a closer look at its **content-level** can reveal information about which times the document is about: Which *periods of time* are explicitly mentioned in the text? When did the *events* mentioned in the text happen? Searching collections at the meta-level is important, and it has been studied in several recent works [16]. In this work, we concentrate on the temporal information present at the content-level. This type of approach requires further steps in the acquisition process of the information contained in the documents. It requires tools to identify and interpret natural language expressions with temporal intent, which are called *temporal expressions* (or *timexes*). More precisely, by looking at the content-level, we are able to make sense of what we call the **temporal scope** of a document. This level of understanding is crucial when we want to search documents based on their temporal content.

Not surprisingly, queries are also rich in temporal features. At the **meta-level** we can identify, for instance, the issue date of a query (which is crucial for query log analysis like Google Trends[1]). But looking at the **content-level** (i.e. the text of queries) can reveal very specific query intents. For example, querying "Obama elections 2008" is very different from querying "Obama elections", for its intent of discriminating among events in a case of clear ambiguity. And the expression "last year" in the query "last year best movies" can be interpreted as a signal for the intent of drifting away from the plausible default behavior of always retrieving the latest information. Temporal expressions are present in a substantial fraction of queries (about 1.9%, as per [5]). Although this is not a very large percentage, when a query contains such a signal, its intent becomes very different from the default one.

As a concrete example, consider a user searching a digital library, an archive, or a crawled portion of the Web. She uses natural language queries to specify her information needs, which pertain not only to the textual content, but also to the temporal scope of the information content. For instance, she queries the phrase "balkan conflicts in 1912 and 1913" to retrieve relevant information about the Balkan Wars. Any information related to those events is relevant to the user, including the causes of the wars and the aftermath. In this scenario, the relevance of a document can be modeled by: (1) a traditional notion of keyword-based similarity (e.g. cosine similarity in the vector space model [17]); (2) a notion of *temporal scope similarity*, which might favor documents regarding time periods that are "close" to the query time period.

---

[1] http://trends.google.com

4   Temporal Scope Similarity Model

One of our major aims is to provide more evidence that a similarity measure based on the temporal scope of documents and queries can lead to improvements in the effectiveness of traditional ranking algorithms based solely on term statistics. With keyword-based similarity models, we can readily identify documents that are textually similar to the query, but we cannot easily distinguish between two documents that are textually too similar to one other. In the search space dictated by keyword similarity, all textually similar documents have similar representations, which means that these models are not rich enough for distinguishing among them. The time dimension, in some cases, can make it possible to have a clearer distinction. Moreover, similarity models based solely on term occurrences are not rich enough for capturing specific time-related aspects of queries and documents. In particular, we identify the following three characteristics of temporal expressions that cannot be modeled with simple keyword-based models, and that can lead to poor results.

Temporal Synonymy. Suppose we treat temporal expressions as in keyword-based models. That is, the expression "2014" is for us a simple term that can occur in a document or a query with a certain frequency. In this model, we would not be able to account for different ways of referring to the year 2014. But there are many. For instance, we could write "last year", or "next year", or even "in a decade", depending on the context.

Temporal Polysemy. Additionally, some temporal expressions can potentially refer to more than just one period of time. Consider for instance the expressions "every Tuesday", "yearly", or the implicitly temporal expression "super bowl".

Structured Domain. Periods of time can be modeled as intervals of numbers, i.e. they can be represented as *time* or *temporal intervals* (as it has been proposed in the context of temporal databases [6]). In the domain of temporal intervals, it is easier to define notions of *overlap*, *containment* and *distance* between them at the semantic level, rather than at the syntactic level.

We frame our model for temporal scope similarity as follows, embedding a notion of distance between time intervals:

1. Different temporal expressions can be mapped to the same temporal interval.
2. A single temporal expression can be mapped to multiple temporal intervals.
3. The temporal space is a metric space (Δ, δ), where δ is a distance function on Δ, modeling a notion of distance between temporal intervals.
4. The temporal scope of documents and queries are subsets of Δ, i.e., $T_D \in \mathcal{P}(\Delta)$ and $T_Q \in \mathcal{P}(\Delta)$, where D and Q are a document and a query, respectively.
5. Let δ^* be a distance function on $\mathcal{P}(\Delta)$ defined in terms of δ. Then, $(\mathcal{P}(\Delta), \delta^*)$ is a metric space for document and query representations, where δ^* models a notion of distance between them in terms of their temporal scope.

The temporal scope similarity can be defined as a similarity in the metric space $(\mathcal{P}(\Delta), \delta^*)$. We formalize these concepts in the next two sections.

5   Modeling the Temporal Scope

In this section, we formally define the domain of temporal intervals Δ, used to represent documents and queries. Our goal is to utilize temporal expressions extracted from the text to model the temporal scope. Temporal expressions are used in natural language texts to express temporality [8,4]. For instance, they might be used to state when a certain event happened (e.g. "two years ago Obama won the elections").

We model the temporal scope by mapping the temporal expressions to a *temporal domain* Δ, which is the set of all possible temporal intervals, represented as ordered pairs of integers. By doing so, we obtain a *temporal scope representation* for each document and each query, that we indicate with T_D and T_Q respectively, and such that $T_D \subseteq \Delta$ and $T_Q \subseteq \Delta$.

Definition 1. CHRONON. *A chronon is the smallest discrete unit of time, i.e., an atomic time. It describes the granularity of the model. Examples of chronons are seconds, days, years, etc.*

Definition 2. TIMELINE. *Let* $t_{min}, t_{max} \in \mathbb{Z} : t_{min} \leq t_{max}$. *The* timeline *is the totally ordered set of numbers*

$$\Gamma = \{t \in \mathbb{Z} \mid t_{min} \leq t \leq t_{max}\}$$

in which each number corresponds to a different chronon, and consecutive numbers correspond to consecutive chronons. Therefore, t_{min} and t_{max} correspond, respectively, to the first and the last chronons that can be captured by the timeline, and $|\Gamma| = t_{max} - t_{min} + 1$ is the cardinality of the timeline.

Definition 3. TEMPORAL DOMAIN. *The* temporal domain *is the set*

$$\Delta = \{[s,t] \mid s, t \in \Gamma \text{ and } s \leq t\} \subseteq \Gamma \times \Gamma$$

that is, the set containing all pairs of timeline elements, internally ordered. It follows that the cardinality of Δ is $|\Delta| = \frac{|\Gamma|(|\Gamma|+1)}{2}$.

Definition 4. TEMPORAL INTERVALS. *Let* TIMEX *be the set of all temporal expressions that can extracted either from documents or queries, and let* Ψ : TIMEX $\rightarrow$ $\mathcal{P}(\Delta)$ *be a function that maps temporal expressions to temporal intervals. Let $e \in$ TIMEX be a temporal expression. The set $\Psi(e)$ is the set of all the temporal intervals of the expression e.*

For convenience, we will also use TIMEX_Q and TIMEX_D to denote the set of expressions extracted from a query Q and a document D, respectively. Further, we will use $[s,t]_Q$ and $[s,t]_D$ to indicate whether the temporal interval $[s,t]$ has been extracted from the query Q or the document D, respectively. Notice that, at this point, the first two points of Sect. 4 are both satisfied.

Definition 5. DOCUMENT/QUERY TEMPORAL SCOPES. *The* document temporal scope *and the* query temporal scope *are the document and the query representations in the temporal domain:*

$$T_D = \{[s,t]_D\} = \{[s,t] \in \Psi(e) \mid e \in \text{TIMEX}_D\} \subseteq \Delta \qquad (1)$$

And similarly for T_Q. Notice that now point 4 of Sect. 4 is also satisfied.

6 Temporal Scope Similarity

To understand the difficulty of modeling a temporal similarity metric, consider the following example. Imagine two textually similar documents, one containing the time expression "during the twentieth century", and one containing the time expression "June 1950", as shown in Fig. 1. The temporal scope of the first document is broad, whereas the second one is narrow. Now, suppose a user formulates the query "between 1940 and 1960", as also shown in the picture. Which of the two documents would the user consider more relevant?

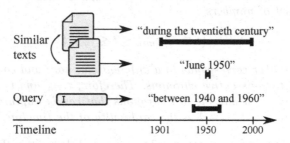

Fig. 1. Example of two textually similar documents with different temporal scopes, and a time-based query

This question cannot be answered without knowing the query semantics and how the user expressed her information needs as a textual query. Some users might consider the broader document more relevant because its temporal scope covers the query scope. Others might think that a broader document is less relevant because it is too generic, and that the narrower document is more relevant because it falls inside the query scope. Perhaps, some other users might think that the best document should have a temporal scope that matches exactly the query scope. Therefore, we propose to use three different *generalized metric spaces* (i.e., metric spaces in which some of the metric properties, in particular symmetry and coincidence, are relaxed) to capture these alternatives.

6.1 Generalized Metric Spaces

The goal of this section is to model δ^*. Since documents and queries are represented as sets of temporal intervals, i.e. T_D and T_Q respectively, as in (1),

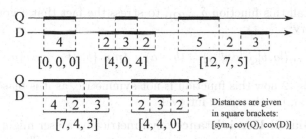

Fig. 2. Five different examples of query and document temporal intervals, and their generalized distances

δ^* is a function between sets of intervals. We define this set-based distance by aggregating the inter-distances between each pair of elements in the two sets, assuming that a ground distance function $\delta : \Delta \times \Delta \to \mathbb{R}$ between pairs of time intervals is provided. Given δ, we define δ^* as the minimum distance between each pair of temporal intervals:

$$\delta^*(T_Q, T_D) = \min_{[s,t] \in T_Q, [u,v] \in T_D} \delta([s,t],[u,v]) \qquad (2)$$

an idea borrowed from hierarchical clustering, and known as *single-link* [18]. Clearly, if δ is a generalized metric, so is δ^*, as the metric properties are preserved by the min function. With this definition, all the burden is placed on the definition of δ. Following our previous discussion, we propose the following three metrics. In all cases, as in (2), we assume that the first interval is a query interval, and the second interval is a document interval.

Manhattan Distance. Recall from Definition 2 that the timeline is a discrete set of integers. One metric that is well-suited for discrete spaces is the Manhattan distance [19] (also known as Taxicab distance, or L_1 distance). Intuitively, the Manhattan distance sums up the distances between the starting and ending times of the two intervals, resulting to zero only when the two intervals are exactly the same. We call this function δ_{sym}, to stress the fact that it is the only symmetric function (and proper metric) we consider:

$$\delta_{sym}([a,b],[c,d]) = |a - c| + |b - d|$$

Figure 2 shows five different possible cases. With δ_{sym}, knowing if an interval is from a query or a document makes no difference, since it is symmetric.

Query-Biased Hemidistance. Recall the example from Fig. 1. A user might consider the broader document more relevant because it covers the query scope. With the query-biased hemidistance, we assign a distance zero to all documents that completely cover the query scope, and a positive distance to documents that do not cover part of the query scope. Furthermore, if the query and the document scopes do not intersect, the gap between the two intervals is also added to their distance. We call this function $\delta_{cov(Q)}$, to stress the fact that a good document covers the query:

$$\delta_{cov(Q)}([a,b]_Q,[c,d]_D) = (b-a) - (min\{b,d\} - max\{a,c\})$$

Notice from Fig. 2 how this function is not symmetric, as it is biased in favor of the first interval, i.e. the query interval.

Document-Biased Hemidistance. Symmetrically, a user might consider the broader document less relevant because it is too generic. She might consider relevant a document which falls inside the query interval or, in other words, which is covered by the query interval. The document-biased hemidistance is the opposite case as the query-biased hemidistance, hence we call this function $\delta_{cov(D)}$:

$$\delta_{cov(D)}([a,b]_Q,[c,d]_D) = (d-c) - (min\{b,d\} - max\{a,c\})$$

Again, Fig. 2 shows that this function is not symmetric, as it is biased in favor of the second interval, i.e. the document interval. It is also interesting to notice that the Manhattan distance is the sum of the two hemidistances, for any given pair of temporal intervals. Depending on the user task, any of these three metrics can be more appropriate than the others.

7 Combining the Rankings

Textual and temporal similarities cannot model the relevance in isolation better than a combination of both. In general, the textual similarity taken in isolation might be more effective than the temporal similarity taken in isolation, which implies that the two measures should be combined with different weights. The method we use is straightforward. We linearly combine the two similarity measures for each document. The resulting combined scores are, in turn, the final ranking. Similar ideas have been proposed in [13,7].

Given a query Q, we compute $sim_{kw}(Q, D_i)$ and $sim_{\delta^*}(Q, D_i)$ for each document D_i, where sim_{kw} is the keyword-only similarity. All scores are in $[0,1]$ (normalized, if necessary) and higher for greater similarity (i.e., lower distance). This process implies transforming the results of δ^*, which are distances, to values indicating similarity. One way for doing this is with an exponential decay function (similarity decreases exponentially with distance), $sim_{\delta^*}(Q, D_i) = e^{-\delta^*(T_Q, T_{D_i})}$, which gives, by definition, scores in $(0,1]$. If $T_Q = \emptyset$ or $T_{D_i} = \emptyset$ we set $sim_{\delta^*}(Q, D_i) = 0$. Modeling the similarity by exponential decay functions has also been studied in psychology [20]. We then compute all combined scores:

$$sim(Q, D_i) = (1-\alpha)sim_{kw}(Q, D_i) + (\alpha)sim_{\delta^*}(Q, D_i) \qquad (3)$$

for a linear combination parameter $\alpha \in [0,1]$. The final ranking is simply given by ordering the resulting set of scores. Setting α to 0 reduces the model to the keyword-only case. Setting α to 1 results in a temporal-only ranking.

8   Experimental Analysis

We evaluated the effectiveness of our proposed method using the TREC Novelty 2004 test collection,[2] consisting of 1808 documents extracted from the AQUAINT corpus, and 50 topics. The documents are news articles from three newswires (New York Times News Service, AP and Xinhua News Service), spanning a period of time of 5 years (from January 1996 through September 2000).

Queries. Topics, numbered N51-N100, comprise a *title*, a *description* and a *narrative* each. While the titles are short and concise descriptions of the query need, the descriptions and narratives are longer and truly natural language texts.

Relevance Assessments. Relevance assessments are given at the finer granularity of sentences. We abstracted from that level by simply considering relevant a document with at least one relevant sentence, obtaining in average 24 relevant documents per query. The "new sentence" assessments introduced in the Novelty track have been ignored in our experiments.

Temporal Features. We extracted temporal expressions (aka timexes) from both documents and queries with state-of-the-art NLP tools. In particular, HeidelTime [10] was used for identification, and TIMEN [11] for normalization. The first tool produced TimeML [21] documents in which timexes were annotated with TIMEX3 tags. The latter step required providing a "dct", i.e. the document (or query) creation time, to solve relative expressions (e.g. "last year"). The collection had dct's for documents but not for queries, hence we used 2013-01-01 for all queries. A set of 13 rules were used to map the normalized strings produced by TIMEN into temporal intervals, strictly following TimeML semantics. They were simple regular expressions capturing references to centuries, decades, years, months, weeks, days, as well as references to past n years, months, weeks, days, and generic past, future and current time references.

Table 1 shows statistics about timex extraction and interpretation. Temporal expressions were less frequent in queries than in documents, as we expected. However, using the descriptions and narratives from the topics gave us enough data to run our tests, resulting in 11 temporal queries.

Table 1. Temporal features in the Novelty collection

	TREC Novelty 2004 Collection		
	Documents	Topic Desc.	Topic Narr.
Number	1808	50	50
Percentage containing timexes	75%	22%	10%
Total number of timexes found	10620	14	6
Percentage of timexes mapped to intervals	81%	100%	100%

---

[2] http://trec.nist.gov/data/t13_novelty.html

8.1   Effectiveness of the Combined Ranking

In the experiment, we compared the effectiveness of combining temporal and non-temporal scores against a non-temporal ranking baseline, and we assessed the impact on temporal queries versus non-temporal queries. We selected $sim_{\delta^*_{cov(D)}}$ to model the temporal scope similarity, since it gave us the best results in terms of *mean average precision* (MAP). Lucene's default similarity[3] (based on the vector space model [17]) was used as the non-temporal, keyword-only baseline.

Sensitivity Varying α. In this test, we compared the MAP of the text-only ranking and the combined ranking, for 50 different combination parameters $\alpha \in [0, 1]$. Results are shown in Fig. 3, computed over the entire Novelty collection (all documents and all queries). Figures 3a and 3b show results when topic titles and topic descriptions, respectively, were used as textual queries. Using narratives as textual queries resulted in the worse keyword rankings, and their results are thus omitted. In all cases, the union of all the temporal expressions extracted from the topic descriptions and narratives were used as temporal queries. From the figures, it is clear how small α's improved the overall effectiveness in all cases.

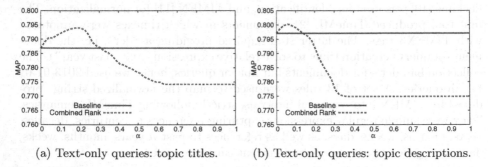

(a) Text-only queries: topic titles.   (b) Text-only queries: topic descriptions.

Fig. 3. Sensitivity of MAP for different combination parameters

Impact on Temporal Queries. We also compared the impact on the 11 temporal queries (i.e., such that $T_Q \neq \emptyset$) versus the entire set of queries.[4] Results are shown in Table 2. In the table, we reported *precision-at-k* (P@k), *recall-at-k* (R@k) and *MAP-at-k* (MAP@k) at different cutoff levels, when topic descriptions are used as textual queries and $\alpha = 0.06$ (the best combination weight from the previous experiment). Several conclusions can be drawn from this table. First of all, it confirms that combining textual with temporal scores improves the baseline in most cases (all the values in bold), both in terms of precision and recall. More importantly, the results obtained on the 11 temporal queries were generally higher than those obtained considering all 50 queries, even when the

---

[3] https://lucene.apache.org/

[4] Considering non-temporal queries alone would not change the rankings since we would get null temporal scores.

baseline was less effective in terms of precision (see values with •), or better in terms of recall and MAP (see values with *). This means that the temporal scope similarity we introduced in this paper has a higher impact on temporal queries, which, in turns, confirms the soundness of our model.

Table 2. Impact of temporal queries. Better than baseline: **bold**; better on temporal queries: *; worse on temporal queries: •.

	Effectiveness over all 50 queries							Effectiveness over 11 temporal queries					
	Baseline			Combined Rank				Baseline			Combined Rank		
k	P@k	R@k	MAP@k	P@k	R@k	MAP@k	k	P@k	R@k	MAP@k	P@k	R@k	MAP@k
5	0.84	0.17	0.16	0.84	0.17	0.16	5	0.83•	0.18*	0.17*	0.81•	0.18*	0.17*
10	0.80	0.33	0.30	**0.81**	0.33	**0.31**	10	0.79•	0.34*	0.31*	**0.81**	**0.35***	**0.32***
20	0.77	0.64	0.57	**0.78**	**0.65**	**0.58**	20	0.76•	0.66*	0.57	**0.79***	**0.69***	**0.60***
	$\alpha = 0$			$\alpha = 0.06$				$\alpha = 0$			$\alpha = 0.06$		

Significance Analysis. Since all score improvements were relatively small, we also performed significance analysis to strengthen our results. We run the Bootstrap Paired Test, as described in [22], using 10,000 bootstrap samples, on the 50 systems from Fig. 3b. The smallest p-value obtained was 0.05, corresponding to the system having $\alpha = 0.06$ (the best-performing one). Re-running the test by only including scores that were better than the baseline resulted in the lowest p-value of 0.04, again for $\alpha = 0.06$. This shows that there is a very low chance that the improvements given by our model are only due to chance.

9 Conclusion and Future Work

In this paper, we have studied temporal aspects of documents and queries, and we have introduced a temporal ranking model based on generalized metrics among the temporal scopes of documents and queries. We have shown that temporal scope similarities lead to effectiveness improvements only when combined with non-temporal similarity measures. This implies a *multi-faceted relevance*, in which time plays an important role. Future work will investigate ways to incorporate other dimensions, such as space, in the ranking model. We will also address the problem of efficiency: we aim at studying properties of the model that can be exploited to allow fast search in the temporal dimension and fast ranking for temporal queries.

Acknowledgments. This work has been supported in part by the Italian Ministry of Education, Universities and Research FIRB project RBFR107725 and OPLON within Smart Cities and Communities and Social Innovation project SCN_00176.

References

1. Alonso, O., Strötgen, J., Baeza-Yates, R., Gertz, M.: Temporal information retrieval: Challenges and opportunities. In: 1st Temporal Web Analytics Workshop at WWW, pp. 1–8 (2011)
2. Jones, R., Diaz, F.: Temporal profiles of queries. ACM Transactions on Information Systems (TOIS) 25(3) (2007)
3. Campos, R., Dias, G., Jorge, A.M., Nunes, C.: Enriching temporal query understanding through date identification: how to tag implicit temporal queries? In: Proceedings of the 2nd Temporal Web Analytics Workshop, pp. 41–48. ACM (2012)
4. Berberich, K., Bedathur, S., Alonso, O., Weikum, G.: A language modeling approach for temporal information needs. In: Advances in Information Retrieval, pp. 13–25 (2010)
5. Nunes, S., Ribeiro, C., David, G.: Use of temporal expressions in web search. In: Advances in Information Retrieval, pp. 580–584 (2008)
6. Snodgrass, R.T.: Temporal databases. IEEE Computer 19, 35–42 (1986)
7. Li, X., Croft, W.: Time-based language models. In: Proceedings of the Twelfth International Conference on Information and Knowledge Management, pp. 469–475. ACM (2003)
8. Alonso, O., Gertz, M., Baeza-Yates, R.: On the value of temporal information in information retrieval. In: ACM SIGIR Forum, vol. 41, pp. 35–41. ACM (2007)
9. Verhagen, M., Gaizauskas, R., Schilder, F., Hepple, M., Moszkowicz, J., Pustejovsky, J.: The tempeval challenge: identifying temporal relations in text. Language Resources and Evaluation 43(2), 161–179 (2009)
10. Strötgen, J., Gertz, M.: Heideltime: High quality rule-based extraction and normalization of temporal expressions. In: Proceedings of the 5th International Workshop on Semantic Evaluation, pp. 321–324. Association for Computational Linguistics, Uppsala (2010)
11. Llorens, H., Derczynski, L., Gaizauskas, R., Saquete, E.: Timen: An open temporal expression normalisation resource. In: Proceedings of the 7th International Conference on Language Resources and Evaluation (2012)
12. Metzler, D., Jones, R., Peng, F., Zhang, R.: Improving search relevance for implicitly temporal queries. In: Proceedings of the 32nd International ACM SIGIR Conference on Research and Development in Information Retrieval, pp. 700–701. ACM (2009)
13. Kanhabua, N., Nørvåg, K.: Determining time of queries for re-ranking search results. In: Lalmas, M., Jose, J., Rauber, A., Sebastiani, F., Frommholz, I. (eds.) ECDL 2010. LNCS, vol. 6273, pp. 261–272. Springer, Heidelberg (2010)
14. Gey, F., Larson, R., Kando, N., Machado, J., Sakai, T.: Ntcir-geotime overview: Evaluating geographic and temporal search. In: NTCIR, vol. 10, pp. 147–153 (2010)
15. Diaz, F., Dumais, S., Efron, M., Radinsky, K., de Rijke, M., Shokouhi, M.: Sigir 2013 workshop on time aware information access (# taia2013). In: Proceedings of the 36th International ACM SIGIR Conference on Research and Development in Information Retrieval, pp. 1137–1137. ACM (2013)
16. Nunes, S.: Exploring temporal evidence in web information retrieval. In: Future Directions in Information Access (FDIA) (2007)
17. Salton, G., Wong, A., Yang, C.: A vector space model for automatic indexing. Communications of the ACM 18(11), 613–620 (1975)
18. Sibson, R.: Slink: an optimally efficient algorithm for the single-link cluster method. The Computer Journal 16(1), 30–34 (1973)

19. Black, P.E.: Manhattan distance. Dictionary of algorithms and data structures. US National Institute of Standards and Technology (2006)
20. Shepard, R.N., et al.: Toward a universal law of generalization for psychological science. Science 237(4820), 1317–1323 (1987)
21. Pustejovsky, J., Castano, J., Ingria, R., Saurí, R., Gaizauskas, R., Setzer, A., Katz, G., Radev, D.: TimeML: Robust specification of event and temporal expressions in text. In: Mani, I., Pustejovsky, J., Gaizauskas, R. (eds.) The Language of time: a Reader. Oxford University Press (2005)
22. Sakai, T.: Evaluating information retrieval metrics based on bootstrap hypothesis tests. Information and Media Technologies 2(4), 1062–1079 (2007)

Local Linear Matrix Factorization for Document Modeling

Lu Bai, Jiafeng Guo, Yanyan Lan, and Xueqi Cheng

Institute of Computing Technology, Chinese Academy of Sciences, BeiJing, China
bailu@software.ict.ac.cn, {guojiafeng,lanyanyan,cxq}@ict.ac.cn

Abstract. Mining low dimensional semantic representations of document is a key problem in many document analysis and information retrieval tasks. Previous studies show better representation mining results by incorporating geometric relationships among documents. However, existing methods model the geometric relationships between a document and its neighbors as independent pairwise relationship; while the pairwise relationship relies on some heuristic similarity/dissimilarity measures and predefined threshold. To address these problems, we propose a Local Linear Matrix Factorization (LLMF), for low dimensional representation learning. Specifically, LLMF exploits the geometric relationships between a document and its neighbors based on local linear combination assumption, which encodes richer geometric information among the documents. Moreover, the linear combination relationships can be learned from the data without any heuristic parameter definition. We present an iterative model fitting algorithm based on quasi-Newton method for the optimization of LLMF. In the experiments, we compare LLMF with the state-of-the-art semantic mining methods on two text data sets. The experimental results show that LLMF can produce better document representations and higher accuracy in document classification task.

Keywords: document modeling, local linear combination, matrix factorization.

1 Introduction

Extracting low dimensional semantic representations of documents has shown great success in wide applications[4][22]. Typically, by representing the corpus as a document-word matrix, matrix factorization can be applied to identify the semantic co-occurrence patterns of words (known as topics), and meanwhile extract low dimensional document representations over the topics [15][6]. Compared to the original document-word matrix, the low dimensional representations exhibit better semantics and achieve higher computation and storage efficiency.

Recent studies suggest that the documents are usually sampled from a non-linear low dimensional subspace which is embedded in the high dimensional ambient space [7][8][14]. Thus, the local geometric structure is essential to reveal the hidden semantics in the corpora, and should be preserved when learning the low dimensional semantic representations. Based on this idea, some works (such

M. de Rijke et al. (Eds.): ECIR 2014, LNCS 8416, pp. 398–411, 2014.

as LapPLSA[7], LTM[8], DTM[14] and GNMF [6]) model the geometric relationship in a manifold way, which requires that the low dimensional representations of two documents to be close if they are neighbors in the original space. The empirical experiments demonstrated these methods can produce better results than the traditional factorization methods.

However, there are two clear drawbacks on only modeling the pairwise geometric relationship between documents. Firstly, the pairwise relationship usually relies on some heuristic similarity/dissimilarity measures, and the local neighborhood structure selected with predefined threshold as well. Unfortunately, it is unclear which measure can well capture the closeness of document pairs and the threshold is often hard to define in practice. Secondly, the pairwise geometric relationships between a document and its neighbors are assumed to be independent, making the rich geometric information among the local pairs lost. Moreover, these models may easily be affected by biased distribution of document pairs (especially when many redundant but less similar document pairs are included in the local neighborhood).

In this work, we propose Local Linear Matrix Factorization (LLMF), a novel low dimensional representation learning method by better exploiting the geometric relationship among documents. Specifically, inspired by the Local Linear Embedding (LLE [17]) method, we capture the geometric relationships among documents through representing a document by a linear combination of its neighbor documents. The linear combination coefficients demonstrate not only the geometric relationships between the document and its neighbors, but also the relationships among the neighbors. Unlike the LLE method, the linear combination coefficients are obtained by solving a regression problem with l_1 constraints [20], which simultaneously build the nearest neighborhood structure of the documents. Therefore, we do not resort to choosing any similarity/dissimilarity measure or predefining a threshold to select neighbors. With the learned combination coefficients, LLMF produces the low dimensional semantic representations by factorizing the document-word matrix with the local linear constraint. The learning process is straight forward, and can be summarized as an iterative process to fit the model in a quasi-Newton way.

We conduct empirical experiments on two benchmark text data sets. The results demonstrate that LLMF can produce better document representations and achieve higher accuracy in document classification task compared to several state-of-the-art semantic representation learning methods.

2 Preliminary Studies

In this section, we briefly introduce some previous studies on matrix factorization and the incorporation of geometric information.

2.1 A Brief View of Matrix Factorization

Matrix factorization or matrix decomposition is a technique that factorizes a source matrix into the production of several matrices based on the discipline of linear algebra. Given a data matrix $W^T = [W_1, W_2, \ldots, W_N]^T \in \mathbb{R}^{N \times M}$, where $W_i, i \in [1, n]$ denotes a M-dimension data vector. A simple factorization of W^T is to produce two related matrices $U \in \mathbb{R}^{N \times K}$, $V \in \mathbb{R}^{K \times M}$ that approximate W^T with the production of U and V:

$$W^T \approx UV \tag{1}$$

The dimension K is usually set as a small value, e.g. $K \ll M$, thus U is a low dimensional and compact representation of W^T with the new basis depicted by V. In some domains, such as document modeling and image processing, U or V are required to be non-negative. Components thus are additive-only to construct data that makes the factorization more interpretable and favorable in practice.

Square Euclidean distance between W^T and UV is a typical objective function for matrix factorization (e.g. in NMF).

$$\min \sum_{i,j} (W_{ij}^T - U_{i\cdot} V_{\cdot j})^2 \tag{2}$$

The problem in (2) is bi-convex, which means the problem is convex on $U(V)$ when $V(U)$ is fixed. We can apply an iterative algorithm for optimization. As suggested in [2], it is convenient to optimize U or V as a linear regression problem alternatively, then stop the iteration when the loss is small enough.

Alternatively, KL-divergence is also a popular measure for the approximation of W^T with U and V, especially when U and V fall into the probabilistic perspective (e.g. PLSA). Usually, the statistic inference methods, e.g. EM or MCMC, are employed to solve the optimization problem.

2.2 Document Modeling with Geometric Constraints

Matrix factorization has been applied in document modeling to find compact representations by minimizing the reconstruction error. Recent studies show that by incorporating geometric information among documents, one can learn better low dimensional representations. A variety of document modeling methods employ the idea of Laplacian Eigenmap(LE) to enhance the geometric properties of learned topics, such as GNMF [6], LapPLSA [7] and LTM [8]. Specifically, a document manifold is first constructed by selecting the neighbors of each document with some similarity measure and threshold. The geometric constraints of LE can then be formulated as the following optimization problem.

$$\min \sum_{i,j} S_{ij} \parallel x_i - x_j \parallel_2^2 \tag{3}$$

where S_{ij} denotes the similarity between i and j. Intuitively, the optimization of (3) is equivalent to make the low dimensional representation x_i, x_j close when i and j are close in the original space.

Obviously, the existing document modeling methods based on LE preserve the geometric information among documents by only modeling the geometric relationships between independent document pairs. Thus, the rich geometric information among the neighbor pairs are lost in this case. Moreover, the pairwise relationship relies on some heuristic similarity measure as well as the predefined threshold for selecting the neighbors. However, it is unclear which measure can well capture the closeness of document pairs and the threshold is often hard to define in practice.

3   Local Linear Matrix Factorization

In this section, we introduce a novel low dimensional representation learning method better exploiting the geometric information among documents, namely Local Linear Matrix Factorization(LLMF). We also provide an effective algorithm for optimization. In addition, we give some detailed discussions about the differences between LLMF and other document modeling methods.

3.1   Model Formalization

Suppose we have N documents over the vocabulary of size M. Let $D^T = [D_1, D_2, \ldots, D_N]^T \in \mathbb{R}_+^{N \times M}$ denote a document-word matrix, where D_{ij}^T is the occurrence number of word j in document i. $\theta \in \mathbb{R}_+^{N \times K}$ is the low dimensional representation of D^T with $K \ll N$. $\beta \in \mathbb{R}_+^{K \times M}$ is the corresponding basis of the latent semantic space. From the perspective of matrix factorization, D^T can be expressed as

$$D^T \approx \theta\beta \qquad (4)$$

By using the square error to measure the approximation in formula (4), we obtain the following non-negative matrix factorization problem for document modeling

$$\mathcal{L}_\theta = \sum_{i=1}^M \sum_{j=1}^N (D_{ij}^T - \theta_{i\cdot}\beta_{\cdot j})^2 + \lambda_\theta \parallel \theta \parallel_2^2 + \lambda_\beta \parallel \beta \parallel_2^2 \qquad (5)$$

$$s.t. \quad \theta \geq 0, \beta \geq 0$$

where λ_θ and λ_β are the weights of l_2 regularizers used to reduce the over-fitting, and the non-negative constraints over θ and β make the learned components interpretable.

Inspired by LLE [17], we consider that local geometric information can be captured by local linear combination relationship (i.e. document can be reconstructed by linear combination of its neighbors), rather than independent pairwise relationships. Specifically, document d can be approximate as

$$D_d^T \approx \phi_d^T \hat{D}^T \qquad (6)$$

where ϕ_d denotes the combination weight vector for document d, and $\hat{D}^T$ denotes the normalized document-word matrix obtained by $\hat{D}_{ij}^T = \frac{D_{ij}^T}{L_i}$, where L_i is the length of document i. We use normalized document-word matrix in local linear combination to avoid the bias of long documents. Note that for the combination weight vector of document d, documents not belonging to the neighbors of d have the value 0 at the corresponding entries of ϕ_d.

The local linear combination constraints can then be expressed by the following objective function

$$\mathcal{L}_\phi = \sum_{i,j}^{N} (D_{ij}^T - \phi_{i\cdot}^T \hat{D}_{\cdot j}^T)^2 + \gamma \parallel \phi \parallel + \lambda_\phi \parallel \phi \parallel_2^2 \qquad (7)$$

where γ and λ_ϕ denotes the weights for l_1 and l_2 norm over ϕ, respectively. Here we put the l_1 norm over ϕ due to the assumption that the number of neighbors of each document is small. It is worth noting that minimizing the formula (7) actually conducts neighbor selection and combination weight learning simultaneously. In this way, we can avoid using heuristic similarity measures and threshold to select neighbors as previous methods.

Unlike previous document modeling methods with geometric constraints (e.g. GNMF, LapPLSA and LTM), it is not straightforward to combine the local linear constraints $\mathcal{L}_\phi$ with the matrix factorization objective $\mathcal{L}_\theta$, since the θ and ϕ is not shared by both optimizations. However, we can bridge θ and ϕ by the normalized basis $\hat{D}$. Based on formula (4), we have that

$$\hat{D}^T = [\hat{D}_1^T, \dots, \hat{D}_N^T] = [\hat{\theta}_1 \beta, \dots, \hat{\theta}_N \beta] = \left([\hat{\theta}_1, \dots, \hat{\theta}_N] \right) \beta = \hat{\theta} \beta \qquad (8)$$

where the entries of $\hat{\theta}_i$ can be derived as $\hat{\theta}_{ik} = \frac{\theta_{ik}}{L_i}$, $k \in [1, \cdots, K]$. In this way, formula (6) can be rewritten as

$$D_d^T \approx \phi_d \hat{D}^T \approx \phi_d \hat{\theta} \beta \qquad (9)$$

Compared formula (9) with formula (4), we obtain that

$$\theta_d \approx \phi_d [\hat{\theta}_1, \dots, \hat{\theta}_N] \qquad (10)$$

Then we can integrate the local linear constraints into the matrix factorization, and the objective function can be expressed as

$$\mathcal{L}(\theta, \beta) = \sum_{i=1}^{N} \sum_{j=1}^{M} (\sum_{k=1}^{K} \theta_{ik} \beta_{kj} - D_{ij})^2 + \lambda_\theta \sum_{i=1}^{N} \sum_{k=1}^{K} \theta_{ik}^2 + \lambda_\beta \sum_{k=1}^{K} \sum_{j=1}^{M} \beta_{kj}^2$$

$$+ \eta \sum_{i=1}^{N} \sum_{k=1}^{K} (\theta_{ik} - \sum_{n=1}^{N} \frac{1}{L_n} \phi_{in} \theta_{nk})^2 \qquad (11)$$

where the coefficient η controls the trade-off between the matrix factorization objective and the local linear constraints.

3.2   Model Fitting

In this section we would show how to infer the latent factor ϕ, θ and β respectively. The inference process can be divided into two optimization problems.

ϕ can be firstly evaluated by optimizing the objective function (7) as a regression problem. Since the basis $\hat{D}^T$ can be pre-computed by normalizing the words count for every documents, the optimization leaves ϕ unknown. It is easy to prove that minimizing the objective function (7) is a convex problem, where the global optimal solution can be found. However, the l_1 regularizer makes the optimization problem in formular (7) not differentiable when some dimension of ϕ is 0. To address this problem, we adopt the OWL-QN [3] algorithm for optimization.

OWL-QN algorithm is based on the famous Quasi-Newton algorithm that leverages the second order information to accelerate the optimization. The algorithm would check the state for each iteration step and revise the value if some dimension cross the orthant boundary. The OWL-QN algorithm requires to calculate the gradient of the function without the l_1 norm part, which is shown as follows

$$\frac{\partial \mathcal{L}'_\phi}{\partial \phi_{in}} \propto \sum_{w=1}^{M} \Big(\sum_{n=1}^{N} \phi_{in} \hat{D}_{nw}^T - D_{iw}^T \Big) \hat{D}_{nw}^T + \lambda_\phi \phi_{in} \tag{12}$$

The iteration stops when the change of objective function (7) is small enough.

After evaluating ϕ, θ and β can be obtained by objective function (5). Since both θ and β are unknown, the optimization problem is bi-convex and we can use the alternative strategy for optimization. Here we solve the problem with gradient method. The gradient of θ and β can be calculate as following:

Algorithm 1. Learning Procedure of LLMF

Require: $D^T, K \in Z_+, \epsilon > 0, \lambda_\theta, \lambda_\phi, \lambda_\beta, \eta$
  $f \leftarrow$ the function to calculate the loss of ϕ as function 7
  $g \leftarrow$ the function to calculate the gradient of ϕ as function 12
  initialize $\phi \in \mathbb{R}_+^{N \times N}$ randomly
  Learn $\phi \leftarrow$ OWL-QN$(D^T, \phi, \eta, \epsilon, f, g)$
  initialize $\theta^{(0)} \in \mathbb{R}_+^{N \times K}$ randomly
  initialize $\beta^{(0)} \in \mathbb{R}_+^{K \times M}$ randomly
  $f \leftarrow$ the function to calculate the loss of ϕ as function 5
  $g_\theta \leftarrow$ the function to calculate the gradient of ϕ as function 14
  $g_\beta \leftarrow$ the function to calculate the gradient of β as function 13
  $l^{old} \leftarrow 0$
  for $t = 1 : T$ **do**
    Learn $\theta^{(t)} \leftarrow$ OWL-QN$(D^T, \theta^{(t-1)}, 0, \epsilon, f, g_\theta)$
    Learn $\beta^{(t)} \leftarrow$ OWL-QN$(D^T, \beta^{(t-1)}, 0, \epsilon, f, g_\beta)$
    $l \leftarrow$ calculate loss function value as function 5
    if $\| l - l^{old} \| < \epsilon$ **then**
      break
    end if
    $l^{old} \leftarrow l$
  end for
  return θ, β, ϕ

$$\frac{\partial \mathcal{L}_\theta}{\partial \beta_{kw}} = \sum_{d=1}^{N} \left(\sum_{k=1}^{K} \theta_{dk}\beta_{kw} - D_{dw}^{T} \right) \theta_{dk} + \lambda_\beta \beta_{kw} \qquad (13)$$

$$\frac{\partial \mathcal{L}_\theta}{\partial \theta_{dk}} = \sum_{w=1}^{M} \left(\sum_{k=1}^{K} \theta_{dk}\beta_{kw} - D_{dw}^{T} \right)\beta_{kw} + \lambda_\theta \theta_{dk} + \eta \sum_{b=1}^{N} \left(\sum_{n=1}^{N} \frac{1}{L_n}\phi_{bn}\theta_{nk} - \theta_{bk} \right)\left(\frac{1}{L_b}\phi_{bd} \right)$$
$$-\eta \left(\sum_{n=1}^{N} \frac{1}{L_n}\phi_{dn}\theta_{nk} - \theta_{dk} \right) \qquad (14)$$

The algorithm to infer ϕ, θ and β are described in the algorithm 1.

3.3 Discussion

In this section, we further provide some discussions on the differences between the proposed LLMF method and existing state-of-the-art document modeling methods.

Compared to NMF, PLSA and LDA, LLMF smoothed the low dimensional representations of documents with its neighbors. The weights for each neighbors are evaluated by solving a least square regression problem that captures the geometric information among documents. Both LDA and NMF smooth the latent representations with a unimodal prior distribution (Dirichlet distribution for LDA and Gaussian distribution for NMF), but the posterior distribution would prefer to fit the most intensive areas globally. In all the three methods (i.e. NMF, PLSA and LDA), the local geometric information among documents is not considered.

When compared with document modeling methods with geometric constraints, like LapPLSA and LTM, the way of capturing the geometric information in these

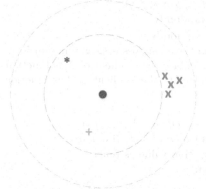

Fig. 1. Biased latent representations when data distribution is unbalanced

methods is quite different from that of our proposed LLMF. Previous methods smooth the latent representations with their neighbors in a pairwise way. It is illustrated in [5] that in these methods the similarity measure and neighborhood threshold should be carefully defined when constructing the local manifold patch, otherwise the improper neighborhood would dramatically bias the latent representations and affect the performance. We explain this phenomenon here and show the advantage of local linear constraints over pairwise constraints. As shown in Fig. 1, one aims to learn the latent representation of red circle point, where the green snow point (*) and yellow plus point (+) are more close than the blue cross points (x). When the blue points are included as neighbors of the target red point, the learned representation of the red circle point would be biased to the blue points due to the pairwise smoothing regularization and unbalanced data distribution. This problem would become more severe when more neighbors are used for pairwise smoothing. However, in our LLMF , the local geometric information is preserved by linear combination of its neighbors, and thus the neighbors are competitive in representing the data. Therefore, the importance of the green and red points in representing the red point would not be affected much even when the blue points are involved, and the weight of each blue point would be reduced due to the redundancy. As a result, LLMF can better preserve the rich geometric relationships among data and learn better low dimensional representations.

4   Experiments

In this section, we demonstrate the results in the task of document classification, with experiments conducted on two widely used benchmark text corpora, i.e. 20newsgroup and la1. Firstly, we introduce the experimental settings. Then we qualitatively evaluate the latent topics learned by LLMF. At last, we evaluate the effectiveness of the proposed LLMF in document classification by comparing with the state-of-the-art semantic learning methods.

4.1   Data Sets and Baseline Methods

Our empirical studies on text semantics learning were conducted on two real-word text corpora, i.e. 20newsgroup dataset and la1 data set.

- The 20newsgroup[1] is a benchmark text collection for topic modeling, which contains almost 18,744 postings to Usenet newsgroups in 20 different categories almost evenly. The vocabulary is pruned by stemming each term to its root and removing the stop words for noise concern.
- la1 is a public dataset in Weka[2] containing 2,850 documents in 5 categories. The vocabulary consists of 13,195 unique words, and is also preprocessed by

---

[1] http://people.csail.mit.edu/jrennie/20Newsgroups/
[2] http://www.cs.waikato.ac.nz/ml/weka/

stemming and stop words removing. Unlike 20newsgroup, la1 is an unbalanced dataset where different categories contain quite different number of documents.

To evaluate the performance of LLMF, we provide comparisons of our methods against several state-of-the-art topic learning methods, including PLSA [13], LDA [4], NMF [15] and LapPLSA [13]. Here we briefly introduce the experimental settings about these methods beside our LLMF.

- **PLSA:** Probabilistic Latent Semantic Analysis(PLSA) is introduced by Hoffman[13] as a probabilistic version of LSI [9]. We use the code from Peter's homepage[3] for experiment.
- **LDA:** Latent Dirichlet Allocation(LDA) is a full Bayes version of PLSA. We user the code from the author's homepage of LDA[4], which is implemented in c-language by Blei.
- **NMF:** Non-negative Matrix Factorization(NMF) is a traditional dimension reduction methods. We adopt the alternating constrained least squares (ACLS) [2] to factorize the document-word matrix. To avoid over-fitting, l_2 norm is added as a constraint over the factorized matrices.
- **LapPLSA:** LapPLSA smooths the topic representations of nearby document pairs. In our experiments, we applied different number of neighbors and weights of the geometric regularizer, and select the best performance to report.
- **LLMF:** In our proposed LLMF, we set the parameters $\lambda_\theta, \lambda_\beta, \lambda_\phi$ as $0.01, 0.1, 1$, respectively. The l_1 norm in the model is weighted by $\gamma \in \{0.001, 0.01, 0.1, 1\}$.

All the above methods are conducted on both dataset several times with random initialization by setting the dimension $K \in \{30, 50, 70, 100\}$. We compare the best results of different methods and demonstrate the results in the following sections.

4.2   Topic Learning

In this section, we qualitatively evaluate the learned semantic information by LLMF. To illustrate the meaning of the inferred semantic factors, we randomly select several column from the matrix β, and re-range the words according to the corresponding weights in that column in descending order.

In Table 1, we list the top 10 important words from the randomly selected 5 learned components over the two datasets, respectively. For better understanding, we manually label each topic according to the meaning of the selected words. It is interesting to see that the selected words are closely related, and show similar semantic meanings expressed by the labels. Therefore, the results show that our LLMF can effectively learn the latent semantic information from the corpus.

---

[3] http://people.kyb.tuebingen.mpg.de/pgehler/code/index.html
[4] http://www.cs.princeton.edu/blei/lda-c

Table 1. Topics Learned by LLMF over the Two Datasets

20news					la1				
image	hardware	hockey	ibm	motor-race	sports	financial	national	market	computer
jpeg	tape	team	ibm	motorcyc	game	fund	west	bank	stor
gif	driver	hockei	hardwar	ride	plai	stock	soviet	compani	disk
compress	adaptec	leag	ram	rec	team	market	chemic	million	electron
viewer	sys	nhl	machin	bmw	season	price	plant	loan	comput
jfif	backup	season	memori	bike	player	invest	weapon	amp	data
convert	memori	game	card	club	coach	trad	german	card	ibm
format	run	player	monitor	time	basketbal	bond	govern	sav	machin
quantis	cdrom	championship	dos	rider	goal	investor	libya	credit	softwar
imag	floppi	wing	cpu	moto	time	exchang	israel	billion	pc
displai	hardwar	vs	bus	biker	win	trade	american	market	user

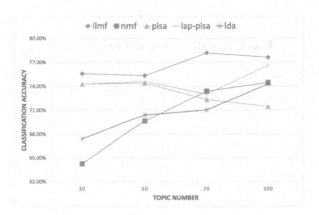

Fig. 2. Classification performance on 20newsgroups

4.3 Document Classification

In this section, we quantitatively evaluate the effectiveness of the learned low dimensional representations. Since the learned low dimensional representation is usually taken as feature engineering in real application tasks such as classification and clustering, we propose to evaluate the representations by comparing the performances of different methods in these tasks.

Here we conduct the task of document classification on the two data sets mentioned above, and compare the classification accuracies of different models. Specifically, we use all the documents of each data set to learn the parameters of different models. Then we randomly select 60% documents with their inferred representations as features to build the multi-class SVM classifier, and the rest 40% documents for test. We adopted the LIBSVM toolbox[5] as our implementation for SVM. Cross validation is conducted to select the parameter C in SVM.

The classification results on the two data sets are reported in Figure 2 and 3. We can see that LLMF consistently achieves better accuracy than all the baseline methods in both data sets. The results indicate that we can learn better

---

[5] http://www.csie.ntu.edu.tw/~cjlin/libsvm/

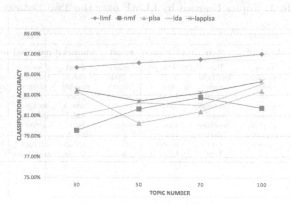

Fig. 3. Classification performance on la1

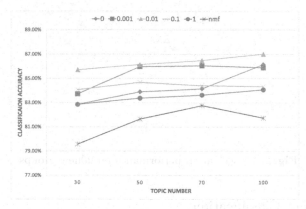

Fig. 4. Classification performance on la1 with the variation of the parameter γ

semantic representation by LLMF. It demonstrates that it is valuable to pre-serve more detailed local geometric information using local linear combination for learning semantic representation. Moreover, we can see that the difference between between LLMF and LapPLSA in Figure 3 is even larger than that in Figure 2. The reason may lie in the imbalance of la1, since the pairwise relation will be highly biased by imbalanced data, as shown in our discussion.

We also study the robustness of our methods under different sizes of neigh-borhood. In our proposed LLMF, different γs represent different sparseness of neighbors. Intuitively, the bigger γ is, the fewer neighbors wil be used in the linear combination, and vise versa. Therefore, we show the variation of classifi-cation accuracy on the unbalanced dataset la1 as the weight γ changes in LLMF in Figure 4. Since LLMF is essentially a generalized non-negative matrix factor-ization method, we take NMF as the baseline for comparison. From the results we can clearly see that with different γs, LLMF are consistently better than the basic NMF method in terms of document classification accuracy. It demonstrates the robustness of our proposed LLMF method.

5   Related Work

Learning an effective semantic representation of data can greatly improve the effectiveness and efficiency of many real applications, such as information retrieval [22], network analysis [1], recommend systems [21] and so on.

For document modeling, matrix factorization is a typical way to learn the low dimensional representations of documents, such as LSI [9] and NMF [15]. These methods directly factorize the document-word matrix into low rank matrices according to different criteria. As an alternative, topic models, such as PLSA [13] and LDA [4], provide a probabilistic view on document modeling. Specifically, each document is taken as a distribution over topics, where each topic is a distribution over words. Through the posterior optimization, [18] provides a probabilistic generative view in interpreting the matrix factorization. [12] and [10] demonstrate the close connections between PLSA and NMF.

Integrating the geometric relationship among documents into the document modeling methods has been proved reasonable and effective. Some studies [11][19][16] leverage the explicit relationship, such as links and citations, in topic learning. Alternatively, several researches employ the geometric relationship, e.g. manifold assumption, to improve the document modeling. For example, Lap-PLSA [7] increases the proximity between the topics of document pairs in neighborhood using Laplacian eigenmap constraints based on PLSA. LTM [8] takes the same assumption as LapPLSA, but leverages the KL-divergence to evaluate the difference of topics instead. GNMF [6] and GraphSC [23] leverage the graph embedding to regularize the latent factor learning. DTM [14] not only enhances the topical proximity between nearby document pairs, but also increases the topical separability between the unfavorable pairs. As far as we known, all these above methods need firstly select the documents' neighbors with heuristic similarity measure and thresholds, and then preserve the geometric relationship by enhancing the pairwise proximity.

6   Conclusions

In this paper, we present a novel method for learning low dimensional representations of document, namely Local Linear Matrix Factorization(LLMF). LLMF exploits the geometric relationships between a document and its neighbors based on local linear combination assumption in document modeling. In this way, LLMF can better capture the rich geometric information among documents than those based on independent pairwise relationships. The experimental results on document classification show LLMF can learn better low dimensional semantic representations than the state-of-the-art baseline methods.

In the future, we would like to extend LLMF to the paralleled and distributed settings for computation efficiency. Moreover, it would also be interesting to apply LLMF to other scenarios, e.g. recommender systems, where dimension reduction has shown benefits for the application.

Acknowledgments. This work was funded by the 973 Program of China under Grants No. 2012CB316303 and No. 2014CB340401, and National Natural Science Foundation of China under Grant No. 61232010, No. 61003166, No. 61203298, No. 61173064 and No. 61272536.

References

1. Airoldi, E.M., Blei, D.M., Fienberg, S.E., Xing, E.P.: Mixed membership stochastic blockmodels. J. Mach. Learn. Res. 9, 1981–2014 (2008)
2. Albright, R., Cox, J., Duling, D., Langville, A.N., Meyer, C.D.: Algorithms, initializations, and convergence for the nonnegative matrix factorization. Matrix (919), 1–18 (2006)
3. Andrew, G., Gao, J.: Scalable training of l1-regularized log-linear models. In: ICML 2007, pp. 33–40. ACM, New York (2007)
4. Blei, D.M., Ng, A.Y., Jordan, M.I., Lafferty, J.: Latent dirichlet allocation. Journal of Machine Learning Research 3 (2003)
5. Cai, D., He, X., Han, J.: Locally consistent concept factorization for document clustering. IEEE Transactions on Knowledge and Data Engineering 23(6), 902–913 (2011)
6. Cai, D., He, X., Han, J., Huang, T.S.: Graph regularized nonnegative matrix factorization for data representation. IEEE Trans. Pattern Anal. Mach. Intell. 33(8), 1548–1560 (2011)
7. Cai, D., Mei, Q., Han, J., Zhai, C.: Modeling hidden topics on document manifold. In: CIKM 2008, pp. 911–920. ACM, New York (2008)
8. Cai, D., Wang, X., He, X.: Probabilistic dyadic data analysis with local and global consistency. In: ICML 2009, pp. 105–112. ACM, New York (2009)
9. Deerwester, S.C., Dumais, S.T., Landauer, T.K., Furnas, G.W., Harshman, R.A.: Indexing by latent semantic analysis. JASIS 41(6), 391–407 (1990)
10. Ding, C., Li, T., Peng, W.: On the equivalence between non-negative matrix factorization and probabilistic latent semantic indexing. Comput. Stat. Data Anal. 52(8), 3913–3927 (2008)
11. Erosheva, E., Fienberg, S., Lafferty, J.: Mixed-membership models of scientific publications 101(suppl. 1), 5220–5227 (2004)
12. Gaussier, E., Goutte, C.: Relation between plsa and nmf and implications. In: SIGIR 2005, pp. 601–602. ACM, New York (2005)
13. Hofmann, T.: Unsupervised learning by probabilistic latent semantic analysis. In: Machine Learning (2001)
14. Huh, S., Fienberg, S.E.: Discriminative topic modeling based on manifold learning. In: KDD 2010, pp. 653–662. ACM, New York (2010)
15. Lee, D.D., Seung, H.S.: Learning the parts of objects by non-negative matrix factorization. Nature 401(6755), 788–791 (1999)
16. Nallapati, R.M., Ahmed, A., Xing, E.P., Cohen, W.W.: Joint latent topic models for text and citations. In: KDD 2008, pp. 542–550. ACM, New York (2008)
17. Roweis, S.T., Saul, L.K.: Nonlinear dimensionality reduction by locally linear embedding. SCIENCE 290, 2323–2326 (2000)
18. Salakhutdinov, R., Mnih, A.: Probabilistic matrix factorization. In: NIPS (2007)

19. Sun, C., Gao, B., Cao, Z., Li, H.: Htm: a topic model for hypertexts. In: EMNLP 2008, Stroudsburg, PA, USA, pp. 514–522. Association for Computational Linguistics (2008)
20. Tibshirani, R.: Regression shrinkage and selection via the lasso. Journal of the Royal Statistical Society, Series B 58, 267–288 (1994)
21. Wang, C., Blei, D.M.: Collaborative topic modeling for recommending scientific articles. In: KDD 2011, pp. 448–456. ACM, New York (2011)
22. Wei, X., Croft, W.B.: Lda-based document models for ad-hoc retrieval. In: SIGIR 2006, pp. 178–185. ACM, New York (2006)
23. Zheng, M., Bu, J., Chen, C., Wang, C., Zhang, L., Qiu, G., Cai, D.: Graph regularized sparse coding for image representation. Trans. Img. Proc. 20(5), 1327–1336 (2011)

Video Clip Retrieval by Graph Matching

Manal Al Ghamdi* and Yoshihiko Gotoh

Department of Computer Science, University of Sheffield, United Kingdom
{m.alghamdi,y.gotoh}@dcs.shef.ac.uk

Abstract. This paper presents a new approach to video clip retrieval using the Earth Mover's Distance (EMD). The approach builds on the many-to-many match methodology between two graph-based representations. The problem of measuring similarity between two clips is formulated as a graph matching task in two stages. First, a bipartite graph with spatio-temporal neighbourhood is constructed to explore the relation between data points and estimate the relevance between a pair of video clips. Secondly, using the EMD, the problem of matching a clip pair is converted to computing the minimum cost of transportation within the spatio-temporal graph. Experimental results on the UCF YouTube Action dataset show that the presented work attained a significant improvement in retrieval capability over conventional techniques.

Keywords: graph matching, Earth Mover's Distance, video retrieval.

1 Introduction

Content-based video processing has been the centre of attention among researchers in recent years. A wide range of applications rely on accurate representation of visual features so that relevant clips can be identified efficiently [1]. Although text is the most commonly used form for presenting contents when retrieving information, text alone may not be sufficient for presenting videos because visual information can be interpreted broadly. Hence alternative representations of visual contents can be explored. Video retrieval has two fundamental issues: how to present video contents and how to measure the similarity. Graphs have drawn attention recently as an effective tool for representation in various computer vision applications [1]. Region of interest in images or videos can be presented by a collection of nodes and edges in a graph. The problem of comparing video clips is reformulated as finding the consistent relation between two sets of features through their graphs. Two graphs are aligned by matching their nodes in a way that conserves the edges of the matched nodes [2].

A number of studies have been conducted to address the graph matching problem for various applications. Zaslavskiy *et al.* proposed a complex approach to detecting many-to-many correspondence between graphs [2], although their solution involved discrete optimisation, it was not very suitable for a graph

---

* The first author would like to thank Umm Al-Qura University, Makkah, Saudi Arabia for funding this work as part of her PhD scholarship program.

M. de Rijke et al. (Eds.): ECIR 2014, LNCS 8416, pp. 412–417, 2014.

matching problem. Zhou and De la Torre proposed the factorised graph matching (FGM) algorithm [3]. They employed the Kronecker product to factorise the affinity matrix into four smaller matrices: a source and a target graph's incidence matrices, a node and an edge affinity matrices. They required large memory to encode the whole affinity matrix without an approximation.

This paper presents a novel video retrieval framework based on graph matching. We formulate the similarity measurement between a pair of clips as a graph matching problem in two stages. Firstly, a graph-based representation is defined with a spatio-temporal neighbourhood. Second, the matching problem is converted to a task of measuring the distance between their graphs using the Earth Mover's Distance (EMD), developed for image retrieval [4]. The contributions of this work are as follows; (1) *Representation:* A high-dimensional representation is mapped to a spatio-temporal graph, where nodes represent frames and edges represent the temporal order. (2) *Matching:* The clip similarity is measured using the distance between their graphs. The EMD finds the minimum cost by solving the transportation problem. (3) *Retrieval:* The unsupervised approach, which does not require prior information or training, to video clip retrieval is presented. Space-time coherence embedded in video sequences is described, then a many-to-many graph mapping technique defines the similarity between clips.

2   The Approach

Given two video clips, the approach is formulated as a graph matching problem in two stages — graph-based video representation (Section 2.1), followed by similarity measurement between the two clips (Section 2.2). We use the following notations in the remainder of this paper: Let $X = \{x_1, \ldots, x_M\} \in \mathbb{R}^{M \times D}$ denote a query clip with M frames and D dimensions, where x_i ($i = 1, \ldots, M$) represents a frame in X. Let $Y_k = \{y_1, \ldots, y_N\} \in \mathbb{R}^{N \times D}$ be the k-th test clip with N frames and D dimensions, where y_j ($j = 1, \ldots, N$) represents a frame in Y_k.

2.1   Graph-Based Construction

The structure in the high-dimensional space is transferred to a spatio-temporal distance graph with nodes representing frame instances in a query X connecting to the related nodes in a test clip Y_k. The method reconstructs the frames order based on their spatio-temporal relationship and recalculates distances along them to ensure the shortest distance. Initially a distance matrix $D_k = \{d_{ij}\}$ between two videos X and Y_k is derived, where d_{ij} is the cost, or the distance, between two frames x_i and y_j, calculated by:

$$d_{ij} = \left(\sum_{d=1}^{D} \|x_{id} - y_{jd}\|^2 \right)^{\frac{1}{2}} \tag{1}$$

where $\| \cdot \|^2$ is the ℓ^2 norm. Then, for each frame instance x_i ($i = 1, \ldots, M, j$):

1. L frames, whose distance is the closest to x_i, are connected. They are referred to as spatial neighbours (sn):

$$sn_{x_i} = \left\{ y_{j1}, \ldots, y_{jL} \mid \underset{j}{\operatorname{argmin}}^L (d_{ij}) \right\} \qquad (2)$$

where $\operatorname{argmin}^L_j$ implies L node indexes j with the shortest distances;

2. Another L frames, chronologically ordered around x_i, are set as temporal neighbours (tn): $tn_{x_i} = \left\{ y_{j-\frac{L}{2}}, \ldots, y_{j-1}, y_{j+1}, \ldots, y_{j+\frac{L}{2}} \right\}$

3. Optimally, (tn_{sn}) is selected from temporal neighbours of spatial neighbours as: $tn_{sn_{x_i}} = \left\{ tn_{y_{j1}} \cup tn_{y_{j2}} \cup \ldots \cup tn_{y_{jL}} \right\} \cap tn_{x_i}$

4. Spatial and temporal neighbours are then integrated, producing spatio-temporal neighbours (stn) for frame x_i as: $stn_{x_i} = sn_{x_i} \cup tn_{sn_{x_i}}$ The above formulation of stn_{x_i} effectively selects x_i's temporal neighbours that are similar, with a good chance, to its spatial neighbours. This means that, suppose x_i is an isolated frame and totally different from the temporal neighbours, only the spatial neighbours will be taken into consideration.

Given the neighbourhood graph, Dijkstra's shortest path algorithm finds the distance between nodes [5]. This forms a new correlation matrix $G_k = \{X, Y_k, E_k\}$ of pairwise geodesic distances with $V_k = X \cup Y_k$ as the vertex set and $E_k = \{\omega_{ij}\}$ as the edge set, where ω_{ij} presents the distance between two frames x_i and y_j.

2.2   Graph Matching

The EMD was initially designed for the image retrieval task [4], which we apply to define the optimal match between two video clips. The approach assumes that the similarity is modelled by a weighted graph, and the EMD determines the minimum cost as similarity between the two weighted graphs. It involves solving the well-known *transportation problem* by determining the minimum amount of 'work' required to achieve the transformation. Given the weighted graph G_k constructed between X and Y_k, the EMD can be calculated as follows:

1. (Transportation problem) Define the flow $F = \{f_{ij}\}$, where f_{ij} is the flow between x_i and y_j, that minimises the following cost function:

$$\text{WORK}(X, Y_k, F) = \sum_{i=1}^{M} \sum_{j=1}^{N} d_{ij} f_{ij} \qquad (3)$$

with respect to the constraints: (1) $f_{ij} \geq 0$, $i = 1, \ldots, M$, $j = 1, \ldots, N$ to limit the moving frames from X to Y_k and not vice versa; (2) $\sum_{j=1}^{N} f_{ij} \leq \omega_{x_i}$, $i = 1, \ldots, M$ to control the number of frames sent by X to their weights; (3) $\sum_{i=1}^{M} f_{ij} \leq \omega_{y_j}$, $j = 1, \ldots, N$ to control the events in Y_k not to receive more

than their weights; (4) and finally $\sum\limits_{i=1}^{M}\sum\limits_{j=1}^{N} f_{ij} = \min\left(\sum\limits_{i=1}^{M}\omega_{x_i}, \sum\limits_{j=1}^{N}\omega_{y_j}\right)$ to allow
the maximum number of frames to move.

2. The EMD is calculated as the resulted work normalised by the total flow:

$$\text{EMD}(X, Y_k) = \frac{\text{WORK}(X, Y_k, F)}{\sum\limits_{i=1}^{M}\sum\limits_{j=1}^{N} f_{ij}} = \frac{\sum\limits_{i=1}^{M}\sum\limits_{j=1}^{N} d_{ij} f_{ij}}{\sum\limits_{i=1}^{M}\sum\limits_{j=1}^{N} f_{ij}} \qquad (4)$$

Finally, $\text{SIM}(X, Y_k) = 1 - \text{EMD}(X, Y_k)$ calculates the similarity between two
clips X and Y_k. Note that $\text{SIM}(X, Y_k)$ ranges between 0 and 1, where X and
Y_k are very similar when the value is close to 1.

3   Experiments

We tested the spatio-temporal approach to graph matching using the UCF
YouTube Action (UCF11) dataset; clips were collected from typical YouTube
videos with large variations in camera move, scale, illumination, *etc.* [6]. The
UCF11 consists of 1600 videos with 11 action categories: basketball shooting,
cycling, diving, trampoline jumping, golf swinging, horse riding, soccer juggling,
volleyball spiking, swinging (by children), walking with dogs and tennis swinging. Each category contains 25 scenes with at least four clips for each scene. For
this experiment, we randomly chose 50 clips to represent each action, two from
each scene, totalling 550 clips in the dataset. From the 550 videos, we picked up
five query clips randomly for each action, making 55 videos in the query set.

For the local features detector we adopted spatio-temporal scale invariant
feature transform (SIFT) combined with the locality-constrained linear coding
(LLC) [7]. Given a video stream the detector identified spatially and temporally invariant interest points, containing the amount of information sufficient to
represent the video content. Spatio-temporal regions around the interest points
were described using the 3-dimensional histogram of Gaussian (HOG) [8]. The
descriptor length was 640-dimensional, determined by the number of bins to
represent the orientation angles in the sub-histograms. In the spatial pyramid
matching (SPM), the LLC codes were computed for each spatio-temporal sub-
region and pooled together using the multi-scale max pooling. We used $4 \times 4, 2 \times 2$
and 1×1 sub-regions. The pooled features were concatenated and normalised
using the ℓ^2-norm. The initial number of frames to be spatially connected in the
manifold appeared dependent on the clip length and was selected manually.

3.1   Results and Analysis

The spatio-temporal graph matching with the EMD (GM/EMD) presented in
this paper was compared with the maximum graph matching (MM), the FGM
and the graph matching with continuous relaxation (GM/CR). The conventional

Table 1. The video clip retrieval task: the average precision (P) and recall (R) using the UCF11 dataset. The spatio-temporal graph matching with the EMD (GM/EMD) was presented along with the maximum graph matching (MM), the factorised graph matching (FGM) and the graph matching with continuous relaxation (GM/CR).

Query category	GM/EMD		MM		FGM		GM/CR	
	P	R	P	R	P	R	P	R
basketball shooting	0.801	0.712	0.375	0.299	0.410	0.527	0.629	0.514
cycling	0.723	0.650	0.429	0.438	0.526	0.491	0.617	0.593
diving	0.622	0.600	0.250	0.461	0.378	0.483	0.522	0.459
trampoline jumping	0.684	0.698	0.382	0.323	0.473	0.400	0.450	0.409
golf swinging	0.733	0.528	0.333	0.255	0.442	0.307	0.603	0.477
horse riding	0.898	0.676	0.500	0.409	0.617	0.496	0.765	0.530
soccer juggling	0.610	0.758	0.155	0.333	0.329	0.485	0.498	0.570
volleyball spiking	0.790	0.754	0.380	0.399	0.501	0.489	0.588	0.565
swinging (by children)	0.821	0.692	0.442	0.220	0.493	0.411	0.592	0.504
walking with dogs	0.875	0.837	0.562	0.491	0.619	0.500	0.694	0.619
tennis swinging	0.661	0.587	0.444	0.206	0.591	0.396	0.537	0.426
average	0.747	0.681	0.387	0.348	0.489	0.453	0.590	0.515

MM did one-to-one mapping [9]; a 'match' was found by computing the maximum cardinality in a weighted bipartite graph. As introduced in Section 1, both FGM [3] and GMCR [2] were state-of-the-art techniques in graph matching[1].

Table 1 presents the video retrieval performance using two classical information retrieval measures, average precision and recall, showing that the GM/EMD outperformed the other techniques. The following observations can be made:

- The GM/EMD proved its ability to represent events in real video sequences with large variations in camera motion, object appearance and pose, scale, viewpoint, cluttered background, illumination conditions, *etc.* It was able to discover the similarity of clips from the same action category and the dissimilarity between clips that shared some common motions but were from different categories (*e.g.*, 'tennis swinging' and 'golf swinging').
- The EMD has many interesting properties — to some extent it imitates the human perception of texture similarities. It allows partial matches and can be applied to general variable-size signatures.
- The GM/CR approach [2] achieved better results than the FGM [3]. The FGM uses the pairwise relation between feature points with unary information. This relationship is rotation-invariant but not scale-invariant nor affine-invariant.
- Many-to-many mapping approaches (GM/EMD, FGM and GM/CR) performed better than the one-to-one approach (MM). One-to-one mapping was too restrictive, while many-to-many mapping allowed flexible matches between the vertices between two graphs. With the many-to-many mapping,

---

[1] We used the publicly available Matlab code from the authors' websites.

object parts can be represented by multiple vertices given noise or various view points. This situation could not be handled through the one-to-one matching.

- The MM [9] presented the lowest performance by enforcing the one-to-one mapping over video clips. One-to-one mapping does not work effectively when clip contents are similar but not identical because one frame can only be mapped to one corresponding frame.
- The 'basketball shooting', 'trampoline jumping', 'soccer juggling' and 'volleyball spiking' are easily mixed as they all involved 'jumping' motion. The 'cycling' and 'horse riding' categories had similar camera motion. All the 'swinging' actions shared a common motion. Some actions were performed with objects such as a horse, a bike or a dog. Despite all this, the GM/EMD was able to match and retrieve the relevant clips in many cases.

4   Conclusions

We presented an approach to graph representation and similarity measurement for the video clip retrieval task using the EMD. A graph-based representation was defined for video sequences with their spatio-temporal neighbourhood. The video clips similarity was then measured using the distance between a pair of graphs. The EMD measured the video clip similarity by computing the minimum transportation cost between two graphs. Experimental results using the real and challenging UCF11 dataset indicated that the approach was more capable of retrieving the relevant video clips than existing techniques.

References

1. Chen, L., Chua, T.S.: A match and tiling approach to content-based video retrieval. In: Proceedings of IEEE International Conference on Multimedia and Expo. (2001)
2. Zaslavskiy, M., Bach, F., Vert, J.-P.: Many-to-many graph matching: a continuous relaxation approach. In: Balcázar, J.L., Bonchi, F., Gionis, A., Sebag, M. (eds.) ECML PKDD 2010, Part III. LNCS, vol. 6323, pp. 515–530. Springer, Heidelberg (2010)
3. Zhou, F., de la Torre, F.: Factorized graph matching. In: Proceedings of IEEE International Conference on Computer Vision and Pattern Recognition (2012)
4. Rubner, Y., Tomasi, C., Guibas, L.J.: The earth mover's distance as a metric for image retrieval. International Journal of Computer Vision (2000)
5. van der Maaten, L.J.P., Postma, E.O., van den Herik, H.J.: Dimensionality reduction: A Comparative Review (2008)
6. Liu, J., Luo, J., Shah, M.: Recognizing realistic actions from videos in the wild. In: Proceedings of IEEE International Conference on Computer Vision and Pattern Recognition (2009)
7. Al Ghamdi, M., Al Harbi, N., Gotoh, Y.: Spatio-temporal video representation with locality-constrained linear coding. In: Fusiello, A., Murino, V., Cucchiara, R. (eds.) ECCV 2012 Ws/Demos, Part III. LNCS, vol. 7585, pp. 101–110. Springer, Heidelberg (2012)
8. Scovanner, P., Ali, S., Shah, M.: A 3-dimensional SIFT descriptor and its application to action recognition. In: Proceedings of ACM Multimedia (2007)
9. Plummer, D., Lovász, L.: Matching theory. Elsevier Science (1986)

EDIUM: Improving Entity Disambiguation via User Modeling

Romil Bansal, Sandeep Panem, Manish Gupta, and Vasudeva Varma

International Institute of Information Technology,
Hyderabad, India

Abstract. Entity Disambiguation is the task of associating entity name mentions in text to the correct referent entities in the knowledge base, with the goal of understanding and extracting useful information from the document. Entity disambiguation is a critical component of systems designed to harness information shared by users on microblogging sites like Twitter. However, noise and lack of context in tweets makes disambiguation a difficult task. In this paper, we describe an Entity Disambiguation system, *EDIUM*, which uses User interest Models to disambiguate the entities in the user's tweets. Our system jointly models the user's interest scores and the context disambiguation scores, thus compensating the sparse context in the tweets for a given user. We evaluated the system's entity linking capabilities on tweets from multiple users and showed that improvement can be achieved by combining the user models and the context based models.

Keywords: Entity Disambiguation, Knowledge Graph, User Modeling.

1 Introduction

Named Entity Disambiguation (NED) is the task of identifying the correct entity reference from the knowledge bases (like DBpedia, Freebase or YAGO), for the given entity. In microblogging sites like Twitter, NED is an important task for understanding the user's intent and for topic detection and tracking, search personalization and recommendations.

In the past, many NED techniques have been proposed. Some utilize contextual information of an entity, while others use candidate's popularity for disambiguation. But tweets being short and noisy, lack sufficient context for these systems to disambiguate precisely. Due to this, the underlying user interests are modeled to disambiguate the entities [1–3]. However, creation of the user models might require some external knowledge (like Wikipedia edits [1]), which are computationally expensive or labor intensive. Many other researchers have also tried to model entity disambiguation through user interest models [2, 3]. Our approach is different from these approaches as it tries to combine contextual models and user models by analyzing the user's tweeting behavior. The user model is built by analyzing user's behavior on the previous tweets.

This approach can be used for modeling users and disambiguating entities in other streaming documents like emails or query logs as well. The next section describes the *EDIUM* system in detail.

M. de Rijke et al. (Eds.): ECIR 2014, LNCS 8416, pp. 418–423, 2014.

2   The *EDIUM* System

EDIUM works by representing user's interests as a distribution over the semantic Wikipedia categories. *EDIUM* has three sub-systems: the context modeling system (Section 2.1), the user modeling system (Section 2.2) and the disambiguation system (Section 2.3). The user's interests are modeled based on the tweet categories by the user. These interests along with the local context are used by the disambiguation system for linking entities in the new tweets. The final results are fed back to the system for improving the user model.

Every tweet has multiple entity mentions and each mention can be aligned to multiple entities. Table 1 represents few notations used while describing the system.

Table 1. Notations used for Describing the System

Symbol	Description
C_i^j	j-th candidate entity for the i-th entity mention in a given tweet
c_i^j	Contextual similarity score for candidate entity C_i^j
$Par(C)$	Parent of C is set of all categories that are the immediate ancestor of category C in Wikipedia Category Network
$G(C)$	Grandparent of C is set of all categories such that $G(C) \subseteq Par(Par(C))$ and $G(C) \cap Par(C) = \emptyset$
$N_r(C)$	Set of all categories in the r-th neighborhood of category C in Wikipedia Category Network
IC_u^i	Set of all categories in the i-th interest cluster for user u
ic_u^i	Score for the interest cluster IC_u^i

2.1   The Contextual Modeling System

Context Model (CM) disambiguates the entities based on the text around the entities. Similarity between the text around the entity mention and the text on Wikipedia page of an entity is compared and an appropriate weightage for disambiguation is given to each candidate entity. The candidate entity with the maximum weightage is considered as the disambiguated entity for the given entity mention.

Many techniques have been proposed to disambiguate the entity mentions in the text [4–6] based on the context. We used existing entity linking systems like DBpedia Spotlight [4] and Wikipedia Miner [5] for linking and disambiguating the entities based on the context.

The contextual score $Score_C(C_i^j)$ is the candidate score normalized based on all the possible candidates for the given entity mention. We improved the referents' disambiguation scores by combining the context based scores with the user's interest based scores in an appropriate manner.

2.2   The User Modeling System

User Model (UM) understands the user's interests and behavior.

UM Creation: We used cluster-weighted models[1] for modeling the user's interests. The following assumptions were made while creating the user models.

- Users only tweet on topics that interest them.
- The amount of interest in a topic is proportional to the information shared by the user on the topic.

Based on these assumptions, we modeled each user into weighted clusters of semantic Wikipedia categories. Each cluster represents the user's interest in specific topic and weight represents the overall interest of the user in that topic.

The UM is updated for user's future tweets based on the categories in the user's current tweet. The tweet categories are extracted using the following steps.

1. Current tweet's entities are discovered via the disambiguation modeling (DM) system (Section 2.3).
2. The entities with sufficiently high confidence are shortlisted to prevent UM from learning incorrect information for the user. We considered only those entities where confidence (ratio of scores of the second ranked entity to the disambiguated entity) is atmost δ.[2]
3. Tweet categories for the shortlisted entities are extracted using Wikipedia. The score of each tweet category is equal to the number of tweet entities belonging to the category.
4. Considering the graph of semantic Wikipedia categories, the tweet categories are smoothed to include the parent categories. Parents are given scores in inverse proportion to their out-degree for each child category. Common parent gets lesser contribution from the child's score as compared to a rare parent.

The UM is created based on the tweet categories. If the category is already present in the model, the score is updated by the normalized sum of the initial and the tweet category score. Otherwise the category and its score is added to the model. As the new tweet category scores get added to the UM, the model evolves to better represent the newly processed tweet.

To find the topics of interest for the user, each category is mapped to a single interest cluster. Although many clustering techniques over the Wikipedia network have been proposed [7], for efficient computations so as to enable streaming scenarios, we formed clusters based on the similar ancestors for a given category. The score of the k-th interest cluster, ic_u^k, for user u, is the sum of the weights of the categories in the cluster k.

Twitter users exhibit different interest behaviors: some users are highly specific while others tweet about almost every category. This can be inferred from the fact that some users tweet about trending hashtags or popular news, while others tweet only about highly specific products or companies. Entity disambiguation depends highly on the behavior of the users. While making use of interest models might be useful in the latter case, it might not be that effective in the former case. To handle this issue,

---

[1] http://en.wikipedia.org/wiki/Cluster-weighted_modeling

[2] The parameter δ depends on the use case and performance of the underlying CM system. While high values ensure the large learning rates, low values ensure the performance of the UM system.

we introduce the concept of relatedness between the User Model (UM) and the Disambiguation Model (DM).

Similarity between the UM and the DM is defined as the cosine similarity between the tweet categories vector obtained when DM is used ($Score_D(C_i)$, Eq. 6) vs. that when only the user model is used ($Score_U(C_i)$, Eq. 5) for disambiguation.

$$sim(UM, DM) = cos(Score_D(C_i), Score_U(C_i)) \tag{1}$$

Similarly, similarity between the CM and the DM is defined as the cosine similarity between the tweet categories vector obtained when DM is used ($Score_D(C_i)$, Eq. 6) vs. that when only CM is used ($Score_C(C_i)$) for disambiguation.

$$sim(CM, DM) = cos(Score_D(C_i), Score_C(C_i)) \tag{2}$$

At time t, let $R^{(t)}$ denote the relatedness which is defined as the weight-normalized similarity of UM and DM with that of CM and DM. Weights are assigned in inverse proportion to the model's contribution while disambiguation.

$$R^{(t)} = \frac{(1 - \alpha^{(t)}) \times sim(UM, DM)}{(1 - \alpha^{(t)}) \times sim(UM, DM) + \alpha^{(t)} \times sim(CM, DM)} \tag{3}$$

The parameter $\alpha^{(t)}$ measures the consistency of the user's behavior relative to the user's learnt model at time t. If the user's interest changes frequently, UM will not be able to disambiguate the entities properly, keeping the $\alpha^{(t)}$ low. The higher the value of $\alpha^{(t)}$, the more precise is the user model. The weight α is used to tradeoff between the contributions of the user model and the context model for disambiguation Eq. 6).

We update $\alpha^{(t)}$ after each tweet based on the relatedness value of the newly disambiguated tweet. So as to obtain a stable estimate of α, we compute α as the average of the last m relatedness values (Eq. 4). We used $m = 20$ for our experiments.

$$\alpha^{(t+1)} = \frac{1}{m} \sum_{k=0}^{m-1} R^{(t-m)} \tag{4}$$

To deal with the changing user's interests, we set $\alpha^{(t+1)}$ to $0.9\alpha^{(t)}$ after each day. This lowers the dependency of the DM on the UM with time. To avoid incorrect UM learning, α is restricted to maximum of 0.7. This resists the UM from making decisions without taking the context into consideration.

The UM is committed to the database after each transaction and is used whenever new tweet from the same user arrives. This helps us track huge number of users and build the streaming disambiguation system for Twitter streams.

Disambiguation: For each category C_i^j in a tweet, the final score given by the UM is

$$Score_U(C_i^j) = \sum_{k=0}^{n} sim(C_i^j, IC_u^k) \times Score(IC_u^k) \tag{5}$$

where $Score(IC_u^i)$ is the normalized score of the i-th interest cluster for user u, $sim(C_i^j, IC_u^i)$ is number of common elements in IC_u^i and $N_3(C_i^j)$ and n is the number of interest clusters of the user u. The $Score_U(C_i)$ is then normalized across all possible candidates for the i-th entity mention.

2.3   The Entity Disambiguation Modeling System (DM)

DM disambiguates the entities based on the textual context as well as the user's interests. The DM system combines both the context based model's score and the user based model's score using the parameter $\alpha^{(t)}$, that relates the stability of the user with respect to the previous tweeted topics. The final score predicted by the DM is

$$Score_D(C_i^j) = \alpha^{(t)} \times Score_U(C_i^j) + (1 - \alpha^{(t)}) \times Score_C(C_i^j) \qquad (6)$$

The model selects the entity that maximizes the $Score_D(C_i^j)$ for the given entity mention i.

$$Entity_i = \arg\max_{C_i^j} Score_D(C_i^j) \qquad (7)$$

3   Results and Discussions

We evaluated the performance of *EDIUM* on a dataset annotated manually by three individuals. The dataset consists of 200 tweets each from randomly selected 20 different Twitter users[3]. α is initialized to 0.001 for each user because UM has no prior information about the user. We performed experiments with $m = 20$ and $\delta = 0.9$. Parameter tuning was performed based on 200 tweets each from 5 different Twitter users. As the UM processes tweets from the user's tweet stream, it improves over time. For such a streaming scenario, Precision at 1 (P@1) score for entity disambiguation is calculated at interval of 20 tweets for each user. The system is evaluated with both DBpedia Spotlight (DS) and Wikipedia Miner (WM) as the context modeling systems. Fig. 1 reports the performance of the system over time when the proposed model is used vs. when just the CM is used or just the UM (built using previous tweets with the proposed model) is used for disambiguation. We observed that *EDIUM* started to outperform the CM after ~60 tweets (of each user) are processed by the system. The maximum performance is achieved when the proposed model is used with the Wikipedia Miner as the CM system.

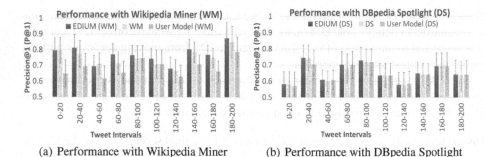

(a) Performance with Wikipedia Miner   (b) Performance with DBpedia Spotlight

Fig. 1. Precision@1 Score of *EDIUM* under Different Configurations

In general, we observed that *EDIUM* performs better with Wikipedia Miner than with DBpedia Spotlight. This is because the system is dependent on the underlying CM

---

[3] We would like to thank Wei Shen [2] for providing the dataset.

for learning the user interests initially. Precise CM leads to faster and more accurate UM. The higher the accuracy of the UM, the better the disambiguation.

Conversely, if the underlying CM has low accuracy for entity disambiguation, the UM usually takes much longer to learn the user's true interests. In that case, UM does not contribute in disambiguation giving insignificant improvement in Precision. However, it is observed that UM alone, after sufficient training, can also disambiguate the entities mentioned in the user's tweets and achieve comparable precision.

4 Conclusion and Future Work

In this paper, we have modeled entity disambiguation based on the user's past interest information. We proposed a way to model the user's interests using the entity linking techniques and then using it later to improve the disambiguation in entity linking systems. The gain in precision is proportional to the accuracy of the underlying entity linking system.

More analysis is required on the user modeling aspect of the system. Experiments on larger datasets are required to test the significance and performance of the system on different categories of the users over longer time duration. Currently user's past tweets are used for building the user model and the model's quality depends significantly on the underlying context model. In the future, we plan to include network and demographic information of the users to improve user modeling and hence the entity disambiguation system.

References

1. Murnane, E.L., Haslhofer, B., Lagoze, C.: RESLVE: Leveraging User Interest to Improve Entity Disambiguation on Short Text. In: Proc. of the 22nd Intl. Conf. on World Wide Web (WWW), Republic and Canton of Geneva, Switzerland, pp. 81–82 (2013)
2. Shen, W., Wang, J., Luo, P., Wang, M.: Linking Named Entities in Tweets with Knowledge Base via User Interest Modeling. In: Proc. of the 19th ACM Conf. on Knowledge Discovery and Data Mining (KDD), pp. 68–76. ACM, New York (2013)
3. Yerva, S.R., Catasta, M., Demartini, G., Aberer, K.: Entity Disambiguation in Tweets Leveraging User Social Profiles. In: Proc. of the 2013 Intl. Conf. on Information Reuse and Integration (IRI), pp. 120–128. IEEE (2013)
4. Mendes, P.N., Jakob, M., García-Silva, A., Bizer, C.: DBpedia Spotlight: Shedding Light on the Web of Documents. In: Proc. of the 7th Intl. Conf. on Semantic Systems, pp. 1–8. ACM, New York (2011)
5. Milne, D., Witten, I.H.: An Open-source Toolkit for Mining Wikipedia. Artificial Intelligence 194, 222–239 (2013)
6. Meij, E., Weerkamp, W., de Rijke, M.: Adding Semantics to Microblog Posts. In: Proc. of the 5th ACM Intl. Conf. on Web Search and Data Mining (WSDM), pp. 563–572. ACM, New York (2012)
7. Qureshi, M.A., O'Riordan, C., Pasi, G.: Short-text Domain Specific Key Terms/Phrases Extraction Using an N-gram Model with Wikipedia. In: Proc. of the 21st ACM Conf. on Information and Knowledge Management (CIKM), pp. 2515–2518. ACM, New York (2012)
8. Michelson, M., Macskassy, S.A.: Discovering Users' Topics of Interest on Twitter: A First Look. In: Proc. of the 4th Workshop on Analytics for Noisy Unstructured Text Data, pp. 73–80. ACM (2010)

A Comparison of Approaches for Measuring Cross-Lingual Similarity of Wikipedia Articles

Alberto Barrón-Cedeño[1], Monica Lestari Paramita[2],
Paul Clough[2], and Paolo Rosso[3]

[1] Talp Research Center, Universitat Politècnica de Catalunya, Barcelona, Spain
albarron@lsi.upc.edu
http://www.lsi.upc.edu/~albarron
[2] Information School, University of Sheffield, Sheffield, UK
{m.paramita,p.d.clough}@sheffield.ac.uk
http://ir.shef.ac.uk/
[3] NLE Lab, PRHLT, Universitat Politècnica de València, Valencia, Spain
prosso@dsic.upv.es
http://users.dsic.upv.es/~prosso

Abstract. Wikipedia has been used as a source of comparable texts for a range of tasks, such as Statistical Machine Translation and Cross-Language Information Retrieval. Articles written in different languages on the same topic are often connected through inter-language-links. However, the extent to which these articles are similar is highly variable and this may impact on the use of Wikipedia as a comparable resource. In this paper we compare various language-independent methods for measuring cross-lingual similarity: character n-grams, cognateness, word count ratio, and an approach based on outlinks. These approaches are compared against a baseline utilising MT resources. Measures are also compared to human judgements of similarity using a manually created resource containing 700 pairs of Wikipedia articles (in 7 language pairs). Results indicate that a combination of language-independent models (char-n-grams, outlinks and word-count ratio) is highly effective for identifying cross-lingual similarity and performs comparably to language-dependent models (translation and monolingual analysis).

Keywords: Wikipedia, Cross-Lingual Similarity.

1 Introduction

Wikipedia, the free online encyclopedia, contains articles on a diverse range of topics written by multiple authors worldwide. It has been used as a source of multilingual data for a range of monolingual and cross-language NLP and IR tasks, such as named entity recognition [14], query translation for CLIR [9], word-sense disambiguation [6] and statistical machine translation [7]. Wikipedia editions have been created in more than 287 languages, with some languages more evolved than others. A proportion of topics appear in multiple Wikipedias

M. de Rijke et al. (Eds.): ECIR 2014, LNCS 8416, pp. 424–429, 2014.
© Springer International Publishing Switzerland 2014

(i.e. in multiple languages), resembling a comparable corpus. However, the similarity and textual relationship between articles written in multiple languages on the same topic (further referred to as *interlanguage-linked articles*) can vary widely. Some articles may be translations of each other; others may have been written independently and cover different aspects of the same topic. In some cases, the articles may even contain contradictory information [3].

Measuring similarity is core to many tasks in IR and NLP. However, few studies have focused on computing Cross-Lingual (CL) similarity in Wikipedia; particularly for under-resourced languages. In this work, we are interested in methods that require few, if any, language resources ("language-independent" approaches). This is important in cases where languages being studied are "under-resourced". We identify various models to compute cross-lingual similarity in Wikipedia (Section 2) and use them to calculate cross-language similarity between 700 Wikipedia article pairs (Section 3). We measure performances of these models using an existing Wikipedia evaluation benchmark [10]. We conclude the paper in Section 4 and provide directions for future research.

2 Cross-Lingual Similarity Models

Multiple models to assess the degree of similarity across languages have been proposed for different tasks, such as the extraction of parallel text fragments for MT [8] and CLIR [2,4]. The purpose is to estimate a similarity value $sim(d, d')$, where $d \in L$ $d' \in L'$ are words, text fragments, or documents written in languages L, L' ($L \neq L'$). The process for CL similarity estimation involves three steps. (*i*) *Pre-processing*: d and d' are passed by a standard normalisation process (e.g. tokenisation, case folding, etc.). (*ii*) *Characterisation*: d and d' are mapped into a common space —some strategies are based on MT techniques, translating d into L', allowing for further monolingual analysis; others break the texts into small chunks (e.g. character n-grams) and exploit syntactic similarities between languages [5,12]; yet other techniques attempt to map concepts in d and d' into a common semantic space on the basis of multilingual thesauri [13] or comparable corpora [11]. (*iii*) *Comparison*: The mappings of d and d' are compared on the basis of a similarity measure (e.g. cosine). In this paper, we focus on a range of methods requiring few, if any, specific language resources. Other models, such as CL-ESA [11], exhibit state-of-the-art performance; however, because we are interested in methods that will operate efficiently for under-resourced languages, we do not consider such resource-demanding approaches. To evaluate performance we compare methods against a language-dependent method which utilises MT followed by a monolingual analysis.

Character n-grams (cng). To calculate char-n-grams, a simplified alphabet $\Sigma = \{a, \ldots, z, 0, \ldots, 9\}$ is considered; i.e. any other symbol, space, and diacritic is discarded and case-folding applied. The text is then codified into a vector of character n-grams ($n = [3, 5]$). This model is an adaptation of McNamee & Mayfield's [5].

Cognateness (cog). This concept was proposed to identify parallel sentences [12]. A token t forms a *cognateness* candidate if: (*a*) t contains at least one digit; (*b*) t contains only letters and $|t| \geq 4$; or (*c*) t is a punctuation mark. t and t' are pseudo-cognates if both belong to (*a*) or (*b*) and are identical, or belong to (*b*) and share the same four leading characters. Hence, we characterise d and d' as follows: if t accomplishes (*a*) , it is maintained verbatim, if it accomplishes (*b*) it is cut down to its first four characters. We neglect (*c*) because we are comparing entire articles. Again, case-folding and removal of diacritics is applied.

Word Count Ratio (wc). This simple measure is computed as the length ratio between the shorter and the longer document (in number of tokens).

Common Outlinks (lnk). This is a model appropriate for analysing Wikipedia articles that has been used on the extraction of similar sentences across languages with encouraging results [1]. It exploits the Wikipedia's internal *outlinks*: if an article in language L (L') links to article a_L ($b_{L'}$) and a and b are about the same topic, the corresponding texts are considered similar. We compute a simplified version where a vector is created representing outlinks in the documents that are mapped to another language using the structure of Wikipedia.

Translation + Monolingual Analysis (trans$_n$). Our baseline is a language-dependent model: we use Google's MT system to translate d into L', generating d_t, which are then compared against d' using a standard monolingual process.

3   Experiments

We selected 7 language pairs: German (de), Greek (el), Estonian (et), Croatian (hr), Latvian (lv), Lithuanian (lt), and Romanian (ro), which are all paired to English (en). These languages were chosen as they exhibit different characteristics, such as writing systems (Greek texts were transliterated using ICU4J: http://site.icu-project.org) and availability of resources: German is highly resourced; the remaining languages are under-resourced.

First, we calculate similarity scores for each document pair using the models described in Section 2. We use an existing Wikipedia evaluation benchmark[1] [10] containing 100 document pairs for each language pair, ranging between 107-1,546 words.The similarities were manually scored between 1–5 by two assessors with appropriate language skills. Overall, 85% of the pairs were given the same score or differ by one, which shows a good agreement between the assessors. For each document pair, we compute the manually-assessed similarity score by averaging assessors' scores, resulting in 9 similarity values (ranging from 1-5). Lastly, we compare the models by calculating Spearman-rank correlation between (combined) automatic to human-assessed similarity scores.

---

[1] We discard Slovenian due to the high number of pairs judged as similar in the evaluation set (>95%) as it would affect the accurate calculation of correlation scores.

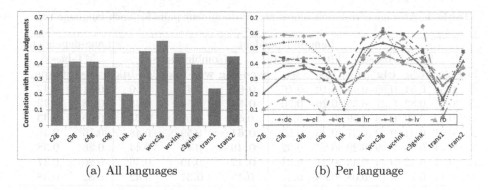

(a) All languages   (b) Per language

Fig. 1. Correlation of different models and human judgements. Note: cng=character n-grams (n=[2-4]); cog=cognateness; lnk=outlinks; wc=word counts; $trans_n$ =translation plus word n-grams comparison.

Spearman-rank correlation coefficient scores are calculated between the automatically- and manually-generated similarity scores for all 700 document pairs and each similarity method (cf. Figure 1(a)). Language-independent models such as char-n-grams ('c2g', 'c3g' and 'c4g') identify cross-lingual similarity with performance comparable to the baseline translation models using bi-gram overlap ('trans2'). The results show that a simplistic language-independent model based on the word count ratio ('wc') correlates higher with human judgements compared to models using MT, which suggests that interlanguage-linked Wikipedia articles with similar lengths are very likely to contain similar content. This correlation, however, can still be improved by using a combination of syntax-based (specifically 'c3g') and structure-based models ('wc' or 'outlinks'). These findings are promising, considering that these models are purely language-independent and can easily be calculated for many language pairs.

Whilst language-independent models perform well overall, their performance may differ for each language pair. Therefore, we also computed correlations in each language. Figure 1(b) shows that whilst char-n-grams perform well on average, their correlations vary significantly across different languages. The simplified outlinks ('lnk') model was less reliable in identifying similarity. However, combining 'lnk' and 'wc' results in a more stable model, performing well for all languages, although slightly lower than the combination of 'wc' and 'c3g'. This combination obtains ρ=0.55; just slightly lower than the combination of 'wc' and 'trans2': ρ=0.57 (not shown in the plot).

We perform a similar analysis for language-dependent models and identify a drastic increase in correlation of using word bi-grams ('trans2') compared to uni-grams ('trans1') overlap. The poor performance for the latter may also be caused by the weighting strategy used: simple tf. A straightforward enhancement of this model would be to remove stopwords and apply tf-idf weighting.

Table 1 shows the various models and their correlation scores (top-performing models highlighted). For all language pairs, the highest correlation to human judgements are achieved using combinations of language-independent methods.

Table 1. Average correlation scores across document pairs for each language pair

	Lang-independent								Lang-dependent	
	Syntax-based				Struct-based		Combinations			MT-based	
lan	c2g	c3g	c4g	cog	lnk	wc	wc+c3g	wc+lnk	c3g+lnk	trans$_1$	trans$_2$
de	0.52	0.54	0.55	0.43	0.10	0.46	**0.63**	0.41	0.50	0.11	0.47
el	0.21	0.32	0.37	0.35	0.26	0.50	**0.54**	0.50	0.38	0.18	0.42
et	0.57	0.59	0.58	0.59	0.34	0.43	0.60	0.52	**0.65**	0.05	0.33
hr	0.47	0.44	0.42	0.37	0.36	0.56	**0.61**	0.59	0.48	0.16	0.48
lt	0.31	0.38	0.39	0.29	0.27	0.32	**0.46**	0.42	0.44	0.26	0.38
lv	0.41	0.42	0.44	0.43	0.21	0.34	**0.47**	0.40	0.36	0.26	0.37
ro	0.11	0.17	0.18	0.08	0.38	0.45	0.45	**0.57**	0.42	0.32	0.38

For five language pairs, 'wc+c3g' is proven to be superior, while 'wc+lnk' and 'c3g+lnk' perform better in Romanian–English and Estonian–English, respectively. These results are promising: by combining language-independent models, it is possible to reliably identify cross-lingual similarity in Wikipedia with better performance (i.e. correlation to human judgement) than using MT systems.

4 Conclusions

This paper compares different methods for computing cross-lingual similarity in Wikipedia. Methods vary from language-independent (syntax-based, structure-based, and a combination of the two), to language-dependent requiring language-specific resources, e.g. an MT system. In contrast to previous work, we investigated the performance of each method for a wide range of under-resourced languages. We analysed correlations between these models and human judgements by making use of an existing evaluation benchmark for Wikipedia.

We conclude that a combination of language-independent models perform better than language-dependent models (i.e. involving translation) of bi-gram word overlap. Word count ratio and char-3-grams perform best in most languages ($\rho = 0.55$), followed by a combination of word count and outlinks ($\rho = 0.47$). A simple translation model using word bi-gram correlates with manual judgements ($\rho=0.45$), followed by the last method combination: char-n-grams and outlinks ($\rho=0.40$). This result is very promising given that these models can be calculated without the need of any translation resources, which will enable these models to be applied to measure cross-lingual similarity to any Wikipedia language pair, possibly after applying transliteration.

As future work we plan to investigate other combinations of language-independent models in order to create a language-independent approach to reliably measure cross-lingual similarity in Wikipedia. We also plan to use these models to further investigate similarity between articles in Wikipedia.

Acknowledgements. The work of the first author was in the framework of the Tacardi research project (TIN2012-38523-C02-00). The work of the fourth author was in the framework of the DIANA-Applications (TIN2012-38603-C02-01) and WIQ-EI IRSES (FP7 Marie Curie No. 269180) research projects.

References

1. Adafre, S., de Rijke, M.: Finding Similar Sentences across Multiple Languages in Wikipedia. In: Proc. of the 11th Conf. of the European Chapter of the Association for Computational Linguistics, pp. 62–69 (2006)
2. Dumais, S., Letsche, T., Littman, M., Landauer, T.: Automatic Cross-Language Retrieval Using Latent Semantic Indexing. In: AAAI 1997 Spring Symposium Series: Cross-Language Text and Speech Retrieval, Stanford University, pp. 24–26 (1997)
3. Filatova, E.: Directions for exploiting asymmetries in multilingual Wikipedia. In: Proc. of the Third Intl. Workshop on Cross Lingual Information Access: Addressing the Information Need of Multilingual Societies, Boulder, CO (2009)
4. Levow, G.A., Oard, D., Resnik, P.: Dictionary-Based Techniques for Cross-Language Information Retrieval. Information Processing and Management: Special Issue on Cross-Language Information Retrieval 41(3), 523–547 (2005)
5. Mcnamee, P., Mayfield, J.: Character N-Gram Tokenization for European Language Text Retrieval. Information Retrieval 7(1-2), 73–97 (2004)
6. Mihalcea, R.: Using Wikipedia for Automatic Word Sense Disambiguation. In: Proc. of NAACL 2007. ACL, Rochester (2007)
7. Mohammadi, M., GhasemAghaee, N.: Building Bilingual Parallel Corpora based on Wikipedia. In: Second Intl. Conf. on Computer Engineering and Applications., vol. 2, pp. 264–268 (2010)
8. Munteanu, D., Fraser, A., Marcu, D.: Improved Machine Translation Performace via Parallel Sentence Extraction from Comparable Corpora. In: Proc. of the Human Language Technology and North American Association for Computational Linguistics Conf (HLT/NAACL 2004), Boston, MA (2004)
9. Nguyen, D., Overwijk, A., Hauff, C., Trieschnigg, D.R.B., Hiemstra, D., de Jong, F.: WikiTranslate: Query Translation for Cross-Lingual Information Retrieval Using Only Wikipedia. In: Peters, C., Deselaers, T., Ferro, N., Gonzalo, J., Jones, G.J.F., Kurimo, M., Mandl, T., Peñas, A., Petras, V. (eds.) CLEF 2008. LNCS, vol. 5706, pp. 58–65. Springer, Heidelberg (2009)
10. Paramita, M.L., Clough, P.D., Aker, A., Gaizauskas, R.: Correlation between Similarity Measures for Inter-Language Linked Wikipedia Articles. In: Calzolari, E.A. (ed.) Proc. of the 8th Intl. Language Resources and Evaluation (LREC 2012), pp. 790–797. ELRA, Istanbul (2012)
11. Potthast, M., Stein, B., Anderka, M.: A Wikipedia-Based Multilingual Retrieval Model. In: Macdonald, C., Ounis, I., Plachouras, V., Ruthven, I., White, R.W. (eds.) ECIR 2008. LNCS, vol. 4956, pp. 522–530. Springer, Heidelberg (2008)
12. Simard, M., Foster, G.F., Isabelle, P.: Using Cognates to Align Sentences in Bilingual Corpora. In: Proc. of the Fourth Intl. Conf. on Theoretical and Methodological Issues in Machine Translation (1992)
13. Steinberger, R., Pouliquen, B., Hagman, J.: Cross-lingual Document Similarity Calculation Using the Multilingual Thesaurus EUROVOC. In: Gelbukh, A. (ed.) CICLing 2002. LNCS, vol. 2276, pp. 415–424. Springer, Heidelberg (2002)
14. Toral, A., Muñoz, R.: A proposal to automatically build and maintain gazetteers for Named Entity Recognition using Wikipedia. In: Proc. of the EACL Workshop on New Text 2006. Association for Computational Linguistics, Trento (2006)

Challenges on Combining Open Web and Dataset Evaluation Results: The Case of the Contextual Suggestion Track

Alejandro Bellogín[1,2], Thaer Samar[1], Arjen P. de Vries[1], and Alan Said[1]

[1] Centrum Wiskunde & Informatica, Amsterdam, The Netherlands
{samar,arjen,alan}@cwi.nl
[2] Universidad Autónoma de Madrid, Madrid, Spain
alejandro.bellogin@uam.es

Abstract. The TREC 2013 Contextual Suggestion Track allowed participants to submit personalised rankings using documents either from the Open Web or from an archived, static Web collection, the ClueWeb12 dataset. We argue that this setting poses problems in how the performance of the participants should be compared. We analyse biases found in the process, both objective and subjective, and discuss these issues in the general framework of evaluating personalised Information Retrieval using dynamic against static datasets.

1 Introduction and Motivation

The Contextual Suggestion TREC Track investigates search techniques for complex information needs that are highly dependent on context and user interests. Input to the task are a set of profiles (*users*), a set of example suggestions (*attractions*), and a set of contexts (*locations*). Each attraction includes a title, a description, and an associated URL. Each profile corresponds to a single user, and indicates the user's preference with respect to each attraction. Two ratings are used: one for the attraction's title and description and another one for its website. Finally, each context corresponds to a particular geographical location (a city and its corresponding state in the United States). With this information, up to 50 ranked suggestions should be generated by each participant for every context and profile pair. Each suggestion should be appropriate to both the user's profile and the context. The description and title of the suggestion may be tailored to reflect the preferences of that user.

As opposed to the 2012 track, where only submissions from the Open Web were allowed [2], in 2013, the organisers introduced the option of submitting rankings from one of the TREC collections. In particular, from the ClueWeb12 dataset, which includes more than 700 million Web pages crawled from the Internet between February and May 2012. This approach would allow further comparison and reproducibility of the results, in agreement with most of the other TREC tracks developed so far.

The Contextual Suggestion track has, however, some characteristics that challenge the standard TREC evaluation setting, and the Cranfield paradigm at large. First, relevance is defined as a bi-dimensional variable, since a document

M. de Rijke et al. (Eds.): ECIR 2014, LNCS 8416, pp. 430–436, 2014.
© Springer International Publishing Switzerland 2014

has to be interesting for the user and appropriate for the given context. Second, it is personalised, hence the typical pooling mechanism where several judges are used and their judgements are aggregated cannot be used. Third, and as a consequence of the second, there is no explicit query in the system, or, as stated in [1], the query "entertain me" is implicitly assumed and fixed during the retrieval/recommendation stage.

In this paper we analyse whether the evaluation results and relevance assessments obtained in this year's track (2013) may be hiding the fact that the ClueWeb12 dataset does not contain as many *interesting* documents as an Open Web dataset. We show that there is an (implicit) bias to receive better judgements by the Open Web submissions when compared with those using ClueWeb12. This result is evidenced by comparing a subset of documents which was assessed as Open Web and ClueWeb12 documents separately, ending up with inconsistent assessments. We finally discuss more general connotations of this work in the broader context of evaluating interactive Information Retrieval with users, combining documents from live and archived web.

2 Experimental Setup

In our analysis, we use ground truth relevance assessments provided by the organisers of the TREC 2013 Contextual Suggestion track. These are provided as two separate categories, depending on whether the relevance is personal (how interesting is this document for the user, in a particular context) or related to the geographical appropriateness of the document in the given context. These judgements also have different scales: subjective judgements range from 0 (strongly uninterested) to 4 (strongly interested) whereas objective judgements go from 0 (not geographically appropriate) to 2 (geographically appropriate). Besides, in both cases, a value of -2 indicates that the document did not load.

Out of the $28,100$ possible combinations of user profiles (562) and contexts (50), only 223 (user, context) pairs were evaluated, roughly 0.8% of the complete set of pairs. For these pairs, the top-5 documents of every submission were judged by the users (profile and geographical relevance) and NIST assessors (geographical relevance). The metrics used to evaluate the submissions were Precision at 5 (P@5), Mean Reciprocal Rank (MRR), and Time-Biased Gain (TBG) [1]. These metrics consider the geographical and profile relevance (both in terms of document and description judgement), taking as thresholds a value of 1 and 3 (inclusive), respectively.

3 Challenges

In this section we present the analysis derived from the data described above. We first compare Open Web and ClueWeb12 in general, to see whether one of them has a higher inherent quality than the other (fair comparison of datasets). Next, we performed a pairwise comparison in a subset of documents shared by the two datasets (consistency of evaluation judgements).

432   A. Bellogín et al.

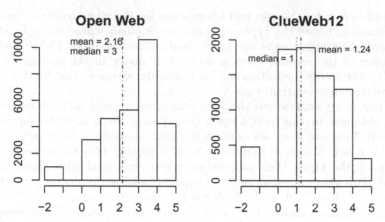

Fig. 1. Histogram of document judgements for documents from the Open Web (left) and from ClueWeb12 (right)

3.1 Fair Comparison of Datasets

We start by analysing how comparable – in terms of inherent quality – the Open Web and ClueWeb12 datasets are for the Contextual Suggestion task. This is important because if, for some reason, one of the datasets is richer, or has a biased view of the world, the documents retrieved will share such a bias. This is, in principle, not desirable [4].

For this analysis, we leave out the user, context, and system variables, and compare the judgements given to documents from the Open Web against those from ClueWeb12. We can see the results in Fig. 1. We observe that the Open Web histogram is slightly skewed towards the positive, relevant judgements. Even though we are not interested in comparing the actual frequencies (this would not be fair, mainly because there were many more Open Web submissions than ClueWeb12 ones[1]), it is still relevant to see the relative frequency of −2's and 5's in each dataset: this is an important difference which will impact the performance of the systems using ClueWeb12 documents, as we shall see next.

This first check may indicate that, although the ClueWeb12 dataset is a snapshot of the Web, it does not contain as many *entertaining* or interesting documents as those datasets specially tailored from the Open Web for this task. In fact, such datasets are typically seeded from candidate sources known for having interesting content, such as Yelp, Google Places, Foursquare, and Trip-Advisor [2], in contrast to ClueWeb12, which aims to cover a more uniform set of topics; even though it was seeded with specific travel sites. This apparent contradiction is explained by the fact that ClueWeb12 is missing Yelp due to very strict indexing rules (see http://yelp.com/robots.txt). It turns out that this site alone provided the highest number of relevant documents in the two editions of the track. In this situation, the fact that one of the datasets lacks a specific bias

---

[1] More specifically, 27 runs submitted URLs from the Open Web, and only 7 used ClueWeb12 documents.

Table 1. Comparison of performance for different methods based on the data used to derive the best ranking. The metrics P@5$_r$ and MRR$_r$ denote that no geographical information is used to account for relevance.

Collection	Method	P@5	MRR	TBG	P@5$_r$	MRR$_r$
Open Web	Oracle + geo	**0.909**	**0.945**	**4.030**	**0.950**	0.957
Open Web	Oracle	0.742	0.845	2.767	**0.950**	**0.962**
ClueWeb12	Oracle + geo	0.509	0.761	2.221	0.700	0.892
ClueWeb12	Oracle	0.413	0.640	1.422	0.700	0.892
ClueWeb12 sub	Oracle + geo	0.418	0.702	1.870	0.551	0.803
ClueWeb12 sub	Oracle	0.393	0.652	1.566	0.557	0.814

which affects the other dataset and, likely, the ultimate user need – "entertain me" – seems to limit its applicability in this context.

As an additional test, we build *oracle* submissions based on the relevance judgements (both subjective and objective). Although we have observed that Open Web documents tend to receive higher ratings, this does not necessarily mean that the *best possible performance* of the different methods should be very different. Table 1 shows the performance values obtained by evaluating the rankings found when the documents are ordered according to subjective and objective assessments (*Oracle + geo*) or only the subjective judgements (*Oracle*). We also present the metrics P@5$_r$ and MRR$_r$ where the geographical (objective) information is not considered. We generated the rankings using three subsets of the assessed documents: those submitted as Open Web, those submitted as ClueWeb12, and among those submitted as ClueWeb12, those coming from a subcollection (ClueWeb12 sub) provided by the organisers. Since each of these methods may potentially generate a different description, in the evaluation of these rankings we do not consider the description judgements (i.e., they are assumed to be always relevant), which should not affect the observed ranking of the methods since the same setting is applied for all of them.

Based on the results presented in Table 1, we find that the performance of rankings using exclusively documents from the Open Web is always the highest. Specifically, the drop in performance seems to be quite significant; for instance, in terms of P@5, the lowest value for Open Web is 0.742, whereas the highest value using ClueWeb12 is 0.509. Furthermore, even when a subcollection from ClueWeb12 specifically tailored for the task[2] is used, the results do not improve at all.

As we discussed before, a possible reason for these values is that ClueWeb12 does not contain as many interesting documents as a tailored subset of the Web. Specifically, we found less than 20% of the assessed Open Web documents appearing in ClueWeb12. Dynamic websites, pages within social networks (e.g., Google+, Facebook), or fresh content (created after the ClueWeb12 was crawled) mostly correspond to these missed documents.

---

[2] As described by the organisers in `https://sites.google.com/site/treccontext/trec-2013/subcollection`: "this subcollection was created by issuing a variety of queries targeted to the Contextual Suggestion track on a commercial search engine."

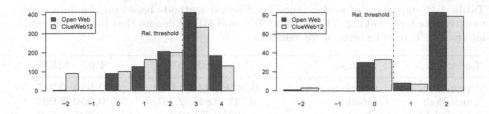

Fig. 2. Histogram of document judgements for the same documents from the Open Web and from ClueWeb12. In the left figure, the ratings show subjective relevance (interest), whereas the right one shows the objective relevance of the document (geographically appropriate).

3.2   Consistency of Evaluation Judgements

While doing the aforementioned analysis, we identified a subset of documents that were submitted as part of the ClueWeb12 dataset whose corresponding URLs were also submitted (by other participants) as Open Web documents. Since both identifiers (URL and ClueWeb12 id) correspond to the same document/webpage, we decided to investigate if we could detect any bias towards one of the datasets based on this sample of the judgements. Hence, out of the 36, 178 available assessments, we ended up with 1, 024 corresponding to the same document retrieved by different systems for exactly the same user and context, and thus, the relevance judgement should be, in principle, the same.

Fig. 2 shows the histograms of the subjective (left) and objective (right) assessments on the sampled documents. We notice how the shape of the objective assessments (geographical relevance) is very similar in both datasets, whereas that of the subjective assessments (profile relevance) has some differences, especially in the number of 3's, 4's, and -2's. To further emphasise this difference, we performed a Wilcoxon signed-rank significance test where the null hypothesis is that both variables have the same mean. We obtained a p-value of 0.5018 for the first case, and a value close to 0 for the second case. This indicates that we cannot reject the null hypothesis in the former, whereas in the latter it is very unlikely that the null hypothesis is true. This justifies a conclusion that the subjective assessments from Open Web and ClueWeb12 are very likely to be different (while this problem does not exist for the geographical relevance case). To provide some context, the consistency of these assessments within each dataset (Open Web and ClueWeb12) as measured by the standard deviation is around 0.05.

An alternative visualisation of this situation is depicted in Fig. 3, where we show in a scatterplot the judgements received by the same document in each dataset. In this figure we can also observe how frequent these combinations occur. For instance, in 245 cases, a document assessed as 3 in the Open Web has also been assessed as 3 in ClueWeb12. It is important to note that ClueWeb12 documents receive a larger amount of -2 judgements (as we could see in Fig. 2 left) almost independently of the corresponding assessment received by the Open Web document. Hence, in the extreme situation, 16 times an Open Web document assessed as 4 received a -2 as a ClueWeb12 document. This

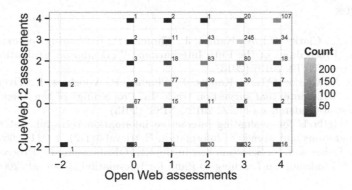

Fig. 3. Scatterplot showing the frequencies of the available assessments combinations for the same document (by the same user in the same context), being judged as a document from the Open Web or from ClueWeb12

behaviour, however, is neglible for low assessed Open Web documents in the inversed situation.

Part of the differences in judgements can be attributed to a different rendering of the document for each dataset[3]. Assessors are influenced by several conditions, one of them is the visual aspect of the interface, but also the response time, the order of examination, the familiarity with the interface, etc. [3]. Therefore, it is important that these details are kept as stable as possible when different datasets are evaluated at the same time.

4   Conclusions

We analysed two concrete challenges for measuring effectiveness of information retrieval systems on a dataset where the documents originate from a mix of sources. A fair comparison would require a similar ratio of potential candidate documents to be relevant. Additionally, when the considered user need is time-sensitive (as in our case), a static, archived dataset is prone to be in disadvantage with respect to a dynamic, live Open Web dataset.

Consistency in the judgements is also an important challenge, not always possible due to incomplete mappings between the datasets at hand. Special efforts are therefore required to design a user experience where the assessor is not aware of the origin of the document being evaluated. Aspects such as timing and fatigue should also be considered [3].

Acknowledgments. Part of this work was carried out during the tenure of an ERCIM "Alain Bensoussan" Fellowship Programme, funded by European Comission FP7 grant agreement no.246016. This research was further supported by the Netherlands Organization for Scientific Research (NWO project 640.005.001) and the Dutch national program COMMIT/.

---

[3] Confirmed via email with the organisers.

References

1. Dean-Hall, A., Clarke, C.L.A., Kamps, J., Thomas, P.: Evaluating contextual suggestion. In: Proceedings of the Fifth International Workshop on Evaluating Information Access (EVIA 2013) (2013)
2. Dean-Hall, A., Clarke, C.L.A., Kamps, J., Thomas, P., Voorhees, E.: Overview of the TREC 2012 contextual suggestion track. In: Proceedings of the Twenty First Text REtrieval Conference (TREC 2012). NIST (2013)
3. Kelly, D.: Methods for evaluating interactive information retrieval systems with users. Foundations and Trends in Information Retrieval 3(1-2), 1–224 (2009)
4. Minka, T., Robertson, S.: Selection bias in the LETOR datasets. In: Proceedings of SIGIR 2008 Workshop on Learning to Rank for Information Retrieval (2008)

Integrating Multiple Resources
for Diversified Query Expansion

Arbi Bouchoucha, Xiaohua Liu, and Jian-Yun Nie

Dept. of Computer Science and Operations Research
University of Montreal
Montreal (Quebec), Canada
{bouchoar,liuxiao,nie}@iro.umontreal.ca

Abstract. Diversified query expansion aims to cover different possible intents of a short and ambiguous query. Most standard approaches use a single source of information, e.g., the initial retrieval list or some external resource like ConceptNet. The coverage is thus limited by that of the resource. To alleviate this issue, we propose the use of multiple resources. More specifically, our framework first automatically generates a list of diversified queries for each resource, and then combines the retrieved documents for all the expanded queries following the Maximal Marginal Relevance principal. We evaluate our framework on several TREC data sets, and demonstrate that our framework outperforms the state-of-the-art approaches, suggesting the effectiveness of incorporating different resources for diversified query expansion.

Keywords: Diversified Query Expansion, Resource Integration.

1 Introduction

Queries in Web search are usually short, ambiguous (e.g., "Java") and under-specified [1]. To address this issue, various search result diversification (SRD) technologies have been proposed (see for example, [1,4,7]). Traditional SRD approaches are based on query expansion (QE) and pseudo-relevance feedback (PRF). One weakness of such approaches is that their performance much depends on the initial retrieval results.

Diversified query expansion (DQE) represents the most recent approach to SRD. One distinguished feature of DQE is the utilization of external resources, e.g., ConceptNet [8], Wikipedia, or query logs, to generate a set of diversified queries, whose retrieval results are then combined into a new list. One representative work of DQE is conducted by Bouchoucha et al. [3], which expands queries using ConceptNet and uses the Maximal Marginal Relevance (MMR) strategy [4] to select diversified terms. Their approach outperforms the state-of-the-art existing SRD methods on TREC data.

Following this work, we propose to combine multiple resources to diversify queries. Our approach is largely motivated by the following observation: there are a large number of queries for which ConceptNet cannot yield good performance but some other resources can suggest good terms. "defender", the #20

M. de Rijke et al. (Eds.): ECIR 2014, LNCS 8416, pp. 437–442, 2014.
© Springer International Publishing Switzerland 2014

query from the TREC 2009 Web track, is such an example. It has six different subtopics[1]. In our experiments, traditional IR models for this query return no relevant documents. Therefore PRF does not work. ConceptNet returns results covering subtopic 2, 3 and 6, while Wikipedia and query logs provide documents covering subtopic 1, 2, 3, 4 and 1, 2, 4, 5, respectively. By integrating all these sources, we obtain a list of documents covering all the subtopics.

We further propose a unified framework to integrate multiple resources for DQE. For a given resource (e.g., ConceptNet), our framework first generates expansion candidates, from which a set of diversified terms are selected following the MMR principle. Then the retrieved documents for any expansion query of any resource are combined and again the MMR principle is used to output diversified results. In this work, we integrate four typical resources, *i.e.*, ConceptNet, Wikipedia, query logs, and initial search results.

It is worth noting that the idea of combining multiple resources has been successfully exploited for other IR tasks. For example, Deveaud et al. [5] combine several external resources to generalize the topical context, and suggest that increasing the number of resources tends to improve the topical representation of the user information need. Bendersky et al. [2] integrate multiple information sources to compute the importance features associated with the explicit query concepts, and to perform PRF. Compared with these studies, our work has two significant differences: 1) these resources are used to directly generate diversified queries; and 2) MMR is used to cover as many aspects as possible of the query.

We evaluate our approach using TREC 2009, 2010 and 2011 Web tracks. The experimental results show that multiple resources do complete each other, and integrating multiple resources can often yield substantial improvements compared to using one single resource.

Our contributions are twofold: 1) we propose the integration of multiple resources for DQE, and a general framework based on MMR for the implementation; 2) we show the effectiveness of our method on several public datasets.

The remainder of this paper is organized as follows. Section 2 presents details of our method. Section 3 describes our experiments. Finally in Section 4, we conclude with future work.

2   Proposed Framework

Our proposed framework consists of two layers. The first layer integrates a set of resources, denoted by R, to generate diversified queries. Given a query Q, it iteratively generates a good expansion term c^* for each resource $r \in R$, which is both similar to the initial query Q and dissimilar to the expansion terms already selected:

$$c^* = argmax_{c \in C_{r,Q}}(\lambda_r \cdot sim(c, Q) - (1 - \lambda_r) \cdot max_{c_i \in S_{r,Q}} sim_r(c, c_i)) \quad (1)$$

Here, $C_{r,Q}$ and $S_{r,Q}$ represent the set of candidate terms and the set of selected terms for r, respectively; the parameter λ_r (in [0,1]) controls the trade-off

---

[1] http://trec.nist.gov/data/web/09/wt09.topics.full.xml

between relevance and redundancy of the selected term; $sim_r(c, c_i)$ returns the similarity score of two terms for resource r; $sim_r(c, Q)$ is the similarity score between term c and the query Q, which is computed using Formula 2, where q is a subset of Q and $|q|$ denotes the number of words of q.

$$sim_r(c, Q) = max_{q \in Q}(sim_r(c, q) \cdot \frac{|q|}{|Q|}) \tag{2}$$

Once c^* is selected, it is removed from $C_{r,Q}$ and appended to $S_{r,Q}$. With the parameter λ_r, initial term candidates $C_{r,Q}$, and the term pair similarity function $sim_r(c, c_i)$, which depend on the particular resource, Formula 1 becomes a generalized version of Maximal Marginal Relevance-based Expansion (MMRE) proposed by Bouchoucha et al. [3]. And by instantiating λ_r, $C_{r,Q}$ and $sim_r(c, c_i)$, our framework can integrate any resource.

We investigate four typical resources in this work: ConceptNet, Wikipedia, query logs, and initial search results, hereafter denoted by C, W, QL and D, respectively. For ConceptNet, we use the same approach introduced in [3] to compute $C_{C,Q}$ and sim_C. For Wikipedia, we define $C_{W,Q}$ as the outlinks, categories, and the set of terms that co-occur with Q or a part of Q [2]; $sim_W(c, c_i)$ is defined by Formula 3, where W (W_i) is the set of vectors containing term c (c_i) obtained by ESA, and $sim(w, w_i)$ is simply the cosine similarity of vector w and w_i.

$$sim_W(c, c_i) = \frac{1}{|W||W_i|} \sum_{w \in W, w_i \in W_i} sim(w, w_i) \tag{3}$$

For query logs, $C_{QL,Q}$ includes the queries that share the same click-through data with Q, as well as the reformulated queries of Q that appear in a user session within a 30 minutes-time window; $sim_{QL}(c, c_i)$ is defined by Formula 4 :

$$sim_{QL}(c, c_i) = \frac{2 \cdot |\{Q'|Q' \in C_{QL,Q}, c \in Q', c_i \in Q'\}|}{|\{Q'|Q' \in C_{QL,Q}, c \in Q'\}| + |\{Q'|Q' \in C_{QL,Q}, c_i \in Q'\}|} \tag{4}$$

For initial search results, we consider top K returned results as relevant documents (K is experimentally set to 50 in our experiments), and use PRF to generate $C_{D,Q}$; $sim_D(c, c_i)$ is computed using Formula 5, where $freq(c, c_i)$ refers to the co-occurrence of term c and c_i within a fixed window of size 15.

$$sim_D(c, c_i) = \frac{2 \cdot freq(c, c_i)}{\sum_{c'} freq(c, c') + \sum_{c'} freq(c_i, c')} \tag{5}$$

The second layer of our framework generates diversified search results in three steps. First, for each resource r, it generates a set of ranked documents $D_{r,Q}$ using the expansion terms $S_{r,Q}$, which are then combined into one unique set D_Q. Finally, it uses again the MMR principal [4] to iteratively select d^* from the current document candidates. Formula 6 defines this process, where DC_Q denotes

---

[2] In cases where no Wikipedia pages match Q or a part of Q, we use *Explicit Semantic Analysis (ESA)*[6] to get semantically related Wikipedia pages, from which to extract the outlinks, categories and representative terms to obtain $C_{W,Q}$.

the document candidates, which is initialized as D_Q; DS_Q denotes the set of selected documents, which is empty at the very beginning; λ is the parameter that controls the tradeoff between relevance and diversity; $rel(d, Q)$ measures the similarity between document d and query Q; $sim(d, d_i)$ denotes the similarity between two documents (we use the cosine similarity in our experiments). The selected document d^* is then removed from DC_Q to DS_Q.

$$d^* = argmax_{d \in DC_Q}(\lambda \cdot rel(d, Q) - (1 - \lambda) \cdot max_{d_i \in DS_Q} sim(d, d_i)) \quad (6)$$

One core element of the second layer is $rel(d, Q)$, which is defined using Formula 7, where $rel(D_{r,Q}, d)$ and $rank(D_{r,Q}, d)$ are the normalized relevance score[3] and the rank of document d in $D_{r,Q}$[4], respectively. This formula captures our intuition that the more a document is ranked on top and with high relevance score, the more relevant it is to the query.

$$rel(d, Q) = \sum_{r \in R} \frac{rel(D_{r,Q}, d)}{rank(D_{r,Q}, d)} \quad (7)$$

3 Experiments

We conduct experiments on the ClueWeb09 (category B) dataset, which contains 50,220,423 documents (1.5 TB), and use the test queries from TREC 2009, 2010 and 2011 Web tracks. Statistics for the query sets being used are reported in [3]. Indri is used as the basic retrieval system. Our baseline is a query generative language model with Dirichlet smoothing (μ=2000), Krovetz stemmer, and stopword removal using the standard INQUERY stopword list.

We consider four typical resources: the last version of ConceptNet[5], the English Wikipedia dumps of July 8th, 2013, the log data of Microsoft Live Search 2006, which spans over one month (starting from May 1st) consisting of almost 14.9M queries shared between around 5.4M user sessions, and the top 50 results returned for the original query.

The evaluation results in the diversity task of the TREC 2009, 2010 and 2011 Web tracks are reported based on five official measures: MAP and nDCG for adhoc performance, α-nDCG (in our experiments, $\alpha = 0.5$) and ERR-IA for diversity measure. We also use S-recall to measure the ratio of covered subtopics for a given query. Using greedy search on each resource (step=0.1), we empirically set $\lambda_C = 0.6$, $\lambda_W = 0.5$, $\lambda_{QL} = 0.4$, $\lambda_D = 0.6$, and $\lambda = 0.3$. For each test query, we generate 10 expansion terms using MMRE, with respect to each resource.

Table 1 reports our evaluation results, from which we make four main observations. Firstly, when each resource is considered separately, using query logs often yields significantly better adhoc retrieval performance and diversity. Maybe, this is because the candidate expansion terms generated from query logs are those

---

[3] exp function is used for normalization, i.e., $x \leftarrow \frac{exp\, x}{\sum_{x'} exp\, x'}$.

[4] For $d \notin D_{r,Q}$, we set $\frac{1}{rank(D_{r,Q}, d)} = 0$.

[5] http://conceptnet5.media.mit.edu

suggested by users (through their query reformulations), which reflect well the user intent. This suggests the important role of query logs for the diversity task.

Secondly, Wikipedia outperforms ConceptNet for TREC 2009 and TREC 2010, but not significantly in general. However, ConceptNet significantly outperforms Wikipedia for TREC 2011 in all the measures. To understand the reason, we manually assess the different queries to see whether they have an exact matching page from Wikipedia. We find that 36/50, 34/48 and 19/50 queries from TREC 2009, TREC 2010 and TREC 2011 respectively, have exact matching pages from Wikipedia (including the disambiguation and redirection pages), and that only when the query corresponds to a known concept (*i.e.* page) from Wikipedia, the candidate expansion terms suggested by Wikipedia tend to be relevant. This means that Wikipedia helps promoting the diversity of the query results, if the query corresponds to a known concept.

Thirdly, the set of feedback documents has the poorest performance among all resources under consideration. Its performance drastically decreases from TREC 2009 to TREC 2010 to TREC 2011 in terms of adhoc retrieval and diversity.

Table 1. Experimental results of different models on TREC Web tracks query sets. BL denotes the baseline model; MMR is the model based on results re-ranking (with λ=0.6) [4]; MMRE$_C$, MMRE$_W$,MMRE$_{QL}$, and MMRE$_D$ refer to the MMRE model based on ConceptNet, Wikipedia, query logs, and search results, respectively; COMB denotes the model combining all the four resources. *, -, +, §,♮, and ‡ indicate significant improvement ($p < 0.05$ in T-test) over BL, MMR, MMRE$_C$, MMRE$_W$, MMRE$_{QL}$, and MMRE$_D$, respectively.

Queries	Model	MAP	nDCG@20	α-nDCG@20	ERR-IA@20	S-recall@20
TREC 2009	BL	0.161	0.240	0.188	0.097	0.367
	MMR	0.166	0.246	0.191*	0.103	0.377
	MMRE$_C$	0.195*-‡	0.293*-‡	0.269*-‡	0.140*-‡	0.482*-
	MMRE$_W$	0.208*-+‡	0.319*-+‡	0.274*-‡	0.146*-‡	0.510*-‡
	MMRE$_{QL}$	0.221*-+§‡	0.340*-+§‡	0.295*-+§‡	0.153*-§‡	0.599*-+§‡
	MMRE$_D$	0.188*	0.276*	0.224*	0.115*	0.435*
	COMB	**0.258*-+§♮‡**	**0.379*-+§♮‡**	**0.328*-+§♮‡**	**0.181*-+§♮‡**	**0.672*-+§♮‡**
TREC 2010	BL	0.103	0.115	0.198	0.110	0.442
	MMR	0.106	0.119	0.209*	0.111	0.459
	MMRE$_C$	0.146*-‡	0.196*-‡	0.293*-‡	0.165*‡	0.664*-‡
	MMRE$_W$	0.149*-‡	0.203*-‡	0.317*-+‡	0.174*-‡	0.683*-‡
	MMRE$_{QL}$	0.158*-+§‡	0.221*-+‡	0.341*-+§‡	0.182*-+§‡	0.694*-‡
	MMRE$_D$	0.117*	0.142*	0.225*	0.148*	0.508*
	COMB	**0.173*-+§♮‡**	**0.239*-+§♮‡**	**0.352*-+§♮‡**	**0.195*-+§♮‡**	**0.703*-+§‡**
TREC 2011	BL	0.093	0.155	0.380	0.272	0.700
	MMR	0.096	0.159	0.382	0.269	0.714
	MMRE$_C$	0.155*-§‡	0.320*-§‡	0.552*-§‡	0.397*-§‡	0.975*-§‡
	MMRE$_W$	0.124*-‡	0.255*-‡	0.449*-‡	0.313*-‡	0.798*-‡
	MMRE$_{QL}$	0.160*-+§‡	0.342*-+§‡	0.578*-+§‡	0.411*-§‡	0.982*-§‡
	MMRE$_D$	0.104*	0.163*	0.397	0.279	0.733
	COMB	**0.167*-+§‡**	**0.359*-+§♮‡**	**0.586*-+§‡**	**0.422*-+§‡**	**0.990*-+§‡**

This may be due to the fact that the topics of TREC 2011 are harder than the topics of TREC 2010, and the topics of the latter are harder than those of TREC 2009 (based on the MAP values). The more the collection contains difficult queries, the more likely the set of top returned documents are irrelevant. Hence, the candidate expansion terms generated from these documents tend to include a lot of noise.

Finally, combing all these resources gives better performance, and in most cases the improvement is significant for almost all the measures. In particular, the diversity scores obtained (for α-nDCG@20, ERR-IA@20, and S-recall@20), are the highest scores. This means that the considered resources are complementary in term of coverage of query subtopics: the subtopics missed by some resources can be recovered by other ones, as demonstrated by "defender" in the introduction. Moreover, our combination strategy promotes the selection of the most relevant documents in the final results set, which explains why higher scores for MAP and nDCG@20 are obtained.

4   Conclusions and Future Work

This paper presents a unified framework to integrate multiple resources for DQE. By implementing two functions, one to generate expansion term candidates and the other to compute the similarity of two terms, any resource can be plugged into this framework. Experimental results on TREC 2009, 2010 and 2011 Web tracks show that combining several complementary resources performs better than using one single resource. We have observed that the degree of the contribution of a resource to SRD depends on the query. In future, we are interested in other approaches to resource integration for DQE, e.g., assigning different resources with different weights that are sensitive to the query.

Acknowledgments. We thank the anonymous reviewers for their valuable suggestions and useful comments.

References

1. Agrawal, R., Gollapudi, S., Halverson, A., Ieong, S.: Diversifying search results. In: Proceedings of WSDM, pp. 5–14 (2009)
2. Bendersky, M., Metzler, D., Croft, W.B.: Effective query formulation with multiple information sources. In: Proceedings of WSDM, pp. 443–452 (2012)
3. Bouchoucha, A., He, J., Nie, J.-Y.: Diversified query expansion using conceptnet. In: Proceedings of CIKM, pp. 1861–1864 (2013)
4. Carbonell, J., Goldstein, J.: The use of mmr diversity-based reranking for documents and producing summaries. In: Proceedings of SIGIR, pp. 335–336 (1998)
5. Deveaud, R., SanJuan, E., Bellot, P.: Estimating topical context by diverging from external resources. In: Proceedings of SIGIR, pp. 1001–1004 (2013)
6. Gabrilovich, E., Markovitch, S.: Computing semantic relatedness using wikipedia-based explicit semantic analysis. In: Proceedings of IJCAI, pp. 1606–1611 (2007)
7. Santos, R.L., Macdonald, C., Ounis, I.: Exploiting query reformulation for web search result diversification. In: Proceedings of WWW, pp. 881–890 (2010)
8. Speer, R., Havasi, C.: Representing general relational knowledge in conceptnet 5. In: Proceedings of LREC, pp. 3679–3686 (2012)

Facet-Based User Modeling in Social Media for Personalized Ranking

Chen Chen[1], Wu Dongxing[1], Hou Chunyan[2,*], and Yuan Xiaojie[1]

[1] Nankai University, Tianjin, China
{nkchenchen,yuanxj}@nankai.edu.cn, wudongxing@dbis.nankai.edu.cn
[2] Tianjin University of Technology, Tianjin, China
houchunyan@tjut.edu.cn

Abstract. Micro-blogging service has grown to a popular social media and provides a number of real-time messages for users. Although these messages allow users to access information on-the-fly, users often complain the problems of information overload and information shortage. Thus, a variety of methods of information filtering and recommendation are proposed, which are associated with user modeling. In this study, we propose an effective method of user modeling, facet-based user modeling, to capture user's interests in social media. We evaluate our models in the context of personalized ranking of microblogs. Experiments on real-world data show that facet-based user modeling can provide significantly better ranking than traditional ranking methods. We also shed some light on how different facets impact user's interest.

1 Introduction

With the rise of social media like Facebook and Twitter in USA, Sina Weibo and Tencent Weibo in China, millions of users are connecting and communicating with each other over rich multimedia content, including text, image and video. Although social media provide a unique opportunity for users to access fresh information, users commonly acknowledge two issues that prevent current social media from being sufficiently relevant. At first, some of users are flooded with a large number of messages from their social network and simply cannot process them in an effective and efficient way. This is known as the problem of information overload. For some others, in contrast, the incoming steam of messages for a user is usually limited to their social circle. Thus, it is very difficult for a user to obtain information outside of their circle, even though it might match their interests. It is called as the problem of information shortage.

To address both these problems, some filtering and recommendation methods [3, 5, 6] are proposed, which are associated with user relevance. The user relevance, namely whether a user is interested in a message, is determined by many factors. Personal interest is an important factor to decide whether information is personally relevant. User modeling is used to represent the personal interest. The basic assumption of user modeling is that personal interests can be inferred from his history preference. Traditional methods of user modeling analyze the content of users' messages in history to

M. de Rijke et al. (Eds.): ECIR 2014, LNCS 8416, pp. 443–448, 2014.

discover the topics of interest for users [1, 2, 5]. However, user modeling in this way is very difficult because the message of social media is short, informal, ungrammatical and noisy. Besides the content of the users' messages, there are other kinds of important information available in social media such as the posted time and social network. To fully mine such information, we propose an effective method of user modeling, facet-based user modeling, to capture the personal interests in social media. We define the facet which is the probability distribution of a user's preference to an attribute. A few facets are proposed as history features in different studies [3, 5, 6] and never compared in the context of personalized ranking. Facet-based user modeling can consider different facets based on history preference. Our approach generalized the traditional methods of user modeling by incorporating users' different preferences to attributes of the microblog such as the content, time and social network so that user's interests can be traceable and understood better.

We conduct our experiments on a real large scale dataset and evaluate our approach in the context of personalized ranking of microblogs. The results suggest that facet-based user modeling can provide the better ranking in social media than traditional ranking methods. In addition, we find that user's forwarding and commenting behaviors are mainly determined by messages posted by the individual social friends and users prefer to read the message of the rich information.

2   Face-based User Modeling in Social Media

We give basic definitions and then present facet-based user modeling.

Definition 1. (Attribute of Message) The main attributes of a message $m \in M$ can be described as $A=\{Publisher, Followee, Content, Source, Time\}$, where *Publisher* is the author of message m, *Followee* is the user who forwards or comments the message m, *Content=\{Words, Entities, Hash Tags, URLs, Image, Video, Mentions\}*, *Source* is the device or news media where message m is edited. *Time* is when the message m is published.

Definition 2. (Facet) Facet is the probability distribution of a user $u \in U$ preference to an attribute $a \in A$ where the distribution function $D(u,a)$ can be computed by a certain data in history.

$$F(u,a) = \{(u,a,D(u,a)) \mid u \in U, a \in A\}$$

where U and A denote the set of users and attributes respectively.

Definition 3. (User Model) The user model of a user $u \in U$ is the set of facets with respect to the given user u for an attribute $a \in A$.

$$U(u) = \{(a, F(u,a)) \mid u \in U, a \in A\}$$

where U and A denote the set of users and attributes respectively.

Facet-based user modeling has advantages over content-based methods. Firstly it is difficult to analyze the content of microblogs because of the 140 characters limitation and the insufficient word co-occurrence. Also, the content-based methods are not aware of this user's interest in other attributes such as mentions (@user), hashtags (#abc#) and followees. From these special attributes of microblogs, we can understand users' personal interests comprehensively rather than only by the content.

Compared with the general feature, the facet relies on the user history and focuses on user modeling from different perspectives so as to capture the user interests and preferences to the microblog. For example, the authority of a publisher is an important feature. If a message is published by an authoritative user, it is more likely to be a popular message. However, this feature is not a facet. In addition, a feature may be nominal or categorical such as weekday, and the facet is able to represent it as a probability, which describes user's preference extent to the attribute. For example, a facet to weekday can shows that a user prefers to forward or comment microblogs with the probability of 30% on Sunday, and 10% on Monday.

We introduce how to compute facet to the publisher by history data as an example and other facets are computed in the same way. Let U be the set of users. For $u \in U$, we can compute u's facet to the publisher p by Eq.(1).

$$F(u, p) = \frac{\#(\text{messages published by } p \text{ in } u\text{'s history})}{\#(\text{messages in } u\text{'s history})} \qquad (1)$$

3   Experiments

3.1   Setup

We conduct our experiments on Sina Weibo, which is the most popular Chinese mi-cro-blogging service and has more than 400 million registered users by the end of the third quarter in 2012. The data set is collected from 2012-04-01 to 2012-06-30, which is through 13 weeks. There are 8 weeks as the history for facet computation from 2012-04-01 to 2012-05-26, and 5 weeks for evaluation from 2012-05-27 to 2012-06-30. To better evaluate our methods, we choose 1711 active users who have at least 5 forwarded messages per week. Finally, we collect 851 users, who have over five followees from active users. As Table 1 shown, there are 5,699,880 messages received by 842 users in five weeks, which are used for evaluation. Of these messages, 24,998 messages (0.44%) are forwarded, and 16,300 messages (0.28%) are not only forwarded but also commented by the users who receive them afterwards. From the statistics, we see that most messages are not forwarded or commented, which shows the problem of information overload.

For entity-based user modeling, we use Baidu Encyclopedia, which is a Chinese language collaborative Web-based encyclopedia provided by the Chinese search engine Baidu. Until October 2012, Baidu Encyclopedia has more than 5.3 million articles, which are organized into the taxonomy with 3 levels deep. There are 1199 categories at the lowest level, 96 at the middle level and 12 top level categories.

User behaviors indirectly reflect users' interests. These behaviors include forwarding and commenting a message. Therefore, we assume that users' reposting behaviors represent their personal judgment of relevance. In the context of personalized ranking, relevance ratings are considered at three levels. Forwarding and commending a message corresponds to a 2 rating, forwarding only to a 1 rating, and others to 0 rating. We use Mean Average Precision (MAP), and Normalized Discounted Cumulative Gain (NDCG), which are popular rank evaluation metrics to evaluate the proposd

Table 1. Statistics of the experimental data. "#User", "#Week" and "#Message" denote the number of users, weeks and messages respectively. "#Fwd" means the number of messages forwarded while "#(Fwd+Cmt)" is the number of messages forwarded and commented.

	#User	#Week	#Message	#Fwd	#(Fwd+Cmt)
Evaluation	842	5	5,699,880	24,998	16,300
Facet	842	8	about 10M	332,714	212,757

methods. Note that these metrics are based on users rather than queries. In our experiment, the significance ($p < 0.05$) is tested using a 2-tailed paired t-test.

Before the evaluation, 8 weeks of history data is used to compute facets for each user. Then, our task is to take the $(n-1)^{th}$ week (past 1 week) of data as training data to predict the ranking of $(n)^{th}$ week (current week) for each user. For example, we have the 1st week of data as training data and the 2nd week of data as testing data. We have 5 weeks of data for evaluation and get experiment results from the 2nd week to the 5th week. We train a logistic regression model for each user, i.e. each user has the own weight of the facet. Once weights of facets have been derived for a training data set, the logistic regression model can be used to predict the probability forwarded or commented for new messages. New messages are ranked by the probability in descending order.

3.2 Comparison of User Modeling Methods

Abel et al. [1] define user model and divide content-based user modeling into hashtag-, topic-, entity-based profile in terms of types of profile. They also find that hashtag-based user modeling fails for some users because these people neither made use of hashtags nor reposted messages that contain hashtags. Therefore, we focus on the topic-based and entity-based user modeling in this study.

Topic-based user modeling uses words in messages and word frequency is regarded as the weight. With respect to entity-based user modeling, we collect named entities from Baidu Encyclopedia and their categories (e.g science, sport, economy, etc.), and identify entities in messages by the Encyclopedia to better understand the semantics of messages. A category of Encyclopedia includes many entities, and in general the more entities in a category are identified, the user will be more interested in this category. For example, when building entity-based user model with 12 categories, we use the 12-dimension vector to represent the user model. For user modeling, the occurrence frequency of a category of entities in historical messages is regarded as the weight of this category. For example, in the entity-based profile, the weight of a category is 50, which means that a user publishes messages in which entities of this category are mentioned for 50 times. New messages can be represented by the vector in this way. Following Abel et al. [1], we use the cosine similarity to rank messages.

We have compared our method against some baselines, which include the chronological, entity-based and topic-based user modeling. The entity-based user modeling is based on 12 categories at top level in Baidu Encyclopedia. As to topic-based user modeling, we use bigrams as words for Chinese characters in messages while forward

Table 2. Evaluation results of chronology, entity-based, topic-based and facet-based user modeling. The result labeled by more stars is statistically significant than that of fewer stars.

		5th week	4th week	3rd week	2nd week	1st week
M	Chrono	0.0081	0.0118	0.0135	0.0173	0.0265
A	Entity*	0.0134	0.0136	0.0169	0.0227	0.0278
P	Topic**	0.0287	0.0299	0.0331	0.0388	0.0454
	Facet***	0.0341	0.0329	0.0412	0.0447	
N	Chrono	0.1779	0.1854	0.1879	0.1985	0.2164
D	Entity*	0.1875	0.1903	0.1961	0.2085	0.2213
C	Topic**	0.2103	0.2141	0.2159	0.2289	0.2437
G	Facet***	0.2187	0.2192	0.2287	0.2380	

maximum matching algorithm is used to identify the entities for entity-based user modeling.

As Table 2 shown, the chronological ranking gets bad performance because whether a user would like to forward or comment a message depends on his interests rather than on time. The entity-based ranking is significantly better than the chronological ranking. Thus, user modeling can benefit from the semantic knowledge because user interests can be understood by the encyclopedia to some extent. To evaluate the different representation of entity-based user model, we conduct experiments in terms of different levels categories of Encyclopedia and find that the difference of performance is not significant. Topic-based user modeling significantly outperforms entity-based user modeling. The reason is that a message is short and lack of entities while bag-of-word model takes all words in messages into account. The facet-based user modeling is significantly better than others, which validates that it is able to understand users' interests comprehensively.

3.3   Facet Influence

For a given facet, we compute Kendall's Tau coefficient between the relevance judgment and a facet for each user. As Table 3 shown, we find that user's forwarding or commenting behavior is mainly influenced by messages posted by social friends. In social media, a user usually sends a microblog to his friend by the mention (i.e. @friend), and the friend gives the comments and/or forwards as a reply. The image and URL is actually the enrichment of microblog message which is limited to the 140 characters. It reveals that users prefer to read the message of the rich information. The facets of the work time, video and weekday have little effect on personalized ranking in social media.

4   Conclusions and Future Work

In this work, we propose facet-based user modeling to deal with the problems of information overload and shortage. Experiments on real-world data show that facet-based user model can improve the ranking performance significantly. Moreover, we find that a user's forwarding or commenting behavior is mainly influenced by

Table 3. Average of Kendall Tau coefficients between facets and relevance over all users

Feature	5th week	4th week	3rd week	2nd week	1st week
followee	0.0507	0.0572	0.0481	0.0521	0.0512
mention	0.0311	0.0332	0.0355	0.0369	0.0376
image	0.0199	0.0181	0.0203	0.0230	0.0235
hashtag	0.0193	0.0135	0.0166	0.0191	0.0259
URL	0.0172	0.0150	0.0010	0.0045	0.0143
source	0.0152	0.0127	0.0143	0.0201	0.0184
word	0.0134	0.0169	0.0166	0.0166	0.0214
publisher	0.0089	0.0111	0.0110	0.0056	0.0107
entity	0.0042	0.0044	0.0047	0.0066	0.0073
worktime	0.0039	0.0047	0.0054	0.0038	0.0066
video	0.0027	0.0072	0.0062	0.0023	0.0097
weekday	0.0017	0.0013	0.0011	0.0003	0.0009

messages posted from social friends and users prefer to read the message of the rich information. Although our work has been done in the context of Sina Weibo, we expect the same results would hold for many other similar social media, such as Twitter and Facebook. For future work, we will take more facets into account such as the geographic location and personal tags.

Acknowledgments. This research was supported by Fundamental Research Funds for the Central Universities, National Natural Science Foundation of China (No. 61170184) and Tianjin Municipal Science and Technology Commission (No. 13ZCZDGX02200).

References

1. Abel, F., Gao, Q., Houben, G.-J., Tao, K.: Analyzing user modeling on twitter for personalized news recommendations. In: Konstan, J.A., Conejo, R., Marzo, J.L., Oliver, N. (eds.) UMAP 2011. LNCS, vol. 6787, pp. 1–12. Springer, Heidelberg (2011)
2. Chen, J., Nairn, R., Nelson, L., Bernstein, M., Chi, E.: Short and tweet: experiments on recommending content from information streams. In: Proceedings of the 28th International Conference on Human Factors in Computing Systems, pp. 1185–1194 (2010)
3. Chen, K., Chen, T., Zheng, G., Jin, O., Yao, E., Yu, Y.: Collaborative personalized tweet recommendation. In: Proceedings of the 35th International ACM SIGIR Conference on Research and Development in Information Retrieval, pp. 661–670 (2012)
4. Hong, L., Bekkerman, R., Adler, J., Davison, B.D.: Learning to rank social update streams. In: Proceedings of the 35th International ACM SIGIR Conference on Research and Development in Information Retrieval, pp. 651–660 (2012)
5. Kapanipathi, P., Orlandi, F., Sheth, A., Passant, A.: Personalized Filtering of the Twitter Stream. In: Proceeding of the 2nd Workshop on Semantic Personalized Information Management: Retrieval and Recommendation, pp. 6–13 (2011)
6. Uysal, I., Croft, W.B.: User oriented tweet ranking: a filtering approach to microblogs. In: Proceedings of the 20th ACM International Conference on Information and Knowledge Management, pp. 2261–2264 (2011)

Spot the Ball: Detecting Sports Events on Twitter

David Corney, Carlos Martin, and Ayse Göker

IDEAS Research Institute, School of Computing & Digital Media,
Robert Gordon University, Aberdeen AB10 7QB
{d.p.a.corney,c.j.martin-dancausa,a.s.goker}@rgu.ac.uk

Abstract. Messages from social media are increasingly being mined to extract useful information and to detect trends. These can relate to matters as serious as earthquakes and wars or as trivial as haircuts and cats. Football remains one of the world's most popular sports, and events within big matches are heavily discussed on Twitter. It therefore provides an excellent case study for event detection.

Here we analyse tweets about the FA Cup final, the climax of the English football season, for 2012 and 2013. We evaluate an automated topic detection system using a ground truth derived from mainstream media. We also show that messages can be associated with different teams' fans, and that they discuss the same events from very different perspectives.

Keywords: topic detection, Twitter, football.

1   Introduction

Twitter has been used to detect and predict events as diverse as earthquakes [1], stock market prices [2] and elections [3] with varying degrees of success. Journalists are increasingly using social media to find breaking news stories, sources and user-generated content [4]. Here, we consider the problem of detecting events from social media within the domain of football. Association Football (or 'soccer') remains the world's most popular sport. The last FIFA World Cup final (2010, South Africa) had a global TV audience of over 500 million, with more than 2 billion people watching at least 30 minutes of football during the tournament [5]. The next World Cup is scheduled for Brazil (2014), which itself is now the country with the second largest number of Twitter users [6].

Twitter is actively increasing its ties to television [7]. Major sporting events, like many live TV shows, occur at pre-specified times; they attract large audiences; and they are fast-paced. These features allow and encourage audience participation such as sharing comments and discussing the event with a focus that moves with the event itself. Facebook and other social media are also competing for access to this valuable "second screen" [8]. Thus there is a substantial and growing demand for linking social media discussions to televised events, including sports, for example to improve targeted advertising.

M. de Rijke et al. (Eds.): ECIR 2014, LNCS 8416, pp. 449–454, 2014.
© Springer International Publishing Switzerland 2014

2   Methods

Our system identifies phrases (word n-grams) that show a sudden increase in frequency (a "burst") in a stream of messages and then finds co-occurring n-grams to identify topics. Such bursts are typically responses to real-world events. Here, we filtered tweets from Twitter's streaming API using the teams' and players' names as keywords.

We want to find n-grams that appear at one point in time more than previously. We measure burstiness using the "temporal document frequency-inverse document frequency", or $df\text{-}idf_t$ (Eq. 1), a temporal variant of the widely-used $tf\text{-}idf$. For each n-gram, we compare its frequency in one slot with that of the preceding slot, where each slot contains x tweets (here, $x = 1500$ based on experiments to optimise recall). We define the document frequency df_{ti} as the number of tweets in slot i that contain the n-gram t. We repeat this for commonly-occurring sequences of two and three words as well as single-word terms.

$$df - idf_{ti} = (df_{ti} + 1)/(\log(df_{t(i-1)} + 1) + 1). \tag{1}$$

This produces a list of terms (i.e. words or 2- or 3-grams) which can be ranked by their $df\text{-}idf_t$ scores. We then group together terms that tend to appear in the same tweets with a standard hierarchical clustering algorithm. We define the similarity between two terms/clusters as the fraction of messages in the same slot that contain both of them, so it is likely that the term clusters whose similarities are high represent the same topic. Therefore, we merge clusters until no two share more than half their tweets at which point we assume that each one represents a distinct topic.

Finally, the clusters are ranked by the maximum df-idf$_t$ score of their terms, so that each cluster is scored according to the most representative term in the cluster. Each of these clusters defines a topic as a list of terms, which we can also illustrate with representative tweets. The whole process is fast enough to be carried out in real-time on a single PC. We recently published more details about this algorithm [9] where we compared it with alternative algorithms and benchmarks such as LDA. We also described evaluation for recall of political news events as well as sporting events.

We also attempt to identify the team that each Twitter user supports (if any). For each user, we count the total number of times they mention each team across all their tweets. Manual inspection suggests that fans tend to use their team's standard abbreviation (e.g. CFC or MCFC) greatly more often than any other teams', irrespective of sentiment. We therefore define a fan's degree of support for one team as how many more times that team's abbreviation is mentioned by the user compared to their second-most mentioned team. Here, we include as "fans" any user with a degree of two or more and treat everyone else as neutral.

3   Related Work

We now briefly review three recent studies of Twitter-based sporting event detection. These all detect sudden increases in the volume of tweets as an indication that an event has occurred; in contrast, our system detects increases in the frequency of words or phrases, which is largely independent of the total number of messages received. Thus even if the total volume of tweets is constant, our system can detect shifts in the topic of conversations in reaction to events.

A recent study also used Twitter detect events during football matches [10]. While we have similar aims, our methods are somewhat different. After collecting tweets using suitable hashtags, their system detects events by finding spikes in the volume of tweets collected. For each spike, they analyse the words of the constituent tweets and use machine learning tools to classify the event. They consider just five classes of event (goals, own-goals, red cards, yellow cards and substitutions) and compare these classifications to the official match data to evaluate their system. Finally, they classify individual tweeters as fans of one team or another by counting the number of mentions of each team over several matches, similar to our approach.

Twitter has also been used to identify events during American football (NFL) matches [11]. In that work, tweets were collected by filtering by team names and by NFL terminology. Events were also detected by finding spikes in the overall volume of collected tweets and each event was assigned to one of a fixed number of classes using lexicographic analysis. They used an adaptive window to find events of varying impact. Their system was very effective at detecting the most significant scoring events (i.e. touchdowns), but was less effective at finding "smaller" events such as interceptions and field goals.

Past work on sporting event detection has often focussed on video (or audio) analysis to discover key moments. One recent study used Twitter to enhance such video annotation of both football and rugby matches [12]. They also detected events using spikes in tweet volumes in a stream filtered using hashtags. They then classified events into a few pre-specified classes and used that to create a video of match highlights.

In contrast to the above studies, our system can find arbitrary events and is not limited to pre-specified classes of event. Nor is it limited to finding just one event at a time, as we do not attempt to classify each "event minute" into a single category (unlike [10]). Our system has the potential to identify a much wider range of events than these systems, even if the events are overlapping.

4 Results and Discussion

Figure 1a shows the relative frequency of tweets from fans during the 2012 final. Both groups are active throughout the match with a number of clear spikes in activity. Chelsea fans are particularly active immediately after their team scores (points A and C) and also at the end of the match in celebration of their victory, as would be expected. Liverpool fans are more active when their team score (D). Both sets of fans are active when Liverpool nearly equalize at the end (E).

Figure 1b shows the frequency of tweets from the 2013 final. One clear feature is the large number of Manchester City tweets compared to Wigan Athletic. At the end of the Premier League season, Manchester City finished 2nd while Wigan

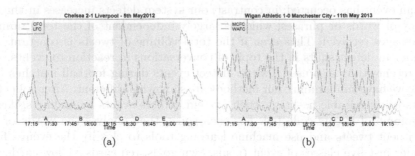

Fig. 1. Tweets from supporters of each team, FA Cup finals of 2012 (left) and 2013 (right). See Table 1 for events key.

Table 1. Key to events of matches shown in Figure 1a (left) and Figure 1b (right)

Key	Team	2012 Final Event	Key	Team	2013 Final Event
A	CFC	Ramires scores	A	WAFC	MacManaman shot, misses
B	CFC	Mikel yellow card for a foul	B	MCFC	Tevez shot, saved
C	CFC	Drogba scores	C	MCFC	Zableta booked
D	LFC	Carroll scores	D	WAFC	MacManaman chance
E	LFC	Carroll shoots, saved	E	MCFC	Rodwell free-kick is saved
			F	WAFC	Watson scores

finished 18th and were relegated. Furthermore, Wigan had an average home attendance of 19,359 compared to City's 46,974 (http://www.soccerstats.com). This discrepancy in recent success is apparently reflected in the number of tweets. These patterns are correlated with the number of followers of the clubs official Twitter accounts. As of 28 October 2013, @LaticsOfficial has 118,512 followers; @MCFC has 1,264,369; @ChelseaFC has 2,943,118 and @LFC 2,072,077.

To evaluate our topic detection algorithm we generated a ground truth using mainstream media descriptions of the matches, primarily the BBC online commentaries. Our ground truth includes not just goals scored but also near-misses, good saves and other events likely to be of interest to fans. We define each event with a list of terms (typically 3-5) and define an event as correctly extracted only if all terms are found. We selected 13 events for the 2012 FA Cup Final (also discussed in [9]) and 15 events for the 2013 Final for this evaluation. Some of these events are listed in Table 1.

Table 2 shows the proportion of correctly recalled topics for each match as we vary the number of topics sought in each slot. As expected, as we increase the number of topics being considered, the total recall increases. For the 2012 match, recall is already nearly perfect even with just one topic, so does not increase further. For the 2013 match, a high recall is achieved when at least 4 topics are identified. This difference is due to the specific topics in the ground truth of the two matches and how many tweets discussed each topic. It also reflects the differing nature of the matches and the corresponding reports.

Table 2. Topic recall for FA Cup 2012 and 2013 sets

Dataset	Top results				
	1	2	3	4	5
FA Cup 2012	0.923	0.923	0.923	0.923	0.923
FA Cup 2013	0.333	0.600	0.667	0.733	0.733

Table 3. Examples of detected topics and example tweets for teams in both the 2012 and 2013 FA Cup finals

Dataset	Topic detected	Sample tweet
Mainstream Media Story (2012): Didier Drogba scores second goal		
FA Cup 2012 Chelsea fans	#cfcwembley #facupfinal sl chelsea goal @chelseafc	RT @chelseafc: Chelsea goal #CFCWembley #FACupFinal (SL)
	king wembley	KING OF WEMBLEY DROGBA #ENOUGHSAID #CFC
	yes drogba	Yes Drogba! #CFC
FA Cup 2012 Liverpool fans	game over #lfc	Game over #LFC sloppy playing, bad passing and conceding easy goals
	fuck #lfc	Fuck Fuck Fuckity Fuck #lfc
	didier drogba 2-0 tackling chelsea	Didier Drogba. 2-0 to Chelsea. Where was the defending? The midfield? The tackling? #LFC #FACupFinal
Mainstream Media Story (2012): Carroll (Liverpool) shoots but Cech just saves		
FA Cup 2012 Chelsea fans	petr cech	petr cech is my hero @chelseafc #CFCWembley #FACupFinal
	goal line	Hahaha! Another case for goal-line technology. #Chelsea 2-1 #Liverpool #FACup
	save #cech	what a save!!#cech#cfc
FA Cup 2012 Liverpool fans	over line ball	Ball is over the line! #ynwa #LFC #liverpoolfc #facupfinal
	goal blind	The ref n linesman is blind!!! That was a goal!!!! #LFC
	robbed #facupfinal	Robbed. Fucking robbed. #FACupFinal
Mainstream Media Story (2013): Wigan goal (Watson)		
FA Cup 2013 MCFC fans	wigan #facupfinal #wigan fucking	Fucking get in Wigan!! #facupfinal #Wigan #mcfc
	#mcfc #ctid utterly deserved goal wigan congratulations worthy winners completely	Completely and utterly deserved goal for Wigan. Congratulations. Worthy winners on the day. #mcfc #CTID
FA Cup 2013 WAFC fans	ben watsoooonnnnnn manchester city 0 1	RT @InfoFootball_: BEN WATSOOOONNNNNN!!!!! Manchester City 0 - 1 Wigan
	win #wafc #facupfinal story delighted dave	Delighted for Dave Whelan. What a man, what a story, what a win! #WAFC #facupfinal
	lead #wafc deserved goal brilliant	GOAL! Brilliant. Deserved lead for #wafc

The high topic recall of Table 2 shows that our system is capable of detecting a variety of topics from live sports events, and is not limited to finding events of a pre-specified class. The good fit between team-specific tweets and team-related events shown in Figure 1 suggests that our classification of tweeters to fans is sufficiently accurate.

Note that after Wigan's late goal ('F' in Figure 1b), there is no spike in Wigan fans' tweets and the City fans become (and remain) almost silent. However, the examples in Table 3 show that our system did successfully identify the goal even without a spike, in contrast to the methods discussed earlier [10–12].

Table 3 gives examples of how different teams' fans discuss the same events in very different ways. Not only does this confirm our fan-team classification is effective, it also shows the potential power of generic topic detection. Journalists, and others seeking news, do not want just the headlines: they also want a variety of perspectives on a story. We have shown that by dividing active tweeters into sets, depending on which team they support, we can find two distinct views. For example, when Chelsea's goalkeeper, Petr Cech, narrowly prevented Liverpool equalizing, Chelsea fans reported this as a great save while Liverpool fans complained that the referee was mistaken and the ball had crossed the line. We believe our methods could be extended to cluster and analyse social media comments in other domains. For example, it may be possible to divide political commentators into groups depending on which party they support, allowing their varied views to be analysed separately, rather than mixed together.

Acknowledgments. This work is supported by the SocialSensor FP7 project, partially funded by the EC under contract number 287975.

References

1. Sakaki, T., Okazaki, M., Matsuo, Y.: Earthquake shakes Twitter users: real-time event detection by social sensors. In: Proc. of WWW 2010, pp. 851–860. ACM (2010)
2. Bollen, J., Mao, H., Zeng, X.: Twitter mood predicts the stock market. Journal of Computational Science 2(1), 1–8 (2011)
3. Tumasjan, A., Sprenger, T.O., Sandner, P.G., Welpe, I.M.: Predicting elections with Twitter: What 140 characters reveal about political sentiment. In: Proceedings of 4th ICWSM, vol. 10, pp. 178–185 (2010)
4. Newman, N.: The rise of social media and its impact on mainstream journalism. Reuters Institute for the Study of Journalism (2009)
5. FIFA TV: 2010 FIFA World Cup South Africa: Television audience report (2010)
6. Semiocast: Brazil becomes 2nd country on Twitter, Japan 3rd Netherlands most active country, Paris, France (January 2012)
7. Twitter: Twitter & TV: Use the power of television to grow your impact, https://business.twitter.com/twitter-tv
8. Goel, V., Stelter, B.: Social networks in a battle for the second screen. The New York Times (October 2, 2013)
9. Aiello, L., Petkos, G., Martin, C., Corney, D., Papadopoulos, S., Skraba, R., Goker, A., Kompatsiaris, I., Jaimes, A.: Sensing trending topics in Twitter. IEEE Transactions on Multimedia 15(6), 1268–1282 (2013)
10. van Oorschot, G., van Erp, M., Dijkshoorn, C.: Automatic extraction of soccer game events from Twitter. In: Proc. of the Workshop on Detection, Representation, and Exploitation of Events in the Semantic Web (2012)
11. Zhao, S., Zhong, L., Wickramasuriya, J., Vasudevan, V.: Human as real-time sensors of social and physical events: A case study of Twitter and sports games. arXiv preprint arXiv:1106.4300 (2011)
12. Lanagan, J., Smeaton, A.F.: Using Twitter to detect and tag important events in live sports. In: Artificial Intelligence, pp. 542–545 (2011)

Exploiting Result Diversification Methods for Feature Selection in Learning to Rank

Kaweh Djafari Naini[1] and Ismail Sengor Altingovde[2]

[1] L3S Research Center, Leibniz University Hannover, Hannover, Germany
naini@l3s.de
[2] Middle East Technical University, Ankara, Turkey
altingovde@ceng.metu.edu.tr

Abstract. In this paper, we adopt various greedy result diversification strategies to the problem of feature selection for learning to rank. Our experimental evaluations using several standard datasets reveal that such diversification methods are quite effective in identifying the feature subsets in comparison to the baselines from the literature.

1 Introduction

Learning to rank (LETOR) is the state of the art method employed by the large-scale commercial search engines to rank the search results. Given the large number of features available in a search engine, which is in the order of several hundreds (e.g., see Yahoo! LETOR Challenge), it is desirable to identify a subset of features that yield a comparable effectiveness to using all the features. Since search engines typically employ a two-stage retrieval where an initial set of candidate documents are re-ranked using a sophisticated LETOR model, a smaller number of features would reduce the feature computation time, which must be done on-the-fly for the query dependent features, and hence overall query processing time. Furthermore, improving the efficiency of the LETOR stage would allow retrieving larger candidate sets and, subsequently, can help enhancing the quality of the search results.

In a recent study, Geng et al. proposed a filtering-based feature selection method that aims to select a subset of features that are both effective and dissimilar to each other [5]. Inspired from this study, we draw an analogy between the feature selection and result diversification problems. In the literature, a rich set of greedy diversification methods are proposed to select both relevant and diverse top-k results for web search queries (e.g., see [1,6,12,9,10]). We apply three representative diversification methods, namely, Maximal Marginal Relevance (MMR) [1], MaxSum Dispersion (MSD) [6] and Modern Portfolio Theory (MPT) [12,9] to the feature selection problem for LETOR. To the best of our knowledge, none of these methods are employed in the context of learning to rank with the standard search engine datasets.

In the next section, we first describe the baseline strategies for the feature selection from the literature, and then discuss how we adopt the result diversification methods for this purpose. In Sections 3 and 4, we present the experimental setup and evaluation results, respectively. Finally, we conclude in Section 5.

M. de Rijke et al. (Eds.): ECIR 2014, LNCS 8416, pp. 455–461, 2014.

2   Feature Selection for LETOR

Feature selection techniques for the classification tasks are heavily investigated in the literature and fall into three different categories, namely, filter, wrapper and embedded approaches [5]. Strategies in the filter category essentially work independently from the classifiers and choose the most promising features in a preprocessing step. In contrast, the strategies following the wrapper approach consider the metric that will be optimized by the classifier whereas those in the embedding category incorporate the feature selection into the learning process. Earlier studies also show that such feature selection methods do not only help improving the accuracy and efficiency of the classifiers, but may also introduce diversity in ensembles of classifiers [3].

For learning to rank, there are only a few recent studies that address the feature selection issue [5,4]. Following the practice in [5], we focus on the feature selection methods that fall into the filter category.

2.1   Preliminaries

For a given feature $f_i \in F$, we obtain its relevance score for a query by ranking the results of a query solely on this feature and computing the effectiveness for the top-10 results. The effectiveness can be measured using any well-known evaluation measure (like MAP, NDCG) or a loss function (as in [5]). In this study, we employ NDCG@10 as the effectiveness measure and denote the *average* relevance score of a feature over all queries by $rel(f_i)$. To capture the similarity of any two features, denoted with $sim(f_i, f_j)$, we compute the Kendall's Tau distance between their top-10 rankings averaged over all queries (as in [5]). The objective is selecting a subset of k features (F_k), where $k < |F|$, such that both the relevance and diversity (dissimilarity) among the selected features are maximized.

2.2   Baseline Feature Selection Methods

Top-k Relevant (TopK): A straightforward method for feature selection is choosing the top-k features that individually yield the highest average relevance scores over the queries [4].

Greedy Search Algorithm (GAS): This is the greedy strategy proposed by Geng et al. in [5]. It starts with choosing the feature, say f_i, with the highest average relevance score into the set F_k. Next, for each of the remaining features f_j, its relevance score is updated with respect to the following equation:

$$rel(f_j) = rel(f_j) - sim(f_i, f_j) \cdot 2c, \tag{1}$$

where c is a parameter to balance the relevance and diversity optimization objectives. The algorithm proceeds in a greedy manner by choosing the next feature with the highest score and updating the remaining scores, until k features are determined.

2.3   Diversification Methods for Feature Selection

As the astute reader would realize, the goal of feature selection as defined in [5,4] is identical to that of the search result diversification techniques: both problems require selecting the most relevant and, at the same time, diverse items. Motivated by this observation, we adopt three different implicit result diversification techniques to the feature selection problem, as follows.

Maximal Marginal Relevance (MMR): This is a well-known greedy strategy originally proposed in [1]. Peng et al. propose a similar idea of minimal-redundancy maximal-relevance in [8]. In a recent study [2], MMR is employed for feature selection in learning to rank in a setup with a limited number of social features, but not evaluated on the standard search datasets, as we do in this paper.

In this study, we adopt a version of MMR described in [11]. The MMR strategy also starts with choosing the feature f_i with the highest relevance score into the F_k. At each iteration, MMR computes the score of an unselected feature f_j according to the following equation:

$$mmr(f_j) = (1 - \lambda)rel(f_j) + \frac{\lambda}{|F_k|} \sum_{f_i \in F_k} 1 - sim(f_i, f_j), \lambda \in [0,1], \quad (2)$$

where λ is again a trade-off parameter to balance the relevance and diversity.

MaxSum Dispersion (MSD): An alternative representation of the diversification (and hence, feature selection) problem is casting it to the facility dispersion problem in the operations research field [6]. In this case, our objective in this paper, i.e., maximizing the sum of relevance and dissimilarity in F_k, can be solved with the greedy 2-approximation algorithm that is originally proposed for the well-known MaxSum Dispersion (MSD) problem. In the MSD solution, a pair of features that maximizes the following equation is selected into F_k at each iteration:

$$msd(f_i, fj) = (1 - \lambda)(rel(f_i) + rel(f_j)) + 2\lambda(1 - sim(f_i, f_j)), \quad (3)$$

where λ is the trade-off parameter.

Modern Portfolio Theory (MPT): This approach is based on the famous financial theory which states that one should diversify her portfolio by maximizing the expected return (i.e, mean) and minimizing the involved risk (i.e., variance). In case of the result diversification, this statement implies that we have to select the documents that maximize the relevance and have a low variance of relevance [12,9]. The latter component has to be treated as a parameter and its best value can be computed by sweeping through the possible values (as in [12]) unless additional data, such as click logs, are available [9].

Fortunately, in case of the feature selection for LETOR, we have adequate data to model both the mean and variance of the relevance of a feature. Obviously, mean relevance of a feature is $rel(f_i)$ as we have already defined. For the variance of a feature ($\sigma^2(f_i)$), we compute the relevance score of f_i for each query q, and then compute the variance for this set of scores in a straightforward manner.

Table 1. Datasets

Dataset	No. of queries	No. of annotated results	No. of features
OHSUMED	106	16,140	45
MQ2008	800	15,212	46
Yahoo! SET2	6,330	172,870	596

Thus, the greedy MPT solution chooses the feature that maximizes the following equation at each iteration:

$$mpt(f_i) = rel(f_i) - [b\sigma^2(rel(f_i)) + 2b\sigma(rel(f_i)) \sum_{f_j \in F_k} \sigma(rel(f_j)) * sim(f_i, f_j)]. \quad (4)$$

Note that, we eliminated the rank position component from the original formula [12,9] as it does not make sense for the feature selection problem. As before, $b \in [0, 1]$ is the trade-off parameter to balance the relevance and diversity.

3   Experimental Setup

Datasets. Our experiments are conducted on three standard LETOR datasets, OHSUMED[1] from Letor3.0, MQ2008 from Letor4.0 and SET2[2] from Yahoo! LETOR Challenge. In Table 1 we summarize the characteristics of each dataset. The Yahoo! SET2 has 596 features and is also the largest dataset with respect to the number of queries and instances. But previous studies have also shown that even for a small number of features, feature selection can improve the ranking [5].

LETOR Algorithm. Our evaluations employ RankSVM [7], which is a very widely used pairwise LETOR algorithm. More specifically, we used SVMRank[3] library implementation. We trained the classifier with a linear kernel with $\epsilon = 0.001$. We report the results with the C values (where $C \in [0.00001, 10]$) that yields the best performance on the test set with all the features.

Evaluation Measures. We evaluate all the feature selection methods using 5-fold cross validation for the OHSUMED and MQ2008 datasets. For Yahoo! SET2, we use the training and test sets as provided. The evaluation measures are MAP and NDCG@10.

4   Experimental Results

In Figure 1, we report the NDCG@10 and MAP scores obtained on the OHSUMED dataset using the baseline and proposed feature selection methods. We observe that when the number of selected features is greater than 10, the performance is comparable or better than using all features (ALL). Furthermore, the methods adapted from the diversity field outperform the baselines (TopK and GAS). In

---

[1] http://research.microsoft.com/en-us/um/beijing/projects/letor
[2] http://webscope.sandbox.yahoo.com/catalog.php?datatype=c
[3] http://www.cs.cornell.edu/people/tj/svm_light/svm_rank.html

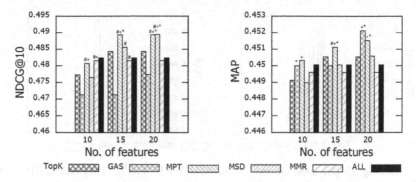

Fig. 1. Ranking effectiveness on OHSUMED: NDCG@10 (left) and MAP (right)

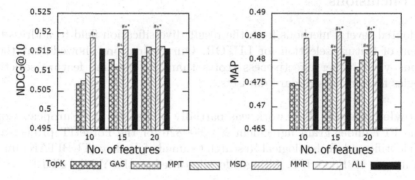

Fig. 2. Ranking effectiveness on MQ2008: NDCG@10 (left) and MAP (right)

particular, MPT is the winner for both evaluation measures when the number of features is set to 15 or 20.

Figure 2 shows the performance for the MQ2008 dataset. In this case, the feature selection algorithms can reach the performance of the ALL only after selecting more than 15 features. For the majority of the cases, the methods adopted from the diversification field are again superior to the baselines, and MSD is the winner method for this dataset.

Finally, in Figure 3, we report the performance for the Yahoo! SET2. As the experiments take much larger time on this dataset, we only present the results for selecting 100 features (out of 596). We observe that, feature selection methods with 100 features cannot beat the all features baseline ALL (not shown in the plots), which is reasonable as we only use one sixth of the available features. MPT is again the best adapted method, and it outperforms TopK baseline for both evaluation measures, and better than or comparable to GAS for MAP and NDCG measures, respectively.

The statistical significance of our methods is verified using the paired t-test with $p < 0.05$. In Figures 1-3, we show the significant differences to the baselines TopK (denoted with +), GAS (denoted with #) and ALL (denoted with *).

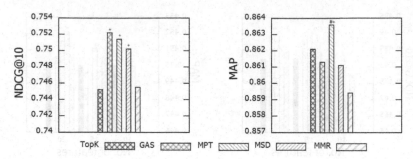

Fig. 3. Ranking effectiveness on the Yahoo! SET2: NDCG@10 (left) and MAP (right)

5 Conclusions

We adopted several methods from the result diversification field to address the problem of feature selection for LETOR. Our evaluations showed that these methods yield higher effectiveness scores than the baseline feature selection strategies for various standard datasets.

Acknowledgments. This work was partially funded by the European Commission FP7 under grant agreement No. 600826 for the ForgetIT project and The Scientific and Technological Research Council of Turkey (TÜBİTAK) under the grant no. 113E065. I. S. Altingovde acknowledges the Yahoo! FREP.

References

1. Carbonell, J.G., Goldstein, J.: The use of MMR, diversity-based reranking for reordering documents and producing summaries. In: Proc. of SIGIR 1998 (1998)
2. Chelaru, S.V., Orellana-Rodriguez, C., Altingovde, I.S.: How useful is social feedback for learning to rank youtube videos? WWW Journal, 1–29 (in press), doi:10.1007/s11280-013-0258-9
3. Cunningham, P., Carney, J.: Diversity versus quality in classification ensembles based on feature selection. In: Lopez de Mantaras, R., Plaza, E. (eds.) ECML 2000. LNCS (LNAI), vol. 1810, pp. 109–116. Springer, Heidelberg (2000)
4. Dang, V., Croft, W.B.: Feature selection for document ranking using best first search and coordinate ascent. In: Proc. SIGIR 2010 Workshop on Feature Generation and Selection for Information Retrieval (2010)
5. Geng, X., Liu, T.-Y., Qin, T., Li, H.: Feature selection for ranking. In: Proc. of SIGIR 2007, pp. 407–414 (2007)
6. Gollapudi, S., Sharma, A.: An axiomatic approach for result diversification. In: Proc. of WWW 2009, pp. 381–390 (2009)
7. Herbrich, R., Graepel, T., Obermayer, K.: Large margin rank boundaries for ordinal regression. In: Advances in Large Margin Classifiers, pp. 115–132 (2000)
8. Peng, H., Long, F., Ding, C.H.Q.: Feature selection based on mutual information: Criteria of max-dependency, max-relevance, and min-redundancy. IEEE Trans. Pattern Anal. Mach. Intell. 27(8), 1226–1238 (2005)

9. Rafiei, D., Bharat, K., Shukla, A.: Diversifying web search results. In: Proc. of WWW 2010, pp. 781–790 (2010)
10. Santos, R.L.T., Castells, P., Altingovde, I.S., Can, F.: Diversity and novelty in information retrieval. In: Proc. of SIGIR 2013, p. 1130 (2013)
11. Vieira, M.R., Razente, H.L., Barioni, M.C.N., Hadjieleftheriou, M., Srivastava, D., Train Jr., C., Tsotras, V.J.: On query result diversification. In: Proc. of ICDE 2011, pp. 1163–1174 (2011)
12. Wang, J., Zhu, J.: Portfolio theory of information retrieval. In: Proc. of SIGIR, pp. 115–122 (2009)

Boilerplate Detection and Recoding

Matthias Gallé and Jean-Michel Renders

Xerox Research Centre Europe

Abstract. Many information access applications have to tackle natural language texts that contain a large proportion of repeated and mostly invariable patterns – called boilerplates –, such as automatic templates, headers, signatures and table formats. These domain-specific standard formulations are usually much longer than traditional collocations or standard noun phrases and typically cover one or more sentences. Such motifs clearly have a non-compositional meaning and an ideal document representation should reflect this phenomenon.

We propose here a method that detects automatically and in an unsupervised way such motifs; and enriches the document representation by including specific features for these motifs. We experimentally show that this document recoding strategy leads to improved classification on different collections.

1 Introduction

These last years, in the Web Information Access community, the meaning of "boilerplate" has shifted from a neutral sense (boilerplate as a standard template) towards a more negative sense (boilerplate as spam). In this paper, we come back to the original definition of boilerplate, considering that identifying them could be a interesting way of capturing unambiguous, highly discriminative features for categories or queries. Indeed boilerplates constitute some kind of signatures made of set of words (with gaps) that are typical of some category, but each word taken individually is highly ambiguous and not at all representative of one class, as shown from this example (the three first from the earn, oil and grain categories of Reuters dataset, and the rest from the Enron e-mail collection):

```
- net •vs •revs •vs •note [typical motif of earn category]
- for •pct sulphur to •dlrs up •[typical motif of oil category]
- pct •vs •pct a year ago and •pct average [typical motif of grain category]
- enron north america •confirm the terms conditions the •the trade date •constitutes confirmation
- the amount set forth •opposite the •credit rating •party the case
- the state •new york •set forth •the meaning set forth section
- the prior written •the other •consent unreasonably withheld •consent required
```

Boilerplate is defined as being "standardized text", "formulaic or hackneyed language" (Merriam-Webster) or "writing that has been used before many times with only very small changes" (Cambridge dictionary). Real-life documents, especially in the business world, contain many of such repeated patterns whose role could be crucial in information access tasks, either because they should be detected and filtered out or, conversely, kept as a single "non-compositional", discriminative feature. Existing work has mostly focused on one instantiation of

M. de Rijke et al. (Eds.): ECIR 2014, LNCS 8416, pp. 462–467, 2014.
© Springer International Publishing Switzerland 2014

this phenomena, namely the cleaning of HTML files by detecting and removing the surplus "clutter" around the main textual content.

The examples from the beginning were detected in the collections we worked with. As we will see later, the • refers to zero or more (up to 5 in this case) words. Due to lack of space we are showing rather small examples, but it should be noted that they can stretch over several sentences. They typically are either automatically generated templates (in emails from customer-care centers for example), legal boilerplate, signatures or table headings.

In order to extract these boilerplates from the text, a formal definition of what is to be considered as boilerplate is needed. The expressiveness of such definitions of *motifs* (using a term from the stringology community) is however strongly related to the size of the resulting set. The most interesting and useful definitions give rise to a potential exponential set of motifs, making the execution time of the extraction algorithms an issue. Therefore, existing motif inference algorithms use heuristics, relying mostly on specific characteristics of the underlying sequences; developed mostly in the bio-informatics community [12,13].

In this paper, we first propose a definition of motif that captures our intuition of what is a boilerplate in a generic text, without making any assumption of the type of text (HTML file, e-mail collection, *etc*). This definition takes into account some user-defined parameters with an obvious interpretation (such as minimum number of occurrences, number of gaps, length, *etc*). Our boilerplate detection algorithm proceeds in two steps. In the first step, we extract all motifs that comply with the requirements given by the user. While exponential in the worst-case, we show how to speed up its execution time by using maximal classes of repeats, using in particular a less known class called largest-maximal repeats [6,5]. In the second step, the extracted motifs are clustered and recoded as new features. Our experiments demonstrate that this two-step approach improves significantly classification on different collections (Reuters news articles and Human Resources forms).

2   Related Work

Most of existing works related to "boilerplate detection" is specific to HTML documents, and tackles specific tasks such as detection of structural features (like tags), style descriptions and visual localization of the content [7,1,11,8]. Our problem and approach are very different from this work, as we look for repeated standard formulations without compositional structure *inside* the text.

The research field of *near-duplicate detection* uses often string-based methods like the one we present here, but its goal is somewhat opposite, as the techniques developed aim at increasing robustness to possible variations in the sequences. In the extreme cases of two forms where only few fields differentiate them, both documents risk to be tagged as duplicated, while our method aims at concentrating the boilerplate of the forms into a single feature.

As far as the the motif extraction problem is concerned, many algorithms have been developed having bio-informatics applications in mind. Several different ways of defining inexact motifs exists but in all of the them their size

464   M. Gallé and J.-M. Renders

soon becomes exponential in some parameter. For the case of gapped motifs, [3] presents a linear solution with the supposition that the number of blocks and the block sizes are fixed. Their solution is based on rewriting the sequence over a bigger alphabet, and the constraint of having a fixed block size derives from there. RISOTTO [12] is a popular alternative for extracting motifs with a given number of mismatches (this is, not necessarily under the form of a gapped motif). Like most of existing works on motif extraction [13,10], this algorithm too is targeted towards genetic application.

Unfortunately, because the underlying problems have a solution set whose size can grow exponentially, most implementations rely on heuristics valid on a certain range of values for the dimensions of the problem (vocabulary size, sequence length, number of repeats, number of gaps or mismatches, *etc*). For instance, current implementations require typically a fixed alphabet of maximal size 4, 16 or 256, while supposing an abundant number of repeats in the sequence. When the working conditions are outside these ranges, the computing performances could degrade dramatically.

3 Methodology

Motif Extraction. Boilerplates will be defined as rigid blocks separated by variable-length gaps. Families of such motifs will be constrained by the (a) minimum number of words of each of these substrings, (b) total length of all blocks, (c) number of different documents they occur in (diversity), (d) number of possible gaps and (e) maximal length of each gap. The last of the examples for instance has 4 blocks (equivalently, 3 gaps) where each block has a maximal length of 3, a total length of 10 and where each gap can be replaced by up to 5 words (this depends on the occurrences in the documents).

Inspired from the bioinformatics approach, we consider a collection as a large string s of words with documents separated by unique symbols. Formally, we define a motif as a tuple $\langle d, N, K, \epsilon, \gamma \rangle$ over s as the sequence $m = B_1 \bullet^{\leq \epsilon} B_2 \bullet^{\leq \epsilon} \ldots B_g$, such that $B_i \in \Sigma^*$ (the *blocks*), $g \leq \gamma$ (number of blocks), $|B_i| \geq d$ (length of each block), $\sum_{i=1}^{g} |B_i| \geq N$ (total length) and $diversity(m) \geq K$ (the diversity of a motif is the number of documents in which it appears). Σ denotes the vocabulary here, and $\bullet$ is a *joker* symbol that matches any symbol in Σ. We denote with $pos_s(m)$ the set of occurrences of motif m. The composition of motif M and B ($composite(M, B, \epsilon)$) is the motif $M \bullet^{\leq \epsilon} B$.

We are particularly interested in maximal motifs: these are motifs such that no other motif covers exactly the same rigid positions by being more specific (being longer or converting a joker into a rigid symbol). For this notion of maximality, a concept we will use in our algorithm is *context diversity*: The left (right) context of a repeat r in s for a subset of its occurrences p is defined as $lc_s(r, p) = \{s[i-1] : i \in p\}$ ($rc_s(r, p) = \{s[i + |r|] : i \in p\}$). For a given r, a subset $p \subseteq pos_s(r)$ is said to be *left-context diverse* ($lcd_s(r, p)$) if $|lc_s(r, p)| \geq 2$ (respectively $rcd_s(r, p)$). If for occurrences p of a repeat r the symbol to the left is always the same (say a), then r is not left-context diverse on this occurrence set p because ar would cover the same as r and r can be considered redundant.

Table 1. Results of the experiments: average accuracy

	R8	R52	HR
BOW	96.95	95.45	71.44
1+2+3-grams	97.05	95.59	71.74
25-grams	97.36	95.68	72.22
50-grams	97.38	95.81	72.24
75-grams	97.08	95.6	72.04
100-grams	96.98	95.47	71.66
Motifs	96.52	95.22	70.86
meta 25-grams	97.78	96.29	72.91
meta 50-grams	97.63	96.15	72.55
meta 75-grams	97.28	95.82	72.22
meta 100-grams	97.09	95.48	71.91
meta-motifs	**98.08**	**96.6**	**73.48**

In order to compute all $\langle d, N, K, \epsilon, \gamma \rangle$-motifs, we start from one block B such that $|B| \geq d$ and try to extend it. The ϵ and γ constraints (on the number and length of the gaps) can then be achieved by construction, while the diversity constraint K can be used to prune possible extensions early on (if a subsequence is not diverse enough, neither super-sequence will be). At the end we filter those motifs that do not satisfy the total length (N) constraint. For this an efficient strategy amounts to define an array occ of size n (the length of the sequence s) that holds for each position i the list of blocks that occur at position i. The sparse number of repeats in natural language makes such an approach feasible, even though the memory requirements for this data structure is linear in the total occurrences of starting blocks. Having pre-computed occ the algorithm that extends a given block is straightforward (Alg. 1): for each starting block B, we find those blocks B' that appear more than once at positions $o + e$, where $o \in pos_s(B)$ and $e \leq \epsilon$. If the motif $composite(B, B', \epsilon)$ satisfies the diversity and maximality constraint (line 5), it is then output and further extended.[1]

Alg. 1 has then to be called once for each starting block. In the most general case, the set of starting blocks is the set of all repeats. This set however is very redundant and several notions of maximality exist to reduce this redundancy, in particular maximal repeats (MR), largest maximal repeats (LMR) and super-maximal repeats (SMR) [6,4]. It can be proved that the motif extraction method gives the same results starting from the set of maximal repeats only or from all standard repeats. We compared the trade-off between speed and coverage of motifs by experimenting with LMR and SMR. We only report here the conclusions of these experiments over different datasets (e-mail and news collections): starting from LMR is in general twice as fast, while covering around 90% of the original motifs, while starting from SMR is three times faster, but covers only 15%. Consequently, we decided to use the LMR starting blocks in all the reported experiments.

---

[1] Strictly speaking, an outputted motif there could still be non-maximal, when one of the gaps has internal positions that cover the same symbol (for example, a motif $B_1 \bullet^3 B_2$ where, for all occurrences, the second position in the gap covers the same symbol). These motifs have also to be extended as there may be no other motif with γ blocks or less that covers any extension. For simplicity's sake, we output also these non-maximal motifs here (they have then to be filtered at the end).

Algorithm 1. Motif Extension

$motif\_extend(B, occ)$

1: **for all** block B' that appears more than once in $\biguplus_{o \in pos_s(B)} \left\{ \bigcup_{0 \le e \le \epsilon} \{occ[o + e]\} \right\}$ **do**
2:     let p be the set of o that fulfill this condition
3:     let p' be the set of $o + e$ that fulfill this condition
4:     $M = \text{composite}(B, B', \epsilon)$
5:     **if** $lcd(B, p) \wedge rcd(B, p) \wedge lcd(B', p') \wedge rcd(B', p') \wedge \text{diversity}(M) \ge K$ **then**
6:         **output** M
7:         **if** $\text{blocks}(M) \le \gamma$ **then**
8:             $motif\_extend(M, occ)$
9:         **end if**
10:     **end if**
11: **end for**

Clustering and Recoding. The described algorithm, even it deals with gaps, fails sometimes in capturing all the variability of a basic pattern. Notably it outputs several motifs whose occurrences strongly overlaps, and which could be considered as one. Using a broader definition of motif would face the challenge of an efficient extraction algorithm. Here we took a different approach, and decided to create *meta-motifs* as clusters of the motifs given by Alg. 1. In our experiments, this was done with a single-link hierarchical clustering algorithm [9] where the similarity between motifs was their Tanimoto coefficient over common bigrams. On different collections, the clustering itself turned out to be very stable with respect to the selected threshold, indicating a clear separation between the clusters. Consequently, we used a threshold of 0.5 and this was not optimized further.

The obtained motifs were added as new features in the extended vector space model representing the documents, while the original words contained in the motifs were removed.

4   Experiments

Our experiments consisted of classifying three different collections: the first is based on a non-public collection of 7 000 OCR'ed forms related to the HR department of a multinational conglomerate, categorized over 35 classes. The other are two subsets (R8 and R52) of the Reuters 21578 collection[2]. In all cases sparse logistic regression (ℓ_2-normalized) classifiers were built.

Besides the baseline of using the simple bag-of-words, we also used the concatenation of uni-, bi- and tri-grams. We also compared against another boilerplate detection approach: we implemented the SPEX algorithm [2] for retrieving n-grams in a memory-efficient (although approximate) way, and processed these n-grams in the same ways as the motifs. We tried two flavors of our motif extraction algorithm: *motifs* stops after the extraction algorithm, while the *meta-motifs*

---

[2] http://csmining.org/index.php/r52-and-r8-of-reuters-21578.html

also performs the clustering. We also experimented with clustering the n-grams (*meta-ngrams*), to see the impact of this step by itself. In all cases, just removing the detected boilerplates decreased accuracy.

We summarize our results in Table 1. The classifiers used *tf-idf*–weighted, ℓ_2-normalised feature vectors. Results are reported as averages over 10 runs, with random splits between training (90%) and test (10%) sets. Wilcoxon signed-rank tests on the performances between the baselines (BOW or multi-grams) and the proposed method ("meta-motifs") gave a p–value of less than 5%. It can be observed that, for both tasks, using n-grams is helpful. However, for large values of n performance decreases. Also, the clustering step by itself alone is also positive. However, the best values are obtained by combining this clustering with the motifs extracted. We believe that this is due to the flexibility offered by our method – not having to fix the length (n) of the motif by focusing on LMR, extending them to motifs and allowing some degree of variations inside meta-motif features – which leverages the usefulness of boilerplates. Note also that we did not tweak the parameters of the motif extraction algorithm and used the same for all datasets.

We believe that the presence of these boilerplate should not be neglected in real-life applications and that a framework like the one we presented here is an important step in an information retrieval pipeline.

References

1. Baroni, M., Chantree, F., Kilgarriff, A., Sharoff, S.: CleanEval: a competition for cleaning webpages. In: LREC (2008)
2. Bernstein, Y., Zobel, J.: Accurate discovery of co-derivative documents via duplicate text detection. Inf. Syst. 31(7), 595–609 (2006)
3. Iliopoulos, C.S., McHugh, J., Peterlongo, P., Pisanti, N., Rytter, W., Sagot, M.: A first approach to finding common motifs with gaps. International Journal of Foundation of Computer Science 16(6), 1145–1155 (2005)
4. Gallé, M.: Searching for Compact Hierarchical Structures in DNA by means of the Smallest Grammar Problem. Université de Rennes 1 (February 2011)
5. Gallé, M.: The bag-of-repeats representation of documents. In: SIGIR (2013)
6. Gusfield, D.: Algorithms on Strings, Trees, and Sequences: Computer Science and Computational Biology. Cambridge University Press (January 1997)
7. Kohlschütter, C., Fankhauser, P., Nejdl, W.: Boilerplate detection using shallow text features. In: WSDM, p. 441. ACM Press, New York (2010)
8. Kohlschütter, C., Nejdl, W.: A Densitometric Approach to Web Page Segmentation Segmentation as a Visual Problem. In: CIKM, pp. 1173–1182 (2008)
9. Manning, C., Raghavan, P., Schütze, H.: Introduction to Inf Retrieval. Cambridge UP (2009)
10. Marsan, L., Sagot, M.-F.: Extracting structured motifs using a suffix tree–algorithms and application to promoter consensus identification. Journal of Computational Biology 7(3/4), 345–362 (2000)
11. Pasternack, J., Roth, D.: Extracting Article Text from the Web with Maximum Subsequence Segmentation. In: WWW, pp. 971–980 (2009)
12. Pisanti, N., Carvalho, A.M., Marsan, L., Sagot, M.-F.: RISOTTO: Fast extraction of motifs with mismatches. In: Correa, J.R., Hevia, A., Kiwi, M. (eds.) LATIN 2006. LNCS, vol. 3887, pp. 757–768. Springer, Heidelberg (2006)
13. Zhang, Y., Zaki, M.: Exmotif: efficient structured motif extraction. Algorithms for Molecular Biology 1(1), 21 (2006)

A Two-level Approach for Subtitle Alignment

Jia Huang[1], Hao Ding[2], Xiaohua Hu[1], and Yong Liu[2]

[1] Drexel University, Philadelphia, PA, USA
{jh645,xh29}@drexel.edu
[2] New York University, New York City, NY, USA
hd510@nyu.edu, yongliu@poly.edu

Abstract. In this paper, we propose a two-level Needleman-Wunsch algorithm to align two subtitle files. We consider each subtitle file as a sequence of sentences, and each sentence as a sequence of characters. Our algorithm aligns the OCR and Web subtitles from both sentence level and character level. Experiments on ten datasets from two TV shows indicate that our algorithm outperforms the state-of-the-art approaches with an average precision and recall of 0.96 and 0.95.

Keywords: sequence alignment, subtitle alignment, dynamic programming.

1 Introduction

Movie subtitles are useful resources for mining interest points in movies. Well-qualified subtitles with high accuracy and no delay can provide better user experience for subtitle related applications. Optical character recognition (OCR) is the most popular approach to extracting real-time subtitles. However, due to the noisy background beneath the displayed subtitle, OCR does not have a high accuracy. This challenge becomes more severe when we need to recognize movie subtitles in Asian languages. Most studies applied NLP and statistical models to correct OCR errors [2]. On the other hand, movie subtitle file retrieved from subtitle website can provide accurate subtitles, but most of them have delays. Therefore, instead of using NLP models, we use the movie subtitle file from the web to correct OCR errors by aligning them together so that the well-qualified subtitle is obtained.

Sequence Alignment is a technique which originated from molecular biology [5] [6] [7]. One of the most popular alignment algorithms is Needleman-Wunsch algorithm [5]. It is a global alignment algorithm which optimizes a global cost function, and tries to align each element in the sequences. The most relevant studies to ours are [3] [4]. In Caroline et al.'s study, in order to construct an English-French parallel corpus, they applied dynamic time warping (DTW) to align 43,013 English and 37,625 French movie subtitle files with a precision of 92.3% [3]. However, they viewed each sentence as a bag of words and aligned the sentence sequences. Sequence alignment has also been applied to matching subtitles with screenplay. The alignment was used to identify characters in the subtitles [4]. However, it broke subtitles and screenplay into words and only aligned two word sequences. In summary,

M. de Rijke et al. (Eds.): ECIR 2014, LNCS 8416, pp. 468–473, 2014.

previous studies aligned subtitles from either the sentence level or the word level, which is insufficient for achieving a meaningful and an accurate alignment result.

In this paper, we propose a two-level Needleman-Wunsch alignment algorithm to align subtitle files. We consider each subtitle file as a sequence of sentences, and each sentence as a sequence of characters. By applying sequence alignment on both levels, we are able to align the OCR and Web files more accurately than the state-of-the-art approaches.

2   Method

In this section, we show how to obtain an optimal alignment between an OCR subtitle file and a Web subtitle file. A subtitle file can be viewed as a two-level structure. Each subtitle file is represented as a sequence of sentences, and each sentence as a sequence of characters. Since global alignment methods such as Needleman-Wunsch algorithm can only give the global optimum on the sentence level, we designed a two-level Needleman-Wunsch algorithm to solve the alignment problem when viewing a subtitle as a two-level structure.

2.1   Sentence Level Sequence Alignment

Suppose the OCR subtitle file A is a sequence of M sentences, i.e., $A = a_1 \cdots a_M$. The web subtitle file B is a sequence of N sentences, $B = b_1 \cdots b_N$. We try to find the optimal alignment between A and B that maximizes alignment score by Needleman-Wunsch algorithm as below:

$$SIM(m,n) = \max \begin{cases} SIM(m-1,n-1) + s(a_m, b_n), & a_m, b_n \text{ aligned} \\ SIM(m-1,n) + g, & a_m \text{ aligned with a null} \\ SIM(m,n-1) + g, & b_n \text{ aligned with a null} \end{cases} \quad (1)$$

$SIM(m,n)$ is the maximum score for sequences $a_1 \cdots a_m$ and $b_1 \cdots b_n$. $s(a_m, b_n)$ represents the similarity between sentences a_m and b_n. g, the gap penalty, is designed to reduce the score when a sentence is aligned with a null, and its value is determined empirically which will be further discussed in Section 3.

The problem then becomes how to compute the similarity $s(a_m, b_n)$ between each sentence a_m in A with each sentence b_n in B. We view each sentence as a sequence of characters, and apply Needleman-Wunsch on the character level. $s(a_m, b_n)$ then equals to the maximum score when the optimal alignment between a_m and b_n is achieved.

2.2   Character Level Sequence Alignment

The similarity between two sentences can be better reflected by the score of character level sequence alignment because it not only increases with the number of common characters of both sentences, but also takes the order of characters into consideration. For instance, the sentences "Here we are!" and "We are here!" have the same words but the word orders are different. Sequence alignment between them can give a high alignment score but still lower than that when two identical sentences are aligned.

Therefore, we utilize Needleman-Wunsch algorithm to assign the highest alignment score between two sentences a_m and b_n as their similarity (a_m, b_n) :

$$s(a_m, b_n)_{i,j} = \max \begin{cases} s(a_m, b_n)_{i-1,j-1} + (a_m(i), b_n(j) \; match ? \; \lambda : 1) \\ s(a_m, b_n)_{i-1,j} \\ s(a_m, b_n)_{i,j-1} \end{cases} \tag{2}$$

$a_m(i)$ is the ith character in the sentence a_m. The term $a_m(i), b_n(j) \; match ? \; \lambda :$ 1 means that, if $a_m(i)$ is identical to $b_n(j)$, match score λ is added onto the alignment score; if they are not identical, mismatch score = 1 is added. The gap penalty on this level is set as 0, which means if $a_m(i)$ or $b_n(j)$ is aligned with a null, no score is added. The formula above indicates that the current optimal alignment comes from three states that $a_m(i)$ and $b_n(j)$ are aligned, $a_m(i)$ is aligned with a null, and $b_n(j)$ is aligned with a null. Hence, the similarity $s(a_m, b_n)$ can be obtained by achieving the final score $s(a_m, b_n)_{I,J}$ where I is the number of characters in a_m and J is the number of characters in b_n. Therefore, by integrating the character level sequence alignment score (2) into the sentence level alignment score (1), the maximum alignment score for subtitle file A and B can be achieved by dynamic programming.

3   Evaluations

3.1   Dataset

Two groups of datasets are used in our experiment. Each dataset consists of a pair of an OCR subtitle file and a web subtitle file. The first group is the subtitle pairs of Episode 12-16 of the TV show "The Big Bang Theory". The second group is the subtitle pairs of Episode 7-11 of the Korean TV show "Miss You" (denoted as TBBT and MissYou). Before running the alignment algorithms, we remove all the empty and repeated subtitles. We then manually align each pair of subtitles to generate the ground truth. There are about 300 OCR and 400 Web subtitles for each TBBT dataset, and 500 OCR and 800 Web subtitles for each MissYou dataset.

3.2   Evaluation Metrics

Precision, recall and F1 score are used as our evaluation metrics. As shown in equation (3), #C is the number of correct alignments. #T is the number of aligning pairs in the algorithms. #A is the number of aligning pairs in the ground truth. Note that all these metrics are based on sentence level alignment.

$$Precision = \frac{\#C}{\#T}, \; Recall = \frac{\#C}{\#A}, \; F1 = 2 \cdot \frac{Precision \cdot Recall}{Precisio + Recall} \tag{3}$$

3.3   Baselines

Existing studies on subtitle alignment do not compare their result with baselines, which makes their conclusions arbitrary. In our experiment, we compare the performance of our algorithm with those of three baselines, F-measure [3], screenplay [4], and Cosine. All three baselines align the subtitles from only one level. F-measure and Cosine aligns the subtitles from the sentence level, while the screenplay aligns the subtitles from the word level. The difference between F-measure with our algorithm lies in the matching scores. Equation (5) presents the definition of sentence similarity in F-measure. $match(e, f)$ is the number of common characters between sentence e and f. $\delta(x, y)$ equals to 1 when x equals to y and 0 otherwise. We see that F-measure view each sentence as a bag of characters. The similarity between two sentences depends on how much common characters they share.

$$S(e,f) = \max \begin{cases} S(e, f-1) + \beta_{FM}(F_M(e,f) + \varepsilon) \\ S(e-1, f-1) + \alpha_{FM}(F_M(e,f) + \varepsilon) \\ S(e-1, f) + \lambda_{FM}(F_M(e,f) + \varepsilon) \end{cases} \tag{4}$$

where

$$F_M(e,f) = 2 \cdot \frac{P(e,f) \cdot R(e,f)}{P(e,f) + R(e,f)} \tag{5}$$

$$P(e,f) = \frac{match(e,f)}{N(f)}, \quad R(e,f) = \frac{match(e,f)}{N(e)} \tag{6}$$

$$match(e,f) = \sum_{i=1}^{N(e)} \sum_{j=1}^{N(f)} \delta(e_i, f_j) \tag{7}$$

The third baseline Cosine differs from F-measure in two aspects. First, it views each sentence as a bag of words, where each word consists of every two consecutive characters. Hence, each sentence is represented as a vector where each entry is the term frequency of the word. Second, the similarity between two sentences is defined as their cosine similarity.

3.4   Parameter Tuning

We tune the parameters to the achieve algorithms' optimal performances. We assume that TBBT and MissYou have distinct characteristics and thus different optimal parameters. For TBBT, we use Episode 17 as the hold-out dataset. For MissYou, we use Episode 12.

For the F-measure algorithm [3], we set ε to 0, β_{FM} and λ_{FM} to 1, and tune α_{FM} from 1 to 9. Results show that for TBBT dataset, precision, recall and F1 increase with α_{FM} until it reaches 3 and then becomes stable. For MissYou dataset, metrics reach optimal value when α_{FM} equals 6 and 9. We thus set α_{FM} to 3 and 6 for two groups of dataset. For the screenplay algorithm, we set the mismatch score as 1 and gap score as 0. Match score a is the only parameter that needs to be tuned. We tune the match score a from 1 to 100. The results show that the optimal value is reached when $a = 100$. Therefore, we set a to 100. For the cosine algorithm, we use the best parameters tuned for the F-measure algorithm. For our own two-level algorithm, the final optimal alignment is decided by the sentence level gap penalty g and character level match score λ. We perform grid search for each feasible pair of (g, λ) and results indicate that optimal value is reached when $g \leq -1000$ and $\lambda \geq 1000$. Hence, we set g to -1000 and λ to 1000 for the two groups of datasets.

4   Results

Figure 1 and 2 show that for both datasets, our algorithm outperforms the baselines on precision and recall. However, there are differences between the results of the two datasets. For TBBT, four algorithms have close performances. Our algorithm outperforms F-measure by only 0.02 in precision. Screenplay performs the worst but is only 0.05 lower than ours. For MissYou, our algorithm shows a more significant advantage. It outperforms F-measure by 0.1 in precision and 0.05 in recall. Screenplay only gets half of our algorithms' precision. Such differences may be due to the dataset characteristics. For TBBT, its OCR and Web subtitles come from the same source and share most characters. Hence, it is easier to align sentences correctly. However, for MissYou, its OCR and Web subtitles use different words to convey the same meaning. As a result, aligning characters, as Screenplay does, would not work, since semantically identical sentences may not share characters. We also notice that the F-measure algorithm has lower precision in our MissYou dataset (around 80%) than in its original dataset (92.3%) [3]. We assume that this is also due to the characteristics of the MissYou dataset.

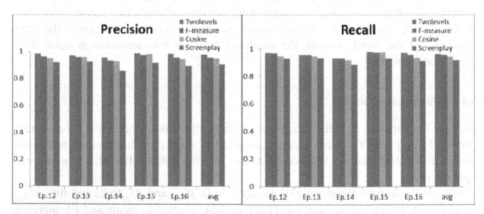

Fig. 1. Performances of alignment algorithms on TBBT dataset

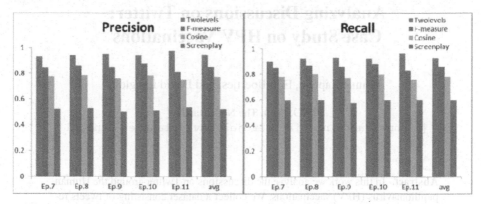

Fig. 2. Performances of alignment algorithms on MissYou dataset

5   Conclusion and Future Work

In our experiments, we consider each subtitle file as a sequence of sentences, and consider each sentence as a sequence of characters. We then use the output of character level alignment as an input for the sentence level alignment. Results show that our two-level algorithm performs the best. Cosine does not have as good performance as we assume, which further indicates that the order of characters is important for accurate sentence alignment. In the future, we will see if one-level alignment algorithms with more complicated word segmentation techniques can outperform our two-level alignment approach.

References

1. Lin, C.-J., Liu, C.-C., Chen, H.-H.: A simple method for Chinese Video OCR and its application to question answering. Computational Linguistics and Chinese Language Processing 6(2), 11–30 (2001)
2. Lopresti, D.: Optical Charater Recognition Errors and Their Effects on Natural Language Processing. In: AND 2008, pp. 9–16 (2008)
3. Caroline, L., Kamel, S., David, L.: Building Parallel Corpora from Movies. In: The 4th International Workshop on Natural Language Processing and Cognitive Science, NLPCS 2007 (2007)
4. Turetsky, R., Dimitrova, N.: Screenplay alignment for closed-system speaker identification and analysis of feature films (2004)
5. Needleman, S.B., Wunsch, C.D.: A general method applicable to the search for similarities in the amino acid sequence of two proteins. Journal of Molecular Biology 48(3), 443–453 (1970), doi:10.1016/0022-2836(70)90057-4, PMID 5420325
6. Mount, D.M.: Bioinformatics: Sequence and Genome Analysis, 2nd edn. Cold Spring Harbor Laboratory Press, Cold Spring Harbor (2004) ISBN 0-87969-608-7
7. Smith, T.F., Waterman, M.S.: Identification of Common Molecular Subsequences. Journal of Molecular Biology 147, 195–197 (1981)

Analyzing Discussions on Twitter:
Case Study on HPV Vaccinations

Rianne Kaptein, Erik Boertjes, and David Langley

TNO, Delft, The Netherlands
{rianne.kaptein,erik.boertjes,david.langley}@tno.nl

Abstract. In this work we analyze the discussions on Twitter around the Human papillomavirus (HPV) vaccinations. We collect a dataset consisting of tweets related to the HPV vaccinations by searching for relevant keywords, by retrieving the conversations on Twitter, and by retrieving tweets from our user group mentioning semi-relevant keywords. We find that by tracking the conversations on Twitter relevant tweets can be found with reasonable precision. Although sentiments and opinions change regularly in a discussion, we find few cases of topic drift.

1 Introduction

Twitter is used by millions of people as a social network to converse with friends, as a professional network to share interesting web pages, as a way to keep up to date on topics, for entertainment and for many other purposes. In this paper we take a closer look at a specific subject of discussion on Twitter: the vaccination of 12 and 13 year old girls in the Netherlands against the Human papillomavirus (HPV) which is the most common sexually transmitted disease and is the leading cause of cervical cancers. The HPV vaccinations were introduced in the Netherlands in 2009, and there has been a lot of public debate about the effectiveness and cost issues. Specifically, there have been a number of online anti-vaccination campaigns which call into question the efficacy and need for the vaccination.

Other countries face similar problems during the roll out of HPV vaccination programs. The debates on HPV takes place in the offline world, but also in the online world, e.g. Keelan et al. [5] analyze the HPV debate on MySpace blogs in the United States.

Organizations increasingly search Twitter messages to form a view of what people are talking about and what different opinions are prevalent. However, such search actions may miss many relevant tweets which form part of longer discussions. For example, only the first of the following two tweets, which form a short discussion, will generally be found when searching for HPV-related opinions:

- I think that the HPV vaccination is important to protect my child's health #hpv #cervicalcancer
- I don't! I've seen research showing it does not really work #justsaying.

In this paper we investigate the discussions on Twitter around the HPV vaccinations in order to answer the following research questions:

M. de Rijke et al. (Eds.): ECIR 2014, LNCS 8416, pp. 474–480, 2014.

1. What is an effective way to retrieve discussions on Twitter?

Most topical datasets are created using a Twitter keyword or hashtag search or stream. In a discussion however not every tweet will mention the search keywords or use the hashtag.

2. What are the characteristics of HPV discussions on Twitter? Do opinions and sentiment change in the discussion?

This may provide us with an important view of the social influence effect taking place during discussions via Twitter.

This paper is organized as follows. The next section discusses related work. Section 3 describes our method to collect Twitter discussions about the HPV vaccinations. In Section 4 we analyze our data, and finally in Section 5 we draw our conclusions.

2 Related Work

Twitter has evolved over the years with new conventions and new possibilities added over time, whereby it is sometimes used as a medium for discussion. Relevant functionality includes directed tweets, retweeting and asking questions. Honeycutt and Herring [4] investigate the use of directed tweets which include the @ sign in conversations on Twitter. They find the @ sign to be used as a marker of addressivity, i.e. to direct a tweet to a specific user, in 91% of the cases. In 5% of the cases the @ sign is used as a reference, i.e. somewhere in the tweet another username is mentioned. The average length of a conversation was 4.62 tweets between 2.5 participants.

A common practice to share information on Twitter is through retweeting. Retweeting started out as a convention without formal grammar, but is now formally integrated in Twitter and most Twitter applications [2].

Morris et al. [8] have investigated what kind of questions people ask their online social networks. Half of the participants in their study uses Twitter to ask questions, ranging from "Why are men so stupid?" to "Should I replace my Blackberry with an iPhone, or just upgrade my Blackberry?". Trust and subjective questions are the main motivations for asking their social network rather than conducting a Web search. For discussions around HPV vaccinations trust and subjectivity are also highly important issues.

In the past few years there have been a number of works that study topic retrieval on Twitter, i.e. how to retrieve all relevant tweets around a topic on Twitter. Bandyopadhyay et al. applies query expansion using external resources as a way to cope with the extreme brevity of tweets [1]. Metzler et al. propose an approach which generates expansion terms based on the temporal co-occurrence of terms [7]. These works focus on query expansion techniques to retrieve more relevant messages, in this work we focus on the context around the tweet, previous messages sent by a user and conversations.

Research on healthcare communication and the HPV vaccinations in particular also takes into account online sources. Zyngier et al. [9] find online sources are scarcely used as a source of health information, and decisions on getting the HPV vaccination depend on trusted authority figures, family and friends. McRee et al. [6] examine associations between parents' Internet information-seeking and their knowledge, attitudes and beliefs about HPV vaccine. A positive influence of accessing information on the Internet about HPV vaccine is found. It was associated with higher knowledge and mostly positive parental attitudes and beliefs. Chou et al. [3] show age is a strong predictor of social networking and blogging activity. Social media is most suited for healthcare communication when the target group is young.

3   Data Set

In this section we first explain the concept of a conversation on Twitter, and then describe how we collect our dataset and use different lists of keywords.

3.1   Conversations

Nowadays on Twitter you have the option to reply to a tweet of another Twitter user. That is, you do not only address a tweet to a user by using the @ sign marker at the beginning of your tweet, but your tweet is a reply to a specific tweet of the other user. In this way conversations between two or more persons can occur. Unfortunately the current Twitter API[1] does not provide an easy way to retrieve complete conversations. To retrieve the complete conversations on Twitter we use a combination of checking for replies to tweets within our user group and screen scraping.

3.2   Data Collection

To create a data set we collect tweets around the topic of the HPV vaccinations. The data is collected during a period of two months: March and April of 2013, which is the period that the first HPV vaccinations are carried out for each cohort. The process for retrieving the data is as follows:

- Using the Twitter streaming API all tweets that match our manually selected relevant keywords are retrieved
- Tweets that are not in the Dutch language are removed

Using the Twitter REST API the following tweets are retrieved:

- For all tweets in the dataset which are a reply on a tweet, the tweets they are replying on are added to the dataset
- Tweets by any user occurring in the dataset so far that are replies to tweets in the dataset collected, are added to the dataset
- Tweets by any user occurring in the dataset so far, mentioning one of the semi-relevant keywords, but none of the unwanted keywords are added to the dataset

Finally, since many conversations were still incomplete we did screen scraping to complete as many conversations as possible. Not all conversations in our dataset are complete, because accounts can be protected, and multiple replies to the same tweet can occur which our method may not detect.

---

[1] Twitter REST API version 1.1 https://dev.twitter.com/docs/api/1.1

3.3 Keyword Selection

Three lists of keywords are used during the data collection: lists of relevant, semi-relevant and non-relevant keywords. The lists of keywords are generated manually using expert judgment and a quick scan of initial search results. When monitoring a topic on Twitter using a stream or a keyword search there is a trade-off between recall and precision. Adding more general keywords to your query will increase recall, but will decrease precision. To tackle this issue we have chosen to keep the list of relevant keywords quite specific e.g. "HPV AND vaccination", and inside the community that is talking about this topic allow for more general keywords e.g. "HPV OR vaccination": the semi-relevant keywords. When a person has already been tweeting about the HPV vaccination before, it is much more likely that this person will tweet about this topic again. For the HPV vaccinations we can distinguish two groups talking about the topic: the 12- or 13-year old girls that are getting the injections, and their parents. We can do this based partly on the users' profile information and partly on the language used. Any girl in this target group talking about getting an injection, is most likely talking about the HPV vaccination.

The list of non-relevant keywords serve as negative relevance feedback, tweets mentioning one of those keywords are considered non relevant. Non relevant keywords include words related to other types of injections, e.g. blood tests, travel vaccinations, and other meanings of the search keywords, e.g. the Dutch translation of stinging eyes ("prikkende ogen") contains the word injection("prik").

4 Twitter Discussion Analysis

Using the strategy described in the previous section we have collected a dataset. We give some statistics on this dataset, analyze the relevance of the collected tweets, as well as make a qualitative analysis of the discussions.

4.1 Data Set Statistics

In total 12.639 tweets are collected. 799 tweets (6%) are retweets, and 421 tweets (3%) contain a URL. The total number of distinct users in the dataset is 4.156, on average each user posted 3.0 tweets. 9.275 (73%) of the tweets are part of one of the 2.165 conversations, making the average conversation length (excluding single tweets not part of a conversation) 4.3 tweets. Due to some shortcomings in our retrieval process not all discussions are completely retrieved, e.g. when more than one person replies to the same initial tweet. The actual average length of the discussions is therefore slightly longer.

The conversations on Twitter are often not just between two persons. Since all your followers can see the public conversations someone is having, other interested people can join in easily. In our dataset 80% of the conversations are between 2 users, 16% between 3 users and 4% of the conversations has 4 or more participants.

The overwhelming majority of the tweets are posted by the young girls getting the vaccinations. This can be seen by the words they use: words related to actually receiving the shots are most frequent in the dataset. Only 50 tweets (0,4%) mention the word

Table 1. Source of the collected HPV tweets

Source	# tweets	% of tweets
Search result mentioning relevant keywords	3.055	24%
Conversation starters without keywords	1.178	9%
Reply to tweet in dataset	2.560	20%
Target user mentioning semi-relevant keywords	2.205	17%
Conversation starters & replies from screen scrape	3.641	29%
Total	12.639	100%

'daughter', a clear indication that a parent is talking. Around 120 tweets are tweets and their retweets of the organisations responsible for the vaccinations. Table 1 shows the sources of the collected tweets. Most results were collected using the screen scrape, the relevant keywords stream accounts for only a quarter of the total number of results. This is, in itself, a highly interesting result: a search using keywords would miss 75% of the relevant tweets.

4.2 Relevance Analysis

A random subset of 222 discussions (10% of all discussions) consisting of 1002 tweets has been manually rated. Each tweet is rated on: relevance (Relevant or Non-Relevant), sentiment (Positive, Neutral, Negative) and opinion on the HPV vaccination (Pro-Vaccination. Anti-Vaccination, Doubt, No opininion). To check interrater agreement a part of the data (234 tweets) has been rated by two raters. Interrater agreement measured by Cohen's kappa is 0.78 for relevance, 0.61 for sentiment and 0.31 for opinions. For tweets rated differently by the two raters, the final judgment is randomly decided between the two judgments. Since tweets are so short, it is hard to evaluate them. Our interrater agreement on sentiment and especially opinions is low. Therefore we concentrate most of our analysis on relevance, for which interrater agreement is high.

Precision, i.e. the fraction of relevant tweets in the collected tweets, for the different sources is shown in Table 2. The search results returned containing relevant keywords are almost all judged as relevant. Tweets before and later in the conversations are not always relevant: on average 59% of these tweets are relevant. Tweets from target users mentioning relevant keywords do not occur very frequently, but from our sample they are relevant in 66% of the cases.

We conclude that with the right list of relevant, semi-relevant and non-relevant keywords, relevant tweets to a certain topic can be retrieved with a good precision. While using only highly relevant keywords lead to the highest precision, you will miss out on a considerable number of relevant tweets. Adding tweets from conversations, as well as adding tweets by users from the target group mentioning semi-relevant keywords substantially improves the number of relevant results, while maintaining an acceptable precision.

4.3 Qualitative Analysis

Analysis of the results shows that there are few in-depth discussions on Twitter. We define a discussion on Twitter as a conversation where multiple opinions are present.

Table 2. Relevance of the collected HPV tweets

Source	# judged	# relevant	Precision
Search result	143	141	0.99
Conversation starters	135	70	0.52
Reply	199	76	0.38
Target user	47	31	0.66
Screen scrape	478	336	0.70
Total	1.002	654	0.65

People might express their own opinion, on which people react with a different opinion, somebody might ask for advice on what to do or somebody might try to convince someone to change their opinion. When we look at our data, in around half of the conversations only one opinion is present, i.e. in 118 out of the 222 conversations. Most users are positive about the HPV vaccination. Even though negative effective effects of the vaccination such as pain and muscle ache are mentioned, there are few girls who decide against the vaccination during the Twitter conversation. Sentiments change most frequently in the conversations. In 126 out of the 222 conversations multiple sentiments can be found.

Looking at topic drift, there are few conversations in which some tweets are relevant, and others are not relevant, i.e. 46 out of the 222 conversations. By far most conversations stick to the topic. Of the judged discussions containing relevant tweets, 128 out of the 176, i.e. 73% of the discussions contain solely relevant tweets.

5   Conclusion

In this paper we propose a new method to track a topic on Twitter, by including discussions which include more relevant tweets than keyword search alone. We used this method to analyze the Dutch discussions around HPV vaccinations on Twitter and show that taking into account the context of relevant tweets substantially increases the number of relevant tweets retrieved around the topic. Considering the types of conversations on the topic, there are few in depth discussions with multiple opinions. On the positive side there is not much topic drift, and conversations containing a tweet with relevant keywords are likely to contain all relevant tweets.

In future work we would like to experiment with creating lists of relevant, semi-relevant and non-relevant words automatically, i.e. further advacance our query expansion method geared at retrieving relevant conversations around a topic. Furthermore, we would like to analyze the characteristics of Twitter behaviour among different types of users.

References

[1] Bandyopadhyay, A., Ghosh, K., Majumder, P., Mitra, M.: Query expansion for microblog retrieval. International Journal of Web Science 1(4), 368–380 (2012)
[2] Boyd, D., Golder, S., Lotan, G.: Tweet, tweet, retweet: Conversational aspects of retweeting on twitter. In: 43rd Hawaii International Conference on System Sciences (HICSS), pp. 1–10. IEEE (2010)

[3] Chou, W.-Y.S., Hunt, Y.M., Beckjord, E.B., Moser, R.P., Hesse, B.W.: Social media use in the United States: implications for health communication. Journal of Medical Internet Research 11(4) (2009)

[4] Honeycutt, C., Herring, S.: Beyond microblogging: Conversation and collaboration via Twitter. In: 42nd Hawaii International Conference on System Sciences (HICSS 2009), pp. 1–10 (2009)

[5] Keelan, J., Pavri, V., Balakrishnan, R., Wilson, K.: An analysis of the human papilloma virus vaccine debate on MySpace blogs. Vaccine 28(6), 1535–1540 (2010)

[6] McRee, A.-L., Reiter, P.L., Brewer, N.T.: Parents internet use for information about hpv vaccine. Vaccine 30(25), 3757–3762 (2012)

[7] Metzler, D., Cai, C., Hovy, E.: Structured event retrieval over microblog archives. In: NAACL HLT 2012, pp. 646–655 (2012)

[8] Morris, M.R., Teevan, J., Panovich, K.: What do people ask their social networks, and why?: a survey study of status message q&a behavior. In: Proceedings of the SIGCHI Conference on Human Factors in Computing Systems, CHI 2010, pp. 1739–1748. ACM, New York (2010)

[9] Zyngier, S., D'Souza, C., Robinson, P., Schlotterlein, M.: Knowledge transfer: Examining a public vaccination initiative in a digital age. In: 44th Hawaii International Conference on System Sciences (HICSS 2011), pp. 1–10. IEEE (2011)

Automatically Retrieving Explanatory Analogies from Webpages

Varun Kumar, Savita Bhat, and Niranjan Pedanekar

Systems Research Lab, Tata Research Development and Design Centre, Pune, India 411013
{kumar.varun1,savita.bhat,n.pedanekar}@tcs.com

Abstract. Explanatory analogies make learning complex concepts easier by elaborately mapping a target concept onto a more familiar source concept. Solutions exist for automatically retrieving shorter metaphors from natural language text, but not for explanatory analogies. In this paper, we propose an approach to find webpages containing explanatory analogies for a given target concept. For this, we propose the use of a 'region of interest' (ROI) based on the observation that linguistic markers and source concept often co-occur with various forms of the word 'analogy'. We also suggest an approach to identify the source concept(s) contained in a retrieved analogy webpage. We demonstrate these approaches on a dataset created using Google custom search to find candidate web pages that may contain analogies.

Keywords: Analogy, Webpages, Information Retrieval, Machine Learning.

1   Introduction

Analogical thinking is believed to be an important contributor to human intelligence and learning [1]. An analogy typically maps relations found among a target concept and its constituents onto similar relations found among a source concept and its constituents [2]. Consider the following analogy for cloud computing found on a forum [3]: *"It's like a Laundromat. You want clean laundry. You can put a washer and dryer in your own apartment [...], but you have to pay for the machines, maintain them, and are limited to one washer and one dryer load at all times unless you buy more machines. [A Laundromat] ... got a lot of machines, so whenever you need laundry done, you can walk over, stuff them into unused machines, and pay by the load. [...] The Laundromat is like a public cloud."* In this analogy, *cloud computing* is the target concept and it is mapped to *Laundromat*, the source concept. We refer to such analogies as explanatory analogies since they explain the target concept by elaborating on certain similarities in the structure or working of the target and the source concept. This is in contrast to metaphors which are more direct and succinct, *e.g. "cloud computing is a Laundromat"*.

M. de Rijke et al. (Eds.): ECIR 2014, LNCS 8416, pp. 481–486, 2014.

1.1 Finding Analogies

Analogies help learners to connect what is familiar to what is new [4]. Multiple studies indicate that use of analogies in explaining concepts enhances the understanding of the concept as well as the recall of this understanding [5-9]. We believe that a searchable repository of analogies for several target concepts, each with multiple source concepts can benefit teachers and learners in classroom learning scenarios as well as in online learning scenarios. Building such a repository is quite a task given the number of concepts and possible analogies. A crowdsourcing website such as Metamia allows crowds to create and search such a repository [10] while a repository such as [11] automatically collects metaphors. In addition, some researchers [12-14] propose approaches to retrieve metaphors from natural language text. But metaphors are expressive in nature while analogies are explanatory. Owing to the explanatory nature of the latter, analogies are employed in learning to a greater extent than metaphors [15]. In this paper, we specifically focus on explanatory analogies.

We observed that several analogies for a variety of concepts are already available on the Internet as parts of articles and forum posts. Users have populated these in order to explain concepts to their audience. A web search engine such as Google may retrieve web pages containing analogies with queries such as *"analogy to explain cache memory"*, but are often accompanied by pages which are not relevant. Sometimes, the pages do not contain any analogies, but only the keywords *'analogy'* and *'cache memory'*. Sometimes, the pages do not contain elaborate explanatory analogies, but only a metaphor such as *"cache memory is like a library"*. Even if search results retrieve analogies with 100% precision, there still remains the need for mining the source concept(s) embedded in these pages for representing them in a repository. We believe that the web search results can be mined to populate a repository of analogies and to identify source concepts contained in them.

1.2 Our Contribution

1. We present, to our knowledge, a first attempt at automatically finding explanatory analogies from webpages for a given target concept.
2. We propose an approach identifying a 'region of interest' in a candidate webpage to utilize linguistic structures typically employed in explanatory analogies.
3. We propose an approach to mine the source concept(s) from the analogies using clustering of concrete nouns found in the pages.
4. We present results obtained by applying our approach to a corpus of web pages found by custom web search queries.

2   Dataset

We employed a custom Google search query such as *"analogy to explain cache memory"* for 60 target concepts (mainly for sciences and computer science) to create the dataset used in this paper. We manually classified the top 10 search results for each query into two classes, *viz.* analogy pages (188 examples) and non-analogy pages

(124 examples). While doing so, we discarded Wikipedia pages, pages without the word *'analogy'*, links to academic papers and links to non-text resources such as videos, PDF files and presentations. We cleaned the webpages by removing irrelevant HTML tags and their content such as text contained in hyperlinks. We manually tagged each analogy webpage with one or more source concepts employed in the analogy.

2.1   Observations about the Dataset

We analyzed the dataset manually and came up with three main observations for pages containing analogies:

1. **Presence of linguistic markers:** Four types of words or phrases are common in analogy webpages, *viz.* explanation markers, similarity markers, pronoun markers and analogy prefix markers. Explanation markers are typically employed while introducing an analogy (*e.g.* "*Suppose there is a rubber band*" or "*Imagine you are at a traffic signal*" or "*This can be explained using...*"). Similarity markers can be present throughout the analogy (*e.g.* "*An eye is similar to a camera*" or "*A cell is like a castle*"). Pronoun markers often accompany other markers or occur on their own (*e.g.* "*Consider that you are on a train*" or "*I buy a new suit*"). Analogy prefix markers are adjectives used to qualify the analogy (e.g. "*A good analogy to understand this is...*"), or a noun (e.g. "*The carousal analogy may work here.*") Besides these, various forms of the word *'analogy'* may also be present. We refer to them as 'analogy markers'.
2. **Presence of text clusters containing source concept(s):** In explanatory analogies, several words or phrases from source concept(s) occur and in many cases occur within a span of a few sentences.
3. **Positions of markers and source concept(s) relative to analogy markers:** The 'Teaching with Analogies' (TWA) model suggests that good analogies introduce the target concept, then initiate the analogy, and then elaborate the relevant features of the source concept [4], [7]. We found that many analogy webpages approximated this pattern. We observed that in webpages containing analogies, linguistic markers as well as source concepts were present in the vicinity of analogy markers.
4. **Size of explanatory analogies:** We found that explanatory analogies were typically longer and elaborate.

3   Classifying Pages with Analogies

We proposed a three-step approach for classification of webpages. The steps are: 1) define a 'region of interest' in each webpage and find its size as well as the counts of linguistic markers contained in it, 2) find candidate source concept clusters from each webpage, and 3) classify webpages into 'analogy' and 'non-analogy' pages using the size of the region of interest, counts of linguistic markers as well as the presence and number of source concept clusters.

3.1   Region of Interest

Based on observation 3 in Section 2.1, we developed a 'region of interest' (ROI) approach to analyze analogy pages. We defined a window of 50 words (approximately 3 to 5 sentences) after the occurrence of an analogy marker as ROI, where linguistic markers are likely to occur and where the source concept is likely to be introduced. We found the ROI for each analogy word in the page and found a union of all such ROIs as the resultant ROI. We found the count of words in this ROI as an indication of it being elaborate and the counts of explanation, similarity, pronoun and analogy prefix markers as an indication of the likelihood of it containing an analogy.

3.2   Finding Source Concept(s)

We observed that many analogy pages had source concepts which were physical objects such as *camera*, *laundry* and *library* rather than abstract entities or physical processes such as *trust*, *gravity* and *lookup*. Existing literature reports a similar observation in case of metaphors [13-14]. To find terms related to source concepts, we found the nouns n-grams in the ROI using the Stanford parser [16]. We then selected the nouns which were likely to be concrete objects by querying their most common word sense using WordNet [17] and selecting those belonging to the *'object'* hypernym category. We also preserved nouns not featuring in WordNet as they could be domain-specific. We further reduced this set to the nouns which had a Wikipedia entry to form a candidate source concept term set. For each pair in this set, we calculated a Wikipedia-based metric indicating relatedness between the two terms [18] as:

$$r\,(t_1, t_2) = \frac{log(max(|T_1|,|T_2|))-log(|T_1 \cap T_2|)}{log(|W|)-log(min(|T_1|,|T_2|))}$$

where T_1 and T_2 are sets of all Wikipedia articles that link to terms t_1 and t_2 respectively and W is the set of all articles in Wikipedia. We then clustered these terms using hierarchical clustering with an empirically determined threshold relatedness of 0.2. We eliminated the clusters containing the target concept term. We also computed the target relatedness of each cluster by averaging relatedness of all terms in the cluster to the target concept. We eliminated clusters with less than 2 terms. We also avoided clusters related to the target term by eliminating above an empirically determined threshold (average target relatedness > 0.4). We thus formed a set of candidate source concept cluster(s) for each file. For example, for the analogy webpage mentioned in Section 1, we found a cluster such as [*Laundromat, dryer, laundry, clothes, apartment, machines, computer, parents*]. We found 494 source concept clusters from 312 files. Out of these, 402 clusters were for files containing analogies.

3.3   Classifying Analogy Webpages

We found that formation of clusters was significantly more in analogy pages than in non-analogy pages. So, we used the number of clusters generated in a page and the presence of a cluster as features along with the size of ROI and count of linguistic markers mentioned in Section 3.1 for a classification experiment. In Table 1, we report the classification performance for three classifiers, *viz.* Naïve Bayes, Support Vector Machines (polynomial kernel) and Random Forests using 10-fold cross-validation with the WEKA data mining toolkit [19].

Table 1. Webpage classification performance for Analogy (A) and Non-analogy (N) classes and its weighted average (W) while using analysis on whole webpage and ROI

Analysis on	Classifier	Class	Precision	Recall	F-measure
Whole webpage	Naive Bayes	A	0.686	0.814	0.745
		N	0.607	0.435	0.507
		W	0.655	0.663	0.650
	SVM	A	0.613	0.979	0.754
		N	0.667	0.065	0.118
		W	0.635	0.615	0.501
	Random Forest	A	0.691	0.819	0.749
		N	0.618	0.444	0.516
		W	0.662	0.670	0.657
ROI	Naive Bayes	A	0.837	0.899	0.867
		N	0.827	0.734	0.778
		W	**0.833**	**0.833**	**0.831**
	SVM	A	0.754	0.814	0.783
		N	0.679	0.597	0.635
		W	0.724	0.728	0.724
	Random Forest	A	0.871	0.936	0.903
		N	0.891	0.790	0.838
		W	**0.879**	**0.878**	**0.877**

In addition to classifying webpages, we also experimented with the validity of the source concept clusters. We manually assessed the clusters formed in analogy webpages and found that 58% of them were valid clusters. We found that if we recommended only the cluster occupying the largest part of the ROI, the precision increased to 76%.

4   Discussion and Outlook

Table 1 shows that the ROI approach increases the classification performance significantly, *e.g.* from an f-measure of 0.657 to that of 0.877 for the Random Forest classifier. This is probably because most explanatory analogies follow the trend of introducing an analogy with analogy markers and then elaborate the source concept using linguistic markers. The ROI approach helps in eliminating noisy text from the rest of the page. We recommend the use of Naïve Bayes classification to avoid overfitting while obtaining an acceptable f-measure of 0.831. We also report a precision of 76% in detecting relevant source concept clusters in analogy pages. This can be further increased by detecting semantic relationships among target and source terms.

We believe that both these results demonstrate the suitability of our approach to creating a repository of analogies.

Our dataset is currently limited in size and we propose to test the approach on a bigger dataset. In future, we plan to detect multiple analogies in a webpage and evaluate the quality of explanatory analogies before recommending them.

References

1. Gentner, D., Colhoun, J.: Analogical processes in human thinking and learning. In: Towards a Theory of Thinking, pp. 35–48. Springer, Heidelberg (2010)
2. Gentner, D.: Structure-mapping: A theoretical framework for analogy. Cognitive Science 7(2), 155–170 (1983)
3. Cloud computing is like a Laundromat or a big office printer,
   http://blog.melchua.com/2011/12/31/cloud-computing-is-like-a-laundromat-or-a-big-office-printer/
4. Glynn, S.: The teaching-with-analogies model. Science and Children 44(8), 52–55 (2007)
5. Harrison, A.G., Treagust, D.F.: Teaching with analogies: A case study in grade ⁻ 10 optics. J. of Research in Science Teaching 30(10), 1291–1307 (1993)
6. Newby, T.J., Ertmer, P.A., Stepich, D.A.: Instructional analogies and the learning of concepts. Educational Technology Research and Development 43(1), 5–18 (1995)
7. Glynn, S.M., Takahashi, T.: Learning from analogy-enhanced science text. J. of Research in Science Teaching 35(10), 1129–1149 (1998)
8. Iding, M.K.: How analogies foster learning from science texts. Instructional Science 25(4), 233–253 (1997)
9. Vosniadou, S., Schommer, M.: Explanatory analogies can help children acquire information from expository text. J. of Educational Psychology 80(4), 524 (1988)
10. Metamia: Analogy as teaching tool, http://www.metamia.com/
11. Narayanan, S.: MetaNet: A Multilingual Metaphor Repository. In: Proc. of the 12th Intl. Cognitive Linguistics Conf. (ICLC 2012), Edmonton, Canada (2013)
12. Neuman, Y., Assaf, D., Cohen, Y., Last, M., Argamon, S., Howard, N., Frieder, O.: Metaphor Identification in Large Texts Corpora. PloS One 8(4), e62343 (2013)
13. Turney, P.D., Neuman, Y., Assaf, D., Cohen, Y.: Literal and metaphorical sense identification through concrete and abstract context. In: Proc. of the 2011 Conf. on the Empirical Methods in Natural Language Processing, pp. 680–690 (2011)
14. Shutova, E., Teufel, S., Korhonen, A.: Statistical metaphor processing. Computational Linguistics 39(2), 301–353 (2013)
15. Gentner, D., Bowdle, B., Wolff, P., Boronat, C.: Metaphor is like analogy. In: The Analogical Mind: Perspectives from Cognitive Science, pp. 199–253 (2001)
16. Klein, D., Manning, C.D.: Accurate unlexicalized parsing. In: Proc. of the 41st Annual Meeting on Assoc. for Computational Linguistics, vol. 1. Association for Computational Linguistics (2003)
17. Miller, G.A.: WordNet: a lexical database for English. Communications of the ACM 38(11), 39–41 (1995)
18. Witten, I., Milne, D.: An effective, low-cost measure of semantic relatedness obtained from Wikipedia links. In: Proc. of AAAI Workshop on Wikipedia and Artificial Intelligence: an Evolving Synergy, pp. 25–30. AAAI Press, Chicago (2008)
19. Hall, M., Frank, E., Holmes, G., Pfahringer, B., Reutemann, P., Witten, I.H.: The WEKA Data Mining Software: An Update. SIGKDD Explorations 11(1), 10–18 (2009)

Geo-spatial Domain Expertise in Microblogs

Wen Li[1], Carsten Eickhoff[2], and Arjen P. de Vries[3]

[1] Delft University of Technology, 2628 CD Delft, The Netherlands
wen.li@tudelft.nl
[2] ETH Zurich, 8092 Zurich, Switzerland
c.eickhoff@acm.org
[3] Centrum Wiskunde & Informatica, 1098 XG Amsterdam, The Netherlands
arjen@acm.org

Abstract. In this paper, we present a framework for describing a user's geo-spatial domain expertise in microblog settings. We investigate a novel way of casting the expertise problem by using *points of interest* (POI) as a possible categorization of expertise. To this end, we study a large-scale sample of geo-tagged tweets and model users' location tracks in order to gain insights into their daily activities and competencies. Based on a qualitative user study among active Twitter users, we present an initial exploration of domain expertise indicators on microblogging portals and design a classification scheme that is able to reliably identify domain experts.

Keywords: Domain expertise, Geo-tagging, Twitter.

1 Introduction

Empowered by affordable Internet-enabled mobile devices, many online services such as social networks or microblogging portals allow users to share their current geo-spatial context. The resulting data traces are a unique combination of digital and real-world activity that allow for a range of interesting academic and industrial applications including location prediction [6], localized search personalization [4], or contextual advertisement [2].

Expert finding is concerned with identifying those individuals that are most knowledgeable about a given topic. This task was originally applied for locating expertise holders in corporate settings [1] and has since then been extended to a wide range of scenarios. Expertise is typically modelled along a number of pre-defined topics and is estimated based on the users' historic activity in the form of document authorship or project participation.

In this paper, we investigate expertise in terms of a user's knowledge about a place or a class of places. Previous work has found a strong connection between places and their function [7], making a *point of interest* (POI) a proxy for the typical range of activities that are carried out there. Based on this finding, we hypothesise that a user's location tracks constitute evidence for expertise towards the place's function. In this way, we model which of your friends to ask

M. de Rijke et al. (Eds.): ECIR 2014, LNCS 8416, pp. 487–492, 2014.
© Springer International Publishing Switzerland 2014

for advice on "historic museums in Shanghai" or who to best turn to for a menu recommendation at that "new tapas place that opened down town". We rely on Twitter and Foursquare as data sources for investigation and experimentation.

Our work makes 3 novel contributions beyond the state of the art in domain expertise modelling and expert finding. (1) We propose a novel domain expertise framework based on the topology of points of interest. (2) We conduct a survey among active Twitter users in order to better understand their usage of geo-tagged tweets. (3) Based on the insights gained from an initial user study, we design and evaluate an automatic method that is able to reliably identify domain experts.

2 Related Work

The task of expertise retrieval originated in the domain of enterprise search, in which authorship of documents or group affiliations are used as evidence to determine an individual's topical expertise [1,5]. For social networks like Twitter, Wagner *et al.* [8] suggested using external resources other than tweet content for identifying users' expertise in order to overcome the high amounts of noise pertaining to the domain. Bar-Haim *et al.* [3] tried to identify stock experts on Twitter by evaluating their expertise according to stock market events and their tweeted buys and sells. Weng *et al.* [9] combined knowledge from topics (distilled by LDA) and social networks to produce a topic-specific network and used random walk methods to find topic-specific influential users (experts on the topic).

In this paper, we cast the problem in analogy to the structure of the real world rather than document-derived structures and topic-specific influential users by finding experts knowledgeable about a location or a class of locations. In previous works, geo-location-related information is generally under-represented. To capture such knowledge, we rely on the POI-tags on tweets rather than the tweets themselves to profile our candidates. To the best of our knowledge, this is the first attempt to the problem.

3 Methodology

Many social media applications represent location information in the form of so-called geo-tags added to the original content. For example, one may post a tweet, "*I really love sandwiches here*", with a POI-tag *Blue Barn Gourmet* containing detailed information about the place. In the context of Foursquare, such geo-tagged messages are also referred to as check-ins. On Twitter, geo-tagged tweets contain a pair of coordinates, a place name, and an address. On Foursquare, a check-in may also include the category that the place belongs to. Categories on Foursquare are organized in a multi-level hierarchy, effectively forming an ontology of geographic entities. In this paper, we refer to the categories at the top level as top-level-categories, and other categories (at lower levels) just as

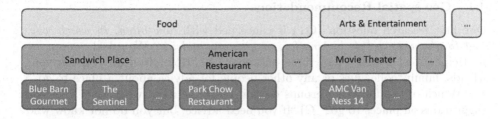

Fig. 1. POI and POI-Category Hierarchy

categories. For example, as shown in Figure 1, Blue Barn Gourmet is a sandwich place in San Francisco, CA, which is categorized as Food at the top level.

Intuitively, an expert knowledgeable in a given topic should have many contact points with it. In previous works, those contact points are modelled as authorships or topical friendships. In this paper, we focus on the check-in activity and postulate three properties a good pair of geo-topic (a location or a class of locations) and geo-expert should satisfy.

Within-Topic Activity. (S_n) The first property is general activity at a given location or category. Intuitively, the more frequently the user interacts with the topic l, the more they will know about it. Since we can only measure check-ins ($C_e = \{(l_c, t_c)\}$) of an expert e at time t, we may not be able to capture all actual physical visits. The check-ins instead represent a lower bound to the number of visits. In this way the expertise can be measured by $S_n(e, l) = \sum_{(l_c, t_c) \in C_e} \mathbf{1}(l_c = l)$, where $\mathbf{1}(.)$ is an indicator function that equals 1 if and only if the condition in the parenthesis is satisfied, 0 otherwise.

Within-Topic Diversity. (S_d) Secondly, we require an expert to know more than just a single instantiation of a category or top-level category. Accordingly, we consider check-ins to a large number of different POIs within a category a stronger indication of expertise than only check-ins at a single location. That is $S_d(e, L) = \sum_{l \in L} \log \sum_{(l_c, t_c) \in C_e} \mathbf{1}(l_c = l)$.

Recency. (S_r) Finally, we require the evidence of expertise to be as fresh as possible. Old check-ins may not represent the user's current range of interests and occupations accurately any more. That is $S_r(e, l) = \sum_{(l_c, t_c) \in C_e} e^{t_c - t} \mathbf{1}(l_c = l)$.

4 Understanding Geo-spatial Expertise

In order to gain a better understanding of how people use geo tags in their tweets, we issued a survey among active users of the microblogging portal Twitter. The survey was distributed via Crowdflower (http://crowdflower.com) and required workers to have a Twitter account to ensure the subjects' familiarity and personal experience with the domain and terminology. A total of 164 forms were received. In the following, we discuss the main findings and implications.

4.1 Geo-spatial Recommendations

Our first research question **(Q1)** is concerned with *how often, in which way, and to whom are people looking for, or giving poi-advice?* We asked our survey participants an initial set of three questions: A) "How often do you ask your friends, family, colleagues or any other people for advice about a place to go?" B) "Which of the following groups do you trust the most when they give you suggestions on places to go?" C) "If you need advice, but you do not know who can help, which of the following channels do you prefer?" For each question, a number of options as well as the possibility to give free-text answers were offered. Participants were invited to select more than one option if applicable. Figure 2 shows the answer distribution for the questions. Most participants stated they rely on location advice from time to time. Only 13% replied that they generally prefer to research places themselves. When it comes to accepting advice, trust in the advisor seems to be a key issue. We observe a clear preference order favouring family and friends over on-line contacts or even unknown review writers. In the case that no friend or family member knows advice, a broadcast to the personal social network and twitter followers is favoured over posting to forums or starting a blog post.

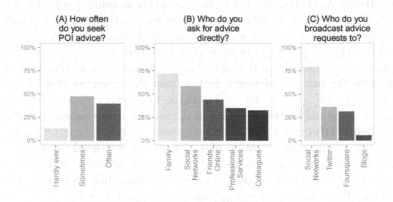

Fig. 2. Geo tag usage on Twitter and Foursquare

4.2 Measures of Expertise

Previously, we established that on-line communication channels are a realistic source of POI-related recommendations and advice. Trust in the advisor appears to be a key issue especially on the Web. Consequently, in our second research question **(Q2)**, we investigate *how to determine the geo-spatial expertise of a person.* In a short experiment, we showed our survey participants examples of anonymous Twitter profiles along with their geo-tagged check-in history. The judges were supposed to determine geo-spatial domain expertise for several profile/topic pairs. Afterwards, the participants were asked to explain their reasoning process and detail which, if any, of the rules presented in Section 3 they used to make their decision. Again, multiple answers were possible. This question resulted in a clear ordering of criteria according to the frequency at which they were chosen by participants: Within-topic activity (70.1%) > Within-topic diversity (48.8%) ≥ Recency (45.1%) > General activity (34.8%).

4.3 Predicting Expertise

To complete off our inspection of geo-spatial notions of expertise, we now aim to make the qualitative insights from our survey usable in an algorithmic manner. Our third and final research question **(Q3)** is therefore, *how well do automatic representations of the survey participants' criteria predict POI-related expertise?*

To achieve this, we algorithmically render different variants of our rules presented in Section 3. Each version aggregates a candidate's check-ins at a location or category. For the variant emphasizing recency, we discounted old check-ins, resulting in candidates with more recent check-ins being ranked at the top. For the variant emphasizing diversity, we discounted repetitive check-ins at the same place, *i.e.*, multiple check-ins at the same place will be given less importance than a check-in at a previously unseen place.

As a first qualitative evaluation of our methods, we approached active users of the Twitter geo-tagging functionality that were dominant in a collection of tweets we crawled between June and August, 2013. A group of 10 such users volunteered to work with us. We presented each of them with our model's predictions of their individual expertise and asked them to judge their actual knowledge about the topic on a 5-point scale from 1 ("I do not know about this") to 5 ("I am an expert"). 6 participants indicated high expertise (Grades 4–5) in the topics predicted by our method. Another 3 reported reasonable competency (Grade 3) towards the predicted topics. Only a single participant indicated mild expertise (Grade 2) towards the predictions. Given the high dimensionality of our problem space (400 categories and thousands of individual POIs), the results of our initial qualitative evaluation study look very promising and encourage future quantitative confirmation in a Web-scale setting.

5 Conclusion

In this paper, we presented a novel categorization of domain expertise along the hierarchy of *points of interest* (POI) as observed on Twitter and Foursquare. We conducted a qualitative user study and investigated the way in which Twitter and other communication channels are used for searching, receiving and giving location-related advice and recommendations. Doing so, we found on-line communication with close friends and family or even the wider social network to be among the major channels for obtaining advice. We also presented participants with examples of Twitter streams and asked them to judge the expertise of the showcased user towards a number of topics, as well as, to explain which criteria influenced their decision. Within-class coverage and diversity turned out to be the most frequently named features. On the basis of these qualitative insights, we designed an automatic classification method that was able to reliably predict domain expertise with high agreement to the profiled persons' self assessment.

This paper describes an ongoing piece of work in progress. There is a substantial amount of envisioned extensions that would have gone beyond the limits of this work. (1) In this paper, we reported the results of a user study at limited scale. In the future, we will take our qualitative insights to Web scale and quantitatively verify their performance on a realistic sample of the Twitter user base. (2) In the future, we will investigate using the textual tweet content as additional evidence for domain expertise. (3) Geo tags are a powerful type of semantic annotation that has been demonstrated to hold significant potential for a wide range of applications. Unfortunately, their coverage amounts to less than 1% of the overall Tweet volume. It is therefore crucial to investigate bootstrapping methods that can help annotate untagged tweets with latent geo tags based on their content or temporal dynamics.

References

1. Balog, K., Azzopardi, L., De Rijke, M.: Formal models for expert finding in enterprise corpora. In: SIGIR 2006, pp. 43–50. ACM (2006)
2. Banerjee, S., Dholakia, R.: Mobile advertising: does location based advertising work? International Journal of Mobile Marketing (2008)
3. Bar-Haim, R., Dinur, E., Feldman, R., Fresko, M., Goldstein, G.: Identifying and following expert investors in stock microblogs. In: EMNLP 2011, pp. 1310–1319 (2011)
4. Bennett, P.N., Radlinski, F., White, R.W., Yilmaz, E.: Inferring and using location metadata to personalize web search. In: SIGIR 2011, pp. 135–144. ACM (2011)
5. Campbell, C.S., Maglio, P.P., Cozzi, A., Dom, B.: Expertise identification using email communications. In: CIKM 2003, pp. 528–531 (2003)
6. Cheng, Z., Caverlee, J., Lee, K.: You are where you tweet: a content-based approach to geo-locating twitter users. In: CIKM 2010, pp. 759–768. ACM (2010)
7. Li, W., Serdyukov, P., de Vries, A.P., Eickhoff, C., Larson, M.: The where in the tweet. In: CIKM 2011, pp. 2473–2476. ACM (2011)
8. Wagner, C., Liao, V., Pirolli, P., Nelson, L., Strohmaier, M.: It's Not in Their Tweets: Modeling Topical Expertise of Twitter Users. In: 2012 International Conference on Privacy, Security, Risk and Trust and 2012 International Confernece on Social Computing, pp. 91–100 (2012)
9. Weng, J., Lim, E.-P., Jiang, J., He, Q.: TwitterRank: Finding Topic-sensitive Influential Twitterers. In: WSDM 20210, pp. 261–270 (2010)

The Impact of Semantic Document Expansion on Cluster-Based Fusion for Microblog Search

Shangsong Liang, Zhaochun Ren, and Maarten de Rijke

ISLA, University of Amsterdam, The Netherlands
{s.liang,z.ren,derijke}@uva.nl

Abstract. Searching microblog posts, with their limited length and creative language usage, is challenging. We frame the microblog search problem as a data fusion problem. We examine the effectiveness of a recent cluster-based fusion method on the task of retrieving microblog posts. We find that in the optimal setting the contribution of the clustering information is very limited, which we hypothesize to be due to the limited length of microblog posts. To increase the contribution of the clustering information in cluster-based fusion, we integrate semantic document expansion as a preprocessing step. We enrich the content of microblog posts appearing in the lists to be fused by Wikipedia articles, based on which clusters are created. We verify the effectiveness of our combined document expansion plus fusion method by making comparisons with microblog search algorithms and other fusion methods.

1 Introduction

Searching microblogs continues to be a challenge, for multiple reasons. For one, the vocabulary mismatch problem takes on a new form. If microblog posts contain only a few words, some of which are misspelled or creatively spelled, the risk of query terms failing to match words observed in relevant posts is large. Additionally, in very short posts, most terms occur only once, making simple operations such as language model estimation difficult [1, 4].

We address these challenges by using data fusion, thereby combining the output of multiple ranking functions, that each combats the unique challenges of microblog search in their own way [2]. That is, instead of searching microblog posts directly, we merge the lists generated by a number of state-of-the-art microblog search algorithms and try to outperform the best component list.

A large number of effective data fusion strategies have been proposed, with the CombSUM family of fusion methods being the oldest and one of the most successful ones in many IR tasks [4, 8]. We are interested in cluster-based fusion [3], as it is a state-of-the-art fusion method that can significantly improve performance in many IR applications [3, 4]. In cluster-based fusion, documents from lists to be fused are clustered and information from the clusters is used to inform the fusion method. As we will see below, in the case of microblog search the contribution of information derived from the clusters is very limited. We hypothesize that this is due to the limited length of posts. Therefore, we propose a document expansion technique and hypothesize that it has a positive impact on the performance of cluster-based fusion for microblog search.

M. de Rijke et al. (Eds.): ECIR 2014, LNCS 8416, pp. 493–499, 2014.
© Springer International Publishing Switzerland 2014

We integrate a specific form of document expansion based on semantic linking into the cluster-based fusion method. We first identify entities in each post appearing in any component list; we then expand each post using the text of Wikipedia articles that the entities link to through semantic linking [6, 7]. We do not utilize the document expansion method proposed by [1], as we lack their explicit information generated from additional datasets. Subsequently, we cluster the posts based both on their content and on the additional information from the Wikipedia articles that the entities link to. We then apply cluster-based fusion, which rewards documents that are (possibly) ranked low by the standard fusion method but that are contained in a cluster where many relevant documents are ranked high in many of the lists. To the best of our knowledge, we are the first to integrate semantic linking with fusion for microblog search.

2   Combining Semantic Linking Based Cluster-Based Fusion

Our fusion methods consists of two steps. The first is semantic linking where each post appearing in the lists to be fused is linked to Wikipedia pages, and the second is cluster-based data fusion proper where we create clusters based on the expanded posts.

2.1   Semantic Linking

We call a sentence in a microblog post a *chunk* so as to be consistent with the literature on semantic linking. In microblog posts, chunks form a sequence $M = \langle m_1, \cdots \rangle$ and our first task is to decide whether a link to Wikipedia should be created for m_i and what the link target should be. A *link candidate* $c_i \in C$ links an *anchor* a in chunk m_i to a target w; an anchor is n-gram in a chunk. Each target w is a Wikipedia article and a target is identified by its unique title in Wikipedia.

In the first step of the semantic linking process, we aim at identifying as many link candidates as possible. We perform lexical matching of each n-gram anchor a of chunk m_i with anchor texts found in Wikipedia, resulting in a set of link candidates C for each chunk m_i that each links to a Wikipedia article w. In the second step, we employ the so-called CMNS method [6] and rank the link candidates in C by considering the prior probability that anchor text a links to Wikipedia article w:

$$CMNS(a, w) = \frac{|C_{a,w}|}{\sum_{w' \in W} |C_{a,w'}|},$$

where $C_{a,w}$ is the set of all links with anchor text a and target w. The intuition is that link candidates with anchors that always link to the same target are more likely to be a correct representation. In the third step, we utilize learning to rerank strategy to enhance the precision of correct link candidates. We extract a set of 29 features proposed in [6, 7], and use a decision tree based approach to rerank the link candidates. Then we obtain three Wikipedia articles that the first three top ranked link candidates link to, and extract the most central sentences from these Wikipedia articles and append them to the microblog post. We call this process *semantic document expansion*.

2.2   Combining Clusters and Semantic Linking

We use "SemFuse" to refer to the integration of semantic document expansion with cluster-based fusion. Let q be the underlying information need expressed as a query, d a microblog post (also called a document), s_d the content of the semantic document expansion for d, L a set of component lists to be fused, C_L be the set of posts that appear in any of the lists, and $F_X(d; q)$ the fusion score for post d computed by a data fusion method X such as CombSUM. We aim at enhancing the ranking effectiveness of the standard fusion method X for microblog search. For each post in the lists, $d \in C_L$, we compute the fusion score $F_{SemFuse}(d; q)$ as:

$$F_{SemFuse}(d; q) := (1 - \alpha)p(d|q) + \alpha \sum_{c \in Cl(C_L)} p(c|q)p(d, s_d|c), \qquad (1)$$

where $Cl(C_L)$ is a set of clusters generated by a simple nearest neighbors based approach that utilizes the content of both the posts and the semantic document expansion, c is a cluster, and $p(d|q)$, $p(c|q)$ and $p(d, s_d|c)$ are the probabilities of d being relevant to q, c being relevant to q and both the post and its semantic expansion content being relevant to c, respectively. To estimate $p(d|q)$ in (1), we use Bayes' theorem so that $p(d|q) = \frac{p(q|d)p(d)}{p(q)}$; let $p(q|d) \propto F_X(d; q)$, $p(q) = \sum_{d' \in C_L} p(q|d')p(d')$ and assume a uniform prior for all documents in C_L so that we obtain:

$$p(d|q) := \frac{F_X(d; q)}{\sum_{d' \in C_L} F_X(d'; q)}. \qquad (2)$$

To estimate $p(c|q)$, we rewrite it as $p(c|q) = p(q|c) \cdot (\sum_{c' \in Cl(C_L)} p(q|c'))^{-1}$, where we use a product-based representation and compute $p(q|c)$ as $p(q|c) = \prod_{d \in c} F_X(d; q)^{\frac{1}{|c|}}$, where $|c|$ is the number of documents in c. Note that here the clusters are obtained by utilizing both microblog posts and the corresponding semantic expansion content. Then we obtain our estimation as:

$$p(c|q) := \frac{\prod_{d \in c} F_X(d; q)^{\frac{1}{|c|}}}{\sum_{c' \in Cl(C_L)} \prod_{d' \in c'} F_X(d'; q)^{\frac{1}{|c'|}}}. \qquad (3)$$

Similar to the above estimations, we can conveniently get $p(d, s_d|c)$ as:

$$p(d, s_d|c) := \frac{\sum_{d' \in c} sim(d', s_{d'}||d, s_d)}{\sum_{d'' \in C_L} \sum_{d' \in c} sim(d', s_{d'}||d'', s_{d''})}, \qquad (4)$$

where $sim(d', s_{d'}||d, s_d)$ is the similarity score between the combination of d' and the semantic expansion $s_{d'}$ and the combination of d and the expansion s_d. We compute this similarity score using symmetric Kullback-Leibler divergence as:

$$sim(d', s_{d'}||d, s_d) := \sum_{t \in d', s_{d'}} p(t|\theta_{d', s'_d}) \log \frac{p(t|\theta_{d', s'_d})}{p(t|\theta_{d, s_d})} + \sum_{t \in d, s_d} p(t|\theta_{d, s_d}) \log \frac{p(t|\theta_{d, s_d})}{p(t|\theta_{d', s'_d})},$$

where t is a token in d or s_d, and $p(t|\theta_{d, s_d})$ is the probability of t given a Dirichlet language model for d and s_d.

After replacing $p(d|q)$, $p(c|q)$ and $p(d, s_d|c)$ by (2), (3) and (4), respectively, in (1), we compute the final SemFuse score and rank documents by $F_{SemFuse}(d; q)$.

3   Experimental Setup

To measure the effectiveness of our fusion approach, we work with the Tweets2011 collection [5].[1] The task studied at the TREC 2011 Microblog track was: given a query with a timestamp, retrieve at least 30 relevant and interesting tweets. In total, 49 test queries were created and 59 groups participated in the TREC 2011 Microblog track, with each team submitting at most four runs, which resulted in 184 runs[2] [5]. The official evaluation metric was precision at 30 (p@30) [5]. The p@30 scores of these 184 runs varied dramatically, with the best run achieving a p@30 score of 0.4551.

In our experiments below, we sample 12 ranked lists based on their p@30 distribution: 6 runs with all of their p@30 scores over 0.40 (Class 1), and another set of 6 runs with p@30 scores between 0.30 and 0.40 (Class 2). The component runs in Class 1 are clarity1, waterlooa3*, FASILKOM02, isiFDL*, DFReeKLIM30* and PRISrun1. The components runs in Class 2 are KAUSTRerank*, ciirRun1, gut, dutirMixFb, normal and UDMicroIDF.[3] Note that in our experiments, the runs in Class 1 are actually the best 6 ones produced by state-of-the-art microblog search algorithms in TREC 2011 Microblog track, and the name of the run marked with a star symbol indicates that this run utilizes expansion information to search posts. In every class, we use run1, run2, run3, run4, run5 and run6 to refer to the runs in descending order of MAP.

We report the performance with the official TREC Microblog 2011 metric, i.e., p@30, plus p@5, 10, 15, and MAP. Statistical significance of observed differences is tested using a two-tailed paired t-test and is denoted using ▲ (or ▼) for significant differences for $\alpha = .01$, or △ (and ▽) for $\alpha = .05$. We make comparisons with recent microblog search algorithms (the runs to be fused), standard fusion methods (CombSUM and CombMNZ) as well as state-of-the-art cluster-based fusion methods (ClustFuseCombSUM and ClustFuseCombMNZ [3]), and λ-Merge [9].

4   Results and Analysis

We report our experimental results of SemFuse and the 5 baselines in Table 1. The performance of our SemFuse and other 5 baselines fusion methods can beat that of the best result list used in the fusion process (run1) in both classes and on most metrics. Many of these improvements are statistically significant. Particularly, in terms of fusing the lists produced by the best individual microblog search algorithms (Class 1), all of the p@30 scores generated by any of the data fusion method are higher than that of the best record in TREC 2011 Microblog track (0.4551), especially for SemFuse, which obtains a score of 0.5259.

It is clear from Table 1 that ClustFuseX (ClustFuseCombSUM or ClustFuseComb-MNZ) cannot beat the standard fusion method (CombSUM and CombMNZ) it integrates in almost all cases, and the performance differences between the two are usually not significant. It is instructive to consider Fig. 1(a), where the optimal value of α in (1) is plotted for ClustFuseCombSUM (black, dotted) and SemFuse (blue, solid); a low

---

[1] http://trec.nist.gov/

[2] The runs can be downloaded from http://trec.nist.gov/

[3] Again, details of the runs can be found at http://trec.nist.gov/

Table 1. Performance on the 12 sample lists. Boldface marks the best result per column; a statistically significant difference between SemFuse and the best baseline fusion method is marked in the upper right hand corner of the SemFuse score. A significant difference with run1 for each fusion method is marked in the upper left hand corner using the same symbols. None of the differences between the cluster-based method and the standard method it incorporates are significant.

	Class 1					Class 2				
	MAP	p@5	p@10	p@15	p@30	MAP	p@5	p@10	p@15	p@30
run1	.2590	.5959	.5796	.5442	.4537	.1886	.4776	.4347	.3878	.3463
run2	.2575	.5673	.4980	.4721	.4211	.1820	.4122	.3796	.3619	.3027
run3	.2318	.5755	.5367	.5034	.4401	.1688	.3878	.3633	.3605	.3136
run4	.2210	.5918	.5673	.5347	.4551	.1525	.4041	.4143	.3878	.3408
run5	.2098	.5469	.5102	.4694	.4095	.1457	.4612	.4143	.3714	.3571
run6	.2058	.5714	.5367	.4939	.4211	.1376	.3959	.3939	.3796	.3218
CombSUM	▲.2659	▲.6245	.5816	.5524	▲.4966	▲.1996	▲.5306	▲.4531	▲.4286	▲.3735
ClustFuseCombSUM	▲.2655	▲.6240	.5802	.5503	▲.4899	▲.1983	▲.5287	△.4500	▲.4213	▲.3686
CombMNZ	▲.2655	▲.6245	.5755	.5524	▲.5020	▲.1963	**.5347**	▲.4592	▲.4354	▲.3789
ClustFuseCombMNZ	▲.2650	▲.6231	.5748	.5502	▲.4987	▲.1956	▲.5330	▲.4523	▲.4311	▲.3731
λ-Merge	△.2548	.5641	▽.5631	.5496	△.4611	▲.1898	.4641	▲.4608	▲.4307	▲.3668
SemFuse	▲**.2822**▲	▲**.6367**▲	▲**.5939**△	△**.5701**▲	▲**.5259**▲	▲**.2122**▲	▲.5306	▲**.4531**▲	▲**.4435**	▲**.4000**▲

optimal value of α means that very little cluster-based information is used in the fusion process, a high value indicates that cluster-based information made a big contribution to the overall performance. We believe that the low optimal value of α for ClustFuse-CombSUM is due to the fact that obtaining reasonably coherent clusters of microblog posts is challenging, due to the limited length of posts.

SemFuse outperforms all baseline fusions method on class 1, on all metrics, and most of the differences are substantial and statistically significant. As shown in Table 1, the performance of λ-Merge is usually a little lower than that of SemFuse, CombSUM, CombMNZ and the ClustFuseX methods when fusing the lists in Class 1 and Class 2 on almost all metrics. This may be due to its overfitting. The results for the runs in Class 2 are not as clear-cut: the higher the quality of the component result lists, the more improvements can be observed for SemFuse in Table 1. For instance, the p@30 scores after fusion are highest in Class 1 compared to those in Class 2, and the quality of Class 1 is better (p@30>0.4) than that of Class 2 (0.3<p@30<0.4).

Finally, we re-consider the distributions of the free parameter α in (1) that governs the optimal weight of cluster information integrated by SemFuse and ClustFuseX (due to space constraints, we only take ClustFuseCombSUM as a representative) in Figure 1. SemFuse tends to use a bigger optimal weight than cluster-based fusion. This shows that cluster information can effectively be used to boost fusion performance when additional document information is available for generating the clusters; clearly, our semantic document expansion is helpful for cluster-based fusion methods. Another observation from Figure 1 is that the weight differences between SemFuse and ClustFuseCombSUM are more obvious in Class 1 that those in Class 2. We believe that this is due the higher quality of the component lists in Class 1.

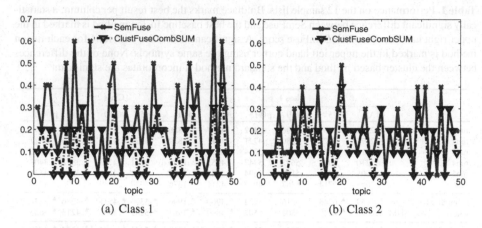

(a) Class 1   (b) Class 2

Fig. 1. Distributions of the weight of parameter α on TREC Microblog track 2011 topics when fusing lists in class 1 (left) and class 2 (right) by SemFuse and ClustFuseCombSUM

5   Conclusion

Microblog search is a challenging IR task because of the special nature of microblog posts. We combine semantic linking with a cluster-based fusion method for searching microblog posts. Our combined fusion method, SemFuse, works with result lists generated by some microblog search algorithms, identifies semantic entities for each post appearing in any of the lists to be fused, appends the most central sentences from the Wikipedia articles that the entities link to to the tweet, and then utilizes the clusters generated from the microblog expansion to enhance cluster-based fusion performance. Our experiments show that data fusion can improve microblog search performance and semantic document expansion can help to enhance cluster-based fusion methods.

Acknowledgments. This research was supported by the European Community's Seventh Framework Programme (FP7/2007-2013) under grant agreement nr 288024 (LiMoSINe project), the Netherlands Organisation for Scientific Research (NWO) under project nrs 640.004.802, 727.011.005, 612.001.116, HOR-11-10, the Center for Creation, Content and Technology (CCCT), the QuaMerdes project funded by the CLARIN-nl program, the TROVe project funded by the CLARIAH program, the Dutch national program COMMIT, the ESF Research Network Program ELIAS, the Elite Network Shifts project funded by the Royal Dutch Academy of Sciences (KNAW), the Netherlands eScience Center under project number 027.012.105 and the Yahoo! Faculty Research and Engagement Program.

References

[1] Efron, M., Organisciak, P., Fenlon, K.: Improving retrieval of short texts through document expansion. In: SIGIR 2012, pp. 911–920. ACM (2012)

[2] Farah, M., Vanderpooten, D.: An outranking approach for rank aggregation in information retrieval. In: SIGIR 2007, pp. 591–598. ACM (2007)

[3] Kozorovitzky, A.K., Kurland, O.: Cluster-based fusion of retrieved lists. In: SIGIR, pp. 893–902 (2011)

[4] Liang, S., de Rijke, M., Tsagkias, M.: Late data fusion for microblog search. In: Serdyukov, P., Braslavski, P., Kuznetsov, S.O., Kamps, J., Rüger, S., Agichtein, E., Segalovich, I., Yilmaz, E. (eds.) ECIR 2013. LNCS, vol. 7814, pp. 743–746. Springer, Heidelberg (2013)

[5] Lin, J., Macdonald, C., Ounis, I., Soboroff, I.: Overview of the TREC 2011 Microblog track. In: TREC 2011. NIST (2011)

[6] Meij, E., Weerkamp, W., de Rijke, M.: Adding semantics to microblog posts. In: WSDM 2012, pp. 563–572. ACM (2012)

[7] Odijk, D., Meij, E., de Rijke, M.: Feeding the second screen: Semantic linking based on subtitles. In: OAIR 2013, pp. 9–16 (2013)

[8] Shaw, J.A., Fox, E.A.: Combination of multiple searches. In: TREC 1992, pp. 243–252. NIST (1993)

[9] Sheldon, D., Shokouhi, M., Szummer, M., Craswell, N.: LambdaMerge: merging the results of query reformulations. In: WSDM 2011, pp. 795–804 (2011)

Towards a Classifier for Digital Sensitivity Review

Graham McDonald, Craig Macdonald, Iadh Ounis, and Timothy Gollins

School of Computing Science
University of Glasgow, G12 8QQ, Glasgow, UK
firstname.lastname@glasgow.ac.uk

Abstract. The sensitivity review of government records is essential before they can be released to the official government archives, to prevent sensitive information (such as personal information, or that which is prejudicial to international relations) from being released. As records are typically reviewed and released after a period of decades, sensitivity review practices are still based on paper records. The transition to digital records brings new challenges, e.g. increased volume of digital records, making current practices impractical to use. In this paper, we describe our current work towards developing a sensitivity review classifier that can identify and prioritise potentially sensitive digital records for review. Using a test collection built from government records with real sensitivities identified by government assessors, we show that considering the entities present in each record can markedly improve upon a text classification baseline.

1 Introduction

Democratic governments are increasingly following policies of openness and transparency by enacting freedom of information legislation that permits anyone to request records from a publicly funded organisation. In the United Kingdom (UK), this is supplemented with regulations that release all government records to the archives after 20 years, subject to some limitations. For example, the UK's Freedom of Information Act 2000 (FOIA)[1] specifies that records containing personal information, or information that might harm international relations, should be withheld for longer periods. In some cases, the requested records may be redacted or *closed / withheld* entirely. To determine such *sensitivities*, all records are reviewed before being released to the archives.

The process of sensitivity review for paper records is a long-established practice[2], however the transition to digital records (including emails) will bring new challenges from several factors: A significant increase in volume (more digital records are created); the structure of digital records is more complex and diffuse than paper; the present resources make the current linear (page-by-page) review procedures infeasible; and there are no commercially available tools to manage digital review procedures. Moreover, the risk associated with the inadvertent release of sensitive records also increases in the digital scenario due to the ubiquitous nature of online search.

---

[1] http://www.legislation.gov.uk/ukpga/2000/36/contents
[2] http://www.nationalarchives.gov.uk/information-management/
our-services/sensitivity-reviews-on-selected-records.htm

M. de Rijke et al. (Eds.): ECIR 2014, LNCS 8416, pp. 500–506, 2014.
© Springer International Publishing Switzerland 2014

We strongly believe that without reliable and efficient sensitivity review tools, policies for transparent government and freedom of information will fail as public bodies will be forced into the precautionary closure of digital records, reducing public scrutiny and negatively impacting social science research. Therefore, we are developing assistive tools to increase the efficiency of digital sensitivity review, building upon information retrieval (IR) technology.

In general, this paper provides an introduction to the IR problems that we believe must be addressed. The contributions of our paper are twofold: (i) We provide an overview of challenges in sensitivity review (ii) We detail an initial empirical investigation of a classification tool to support digital sensitivity review using real government records with real sensitivities.

The remainder of this paper is structured as follows: Section 2 discusses the sensitivity problem; Section 3 describes the classification features that we employ in this work; Section 4 describes our experimental setup; We describe our classification results in Section 5, while concluding remarks follow in Section 6.

2   Sensitivity in Digital Records

Freedom of information laws are enacted within many countries, providing access to public information. For instance, in the European Union, it is enshrined within Article 42 of the Charter of Fundamental Rights, while the USA and other countries have analogous provisions. In this section we discuss the sensitivities relating to public records.

The assumption behind freedom of information is that public records should be open. A common attribute of freedom of information enactments, however, are some standard exemptions limiting what can be released. For instance, information that may prejudice commercial operations is commonly exempt.

The role of sensitivity review is to ensure that all appropriate exemptions are checked before the records are opened. Hence, it is essential that the sensitivity review is efficient and cost-effective so that it does not stop the timely release of records. In this paper, we present work into developing a classifier for digital sensitivity review, so that reviewers can prioritise the records of highest perceived risk.

Our work is framed within the context of the UK, and of a set of government records and assessors that we have access to[3]. However, the notions encapsulated within the UK exemptions transfer easily to the legislation in many other countries.

Table 1 lists exemptions of the FOIA that apply to historical records. As can be seen, each record must be reviewed within the context of fifteen exemptions. Each exemption is assessed against a detailed set of criteria. For example, in assessing Section 27, which aims to limit damage to international relations, evidence of information being passed in confidence is of particular importance for informing a reviewer's decision.

In this work, we focus on two exemptions, namely Section 27 (International Relations) and Section 40 (Personal Information), as we believe these sections to be both representative of the issues we might expect to find in addressing many other exemptions and sufficiently challenging to test our proposed approach. The closest related work that we are aware of addresses tasks such as anonymisation of unstructured

---

[3] Due to the obvious sensitivities involved, the collection is not publicly available.

502   G. McDonald et al.

Table 1. UK Freedom of Information Act 2000: Exemptions that apply to historical records

Section 21: Information Accessible by Other Means	Section 34: Parliamentary Privilege
Section 22: Information Intended for Future Publication	Section 37: Certain Aspects Relating to the Royal Family and Honours
Section 23: Bodies Dealing with Security Matters	Section 38: Health and Safety
Section 24: National Security	Section 39: Environmental Information
Section 26: Defence	Section 40: Personal Information
Section 27: International Relations	Section 41: Information Provided in Confidence
Section 29: The Economy	Section 44: Prohibitions on Disclosure
Section 31: Law Enforcement	

data [1], or data-loss prevention classifiers [2]. However, we believe that no other work has directly addressed the task of sensitivity review. In the next section, we describe our approach to sensitivity review.

3 Features for Sensitivity Review

Our aim in this work is to study appropriate techniques to classify a record's likely sensitivities, focusing upon Section 27 and Section 40. We believe that such automatic classification goes beyond textual/topic classification, as addressed in classical text classification test collections such as 20 Newsgroups[4]. Hence, in the following, we propose several features that we postulate can aid in effective sensitivity classification.

Firstly, a record's sensitivities are likely to be anchored by topical entities, such as people or countries. For example, Personal Information is intrinsically linked to a person and International Relations link to one or more countries. These links can be implicit within a record, which makes the task of identifying sensitivity-entity links very challenging. Moreover, for Section 27, expressed sentiment relating to an entity may help in deciding if the information is sensitive. For these reasons, we chose to focus on the identification of named entities and subjectivity within records.

For entity identification, firstly, we use a dictionary of 43,286 named entities of interest (Politicians, Prime Ministers, Presidents, Royals, Monarchs and Dictators), constructed from the DBpedia[5] knowledge base. We also use a dictionary of 131,232 person names, constructed from the Drupal Name Database[6] and from the lists of unambiguous names supplied with *deid* [3], removing duplicates and non-Latin names (because they do not appear in the corpus), to extract generic instances of person entities from the records. We use LingPipe[7] to efficiently match dictionary entries with record instances and for each record r we define the number of extracted named and generic person entities, as the *nEntity* and *pCount* features respectively.

---
[4] http://qwone.com/~jason/20Newsgroups/
[5] http://dbpedia.org/
[6] https://drupal.org/project/namedb
[7] http://alias-i.com/lingpipe/

Country entities are significant for certain sensitivities, for example Section 27. Relations between countries are not all on par, therefore, the accidental release of records has varying potential for damaging the international relations between a country producing the records and a referenced country or a third-party. The real nature of these relations is privileged information and in flux. Therefore, we model this fragility using *our* perception of current international relations and supply a country-risk map as a system parameter. We define the country risk score of a record r as follows:

$$cRisk(r) = \sum_{c \in r} countryRisk(c) \qquad (1)$$

where c is a country occurring in record r and *countryRisk* is the risk score from the set {1:None, 2:Moderate, 3:High} associated with country c.

Next, we hypothesise that subjective expressions of sentiment might be correlated with sensitivity. For instance, a negatively phrased discussion about another country might be closed under Section 27. For this reason, and inspired by previous work on sentiment analysis (e.g. identifying opinionated content in blog posts [4]), we use the Opinion Finder (OF) sentiment analysis toolkit [5] to detect opinionated sentences within a record and score the record on its subjectivity. Following the work of [4] in finding overall opinionated scores for documents, each record r is scored as follows:

$$subjConf(r) = summConf \cdot \frac{\#subjective}{\#sentences} \qquad (2)$$

where *#subjective* is the number of subjective sentences, *#sentences* is the total number of sentences in the record and *summConf* is the sum of the confidence score from OF's precision-oriented subjective sentence classifier.

4 Experimental Setup

The research question that we address is the following: Can we improve upon a text classification baseline for identifying sensitive records? In this section, we describe the experimental setup and test collection used to address this research question.

The test collection comprises 1111 government records, sampled from a larger corpus addressing international activities. The sampled records were split between seventeen assessors, twelve of whom are from government departments and experienced in sensitivity review. As discussed in Section 2, we consider only two areas of sensitivity relating to the FOIA, namely Section 27 and Section 40. Each assessor conducted at least 50 initial judgements to gain familiarity with the collection.

Assessors were supplied with a guidance document and were asked to judge whether each record contained sensitive information, that would be withheld by the government, under the sensitivities of interest. Four judgement options were provided, *Not Sensitive*, *Sensitive (Section 27)*, *Sensitive (Section 40)*, or *Sensitive (Both)*. Of the 1111 judged records, 104 were sensitive for Section 27, and 86 for Section 40.

To assess inter-assessor agreement, 150 records were judged by two assessors and 50 records were judged by four assessors. Agreement was found to be 0.5525 measured by Cohen's κ [6] for the double-judged records and a Fleiss' κ [7] score of 0.4414 for

504   G. McDonald et al.

Table 2. Results for sensitivity analysis for Sections 27 & 40

	Section 27 International Relations				Section 40 Personal Information			
	Precision	Recall	F-measure	BAC	Precision	Recall	F-measure	BAC
Text Classification	0.3360	0.2972	0.3125	0.6197	0.3756	0.5595	0.4020	**0.7372**
+ *pCount*	0.2224	0.2205	0.2154	0.5689	0.3698	0.3961	0.3188	0.6352
+ *cRisk*	0.3157	0.3067	0.3089	0.6201	0.3549	0.5719	0.3844	0.7088
+ *nEntity*	**0.3605**	**0.3072**	**0.3282**	**0.6255**	**0.3901**	0.5595	**0.4107**	0.7186
+ *subjConf*	0.3123	0.2872	0.2938	0.6125	0.2684	**0.5941**	0.3150	0.6868

records which received four judgements each. While these values indicate a moderate agreement [8], we note that levels of agreement in the current paper-based review process are unknown, as only one assessor will routinely judge each record. Record labels were assigned based on the judgements, using a majority vote where appropriate.

As a baseline, we deploy a text classification approach, where a record is represented by a term frequency vector, over all terms in the collection. This is intuitive as there may exist inherent underlying topical patterns to sensitivities within genres [9] of a collection - e.g. non-sensitive press releases may have various co-occurring terms. Text classification features are extracted and scored using the Terrier IR platform [10]. We then extend the text classification approach by individually applying each of our identified features: *pCount*, *cRisk*, *nEntity* and *subjConf*.

As a classifier, we use SVMLight [11] with a linear kernel. We measure our results using several classification measures, namely: Precision, Recall, and F-measure, as well as Balanced Accuracy (BAC), which is the arithmetic mean of true positive and true negative rates with a BAC of 0.5 indicating a random prediction. All measures are reported over a 5-fold cross validation. Moreover, there is a bias in the distribution of sensitive and non-sensitive records throughout the collection, with over 80% of records being non-sensitive. Hence, due to SVM's sensitivities to imbalanced training data, we up-sample the training sets in each fold by repeating sensitive records until the number of sensitive and non-sensitive records match. The test sets in each fold retain their observed distribution of sensitivities.

5   Results

Table 2 reports the results of the Section 27 and Section 40 classification tasks. Focusing on BAC, as it accounts for imbalanced test sets, we observe that the text classification baseline achieves a BAC of 0.6197 and 0.7372 for Sections 27 and 40, respectively, which are markedly above random. Next, we find that the *cRisk* and *nEntity* features improve the classifier's performance for Section 27 to 0.6201 and 0.6255 respectively. Identifying the presence of entities of interest and countries risk factors, intrinsic to the notion of International Relations, appear to be promising future research directions.

Conversely, *nEntity* and *cRisk* are not fundamental to the notion of Personal Information and we see that these features are detrimental to the baseline classifier's BAC. The detrimental effects of applying the selected feature sets across Sections 27 and 40, illustrate the need for individual feature sets for different aspects of sensitivities.

The *pCount* feature, a simple count of person names occurring in a record, performs poorly for both areas of sensitivity, reducing the performance of the text classification baseline. This is likely due to an over-aggressive selection process, as not all mentions of people's names are in fact personal information.

Finally, it is surprising that *subjConf* leads to the classifier's degradation for Section 27. Opinions within the records have been tagged by the judges as indicative of that exemption, and we believe that this feature is worthy of future investigation. Overall, in addressing our research question, we find that our proposed domain-specific features for sensitivity review, *cRisk* and *nEntity*, can provide benefit in enhancing the accuracy of a text classification approach for digital sensitivity review, especially for Section 27.

6   Conclusions and Future Work

We have provided an overview of the challenges faced by government departments from the imminent switch to digital records sensitivity review. Moreover, we presented some work to develop a classification tool to assist the review process. The challenges discussed in this paper will inevitably be of increasing importance for governments obliged to be transparent and open, while securing the safety of individuals and countries.

We found that two features, namely the number of people in specific roles of interest and a risk score for countries identified within a record, can help to identify sensitive records that risk damaging international relations, by improving on a text classification baseline. We further found that these features did not help to improve BAC for personal information sensitivities. This illustrates the need for individual feature sets to identify different aspects of sensitivity.

As future work, we intend to conduct a study of assessor disagreement, having assessors revisit the disputed judgements to construct a gold standard test collection. We also intend to develop our feature usage. For example, determining term specificity using the Z-score statistical measure [12], investigating feature co-occurrence such as subjective sentences containing named entities and applying automatic features selection.

Acknowledgements. The authors would like to thank Michael Moss, Norman Gray and James Girdwood for their valuable comments, as well as, the assessors for their help in constructing the test collection.

References

1. Nguyen-Son, H.Q., Nguyen, Q.B., Tran, M.T., Nguyen, D.T., Yoshiura, H., Echizen, I.: Automatic anonymization of natural languages texts posted on social networking services and automatic detection of disclosure. In: Proc. ARES (2012)
2. Hart, M., Manadhata, P., Johnson, R.: Text classification for data loss prevention. In: Fischer-Hübner, S., Hopper, N. (eds.) PETS 2011. LNCS, vol. 6794, pp. 18–37. Springer, Heidelberg (2011)
3. Neamatullah, I., Douglass, M.M., Li-wei, H.L., Reisner, A., Villarroel, M., Long, W.J., Szolovits, P., Moody, G.B., Mark, R.G., Clifford, G.D.: Automated de-identification of free-text medical records. BMC Medical Informatics and Decision Making 8(1), 32 (2008)

4. He, B., Macdonald, C., Ounis, I.: Ranking opinionated blog posts using opinionfinder. In: Proc. SIGIR (2008)
5. Wilson, T., Hoffmann, P., Somasundaran, S., Kessler, J., Wiebe, J., Choi, Y., Cardie, C., Siddharth Patwardhan, E.R.: Opinionfinder: a system for subjectivity analysis. In: Proc. HLT/EMNLP (2005)
6. Cohen, J., et al.: A coefficient of agreement for nominal scales. Educational and Psychological Measurement 20(1), 37–46 (1960)
7. Fleiss, J.L.: Measuring nominal scale agreement among many raters. Psychological Bulletin 76(5), 378 (1971)
8. Landis, J.R., Koch, G.G.: The measurement of observer agreement for categorical data. Biometrics 33(1), 159–174 (1977)
9. Orlikowski, W.J., Yates, J.: Genre repertoire: The structuring of communicative practices in organizations. Administrative Science Quarterly 39(4), 541–574 (1994)
10. Ounis, I., Amati, G., Plachouras, V., He, B., Macdonald, C., Lioma, C.: Terrier: A high performance and scalable information retrieval platform. In: Proc. OSIR (2006)
11. Joachims, T.: Learning to Classify Text Using Support Vector Machines – Methods, Theory, and Algorithms. Kluwer/Springer (2002)
12. Savoy, J.: Authorship attribution based on specific vocabulary. ACM Transactions on Information Systems (TOIS) 30(2), 12 (2012)

Unsupervised Approach for Identifying Users' Political Orientations

Youssef Meguebli[1], Mouna Kacimi[2], Bich-Liên Doan[1], and Fabrice Popineau[1]

[1] SUPELEC Systems Sciences (E3S), Gif sur Yvette, France
{youssef.meguebli,bich-lien.doan,fabrice.popineau}@supelec.fr
[2] Free University of Bozen-Bolzano, Italy
mouna.kacimi@unibz.it

Abstract. Opinions, in news media platforms, provide a world wide access to what people think about daily life topics. Thus, exploiting such a source of information to identify the trends can be very useful in many scenarios, such as political parties who are interested in monitoring their impact. In this paper, we present an unsupervised technique to classify users based on their political orientations. Our approach is based on two main concepts: (1) the selection of the aspects and the sentiments users have expressed in their opinions, and (2) the creation of knowledge base from Wikipedia to automatically classify users according to their political orientations. We have tested our approach on two datasets crawled from CNN and Aljazeera. The results show that our approach achieves high quality results.

Keywords: Political leaning, Political Opinion Mining, Sentiment analysis.

1 Introduction

Political views are freely and explicitly expressed through opinions in news media platforms. These opinions represent an interesting sample about political trends and orientations of users. Extracting such type of knowledge would allow news portal publishers to have an idea about the orientation of their commenters, the main issues related to each orientation, and the possible political persuasions and ideological viewpoints for all topics. The opinions expressed by users are not restrained by journalism values such as fairness or balance, and do not go through a formal editorial process. Moreover, the number of opinions about a given topic might continuously increase. The unstructured and the dynamic nature of opinions, provided in news media platforms, call for effective and efficient techniques for identifying political trends.

Several approaches have been proposed to classify political positions from texts. One line of work focused on using SVM with optimization of text feature selection [7][14] [18] [6], as well as complementing with sentiment analysis [4] [13] [10] [2]. Another line of work used word frequencies, Bayesian statistical models, and topic models [8] [11] [9] [12] [16]. Most of these approaches use

M. de Rijke et al. (Eds.): ECIR 2014, LNCS 8416, pp. 507–512, 2014.
© Springer International Publishing Switzerland 2014

supervised techniques which can be expensive as they require training. Moreover, they mainly use semi-structured data to classify users. Examples include data extracted from twitter and microblogs which is characterized by short fragments (tweets, short messages), where each fragment covers a known and a unique aspect. More specifically, approaches based on twitter samples use hashtags of controversial topics such as *'USElection'* or *'Arabspring'* and a set of stakeholders such as actors or politicians, to classify the stakeholders opinions into pro or con categories for the respective topics [1]. Comments published in microblogs are frequently short and do not contain more than one aspect whereas, in news media platforms, users publish long opinions covering more than one aspect.

In this paper, we propose an unsupervised technique for defining the political orientation of users based on their opinions in news media platforms. To the best of our knowledge, we are the first to propose an unsupervised approach on such unstructured and dynamic data. Our contribution is twofold (1) we generate user profile based on the aspects he has discussed in his opinions and their sentiments and (2) we construct a knowledge base of political orientations, using Wikipedia, to automatically classify users based on their profiles. We have conducted extensive experiments with US and Egypt user groups crawled from CNN and Aljazeera. The experiments showed that our approach provides high quality results to classify US users into Republican/Democrat leanings and Egypt users into secular/Islamist leanings.

2   Generating User Profile

To define the political profile of a given user U, we collect the opinions he has expressed, in a given media platform, during a period of time T. Then, we analyze the opinions and extract from them all the aspects that user U has discussed. For each aspect, we define the sentiment expressed by the user U. For example, a user can discuss the aspect of *abortion rights* and be *negative* about it. As a result, the user U is described by a set of aspects $\{a_1, ...a_n\}$ and their related sentiments $\{s_1, ...s_n\}$. To this end, we proceed in three main steps.

Step1. Extraction of Opinionated Sentences. We first identify the sentences[1] expressed in all the opinions of user U. For each sentence, we extract all its contained terms and we assign to each term a sentiment that can be positive, negative, or neutral using the lexicon provided by Ding et.al., [3]. We count the number of positive and negative terms in each sentence. If the number of positive terms is higher than the number of negative terms, we classify the sentence as positive, otherwise it is classified as negative.

Step2. Generation of Candidate Aspects. We take all the opinionated sentences extracted from the previous step, and we rank their contained terms using $tf * idf$ scoring function. In our work, tf represents the term frequency in the set of opinionated sentences of user U, and idf represents the inverted

---

[1] Using OpenNLP http://opennlp.sourceforge.net/

document frequency in the set of opinionated sentences of all users. The idea is to select highly scored unigrams as a base for generating candidate aspects. From these unigrams, we generate bi-grams, then we take the bi-grams as input and we build a set of n-grams by concatenating bi-grams that share an overlapping word. At each step we take the topk n-grams based on the score of their composed unigrams[2]. We check the redundancy of the generated candidates, using Jaccard similarity [15]. If two n-grams have a similarity higher than a defined threshold, we would discard one of them. In our work, we have set the maximum length of the n-grams to 5 since there were no meaningful n-grams of a higher length.

Step3. Selection of Promising Aspects. Generating n-grams that have high $tf * idf$ scores is not enough for identifying the aspects discussed in users' opinions. It is important for the words in the generated n-grams to be strongly associated within a sentence in the original text to avoid covering incorrect information. To capture this association, we use *pointwise mutual information* [17] (PMI) of words in n-grams.Formally, suppose $m_i = w_1...w_n$ is a generated n-grams. We define the $Score_n$ as follows:

$$S_{PMI}(w_1...w_n) = \frac{1}{n} \sum_{i=1}^{n} pmi_{local}(w_i)$$ (1)

where $pmi_{local}(w_i)$ is a local pointwise mutual information function defined as:

$$pmi_{local}(w_i) = \frac{1}{2C} \sum_{j=i-C}^{i+C} pmi'(w_i, w_j), i \neq j$$ (2)

where C is a contextual window size. The $pmi_{local}(w_i)$ measures the average strength of association of a word w_i with all its C neighboring words (on the left and on the right). For example, in *gun control law* phrase, assuming $C = 1$, for *gun* we would obtain the average PMI score of *gun* with *control* and for *control* we would obtain the average PMI of *control* with *gun* and *control* with *law*. When this is done for each $w_i \in m$, this would give a good estimate of how strongly associated the words are in m. We used a modified PMI scoring [5] referred to as *pmi* where the *pmi* between two words, w_i and w_j, is defined as:

$$pmi'(w_i, w_j) = \log_2 \frac{p(w_i, w_j) \cdot c(w_i, w_j)}{p(w_i) \cdot p(w_j)}$$ (3)

where $c(w_i, w_j)$ is the frequency of two words co-occurring in a sentence from the original text within the context window of C (in any direction) and $p(w_i, w_j)$ is the corresponding joint probability. The co-occurrence frequency, $c(w_i, w_j)$, which is not part of the original PMI formula, is integrated into our PMI scoring to reward frequently occurring words from the original text. By adding $c(w_i, w_j)$ into the PMI scoring, we ensure that low frequency words do not dominate and moderately associated words with high co-occurrences have relatively high scores.

---

[2] In this work we have set k=500.

3   Defining Users' Political Orientations

An unsupervised technique of identifying the political orientation of users based on their profiles calls for the use of a knowledge base. To this end, we have created a knowledge base of political orientations from Wikipedia. For a given political orientation, we start from a Wikipedia seed page. We extract from text part of the page all outgoing links that point to other Wikipedia articles. Then, we select the anchor text of these links as aspects related to the political orientation. Each aspect occurs in a sentence that have a sentiment orientation. Thus, in a similar way to user profile, we identify each political orientation to be described by a set of aspects $\{a_1, ... a_m\}$ and their related sentiments $\{s_1, ... s_m\}$. Table 1 shows an example of Liberal and Conservative orientation and their relevant aspects extracted from Wikipedia. To cover more aspects and enrich the knowledge base, we also include the Wikipedia pages pointed by the seed page of the political orientation. For example, a seed page reports that liberals are in favor of universal health care. We take the Wikipedia page of universal health care and add its aspects to the favorite list of Liberals.

Table 1. The structure of the orientation knowledge base

Orientation	Some Aspects Extracted from Wikipedia	
Liberals	favor	universal health care, strict gun control, diplomacy, stem cell research, same-sex marriage, abortion rights
	against	increased military spending The Ten Commandments display in public buildings
Conservatives	favor	small government, low taxes, limited regulation free enterprise, school prayer, capital punishment
	against	same-sex marriage, abortion rights, multiculturalism

To identify the political orientation of user U, we compute the similarity between its profile and the description of all the political orientations that exist in the knowledge base. The most similar description is assigned to the user as its political orientation.

4   Experimental Results

We have crawled 2 datasets from CNN and Aljazeera English news portals. From CNN, we have extracted $11,322$ users, and their $684,058$ opinions about $15,365$ news articles. From Aljazeera, we have extracted 539, and their users $24,826$ about $2,773$ news articles. For each user, we have extracted all his opinions that concern politics from *October 2009* to *September 2013*. We have run our experiments on 500 users: 290 from US (CNN) and 210 from Egypt (Aljazeera). We have shown the list of opinions of each user and asked human assessors, who were students not involved in this project, to analyze the users opinions and classify them into the following categories: *Democrat/Republican* for US users and *Secular/Islamist* for Egypt users. The result of the human assessment is the ground truth for our evaluation.

We applied our approach on the same 500 users selected before. The outcome of the classification was then compared to our golden standard. To measure the effectiveness of our approach, we have computed the accuracy which represents the fraction of users that were correctly classified. We have compared different variations of our approach. The first one uses the $top100$ unigrams, based on $tf * idf$ scoring function, to classify the user. The second approach uses n-grams of length between 1 and 5. The $top100$ n-grams, based on $tf * idf$ scoring function, are selected to classify the user. In the third approach, the $top100$ n-grams, based on PMI, are selected to classify the user.

Table 2. Accuracy of User classification

	US		Egypt	
	Democrats	Republicans	Islamists	Seculars
Unigrams (tf*idf)	50%	16,66%	72,72%	25%
N-grams (tf*idf)	67,60%	50%	70,70%	51,56%
N-grams (PMI)	95,07%	79,41%	85,85%	84,37%

The results are shown in Table 2. We can see the impact of the different steps of our approach on the accuracy of our technique. Using only unigrams generates incomplete information about the aspects discussed by users and thus provides very inaccurate results. We can see that using n-grams improves the results in most cases, however they still have a low accuracy. Using PMI to select the aspects of opinions is the best providing an accuracy that goes up to $95,07\%$.

5   Conclusion and Future Work

We have proposed a new technique for defining the political orientation of users based on their opinions around news articles. The proposed approach is promising as it provides means for dealing with unstructured source of information. Moreover, it is completely unsupervised which makes it flexible to be applied on any kind of dynamic knowledge such as opinions. As future work, we plan to extend the knowledge base to other types of orientations in other domains and propose a general approach for extracting the main aspects of daily life topics and their main trends.

References

1. Awadallah, R., Ramanath, M., Weikum, G.: Harmony and dissonance: organizing the people's voices on political controversies. In: Proceedings of the Fifth ACM International Conference on Web Search and Data Mining, WSDM 2012, pp. 523–532. ACM, New York (2012)

2. Conover, M.D., Gonçalves, B., Ratkiewicz, J., Flammini, A., Menczer, F.: Predicting the political alignment of twitter users. In: 2011 IEEE Third International Conference on Privacy, Security, Risk and Trust (passat), and 2011 IEEE Third International Conference on Social Computing (socialcom), pp. 192–199. IEEE (2011)
3. Ding, X., Liu, B., Yu, P.S.: A holistic lexicon-based approach to opinion mining. In: WSDM, pp. 231–240 (2008)
4. Durant, K.T., Smith, M.D.: Mining sentiment classification from political web logs. In: Proceedings of Workshop on Web Mining and Web Usage Analysis of the 12th ACM SIGKDD International Conference on Knowledge Discovery and Data Mining (WebKDD 2006), Philadelphia, PA (2006)
5. Ganesan, K., Zhai, C., Viegas, E.: Micropinion generation: an unsupervised approach to generating ultra-concise summaries of opinions. In: Proceedings of the 21st International Conference on World Wide Web, WWW 2012, New York, NY, USA, pp. 869–878 (2012)
6. Hirst, G., Riabinin, Y., Graham, J.: Party status as a confound in the automatic classification of political speech by ideology. In: Proceedings of JADT 2010 (2010)
7. Jiang, M., Argamon, S.: Political leaning categorization by exploring subjectivities in political blogs. In: DMIN, Citeseer, pp. 647–653 (2008)
8. Laver, M., Benoit, K., College, T.: Extracting policy positions from political texts using words as data. American Political Science Review, 311–331 (2003)
9. Lin, F., Cohen, W.W.: The multirank bootstrap algorithm: Self-supervised political blog classification and ranking using semi-supervised link classification. In: ICWSM (2008)
10. Malouf, R., Mullen, T.: Graph-based user classification for informal online political discourse. In: Proceedings of the 1st Workshop on Information Credibility on the Web (2007)
11. Martin, L.W., Vanberg, G.: A robust transformation procedure for interpreting political text. In: SPM-PMSAPSA, vol. 16, pp. 93–100 (2008)
12. Monroe, B.L., Colaresi, M.P., Quinn, K.M.: Fightin'words: Lexical feature selection and evaluation for identifying the content of political conflict. In: SPM-PMSAPSA, vol. 16, pp. 372–403 (2008)
13. Mullen, T., Malouf, R.: A preliminary investigation into sentiment analysis of informal political discourse. In: AAAI Spring Symposium: Computational Approaches to Analyzing Weblogs, pp. 159–162 (2006)
14. Oh, A.H., Lee, H.-J., Kim, Y.-M.: User evaluation of a system for classifying and displaying political viewpoints of weblogs. In: ICWSM (2009)
15. Real, R., Vargas, J.M.: The probabilistic basis of jaccard's index of similarity, vol. 45, pp. 380–385. Oxford University Press (1996)
16. Slapin, J.B., Proksch, S.-O.: A scaling model for estimating time-series party positions from texts, vol. 52, pp. 705–722. Wiley Online Library (2008)
17. Terra, E., Clarke, C.L.A.: Frequency estimates for statistical word similarity measures. In: Proceedings of the 2003 Conference of the North American Chapter of the Association for Computational Linguistics on Human Language Technology, NAACL 2003, vol. 1, pp. 165–172. Association for Computational Linguistics, Stroudsburg (2003)
18. Yu, B., Kaufmann, S., Diermeier, D.: Classifying party affiliation from political speech, vol. 5, pp. 33–48. Taylor & Francis (2008)

User Perception of Information Credibility of News on Twitter

Shafiza Mohd Shariff, Xiuzhen Zhang, and Mark Sanderson

School of Computer Science and IT, RMIT University, Australia
{shafiza.mohdshariff,xiuzhen.zhang,mark.sanderson}@rmit.edu.au

Abstract. In this paper, we examine user perception of credibility for news-related tweets. We conduct a user study on a crowd-sourcing platform to judge the credibility of such tweets. By analysing user judgments and comments, we find that eight features, including some that can not be automatically identified from tweets, are perceived by users as important for judging information credibility. Moreover, distinct features like link in tweet, display name and user belief consistently lead users to judge tweets as credible. We also find that users can not consistently judge or even misjudge the credibility for some tweets on politics news.

1   Introduction

As of May 2013, an average of 58 million tweets are posted per day on Twitter.[1] Currently Twitter not only acts as a social medium, it is also becoming a news media source. Twitter citizens not only share news headlines from newswires, but also report real time events before they reach the press [5]. News on Twitter comes from a wide variety of sources: some from well known news organisations and government departments, while most from members of the public. Consequently twitterers often need to judge the credibility of tweets. Morris et al. (2012) discovered that twitterers have poor judgement on the truth of information on Twitter. Features such as the number of retweets, information on users who post tweets and their relationship network (number of followers and followees) help little in determining the level of information credibility on Twitter.

Spammers exploit the anonymity feature of Twitter to propagate their messages, retweeting them to increase their popularity rating [10]. In a Twitter dataset analysed by Gupta and Kumaraguru (2012), nearly half of the tweets about an event were found to be spam. In the work by Castillo et al. (2011), it was discovered that the credibility of information on Twitter is determined mainly by four types of features: message-based, content-based, user-based, and propagation-based. In most existing work, the features may need to be compiled by crawling the Twitter space and extracting the link relationship between twitterers. The purpose of these features are for automatic prediction and may not necessarily be users' perception of important signals for credibility.

---

[1] http://www.statisticbrain.com/twitter-statistics/

M. de Rijke et al. (Eds.): ECIR 2014, LNCS 8416, pp. 513–518, 2014.

In this paper, we focus on studying the tweet-based features that the general public mostly use to determine the credibility level of newsworthy tweet messages. The research questions that we will cover in this current work are:

1. What features do users use to judge the credibility of tweets?
2. How do users use tweet features to make their credibility judgment?
3. Does the tweet topic have an effect on a user's credibility judgment?

Related Work: Information credibility on Twitter has attracted significant research recently [1–4, 6, 7], where most focuses on automated approaches predicting the credibility of topics [1] and events [2, 3] by engineering complex features based on data and meta-data for tweets as well as their social structures. Morris et al. [6] study user perception of information credibility on Twitter and show that users rely on tweet contents and other heuristics for credibility judgments. On the other hand, it is established that Twitter posts report real-time news overlapping with the reported news in newswire with the addition of minor and local news not reported by other sources [8]. Despite the real-time news post, Twitter users are more concerned with the credibility of tweets relating to breaking news, politics, and disaster events [6].

2   Methodology

We design a user study based on the CrowdFlower[2] crowd source platform to examine user perception of the credibility of tweets for news events. We select news event topics, and their relevant tweets for our study. We recruit crowd source evaluators to judge the credibility level of tweets, and leave comments on their judgments. Through their comments, we extract the features and apply predictive association rule analysis [9] to establish the associations between features and credibility levels.

In this research, credibility is defined as *"the quality of being believed or accepted as true, real, or honest"*. [3] The criteria to determine credible tweets [1] are that they must affirm a fact, be informative for the public, not be self opinionated, and not be a chat between friends. To ensure that relevant and truthful news is used in our dataset, we selected twenty news event-related topics (judged by the authors) based on major news recently reported on on-line newswire including BBC, Reuters, CNN, Guardian, and The New York Times. The news events occurred between 1 June 2013 and 15 October 2013. Table 1 describes the twenty topics we selected. Tweets were collected based on the search API of Twitter, using the news event topics shown in the left column of Table 1 as query terms. To ensure that we do not include redundant tweets, directly retweeted messages are excluded. In total, 400 credible tweets in English for twenty news events were presented to CrowdFlower evaluators to judge.

---

[2] https://crowdflower.com

[3] http://www.merriam-webster.com/dictionary/credibility

Table 1. Twenty news event topics

Topic	News event description
US government shutdown	US Government heads toward a shutdown
Iran-US relationship	Iranian President takes steps to thaw relations with the West
Sarin attack in Syria confirmed	United Nations confirms use of chemical weapons in Syria
Shipwrecked at Europe	Boat sinks in the Mediterranean, killing dozens
Egypt state of emergency	Egypt declares state of emergency
Train kills dozens in India	Train kills dozens of religious pilgrims in India
Navy Yard shooting	Gunman and 12 victims killed in Washington D.C. Navy Yard shooting
Earthquake in Pakistan	Magnitude 7.7 earthquake kills at least 327 in Pakistan
Terrorist attack mall	Somalian militants terrorize luxury mall
Military ousted president	President Morsi deposed by military after one year in office
NSA whistle blower	Edward Snowden: whistle-blower behind NSA surveillance revelations
UK new prince	The Duchess of Cambridge gives birth to a baby boy
Oil train derails	A train in Quebec derails and explodes
Colorado flood	Colorado flood 2013 tragedy
Australia's new prime minister	Australia's new Prime Minister Tony Abbott
Iraq suicide attacks	Suicide bomb attacks on Iraqi school, Shi'ite pilgrims, kill 29
Mexico storm disaster	Mexico storms death toll rises, crop lands damaged
Cyclone hits India	Many evacuated as Powerful Cyclone Hits India
Protest in Egypt	More than 50 people are killed as pro-Morsi protest
Riot in Moscow	Rioting erupts in Moscow after killing blamed on migrant

In the crowd source evaluation, the date, topic and topic description of each tweet are given to the evaluators to help them distinguish the credibility level of tweets. The credibility definition and criteria are also presented to the evaluators. To trap unreliable evaluators, gold questions are set up, which are credible news tweets mingled with not credible tweets containing opinions or social chats. For each of the twenty topics, two gold questions are randomly inserted into the credible tweets. Only evaluators that judge the gold questions correctly are considered reliable, and their judgments are accepted.

To further elicit the features the public uses to judge the tweet credibility, we also ask CrowdFlower evaluators to leave textual comments to explain their judgements. We manually examine the comments to ensure quality comments are used to analyse user perception. To this end, we remove nonsensical comments, such as those containing the word "none", numbers or words that are out of context for the topic.

Table 2. Distribution of credibility ratings for 400 tweets

Credibility level	#comments
Definitely credible	342 (85.5%)
Seems credible	2 (0.5%)
Not credible	35 (8.75%)
Can't decide	0 (0%)
No consensus rating	21 (5.25%)

Table 3. Features derived from user comments for credibility rating

Category	Feature	#cmts
Topic-based	Topic keyword - *e.g. Prince (UK new prince topic)*	315 (54%)
Message-based	Link in tweet - *URLs, URL shortener, image links*	95 (16.3%)
User-based	Display name - *Twitter ID e.g. BBCNews, Anonymous*	88 (15%)
User-based	User belief of the topic - *e.g. plausible, professional, it actu- ally happened, facts, informative*	44 (7.5%)
Message-based	Credibility keyword - *e.g. Update, Breaking, Liveupdates*	26 (4.5%)
Message-based	Hashtag - *e.g. #Lampedusa, #Egypt*	8 (1.4%)
Message-based	Retweet - *Contains the letters 'RT' in the tweet messages*	6 (1%)
User-based	User mention - *e.g. @OMBPress, @cctvnewsafrica*	2 (0.3%)

3   Deriving and Analysing Features for Credibility

In our user study, evaluators were asked to judge the credibility level for each tweet as "Definitely credible", "Seems credible", "Not credible", or "Can't decide". At the conclusion of our user study, a total of 2,005 judgements by 98 evaluators for 400 tweets were collected, where five out of 400 tweets received six judgments and the rest received five judgments each. The consensus rule was used to assign credibility rating for tweets. If a tweet receives three out of five or four out of six votes for a credibility level, the message is assigned the corresponding credibility rating; otherwise no consensus credibility rating (recall that there are three credibility levels) can be reached for the tweet. Table 2 lists the distribution of credibility ratings for all tweets. Note that none of the tweets received the judgment of "Can't decide". Our results confirm that users generally trust the information disseminated on Twitter, which mirrors the findings in [1].

3.1   Analysing User Comments for Credible Tweets

We analyse user comments for 342 and two tweets received "Definitely credible" and "Seems credible" ratings to derive features users use for their credibility judgments. The comments collected from the user study consist of 558 valid comments from 22 evaluators, which describe features they feel important for their judgment of the truth and falseness for tweets. Following the categorisation in [1] and [3] we manually summarise the comments into three categories of eight features, as shown in Table 3.

Table 4. Top association rules

Association Rules	Accuracy
Link in Tweet=available 74 => Credible 72	97.7%
Hashtag=yes 8 => Credible 8	97.6%
Retweets=yes 6 => Credible 6	97.2%
Twitter display name=yes, User belief=yes 3 => Credible 3	96.2%
Twitter display name=yes 88 => Credible 81	91.0%
User belief=yes, Topic keyword=yes 36 => Credible 27	77.4%
User belief=yes 44 => Credible 33	76.7%

Note that "User belief of the topic" refers to user's prior belief on the relevant topic and is external to Twitter, while in [1] all features are derived based on Twitter. Table 3 shows that users perceive these features in general with significantly different weights, where Topic keyword is commonly used and User mention is rarely used. In contrast the carefully engineered tens of features in [1] are used collectively by machine learning models for predicting topic credibility.

3.2   Analysing Misjudged and Difficult-to-Judge Tweets

We analyse the 35 tweets with the "Not credible" rating in Table 2. These tweets are misjudged by evaluators, as all tweets in our study have been manually verified as credible. The politics news topics 'Iran and US relationship' and 'US Government shutdown' have the largest number of misjudged tweets. We observe that these tweets are often questions, which may be why users have misperception of their credibility; indeed they are titles for news articles from reliable news agencies with short url links. Although Link in tweet is an important feature for users to judge credible tweets (See Table 4), the language features of tweets also play important roles for user perception of credibility.

We also analyse the 21 difficult-to-judge tweets where users could not reach consensus ratings. 95.6% of these difficult tweets are breaking news (42.8%) and politics news (42.8%). We observe that these tweets mostly lack links to external sources, which result in that users can not consistently judge their credibility. This Link in tweet and tweet credibility association is also shown in Table 4.

3.3   Feature and Credibility Association Analysis

To uncover relationships between features and tweet credibility, we apply association rule mining to the 379 tweets in Table 2 with consensus ratings of Definitely credible, Seems credible and Not credible based on the features in Table 3. We use the WEKA Predictive Apriori package [9][4] to mine for the best 100 association rules of the form "feature set => credibility" with an accuracy threshold of 70%. Table 4 lists the top association rules, where numbers of comments supporting the left and right hand sides are shown. According to the table,

---

[4] http://www.cs.waikato.ac.nz/ml/weka/

for all top rules the right hand side is always the Credible rating – users tend to believe in the information conveyed in tweets yet can not reach the Not credible rating consistently. Moreover, Link in tweet, Display name and User belief are important features often leading users to the Credible rating for tweets. From Tables 3 and 4 it can be seen that the Topic keyword feature, the most important feature commented by evaluators, does not form a strong association rule; only when combined with User belief it gives high accuracy for predicting credible tweets. The Link in tweet feature is used as an indicator for credibility.

4   Conclusions

We have studied the user perception of credibility for news twects on Twitter via a user study on the CrowdFlower platform. Through analysing user credibility judgements and comments, eight features have been identified, where display name, link in tweet and user belief in the tweet topic are most important. By feature and credibility association analysis, we find strong associations between features and tweet credibility. We further find that politics and breaking news are more difficult for users to consistently reach credibility rating.

Acknowledgments. This research is partially supported by Universiti Kuala Lumpur (UniKL), Majlis Amanah Rakyat (MARA), and by the QNRF project "Answering Real-time Questions from Arabic Social Media" NPRP grant number 6-1377-1-257.

References

1. Castillo, C., Mendoza, M., Poblete, B.: Information credibility on Twitter. In: Proc. WWW, pp. 675–684 (2011)
2. Gupta, A., Kumaraguru, P.: Credibility ranking of tweets during high impact events. In: Proc. PSOSM, pp. 2–9 (2012)
3. Gupta, M., Zhao, P., Han, J.: Evaluating event credibility on twitter. In: Proc. SDM, pp. 153–164 (2012)
4. Kang, B., O'Donovan, J., Höllerer, T.: Modeling topic specific credibility on twitter. In: Proc. IUI, pp. 179–188 (2012)
5. Kwak, H., Lee, C., Park, H., Moon, S.: What is twitter, a social network or a news media? In: Proc. WWW, pp. 591–600 (2010)
6. Morris, M., et al.: Tweeting is believing?: understanding microblog credibility perceptions. In: Proc. ACM CSCW, pp. 441–450 (2012)
7. O'Donovan, J., et al.: Credibility in context: An analysis of feature distributions in twitter. In: PASSAT and SocialCom, pp. 293–301 (2012)
8. Petrovic, S., et al.: Can twitter replace newswire for breaking news? In: Proc. AAAI (2013)
9. Scheffer, T.: Finding association rules that trade support optimally against confidence. In: Siebes, A., De Raedt, L. (eds.) PKDD 2001. LNCS (LNAI), vol. 2168, pp. 424–435. Springer, Heidelberg (2001)
10. Wang, A.: Don't follow me: Spam detection in twitter. In: Proc. SECRYPT, pp. 1–10 (2010)

Sentiment Analysis and the Impact
of Employee Satisfaction on Firm Earnings

Andy Moniz[1] and Franciska de Jong[2,3]

[1] Rotterdam School of Management, Rotterdam, The Netherlands
moniz@rsm.nl
[2] Erasmus Studio, Erasmus University, Rotterdam, The Netherlands
fdejong@ese.eur.nl
[3] Human Media Interaction, University of Twente, Enschede, The Netherlands
f.m.g.dejong@utwente.nl

Abstract. Prior text mining studies of corporate reputational sentiment based on newswires, blogs and Twitter feeds have mostly captured reputation from the perspective of two groups of stakeholders – the media and consumers. In this study we examine the sentiment of a potentially overlooked stakeholder group, namely, the firm's employees. First, we present a novel dataset that uses online employee reviews to capture employee satisfaction. We employ LDA to identify salient aspects in employees' reviews, and manually infer one latent topic that appears to be associated with the firm's outlook. Second, we create a composite document by aggregating employee reviews for each firm and measure employee sentiment as the polarity of the composite document using the *General Inquirer* dictionary to count positive and negative terms. Finally, we define employee satisfaction as a weighted combination of the firm outlook topic cluster and employee sentiment. The results of our joint aspect-polarity model suggest that it may be beneficial for investors to incorporate a measure of employee satisfaction into their method for forecasting firm earnings.

1   Introduction

This study intends to contribute to the growing literature about applications of text mining within the field of finance. Our approach towards employees' sentiment analysis starts from the assumption that employees are organizational assets. Management studies [1] suggest that corporate culture influences organizational behavior, especially in the areas of corporate efficiency, effectiveness and employee commitment. Indeed, according to the former CEO of IBM, *"culture is not just one aspect of the game, it is the game"* [2].

From an applications stance, our results may be of interest to investors seeking to predict firm earnings. Prior accounting research suggests that such information is not properly incorporated by the stock market due to its intangible nature, hindering the ability to measure the construct itself. To provide evidence in support of this Edmans [1] tracks the "100 Best Companies to Work for in America" published in Fortune magazine. The study posits a link between current employee satisfaction and future

M. de Rijke et al. (Eds.): ECIR 2014, LNCS 8416, pp. 519–527, 2014.
© Springer International Publishing Switzerland 2014

firm earnings that is not immediately visible to investors. We seek to complement Edmans' work and find evidence to suggest that the forecasting power of our model is incremental to the Fortune study. We extend the regression-based approach adopted by [1] to denote the properties of an object that proxies firm outlook.

The rest of this study is structured as follows: Section 2 provides an overview of the online employee reviews dataset and highlights its advantages over the Fortune dataset. Section 3 defines employee satisfaction by developing the concepts of polarity and aspect. Throughout this paper we use the term sentiment to denote the polarity of employees' reviews and aspect to denote the properties of an object that are commented on by reviewers. We then describe our approach to determine the classification of employee satisfaction via its impact on future firm earnings. In Section 4 we develop a polarity-only and a joint polarity-aspect model to predict firm earnings. Section 5 provides an empirical evaluation of the proposed model. We conclude in Section 6 and provide suggestions for future research.

2   The Dataset

We collected employee reviews from the career community website Glassdoor.com. The platform covers more than 250,000 global companies and contains almost 3 million anonymous salaries and reviews from 2008 onwards [3]. Reviewers provide an *Overall Score* on a scale of 1-5 and rate companies across five dimensions: *Culture & Values, Work/Life Balance, Senior Management, Comp & Benefits* and *Career Opportunities*. Many of these ratings only begin in 2012. We extract employees' full reviews, including their perceived pros and cons of the company [4] and their 'Advice to Senior Management'. The opening sentence of reviewers' text follows a structured format, identifying whether the reviewer is a current or former employee together with the number of years' service. Comments are reviewed by website editors before publically posted. This prevents reviewers from posting defamatory attacks and from drifting off-topic that may otherwise hinder topic modelling and sentiment analysis [5] [6].

As a means to aide comparability to [1], we restrict our analysis to publically traded companies that are published in Fortune magazine's "100 Best Companies to Work for in America" list. Our corpus comprises 41,227 individual reviews, two-thirds of which were written by current employees and the remainder by former employees. The median number of reviews per company is 340, with 84% of company reviews starting in 2008.

Unlike the Fortune dataset which suffers both from untimely (annual) updates and limited data coverage, we believe that employee website comments mitigate such issues, provide a richer source of information and a novel way to look inside a company's culture [3]. Our research employs sentiment analysis using a non-proprietary dataset that we make available in open access to encourage further research[1].

---

[1] https://dl.dropboxusercontent.com/u/
57143190/ECIR2014/employee_reviews.zip

3   Classification of Employee Satisfaction

The approach towards employees' sentiment analysis presented here starts from the assumption that employees are organizational assets and comprises of three steps. First, we employ Latent Dirichlet Allocation (LDA) to identify the aspects in employees' reviews and manually infer one latent topic that appears to be associated with firm outlook. Second, we measure employee sentiment as the polarity of a composite document, defined by aggregating employee reviews for each firm over each fiscal quarter. We use the *General Inquirer* dictionary to count positive and negative terms. In line with [9], our goal is not to show that a term counting method can perform as well as a Machine Learning method, but to provide a methodology to measure the impact of employee sentiment on firm earnings. Finally we define employee satisfaction as a weighted combination of firm outlook and employee sentiment. We develop a regression-based model [8][10] to forecast firm earnings by placing greater weight on documents that emphasize firm outlook.

3.1   Document

We start by defining a document as a single employee review. As the title of each document tends to summarize the review, the title and text are merged. We apply a shallow pre-processing over the text, including removal of stopwords, high frequency terms, company names and company advertisements. We use this definition of a document to train and extract the global aspects [11] of our corpus as described in Section 3.2.

We then redefine the concept of a document by combining all employee reviews written about a company into a composite document. This is because our primary goal is to evaluate the impact of aggregated employee satisfaction on firm earnings. As firms report earnings quarterly, we amalgamate[2] employee reviews posted during the three months' between successive quarterly earnings announcement dates. An analogous approach is adopted by [12].

3.2   Aspect

To infer salient aspects, we employ a standard implementation of LDA [13] using collapsed Gibbs sampling. Probabilistic topic models provide an unsupervised way to identify the hidden dimensions within a document and explain how much of a word in a document is related to each topic. We implement standard settings for LDA hyperparameters, $\alpha = 50/K$ and $\beta=.01$ where K is the number of topics [14]. Table 1 presents the aspects inferred by the LDA model.

---

[2] We require a minimum of 30 reviews [7] to form a document as a way to avoid making statistical inference on a small, potentially biased sample dataset [8].

Table 1. Topic clusters and top words identified by LDA

Representative words are the highest probability document terms for each topic cluster. The inferred aspect titles are manual annotations associated with the topic clusters.

firm outlook	development opportunties	salaries	skillset	interview tips
outlook	learn	raise	innovate	interviews
recommend	stretched	professional	individual	employers
learning	contribute	implement	specialization	private
career	ensure	costsaving	cosmetics	reviews
future	chances	solutions	skill	instructions
opportunities	career	salaries	peers	sent

Our interest lies in the first topic cluster, that we manually annotate as *firm outlook*.

3.3   Determining Sentiment

Our main resource to identify polarity is the *General Inquirer* dictionary[3] [27]. The *General Inquirer* classifies words according to multiple categories, including positive and negative. This dictionary contains 1,915 positive words and 2,291 negative words. We measure polarity by counting the number of positive (P) versus negative (N) terms of a firm's composite document [12]:

$$Polarity = (P \quad N)/(P + N)$$

Since former/older employees may be perversely incentivized [16] to provide negative feedback, we first statistically test for differences across different cohorts in the dataset. We compare the sentiment scores across four groups of employee reviews, distinguishing between former and current employees, junior (<5 years work experience) and senior staff (5+ years) and conduct a multivariate t-test [8] on the average sentiment scores across the four groups. We do not find a statistically significant difference in mean sentiment scores. This provides comfort that all reviews can be amalgamated into a composite document without hindering statistical inference.

3.4   Combined Approach

We adopt a statistical regression-based technique by creating a multiplicative interaction term [17] that combines *firm outlook* with sentiment. Specifically, we define the variable:

$$Outlook\_sentiment_{it} = firm\ outlook_{it} \times Tone_{it}$$

---

[3] http://www.wjh.harvard.edu/~inquirer/homecat.htm

The inclusion of *Outlook_sentiment* within a regression model provides a means to test that it is specifically employee sentiment related to the *firm outlook* topic cluster that is correlated to firm earnings. Our method is aligned with [18], treating positive and negative sentiment as additional topics within a LDA model.

3.5   Measuring the Impact of Employee Satisfaction on Firm Earnings

Classification of employee satisfaction is challenging due to the lack of an obvious outcome to evaluate model performance [19][20][21]. The approach we take is to classify employee sentiment as positive/negative by measuring its ex-post impact on firm earnings using the concept of earnings' surprises adopted by the financial literature [1] [10]. We first define unexpected earnings [1] for firm i during the financial quarter t as the difference between realized firm earnings (EPS_{it}) and the consensus broker estimate $E(EPS_{it})$ prior to the company's earnings announcement. These differences are then divided by the standard deviation of broker forecasts (σ^{EPS}_{it}), so that the resulting SUE_{it} measure can be compared in the same units across all firms:

$$SUE_{it} = 1/\sigma^{EPS}_{it} \times [EPS_{it} - E(EPS_{it})]$$

The Standardized Unexpected Earnings of a firm, SUE_{it}, measures the number of standard deviations that realized earnings are above or below the consensus estimate and can be viewed as an outcome of employee satisfaction [1].

4   Model for Firm Earnings

Our primary means to evaluate the impact of employee satisfaction on firm earnings is via an ordinary least squares regression [8]. This is the standard approach adopted in financial accounting research [1] [10] [22] as a means to isolate the impact of employee satisfaction after controlling for other firm attributes. We adopt this methodology rather than more sophisticated Machine Learning techniques to aide comparability to [1]. In contrast to SVMs and neural networks, the main appeal of a regression-based approach is that the incremental forecasting power of features can readily be determined.

For a baseline, we create a naïve model that forecasts company $i's$ earnings surprise at time $t+1$ (the subsequent quarter) as a linear function of the company's most recent earnings surprise at time t [22]:

$$SUE_{it+1} = \beta_0 + \beta_1 SUE_{it} + \varepsilon_{it}$$

Our polarity-only model incrementally adds *Tone* to the naïve model forecast:

$$SUE_{it+1} = \beta_0 + \beta_1 SUE_{it} + \beta_2 Tone_{it} + \varepsilon_{it}$$

Finally, our joint polarity-aspect model combines both *firm outlook* and *Tone* via the multiplicative interaction term *Outlook_sentiment*. The identification of a statistically significant regression coefficient serves to test the hypothesis that a positive outlook is associated with higher than expected firm earnings over the subsequent quarter and that the feature adds incremental forecasting power to the information contained in *Tone*.

$$SUE_{it+1} = \beta_0 + \beta_1 SUE_{it} + \beta_2 Tone_{it} + \beta_3 Outlook\_Sentiment_{it} + \varepsilon_{it}$$

Table 2 documents the regression results over the full sample for each model.

Table 2. Regression analysis of the models defining SUE_{it+1} as the forecast variable

Model	Intercept	SUE_{it}	$Tone_{it}$	$Outlook\_Sentiment_{it}$
Naïve	-1.393	0.230		
	(-1.59)	*(4.90)****		
Polarity-only	-3.338	0.225	4.672	
	(-2.44)	*(4.79)****	*(1.85)*	
Joint polarity-aspect	-3.026	0.213	4.864	1.435
	(-2.23)*	*(4.57)****	*(1.94)*	*(3.00)****

Numbers in brackets provide the test statistics. The asterisks provide the level of significance where * indicates the variable is statistically significant at the 5% level, ** at the 1% level and *** at the 0.1% level. All test statistics are based on robust standard errors [23].

Following prior financial accounting studies [24] [25], we include control variables in the regression to account for known firm attributes that may otherwise influence earnings. We include the log book-to-market ratio and the log market capitalization and the firm's prior 12 month price return. For presentation purposes only, we omit the estimated coefficients from Table 2.

The polarity-only model appears to be mildly incremental to the baseline, while the joint polarity-aspect model indicates that the interaction term is highly significant as a predictor of firm earnings.

5   Model Evaluation and Analysis

For evaluation, we select the root-mean-square error (RMSE) as a measure of the difference between the predicted model values (E_i) and the firm values actually observed (O_i):

$$RMSE = \left[\frac{1}{N} \sum_{i=1}^{n} (E_i - O_i)^2 \right]^{\frac{1}{2}}$$

Our choice is deemed appropriate since firm earnings are continuous rather than binary variables. We implement cross-validation using a Jack-knife approach [26] due to the limited size of our dataset (288 observations). We draw 1,000 bootstrapped samples (with replacement) using n-1 observations, and estimate the parameters for the regression models to predict the earnings surprise for the out-of-sample observation. The performance of the two sentiment systems are compared to the baseline. We separately identify the RMSE for positive and negative outcomes of earnings surprises.

Table 3. Comparison of RMSE across models

Model	Positive earnings surprises	Negative earnings surprises
Naïve baseline	1.823	2.952
Polarity-only	1.820	2.910
Joint polarity-aspect	1.817	2.624

The results in Table 3 show that the difference in RMSE for positive earnings surprises is negligible across the three forecast models, while RMSE for negative surprises monotonically decreases along each row and is considerably lower for joint polarity-aspect model (-11% below the Naïve baseline model). One interpretation of this result is that employee sentiment has an asymmetric effect on firm earnings. Companies with poor sentiment see negative earnings surprises during the following quarter, while companies with high employee sentiment do not see a noticeable improvement.

6   Conclusion and Future Research

To our knowledge, previous studies have only measured the impact of corporate reputation from the perception of the media and consumers. In this study, we identify a potentially neglected yet primary stakeholder of the firm and suggest that automated sentiment analysis based on employee reviews can provide a novel insight into company culture. Our findings indicate that the interaction of employee sentiment with the *firm outlook* topic cluster contains predictive power for firm earnings. This effect appears to be asymmetric, adversely affecting those companies that do not exhibit positive sentiment related to *firm outlook*.

In future work, we plan to extend our online corpus to include additional jobs and community websites and to extend coverage of companies globally. Interestingly, in an unreported principal components analysis we noticed that *firm outlook* appears to capture different dimensions to those scored by reviewers themselves. Identifying the reasons for this may be an interesting area for future classification research.

Acknowledgement. The research leading to these results has partially been supported by the Dutch national program COMMIT. The authors wish to thank Hubert Jeaneau and Julie Hudson at UBS Investment Bank for their insightful comments, and gratefully acknowledge the support of APG Asset Management.

References

[1] Edmans, A.: Does the Stock Market Fully Value Intangibles? Employee Satisfaction and Equity Prices. Journal of Financial Economics 101(3) (2011)

[2] Jeaneau, H., Hudson, J., Zlotnicka, E.: ESG Keys: Human Capital – Looking for questions (2013)

[3] Jeaneau, H., Hudson, J., Zlotnicka, E.: Corporate culture: Relevant to investors? UBS Investment Research (2013)

[4] Kim, S.M., Hovy, E.: Determining the sentiment of opinions. In: Proceedings of the 20th International Conference on Computational Linguistics (2004)

[5] Pang, B., Lee, L.: A sentimental education: Sentiment analysis using subjectivity summarization based on minimum cuts. In: Proceedings of the 42nd Annual Meeting on Association for Computational Linguistics (2004)

[6] Hussaini, M., Kocyigit, A., Tapucu, D., Yanikoglu, B., Saygin, Y.: An aspect-lexicon creation and evaluation tool for sentiment analysis researchers. In: ECMLPKDD (2012)

[7] Hogg, R., Tanis, E.: Probability and Statistical Inference, 8th edn. (2012)

[8] Mardia, K.V., Kent, J.T., Bibby, J.M.: Multivariate Analysis. Academic Press (1979)

[9] Pang, B., Lee, L., Vaithyanathan, S.: Thumbs up? Sentiment classification using machine learning techniques. In: Proceedings of EMNLP 2002 (2002)

[10] Brown, L.D.: Earnings forecasting research: Its implications for capital markets research. International Journal of Forecasting 9, 295–320 (1993)

[11] Titov, I., McDonald, R.: A Joint Model of Text and Aspect Ratings for Sentiment Summarization. In: Proceedings of the 46th ACL, pp. 308–316 (2008)

[12] Tetlock, P.C.: Giving content to investor sentiment: The role of media in the stock market. Journal of Finance 62, 1139–1168 (2007)

[13] Blei, D.M., Ng, A., Jordan, M.I.: Latent Dirichlet Allocation. Journal of Machine Learning Research 3, 993–1022 (2003)

[14] Griffiths, T.L., Steyvers, M.: Finding scientific topics. Proceedings of the National Academy of Science 101, 5228–5235 (2004)

[15] Kennedy, A., Inkpen, D.: Sentiment Classification of Movie Reviews using Contextual Valence Shifters. Computational Intelligence 22(2), 110–125 (2006)

[16] Tversky, A., Kahneman, D.: Availability: A Heuristic for Judging Frequency and Probability. Cognitive Psychology 5(2) (1973)

[17] Brambor, T., Clark, W.R., Golder, M.: Understanding Interaction Models: Improving Empirical Analyses. Political Analysis 14, 63–82 (2006)

[18] Mei, X.S., Zhai, C.: Automatic labelling of multinomial topic models. In: SIGKDD (2007)

[19] Turney, P.: Thumbs up or thumbs down? Semantic orientation applied to unsupervised classification of reviews. In: Proceedings of the 40th Annual Meeting of the Association for Computational Linguistics (2002)

[20] Wilson, T., Wiebe, J., Hoffmann, P.: Recognizing contextual polarity in phrase-level sentiment analysis. In: Proceedings of the Conference on Human Language Technology and Empirical Methods in Natural Language Processing (2005)

[21] Ku, L.W., Lo, Y.S., Chen, H.H.: Test collection selection and gold standard generation for a multiply-annotated opinion corpus. In: Proceedings of the 45th Annual Meeting of the ACL on Interactive Poster and Demonstration Sessions (2007)

[22] Bernard, V., Thomas, T.: Evidence that stock prices do not fully reflect the implications of current earnings for future earnings. Journal of Accounting and Economics 13, 305–340 (1990)

[23] White, H.: A Heteroskedasticity-Consistent Covariance Matrix Estimator and a Direct Test for Heteroskedasticity. Econometrica 48, 817–838 (1980)

[24] Fama, E.F., French, K.R.: The cross-section of expected stock returns. Journal of Finance 47 (1992)

[25] Carhart, M.M.: On persistence in mutual fund performance. Journal of Finance 52, 57–82 (1997)

[26] Efron, B., Tibshirani, R.J.: An Introduction to the Bootstrap. Chapman & Hall, New York (1993)

[27] Stone, P., Dumphy, D.C., Smith, M.S., Ogilvie, D.M.: The General Inquirer: A Computer Approach to Content Analysis. The MIT Press (1966)

Entity Tracking in Real-Time
Using Sub-topic Detection on Twitter

Sandeep Panem, Romil Bansal, Manish Gupta, and Vasudeva Varma

International Institute of Information Technology, Hyderabad, India

Abstract. The velocity, volume and variety with which Twitter generates text is increasing exponentially. It is critical to determine latent sub-topics from such tweet data at any given point of time for providing better topic-wise search results relevant to users' informational needs. The two main challenges in mining sub-topics from tweets in real-time are (1) understanding the semantic and the conceptual representation of the tweets, and (2) the ability to determine when a new sub-topic (or cluster) appears in the tweet stream. We address these challenges by proposing two unsupervised clustering approaches. In the first approach, we generate a semantic space representation for each tweet by keyword expansion and keyphrase identification. In the second approach, we transform each tweet into a conceptual space that represents the latent concepts of the tweet. We empirically show that the proposed methods outperform the state-of-the-art methods.

Keywords: Sub-Topic Detection, Clustering, Entity Tracking, Text Mining.

1 Introduction

In the recent past, Twitter has been widely used for spreading the social pulse about real world entities. Mining sub-topics from entities helps in trend analysis, social monitoring, topic tracking and reputation mining. "Topics" on Twitter relate to major events in the real world; "sub-topics" on the other hand are fine-grained aspects of such events. For example, consider the tweet, "Recently listed on MLS: 2003 Volvo VHD64B200 #mixer from Transport Truck Sales in Kansas City, KS". Here the sub-topic is "buying or selling of trucks" and the topic is "Volvo". Existing topic detection methodologies are generally based on probabilistic language models, such as Probabilistic Latent Semantic Analysis (PLSA) [7] and Latent Dirichlet Allocation (LDA) [4]. By exploiting tweet contents, LIA at CLEF 2013 [2] applied a large variety of machine learning methods for clustering of tweets. However, these methods require the number of topics as an input, and assume that a single document contains rich information, which is not applicable to microblogs. REINA at CLEF 2013 [2] used similarity matrix and community detection techniques for topic detection. UNED_ORM at CLEF 2013 [1] experimented with approaches like agglomerative clustering based on term co-occurrences and clustering of wikified concepts. Few limitations of these approaches are (1) they often mix multiple incoherent sub-topics together, and (2) they cannot find novel topics in streaming scenarios as they need all the data at once. Our two phase clustering approach as described in Section 2 deals with the above mentioned issues by mining sub-topics from streaming text. The proposed clustering is based on a novel representation of tweets using

M. de Rijke et al. (Eds.): ECIR 2014, LNCS 8416, pp. 528–533, 2014.
© Springer International Publishing Switzerland 2014

various combinations of concepts, keyphrases and keywords with appropriate weights assigned to them. We improve the accuracy of detecting sub-topics by discarding less frequent concepts. Also, *batched* updates ensure that our system is efficient enough to be practical.

2   Approach

In this paper, we propose the following two approaches to tackle the problem of dynamic clustering by exploiting the semantic structure of tweets.

1. Semantic Space Representation (*SSR*) based approach
2. Concept Space Representation (*CSR*) based approach

Both the approaches consist of two phases: an offline cluster generation phase and an online cluster maintenance phase. In the offline phase, a graph-based clustering algorithm is used to obtain the initial clusters from a few tweets. These initial clusters are later used in the online phase to cluster new tweets from the tweet stream and also to update the clusters themselves. Figures 1 and 2 illustrate the offline and online phases for the two approaches respectively.

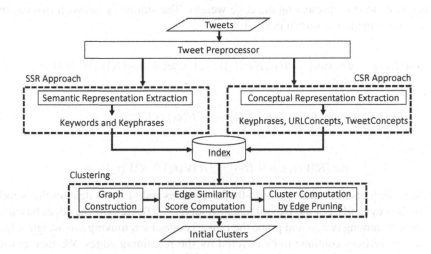

Fig. 1. Offline Initial Cluster Generation Phase for Both Approaches

Semantic Space Representation (*SSR*) Based Approach
In the offline phase, we first preprocess the tweets by removing stopwords and performing POS tagging and URL extraction. We consider only English tweets across multiple domains maintaining the heterogeneity. We extract the longest sequence of nouns as well as proper nouns and keyphrases using POS chunking [5]. We consider nouns, hashtags and proper nouns as keywords and enhance them by finding the synonyms using WordNet and extracting top n words from the page content of the URL contained in

530   S. Panem et al.

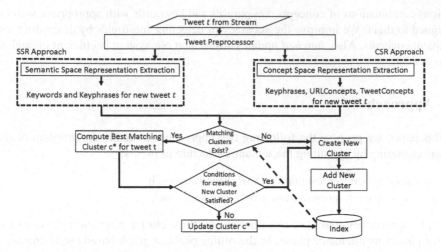

Fig. 2. Online Cluster Maintenance Phase for Both Approaches

the tweet. Thus, the semantic representation of a tweet t consists of keywords ($KW(t)$) and keyphrases ($KP(t)$). Next, we build a graph with tweets as nodes and the similarity between two tweets representing the edge weight. The similarity between two tweets, t_1 and t_2 is computed as shown in Eq. 1.

$$sim(t_1, t_2) = w \times sim(KP(t_1), KP(t_2)) + (1 - w) \times sim(KW(t_1), KW(t_2)) \quad (1)$$

where,

$$sim(KP(t_1), KP(t_2)) = |KP(t_1) \cap KP(t_2)| \quad (2)$$

and

$$sim(KW(t_1), KW(t_2)) = |KW(t_1) \cap KW(t_2)| \quad (3)$$

Here, w denotes the weight given to the keyphrases and $(1 - w)$ denotes the weight given to the keywords. For our experiments we set $w = 0.6$. We rank the edges based on the similarity among tweets and prune the ranked edges by removing low weight edges until all the vertices continue to be covered by the remaining edges. We then cluster the tweets based on nearest-neighbors similarity. Each cluster is stored along with its keywords and keyphrases in the index.

In the online phase, we process each new tweet t by first extracting its keywords and keyphrases. We then query the index for clusters containing the tweet's keywords and retrieve the cluster c^* with the highest score. Score of each cluster is based on the similarity between the tweet and the cluster in terms of their keywords and keyphrases. Tweet t is assigned to the cluster c^* or to a new cluster based on the following intuitive rules. If no clusters in the index match with the tweet keywords, tweet is assigned to a new cluster. Sometimes even though there is a match in the set of keywords, the tweet may belong to a very different sub-topic compared to the sub-topic of the most similar cluster c^*.

To avoid incorrect cluster assignment for the new tweet, we used "Wikipedia title matching" to distinguish sub-topics related to a particular topic. We compare tweet's keyphrases with Wikipedia titles by performing a substring match. If at least one keyphrase occurs in one of the titles of a Wikipedia page and if none of the tweet's keyphrases match with the keyphrases of the matched cluster c^*, then we create a new cluster. Otherwise we assign the tweet to the cluster c^*. For example, consider the tweet "The Volvo ocean race has started this year on a high range in England". Based on the simple keyword match, this tweet would get assigned to the cluster containing "Volvo" and "England", but in reality it should form a new cluster namely "Volvo ocean race".

Concept Space Representation (*CSR*) Based Approach

A lexical mismatch is caused by occurrence of different words in the tweet which are otherwise semantically related. Transformation of tweets into concept space can help reduce the lexical mismatch. This can enable matching even those tweets that are semantically relevant to each other but do not have any overlapping words. We obtain the semantic representation of a tweet by extracting concepts for a tweet using the TagMe [6] API. The API takes short text snippets as input, disambiguates the entities in the text, and maps these entities to Wikipedia pages. Using this API, we represent each tweet t conceptually as a combination of keyphrases ($KP(t)$), URLConcepts ($UC(t)$) and TweetConcepts ($TC(t)$). The offline phase for the *CSR* based approach is similar to the one for the *SSR* based approach.

In the online phase, we query the index for clusters containing the tweet's keywords and retrieve the k nearest clusters. Similarly, we retrieve the top k nearest clusters for URLConcepts and TweetConcepts. For our experiments, we set k as 10. We then assign a score $R(c)$ to cluster c as shown in Eq. 4.

$$R(c) = \frac{\alpha}{R_{KP}(c)} + \frac{\beta}{R_{UC}(c)} + \frac{\gamma}{R_{TC}(c)} \tag{4}$$

Here, α, β and γ are the weights given to keyphrases, URLConcepts and TweetConcepts respectively such that $\alpha + \beta + \gamma = 1$. Here $R_{KP}(c)$, $R_{UC}(c)$, $R_{TC}(c)$ are the ranks of the cluster retrieved when queried with keyphrases, URLConcepts and TweetConcepts respectively. A new tweet t is either assigned to the cluster c^* with the highest score or to a new cluster. If no clusters in the index match with the tweet keywords, the tweet is assigned to a new cluster. The tweet is assigned to the cluster c^* if at least one keyphrase matches between cluster c^* and tweet t, or at least one concept matches other than the entity name. Otherwise the tweet t is assigned to a new cluster.

Maintaining Cluster Purity and Cluster Labels

We preserve the purity of clusters, by removing the irrelevant concepts from the clusters at regular intervals of tweet arrivals into the clusters. After every m (we set m to four) consecutive tweets a cluster receives, we update the cluster by storing only the most frequently occurring concepts. The concepts of tweets in the cluster, and their frequency values are updated. In *SSR*, we label the clusters using the top occurring keyphrase and keyword. In CSR, we define cluster label using the top occurring keyphrase and concept.

As the labels of the cluster change frequently with the incoming tweets, we update the labels of the clusters at regular intervals, δt. For our experiments, we set δt to fifty tweets.

3 Experiments

We focus on the task of entity tracking using sub-topic detection. For the offline phase, tweets can be collected by querying for the entity name on Twitter. If there are no initial tweets related to an entity then the online phase starts with an empty cluster set. For our evaluation, we use RepLab (CLEF) 2013 [2] dataset. The dataset contains tweets for 61 entities. Each entity has about 700 tweets for training and 1500 tweets for testing. In the offline phase, we use the training tweets to obtain seed clusters, which are then used in the online phase to cluster test tweets. The data sets are manually labeled by expert annotators.

For evaluation we use two complementary measures, Reliability and Sensitivity as defined by Amigó et al. [3]. Let us consider a system output X and a gold standard $\mathcal{G}$, which are both a set of document relationships $r(d, d')$. The Reliability (R) of relationships in the system output is the probability of finding them in the gold standard. The Sensitivity (S) of predicted relationships is the probability of finding them in the system output when they appear in the gold standard.

Table 1 shows the system performance with various values of α, β, and γ. The setting $\alpha = 0.3, \beta = 0.2$ and $\gamma = 0.5$ gave the best results. We infer that both the keyphrases and TweetConcepts features are equally important. Because many tweets do not contain a URL, low weight is assigned to the URLConcepts. Table 2 compares the performance of various methods. Here *Baseline* is the memory-based learning baseline supplied by RepLab. We compare our results with *UNED_ORM, REINA, LIA* (teams that participated in RepLab 2013 [2]) whose approaches are described in Section 1. The total number of sub-topics in the dataset were 9570. The proposed *SSR* and *CSR* based approaches detected 7545 and 8633 sub-topics respectively. We observe that for a few tweets, the concepts retrieved from TagMe were relatively inaccurate, and this resulted in lower reliability for *CSR*. *SSR* has higher reliability compared to *CSR* because most of the relationships predicted by the system are also found in gold standard. However, the sensitivity is lower for *SSR* because the number of discovered relationships in the system are less, as excessive keyword matching (probably because of WordNet usage) caused some sub-topics to merge into a single sub-topic. The higher sensitivity of *CSR* as compared to that of *SSR* is because *CSR* has higher coverage of relationships over the gold standard.

CSR based approach maintains the consistency in both, identifying the number of sub-topics as well as their presence in the gold dataset for each entity. The $F1$ measure values are calculated for each topic individually and averaged over all the topics. As shown in Table 2, the proposed approach achieves the highest $F1$ measure value. The increase in the $F1$ measure value by **16.9%** as compared to the *Baseline* and $\sim$**1%** as compared to the best system in RepLab 2013 indicates that the proposed *CSR* approach performs better than the state-of-the-art methods.

Table 1. Accuracy Results for Various Values of α, β and γ

α	β	γ	R	S	$F1$	#Sub-Topics
0.5	0.2	0.3	0.303	0.521	0.338	8586
0.3	**0.2**	**0.5**	**0.304**	**0.516**	**0.339**	**8633**
0.4	0.3	0.3	0.304	0.522	0.338	8794
0.6	0.2	0.2	0.303	0.512	0.335	8785
0.2	0.3	0.5	0.305	0.516	0.337	8760
0.2	0.4	0.4	0.303	0.529	0.338	8685

Table 2. Performance Comparison of Various Methods

Method	R	S	F
CSR	0.304	**0.516**	**0.339**
UNED_ORM	0.460	0.320	0.330
REINA	0.320	0.430	0.290
SSR	**0.496**	0.203	**0.259**
LIA	0.220	0.350	0.250
Baseline	0.150	0.220	0.170

4   Conclusions

In this paper, we have explored a novel approach by exploiting the semantic and concept based representations of tweets for sub-topic clustering. In the *SSR* based approach, we used keywords (WordNet synonyms, URL keywords), keyphrases and Wikipedia title matching (as criterion for creating new cluster) as features. To handle the issue of over-matching of keywords due to WordNet usage, we propose the *CSR* based approach. In the *CSR* based approach, we used TweetConcepts, URLConcepts and keyphrases as features. We maintain the purity of clusters by periodically cleaning up the clusters. Experiments on the RepLab (at CLEF 2013) dataset showed that the proposed approach achieves significant performance gains over the baseline and other systems using metrics like Reliability, Sensitivity and $F1$ measure. In the future, we would like to extend this study by incorporating the similarity between concepts in the *CSR* based approach.

References

1. Amigó, E., Carrillo de Albornoz, J., Chugur, I., Corujo, A., Gonzalo, J., Martín, T., Meij, E., de Rijke, M., Spina, D.: Overview of RepLab 2012: Evaluating Online Reputation Management Systems. In: Forner, P., Müller, H., Paredes, R., Rosso, P., Stein, B. (eds.) CLEF 2013. LNCS, vol. 8138, pp. 333–352. Springer, Heidelberg (2013)
2. Amigó, E., de Albornoz, J.C., Chugur, I., Corujo, A., Gonzalo, J., Martín-Wanton, T., Meij, E., de Rijke, M., Spina, D.: Overview of RepLab 2013: Evaluating Online Reputation Monitoring Systems. In: Proc. of the 4th Intl. Conf. of the CLEF Initiative, pp. 333–352 (2013)
3. Amigó, E., Gonzalo, J., Verdejo, F.: A General Evaluation Measure for Document Organization Tasks. In: Proc. of the 36th Intl. ACM SIGIR Conf. on Research and Development in Information Retrieval (SIGIR), pp. 643–652 (2013)
4. Blei, D.M., Ng, A.Y., Jordan, M.I.: Latent Dirichlet Allocation. Journal of Machine Learning Research 3, 993–1022 (2003)
5. Dave, K.S., Varma, V.: Pattern Based Keyword Extraction for Contextual Advertising. In: Proc. of the 19th ACM Intl. Conf. on Information and Knowledge Management (CIKM), pp. 1885–1888 (2010)
6. Ferragina, P., Scaiella, U.: TagMe: On-the-fly Annotation of Short Text Fragments (by Wikipedia Entities). In: Proceedings of the 19th ACM International Conference on Information and Knowledge Management (CIKM), pp. 1625–1628. ACM (2010)
7. Hofmann, T.: Probabilistic Latent Semantic Indexing. In: Proc. of the 22nd Annual Intl. ACM SIGIR Conf. on Research and Development in Information Retrieval (SIGIR), pp. 50–57 (1999)

Time-Aware Focused Web Crawling

Pedro Pereira[1], Joaquim Macedo[1], Olga Craveiro[2,3], and Henrique Madeira[3]

[1] Centro Algoritmi/Dep. of Informatics, University of Minho,Portugal
[2] ESTG, Polytechnic Institute of Leiria, Portugal
[3] CISUC/Dep. of Informatics Engineering, University of Coimbra, Portugal

Abstract. There is a plethora of information inside the Web. Even the top commercial search engines can not download and index all the available information. So, in the recent years, there are several research works on the design and implementation of focused topic crawlers and also on geographic scope crawlers.

Despite other areas of information retrieval, research on Web crawling is not using the temporal information extracted from Web pages in the used crawling criteria. Therefore, our research challenge is the use of temporal data extracted from Web pages as the main crawling criteria to satisfy a given temporal focus. The importance of the time dimension is quite amplified when combined with topic or geography, but now we want to study it isolated. The used approach is based on temporal segmentation of Web pages text. It only follows links within segments tagged with dates in the scope of restriction. A precision around 75% was achieved in preliminary experimental results.

Keywords: Web Crawling, Temporal Text Segmentation, Temporal Information Extraction, Temporal Information Retrieval.

1 Introduction

Crawling the Web is an old problem that it was subject of extensive research [1] and is widely used today, mainly by search engines like Google, Yahoo! or Bing.

These crawlers try to cover a significant part of the Web, looking for information to build their indexes.Whenever users submit queries, these indexes are processed and the search engine returns responses that include URLs pointing to supposed relevant documents.

The huge size of the Web points to the need of the decentralization of the crawling process, based for instance in geographical partition of the Web [2]. Additional criteria are used to drive the crawling process like Web page rank, freshness, topic focus, geographic scope for overcome such scalability problems. This work introduces the temporal scope, based on dates extracted from Web pages content, as a new crawling criteria.

Temporal information is also be used for many different purposes, such as searching future events [3], clustering search results with timelines [4], defining snippets based on temporal information [5]. Nunes et al. [6] proposed an approach

M. de Rijke et al. (Eds.): ECIR 2014, LNCS 8416, pp. 534–539, 2014.

of web pages timestamping, to determine the publication or modification date of a web page. Yu et al. [7] modified the Page Rank algorithm, using the date of citation to improve the quality of the search results. Dai and Davidson [8] proposed a link-based ranking method considering the freshness of the page content.

In section 2, a strategy of temporal segmentation of Web pages text is presented. Later, this segmentation is used by a crawler to download Web pages, using temporal constraints (section 3). Finally (sections 4 and 5) ,some promising preliminary results and directions for future work are presented.

2 Temporal Segmentation of Text

The main objective of the text segmentation is to divide text into smaller units according to many different criteria. The proposed approach follows a segmentation algorithm based on temporal discontinuities identified in the text [9]. A temporal segment is a set of contiguous sentences or paragraphs sharing the same temporal focus. The identification of the temporal boundaries is given by the temporal information found in the text which could be mapped in chronons with normalized dates anchored in a calendar/clock system [10]. If two adjacent sentences do not have the same chronons, then there is a temporal discontinuity. In this case, these sentences must belong to different segments. Thus, adjacent sentences with the same chronons must belong to the same segment. In the absence of this normalized temporal information, this algorithm follows the approach of traditional topic segmentation. Figure 1 shows an example of temporal segmentation. The thick rectangle shows the document timestamp.

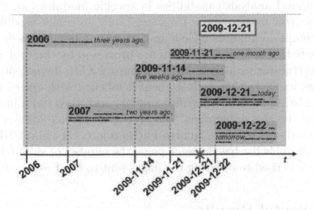

Fig. 1. Temporal Segmentation of a Text [9]

3 Temporal Crawler

A time-aware crawler must traverse the Web in search of informations which are within a given temporal scope. The downloaded pages are analyzed temporally to verify if they are within the temporal scope. The start hypothesis is that a Web page (or a segment of a page) within a given temporal scope has, with high probability, links to to Web pages with the same temporal scope.

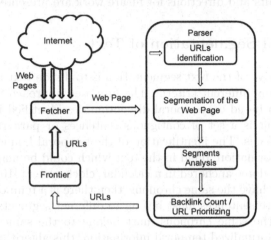

Fig. 2. Temporal Crawler Architecture

The developed time-aware crawler is no more than a general crawler with embedded temporal analysis capabilities in specific modules (see Figure 2). Its implementation was based on the *Crawler4j* [11], an easily modifiable generic crawler. The temporal analysis is done in two steps. First, the web page is temporally segmented. The Web documents are processed by the tool mentioned in section 2. This transforms a Web document into a XML document divided into temporal segments. Note that each segment is labelled with one or more dates[1].

Then, the existing links are classified in two categories: those inside the temporal focus and the remaining ones. The first set is eligible for download (valid URLs) and the others URLs are ignored.Furthermore, the valid URLs are added to a backlink list. This list has, for each page, the number of valid URLs pointing to it. This count is used to ordering the pages sent to the Frontier for download.

4 Experimental Results

To evaluate this crawler we will use precision and recall. In this preliminary evaluation we used only pages from Portuguese Wikipedia domain

---

[1] Most segments have a single date.

(http://pt.wikipedia.org/). This choice may be considered inappropriate[2], but we run the risk for this first proof of concept. Wikipedia is a domain which offers more guarantees in terms of quality of the pages with temporal data. Also, we selected two important events in our recent history including the Second World War (1939-1945) and the terrorist attacks at 9/11/2001.

We used Google to select 10 seeds for each scope. Along the topic, the queries include the word Wikipedia for only Portuguese language documents. Our segmmenter only supports Portuguese language. The used seeds are omitted due to space restrictions. This technique will be used normally combined with a topic or even a geographical scope. So, it is expected to avoid the consideration that such seeds cause a bias on the results.

As output the crawler generates a list with URLs, ordered by download time. For each event, we perform two runs. In the first one, the general crawler downloaded 5000 pages without any time restriction. In the second one, we used the time-aware crawler with the temporal scope (1939-1945 for WW2 and 2001 for 9/11/2001). At end, each run provides 5000 fully segment files and a URL crawl list. For general crawler the segmentation was a post-processing operation.

A file is considered valid, when at least one segment can be placed inside the temporal scope. To compute the total amount of positive files (required to calculate recall), we considered total valid files crawled by the temporal crawler plus the ones that were crawled by the general crawler. So, the recall@N is the quotient between the actual valid files and the total ones. For the precision@N are used the percent of valid files crawled.

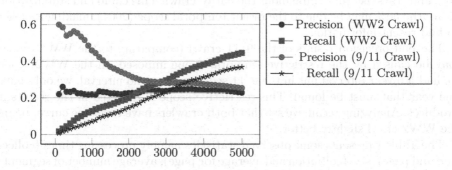

Fig. 3. Precision and Recall of the General Crawler

The Figure 3 shows that the precision for generic crawl using the WW2 seeds is very low throughout the entire crawl. The seeds do not give to the crawler a good start because, as we can see, the precision value at merely 100 files is very low. In the 9/11 crawl, for the first 800 pages the precision is above the 50%. The fairly good results in the first few hundred files do not mean anything. After there is a steep decline in the precision results. Both crawls finish around

---

[2] Wikipedia is a good representation for common web page?

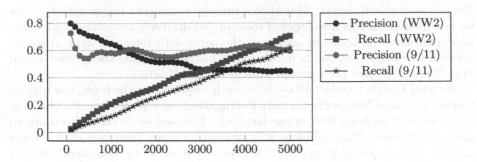

Fig. 4. Precision and Recall of the Temporal Crawler

the 25% precision. The recall value is similar for both crawls, finishing a around 40% for the 5000 files.

As we can see in the Figure 4, for the WW2 crawl, we begin with a precision of exactly 80% at the 100 files mark then it starts to descend. After that descent, we see just a little climb at the 2000 files stabilizing at the 50% mark but then the precision begins to fall and reaches the 50% precision at the 2600 files mark, stabilizing around the 45% precision mark.

If we look to the 9/11 crawl we do not see the same slow descent. The precision results at the 100 files mark is over 70% but then the precision begins to drop. At around 400 files the precision hits the lowest point and then climbs to around 60% and stays like that throughout the entire crawl. This climb and stabilization above the 50% may be due to the recent temporal scope (2001), making it easier to find recent information.

The bad precision results of the 9/11 crawl (comparing to the WW2 crawl) may be caused by the restrictive temporal scope imposed. In the WW2 scope we had a 6 year interval but in the 9/11 scope we have no interval, we only have one year that must be found. This restrictive scope is reflected in the results it produces. Analyzing recall, we see that both crawlers have the same curve, being the WW2 crawl slightly better.

The Table 1 presents some observed statistics namely processing times (collection and page), size (collection and average for page), average number of segments and chronons per page, and number of chronons in temporal scope (TS).

Table 1. Statistics of the Crawled Collections

	Total Proc. Time (m)	Page Proc. Time (s)	Collection size (MB)	Page Size(KB)	Seg/ Page	Seg. in TS	Chronon /Page
WW2	85	61.2	441.0	90.0	82.6	32.6	192.3
9/11	90	66.2	430.0	88.0	74.9	71.8	112.0

5   Conclusions

The main objective of this work is the creation of a time-aware crawler architecture that analyzes temporally the documents, using time as crawling criteria.

Even using as evaluation Web pages from Portuguese Wikipedia, we get promising results. The results from the two different crawls exhibit some variance. This means that more exhaustive experiments are necessary and preferably with the most common web pages.

Either way, it was confirmed our hypothesis, albeit with still preliminary results. Using pages or portions of pages (segments) within a given temporal scope as a starting point, and using the URLs included in them, it is most likely to find new pages with the same temporal scope.

Acknowledgments. This work is partially supported by Algoritmi and CISUC, financed by FCT - Fundação para a Ciência e Tecnologia, within the scope of the projects PEst-OE/EE/UI0319/2014 and PEst-OE/EE/UI0326/2014.

References

1. Olston, C., Najork, M.: Web crawling. Foundations and Trends in Information Retrieval 4(3), 175–246 (2010)
2. Exposto, J., Macedo, J., Pina, A.: Geographical partition for distributed web crawling. In: GIR (2005)
3. Baeza-Yates, R.: Searching the future. In: SIGIR Workshop (2005)
4. Alonso, O., Gertz, M., Baeza-Yates, R.: Clustering and exploring search results using timeline constructions. In: Proceedings of the 18th ACM Conference on Information and Knowledge Management, CIKM 2009, pp. 97–106. ACM, New York (2009)
5. Alonso, O., Baeza-Yates, R., Gertz, M.: Effectiveness of temporal snippets. In: WSSP Workshop, WWW 2009 (2009)
6. Nunes, S., Ribeiro, C., David, G.: Using neighbors to date web documents. In: WIDM, pp. 129–136 (2007)
7. Yu, P.S., Li, X., Liu, B.: On the temporal dimension of search. In: Proceedings of the 13th International World Wide Web Conference on Alternate Track Papers & Posters, WWW 2004, pp. 448–449 (2004)
8. Dai, N., Davison, B.D.: Freshness matters: in flowers, food, and web authority. In: Proceedings of the 33rd International ACM SIGIR Conference on Research and Development in Information Retrieval, SIGIR 2010, pp. 114–121 (2010)
9. Craveiro, O., Macedo, J., Madeira, H.: It is the time for portuguese texts! In: Caseli, H., Villavicencio, A., Teixeira, A., Perdigão, F. (eds.) PROPOR 2012. LNCS, vol. 7243, pp. 106–112. Springer, Heidelberg (2012)
10. Craveiro, O., Macedo, J., Madeira, H.: Leveraging temporal expressions for segmented-based information retrieval. In: ISDA 2010, pp. 754–759 (2010)
11. Crawler4j website, https://code.google.com/p/crawler4j/ Technical report

Temporal Expertise Profiling

Jan Rybak[1], Krisztian Balog[2], and Kjetil Nørvåg[1]

[1] Norwegian University of Science and Technology, Trondheim, Norway
[2] University of Stavanger, Stavanger, Norway
{jan.rybak,kjetil.norvag}@idi.ntnu.no, krisztian.balog@uis.no

Abstract. We introduce the temporal expertise profiling task: identifying the skills and knowledge of an individual and tracking how they change over time. To be able to capture and distinguish meaningful changes, we propose the concept of a hierarchical expertise profile, where topical areas are organized in a taxonomy. Snapshots of hierarchical profiles are then taken at regular time intervals. Further, we develop methods for detecting and characterizing changes in a person's profile, such as, switching the main field of research or narrowing/broadening the topics of research. Initial results demonstrate the potential of our approach.

1 Introduction

Expertise retrieval refers to the general area of linking humans to knowledge areas, and vice versa [2]. Thanks to the increasing amount of information available online that can be traced and mined for evidence of expertise, there has been a great deal of work in this area within the IR community over the past decade. Specifically, two main expertise retrieval tasks have been investigated: expert finding (*"Who are the experts on topic X?"*) and expert profiling (*"What topics does person Y know about?"*), where the former received considerably more attention than the latter.

In this paper, we focus on the expert profiling task with the ultimate goal of identifying and characterizing changes in expertise of individuals over time. To the best of our knowledge, we are the first to propose this task. To be able to capture and distinguish meaningful changes, we first introduce the concept of a *hierarchical expertise profile*, where topical areas are organized in a taxonomy and expertise is represented as a weighted tree (Section 3.1). *Temporal expertise profile* is then defined as a series of timestamped hierarchical profiles (Section 3.2). Next, we develop methods for detecting and characterizing changes in a person's profile. The core idea of our approach is the identification of so-called *focus nodes*: a single node or small set of nodes that accumulate the majority of the node weights, with respect to a given parent node (Section 4.1). A change occurs if there is a difference in the set of focus nodes between two points in time; the change is then interpreted depending on which level of the topic hierarchy is affected (Section 4.2). We illustrate our approach for a selected person (Section 5).

2 Related Work

Existing work on expert profiling has primarily focused on identifying [5] and ranking [1, 3] topics for a given expert. De Rijke et al. [4] consider hierarchical profiles for the more general task of entity profiling, however, their work concentrates on evaluation

M. de Rijke et al. (Eds.): ECIR 2014, LNCS 8416, pp. 540–546, 2014.

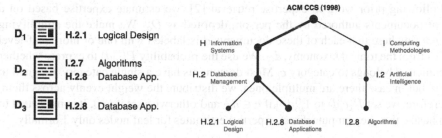

Fig. 1. Example hierarchical expertise profile, constructed from documents shown on the left. Node sizes are set proportional their weight (note that edges are not weighted according to our definition, thickness is only applied here for presentation purposes).

aspects and not on the actual construction of such profiles. Berendsen et al. [3] provide a critical assessment and analysis for the evaluation of expert-profiling systems. Sun et al. [6] present the BibNetMiner system, a system for visualizing bibliographic databases, with a focus on clustering and ranking of conferences. This is subsequently used for author, venue, and research area profiling. None of these works consider the temporal aspects of expertise. Tsatsaronis et al. [7] study the evolution of power graphs for authors over time, based on co-authorship information, the volume of published papers, and impact factors of the respective venues. Albeit they consider temporal aspects, the focus is on classifying authors into 4 predefined types, and not on topical expertise.

3 Temporal Expertise Profile

The purpose of expert profiling is to answer the following question: *"What topics does a person know about?"* In [1] the topical profile of an individual is defined as "a record of the types and areas of skills and knowledge of that individual, together with an identification of levels of 'competency' in each." Based on this definition, we extend the notion of a topical profile to a hierarchical case, where topical areas are not treated as a flat list, but are organized in a taxonomy (where parent-child relationships between topics define a hierarchy.) We represent a person's *hierarchical expertise profile* as a weighted tree, where the weights on the nodes reflect the person's expertise in the given topic. Finally, we define *temporal expertise profiles* as a series of timestamped hierarchical profiles. This allows us to track changes in a person's expertise over time.

3.1 Hierarchical Expertise Profile

The *hierarchical expertise profile* of person a is defined as a weighted tree $T_a = (C, E)$ where tree nodes $C = \{c_1, \ldots, c_n\}$ represent topics and edges E represent hierarchical relationships between topics. We write $e(c_i, c_j)$ to denote that c_j is a sub-topic of c_i. The weights on nodes $\{w_1, \ldots, w_n\}$ indicate the person's expertise on the corresponding topics. We assume that we are given some taxonomy that defines the topics and their hierarchical relationships (such as the ACM Computing Classification System that we will use in our experiments), that is, C and E. Our task, then, is to estimate the node weights $\{w_1, \ldots, w_n\}$.

Following prior work in expertise retrieval [2], we estimate expertise based on a set of documents authored by the person, denoted as D_a. We make the simplifying assumption here that each of these documents d is labeled with one or more leaf-level nodes from the topical taxonomy, d_C. We use the probability $P(c|d)$ to express whether document d belongs to category c. Most documents have a single category assigned to them, but in case there are multiple ones, we distribute the weight evenly across them. Therefore, we set $P(c|d)$ to $1/|d_C|$ if $c \in d_C$ and otherwise set it to 0. It is important to emphasize that we compute direct expertise estimates for leaf nodes only. Formally,

$$w_i = \sum_{d \in D_a} P(c_i|d)P(d). \tag{1}$$

The formula also includes a document prior $P(d)$ which can be used to express the importance of documents; for example, one could assign more importance to articles published at top-tier venues (following the intuition that these might constitute stronger evidence of expertise). However, we leave that to future work and set $P(d) = 1$ for all documents. Note that while we use probabilities in Eq. 1, w_i is not a probability; it is simply a weighted sum of publications that are labeled with a given topic.

For non-leaf nodes we sum up the weights of direct descendants:

$$w_i = \sum_{\{c_j | e(c_i, c_j)\}} w_j. \tag{2}$$

Effectively, weights are calculated in a bottom-up fashion, starting with the leaf nodes and then propagating weights to the upper levels until the root of the tree is reached. An example hierarchical profile is displayed on Figure 1.

3.2 Temporal Expertise Profile

We define the *temporal expertise profile* of a person as a series of hierarchical expertise profiles $\mathcal{T}_a = \{T_a^{t_1}, \ldots, T_a^{t_m}\}$ computed at different points in time, $t_1, \ldots, t_m$. We refer to $T_a^{t_j}$ as the *profile snapshot* taken an t_j. In this work, we assume regular time intervals, but our approach could also be applied to non-regular intervals.

We estimate the weights for leaf nodes $w_i^{t_j}$ for the profile snapshot $T_a^{t_j}$ as a mixture of two components: (1) expertise acquired in the corresponding time period (i.e., in (t_{i-1}, t_i)) based on the authored documents, and (2) expertise "carried over" from the past. The first component is the same as in Eq. 1, the only difference being that we restrict ourselves to documents originating from the given time period. As for (2), we use a decay function τ to capture the notion of expertise "fading away" over time.

$$w_i^{t_j} = \lambda \sum_{\substack{d \in D_a \\ d \in (t_{j-1}, t_j]}} P(c_i|d)P(d) + (1 - \lambda) \sum_{k=1}^{j-1} \tau(j - k) w_i^{t_k}. \tag{3}$$

This might also be viewed as "smoothing with the past" controlled by parameter λ; for the sake of simplicity, λ is set to 0.5 in our experiments. There are many possibilities for setting the decay function $\tau(t)$, where t denotes the distance in time. We employ linear decay based on time distance with two additional constraints: (1) distances below

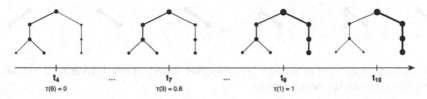

Fig. 2. Example temporal expertise profile. Decay function values $\tau(t)$ are displayed with respect to distances from the rightmost node.

δ_b are still considered "the present" where there is no decay applied, i.e., $\tau(t) = 1$ if $t \leq \delta_b$; (2) distances beyond δ_e are considered "distant past" that does not have any influence anymore, i.e., $\tau(t) = 0$ if $t \geq \delta_e$. For $\delta_b < t < \delta_e$ a linear decay is applied: $\tau(t) = \frac{\delta_e - t}{\delta_e - \delta_b}$. In our experiments we create profiles at yearly regularity and set $\delta_b = 1$ and $\delta_e = 6$. Figure 2 shows an example of a temporal profile with these settings.

It is important to note that the computations described above are applied only to leaf nodes. The weights for non-leaf nodes are calculated as before, i.e., according to Eq. 2.

4   Detecting Changes

Our general strategy for identifying and characterizing changes in temporal expertise profiles works as follows. First, we pin down a single node or small set of nodes that accumulate the majority of the node weights, with respect to a given parent node c_p, called the *set of focus nodes* $F_p^{t_i}$. This is done for each profile snapshot, where t_i in the superscript indicates the timestamp. Next, we say that a change has occurred if there is a difference in the set of focus nodes between two timestamps t_i and t_j, that is, $F_p^{t_i} \neq F_p^{t_j}$. Finally, we characterize the change based on how exactly $F_p^{t_i}$ and $F_p^{t_j}$ differ; the interpretation depends on the parent node's placement in the topic hierarchy. Specifically, we distinguish between changes depending on whether the top level or lower levels of the topic hierarchy are concerned.

4.1   Identifying Focus

We identify the set of focus nodes with respect to a given parent node as follows. First, we rank nodes by their weight (that is, the node with the highest weight comes first)

Algorithm 1. FINDFOCUS identifies the set of focus nodes

Require: profile snapshot T^{t_i}, parent node c_p, weight threshold $\eta \in [0,1]$
Ensure: F
1: $F \leftarrow \emptyset, W_F \leftarrow 0$
2: $r \leftarrow sort(\{c_i | e(c_p, c_i)\})$ ▷ r holds indices of nodes sorted by their weight
3: $i \leftarrow 0$
4: **while** $W_F < \eta \cdot w_p$ **do**
5:     $F \leftarrow F \bigcup \{c_{r[i]}\}, W_F \leftarrow W_F + w_{r[i]}$
6:     $i \leftarrow i + 1$
7: **end while**

Fig. 3. Changes in the field of research (i.e., top-level nodes). (Left): leaving field, (Middle): moving into field, (Right): switching field.

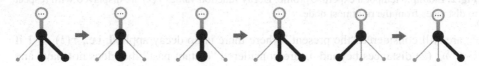

Fig. 4. Changes in the topic of research (i.e., lower level nodes). (Left): narrowing topics, (Middle): broadening topics, (Right): topic switch.

and set the focus nodes to an empty set. Then, we add nodes iteratively, in a rank-based order, until the weight accumulated in the set, relative to the total weight (that is, the parent node's weight), reaches a threshold. Algorithm 1 details our method.

4.2 Characterizing Changes

We distinguish between three types of changes, depending on how $F_p^{t_i}$ and $F_p^{t_j}$ differ:

- **F-** One of the focus nodes is removed: $|F_p^{t_i}| > 1, \exists c_k : c_k \in F_p^{t_i} \land c_k \notin F_p^{t_j}$.
- **F+** New focus node is added: $|F_p^{t_j}| > 1, \exists c_k : c_k \in F_p^{t_j} \land c_k \notin F_p^{t_i}$.
- **Fx** Exchanging a single focus node for another: $|F_p^{t_i}| = |F_p^{t_j}| = 1, F_p^{t_i} \neq F_p^{t_j}$.

Changes in the field of research. The top-level nodes of the hierarchy correspond to the *fields of research*. Focus detection performed on the root of the topic tree (as the parent node), therefore, results in the main fields of research of the person. Specific changes are interpreted as follows (see Figure 3 for an illustration).

- **Leaving field** (F-) The person leaves one of multiple main research fields.
- **Moving into field** (F+) The person takes on a new main field of research.
- **Switching field** (Fx) There is a single main field of research and it changes.

Changes in the topics of research. When the parent node used in the change detection method is not the root of the tree, the nodes affected by the changes are at least on level 2 of the hierarchy and correspond to *research topics*. Therefore, changes in the focus set should be interpreted differently; Figure 4 displays some illustrative examples.

- **Narrowing topics** (F-) The focus is distributed between multiple research topics, one of which gets removed.
- **Broadening topics** (F+) A new research topic gets into the focus.
- **Topic switch** (Fx) Research is focused on a single topic and it changes.

5   Results

We use DBLP[1] as our data collection and generated temporal profiles with yearly steps. Each paper was classified according to the 1998 ACM Computing Classification System using an automated approach. Due to space limitations, we display the temporal profile of a single person, Dutch computer scientist Maarten de Rijke. On Figure 5 we can find examples for all 6 types of change introduced in the previous section: (1) leaving field, 2004 vs. 2008; (2) moving into field, 2002 vs. 2004; (3) switching field, 2002 vs. 2008; (4) narrowing topics, 2000 vs. 2002, denoted with $\ominus$; (5) broadening topics, 2002 vs. 2004, denoted with $\oplus$; (6) topic switch, 2010 vs. 2012, denoted with $\otimes$.

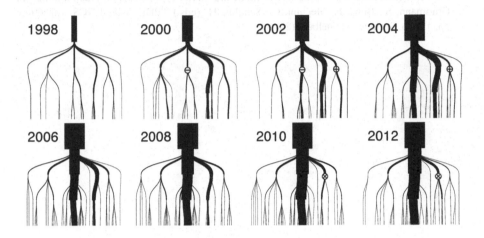

Fig. 5. Temporal expertise profile for a selected person

6   Conclusions

We have presented the task of temporal expert profiling and an approach for constructing temporal profiles based on documents labeled with leaf-level categories from a topic taxonomy. Further, we developed methods for identifying and explaining changes in a person's profile. We illustrated our ideas using a collection of computer science papers from DBLP classified according to the ACM taxonomy, but our approach is not limited to this setting; it could be applied, for example, on PubMed using the MeSH concept hierarchy. The evaluation of temporal expertise profiles remains an open question.

References

[1] Balog, K., de Rijke, M.: Determining expert profiles (with an application to expert finding). In: Proceedings of IJCAI 2007 (2007)
[2] Balog, K., Fang, Y., de Rijke, M., Serdyukov, P., Si, L.: Expertise retrieval. Foundations and Trends in Information Retrieval 6(2-3) (2012)

---

[1] http://www.informatik.uni-trier.de/~ley/db/

[3] Berendsen, R., de Rijke, M., Balog, K., Bogers, T., van den Bosch, A.: On the assessment of expertise profiles. J. Am. Soc. Inf. Sci. Technol. 64(10) (October 2013)

[4] de Rijke, M., Balog, K., Bogers, T., van den Bosch, A.: On the evaluation of entity profiles. In: Proceedings of CLEF 2010 (2010)

[5] Serdyukov, P., Taylor, M., Vinay, V., Richardson, M., White, R.W.: Automatic people tagging for expertise profiling in the enterprise. In: Clough, P., Foley, C., Gurrin, C., Jones, G.J.F., Kraaij, W., Lee, H., Mudoch, V. (eds.) ECIR 2011. LNCS, vol. 6611, pp. 399–410. Springer, Heidelberg (2011)

[6] Sun, Y., Wu, T., Yin, Z., Cheng, H., Han, J., Yin, X., Zhao, P.: BibNetMiner: mining bibliographic information networks. In: Proceedings of SIGMOD 2008 (2008)

[7] Tsatsaronis, G., Varlamis, I., Torge, S., Reimann, M., Nørvåg, K., Schroeder, M., Zschunke, M.: How to become a group leader? or modeling author types based on graph mining. In: Gradmann, S., Borri, F., Meghini, C., Schuldt, H. (eds.) TPDL 2011. LNCS, vol. 6966, pp. 15–26. Springer, Heidelberg (2011)

Assessing Quality of Unsupervised Topics in Song Lyrics

Lucas Sterckx, Thomas Demeester,
Johannes Deleu, Laurent Mertens, and Chris Develder

Ghent University - iMinds, Belgium
firstname.lastname@intec.ugent.be

Abstract. How useful are topic models based on song lyrics for applications in music information retrieval? Unsupervised topic models on text corpora are often difficult to interpret. Based on a large collection of lyrics, we investigate how well automatically generated topics are related to manual topic annotations. We propose to use the kurtosis metric to align unsupervised topics with a reference model of supervised topics. This metric is well-suited for topic assessments, as it turns out to be more strongly correlated with manual topic quality scores than existing measures for semantic coherence. We also show how it can be used for a detailed graphical topic quality assessment.

1   Introduction

This paper presents an analysis of how well topic models can be used to detect lyrical themes for use in Music Information Retrieval (MIR), an interdisciplinary science developing techniques including music recommendation.

Probabilistic topic models are a tool for the unsupervised analysis of text, providing both a predictive model of future text and a latent topic representation of the corpus. Latent Dirichlet Allocation (LDA) is a Bayesian graphical model for text document collections represented by bags-of-words [1]. In a topic model, each document in the collection of documents is modeled as a multinomial distribution over a chosen number of topics, each topic is a multinomial distribution over all words. We evaluate the quality and usefulness of topic models for new music recommendation applications.

Although lyricism and themes are undeniably contributing to a musical identity, they are often treated as mere secondary features, e.g., for obtaining music or artist similarity, which are dominantly determined by the audio signal. Nevertheless, previous works have analyzed lyrics, mainly aimed at determining the major themes they address. Mahadero et al. [2] performed a small scale evaluation of a probabilistic classifier, classifying lyrics into five manually applied thematic categories. Kleedorfer et al. [3] focused on topic detection in lyrics using an unsupervised statistical model called Non-negative Matrix Factorization (NMF) on 32,323 lyrics.

After clustering by NMF, a limited evaluation was performed by a judgment of the most significant terms for each cluster. We expand on this work by performing a large-scale evaluation of unsupervised topic models using a smaller dataset of labeled lyrics and a supervised topic model.

M. de Rijke et al. (Eds.): ECIR 2014, LNCS 8416, pp. 547–552, 2014.

While state-of-the-art unsupervised topic models lead to reasonable statistical models of documents, they offer no guarantee of producing results that are interpretable by humans and require a thorough evaluation of the output. When considering lyrics, there is no general consensus on the amount and nature of the main themes, as opposed to news-corpora (sports, science,...). A useful topic model for MIR, appends the music with a representation of the thematic composition of the lyrics. For use in applications like music recommendation, playlist generation, ..., the topics should be interpretable. Evaluation methodologies based on statistical [1] or coherence [4] measures are not optimal for this purpose since they do not account for interpretability and relevance to the application. Chuang et al. [5] introduced a framework for the large-scale assessment of topical relevance using supervised topics and alignment between unsupervised and supervised topics. Our contributions presented in this paper apply and build on aforementioned work, by assessing quality of unsupervised topics for use in MIR, and by introducing a new method for measuring and visualizing the quality of topical alignment, based on the kurtosis of the similarity between unsupervised topics and a reference set of supervised topics.

In Section 2, we present the data and our experimental set-up. The main topic model analysis is presented in Section 3, followed in Section 4 by conclusions.

We used the results presented below to create an online demo that demonstrates the use of high-quality topics for MIR with an application which automatically generates playlists based on preference of lyrical themes. This demo can be found at http://users.ugent.be/~lusterck (Login = demo:ldamir).

2   Experimental Setup

The main dataset used for this research is the 'Million Song Dataset'(MSD) [6], with metadata for one million songs, and lyrics as bags-of-words for a subset of 237,662 songs from a commercial lyrics catalogue, 'musiXmatch'.

LDA was applied on the set of lyrics, using the java-based package MALLET [7]. Three topic models were inferred from the subset of 181,892 English lyrics for evaluation[1], one with 60 (T60), 120 (T120) and 200 (T200) topics. A manual quality assessment of all of these topics was performed, with scores ranging from 1 (useless) to 3 (highly useful). As an additional resource, a clean dataset of labels was provided by the website, 'GreenbookofSongs.com®'[2] (GOS), a searchable database of songs categorized by subject. This dataset contains 9,261 manually annotated song lyrics matched with the MSD (a small subsample of the GOS' complete database), with multiple labels from a large class-hierarchy of 24 super-categories with a total of 877 subcategories. Labeled Latent Dirichlet Allocation (L-LDA) is a variation of LDA for labeled corpora by incorporating user supervision in the form of a one-to-one mapping between topics and labels [8]. An L-LDA model with 38 supervised topics was inferred from the much smaller set of GOS-labeled lyrics, based on the GOS super-categories (but with the omission of minor

---

[1] Track-id's can be provided upon request.

[2] http://www.greenbookofsongs.com, the authors would like to thank Lauren Virshup and Jeff Green for providing access to the GOS-database.

categories like 'Tools', and splitting up of major categories like 'Love'). These are high-quality topics, but because of the limited size of the GOS data set, less representative for the entire scope of themes in the complete MSD lyrics collection.

3   Topic Model Assessment

The suitability of topic models for use in MIR is determined by the amount of relevant and interpretable topics they produce. We first introduce suitable metrics to evaluate to what extent unsupervised topics can be mapped to supervised topics obtained from tagged documents. We then show how these can be used as a better measure for the interpretability of topics than an existing alternative, and provide a visual analysis of topical alignment.

3.1   Measuring Topical Alignment

We define high-quality topics as topics for which a human judge finds a clear semantic coherence between the relevant terms in relation to an underlying concept (such as 'Love', or 'Christmas'). Such concepts are made explicit by an L-LDA model based on tagged documents, and we detect high-quality LDA-topics as those that bear a strong resemblance with L-LDA topics. For an unsupervised topic to represent a highly distinctive theme, ideally it should be highly similar to only a single supervised topic. For each of the unsupervised LDA-topics, the cosine similarity between the word-topic probability distribution is calculated with the distribution of each L-LDA topic. We introduce two metrics to assess the distribution of these similarities per LDA-topic, which measure how strongly the variance of the mean cosine similarity depends on extreme values (in this case, because of similarities that are much higher than the average). The first is the excess kurtosis (γ_2), traditionally used to detect peakedness and heavy tails. The second is the normalized maximum similarity ($z_{\max}$), used in several outlier detection strategies.

$$\gamma_2 = \frac{\mu_4}{\sigma^4} - 3, \qquad z_{\max} = \frac{X_{\max} - \mu}{\sigma} \tag{1}$$

with μ_4 the fourth moment about the mean μ, σ the standard deviation, and $X_{\max}$ the maximum similarity. Figure 1 shows the similarities with the unsupervised topics for the high-quality LDA-topic 29, with a clearly matched supervised topic (and high values for γ_2 and $z_{\max}$), and for the low-quality LDA-topic 39 (with low γ_2 and $z_{\max}$). The insets show the histograms of the similarities. Various other metrics were evaluated as well, but with a lower ability of detecting the interesting distributions.

3.2   Semantic Coherence

A second evaluation was performed using metrics presented in [4], where the authors show that measures for semantic coherence are highly correlated with human quality judgments. These metrics use WordNet, a lexical ontology, to score topics by measuring the average distance between words of a topic using a variety of distance metrics based on the ontology. The best performing metric was reported to be the LESK-metric [9], based on lexical overlap in dictionary definitions. Table 1 shows the Spearman rank correlation between the LESK score for each topic and the manually assigned quality scores. For comparison, the rank correlation between the manual quality scores and γ_2 and z_{max} (as calculated in Section 3.1) are shown as well, and lead to significantly higher correlation values than with the LESK metric.

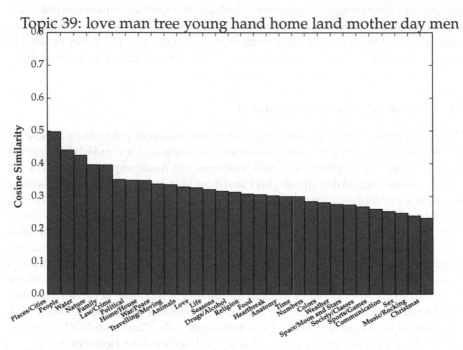

Topic 39: love man tree young hand home land mother day men

Fig. 1. Kurtosis measure and Normalized Maximum Similarity for topic evaluation

Table 1. Spearman correlations with manual quality-scores for the three topic models

Evaluation Metric	T60	T120	T200
Semantic Coherence using Wordnet (LESK)	0,35	0,23	0,31
Kurtosis (γ_2)	0,49	0,49	0,56
Normalized Maximum Similarity (z_{max})	0,49	0,50	0,53

3.3   Graphical Alignment of Topics

We can visualize the alignment between the supervised and unsupervised topics by calculating the kurtosis on the similarities between both topic sets. These are shown in Fig. 2, a *correspondence chart* similar to the one presented in [5], for the 60 topics LDA-model (T60). Our chart differs from the one presented in [5] in that it uses topics from an L-LDA model for the matching of unsupervised topics instead of a list of words generated by experts, and uses bar-charts

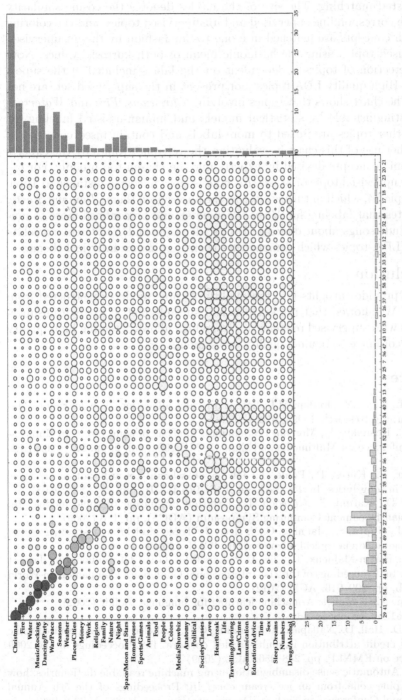

Fig. 2. Correspondence Chart between topics. The size of circles depicts cosine similarity between corresponding supervised and unsupervised topics. Bars on the sides of the graph show the kurtosis scores for their corresponding topics. High scores show that topics are aligned, low scores mean topics are not matched or junk in the case of LDA-topics. Circle coloring means a strong match between LDA and L-LDA topics, thus a useful unsupervised topic.

to display the automatically calculated kurtosis scores instead of likelihoods of human-assisted matching. The size of the circles denotes the cosine similarity between the corresponding supervised and unsupervised topics, and the coloring shows which concepts are matched in a one-to-one fashion by the unsupervised and supervised topics using the harmonic mean of both kurtosis' values. Note that the detection of topics is dependent on the labels included in the supervised data. High-quality LDA-topics, not present in the supervised set, are not detected. The chart shows that topics involving *Christmas, Fire* and *Water* are all very distinguishable by statistical models and human-assisted labeling, or resolved. Other topics are linked to more labels and contain fused concepts or junk. Another use of this chart is evaluating the reference topics by the experts of GOS. Some concepts devised by experts may be chosen too broadly. For example, the supervised topic of *Music/Rocking* is close in cosine-distance to Topic 6 and to Topic 54, which in turn is close to the supervised theme *Dancing/Party*. This indicates that labeling for *Music/Rocking* should be confined more to music and exclude songs about dancing. Topics like *Love* and *Heartbreak* correlate with many LDA-topics which demonstrate their dominance in lyrical themes.

4 Conclusion

This paper provides insights into the quality of topic models constructed from song lyrics. We showed that the kurtosis is a suitable metric to align unsupervised topics with supervised reference topics, which allows detecting high-quality topics in accordance to manual quality assessments.

References

1. Blei, D.M., Ng, A.Y., Jordan, M.I.: Latent dirichlet allocation. The Journal of Machine Learning Research 3, 993–1022 (2003)
2. Logan, B., Kositsky, A., Moreno, P.: Semantic analysis of song lyrics. In: International Conference on Multimedia and Expo, ICME 2004, vol. 2, pp. 827–830. IEEE (2004)
3. Kleedorfer, F., Knees, P., Pohle, T.: Oh oh oh whoah! towards automatic topic detection in song lyrics. In: Proceedings of the 9th ISMIR, pp. 287–292 (2008)
4. Newman, D., Karimi, S., Cavedon, L.: External evaluation of topic models. In: Australasian Document Computing Symposium (ADCS), pp. 1–8 (2009)
5. Chuang, J., Gupta, S., Manning, C.D., Heer, J.: Topic model diagnostics: Assessing domain relevance via topical alignment. In: ICML (2013)
6. McFee, B., Bertin-Mahieux, T., Ellis, D.P., Lanckriet, G.R.: The million song dataset challenge. In: Proceedings of the 21st International Conference Companion on World Wide Web, pp. 909–916. ACM (2012)
7. McCallum, A.K.: Mallet: A machine learning for language toolkit (2002), http://mallet.cs.umass.edu
8. Ramage, D., Hall, D., Nallapati, R., Manning, C.D.: Labeled lda: A supervised topic model for credit attribution in multi-labeled corpora. In: Proceedings of the 2009 Conference on EMNLP, pp. 248–256. ACL (2009)
9. Lesk, M.: Automatic sense disambiguation using machine readable dictionaries: how to tell a pine cone from an ice cream cone. In: Proceedings of the 5th Annual International Conference on Systems Documentation, pp. 24–26. ACM (1986)

An Information Retrieval-Based Approach to Determining Contextual Opinion Polarity of Words

Olga Vechtomova[1], Kaheer Suleman[2], and Jack Thomas[2]

[1] Department of Management Sciences, University of Waterloo, Waterloo, ON, Canada
ovechtom@uwaterloo.ca
[2] Cheriton School of Computer Science, University of Waterloo, Waterloo, ON, Canada
{ksuleman,j26thoma}@uwaterloo.ca

Abstract. The paper presents a novel method for determining contextual polarity of ambiguous opinion words. The task of categorizing polarity of opinion words is cast as an information retrieval problem. The advantage of the approach is that it does not rely on hand-crafted rules and opinion lexicons. Evaluation on a set of polarity-ambiguous adjectives as well as a set of both ambiguous and unambiguous adjectives shows improvements compared to a context-independent method.

1 Introduction

Opinion detection has been an active research area in recent years. There exist a large number of approaches that attempt to identify a static sentiment polarity of words (e.g. [1-3]). It has, however, been recognized that while certain words have an unambiguous polarity, e.g. "amazing", "distasteful", others change their polarity depending on the context, e.g., "pizza was cold" vs. "beer was cold". A number of methods have been proposed to address this problem [4-7]. In [4] a supervised method was proposed to determine contextual polarity of phrases. In [5] a number of rules were used, such as conjunctions and disjunctions, manually created syntactic dependency rule templates, automatically derived morphological relationships and synonymy/antonymy relationships from WordNet. Another approach [6] used an existing opinion lexicon and a number of rules (e.g. negation rule, intra- and inter- sentence conjunction rules, synonym and antonym rules). An approach in [7] used conjunctions of ambiguous adjectives with unambiguous ones with known polarity from an opinion lexicon, and also extracted groups of related target words from Wikipedia. All of the above methods rely on rules and/or existing resources, such as WordNet or opinion lexicons. In this paper we propose an extensible framework for context-dependent polarity determination. To our knowledge this is the first method for this task, which does not rely on hand-crafted or automatically generated rules and does not utilize any pre-existing opinion vocabulary. The task of categorizing an opinion word instance into positive or negative is cast as an information retrieval problem. We build one vector of all contexts of the word a in the positive document set (e.g. reviews with high ratings) and another vector – of its contexts in the negative set. These vectors are treated as

M. de Rijke et al. (Eds.): ECIR 2014, LNCS 8416, pp. 553–559, 2014.

documents. We then build a context vector for the specific instance of a that we want to categorize, which is treated as the query. An IR model is then applied to calculate the query's similarity to each of the two "documents". As contexts we use dependency triples containing a. The approach utilizes automatically extracted lexico-syntactic contexts of the word's occurrences and their frequencies without the need to build hand-crafted rules or patterns or to use pre-existing opinion lexicons. For instance, the method in [6] has an explicit rule for conjunctives. In contrast, in our approach any conjunctives (e.g. "nice and cold"), that a word co-occurs with, say, in positive reviews, are automatically added with all other dependency triples to the positive vector of the word. In this way, the method captures a wide range of lexico-syntactic polarity clues, such as adverbial modifiers (e.g., "barely"), nouns that are targets of the opinion words, and miscellaneous syntactic constructs, such as "but" and negations. The proposed framework is extensible in a number of ways: features could be expanded (e.g., by adding other dependency triples in the sentence), filtered (e.g. by dependency relation type), or grouped by similarity. The method is evaluated on a set of adjectives with ambiguous polarity, and on another set of both ambiguous and unambiguous adjectives.

2   Methodology

Most of the product and business review sites let users assign a numerical rating representing their level of satisfaction with a product or business. In our experiments, we used a dataset of restaurant reviews, where each review has an associated rating on a scale from 1 to 10. All reviews with a rating of 10 were used as a positive training set, and all reviews with ratings 1 and 2 as negative. During the preparatory stage two vectors of context features are created for each adjective a. One vector $posV$ is built based on the adjective's occurrences in the positive set, and the second vector $negV$ is built based on its occurrences in the negative set. At the next stage, polarity of an adjective occurrence a in a previously unseen document d is determined as follows: vector $evalV$ is built for this adjective based on its context within its sentence of occurrence in document d only. Then, a pairwise similarity of $EvalV$ to the vector of the same adjective in the positive set (vector $posV$) and in the negative set (vector $negV$) is calculated.

2.1   Context Feature Vector Construction

The following steps are performed on each of the two training sets: positive and negative. Each document in a training set is processed by using a dependency parser in the Stanford CoreNLP package. In each document, we first locate all nouns that appear as governing words in at least one dependency relation. At this stage in the algorithm, we can optionally apply a filter to process only those nouns that belong to a specific list, e.g. words denoting a specific category of review aspects (e.g. food in restaurant reviews). In our experiments we filtered the list by 456 food names which were created by using a clustering method from another project in progress. Then, for each governing word, its dependency triples with adjectives are extracted, where the dependency relation is either an adjectival modifier (amod), nominal subject (nsubj) or relative clause modifier (rcmod). An example of a dependency triple is *nsubj(pizza, hot)*, where "pizza" is a governor, while "hot" is a dependent word. For each adjective instance we extract all triples, in which they occur as dependent words. If one of the triples represents negation dependency relation (neg), we record that the adjective is negated. For each adjective occurrence, the following information is recorded:

- negation (1 – adjective is negated; 0 – adjective is not negated);
- dependency relation of adjective with its governing noun (amod, nsubj or rcmod);
- adjective lemma (output by Stanford CoreNLP).

These three pieces of information form adjective pattern (*AdjP*), e.g., "negation=0; amod; better". A context feature vector is built for these patterns. The reason for building vectors for lexico-syntactic adjective patterns as opposed to just adjective lemmas, is that, firstly, we want to differentiate between the negated and non-negated instances, and, secondly, between various syntactic usages of the adjective. For instance, adjectives occurring in a post-modifier position (e.g., in "nsubj" relationship to the noun) tend to be used more in evaluative manner compared to those used in pre-modifier position (c.f: "tea was cold" and "cold tea"). While "cold tea" usually refers to a type of drink, "tea was cold" has an evaluative connotation. Also, the types of dependency relations they occur in can be different, e.g. adjectives in post-modifier position occur more with certain adverbial modifiers, which can give clues as to the adjective's polarity, such as "barely", "too", "overly", "hardly".

Next, for each adjective instance, represented as "negation; dependency relation; lemma" adjective pattern, we extract all dependency relations that contain it. Each of them is transformed into a context feature f of the form: "lemma; Part Of Speech (POS); dependency relation". For instance, if adjective "hot" occurs in dependency triple nsubj(tea, hot), the following feature is created to represent "tea" and its syntactic role with respect to the adjective: "tea, NN, nsubj". For each feature we record its frequency of co-occurrence with the adjective pattern (used as *TF* in Eq. 1). More formally, the algorithm is described below:

Table 1. Algorithm 1: Construction of feature vectors for adjective syntactic patterns

1: For each document $d \in T$
2:    For each valid noun n
3:       For each adjective a, dependent of n
4:          If *DepRel(n,a)* $\in$ {amod, rcmod, nsubj}
5:             If any *DepRel(a,w)* = *"neg"*
6:                *negation(a)* = 1
7:             Else
8:                *negation(a)* = 0
9:             End If
10:            Create adjective pattern *AdjP* as *"negation(a); DepRel(n,a); lemma(a)"*
11:            For each *DepRel(a,w)*
12:               Create feature f as *"lemma(w); POS(w); DepRel(a,w)"*
13:               Add f to V_{AdjP}; Increment frequency of $f \in V_{AdjP}$

Where: *valid noun n* – noun that occurs in the list of nouns belonging to a specific category of review aspects (optional step); *T* – training document set, either with positive or negative review ratings (the algorithm is run separately for positive and negative document sets); *DepRel(n,a)* – dependency relation between noun *n* and adjective *a*; *DepRel(a,w)* – dependency relation between adjective *a* as either governor or dependent and any other word *w*; *POS(w)* – part of speech of *w*. V_{AdjP} – feature vector for adjective pattern *AdjP*.

Algorithm 1 is used to generate vectors for all *AdjP* patterns extracted from the positive set and, separately, from the negative set during the preparatory stage. The same algorithm is also used at the stage of determining the polarity of a specific adjective occurrence. At that stage, only the sentence containing this adjective occurrence is used to generate the vector $Eval_{AdjP}$. The pairwise similarity of $Eval_{AdjP}$ with $posV_{AdjP}$ and $Eval_{AdjP}$ with $negV_{AdjP}$ is computed. If similarity with $posV_{AdjP}$ is higher, it is categorized as positive, and as negative if similarity with $negV_{AdjP}$ is higher.

2.2   Computing Similarity between Vectors

We view the problem of computing similarity between vectors as a document retrieval problem. The vector ($EvalV_{AdjP}$) of a specific adjective occurrence *AdjP*, whose polarity we want to determine, is treated as the query, while the two vectors of *AdjP* ($posV_{AdjP}$ and $negV_{AdjP}$) created from the positive and negative training sets respectively, are treated as documents. For the purpose of computing similarity we use BM25 Query Adjusted Combined Weight (QACW) document retrieval function [8]. In [9] it was proposed to use it as a term-term similarity function. The $EvalV_{AdjP}$ is treated as the query, while $posV_{AdjP}$ and $negV_{AdjP}$ as documents (V_{AdjP} in Eq. 1)

$$Sim(EvalV_{AdjP}, V_{AdjP}) = \sum_{f=1}^{F} \frac{TF(k_1 + 1)}{K + TF} \times QTF \times IDF_f \qquad (1)$$

Where: *F* – the number of features that $EvalV_{AdjP}$ and V_{AdjP} have in common; *TF* – frequency of feature *f* in V_{AdjP}; *QTF* – frequency of feature *f* in $EvalV_{AdjP}$; $K = k_1 \times ((1-b)+b \times DL/AVDL)$; k_1 – feature frequency normalization factor; *b* – V_{AdjP} length normalization factor; *DL* – number of features in V_{AdjP}; *AVDL* – average number of features in the vectors *V* for all *AdjP* patterns in the training set (positive or negative). The *b* and k_1 parameters were set to 0.9 and 1.6 respectively, as these showed best performance in computing term-term similarity in [9]. The *IDF* (Inverse Document Frequency) of the feature *f* is calculated as $IDF_f = \log(N/n_f)$, where, n_f – number of vectors *V* in the training set (positive or negative) containing feature *f*; *N* – total number of vectors *V* in the training set. A polarity score of *AdjP* is then calculated for both positive and negative sets as follows:

$$PolarityScore = \alpha \times Sim(EvalV_{AdjP}, V_{AdjP}) + (1 - \alpha) \times P(AdjP) \qquad (2)$$

Where: $P(AdjP)$ is calculated as number of occurrences of $AdjP$ in the set (positive or negative) / total number of occurrences of all $AdjP$ patterns in this set; the best result for α was 0.5. If *PolarityScore* is higher for the positive set, the polarity is positive, and if lower – negative.

3   Evaluation

For evaluation we used a corpus of 157,865 restaurant reviews from one of the major business review websites, provided to us by a partner organization. The collection contains reviews for 32,782 restaurants in the U.S. The average number of words per review is 64.7. All reviews (63,519) with the rating of 10 were used as positive training set, and all reviews with the ratings of 1 or 2 (18,713) as negative.

3.1   Evaluation on Ambiguous Adjectives

For this evaluation we specifically chose four adjectives (cold, warm, hot and soft) that can have a positive or negative meaning depending on the context. From reviews with ratings 3-9, we extracted all dependency triples, containing one of these adjectives in "nsubj" dependency relation with a noun representing a food name. The reason why we used "nsubj" is that post-modifier adjectives are more likely to be opinionated than pre-modifiers (i.e. related with "amod"). To select food nouns only, we applied a filter of 456 food names, created by a clustering method from another project in progress. For this experiment, we focused only on those cases that are not negated, i.e. do not occur in a dependency triple with "neg" relation. Two annotators read 888 original sentences containing these adjectives, and judged the adjective occurrences as "positive", "negative" or "objective" when they refer to food, and as "non-food modifier" for cases not referring to food. The inter-annotator agreement (Cohen's Kappa) is 0.81. There were only 2 objective cases agreed upon by the annotators, which are not included in the evaluation. The evaluation set consists of 519 positive and negative cases agreed upon by the two annotators. The cases are in the following format: "document ID; noun token; negation; dependency relation; adjective lemma; polarity". The number of positive/negative cases for "cold" is 34/180, for "warm": 29/25, for "hot": 196/10, and for "soft": 31/14.

As the baseline a context-independent method was used based on the Kullback-Leibler Divergence (KLD). KLD is used widely in IR, e.g. as a term selection measure for query expansion [10] and as a measure for weighting subjective words [11]. Polarity for each $AdjP$ pattern is calculated as $P_{pos}(AdjP)*\log(P_{pos}(AdjP)/P_{neg}(AdjP))$. $P_{pos}(AdjP)$ is calculated as $F_{pos}(AdjP))/N$, where $F_{pos}(AdjP)$ is frequency of $AdjP$ in the positive set, N is the total number of occurrences of all $AdjP$ pattern in the positive set. $P_{neg}(AdjP)$ is calculated in the same way. Cases with $KLD>0$ are considered as positive, and with $KLD<0$ as negative. Table 2 shows Precision, Recall and F-measure for the context-based method (ContextSim) and KLD.

558   O. Vechtomova, K. Suleman, and J. Thomas

3.2   Evaluation on a Larger Set of Adjectives

A larger scale evaluation was done on 606 "nsubj" and "amod" adjective patterns (482 positive and 124 negative) from 600 restaurant reviews. The dataset contains 164 distinct adjectives. The results are presented in Table 3.

While the overall improvement (F-measure) is higher for ContextSim, the precision is somewhat lower than KLD. Since the method demonstrates a much better performance on ambiguous adjectives, it makes sense to apply it only to such adjectives. We need, therefore, a method for detecting unambiguous adjectives (e.g. *excellent*) with static polarity. This is left for future work.

Table 2. Results based on a set of ambiguous adjectives

Method	Precision	Recall	F-measure
ContextSim	0.9114	1	0.9536
KLD	0.8324	1	0.9085

Table 3. Results based on adjectives from 600 reviews

Method	Precision	Recall	F-measure
ContextSim	0.8874	0.967	0.9255
KLD	0.9185	0.9109	0.9147

4   Conclusion

The paper described a framework for determining contextual polarity of ambiguous adjectives. The advantage of the proposed approach is that it does not rely on hand-crafted rules of opinion lexicons. Performance on a number of ambiguous adjectives is promising compared to a context-independent method using KLD. The proposed framework is extensible in a number of ways: features could be expanded to include, for instance, other dependency triples in the sentence or document, or on the contrary, filtered by the dependency relation type. Currently, we are working on various extensions of this framework, in particular, feature grouping, and are performing a larger scale evaluation on different corpora.

References

1. Esuli, A., Sebastiani, F.: Determining Term Subjectivity and Term Orientation for Opinion Mining. In: Proc. of EACL (2006)
2. Hu, M., Liu, B.: Mining and summarizing customer reviews. In: Proc. of KDD (2004)
3. Hatzivassiloglou, V., McKeown, K.R.: Predicting the semantic orientation of adjectives. In: Proc. of ACL, pp. 174–181 (1997)
4. Wilson, T., Wiebe, J., Hoffman, P.: Recognizing Contextual Polarity in Phrase-Level Sentiment Analysis. In: Proc. of EMNLP (2005)

5. Popescu, A., Etzioni, O.: Extracting Product Features and Opinions from Reviews. In: Proc. of EMNLP (2005)
6. Ding, X., Liu, B., Yu, P.: A holistic lexicon-based approach to opinion mining. In: Proc. of WSDM (2008)
7. Fahrni, A., Klenner, M.: Old Wine or Warm Beer: Target-specific Sentiment Analysis of Adjectives. In: Proc. of the Symposium on Affective Language in Human and Machine, AISB 2008 Convention (2008)
8. Spärck Jones, K., Walker, S., Robertson, S.E.: A probabilistic model of information retrieval: Development and comparative experiments. Information Processing and Management 36(6), 779–808 (pt. 1); 809–840 (pt. 2) (2000)
9. Vechtomova, O., Robertson, S.E.: A Domain-Independent Approach to Finding Related Entities. Information Processing and Management 48(4), 654–670 (2012)
10. Carpineto, C., De Mori, R., Romano, G., Bigi, B.: An information-theoretic approach to automatic query expansion. ACM ToIS 19(1), 1–27 (2001)
11. Vechtomova, O.: Facet-based Opinion Retrieval from Blogs. Information Processing and Management 46(1), 71–88 (2010)

Query Term Suggestion in Academic Search

Suzan Verberne[1], Maya Sappelli[1,2], and Wessel Kraaij[1,2]

[1] Institute for Computing and Information Sciences,
Radboud University Nijmegen, Nijmegen, The Netherlands
[2] TNO, Delft, The Netherlands

Abstract. In this paper, we evaluate query term suggestion in the context of academic professional search. Our overall goal is to support scientists in their information seeking tasks. We set up an interactive search system in which terms are extracted from clicked documents and suggested to the user before every query specification step. We evaluated our method with the iSearch collection of academic information seeking behaviour and crowdsourced term relevance judgements. We found that query term suggestion can significantly improve recall in academic search.

Keywords: Professional search, user interaction, query term suggestion.

1 Introduction

Academic search is a form of professional search that is carried out by scientists. It has the following characteristics [6]: (1) it is interactive: multiple queries are needed to satisfy one information need; (2) it takes place in a specific domain and (3) it tends to be recall-oriented. In our project, we aim to support academic searchers in their information seeking tasks. In the current paper, we focus on support in the form of query term suggestion. Query formulation is an interactive process in which the user adapts the initial query after inspection of the result list. We focus on query *specification* by adding terms. We propose a set-up for academic information seeking in which the user gets suggestions for additional query terms (both single-word and multiword) after inspection of the result list for his previous query. For our experiments, we use the iSearch data collection [8]. This collection contains elaborate descriptions of real search tasks by academic researchers together with a domain-specific corpus and relevance judgements. Information on actual user interaction is not available. To compensate for this, we use a click model for simulating user interaction with the system. The main contribution of the current paper is that we show the potential value of query term suggestion in academic search.

2 Related Work

In previous work on query term suggestion, three different sources for query terms are used: search engine query logs, documents in the retrieval corpus or external knowledge sources. Query logs are especially useful when the queries of

M. de Rijke et al. (Eds.): ECIR 2014, LNCS 8416, pp. 560–566, 2014.

other users can be reused by the current user, for example because they occur in similar sessions [5]. For personalization purposes, the user's own previous queries are sometimes used as a source for query term recommendation, but this data is sparse and topic-dependent [2].

Documents in the corpus are an often-used source for query term suggestion if there are no relevant query logs available. The idea is to extract terms from the documents in the corpus that are most relevant to the user's current query. Relevance can either be defined by the search engine itself, using the top-n highest ranked documents ('pseudo-relevance feedback'), or by the user's clicks, extracting terms from the documents that are clicked by the user ('relevance feedback') [1]. For a topic domain with a controlled vocabulary, terms from a domain-specific thesaurus prove to be a good additional source for query term suggestion [3]. In all cases, the extracted terms are recommended to the user in a short suggestion list.

The current work is — to our knowledge — the first to evaluate query term suggestion for academic search. In large-scale academic search (such as Google Scholar or Microsoft Academic Search), query logs may be helpful for frequent queries, but for the long tail of low-frequency queries we need other sources of query terms. We experiment with clicked documents as source for query terms in a search session. We combine query terms into free text queries in a Language Modelling (LM) framework. The previous work that is most similar to the current work is that of Kim et al. (2011) [6], who evaluate boolean query term suggestion for patent retrieval and medical information search, tasks in which boolean operators are considered very important for reasons of reproducibility and full control. Like Kim et al., we apply user simulation to the evaluation of our approach, but instead of assuming that the user always selects the highest-ranked selected term, we use human judgments for term selection.

3   Methodology

Data Collection. The iSearch collection of academic information seeking behaviour [8] consists of 65 natural search tasks (topics) from 23 researchers and students from university physics departments. The topic owners were given a search task description form with five fields:

(a) What are you looking for? (information need)
(b) Why are you looking for this? (work task context)
(c) What is your background knowledge of this topic? (knowledge state)
(d) What should an ideal answer contain to solve your problem or task? (ideal answer)
(e) Which central search terms would you use to express your situation and information need? (search terms)

A collection of 18K book records, 144K full text articles and 291K metadata records from the physics field is distributed together with the topics. For each topic, 200 documents were manually assessed on their relevance for the information need using a 4-point scale.

For the purpose of evaluation and the simulation of term selection by independent human users, we set up a Human Intelligence Tasks (HIT) on Amazon Mechanical Turk to judge the relevance of automatically generated terms. The subjects were presented with the information need, work task context, background knowledge and ideal answer of a topic from the iSearch data, and a list of terms that were automatically extracted from clicked documents in inital system runs. The top-15 highest scoring terms (see the next section for a description of the term extraction and scoring method) were presented in alphabetical order to the participants and we asked them to select the relevant terms. The workers were told that their queries were reviewed by an expert before their task was approved and they would receive their payment in order to prevent spam submissions. Each topic was assessed by two workers in order to be able to calculate inter-rater agreement. 17% of the suggested terms were judged as relevant. The two workers agreed in their judgment for 78% of the terms. Cohen's κ for inter-rater agreement is 0.50, which implicates a moderate agreement. The workers were also asked to indicate their level of familiarity with the topic on a three-level scale. All workers indicated that they were unfamiliar with the topic. This implicates that we cannot claim that we have collected expert judgements, but the fact that they come from human workers with knowledge of the original user's information need and background knowledge, makes them more reliable than automatically estimated judgements.

Retrieval Process and Query Term Suggestion. We indexed the iSearch collection with the Indri search engine[1]. We used the Indri API to set up a query interface to the combined index of Metadata, Book and Article records. All characters that are not alphanumeric, no hyphen or whitespace[2] are removed from the query terms. Multiple query terms are concatenated and combined using the `combine` function in the Indri query language. For example, the two terms 'ZNO' and 'Transparent conductive oxides' together form the Indri query `#combine(zno transparent conductive oxides)`. As ranking model, we use the Indri LM with Dirichlet smoothing. Per query, we retrieve 100 results from the combined index.

The first user query is the first term from field e 'search terms'. A result list of 100 documents from Indri is presented to the user. In the simple baseline setting without query term suggestion, each follow-up query is the previous query expanded with the next term from the search terms field, until all the terms have been added. For example, the initial query 'zno' is first expanded to 'zno transparent conductive oxides' and then to 'zno transparent conductive oxides magnetron sputtering'[3]

---

[1] http://www.lemurproject.org/indri/
[2] Note that a term can consist of multiple words.
[3] Although the queries get longer and longer, no empty result sets will occur because in Indri not all search terms necessarily have to be present in the result.

Table 1. An example of the queries that the simulated user builds with and without term suggestion. In the first row, all consecutive terms that are added come from field e in the iSearch data; after 4 queries the user is out of terms. In the second row, the first three terms come from the list of query suggestions provided by the system. After these three, the user considers the suggested queries to be not relevant and he adds the remaining terms from field e.

setting	initial query	terms added in consecutive queries, comma separated (Recall for query after the addition of this term)
without term suggestion	zno (0.106)	transparent conductive oxides (0.126), magnetron sputtering (0.136), doping (0.136)
with term suggestion	zno (0.106)	ferromagnetic (0.131), doped (0.162), ferromagnetism (0.162), transparent conductive oxides (0.172), magnetron sputtering (0.177), doping (0.177)

In the experimental setting, the simulated user gets 10 suggestions for query terms to be added to the next query. These terms have automatically been extracted from the documents that the user clicked on in the current search session.[4] All n-grams with $n = 1, 2, 3$ in these documents were considered candidate terms. We scored them with Kullback-Leibler divergence for informativeness and phraseness (KLIP) [9] in order to rank the terms. For the informativeness component of KLIP, KL divergence is calculated between the probability distributions for a term in a foreground collection and in a background collection. In our case, the foreground collection is the collection of documents clicked by the user and as background collection we use the Corpus of Contemporary American English.[5] In a comparison with two other term scoring methods [10], KLIP was found to generate the most descriptive terms for a given document set according to the owner of the document set.

We use a click model to decide whether a user clicks on a document. The click model that we use is the perfect click model from [4][6]: it assumes that a user never clicks on an irrelevant document and the user does not stop inspection of the result list before he has seen all 100 results. When the user has finished inspection of the result list, the system presents a ranked list of terms extracted from the clicked documents. The user selects the highest ranked term that was judged as relevant by at least one human judge in the crowdsourcing task and that has not been added to the query before. If none of the terms is relevant, the user adds the next term from the search terms field to the query. With the additional term, the query is issued again. This process is repeated until the user is out of query terms. An example is shown in Table 1.

---

[4] In the case of metadata and book records we took the fields title and description to extract the terms from; in the case of articles in PDF, for which no metadata is available, we extracted the terms from the first 200 words of the document.

[5] We also experimented with the iSearch corpus as background collection but this gave significantly poorer results.

[6] We adapted the model from a 3-level relevance scale to a 4-level relevance scale: the probability that the user clicks on a document with relevance 0 is 0, with relevance 1 it is 0.33, with relevance 2 it is 0.67 and with relevance 3 it is 1.0.

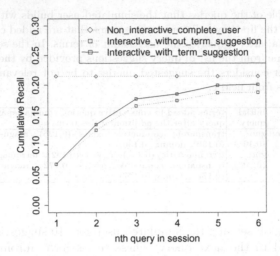

Fig. 1. Cumulative recall over session, averaged over all topics

4   Results

We evaluate our term suggestion in two ways: First, we evaluate the relevance of the suggested terms using the crowdsourced human judgments. The measures that we use are precision (what proportion of the terms that are suggested during the interactive retrieval process have been judged as relevant) and success rate (the proportion of term suggestion lists from which the user chose a suggested term instead of a term from the search terms field). Second, we compare the effectiveness of the retrieval process when query terms are suggested by the system to the effectiveness of the retrieval process when the search terms from the iSearch data are used for all query specifications. Since the search terms have been formulated by the owner of the information need, they can be considered a reference for the automatically extracted terms. We measure effectiveness in terms of cumulative recall (over the session).

We obtained a success rate for query term suggestion of 12.8%. The precision of the suggested query terms was 17.0% according to the MTurk workers. These results seem poor at first sight. The only way to judge their real value is to measure the effect of query term suggestion on the retrieval process. This is shown in Figure 1. We show the average results over the course of the session up until the sixth query, because very few topics have more than six queries. As a reference, we compare our results to the results for the *non-interactive complete user*: This user issues one 'complete' query per topic, consisting of all terms from the fields 'information need', 'work task context' and 'search terms' from the iSearch data (average query length: 106 words) [7].

According to a two-sided paired t-test ($n = 66$ topics), the difference between the best-scoring interactive setting and the non-interactive complete user is highly significant with $P < 0.00001$, meaning that the non-interactive complete user achieves significantly better recall than both the interactive users.

The difference in the final recall reached with and without term suggestion is significant on the 0.05 level ($P = 0.019$). The results show that the use of query term suggestion can lead to a higher recall compared to using the search terms that were formulated by the topic owner without the help from query term suggestion. This is surprising given the finding that only in 12.8% of the query specifications, the user picked a term from the suggestion list.

5   Conclusion and Future Work

We showed that query term suggestion with terms extracted from clicked documents can have a positive effect on the effectiveness of interactive academic search. Our next steps include improvement of the term extraction algorithm, the application of other (non-perfect) click models, comparing our term suggestion method to pseudo-relevance feedback (PRF), and the evaluation of our query term suggestion method on real users.

Acknowledgements. This publication was supported by the Dutch national program COMMIT (project P7 SWELL).

References

1. Belkin, N.J., Cool, C., Kelly, D., Lin, S.J., Park, S., Perez-Carballo, J., Sikora, C.: Iterative exploration, design and evaluation of support for query reformulation in interactive information retrieval. Information Processing & Management 37(3), 403–434 (2001)
2. Feild, H., Allan, J.: Task-aware query recommendation. In: Proceedings of the 36th International ACM SIGIR Conference on Research and Development in Information Retrieval, SIGIR 2013, pp. 83–92. ACM, New York (2013)
3. Hienert, D., Schaer, P., Schaible, J., Mayr, P.: A novel combined term suggestion service for domain-specific digital libraries. In: Gradmann, S., Borri, F., Meghini, C., Schuldt, H. (eds.) TPDL 2011. LNCS, vol. 6966, pp. 192–203. Springer, Heidelberg (2011)
4. Hofmann, K., Schuth, A., Whiteson, S., de Rijke, M.: Reusing historical interaction data for faster online learning to rank for ir. In: Proceedings of the Sixth ACM International Conference on Web Search and Data Mining, WSDM 2013, pp. 183–192. ACM, New York (2013)
5. Huang, C.K., Chien, L.F., Oyang, Y.J.: Relevant term suggestion in interactive web search based on contextual information in query session logs. Journal of the American Society for Information Science and Technology 54(7), 638–649 (2003)
6. Kim, Y., Seo, J., Croft, W.B.: Automatic boolean query suggestion for professional search. In: Proceedings of the 34th International ACM SIGIR Conference on Research and Development in Information Retrieval, SIGIR 2011, pp. 825–834. ACM, New York (2011)
7. Lioma, C., Larsen, B., Ingwersen, P.: Preliminary experiments using subjective logic for the polyrepresentation of information needs. In: Proceedings of the 4th Information Interaction in Context Symposium, pp. 174–183. ACM (2012)

8. Lykke, M., Larsen, B., Lund, H., Ingwersen, P.: Developing a test collection for the evaluation of integrated search. In: Gurrin, C., He, Y., Kazai, G., Kruschwitz, U., Little, S., Roelleke, T., Rüger, S., van Rijsbergen, K. (eds.) ECIR 2010. LNCS, vol. 5993, pp. 627–630. Springer, Heidelberg (2010)
9. Tomokiyo, T., Hurst, M.: A language model approach to keyphrase extraction. In: Proceedings of the ACL 2003 Workshop on Multiword Expressions: Analysis, Acquisition and Treatment, vol. 18, pp. 33–40. Association for Computational Linguistics (2003)
10. Verberne, S., Sappelli, M., Kraaij, W.: Term extraction for user profiling: evaluation by the user. In: Late-Breaking Results, Project Papers and Workshop Proceedings of the 21st Conference on User Modeling, Adaptation, and Personalization, CEUR, pp. 51–57 (2013)

The Impact of Future Term Statistics
in Real-Time Tweet Search

Yulu Wang[1,2] and Jimmy Lin[1,2,3]

[1] Dept. of Computer Science,
[2] Institute for Advanced Computer Studies, [3] The iSchool
University of Maryland, College Park, USA
ylwang@cs.umd.edu, jimmylin@umd.edu

Abstract. In the real-time tweet search task operationalized in the TREC Microblog evaluations, a topic consists of a query Q and a time t, modeling the task where the user wishes to see the most recent but relevant tweets that address the information need. To simulate the real-time aspect of the task in an evaluation setting, many systems search over the entire collection and then discard results that occur after the query time. This approach, while computationally efficient, "cheats" in that it takes advantage of term statistics from documents not available at query time (i.e., future information). We show, however, that such results are nearly identical to a "gold standard" method that builds a separate index for each topic containing only those documents that occur before the query time. The implications of this finding on evaluation, system design, and user task models are discussed.

1 Introduction

In this paper we tackle a simple but substantive question: do term statistics from future documents impact retrieval effectiveness in the real-time tweet search task? The answer to this question holds important implications for the design of retrieval systems and how they are evaluated. We explore this question in the context of the TREC 2011 and 2012 Microblog evaluations and conclude that simple term statistics from future documents (i.e., tweets posted after the query time) do not have a significant impact on retrieval effectiveness.

We adopt the definition of the real-time tweet search task operationalized in TREC [1]. A topic consists of a query Q and a time t, modeling the task where the user wishes to see the most recent but relevant tweets that address the information need. Operationally, the "real-time" nature is *simulated* in the evaluation, as participants are able to acquire the entire collection of tweets all at once. One common approach is to treat the problem as standard *ad hoc* retrieval over the *entire* collection, and then filter any hits that occur after the query time. This widely-adopted approach, however, violates the real-time restriction, because by performing retrieval over the entire collection, the model gains access to future information—for example, term statistics from documents that were created after the query time. To meet the real-time criterion, participants must

M. de Rijke et al. (Eds.): ECIR 2014, LNCS 8416, pp. 567–572, 2014.
© Springer International Publishing Switzerland 2014

ensure that all features used in the ranking model are available at the time of the query. The easiest way to accomplish this is to build a separate inverted index for each topic, and in each index ingest only those tweets that appear before the query time. Needless to say, this "gold standard' approach is awkward, time consuming, and impractical for large collections.

Our study asks a simple question: does it matter? If we compare the "search the entire collection and filter" approach vs. the "one index per topic" approach, are the results different? In other words, does information about future documents such as term statistics affect retrieval effectiveness? We experimentally compare these two approaches and conclude that the answer is *no*—results from both are nearly identical. This finding has implications for system and evaluation design: we can adopt the much simpler approach without "cheating".

2   Methodology

Our study used test collections created from the recent Microblog tracks at TREC [1,2]. The 2011 and 2012 evaluations used the Tweets2011 collection, which consists of, after some spam removal, an approximately 1% sample of tweets from January 23, 2011 to February 7, 2011 (inclusive), totaling about 16 million tweets. Major events that took place within this time frame include the massive democracy demonstrations in Egypt as well as the Super Bowl in the United States. There are 49 topics for TREC 2011[1] and 60 topics for TREC 2012. Each topic consists of a query and an associated timestamp, which indicates when the query was issued. Using a standard pooling strategy, NIST assessors evaluated a total of 114K tweets and assigned one of four judgments to each: spam, not relevant, relevant, and highly-relevant. For the purpose of our experiments, we considered both relevant and highly-relevant tweets "relevant".

Our experiment compared two conditions: In the "complete index" condition, we built a standard inverted index over all 16 million tweets. In the "per-topic index" condition, we built 108 separate indexes, one for each topic. Each index contains only tweets that were created before the query time, making it impossible to inadvertently use future information not available at query time. For all experiments we used the open-source tools that were released as part of the TREC 2013 Microblog evaluation,[2] which are based on the Lucene search engine. For retrieval we used the query-likelihood implementation in Lucene, which serves as the baseline for the 2013 evaluation.

In the complete index condition, the query associated with each topic was treated as a bag of words and run over the index for the entire collection; tweets after the query time were then discarded. In the per-topic index condition, each query was run over its dedicated index. The top 1000 hits in both cases were evaluated. We compared mean average precision (MAP) and precision at rank cutoff 30 (P30), which were the two metrics used in the TREC evaluations.

---

[1] One topic had no relevant documents and was discarded from our analyses.
[2] http://twittertools.cc/

Table 1. Comparing indexing approaches on TREC 2011 and 2012 topics

Condition	TREC 2011		TREC 2012	
	MAP	P30	MAP	P30
Complete index	0.3042	0.3476	0.1818	0.2932
One index per topic	0.3062	0.3497	0.1816	0.2932

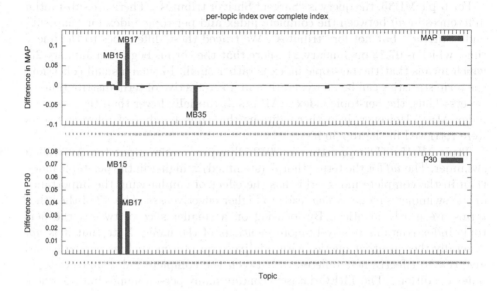

Fig. 1. Per-topic absolute differences in MAP (top) and P30 (bottom) for TREC topics. Bars above the origin denote cases where the per-topic index is more effective.

3 Experimental Results

Our evaluation results are shown in Table 1, comparing the "complete index" approach vs. the "one index per topic" approach. Numbers for the TREC 2011 and 2012 topics are shown separately. Per-topic absolute differences between the two experimental conditions are presented in Figure 1. The x-axis shows each topic, with sets of bars denoting absolute differences in MAP and P30. Across the 2011 and 2012 test collections, if we measure by MAP, the effectiveness of both approaches is identical for 77 topics; the effectiveness is better using the "one index per topic" approach for 14 topics and the effectiveness of the "complete index" approach is better for 17 topics. If we measure by P30, the effectiveness of the two approaches is identical for 106 topics; for the remaining two topics the "one index per topic" approach is better. The overall differences between the two approaches are not statistically significant.

For each topic, we examined the inverse document frequency (*idf*) of the query terms in both the complete index and per-topic index.[3] We found that the *idf* in both conditions were nearly identical across all query terms—this is not surprising considering that *idf* is on a log scale, and it takes substantial variations in document frequencies to have a noticeable affect on the value. However, Figure 1 shows that there is a large difference in effectiveness for a few topics: MB15, MB17, and MB35. We explore these topics below.

For topic MB35, the query is "Sargent Shriver tributes". There are substantial differences in *idf* between the complete index and per-topic index for "Sargent" and "Shriver", but not for "tributes". We traced these differences to the query time, which is 07:18 on January 24—note that the corpus begins on January 23, which means that the per-topic index is rather small. In such a small collection, the term statistics can be idiosyncratic and not reflective of term distributions in tweets. Thus, the per-topic index MAP is substantially lower than the complete index MAP. However, in both conditions the same number of relevant tweets were retrieved in the top 30 hits.

Topic MB15, "Thorpe return in 2012 Olympics", asks about the Australian swimmer. The *idf* for the term "thorp" (stemmed) is higher in the per-topic index than in the complete index—this has the effect of emphasizing the importance of the swimmer's name, which leads to higher effectiveness. The *idf* of the other terms are nearly identical. By focusing on a smaller slice of tweets, the per-topic index contains relatively more mentions of the name. Note that it is *not* generally the case that queries with relatively rare terms such as person names give rise to effectiveness differences between the complete index and per-topic index conditions. The TREC datasets contain many person names but no others exhibit the effectiveness difference we observe in MB15.

Topic MB17, "White Stripes breakup", asks about The White Stripes, an American rock duo. The term "stripe" has a higher *idf* in the per-topic index than the complete index; the other terms have nearly identical *idf* values. Since the name of the group consists of common words in English, the effectiveness difference appears to be idiosyncratic.

It is well known that term occurrences can change rapidly on Twitter—this is the basis of Twitter trends. However, most rapid increases in term occurrences are relatively short lived and correspond to "what's hot" at the moment (e.g., a breaking news story, a TV show, a sports contest, etc.). For the most part, the global Twitter conversation shifts to some other topic within a short span of time [4]. These bursty behaviors do not appear to contain sufficient "mass" to substantially impact term statistics such as *idf*, which has built in "damping" since it is on a log scale. Furthermore, changes in *idf* values alone are not sufficient to yield different rank orders in retrieved hits—if all the query terms become more frequent to the same extent, the document scores would change but not

---

[3] We are aware that collection frequency, rather than document frequency, is used in query likelihood. However, i) *idf* is a general concept that applies broadly to other scoring models, and ii) previous work has confirmed that *df* and *cf* are nearly identical for Twitter search since terms almost always have $tf = 1$ in tweets [3].

the ranking. Since user queries often ask about a coherent concept or terms that are otherwise correlated, changes in term statistics are often correlated as well. To yield a different rank order of results, two conditions must be met: i) the term statistics must change enough to have an impact on a log scale, and ii) the relationship between the term statistics must be different. We have empirically shown that these conditions are rarely met in practice, and thus results from the complete index and per-topic indexes are mostly indistinguishable.

4   Discussion

One potential objection to these results is that we use a simple baseline query-likelihood model for comparing the alternative approaches. We justify this setup by considering modern search architectures, which typically break ranking into three stages: candidate generation, feature extraction, and document reranking. There are two main reasons for this multi-stage design. First, there is a general consensus that learning to rank provides the best solution to document ranking [5]. As it is difficult to apply machine-learned models over the *entire* collection, in practice a candidate list of potentially-relevant documents is first generated. Thus, learning to rank is actually a *reranking* problem (hence the first and third stages). Second, separating candidate generation from feature extraction has the advantage of providing better control over cost/quality tradeoffs. In this architecture, our experiments consider the candidate generation stage. Additional work has shown that end-to-end retrieval effectiveness is insensitive to the candidate generation algorithm [6,3], which means that our experiments using simple query-likelihood accurately reflect real-world conditions.

Examples of this search architecture abound in industry [7] and academia [8,6]. In fact, the TREC 2013 Microblog evaluation is exactly set up along these lines: participants do not have access to the raw collection—instead, they must complete the task via a search API that returns candidate results. Note that in the feature extraction and reranking stages, features manipulated by a system are properties of the candidate documents (which are guaranteed to occur before the query time by filtering). Thus, it is more difficult to inadvertently include future information not available at query time.

Why is our experimental finding important? These results show that for evaluation purposes, we can deploy a *single* complete index to *simulate* the real-time search task, provided that we filter results that occur after the query time. It is not necessary to build a separate index for each topic, which is cumbersome, time-intensive, space-wasting, and simply impractical for large collections.

Finally, our results hold implications for user task models in searching tweet collections. Consider a scenario where a journalist is investigating a sports scandal that has been brewing for the past several weeks. She just got news of a breaking development, and turns to searching tweets to find out more details. Since this particular news story has been developing for several weeks, any keyword search involving the athlete's name might bring up results from many different points in time. It would be desirable if the journalist could specify that she is only interested in the most recent tweets. This is the real-time search scenario we have been considering. However, consider another scenario in which a journalist is searching an archive of tweets as part of a retrospective piece on the impact of social media on the course of the Egyptian revolution. Let's say she is particularly interested in activists using Twitter for "on the ground" reporting purposes. In this case, the journalist has a concrete idea when the relevant tweets should occur (e.g., when protesters were gathered in Tahrir Square) and would like to include this external knowledge as a query time.

These are two distinct task models and they roughly correspond to the per-topic index and complete index conditions, respectively. Our results suggest that a single backend might be sufficient for developing systems that are specialized to either task. This does not necessarily mean that we can use the same algorithms—simply that they might be able to share the same infrastructure, thus simplifying development.

5 Conclusions

We examine a question that arose from recent TREC Microblog evaluations—whether the computational expedient of searching the entire collection with post-hoc filtering unfairly takes advantage of future information. The answer is *no*, which validates the empirical approach that many researchers had been taking all along. Now we affirm that they really weren't "cheating". Whew!

Acknowledgments. This work has been supported by NSF under awards IIS-1144034 and IIS-1218043. Any opinions, findings, or conclusions are the authors' and do not necessarily reflect those of the sponsor. The second author is grateful to Esther for her loving support and dedicates this work to Joshua and Jacob.

References

1. Ounis, I., Macdonald, C., Lin, J., Soboroff, I.: Overview of the TREC-2011 Microblog Track. In: TREC (2011)
2. Soboroff, I., Ounis, I., Macdonald, C., Lin, J.: Overview of the TREC-2012 Microblog Track. In: TREC (2012)
3. Asadi, N., Lin, J.: Fast candidate generation for real-time tweet search with Bloom filter chains. ACM Transactions on Information Systems 31(3), article 13 (2013)
4. Lin, J., Mishne, G.: A study of "churn" in tweets and real-time search queries. In: ICWSM, pp. 503–506 (2012)
5. Li, H.: Learning to Rank for Information Retrieval and Natural Language Processing. Morgan & Claypool Publishers (2011)
6. Asadi, N., Lin, J.: Effectiveness/efficiency tradeoffs for candidate generation in multi-stage retrieval architectures. In: SIGIR, pp. 997–1000 (2013)
7. Cambazoglu, B.B., Zaragoza, H., Chapelle, O., Chen, J., Liao, C., Zheng, Z., Degenhardt, J.: Early exit optimizations for additive machine learned ranking systems. In: WSDM, pp. 411–420 (2010)
8. Macdonald, C., Santos, R.L., Ounis, I.: The whens and hows of learning to rank for web search. Information Retrieval 16(5), 584–628 (2013)

Exploring Adaptive Window Sizes
for Entity Retrieval

Fawaz Alarfaj, Udo Kruschwitz, and Chris Fox

School of Computer Science and Electronic Engineering
University of Essex Colchester, CO4 3SQ, UK
{falarf,udo,foxcj}@essex.ac.uk

Abstract. With the continuous attention of modern search engines to
retrieve entities and not just documents for any given query, we introduce
a new method for enhancing the entity-ranking task. An entity-ranking
task is concerned with retrieving a ranked list of entities as a response to
a specific query. Some successful models used the idea of association discovery in a window of text, rather than in the whole document. However,
these studies considered only fixed window sizes. This work proposes a
way of generating an adaptive window size for each document by utilising
some of the document features. These features include document length,
average sentence length, number of entities in the document, and the
readability index. Experimental results show a positive effect once taking these document features into consideration when determining window
size.

1 Introduction

In an organisational setting, search engines have become mandatory for aiding
knowledge workers with their day-to-day information needs. Traditionally, information retrieval systems function by returning a list of documents in response
to the user's query, although, the needed information may not be necessarily in
the form of documents. In fact, users more often search for specific things or
entities which include people, organisations, or products.

One special type of entity-search is concerned with finding people who have
specific knowledge; this type of entity-search is called expert-finding, i.e. identifying experts who have the relevant skills and knowledge on a given topic [4].

With a search topic input, state-of-the-art expert-finding systems will measure
the knowledge of any candidate expert using the content of the highest ranking documents by highlighting associations based on co-occurrences between
the search topic and the candidate evidence [2]. Evidence of expertise could be
considered to be highlighted with search terms, furthermore the number and frequency is used to ascertain the likelihood of an individual being considered an
expert. There are two main assumptions, firstly, the more a candidate is located
within a document including terms of description the more likely they are to
be an expert on the subject and secondly, a stronger association is seen when
the identifiers are closer to the search terms. With these assumptions in mind,

M. de Rijke et al. (Eds.): ECIR 2014, LNCS 8416, pp. 573–578, 2014.
© Springer International Publishing Switzerland 2014

some research has used fixed-size windows to measure the proximity between candidate identifiers and search terms. Zhu *et al.* tested 31 window sizes on the W3C collection. They found the best window size to be around 200 words. According to Zhu *et al.*, small window sizes could lead to high precision but low recall. Conversely, large window sizes lead to high recall but low precision [7]. Therefore, other studies consider multiple levels of associations in documents by combining multiple fixed window sizes [3].

In this paper, we consider the idea of an *adaptive* window size, where the size of the window is a function of various document features. We argue that, in general, each document has distinct features differing from other documents in the collection. The proposed idea is to use these features to set the window size in order to improve the overall ranking function, while many document features could be examined. The focus in this work is aimed at four main features: document length; candidate frequency (i.e., number of candidates that appear in a document); average sentence length; and readability index. To the best of our knowledge, no existing work has dealt with using the document features to determine the optimal window size for the proximity function, apart from our earlier work [1]. It is important to note that the adaptive window size approach could be applied to any proximity search, in particular for an entity-oriented search that generalises expert search. The study is performed in the expert-search domain due to the availability of expert-search benchmarks. The main research question considered is whether an adaptive window size leads to improvements over fixed window size methods.

2   Adaptive Window Size for Proximity Ranking

The window size for the proximity function will be determined for each document based on the following features. **Document Length**: according to Miao *et. al.* [5], in large documents, it is more likely to find more occurrences of a query topic. It is also more likely to have irrelevant words (noise) in such documents. Thus, in order to minimise the negative influence of noise, the window size should be relatively smaller as the document gets bigger. **Candidate Frequency**: refers to the number of candidates found in a document. When a document has more occurrences of candidates' evidence, the window size should be relatively larger to accommodate more occurrences. **Average Sentence Length**: the window size is adjusted in proportion to the average sentence length (in tokens) in the document. **Readability Index**[1]: the window size is adjusted using the readability index where the window gets bigger whenever the index gets smaller. These features are combined in the following equation:

$$WindowSize = \frac{\sigma}{4} * (\log(\frac{1}{DocLength}) * \beta_1 + CanFreq * \beta_2$$
$$+ AvgSentSize * \beta_3 + ReadabilityIndex * \beta_4) \quad (1)$$

---

[1] FleschKincaid test is used to calculate the Readability Index in this experiment.

The variable σ allows to scale the window size. The weighting factors β, which determine each feature's contribution in the equation are determined empirically. Once the size of the window has been identified, it can be applied to all search terms found in the document, enabling the extraction of the candidates evidence accompanying the search term. Each of these are given a weight in the window depending upon their proximity to the search query. The proximity weight is calculated using Gaussian kernel function, which according to previous work [6], produces the best results in this context.

3   Experiments

Improving on our earlier work [1], we have added the new feature, readability index, to the set of features. Moreover, we have applied this method on extra test collections. In this work, two datasets are used to test the proposed approach: W3C corpus and CSIRO corpus, and the four test collections of TREC Enterprise Track between 2005-2008[2] see (Table 1). We used 10 training topics to train our variables, thus having a clear distinction between test and training data.

Table 1. TREC Expert Finding Test Collections

	W3C		CSIRO	
	TREC 2005	TREC 2006	TREC 2007	TREC 2008
Documents	331,1037	331,1037	370,715	370,715
Candidates	1,092	1,092	$\approx$3,000	$\approx$ 3,000
Size	5.7 GB	5.7 GB	4.2 GB	4.2 GB
Topics	50	49	50	77
Qrels	1509	8351	152	2710

Stopwords and HTML markup were eliminated prior to processing. Lucene[3] was used as a retrieval engine. For evaluation, we applied a range of standard IR measures, but in our discussion, we focus on mean average precision (MAP). In this work, we use the two-stage model [2] for the initial candidate ranking as follows:

$$p(ca|q) = \sum_{i=1}^{|D|} p(d_i|q) \cdot p(ca|d_i, q) \tag{2}$$

where $p(d|q)$ is the document relevance to the query, which is calculated by the underlying search engine.

$p(ca|d_i, q)$ is calculated using the two assumptions mentioned earlier:

$$p(ca|d, q) = \frac{P_{occu}(ca|d) + P_{kernel}(ca|d)}{\zeta} \tag{3}$$

---

[2] http://www.ins.cwi.nl/projects/trec-ent/wiki/index.php/Main_Page
[3] http://lucene.apache.org

where $P_{occu}(ca|d)$ represents the first assumption (i.e., the more often the candidate appears in relevant documents, the more likely he/she is an expert), and $P_{kernel}(ca|d)$ represents the second assumption (i.e., the closer the candidate appears to relevant terms, the more likely he/she is an expert). In this work, the two probabilities are considered as independent, hence the summation. The value of the constant ζ is chosen to ensure that $p(c|d, q)$ is a probability measure. The value of ζ is computed as follows:

$$\zeta = \sum_{i=1}^{N} (P_{occu}(ca_i|d) + P_{kernel}(ca_i|d))$$

where N is the total number of candidates in the document d. For the co-occurrence part, (i.e., $P_{occu}(ca|d)$), a $TF - IDF$ weighting scheme is applied [3]:

$$P_{occu}(ca|d) = \frac{n(ca, d)}{\sum_i n(ca_i, d)} \cdot \log \frac{|D|}{|\{d' : n(ca, d') > 0\}|} \qquad (4)$$

where $n(ca, d)$ is the number of times the candidate appears in the document. $\sum_i n(ca_i, d)$ is the number of times any candidate appears. $|D|$ is the number of documents in the collection. $d' : n(ca, d') > 0$ is number of documents where the candidate appears. Finally, $P_{kernel}(ca|d)$ is defined as follows:

$$P_{kernel}(ca|d) = \frac{k(t, c)}{\sum_{i=1}^{N} k(t, ca_i)} \qquad (5)$$

As mentioned above, non-uniform Gaussian kernel functions have been used to calculate the candidate's proximity:

$$k(t, c) = \frac{1}{\sqrt{2\pi\sigma^2}} \exp(\frac{-u^2}{2\sigma^2}), \quad u = \begin{cases} |c - t|, & \text{if } |c - t| \leq w \\ \infty, & \text{otherwise} \end{cases} \qquad (6)$$

where c is the candidate position in the document, t is the topic position, and w is the window size for the current document.

For further elucidation, Figure 1 shows a simple illustrative example of how $p(ca|q)$ is measured. The example topic returned three relevant documents, which used to rank three candidates. In this example, each candidate c_i has n = number of times he/she appears in the document and k = the result of the kernel function. The two ranking models ($P_{occu}(ca|d)$ and $P_{kernel}(ca|d)$) are combined to determine the final candidate rank.

To test the effect of each document feature separately, using training topics, we first generate the adaptive window size with a single feature. Figure 2 shows the MAP at sigma values between 0 and 1000. The analysis of variance, ANOVA, test at $p < 0.05$ suggested a statistical difference between the features. This is true for all datasets. It is clear from the figure that the second feature (i.e., the number of candidates in the document) appears to score the highest in all datasets.

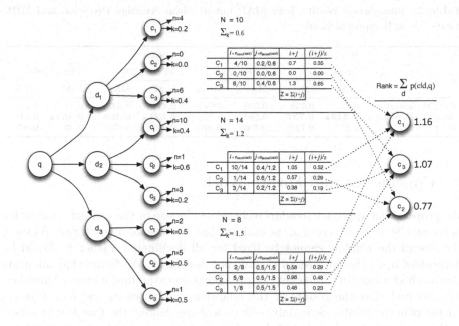

Fig. 1. An example for the system framework

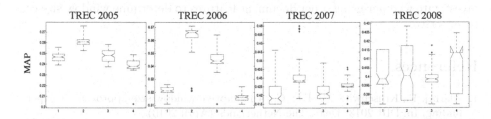

Fig. 2. The results with an adaptive window using a single feature, where 1 is the document length, 2 is the number of candidates in the document, 3 is the average sentence length, and 4 is the readability index

In order to compare the proposed adaptive-window method to a strong baseline, a fixed window size of 200 words is used as suggested by [7] with Gaussian proximity functions. For comparison, we also added the highest-scoring result from the TREC Enterprise track. From the results table, (Table 2), it can be seen that the use of the proposed method resulted in an improvement ranging from 10% to 20% over the fixed window baseline. Using paired t-test on average precision values, we found the difference between our best run and the corresponding baseline to be statistically significant. We indicate $p < 0.01$ using ▲ and $p < 0.05$ using △. The significant improvement is reported for MAP only.

Table 2. Summarised results, here MAP means Mean Average Precision and MRR means Mean Reciprocal Rank

	W3C				CSIRO			
	TREC 2005		TREC 2006		TREC 2007		TREC 2008	
	MAP	MRR	MAP	MRR	MAP	MRR	MAP	MRR
Fix200 Gaussian	0.290	0.748	0.536	0.704	0.474	0.643	0.411	0.791
AdaptiveGaussian	**0.349**▲	**0.797**	0.591 △	0.765	**0.521**▲	**0.763**	**0.457**▲	0.816
Best TREC	0.275	0.727	**0.643**	**0.961**	0.463	0.633	0.449	**0.872**

4   Conclusions and Future Work

We proposed an approach to adaptively select the size of the context window for boosting the retrieval scores of the entities that are close to query terms. As such, the size of the window cannot be fixed for all documents, rather it should be dependent upon the features of the current document. We found that adopting this method results in significant improvements over standard metrics. Moreover, we also find that the results of the adaptive-window using the four features outperform the results using only a single feature. Among the four features used in this study, *the number of candidates* feature appears to be the most important. Going forward, we intend to put the adaptive window size method into practice on other TREC benchmarks and expert-finding collections. Furthermore, we will investigate whether using other document features to determine window size can be effective.

References

1. Alarfaj, F., Kruschwitz, U., Fox, C.: An adaptive window-size approach for expert-finding. In: DIR 2013, Delft, The Netherlands (April 2013)
2. Balog, K., Fang, Y., de Rijke, M., Serdyukov, P., Si, L.: Expertise retrieval. Foundations and Trends in Information Retrieval 6(2-3), 127–256 (2012)
3. Balog, K., Azzopardi, L., de Rijke, M.: A language modeling framework for expert finding. Information Processing and Management 45(1), 1–19 (2009)
4. Macdonald, C., Ounis, I.: Searching for expertise: Experiments with the voting model. The Computer Journal 52(7), 729–748 (2009)
5. Miao, J., Huang, J.X., Ye, Z.: Proximity-based rocchio's model for pseudo relevance. In: SIGIR 2012, Portland, Oregon, pp. 535–544 (2012)
6. Petkova, D., Croft, W.: Proximity-based document representation for named entity retrieval. In: CIKM 2007, pp. 731–740. ACM, New York (2007)
7. Zhu, J., Song, D., Rüger, S.: Integrating multiple windows and document features for expert finding. JASIST 60(4), 694–715 (2009)

Score Normalization Using Logistic Regression with Expected Parameters

Human Media Interaction, University Twente, 7522AE Enschede, The Netherlands
r.aly@utwente.nl

Abstract. State-of-the-art score normalization methods use generative
models that rely on sometimes unrealistic assumptions. We propose a
novel parameter estimation method for score normalization based on logistic regression, using the expected parameters from past queries. Experiments on the Gov2 and CluewebA collection indicate that our method is
consistently more precise in predicting the number of relevant documents
in the top-n ranks compared to a state-of-the-art generative approach
and another parameter estimate for logistic regression.

1 Introduction

Search engines rank documents by scores that were designed to express "higher
is better" and vary wildly between queries. Some applications however benefit
from scores reflecting the probability of relevance, e.g. to model the number of
documents a user should read from a ranked list [4] or to fuse results of different search engines [5]. The main challenge for score normalization methods that
transform scores into such probabilities is to find precise and robust transformation functions for the scores of a given query. In this paper we propose a novel
score normalization method based on logistic regression that uses results from
previous queries.

State-of-the-art score normalization methods model the probability of relevance by making assumptions about the shapes of the density functions of scores
in relevant and non-relevant documents [3]. These methods face the challenge
that the actual scores sometimes violate the assumed shapes [2]. We normalize
scores using logistic regression, which model the relationship between scores and
probabilities as a sigmoid function, i.e. without prior modeling of density functions. Logistic regression models make weaker assumptions about scores and use
less parameters, which might be the reason why Arampatzis and Robertson [3]
recently referred to them as "an under-explored avenue worth pursuing".

Nottelmann and Fuhr [13], who were among the first to use logistic regression
in score normalization, learn parameters from relevance judgments of a set of
training queries. This approach has the disadvantage that it learns a transformation between scores to the probability of relevance to *any* of these queries. In [1],
we found that such probabilistic models do not conform to the probability of relevance principle because they are not specific to the current query. We therefore

M. de Rijke et al. (Eds.): ECIR 2014, LNCS 8416, pp. 579–584, 2014.
© Springer International Publishing Switzerland 2014

approach parameter estimation by learning parameters for each training query individually by using the expected parameter values from training queries. Our experiments show that this method improves accuracy even with limited training material. Additionally, the expected parameter values can be seen as a naive query specific estimate and therefore have potential to be improved in future work.

The rest of this paper is structured as follows: in Section 2 we describe related work to this paper. Section 3 describes our ranking framework for news items. The experiments which conducted to evaluate our approach is described in Section 4. We finish this paper with conclusions and proposals for future work in Section 5.

2 Related Work

Arampatzis and Robertson [3] recently provide an exhaustive survey over the vast body of literature on score normalization methods that produce probabilities of relevance: most normalization methods use generative approaches based on score distributions in relevant documents and non-relevant documents. For example, Gamma distributions [11] or an exponential distribution for scores in non-relevant documents and a normal distribution for scores in relevant documents [12] have been tried. [4] consider a truncated normal distribution for scores in relevant documents, which is more realistic than normal distributions, which have infinity support. Arampatzis and Robertson [3] conclude that a pair of exponential-normal distributions is currently the best performing. We use logistic regression models for score normalization, which only assume that the probability of relevance increases monotonically with scores. Additionally, logistic regression models are mathematically less complex compared to generative methods and require less parameters (two parameters versus four in generative approaches).

Nottelmann and Fuhr [13] are among the first to investigate logistic regression for score normalization, which has been considered before as a ranking function, see for example Cooper et al. [6]. Nottelmann and Fuhr assume a single logistic regression model that transforms scores of arbitrary queries to probabilities of relevance. They estimate the corresponding parameter setting from the scores of judged documents for a set of training queries. As a result, the transformation function for typical scores over multiple queries. However, scores have been found to differ between queries and our experiments indicate that this approach sometimes produces poor performance. Our estimation approach generates for each query a parameter setting, which we combine to the expected parameters assuming queries are random samples from a population. Our estimate is therefore similar to a prior estimate for the parameters for a query at hand that does not consider any query-specific data.

Note that there are also score normalization methods that produce other quantities than probabilities of relevance. For example, [10] consider the commutative score distributions of historical queries and Arampatzis and Kamps [2] consider the part of a signal in a noisy channel as normalized scores. Although these methods do sometimes achieve strong performance, they are only distantly related to our method and will not be treated further here.

3   Estimating Logistic Models

We now describe our method to transform scores into probabilities of relevance using logistic regression. Because logistic regression models are seldom used in information retrieval, we first provide a brief introduction and we refer the interested reader to [13] for a more in-depth discussion. In the score normalization scenario, logistic regression assume that the probability of relevance R for a document with score s is defined as:

$$P(R|s) = \frac{1}{1 + exp(-w_1 - w_2\ s)} \tag{1}$$

where $w = (w_1, w_2)$ are the model parameters. This notation does not specify the exact probabilistic model used. As explained in [1], the model could belong to Model 2, which consider the relevance of documents to a specific queries, or to Model 0, which considers the relevance between multiple queries and documents. Nottelmann and Fuhr [13] model probabilities of relevance for Model 1 because they use as training data a set of relevance judgments between a set of queries Q and the documents in the collection $X = \{(r_{q,d}, s_{q,d})\}\forall q \in Q$, where $r_{q,d} \in \{0, 1\}$ is the relevance status and $s_{q,d}$ is the scores between query q and document d. They use the maximum likelihood estimate w_n given the data X:

$$w_n = \underset{w}{\operatorname{argmax}}\, P_Q(X|w)$$

where the P_Q is the probability measure for the probability of relevance to *any* query in Q with a score s. For a new query q these parameters are used in equation Eq. (1) to calculate the probability of relevance.

In contrast to this estimation procedure, our method considers for each training query $q \in Q$ the data $X_q = \{(r_{q',d}, s_{q',d}) \in X | q' = q\}$, where X is defined above. For each training query, we calculate the maximum likelihood parameter estimate w_q given the data X_q:

$$w_q = \underset{w}{\operatorname{argmax}}\, P_q(X_q|w) \tag{2}$$

where P_q is the measure for the probability of relevance for the query q. We consider w_q as optimal values of a random variable W defined on the sample space of queries Q^+. Now, when a retrieval system encounters a new query $\hat{q}$, we cannot calculate the parameters $w_{\hat{q}}$ by Eq. (2) as there are no relevance judgments available. Instead, we assume that $\hat{q}$ is random sample from Q^+, and use the expected parameter values in query space Q^+, $E_{Q^+}[W]$, as an unbiased estimate of for $w_{\hat{q}}$. This expectation can be estimated by the mean of the parameters in the training queries:

$$w_{\hat{q}} := E_{\mathcal{Q}^+}[\boldsymbol{W}] \simeq E_{\mathcal{Q}}[\boldsymbol{W}] = \left(\frac{\sum_q w_{q,1}}{|\mathcal{Q}|}, \frac{\sum_q w_{q,2}}{|\mathcal{Q}|} \right) \tag{3}$$

where the last last term are the mean in the training query set and $(w_{q,1}, \boldsymbol{w}_{q,2})$ are the parameters of training query q, which we determined through Eq. (2). Note that although this estimate is clearly coarse and equal for all new queries, it estimates parameters for individual queries and therefore estimates Model 2 probabilities.

Table 1. Results: mean absolute error in the given evaluation query set $\mathcal{Q}^t$. Column heads indicated the cut-off value n. $*$ indicates statistical significance compared to TruncExpNorm according to a paird t-test with $p = 0.05$.

Estimator Type	10	30	50	100	1000
TruncExpNorm	4.29	13.29	21.64	40.85	323.63
LogNottelman	4.70	14.79	26.42	58.37	812.87
LogExpectation	3.11*	8.66*	14.52*	26.72*	118.18*

(a) Gov2 $\mathcal{Q}^t = 701 - 750$

Estimator Type	10	30	50	100	1000
TruncExpNorm	3.77	11.65	19.18	38.26	329.92
LogNottelman	3.85	13.29	24.42	56.09	809.29
LogExpectation	2.62*	9.30	16.11	32.84	163.03*

(b) Gov2 $\mathcal{Q}^t = 751 - 800$

Estimator Type	10	30	50	100	1000
TruncExpNorm	4.94	13.65	20.25	34.57	317.10
LogNottelman	4.80	14.80	26.20	59.90	869.60
LogExpectation	3.04*	7.95*	12.60*	25.15*	130.65*

(c) Gov2 $\mathcal{Q}^t = 801 - 850$

Estimator Type	10	30	50	100	1000
TruncExpNorm	3.58	9.20	13.96	30.77	327.34
LogNottelman	6.76	19.96	35.72	75.21	701.30
LogExpectation	2.39*	7.44	10.19*	17.48*	70.85*

(d) CluewebA $\mathcal{Q}^t = 001 - 50$

Estimator Type	10	30	50	100	1000
TruncExpNorm	3.09	9.74	15.25	28.17	347.08
LogNottelman	7.03	20.65	35.32	73.92	662.67
LogExpectation	2.54	7.97	12.90	20.28	60.56*

(e) CluewebA $\mathcal{Q}^t = 051 - 100$

Estimator Type	10	30	50	100	1000
TruncExpNorm	3.31	8.57	16.83	32.12	324.84
LogNottelman	4.88	16.12	30.12	68.40	893.46
LogExpectation	3.84	9.51	16.94	32.18	156.66*

(f) CluewebA $\mathcal{Q}^t = 101 - 150$

4   Experiments

We evaluated our approach on two web datasets: Gov2 (25M documents) using the TREC terrabyte track queries 1-150 and CluewebA (250M documents after removal of the documents with an above median spam score [7]) using the TREC web track queries 700-850. To balance between the number of training queries and evaluation queries, we consider queries in batches of 50 as test queries ($\mathcal{Q}^t$) and the remaining 100 queries as training queries ($\mathcal{Q}$) according to the year that they were used in TREC. In order to compare our method against the state-of-the-art, we implemented the generative model using truncated exponential and normal distributions by Arampatzis et al. [4] (TruncExpNorm) and the logistic regression model with the estimation method by Nottelmann and Fuhr [13] (LogNottelmann). We refer to the method presented here as LogExpectation. As the score function s we used Indri with default settings [1], which produces negative scores. As all methods require positive scores, we followed [4] and cut the original ranking at rank 1000 and added the smallest score to the scores of each document. Because LogNottelmann showed poor performance with these scores, we also divided the scores for this method by the maximum score, ensuring that

---

[1] http://www.lemurproject.org/indri/

they were between 0 and 1. For logistic regression training we used the libLinear software package [9]. Similar to the evaluation of the TREC Legal Track [8], we used the mean absolute error ME_n of the expected number of relevant documents and the actual number, which can be defined as:

$$ME_n = \frac{1}{|\mathcal{Q}^t|} \sum_{q \in \mathcal{Q}^t} \left| R_q^n - \sum_{i=1}^{n} P(R|s(d_i)) \right|$$

where R_q^n is the number of relevant documents in the top n of query q and $P_q(R|s(d_i))$ is the probability of relevance of the ith document d_i using the parameter estimates from Eq. (3). Note that a lower mean error is better.

Table 1 shows the results of our experiments. Our method has a lower mean error when estimating the number of relevant documents at almost all given cut-off values. The improvements are stronger for higher cut-off values. Due to space limitations, we refer the reader to the table for more detailed results. Note that we considered

5 Conclusions

We proposed a logistic regression model for score normalization, which requires fewer parameters and makes milder assumptions than state-of-the-art generative models. Compared to the method by Nottelmann and Fuhr [13], which uses logistic regression models to estimate parameters of a probability measure for all queries, our method uses the expected weights from sample queries as a constant estimate for a probability measure for individual queries. Because we recently found in [1] that probability measures for all queries do not conform to the probability ranking principle, our estimation method is therefore a first step towards solutions that obey this principle.

Using the Gov2 and Cluweb9A datasets with three query batches of 50 queries per dataset, we evaluated our method against a state-of-the-art generative model and the logistic regression model by Nottelmann and Fuhr [13]. Similar to the TREC Legal Track we used the mean error when estimating the number of relevant documents in the top-n ranks as an evaluation measure. Our method consistently improved the evaluation measure in both datasets for all considered query batches and was often statistically significant.

This work is among the first to use logistic regression models for score normalization. While the proposed method already achieves improvements compared to strong baselines, we believe that future work can leverage further improvements by adapting parameter estimates for each query.

Acknowledgments. The work reported in this paper was partially funded by the EU Project AXES (FP7-269980).

References

[1] Aly, R., Demeester, T., Robertson, S.: Probabilistic models in ir and their relationships. Information Retrieval, 1386–4564 (2013) ISSN 1386-4564, http://dx.doi.org/10.1007/s10791-013-9226-3, doi:10.1007/s10791-013-9226-3

[2] Arampatzis, A., Kamps, J.: A signal-to-noise approach to score normalization. In: CIKM 2009, USA. ACM (2009) ISBN 978-1-60558-512-3

[3] Arampatzis, A., Robertson, S.E.: Modeling score distributions in information retrieval. Information Retrieval 14, 1–21 (2010) ISSN 1386-4564

[4] Arampatzis, A., Kamps, J., Robertson, S.: Where to stop reading a ranked list?: threshold optimization using truncated score distributions. In: SIGIR 2009, USA, pp. 524–531. ACM (2009) ISBN 978-1-60558-483-6

[5] Callan, J.: Distributed information retrieval. In: Croft, W. (ed.) Advances in Information Retrieval. The Information Retrieval Series, vol. 7, pp. 127–150. Springer US (2000) ISBN 978-0-7923-7812-9, http://dx.doi.org/10.1007/0-306-47019-5_5, doi:10.1007/0-306-47019-5_5

[6] Cooper, W., Chen, A., Gey, F.C.: Experiments in the probabilistic retrieval based on staged logistic regression. In: TREC 1994. NIST (1994)

[7] Cormack, G.V., Smucker, M.D., Clarke, C.L.A.: Efficient and effective spam filtering and re-ranking for large web datasets. CoRR, abs/1004.5168 (2010)

[8] Cormack, G.V., Grossman, M.R., Hedin, B., Oard, D.W.: Overview of the trec 2011 legal track. In: TREC 2011, p. 1 (2011)

[9] Fan, R.-E., Chang, K.-W., Hsieh, C.-J., Wang, X.-R., Lin, C.-J.: Liblinear: A library for large linear classification. Journal of Machine Learning Research 9, 1871–1874 (2008)

[10] Fernández, M., Vallet, D., Castells, P.: Using historical data to enhance rank aggregation. In: SIGIR 2006, USA. ACM (2006) ISBN 1-59593-369-7

[11] Kanoulas, E., Dai, K., Pavlu, V., Aslam, J.A.: Score distribution models: assumptions, intuition, and robustness to score manipulation. In: SIGIR 2010, USA, pp. 242–249. ACM (2010) ISBN 978-1-4503-0153-4

[12] Manmatha, R., Rath, T., Feng, F.: Modeling score distributions for combining the outputs of search engines. In: SIGIR 2001, USA, pp. 267–275. ACM (2001) ISBN 1-58113-331-6

[13] Nottelmann, H., Fuhr, N.: From uncertain inference to probability of relevance for advanced ir applications. In: Sebastiani, F. (ed.) ECIR 2003. LNCS, vol. 2633, pp. 235–250. Springer, Heidelberg (2003)

More Informative Open Information Extraction via Simple Inference

Hannah Bast and Elmar Haussmann

Department of Computer Science
University of Freiburg, 79110 Freiburg, Germany
{bast,haussmann}@informatik.uni-freiburg.de

Abstract. Recent Open Information Extraction (OpenIE) systems utilize grammatical structure to extract facts with very high recall and good precision. In this paper, we point out that a significant fraction of the extracted facts is, however, not *informative*. For example, for the sentence *The ICRW is a non-profit organization headquartered in Washington*, the extracted fact *(a non-profit organization) (is headquartered in) (Washington)* is not informative. This is a problem for semantic search applications utilizing these triples, which is hard to fix once the triple extraction is completed. We therefore propose to integrate a set of simple inference rules into the extraction process. Our evaluation shows that, even with these simple rules, the percentage of informative triples can be improved considerably and the already high recall can be improved even further. Both improvements directly increase the quality of search on these triples.[1]

1   Introduction

Information extraction (IE) is the task of automatically extracting relational tuples, typically triples, from natural language text. In recent years, the trend has been towards Open Information Extraction (OpenIE), where identifying the predicate and hence the relation is part of the problem. For example, from the sentence

The ICRW is a non-profit organization headquartered in Washington.

the following triples might be extracted:

#1: *(The ICRW) (is) (a non-profit organization)*
#2: *(a non-profit organization) (is headquartered in) (Washington)*

Extracted triples are an important source of information for many information retrieval (IR) systems, in particular in the area of semantic search. For example, systems for the semantic search challenges in the SemSearch 2010/2011 [1] and TREC Entity Track 2009/2010 [2] perform search on triples. A public demo of a search on triple extractions of the ReVerb OpenIE system [3] is available at: http://openie.cs.washington.edu/. Semantic search systems like Broccoli [4] search in triples or triple-like excerpts extracted from the full text as well. All these approaches rely on the usefulness of extracted triples, usually indicated by how much facts were extracted (*recall*) and whether they are correct (*precision*).

---

[1] A demo of our system is available via
http://ad.informatik.uni-freiburg.de/publications

M. de Rijke et al. (Eds.): ECIR 2014, LNCS 8416, pp. 585–590, 2014.

Early approaches to OpenIE focused on extracting triples with high precision but comparably low recall [5]. Later systems focused on improving recall at best possible precision. Newer systems also addressed other important quality aspects, for example in [3] *incoherent* extractions were addressed and [6] considers the *context* of triples. The by far highest recall (at reasonably good precision) was recently achieved with rule-based approaches utilizing grammatical structure, namely ClausIE [7] and CSD-IE [8]. They rely on grammatical rules based on deep parses of a sentence to extract *direct* facts, as e.g., triples #1 and #2 above. The facts are direct in the sense that subject, predicate (possibly implicit) and object are in some form directly connected via a grammatical relation. With respect to these direct facts, the systems achieve almost perfect recall which makes them suitable for a wide range of applications in IR. However, these systems ignore various quality aspects from earlier work.

In this paper, we show that a significant amount of the extracted facts is not *informative*. In the example above, triples #1 and #2 can both be considered correct, but only triple #1 is, by itself, informative. For all practical purposes, the fact that some non-profit organization is headquartered in Washington is useless. This is a serious problem for systems utilizing these triples for search. For example, a search for where the ICRW is headquartered would not be possible to answer from the extracted triples above. An informative extraction that instead can be inferred is:

#3: *(The ICRW) (is headquartered in) (Washington)*

With this, the extracted triple #2 becomes superfluous - all information of the sentence is covered in a precise form in triples #1 and #3. Note that inferring this is only possible while processing the sentence, when individual subjects, objects and their relations are uniquely identified. Afterwards, multiple facts extracted from different sentences mentioning (say) *a non-profit organization* cannot be guaranteed to refer to the same organization.

Based on the observation that this phenomenon is frequent we propose to integrate simple inference into the extraction process of an OpenIE system. Our approach utilizes some of the generic rules used in large scale inference systems, see Section 3. The process is simple and fast, only uses few inference rules and already shows good results. We provide a brief overview of related work in the next section, describe our approach in Section 3 and provide an evaluation in Section 4.

2 Related Work

For an elaborate overview of recent OpenIE systems we refer to [7]. To the best of our knowledge no existing OpenIE system addresses the issue of inference during its extractions process. Some earlier systems, e.g. [9], extract "indirect" facts, similar to inferred facts, using learned patterns. This only works if the text pattern learned for extraction is part of the used training set. In contrast, our inference rules are generic and independent of the exact text surface of a relation.

A lot of work on inferring new information from triples or knowledge bases exists, e.g. [10,11]. The goal is usually to infer facts from triples extracted from different sources in order to, for example, extend knowledge bases or perform question answering. Our goal is to improve the informativeness of extracted triples in the first place, not to perform elaborate inference. Furthermore, as we argued in Section 1, some information can only be inferred while extracting triples, and is irrevocably lost afterwards.

Informativeness of extracted triples has previously been addressed in [3]. Triples were considered uninformative if they omit critical information, for example, when relation phrases are underspecified. The informativeness of extracted triples was evaluated as part of correctness, i.e. uninformative triples were labeled incorrect. We take this one step further and consider a triple uninformative if there is a more precise triple that should be extracted instead, e.g., if the subject should be different (as in the example in Section 1). In our evaluation we explicitly label informativeness as well as correctness of extracted triples.

3   Simple Inference for OpenIE

Our approach consists of three straightforward steps, which are comparably simple yet effective (as our quality evaluation in section 4 shows). The steps are performed after subjects, predicates and objects of all triples in a sentence have been identified, but when all other information (underlying parse tree, supporting data structures etc.) is still available. Given the predicate of each triple in a sentence we first classify the predicate into one of several *semantic relation classes*. Based on the semantic relation we apply a set of inference rules to derive new triples. In the final step we remove existing triples that we consider uninformative depending on whether and how they were used to derive new triples. The next subsections each describe one of the three steps.

3.1   Identifying Semantic Predicate Class

We first classify the predicate of each triple into one of five semantic relation classes shown in Table 1. The relations have previously been successfully used for inference [10] and allow deriving generic, domain-independent inference rules.

To identify the relations we match simple indicator words and patterns. The patterns are implemented as regular expressions over text or parse tree fragments using Tregex [12].

Table 1. Semantic relation classes and patterns for identification

	Semantics	Pattern
SYN	synonymy	*is, was, has/have been, are, nicknamed, known as*
IS-A	hyponymy	*(has/have been, are, is) a/an*
PART-OF	meronymy	*part of, consist* of*
IN	containment or placement	** in*
OTHER	all other relations	***

3.2   Inferring New Triples

Given the triples with identified semantic predicates we infer new triples using a set of generic inference rules. Table 2 shows the rules used. For our example from the introduction, the last rule matches because the semantic relation IS-A holds between *The ICRW* (*A*) and *a non-profit organization* (*B*) and *C* can be bound to *Washington*. As a result, it is inferred that *The ICRW is headquartered in Washington*.

These rules are similar to the up-ward monotone rules from [10], but have been extended with an additional rule to reason over IS-A relations. The implementation differentiates between lexically identical subjects and objects that occur in different places of a sentence. This is a fundamental difference to approaches inferring information after triple extraction, where this information is no longer available.

Table 2. Inference rules for new triples

Table 3. Rule for deleting triples

$$OTHER(A', B) \leftarrow OTHER(A, B) \wedge SYN(A, A')$$
$$OTHER(A', B) \leftarrow OTHER(A, B) \wedge SYN(A', A)$$
$$OTHER(A, B') \leftarrow OTHER(A, B) \wedge SYN(B, B')$$
$$OTHER(A, B') \leftarrow OTHER(A, B) \wedge SYN(B', B)$$
$$IN(A, C) \quad\quad \leftarrow IN(A, B) \wedge PART\text{-}OF(B, C)$$
$$IN(A, C) \quad\quad \leftarrow IN(A, B) \wedge IS\text{-}A(B, C)$$
$$OTHER(A, C) \leftarrow IS\text{-}A(A, B) \wedge OTHER(B, C)$$

$$IS\text{-}A(A, B)$$
$$remove(OTHER(B, C)) \leftarrow \wedge OTHER(B, C)$$
$$\wedge OTHER(A, C)$$

3.3 Removing Uninformative Triples

As described in Section 1 some triples become redundant after they were used to infer additional information. These triples should not be part of the output of the system. This is often the case for IS-A relations and we use a single rule shown in Table 3 to remove triples from our result list.

4 Evaluation

We evaluate the quality of extracted triples with respect to correctness and informativeness. A system similar to the OpenIE system in [8] was used to integrate inference as described above. We compared it against the OpenIE system without inference.

As dataset we used 200 random sentences from Wikipedia. The sentences contain only few incorrect grammatical constructions and cover a wide range of complexity and length. This is the exact same dataset that has already been used in [7].

For each extracted triple we manually assigned two labels: one for correctness (yes or no) and one for informativeness (yes or no). We follow the definition of [5] and consider a triple correct if it is consistent with the truth value of the corresponding sentence. A correct triple is considered informative if there is no extraction that is more precise, according to the sentence it was extracted from. For example, in the sentence from the introduction, triples #1 and #2 would be considered correct, but only triple #1 would be considered informative and triple #3 would be considered both, correct and informative. From the labeled triples we calculated precision of correct triples and estimate recall using the number of extracted correct triples. We also calculated corresponding breakdown statistics for triples that are informative (inf.) as well as correct (corr.). Table 4 shows overall results and Table 5 provides detailed information about the inferred triples.

We first discuss the results in Table 4. Without inference, a large fraction of 10% of correct triples is not informative (prec-corr. inf.). This means that, on average, every 10th extracted correct triple is more or less useless. Using inference the overall number of extracted facts increases from 649 to 762, a relative increase of 17%. The number of correct facts (#facts corr.) also increases: from 429 to 484, corresponding to a relative increase of 13%. The relative increase in correct triples is smaller, because a small number of incorrect triples are inferred (see next paragraph). This is also the reason for the small decrease in the percentage of correct triples (prec corr.) from 66% to 64% (at a 13% higher recall, however). Overall, the number of triples that are both correct and informative (#facts corr. + inf.) increases from 385 to 444: a 15% increase. This is a major improvement, caused by the large number of correct informative triples that were inferred and the uninformative triples removed. Correspondingly, the percentage of correct triples that are also informative (prec-corr. + inf.) increases from 90% to 92%.

Table 4. Quality evaluation results with inference (top row) and without inference (bottom row) over the labels correct (corr.) and informative (inf.). *prec corr.* refers to the percentage of all triples labeled correct, *prec-corr. inf.* to the percentage of correct triples labeled informative.

	#facts	#facts corr.	#facts corr. + inf.	prec corr.	prec-corr. inf.
No Inference	649	429	385	66%	90%
Inference	762	484	444	64%	92%

Table 5. Detailed statistics for inferred triples. *prec inf.* refers to the percentage of inferred triples labeled correct and *prec-corr. inf.* to the percentage of inferred correct triples also labeled informative.

#inferred	#inferred corr.	prec inf.	#inferred corr. + inf.	prec-corr. inf.
127	69	54%	59	85%

Table 5 shows the statistics for inferred triples. Note that, as described in Section 3.3, during inference previously extracted triples may be removed. Therefore, the number of extracted facts with inference does not equal the sum of facts extracted without inference and inferred facts (see Table 4). Overall, about 54% of inferred triples are correct (prec. inf.). A preliminary investigation shows that this is mainly caused by mistakes in preceding phases, in particular wrong parses, wrong identification of objects or predicates and wrong mapping of predicates to their semantic class (see Section 3.1). Eliminating these errors should be part of our next steps (see Section 5). About 85% of correct inferred triples are also informative (prec-corr. inf.). Closer analysis shows that, due to inference, 32% of the triples that were correct but uninformative were removed and replaced with informative triples. Together with the inferred triples, this causes the increase in the percentage of correct triples that are also informative (prec-corr. + inf. in Table 4).

In some cases uninformative triples were inferred. For example, from the sentence *She joined WGBH-TV, Boston's public television station* it is inferred that *(She) (joined)*

(Boston's public television station). Given that our current approach does not differentiate between concrete and abstract subjects and objects (and therefore the "direction" of inference) the high percentage of informative triples derived is remarkable. Further work should, however, try to prevent these extractions.

5   Conclusions

We have presented a simple yet effective way to increase the informativeness of extracted triples for a recent OpenIE system. Using only a few simple inference rules integrated into triple extraction can increase the number of extracted informative triples by 15%. There are a lot of promising directions to improve our work.

A preliminary error analysis shows that most mistakes happen in preceding extraction stages, in particular the precise identification of predicates and objects. Improvements in these areas will likely obviate the small negative effect on precision. To improve the recognition of semantic relations, utilizing existing collections of semantic patterns, such as provided in [13], seems promising. Our manually designed inference rules could be exchanged for automatically derived rules, e.g., as suggested in [14]. Finally, to derive additional facts and distinguish between abstract and concrete facts utilizing information from named entity recognition seems promising.

References

1. Tran, T., Mika, P., Wang, H., Grobelnik, M.: Semsearch 2011: the 4th Semantic Search Workshop. In: WWW (2011)
2. Balog, K., Serdyukov, P., de Vries, A.P.: Overview of the TREC 2010 Entity Track. In: TREC (2010)
3. Fader, A., Soderland, S., Etzioni, O.: Identifying relations for open information extraction. In: EMNLP, pp. 1535–1545 (2011)
4. Bast, H., Bäurle, F., Buchhold, B., Haussmann, E.: Broccoli: Semantic full-text search at your fingertips. CoRR (2012)
5. Banko, M., Cafarella, M.J., Soderland, S., Broadhead, M., Etzioni, O.: Open information extraction from the web. In: IJCAI 2007, pp. 2670–2676 (2007)
6. Mausam, S.M., Soderland, S., Bart, R., Etzioni, O.: Open language learning for information extraction. In: EMNLP-CoNLL, pp. 523–534 (2012)
7. Corro, L.D., Gemulla, R.: ClausIE: clause-based open information extraction. In: WWW, pp. 355–366 (2013)
8. Bast, H., Haussmann, E.: Open information extraction via contextual sentence decomposition. In: ICSC (2013)
9. Banko, M., Etzioni, O.: The tradeoffs between open and traditional relation extraction. In: ACL, pp. 28–36 (2008)
10. Schoenmackers, S., Etzioni, O., Weld, D.S.: Scaling textual inference to the web. In: EMNLP, pp. 79–88 (2008)
11. Lao, N., Mitchell, T.M., Cohen, W.W.: Random walk inference and learning in a large scale knowledge base. In: EMNLP, pp. 529–539 (2011)
12. Levy, R., Andrew, G.: Tregex and Tsurgeon: tools for querying and manipulating tree data structures. In: LREC, pp. 2231–2234 (2006)
13. Nakashole, N., Weikum, G., Suchanek, F.M.: PATTY: A taxonomy of relational patterns with semantic types. In: EMNLP-CoNLL, pp. 1135–1145 (2012)
14. Schoenmackers, S., Davis, J., Etzioni, O., Weld, D.S.: Learning first-order horn clauses from web text. In: EMNLP, pp. 1088–1098 (2010)

Learning a Theory of Marriage
(and Other Relations) from a Web Corpus

Sandro Bauer[1], Stephen Clark[1], Laura Rimell[1], and Thore Graepel[2]

[1] University of Cambridge, Cambridge, United Kingdom
firstname.lastname@cl.cam.ac.uk
[2] Microsoft Research Cambridge, Cambridge, United Kingdom
thore.graepel@microsoft.com

Abstract. This paper describes a method for learning which relations are highly associated with a given seed relation such as *marriage* or *working for a company*. Relation instances taken from a large knowledge base are used as seeds for obtaining candidate sentences expressing the associated relations. Relations of interest are identified by parsing the sentences and extracting dependency graph fragments, which are then ranked to determine which of them are most closely associated with the seed relation. We call the sets of associated relations *relation theories*. The quality of the induced theories is evaluated using human judgements.

1 Introduction

Information Retrieval, and related areas such as Information Extraction and Question Answering, are starting to move away from "shallow" approaches based on keywords to more semantically informed approaches. One example is the use of entailment rules, allowing "Tom Cruise divorced Katie Holmes", for example, to be recognized as providing the answer to the question "Who did Tom Cruise marry?" (through the rule "X divorces Y" implies "X was married to Y"). Taking this example further, suppose the text says that "X has given birth to her fifth child with Y". Whilst not entailing that X is married to Y, there is some likelihood that X and Y are indeed married. Knowledge such as this constitutes what we are calling a *relation theory*. Inducing such a theory is in line with the more general vision of Machine Reading [1], which aims to allow language processing systems to make many of the inferences that humans make when processing text.

The aim of our work is to infer these tacit associations automatically from text, with a score for each associated relation indicating the strength of the association. More concretely, it is a set of such (*relation, score*) pairs for the marriage relation that we call a *theory of marriage* in this paper. The associated relations are in the form of dependency graph fragments. Our method is inspired by the *distant supervision hypothesis* [2], which assumes that all sentences containing instances of a relation do express that relation (e.g. that all sentences containing *William* and *Kate* express the fact that Prince William and Kate Middleton are married). However, rather than maintaining this hypothesis, we argue that its

M. de Rijke et al. (Eds.): ECIR 2014, LNCS 8416, pp. 591–597, 2014.

apparent *failure* in many contexts can be used to our advantage, because many inaccurate examples are semantically close to the seed relations. For example, a sentence stating that William gave Kate an engagement ring can be used as evidence that *giving an engagement ring* is associated with the marriage relation.

2 Methodology

Our experiments exploited three freely-available resources. Freebase[1], a crowd-sourced knowledge base, was the source of entity pairs standing in the seed relation. ClueWeb09[2], a corpus of 500 million English web pages, provided text from which candidates for the associated relations were extracted. Finally, a large background corpus of parsed sentences from Wikipedia was used to rank candidate relations. We will use the marriage relation as a running example.

First, the ClueWeb09 corpus was processed using the boilerplate removal tool in [3], together with some additional simple pre-processing steps such as removing overly long or short sentences. Next, all ClueWeb sentences were extracted which contained references to any pair of entities standing in the marriage relation in Freebase. This resulted in a set of 1,022,271 sentences, which provided the source from which to extract the additional relations associated with marriage.

In terms of how to extract the additional relations, we considered two alternatives. First, we experimented with Open IE tools, in particular ReVerb [4] and OLLIE [5], an extension to ReVerb based on dependency paths. These tools often correctly detected associated relations, but missed some examples, particularly those involving long-range dependencies. Also, the flat output in the form of triples makes it difficult to generalise over syntactically similar relations (e.g. "be girlfriend of" and "be long-time girlfriend of"). We therefore decided to build our own extraction tool based on full dependency graphs. Our representation, like that of OLLIE, is an extension to the notion of dependency path in [6].

First, all the extracted sentences from ClueWeb were parsed with the C&C parser [7], which produces typed dependency graphs. Then all sentences with no dependency path between the two entities were removed. Using a parser in this way we obtain around twice as many extractions as with ReVerb. For the remaining sentences, the direct dependency path between the two married entities is extracted, and a number of heuristics are used to add side nodes from the dependency path, i.e. nodes connected to one of the nodes on the dependency path through a single edge. Specifically, we add common-noun direct objects of verbs (where an indirect object or the second object of a ditransitive verb is on the path) and prepositions with no further outlinks. This way, we capture phrasal verbs such as "X puts up with Y" (which would otherwise be reduced to "X puts with Y") and verbal constructions such as "X ties knot with Y", where the direct object "knot" is crucial for the meaning of the relation.

This process results in dependency graph fragments such as the one shown in Figure 1 (from the sentence *Mr Sauer is known for his long marriage to Merkel*):

---

[1] http://www.freebase.com/
[2] http://lemurproject.org/clueweb09/

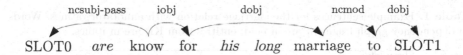

Fig. 1. Example of a dependency graph fragment, with lemmas as tokens

The tokens making up the dependency graph fragment are linked through typed dependency edges, and the SLOTs represent the married entities from the KB.

In order to induce the tacit associations, we employ association metrics typically used for finding collocations [8]. The idea is that we want to rank highly those graph fragments which appear more often in the marriage corpus (the set of sentences containing instances of married entities) than would be expected in a general corpus. We used a parsed version of Wikipedia as the background corpus, and calculated the number of times that the relevant graph fragments occur in this corpus. We tried several association measures, and found that the standard t-test produced promising results. As a final filter, graph fragments which occured less than 5 times in the corpus, or which appeared with less than 5 entity pairs, were removed. Note that the induced relations are not canonical, in the sense that more than one graph fragment can denote the same relation, and we leave clustering of the fragments for future work.

3 Human Evaluation

In order to verify that the ranked lists of dependency graph fragments capture relations that humans associate with the given seed relations, we asked annotators to judge the quality of the induced rankings. Thirteen annotators, all computational linguists, participated, and each annotator was asked to evaluate a total of 60 graph fragments taken from two different seed relations. Participants were given two example sentences per fragment, with the entities and the words in the graph fragment marked with different colours. The task was to assess whether the relation represented by the fragment is highly, somewhat or not at all associated with the seed relation. There was a fourth option for cases where there is an associated relation in at least one of the sentences, but our heuristics failed to capture all the words representing it (see Table 1 for examples).

Our method was evaluated on four Freebase relations: *marriage, parent, employment by a company*, and *birthplace*. For each seed relation, we evaluated the 100 most highly ranked graph fragments and the first 10 of every 100 fragments outside the top 100 ("less highly ranked fragments"). The total number of fragments varied from 429 to 1,084 depending on the seed relation. Each annotator was presented with, for each of two seed relations, 20 graph fragments from the top 100 and 10 fragments from the less highly ranked fragments, in random order. A subset of the data (10 of the 39 chunks) was presented to a second set of annotators to measure inter-annotator agreement. We calculated percentage agreement per rank, averaged across all relations, as well as the κ coefficient [9].

Table 1. Example sentences for the marriage relation with annotator ratings. Words in dependency graph fragment are in bold; entities from KB are in italics.

rating	example
highly	*Sonia Gandhi* **is** the **widow of** *Rajiv Gandhi* who was assassinated in 1991.
somewhat	In October 2007 *Miranda Kerr* **started dating** English hottie *Orlando Bloom*.
not at all	*Nicoletta Braschi* **worked with** *Roberto Benigni* in a lot of his films.
wrong words	UsMagazine.com is reporting that *Amy Winehouse* **did** marry boyfriend *Blake Fielder-Civil* in Florida yesterday.

Table 2. Examples from the top 100 graph fragments for the four relations. Words in dependency graph fragments are in bold; entities from KB are in italics.

relation	examples
marriage	Baywatch hottie *Pamela Anderson* **tied** a **knot with** fiance *Kid Rock* this weekend.
	Lauryn Hill **gave birth** over the weekend **to** her fifth **child with** *Rohan Marley*.
parenthood	As God had commanded, *Abraham* **circumcised** *Isaac* when he was eight days old.
	... and *Liev Schreiber* have **welcomed** their second son, *Samuel Kai Schreiber*.
birthplace	*Harold Washington* (1922-1987) **was** the first African-American **mayor of** *Chicago*.
	Han Hoogerbrugge is a digital artist **living in** *Rotterdam*, Netherlands.
employment	*Buzz Aldrin* **retired from** *NASA* as long ago as 1972.
	Scott Gnau **is** vice president and general **manager** ... **at** *Teradata*.

Tables 3a to 3d give the percentages at each rank for the four relations (e.g. 42% of the top 100 graph fragments for the marriage relation were judged as highly relevant). For all four relations, more than half of the top 100 fragments in the lists were judged highly or somewhat associated. Table 2 gives some examples from the top 100 graph fragments for each of the relations. In contrast, the less highly ranked fragments are overwhelmingly rated as somewhat associated or below. Some relations show an especially quick drop-off: the *marriage* relation has many examples such as *write letter to* and *save* in the less highly rated associated fragments, which could apply to a married couple, but are arguably not part of a core theory of marriage.

Figure 2 shows the precision at rank summed over all four relations, where a point is awarded when a fragment is judged highly or somewhat associated. For the agreement experiment, we obtained an overall percentage agreement score of 67.5% and a κ coefficient of 0.55, indicating a fair level of agreement.

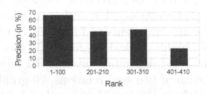

Fig. 2. Precision at rank across all relations

4 Related Work

There is a large body of broadly related work attempting to induce general entailment rules, such as X *writes* Y implies X *is the author of* Y [6]; or if a company is based in a city, and the city is located in a state, then the company is headquarted in that state [10,11]. Whilst our work can be situated in this broad area of IE, it is more closely associated with attempts to derive "domain theories" from text [12], which might contain rules stating that, for example, flights do not start and end in the same city; or attempts to induce rules between relations in a KB, such as: if two people have children in common, then they are often married [13]. Chambers et al.[14] exploit co-reference in documents to extract narrative schemas in an unsupervised setting, for example the events associated with a criminal being arrested. In some ways the failure of the distant supervision hypothesis could also be seen as important for their system, since it relies on the semantic association of sentences involving the same actors; however, in our system we use seed instances from a KB in order to test whether richer representations of relations can be bootstrapped from a KB.

Table 3. Scores for the different relations

rank	highly	some-what	not at all	wrong words
1-100	42	29	23	6
201-210	0	60	30	10
301-310	20	10	40	30
401-410	0	10	90	0
501-510	0	10	60	30

(a) *marriage*

rank	highly	some-what	not at all	wrong words
1-100	19	33	42	6
201-210	10	20	50	20
301-310	30	40	30	0
401-410	10	10	70	10
501-510	10	20	70	0
601-610	10	40	40	10
701-710	0	0	90	10
801-810	0	20	80	0
901-910	0	10	90	0
1001-1010	0	20	80	0

(b) *birthplace*

rank	highly	some-what	not at all	wrong words
1-100	43	8	31	18
201-210	10	0	70	20
301-310	0	10	80	10
401-410	0	10	50	40

(c) *parenthood*

rank	highly	some-what	not at all	wrong words
1-100	70	22	2	6
201-210	40	40	10	10
301-310	40	40	10	10
401-410	30	20	40	10

(d) *employment*

5 Conclusion

We have described a novel method for inducing theories of relations that enhance the information present in KBs, which we believe is a valuable ingredient for more semantically informed IR systems. The theories presented here are simplified versions of what an ideal relation theory would contain. Obvious extensions include making the theories culture- and era-dependent; for example, divorce is much more associated with marriage in the modern era than in the past. Similarly, we could include temporal ordering; for example the fact that engagement happens before and divorce occurs after and is the end of marriage.

As a step in this direction, we performed a small pilot study making use of the birthdates of people available in Freebase. We induced theories of marriage for different time periods, including the biblical era and the 20th century, by seeding the extraction with only couples born in the relevant era. The results reflect how marriage depends on historical context. Relations such as *commit sin with* or *bury one's partner* are more frequent for protagonists in the bible, and are ranked highly for the biblical theory, while *suing one's partner* and *being spotted with one's partner* are more highly ranked for the contemporary theory.

Acknowledgments. The first author was supported by Microsoft Research through its PhD Scholarship Programme.

References

1. Etzioni, O., Banko, M., Cafarella, M.J.: Machine Reading. In: Proceedings of the Twenty-First National Conference on Artificial Intelligence, AAAI 2006, Boston, Massachusetts, USA, pp. 1517–1519 (2006)
2. Mintz, M., Bills, S., Snow, R., Jurafsky, D.: Distant supervision for relation extraction without labeled data. In: Proceedings of the Joint Conference of the 47th Annual Meeting of the ACL and the 4th International Joint Conference on Natural Language Processing of the AFNLP, ACL 2009, Suntec, Singapore, vol. 2, pp. 1003–1011 (2009)
3. Kohlschütter, C., Fankhauser, P., Nejdl, W.: Boilerplate Detection using Shallow Text Features. In: Proceedings of the Third ACM International Conference on Web Search and Data Mining, WSDM 2010, New York, USA, pp. 441–450 (2010)
4. Fader, A., Soderland, S., Etzioni, O.: Identifying Relations for Open Information Extraction. In: Proceedings of the Conference on Empirical Methods in Natural Language Processing, EMNLP 2011, Edinburgh, United Kingdom, pp. 1535–1545 (2011)
5. Mausam, S.M., Bart, R., Soderland, S., Etzioni, O.: Open Language Learning for Information Extraction. In: Proceedings of the 2012 Joint Conference on Empirical Methods in Natural Language Processing and Computational Natural Language Learning, EMNLP-CoNLL 2012, Jeju Island, Korea, pp. 523–534 (2012)
6. Lin, D., Pantel, P.: DIRT – Discovery of Inference Rules from Text. In: Proceedings of ACM SIGKDD Conference on Knowledge Discovery and Data Mining 2001, Location Unknown, pp. 323–328 (2001)
7. Clark, S., Curran, J.R.: Wide-Coverage Efficient Statistical Parsing with CCG and Log-Linear Models. Computational Linguistics 33(4), 493–547 (2007)
8. Manning, C.D., Schütze, H.: Foundations of statistical natural language processing. MIT Press, Cambridge (1999)
9. Cohen, J.: A Coefficient of Agreement for Nominal Scales. Educational and Psychological Measurement 20, 37–46 (1960)
10. Schoenmackers, S., Etzioni, O., Weld, D.S., Davis, J.: Learning first-order horn clauses from web text. In: Proceedings of the 2010 Conference on Empirical Methods in Natural Language Processing, EMNLP 2010, Cambridge, Massachusetts, USA, pp. 1088–1098 (2010)

11. Berant, J., Dagan, I., Goldberger, J.: Global learning of typed entailment rules. In: Proceedings of the 49th Annual Meeting of the Association for Computational Linguistics: Human Language Technologies, HLT 2011, Portland, Oregon, USA, vol. 1, pp. 610–619 (2011)
12. Liakata, M., Pulman, S.: Learning theories from text. In: Proceedings of the 20th International Conference on Computational Linguistics, COLING 2004, Geneva, Switzerland (2004)
13. Galárraga, L., Teflioudi, C., Suchanek, F., Hose, K.: AMIE: Association Rule Mining under Incomplete Evidence in Ontological Knowledge Bases. In: Proceedings of the 22nd International World Wide Web Conference (WWW 2013), Rio do Janeiro, Brazil (2013)
14. Chambers, N., Jurafsky, D.: Unsupervised Learning of Narrative Schemas and their Participants. In: Proceedings of the Joint Conference of the 47th Annual Meeting of the ACL and the 4th International Joint Conference on Natural Language Processing of the AFNLP, ACL 2009, Suntec, Singapore, vol. 2, pp. 602–610 (2009)

Learning from User Interactions
for Recommending Content in Social Media

Mathias Breuss and Manos Tsagkias

ISLA, University of Amsterdam,
Amsterdam, The Netherlands
mathiasbreuss@gmx.at, e.tsagkias@uva.nl

Abstract. We study the problem of recommending hyperlinks to users in social media in the form of status updates. We start with a candidate set of links posted by a user's social circle (e.g., friends, followers) and rank these links using a combination of (i) a user interaction model, and (ii) the similarity of a user profile and a candidate link. Experiments on two datasets demonstrate that our method is robust and, on average, outperforms, a strong chronological baseline.

1 Introduction

We have become information overloaded. Our online social circles generate a continuous flood of information which is difficult to keep up with. Anecdotal evidence suggests that a typical Twitter user who follows 80 users [6] needs to sift through more than a few hundreds tweets a day, which renders finding interesting and relevant information difficult. One way to tackle this problem is via content personalization. Central here is to infer a user's preferences from their activities. Our idea builds on previous successful research [2, 3] that encodes signals from how the user interacts with content from their social network for building a content recommender system. Xu et al. [9] established that people's posting behavior is influenced by the news, what their friends have posted, and their intrinsic interests [10]. Chen et al. [4] explored the effect of content sources, topic interest models for users, and social voting for recommending (ranking) hyperlinks to users. While their work focuses on discovering relevant popular hyperlinks, we also consider the underlying users and their relation to the target user of the recommendations. Abel et al. [1] investigated the effects of topic modeling, semantic enrichment and time awareness in user modeling for hyperlink recommendation. Our work is parallel, with the difference on how we generate, rank and combine the candidate set of hyperlinks, and in the quantitative evaluation method we use.

Our system learns from a user's interactions with the items posted from their social circles. When two users have similar behavioral patterns, we posit that they are likely to find items posted from each other relevant. For example, two users are more similar if they tweet in similar times, if they tend to tweet similar sets of hyperlinks, or if they tend to use mentions or not. We define and model user behavior using six dimensions (see §2.1 for details), e.g., commonly shared items, activity patterns, which we use for gauging user similarity. Instead of making heuristic decisions on the weight of these dimensions, we use a semi-supervised method for learning the optimal weights. On top of

M. de Rijke et al. (Eds.): ECIR 2014, LNCS 8416, pp. 598–604, 2014.

the user models, we profile user interests using the content of their tweets that include a hyperlink, and the content of the hyperlinks themselves (§2.2). The final recommendations are a combination of the recommendations from the user-based model, and the recommendations from the content similarity between a hyperlink and the user profile using late data fusion.

The research questions we aim to answer are: (RQ1) What is the recommendation effectiveness of models that learn from user interactions? (RQ2) Are content-based approaches effective? and (RQ3) What is the effectiveness of combining user- and content-based models? Our main contribution is a recommender system that infers user preferences from users' interactions with their social circles.

2   Models for Recommendation in Social Media

We begin with defining our terminology, and an overview of our recommendation pipeline. We refer to users for whom we want to provide recommendations as *target users*. For each target user we collect the lists of their friends, followers and their tweets. We refer to the friends and followers of a target user as *associate users*. We work on a streaming setting and move in time in steps of t. The recommendation system uses the entire history of a target user's associate users and their items up to t to build a candidate set of items to recommend to the target user. Below, we present three recommendation models: (i) user-based, (ii) content-based, and (iii) hybrid.

2.1   User-Based Models

Items are recommended by selecting the associate user that is most similar to the target user, using the associate user's items and then proceed to the second most similar associate user. We consider the following six dimensions for assessing the similarity between a target user and an associate user based on the target user's interaction patterns.

Common items: Users who have shared the same items in the past might do so in the future. We record the similarity of a target user and an associate user as the Jaccard similarity between the set of items they have shared. We expect this model to provide strong evidence for user similarity, however, due to data sparsity its use might be limited.

Mentions: Interaction between users via mentions hints a stronger tie between them, and a higher likelihood for sharing common interests. The more often a user mentions another user the stronger this bond becomes. We model this interaction by counting how often a target user mentions an associate user, and normalize this by the maximum number of mentions of any associate user.

Activity patterns: Users who share the same temporal activity patterns, e.g., they tweet similar times in the day, may originate from the same location, culture, or demographics group and can potentially share the same interests. We represent each user with two vectors, one with 24 dimensions to denote their activity within a day, and one with 7 dimensions for their activity within a week, and count the number of tweets posted by the user in each time unit. The similarity between two users is the cosine similarity between the corresponding vectors.

Intrinsic features: A user's writing style can be potential good surrogate for user similarity [7]. We consider three features, the number of mentions (nmentions), the number of hashtags (nhashtag) and the number of hyperlinks (nurl) within a tweet. E.g., we count the number of tweets that contain three hashtags.

Social features: We follow [8] and represent our users as a vector of the number of tweets, favorites, followers, friends, times listed, the UTC offset, the location of the user, the account language, and whether the user's account is verified. The similarity of two users is the cosine similarity between their respective vectors. Unlike the other dimensions, we consider social features stable over time, and therefore we do not compute them for every time bin.

Item popularity: Generally popular items are likely to be relevant items. We gauge the popularity of a hyperlink via Facebook's popularity indicators, i.e., the number of shares, likes, and comments. The score for an item is the sum of shares, likes and comments divided by the maximum sum in our collection. The score for an associate user is, then, the sum of popularity scores of their items.

We combine the above dimensions using a Naive Bayes classifier trained at every time step t in a semi-supervised manner. A training example corresponds to an associate user, and consists of six features each of which corresponds to one of the above dimensions at $t - 1$. If the target user has retweeted at least one item from an associate user at t, then the associate user is assigned to the positive class, otherwise to the negative. To rank associate users in t, we classify them using the model trained on $t - 1$, and rank the classified associate users by the classifier's confidence score for the positive class. Items are then recommended first from the user with the highest confidence, followed by the user with the second highest confidence.

2.2 Content-Based Models

In addition to recommending items from a ranked list of associate users, we can also recommend items directly, based on the contents of the items. In this respect, the recommendation problem can be cast as an information retrieval problem: the terms of a candidate item from an associate user are considered the terms of a document, and the terms of the items of a target user are considered the query. We consider two methods for representing candidate items: (i) using the terms found in the tweet of the item (tweet), (ii) using the terms in the website referred to by a hyperlink of the item (hyperlink). The resulting recommendation list is then ordered by the rank of an item in the underlying information retrieval system's result list. For ranking candidate items given a target user profile, we use Apache Lucene.

2.3 Hybrid Model

Our hybrid model combines the recommendation lists from user- and content-based models into one single list using late data fusion. Our fusion method is a variant of

WcombSUM [5], a weighted late data fusion method.[1] Given a set L of recommendation lists r and their weights w_r with $\sum_r w_r = 1$, the final score of an item i is $\left[\sum_{r \in L} \frac{rank_r(i)}{w_r}\right]^{-1}$, where $rank_r(i)$ is the rank of i in r.

3 Experimental Setup

We report on recommendation experiments using the following experimental conditions: individual user-based models, semi-supervised combination of user-based models, content-based models, and hybrid models.

Datasets: We report results on two datasets: (i) the average user dataset, and (ii) the influential user dataset; see Table 1 for a summary. The *average user dataset* consists of 120 users, 124,828 associate users (friends and followers), and 31M tweets with 6.3M hyperlinks in the period Aug 19–Sep 17, 2012. We sampled users who meet the following criteria after analyzing a random sample of users from Twitter's sample stream: they tweet in English, their number of friends is within 10% of either the 1st Qu., Median and Mean of the random sample, their number of followers is within 10% of the median (given the previous constrain), and their number of tweets is within 20% of the median or the mean. The *influential user dataset* consists of 240 target users, 101,573 associate users (friends only), and 21M tweets with 6.8M hyperlinks in the period Jan 8–Feb 15, 2013. We sampled users who ranked highly influential at particular topic groups (e.g., arts, blogger) according to Peerreach.com, a service that analyzes user influence in Twitter. Due to the high number of followers of these users, we were unable to crawl them due to Twitter's rate limits. To cope with possible changes in a target user's associate users, we regularly poll and update all target users in both datasets. For each dataset, we set aside 30 target users as a development set for parameter tuning, and use the remaining for testing.

Table 1. Summary of the average and influential user datasets

Counts	Average user dataset					Influential user dataset				
	1st Qu.	Median	Mean	3rd Qu.	Max.	1st Qu.	Median	Mean	3rd Qu.	Max.
Friends	112.8	198.0	643.2	1380.0	2,009	66.5	197.0	463.8	557.5	4,377
Followers	123.0	225.0	414.7	788.2	1,283	—	—	—	—	—
Tweets	193.0	424.0	741.9	974.0	3,500	72.5	220.5	432.0	671.2	2,097
Hyperlinks	8.5	24.5	126.8	68.8	3,275	24.5	77.0	163.5	237.5	1,941

Evaluation and tuning: We choose mean average precision (MAP) as our evaluation metric. We run multiple tests for a specific target user using candidate and test sets from different time bins. We assemble our ground truth per target user as follows. An item from a candidate set at time $t - 1$ is deemed relevant if it is posted by the target user in the test set at time t. We use the mean of the average precision over all target

---

[1] Due to the poor performance of WcombSUM in our experiments, we omit the details.

users and all time bins to evaluate system performance. We compare our models against two baselines: (i) random, and (ii) chronological. In the *random baseline*, given a target user, we assign a random score between 0 and 1 to all their associate users. To increase stability, we repeat the process 100 times, and report on the average score for each associate user. In the *chronological baseline*, candidate items are ordered chronologically with the most recent one showing first. Statistical significance is tested using a two-tailed t-test and is marked as ▲ (or ▼) for significant differences for $\alpha = .01$, or △ (and ▽) for $\alpha = .05$. The parameters of our late data fusion model are tuned by performing a parameter sweep using 30 randomly selected target users per dataset. We find that for the average users an optimal performance is achieved using the tweet content (weight of 0.01) and for the influential users using the hyperlink content (weight of 0.5).

4 Results and Analysis

In our first experiment, we answer RQ1. In Table 2 (top) we report on improvement of MAP relative to baselines for individual features extracted from six interaction dimensions, and their combination using a semi-supervised method. The chronological baseline is stronger than random, as expected, however, both yield very low MAP score

Table 2. Improvement in MAP for user-based, content-based, and hybrid recommendation models over random (Rand) and chronological (Chron) baselines for both average user dataset (AUD) and influential user dataset (IUD). For user-based models, we list the effectiveness of the individual components, and their combination (semi-supervised). For our hybrid approach, we provide the weight w for the content-based method. Boldface indicates best performance.

Model	Dimension	AUD (%) Rand	Chron	IUD (%) Rand	Chron
User-based					
item	Common	697▲	91▲	227▲	29△
mention	Mentions	56	-63▼	138▲	-6
hod	Activity	283▲	-8	35	-47▼
dow	Activity	185▲	-32	42△	-44▼
nurl	Intrinsic	-33	-84▼	-29▽	-72▼
nmentions	Intrinsic	-27	-82▼	55△	-39▼
nhashtag	Intrinsic	0	-76▼	61▲	-36▼
social	Social	14	-73▼	52▲	-40▼
fbitem	Popularity	-67▼	-92▼	-13	-65▼
semi-supervised		1230▲	219▲	882▲	288▲
Content-based					
tweet		-44▼	-86▼	233▲	32△
hyperlink		-32	-84▼	82▲	-28▼
Hybrid					
tweet $w = 0.01$		1608▲	310▲		
hyperlinks $w = 0.5$				737▲	231▲

mainly due to data sparsity; on AUD, Rand and Chron score 0.0010 and 0.0027 respectively, and on IUD they score 0.0035 and 0.0126. Commonly shared items (item) proves the strongest individual feature. Our semi-supervised method outperforms all individual features and supports our hypothesis that learning a combination of interaction dimensions is useful. After performing feature selection, fbitem and mention showed to be the most discriminative features. In our second experiment, we answer RQ2. We find that content-based methods are not as effective as user-based methods; see Table 2 (middle). The reason is related to the kind of content the system has to deal with. The performance of content-based models is better for influential users because they often share newspaper articles and tend to create quality text, and drops for average users who tend to share more links to multimedia sites. In our third experiment, we answer RQ3. Our late data fusion method outperforms semi-supervised in the average user dataset, but not in the influential user dataset. On average, it outperforms the chronological baseline by 270%; see Table 2 (bottom). The performance difference between datasets hints that the weights are not optimal, possibly due to the small size of our development set.

Next, to better understand our results, we investigate the effect of the number of items shared by a target user on recommendation effectiveness. We plot MAP as a function of the average number of hyperlinks posted by a target user for the hybrid model on the average user dataset, and find a moderate linear correlation between the two (Pearson's $\rho = 0.6252$). This correlation means that our model performs better for users who post more hyperlinks, and can be used as indicator for deciding whether the system should provide recommendations to a given user. During manual investigation we found many users who made use of services that tweet on behalf of the user, e.g., the number of a user's new followers, and quantitative summaries of a user's tweet activity. This type of automatically generated content is likely to have adversely affected recommendation performance. Future work should pay special attention to automatically generated content generation or bots.

5 Conclusions

We focused on recommending hyperlinks to users in Twitter in the form of status updates. Our main contribution is a model that infers user preferences from how a user interacts with items posted in their social circles, and we combine this with a content-based model. We tested our models on two datasets of different user demographics and found similar patterns in performance. This demonstrates the robustness of our methods. In the future, we envisage an adaptive weighting scheme for our late data fusion method where the recommendation list from individual models are weighted per user.

Acknowledgments. This research was partially supported by the European Community's 7th Framework Programme (FP7/2007-2013) under grant agreement nr 288024 (LiMoSINe project).

604   M. Breuss and M. Tsagkias

References

[1] Abel, F., Gao, Q., Houben, G.-J., Tao, K.: Semantic enrichment of Twitter posts for user profile construction on the social web. In: Antoniou, G., Grobelnik, M., Simperl, E., Parsia, B., Plexousakis, D., De Leenheer, P., Pan, J. (eds.) ESWC 2011, Part II. LNCS, vol. 6644, pp. 375–389. Springer, Heidelberg (2011)
[2] Abel, F., Gao, Q., Houben, G.-J., Tao, K.: Analyzing user modeling on twitter for personalized news recommendations. In: Konstan, J.A., Conejo, R., Marzo, J.L., Oliver, N. (eds.) UMAP 2011. LNCS, vol. 6787, pp. 1–12. Springer, Heidelberg (2011)
[3] Anagnostopoulos, A., Kumar, R., Mahdian, M.: Influence and correlation in social networks. In: KDD 2008, pp. 7–15. ACM (2008)
[4] Chen, J., Nairn, R., Nelson, L., Bernstein, M., Chi, E.: Short and tweet: experiments on recommending content from information streams. In: CHI 2010, pp. 1185–1194. ACM (2010)
[5] He, D., Wu, D.: Toward a robust data fusion for document retrieval. In: NLP-KE 2008, pp. 1–8 (2008)
[6] Huberman, B.A., Romero, D.M., Wu, F.: Social networks that matter: Twitter under the microscope. CoRR (2008)
[7] Massoudi, K., Tsagkias, M., de Rijke, M., Weerkamp, W.: Incorporating query expansion and quality indicators in searching microblog posts. In: Clough, P., Foley, C., Gurrin, C., Jones, G.J.F., Kraaij, W., Lee, H., Mudoch, V. (eds.) ECIR 2011. LNCS, vol. 6611, pp. 362–367. Springer, Heidelberg (2011)
[8] Petrovic, S., Osborne, M., Lavrenko, V.: RT to win! Predicting message propagation in Twitter. In: AAAI 2011 (2011)
[9] Xu, Z., Zhang, Y., Wu, Y., Yang, Q.: Modeling user posting behavior on social media. In: SIGIR 2012, pp. 545–554. ACM (2012)
[10] Yan, R., Lapata, M., Li, X.: Tweet recommendation with graph co-ranking. In: ACL 2012, pp. 516–525. ACL (2012)

Towards an Entity–Based Automatic Event Validation

Andrea Ceroni and Marco Fisichella

L3S Research Center, Germany
{ceroni,fisichella}@L3S.de

Abstract. Event Detection algorithms infer the occurrence of real–world events from natural language text and always require a ground truth for their validation. However, the lack of an annotated and comprehensive ground truth makes the evaluation onerous for humans, who have to manually search for events inside it. In this paper, we envision to automatize the evaluation process by defining the novel problem of Entity–based Automatic Event Validation. We propose a first approach which validates events by estimating the temporal relationships among their representative entities within documents in the Web. Our approach reached a Kappa Statistic of 0.68 when compared with the evaluation of real–world events done by humans. This and other preliminary results motivate further research effort on this novel problem.

Keywords: Event Detection, Automatic Event Validation.

1 Introduction and Related Work

Inferring the magnitude and occurrence of real-world events from natural language text is a crucial task in various domains. A large body of previous work focuses on the problem to identify real events using Web sources as part of a broader initiative named topic detection and tracking [1]. In general, two main approaches have been considered to solve the problem of event detection. Document-based approaches [1,8] exploit generative models to cluster documents based on semantics and timestamps. Entity-based approaches [4,5] discover events by studying distributions of their entities over documents and time.

In order to assess the retrieved events, all the aforementioned methods require a validation methodology, always involving the presence of a ground truth. However, to the best of our knowledge, there are no existing resources for a comprehensive list of real-world events. A solution is to strive a human assessment. In [3,5], authors start from the idea that entities can appear in various data sources, such as search engine query logs, news and blog articles. Given an entity, annotators examined various Web-based sources (e.g., Wikipedia, news search, and Web search in general) to identify events involving the specific entity.

So far several and promising attempts for building collections of real-world events have been conducted: one example is Wikipedia Current Event Portal[1]

---

[1] http://en.wikipedia.org/wiki/Portal:Current_events

M. de Rijke et al. (Eds.): ECIR 2014, LNCS 8416, pp. 605–611, 2014.
© Springer International Publishing Switzerland 2014

which tries to collect human generated event descriptions; another example is the YAGO2 Knowledge Base [6], where entity relationships are modeled through facts. However, these approaches are limited in term of number of events, complexity (e.g., YAGO2 facts only involves two entities), and granularity (i.e. "small events" such as the loan of a football player are often ignored and not reported in favor of "big events" such as the Arab Spring).

One way to overcome the aforesaid limitations is to automate the validation of event detection algorithms, starting from a no annotated ground truth. Although any document collection can be used in principle as ground truth, the choice for a collection taken from the Web rises from the intuition that events on the Web cover any granularity, are continuously updated, and are freely available. We cast a new light on this task. Similar to our intent is the work conducted in [9], but authors'approach is oriented to event detection instead of event validation. Finally, our research is a preliminary investigation in this direction with the main purpose to support the human validation and to provide in the next future a unique tool for event validation to the scientific community.

2   Problem Statement

Event Model. We define an *event* e as a tuple $e := (C_e, t_e^0, t_e^f)$, where $C_e = \{c_e^1, \ldots, c_e^{n_e}\}$ is the set representing the *entities* that participated to e, t_e^0 is the time instant when the event started, and t_e^f is the time instant when the event ended. We do not assume that entity names are linked to a Knowledge Base and to possible other surface forms: the only assumption required by our model is that each entity c is identified by one or more words.

Ground Truth. A *ground truth* is defined as a document collection G, where each document $d_i \in G$ can be representative of one or more events.

Event Validation. Given an event set E and a ground truth G, an *event validation* is a procedure that checks if any event $e \in E$ has evidence in G. If that is the case, then e is labeled as *true*, otherwise it is labeled as *false*.

Automatic Event Validation. Given an event set E and a ground truth G, an *automatic event validation* is an event validation which automatically labels events through an *evaluation policy*, i.e. a criterion to state if an event is true.

3   Approach

Our approach is depicted in Figure 1. For every event, the *query formulation* component generates one or more queries by taking into account the participating entities, the starting time, and the ending time. Then, the *document retrieval* component interacts with the *ground truth* by performing the queries and retrieving the results, i.e. documents. Note that, in order to make queries, the ground truth is supposed to be indexed. The *document processing* component extracts features from the retrieved documents, which are used by the *validation policy* to validate the event. The problem solution sketched above is quite general and

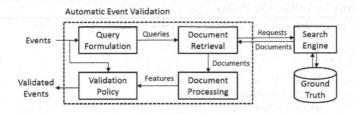

Fig. 1. Architecture of the automatic event validation

could be applied in principle to different setups having different ground truths, search engines, and supported queries. In the following we describe our choices regarding each component in our solution.

Query Formulation. For every event $e \in E$, a query q is generated from its participating entities C_e, its starting time t_e^0, and its ending time t_e^f. Every query q is defined as a tuple $q := (q_{text}, q_{time})$, where q_{text} is a *textual query* representing the keywords of the query, and q_{time} is a *temporal query*, which represents the time period that q refers to. We define q_{text} as a bag of words of the entities appearing in C_e, i.e. $q_{text} = (c_e^1, \dots, c_e^{n_e})$, while q_{time} is a time interval whose start and end time are respectively t_e^0 and t_e^f, i.e. $q_{time} = [t_e^0; t_e^f]$. Since temporal queries are not assumed to be supported, a set Q_{text} of textual queries $\hat{q}_{text}$ is built from q as follows. A set of dates is created by exploding the time interval represented by q_{time}, according to a given granularity. For instance, if $q_{time} = [2013\text{-}05\text{-}18; 2013\text{-}06\text{-}18]$ and the granularity is monthly, then the set will be $\{\text{may } 2013, \text{june } 2013\}$. For every date, a textual query $\hat{q}_{text}$ is then created by concatenating the original textual query q_{text} with the date itself. Note that the granularity is an input parameter, and it should be coherent with the duration of the event.

Document Retrieval. The document retrieval component interacts with the ground truth G by performing queries and by retrieving documents. For every $\hat{q}_{text} \in Q_{text}$, a document set $D_{\hat{q}} \subset G, |D_{\hat{q}}| = k$, is filled with the top–k documents retrieved by the search engine in response to $\hat{q}$. According to this, the document set containing all the documents associated to the original query q is $D_q := D_{\hat{q}_1} \cup \dots \cup D_{\hat{q}_{|Q_{text}|}}$. Note that the value of k is an open parameter.

Ground Truth. We chose the Web as ground truth because it offers a wide event coverage and it is globally developed and updated by many different sources.

Feature Extraction. Two types of *features*, namely *terms* and *temporal expressions*, are extracted from each document d in order to determine whether an event e has evidence in d. Each *term* is a tuple $w := (v_w, p_w)$ where v_w is a non stop word and p_w is its position within the text. Each *temporal expression* consists in a tuple $t := (v_t, p_t)$, where v_t is the temporal expression value and p_t is its position within the text. Thus, for each $d \in D_q$, a set of features $F_d := \{W_d, T_d\}$ is extracted, where $W_d = \{w_1, \dots, w_{n_W}\}$ is a set of terms and $T_d = \{t_1, \dots, t_{n_T}\}$ is a set of temporal expressions.

Algorithm 1. Validation Policy

Input   : document features $F_d = \{W_d, T_d\}$, event $e = (C_e, t_e^0, t_e^f)$
Output: event confidence γ_e
Set $c_{max} = 0$
for each $t \in T_d$ **do**
   | **if** $t \subset [t_e^0; t_e^f]$ **then**
   | | Set $c_t = 0, p_t^{start} = 0, p_t^{end} = 0$
   | | $[p_t^{start}, p_t^{end}]$ = getRegionBoundaries(t)
   | | W_t = getTermsInRegion($p_t^{start}, p_t^{end}, W_d$)
   | | **for** each $c \in C_e$ **do**
   | | | **if** $c \subset W_t$ **then**
   | | | | $c_t = c_t + 1$
   | | | **end**
   | | **end**
   | | $c_{max} = \max(c_{max}, c_t)$
   | **end**
end
return $\gamma_e = \frac{c_{max}}{|C_e|}$

Validation Policy. The validation policy, described in Algorithm 1, determines if an event e has evidence in a document d or not. It takes as input the document features F_d and the event e, and returns as output the confidence $\gamma_e \in [0; 1]$ for the event to be true, calculated as follows. For every temporal expression $t \in T_d$ included in the event time span $[t_e^0; t_e^f]$, the boundaries of the document region associated to t are computed by considering $p_t^{start} = p_t$ as starting position and $p_t^{end} = p_{\tilde{t}}$ as ending position, where $\tilde{t}$ is the first temporal expression after t which is not included in $[t_e^0; t_e^f]$. Matching of entities are then checked inside the set W_t of terms whose position is included in $[p_t^{start}; p_t^{end}]$. The number c_t of entities in C_e having at least one matching in W_t is computed for every t, and the maximum value c_{max} is used to compute the event confidence $\gamma_e = \frac{c_{max}}{|C_e|}$. The automatic validation of a give event e is done by applying the evaluation policy described above for every document $d \in D_q$.

4   Experiments

We tested our approach by measuring the level of agreement between a human evaluator and our automatic event validation in evaluating events.

Dataset. We used a set E of 258 events[2], detected by an event detection algorithm between 18 January 2011 and 7 February 2011, where entities are defined as Wikipedia pages representing persons, locations, artifacts, and groups. Every event e is associated to a label $l_e \in [0; 1]$ given by a human evaluator, indicating the percentage of entities that truly participated to the event.

We also focused our analysis on different subsets : the *Clean Label Set* (E_{cl}), containing only those events whose l_e is either equal to 0 or to 1; the *Clean Confidence Set* (E_{cc}), containing those events for which the automatic evaluation returned either $\gamma_e = 0$ or $\gamma_e = 1$; the *Clean Set* ($E_c = E_{cl} \cap E_{cc}$), including events

---

[2] http://fs3.l3s.uni-hannover.de/~ceroni/ECIR_2014/dataset.xlsx

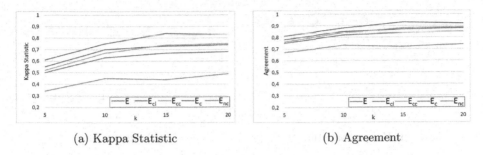

(a) Kappa Statistic   (b) Agreement

Fig. 2. Kappa Statistic and Agreement for different datasets and values of k

that have integer labels and confidences, i.e. $l_e = \{0,1\}$ and $\gamma_e = \{0,1\}$; the *Not Clean Set* (E_{nc}) containing all the events that are not contained in E_c.

Implementation Details. We used the Bing Search API to perform queries and to retrieve websites in terms of URLs. Document features have been extracted through the Stanford CoreNLP parser[3].

Evaluation Criteria. We measured the Kappa Statistic [2] (hereafter called KS) between human evaluator and automatic validation. It determines the level of agreement between two judges by taking into account also the probability that they agree by chance. We also report the rough percentage of Agreement (A) for sake of completeness. In order to assess the performances when either l_e or γ_e are strictly greater than 0 and lower than 1, we binarized their values by applying a threshold $\tau = 0.5$, i.e. $\gamma_e, l_e = 1$ if $\gamma_e, l_e > \tau$; $\gamma_e, l_e = 0$ otherwise.

We computed KS and A for our different datasets, for different values of k, and we repeated the validation at different time instants to experiment how variations in the document collection and in the ranking affect our approach.

Results. The performances of our approach for different datasets and different values of k are reported in Figure 2a (KS) and Figure 2b (A). The KS generally increases with increasing values of k, which means that the validation is more reliable if more documents are considered. The following discussion is done by considering the performances achieved with $k = 20$. The sizes of the different subsets are: $|E_{cl}| = 206$, $|E_{cc}| = 178$, $|E_c| = 152$, $|E_{nc}| = 106$.

Our approach achieves a KS equal to 0.68 over the entire event set E. According to the KS ranges introduced in [7], 0.68 can be considered as a *substantial* level of agreement. Higher values of KS are achieved over E_{cl} (0.74) and E_{cc} (0.75). The performance over dataset E_{cc} is particularly relevant because it measures how one can rely on the automatic validation when it returns a *strong* judgment regarding an event, i.e. strictly true or false. The highest KS value of 0.83, categorized as *almost perfect* agreement in [7], is achieved over E_c. On the contrary, the lowest KS value (0.49) is achieved over the dataset E_{nc}.

Finally, we performed the automatic validation for 4 different times in September 2013 and October 2013. The results, achieved by fixing $k = 20$, are shown

---

[3] http://nlp.stanford.edu/software/corenlp.shtml

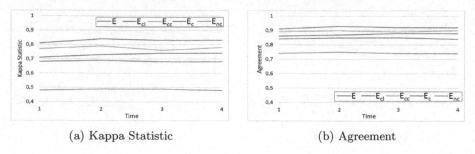

(a) Kappa Statistic   (b) Agreement

Fig. 3. Kappa Statistic and Agreement for different datasets over time ($k = 20$)

in Figure 3. It is possible to appreciate small variations for both KS (Figure 3a) and A (Figure 3b), although the overall trend can be recognized as constant in all the datasets. Variations can be due to different reasons, e.g. unavailability of Web pages, changes in the ranking parameters of the search engine, etc.

5   Conclusions and Future Work

In this paper we introduced the novel problem of Entity–based Automatic Event Validation as an attempt to automatize the event validation process. We also proposed a first approach which validates events by estimating the temporal relationships among their representative entities within documents on the Web. The preliminary results achieved in our work are promising, but also show that a consistent research effort still has to be spent on this problem.

The main future works that we envision are the development of other validation policies, a comparison over different ground truths, and the release of a reliable tool to be used by the research community to perform event validation.

Acknowledgments. This work was partially funded by the European commission in the context of the FP7 projects ForgetIT (grant no: 600826) and DURAARK (grant no: 600908).

References

1. Allan, J., Papka, R., Lavrenko, V.: On-line new event detection and tracking. In: ACM SIGIR (1998)
2. Carletta, J.: Assessing agreement on classification tasks: the kappa statistic. Comput. Linguist. (1996)
3. Das Sarma, A., et al.: Dynamic relationship and event discovery. In: WSDM (2011)
4. Fisichella, M., Stewart, A., Denecke, K., Nejdl, W.: Unsupervised public health event detection for epidemic intelligence. In: CIKM (2010)
5. Fung, G.P.C., Yu, J.X., Yu, P.S., Lu, H.: Parameter free bursty events detection in text streams. In: VLDB (2005)

6. Hoffart, J., Suchanek, F., Berberich, K., Weikum, G.: Yago2: A spatially and temporally enhanced knowledge base from Wikipedia. Artificial Intelligence (2012)
7. Landis, J.R., Koch, G.G.: The measurement of observer agreement for categorical data. Biometrics 33(1), 159–174 (1977)
8. Li, Z., Wang, B., Li, M., Ma, W.-Y.: A probabilistic model for retrospective news event detection. In: ACM SIGIR (2005)
9. Shan, D., et al.: Eventsearch: a system for event discovery and retrieval on multi-type historical data. In: SIGKDD 2012 (2012)

Query Expansion with Temporal Segmented Texts

Olga Craveiro[1,2], Joaquim Macedo[3], and Henrique Madeira[2]

[1] School of Technology and Management, Polytechnic Institute of Leiria, Portugal
[2] CISUC, Department of Informatics Engineering, University of Coimbra, Portugal
{marine,henrique}@dei.uc.pt
[3] Algoritmi, Department of Informatics, University of Minho, Portugal
macedo@di.uminho.pt

Abstract. The use of temporal data extracted from text, to improve the effectiveness of Information Retrieval systems, has recently been the focus of important research work. Our research hypothesis is that the usage of the temporal relationship between words improves the Information Retrieval results. For this purpose, the texts are temporally segmented to establish a relationship between words and dates found in texts. This approach was applied in Query Expansion systems, using a collection with Portuguese newspaper texts. The results showed that the use of the temporality of words can enhance retrieval effectiveness. In particular for time-sensitive queries, we achieved 9.5% improvement in Precision@10. To our knowledge, this is the first work using temporal text segmentation to improve retrieval results.

Keywords: Query Expansion, Temporal Information Retrieval, Temporal Text Segmentation.

1 Introduction

The importance of time in Information Retrieval has been studied from a various perspectives including document clustering, searching in time, Web archiving, etc. However, there is still much to do to achieve its full integration in the most popular retrieval models [1]. Some approaches only use the documents' metadata, such as document's timestamp, while some others also consider the temporal information found in the documents' content. However, none of them establishes an association between the words and the time, as it is proposed in our approach.

Establishing a relationship between words and time, we can temporally relate the words and make our interpretation of texts richer, allowing the identification of relations between entities, facts or events described by documents.

One of the possible applications for which words temporality could be of benefit is automated query expansion, henceforth QE. Traditionally, in QE, the relevant retrieved documents of an initial query are the top-ranked documents. Some approaches also incorporate the time dimension to improve traditional QE, although they only consider the documents' timestamps, unlike our approach which also uses the temporality of each word found in documents.

M. de Rijke et al. (Eds.): ECIR 2014, LNCS 8416, pp. 612–617, 2014.
© Springer International Publishing Switzerland 2014

The work of Amodeo et al. [3] relates the publication date of documents and their relevance for a given query to select the pseudo-relevant documents. The proposal of Whiting et al. [10] is to get the terms selection better, considering the temporal profile correlation between terms from the pseudo-relevant documents and temporally significant terms.

In the following, we present a brief description of the temporal segmentation, which is the base of this work. Section 3 describes three methods to incorporate temporal information into QE systems. Experiments and results are reported in Section 4 and we conclude in Section 5.

2   Temporal Segmentation of Texts

The temporal segmentation of texts has a crucial importance in this work. So, we present a brief description of the followed approach, detailed on our prior work [7].

The used segmentation is based on temporal discontinuities found in texts. For a given segment, dates and words establish relationships, later used on our QE methods.

The temporal segmentation partitions the text into coherent segments as in topic segmentation, but these segments are tagged with timestamps. In this way, the words of the segment become associated to the timestamps of the same segment.

In fact, the timestamps are the *chronons* collected from the text which belongs to the segment. A *chronon* is a normalized date which is anchored in a calendar/clock system by timelines defined by points. The timeline can have different levels of granules, such as year, month, week, day, and hour. To obtain the *chronons*, we developed a system which extracts the temporal information found in Portuguese texts, putting it into the *chronon* format. A detailed description of this system can be found in [5, 6].

A temporal segment is defined as a set of adjacent sentences that shares the same temporal focus. The segment length ranges from a single sentence to a multi-paragraph text. Thus, adjacent sentences with the same *chronons* must belong to the same segment.

The boundaries of a segment are firstly defined, by the temporal information found in the segment, unless this information does not exist in the text. In that case, the boundaries are identified by the topic changes using the cosine similarity formula.

Fig.1 presents two examples of temporal segmentation. In the first example, the segment "*A tempestade de Domingo causou alguns problemas nas redes de energia eléctrica. A empresa de energia eléctrica recebeu cerca de 31 mil chamadas.*"[1] is composed of two sentences. The first one has a date (*Domingo*) which was resolved by our system of temporal information extraction. In the next sentence, the topic stays the same, so that sentence will also belong to this segment.

The other segment "*Choveu na sexta-feira e no sábado.*"[2] is tagged with two *chronons*, because these two normalized dates are in the same sentence.

---

[1] English version: **Sunday**'s storm caused some problems in electricity networks. The company of electricity received about 31,000 calls.

[2] English version: It rained on **Friday** and **Saturday**.

(1) <SEGMENT DN="2011-10-31">A tempestade de **Domingo** causou alguns problemas nas redes de energia eléctrica. A empresa de energia eléctrica recebeu cerca de 31 mil chamadas.</SEGMENT>

(2) <SEGMENT DN="2011-11-10 2011-11-11">Choveu na **Sexta-feira** e no **Sábado**.</SEGMENT>

Fig. 1. Two examples of the temporal segmentation

3   Time-Aware Query Expansion

Since it is consensual that the time dimension is important in Information Retrieval systems, the dependence existent between words and dates given by the temporal segmentation of texts can be used to improve these systems. Considering the automated query expansion models as one of the possible applications, our proposal is to incorporate the word temporality in these models.

Starting from the general algorithm of QE, we propose three methods that exploit the temporal links between words in the different stages of the algorithm.

QE is used to increase the quality of the results retrieved from a user's request. The original query is expanded with more terms that have a similar meaning to define the query with more clarity and less ambiguity. In our proposal the new terms must have a temporal relationship with the terms of the original query. In other words, our methods promote the terms that occur in the same temporal segments of the query terms, by giving them more importance in the different steps of the query reformulation.

Temporal Filtering (TmpF). The main objective of this method is to decrease the number of candidate terms to be included in the reformulated query. The initial set of candidate terms is composed of all the existent terms in the content of the pseudo-relevant documents. Usually, a great number of systems apply stopwords removal and stemming to this set. After that, our method also applies a temporal filtering which only keeps in the candidates set, the terms that co-occur with the original query terms in the same temporal segments; all other terms are removed from the set.

Temporal Weighting (TmpW). This method penalizes the score assigned to the candidate terms that do not belong to the same temporal segments of the query terms. Indeed, the terms are not excluded, but they will be in different position of the ranking. So, the terms with a temporal relationship with the query terms are promoted in the ranking, obtaining more probability to be picked-up. Formula (1) computes the new score of the candidate terms. This formula uses a temporal distance $td(t)$ between the term t and the query terms, following a discrete approach.

$td(t)$ is 1 when t and the query terms are not in the same segments; otherwise 0.

$$score^*(t) = score(t) \times \frac{1}{1 + td(t)} \qquad td(t) \in [0,1] \qquad (1)$$

Temporal Reweighting (TmpR). This method is focused on the query reformulation without interfering with the terms selection, unlike the others. The set of terms for the reformulated query is the same. However, the weight of a term that co-occurs in the same temporal segments of the original query terms is increased, by using a weight δ. Formula (2) is based on the Rocchio's formula [9], where α and β are positive weights associated to the initial and the new query terms, respectively. The formula gives the weight to apply to each term, considering that $\beta+\delta\leq1$. $score(t)$ is the weight of the term t given by a term-weighting traditional formula.

$$w^*(t \mid q) = \alpha \times w(t \mid q) + (\beta + \delta) \times score(t) \qquad \beta+\delta\leq1 \qquad (2)$$

4   Experiments and Results

In order to evaluate the effectiveness of our temporal methods for improving QE, we carried out some experiments with the CHAVE collection created by Linguateca[3], which is the only test collection in Portuguese. The collection is composed of full-text from two major daily Portuguese and Brazilian newspapers, namely PUBLICO[4] and Folha de São Paulo[5], from complete editions of 1994 and 1995. This collection has a total of 210,734 documents.

The experiments were performed with 2 sets of topics: *TopicsSet1*, which is composed of all 100 topics of the collection, from C251 to C350, and *TopicsSet2*[6], which is only composed of time-sensitive topics with explicit temporal information in their title and/or their description, a total of 18 topics.

Documents were indexed using Terrier 3.5[7] [8] with stopwords removal and *PortugueseSnowballStemmer* applied for stemming. The documents were also temporally segmented. The top 1000 documents were retrieved for each topic using TF-IDF. The results are used as the baseline noQE to compare with our methods. We also report a run with the query expansion model Bo1, as the stronger baseline, named as Bo1QE. Bo1, developed by Amati [2], was the base model of our methods. In the experiments, the top *10* terms of the top *3* retrieved documents were used to expand the query. The weight of the terms in the expanded queries was calculated by the parameter free expansion option provided by Terrier, computing by the Amati's formula [2].

The retrieval effectiveness was measured by Mean Average Precision (MAP) for the top 1000 documents and Precision@10. For the robustness evaluation [4], we calculated the robustness index (RI) with the difference between the number of improved (IMP) and the number of penalized (PQ) queries based on MAP, dividing by the total number of queries.

---

[3] http://www.linguateca.pt
[4] http://www.publico.pt
[5] http://www.folha.com.br
[6] TopicsSet2: C326, C327, C332, C334, C335, C336, C337, C339, C340, C341, C343, C344, C345, C346, C347, C348, C349, and C350.
[7] http://ir.dcs.gla.ac.uk/terrier/

Two sets of *chronons* are created during the processing of a query: *QueryChronons*, and *CandidateTermsChronons*. They are composed of the segments timestamps where the query terms and the candidate term occur in pseudo-relevant documents, respectively. So, we study the number of *chronons* (j) associated to a candidate term that matches the set *QueryChronons*, by the variation of the parameter j between 1 and 10. We verified that the best results were obtained when $j<4$.

In TmpR evaluation, α was set to 1, given the most importance to the initial query terms. β and δ were ranged between 0.1 and 0.9, considering that $\beta+\delta\leq1$.

Table 1 shows the results obtained by our methods using the 2 topics sets. When the *TopicsSet1* is used, the methods achieved similar results, but all the results are better than both the baseline noQE and Bo1QE. The method TmpR with $\beta=0.4;\delta=0.3$ achieved the best MAP 0.363, although it has penalized one more query than the run Bo1QE100. The best Precision@10 and RI were achieved by TmpW.

The best results were achieved by our methods when the set of the 18 time-sensitive topics with explicit temporal information was used. Keeping the same behavior of the other topics set, TmpR and TmpW achieved the best results with a significant difference. In this case, TmpW achieved the best RI, and TmpR with $\beta=0.5;\delta=0.4$ obtained the best MAP and Precision@10.

With the results reported, we can conclude that our proposals methods can improve the retrieval effectiveness. Although, we also observed that our approach promote better results in time-sensitive queries, since the best results were obtained with *TopicsSet2_18*. We verified that TmpF becomes very restrictive in the process of terms selection. TmpW can enhance more queries, although TmpR improves the average precision, promoting the relevant documents in the ranking.

Table 1. Results obtained by the methods Temporal Filtering, Temporal Weighting and Temporal Reweighting using the CHAVE collection with 100 topics and 18 topics

TopicsSet1_100	noQE100	Bo1QE100	TmpF	TmpW	TmpR $\beta=0.4;\delta=0.3$
MAP	0.322	0.361	0.362	0.362	**0.363**
P@10	0.484	0.498	0.501	**0.502**	0.500
IMP; PQ		69; 29	70; 29	71; 28	69; 30
RI		0.400	0.410	**0.430**	0.390
TopicsSet2_18	noQE18	Bo1QE18	TmpF	TmpW	TmpR $\beta=0.5;\delta=0.4$
MAP	0.431	0.441	0.456	0.453	**0.457**
P@10	0.444	0.472	0.500	0.500	**0.517**
IMP; PQ		11; 5	14; 3	15; 3	13; 5
RI		0.333	0.611	**0.667**	0.444

5   Conclusions and Future Work

Unlike other proposals for Temporal Information Retrieval, our approach keeps the dependence between words and time given by the temporal segmentation. This work launches the foundations for future research on taking advantage of this dependence to improve the effectiveness of retrieval systems. In this paper, we presented the application of this approach to QE, although it can be applied to other issues and components of retrieval systems. We obtained auspicious results in the experiments carried out. We conclude that our approach provides a significant improvement in time-sensitive queries. For future work, we intend to combine words temporality with the words occurrence in the term-weighting formulas of the QE systems.

Acknowledgements. We are especially grateful to Catarina Reis for her collaboration, and the reviewers of the paper for their insightful comments and suggestions. This work is partially supported by Algoritmi and CISUC, financed by FCT - Fundação para a Ciência e Tecnologia, within the scope of the projects PEst-OE/EE/UI0319/2014 and PEst-OE/EE/UI0326/2014.

References

1. Alonso, O., Strötgen, J., Baeza-Yates, R., Gertz, M.: Temporal Information Retrieval: Challenges and Opportunities. In: 1st International Temporal Web Analytics Workshop (TWA-WWW 2011), pp. 1–8 (2011)
2. Amati, G.: Probability Models for Information Retrieval based on Divergence from Randomness. PhD thesis, Department of Computing Science, University of Glasgow (2003)
3. Amodeo, G., Amati, G., Gambosi, G.: On relevance, time and query expansion. In: Proceedings of the 20th ACM International Conference on Information and Knowledge Management, CIKM 2011, pp. 1973–1976. ACM, New York (2011)
4. Carpineto, C., Romano, G.: A Survey of Automatic Query Expansion in Information Retrieval. ACM Comput. Surv. 44(1), 1 (2012)
5. Craveiro, O., Macedo, J., Madeira, H.: Use of Co-occurrences for Temporal Expressions Annotation. In: Karlgren, J., Tarhio, J., Hyyrö, H. (eds.) SPIRE 2009. LNCS, vol. 5721, pp. 156–164. Springer, Heidelberg (2009)
6. Craveiro, O., Macedo, J., Madeira, H.: Leveraging temporal expressions for segmented-based information retrieval. In: ISDA, pp. 754–759. IEEE (2010)
7. Craveiro, O., Macedo, J., Madeira, H.: It is the time for Portuguese texts! In: Caseli, H., Villavicencio, A., Teixeira, A., Perdigão, F. (eds.) PROPOR 2012. LNCS, vol. 7243, pp. 106–112. Springer, Heidelberg (2012)
8. Ounis, I., Amati, G., Plachouras, V., He, B., Macdonald, C., Lioma, C.: Terrier: A High Performance and Scalable Information Retrieval Platform. In: Proceedings of ACM SIGIR 2006 Workshop on Open Source Information Retrieval (OSIR 2006), Seattle, Washington, USA (August 10, 2006)
9. Rocchio, J.: Relevance Feedback in Information Retrieval. In: Salton, G. (ed.) The SMART Retrieval System—Experiment in Automatic Document Processing, pp. 313–323. Prentice-Hall, New Jersey (1971)
10. Whiting, S., Moshfeghi, Y., Jose, J.M.: Exploring term temporality for pseudo-relevance feedback. In: Proceedings of the 34th International ACM SIGIR Conference on Research and Development in Information Retrieval, SIGIR 2011, pp. 1245–1246. ACM, NY (2011)

On Clustering and Polyrepresentation

Ingo Frommholz and Muhammad Kamran Abbasi

Institute for Research in Applicable Computing
University of Bedfordshire, UK
{ingo.frommholz,muhammad.abbasi}@beds.ac.uk

Abstract. Polyrepresentation is one of the most prominent principles in a cognitive approach to interactive information seeking and retrieval. When it comes to interactive retrieval, clustering is another method for accessing information. While polyrepresentation has been explored and validated in a scenario where a system returns a ranking of documents, so far there are no insights if and how polyrepresentation and clustering can be combined. In this paper we discuss how both are related and present an approach to integrate polyrepresentation into clustering. We further report some initial evaluation results.

1 Introduction

Polyrepresentation is one of the most prominent principles in a cognitive approach to interactive information seeking and retrieval [1,2]. The idea behind polyrepresentation is to make deliberate use of a variety of different representations to allow for the consideration of the different contexts or interpretations. Different cognitively (i.e. created by a different actor) and functionally different representations may exist on different levels, for instance by providing different views on information objects or different representations of the user's information need. In the literature the principle of polyrepresentation is applied and explored in various experimental settings and contexts ranging from information need cognitive perspectives [1,3], query expansion [4,5], polyrepresentation based implicit feedback [6], query representations [7], inter and intra document contexts [8] to information seeking strategies [9]. Besides this a good account of tested and expected possibilities regarding polyrepresentation is given in [10].

The principle of polyrepresentation states that if an information object (e.g. a document) is relevant with respect to more representations, the more likely it is relevant to the user's information need. The situation is depicted in Figure 1. In this example we assume 3 different representations. $\mathcal{R}$ denotes the relevance of representations. $\mathcal{R}_1$ is the set of documents relevant with respect to representation 1 (but not w.r.t. representation 2 and 3), $\mathcal{R}_{12}$ is the set of documents where representations 1 and 2 are relevant, but not representation 3, etc. $\mathcal{R}_0$ is the set of documents that are totally irrelevant w.r.t. any representation. Following this notation, $\mathcal{R}_{123}$ is the so-called *cognitive overlap* – the set of documents where all representations are relevant. According to the polyrepresentation principle many relevant documents can be found here, which has been confirmed in several experiments (see, e.g.[8,10,4]).

M. de Rijke et al. (Eds.): ECIR 2014, LNCS 8416, pp. 618–623, 2014.

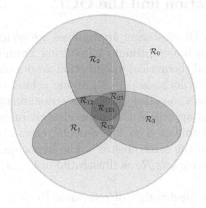

Fig. 1. Polyrepresentation-based relevance

However, to exploit the full potential of polyrepresentation we need to look beyond the cognitive overlap [11]. An example, here for the polyrepresentation of information objects, should illustrate this. Assume a user seeks for "good introductions to quantum mechanics". Certainly the content of a book (quantum mechanics) helps us to estimate relevance, but also other representations (e.g. reviews telling that this book is of introductory nature as well as ratings saying that a book is a good one) need to be involved. If a user just seeks for "good books about quantum mechanics", reviews may be less important while ratings and the content still are. If we only look at the cognitive overlap of all 3 representations we may fail to retrieve relevant documents for the latter query as it would ignore documents that are relevant by content and reviews only. An immediate problem arises, provided we have little to no information about the user's cognitive space: how should we rank documents outside the cognitive overlap? Should we consider $\mathcal{R}_{12}$ before $\mathcal{R}_{23}$? So far polyrepresentation has mostly been used as a means to rank documents. However, when it comes to interactive retrieval, *clustering* is another method for accessing information. Therefore our basic idea is to create clusters that correspond to the different sets $\mathcal{R}$ – polyrepresentation, like clustering, creates a partitioning of the document set based on which representations are relevant and clustering can be used to match these partitions $\mathcal{R}$. For instance, the cognitive overlap $\mathcal{R}_{123}$ would ideally correspond to a cluster C_{123} that contains documents that are highly relevant to all representations. Instead of producing a ranked list of documents, an interactive polyrepresentative IR system could present the user the cognitive overlap cluster first and cluster representations of other sets $\mathcal{R}$ as alternatives.

The rest of this paper is structured as follows. In the next section we briefly discuss how the OCF can be applied in a polyrepresentation context. In Section 3 we present some first evaluations before we conclude in Section 4.

2   Polyrepresentation and the OCF

As many contemporary IR clustering approaches are heuristic, recently a prob-abilistic framework known as Optimum Clustering Framework (OCF) [12] has been proposed as a sound theoretical justification for document clustering, which is based on the cluster hypothesis [13]. Applying a kind of "reverse cluster hypothesis" the OCF utilises so-called *query sets* – documents that are relevant to the same queries should appear in the same cluster. In our polyrepresentation scenario this can be reformulated as "documents relevant to the same representations should appear in the same cluster". The hypothesis is that this way the different polyrepresentative sets $\mathcal{R}$, as illustrated in Fig. 1, can be reconstructed by means of clustering.

The OCF utilises the probability of relevance $\Pr(R|q,d)$ for each document d and query q in the query set. Here we will apply this framework for polyrepresentation by creating a query set that is based on the different information need representations $r_i \in REP$ that the system considers. To apply OCF-based clustering we need to calculate $\Pr(R|d,r_i)$ for each document d first. After computing $\Pr(R|d,r_i)$ for each document d and each of the n representations, d is represented in OCF as a vector

$$\tau^T(d) = (\Pr(R|d,r_1),\dots,\Pr(R|d,r_n)).$$

Clustering algorithms may now apply for instance the cosine similarity of the $\tau(d)$ vectors as document similarity metrics.

3   Evaluation

3.1   Test Collection, Metrics and Evaluation Goal

The experiments were carried out on the PF subcollection of iSearch[1] [14], as this contains full texts. Similar to [15] we focused on the polyrepresentation of information needs, hence the selection of this collection. iSearch provides 65 search tasks with five representation for each task, which are Search Task, Work Task, Current Information Need, Background Knowledge and Ideal Answer. These five representations are used as the polyrepresentative description of an information need.

The goal of our evaluation, which is highly preliminary, is to gain initial insights about clustering and polyrepresentation. Here we focused on one task: can clustering reveal a cluster that is potentially the *real* cognitive overlap (the set of documents relevant w.r.t. all representations)? The problem we face is that we do not have an indication about the real cognitive overlap as the iSearch collection only provides relevance judgements for whole documents but not for single representations. Therefore, based on the relevance judgements, we identified for each iSearch search task 3 possible clusters that could initially be presented to

---

[1] http://itlab.dbit.dk/~isearch/

the user: C_{prec}, the cluster with the highest cluster precision (i.e. number of relevant document in the cluster divided by the number of documents in the cluster); C_{pair}, the cluster with the highest pairwise precision (see below) and C_{rep}, the cluster where all representations r_i highly contributed (i.e. $\Pr(R|d, r_i)$ is high for each r_i). By its definition the latter one would be the *computed* cognitive overlap and not be based on actual relevance judgements. Our question to investigate is: can clustering identify one of these 3 kinds of clusters? This translates into a cluster ranking task where we have to take into account the position in the ranking of the cluster under observation.

Our *pairwise precision* is derived from the pairwise precision used in [12] as a measure for cluster validity. The basic idea is to divide the pairs of relevant documents occurring in the same cluster by the total number of pairs in the cluster. In contrast to [12] we do not have relevance judgements for each of our representations. We therefore arrive at a simpler expression of the pairwise precision of a cluster that we use in our evaluation: $P_P(C_i) = |C_i| \frac{r_i(r_i-1)}{|C_i|(|C_i|-1)}$ with r_i the number of relevant documents in cluster C_i. C_{pair} is then the cluster so that $P_P(C_{pair}) \geq P_P(C_i) \ \forall C_i \in \mathcal{C}$.

3.2 Ranking Clusters

To match the different subsets from Figure 1 we computed $2^{|REP|}$ clusters, with $|REP|$ being the number of representations used. To estimate $\Pr(R|d, r_i)$ terms were weighted with BM25, as it was done in [12]. The documents were indexed and retrieved with Terrier 3.5[2] [16].

The maximum number of retrieved documents was set to the number of documents in the collection (143569). The documents were clustered using *k-means* with Euclidean distance. The mean cluster size was 4486.53 documents per cluster with a minimum and maximum standard deviation of 1186 and 9020 for the single search tasks. The minimum and maximum global cluster sizes were 133 and 44565 documents, respectively.

We ranked the clusters according to different criteria. One criterion was the *arithmetic mean* of each cluster [17] computed as $arith(C) = \frac{1}{|C|} \sum_{d \in C} \sum_{i=1}^{n} \frac{\Pr(R|d, r_i)}{n}$. The other criterion is derived from the notion of expected F-measure in the context of the OCF [12]. For a cluster C in the clustering $\mathcal{C}$ let $\sigma(C) = \frac{1}{|C|-1} \sum_{(d_l, d_m) \in C_i \times C_i} \tau(d_l)^T \times \tau(d_m)$ $(l \neq m)$ if $|C| > 1$, and 0 otherwise. Then the *expected pairwise precision* of C is defined as $\pi(C) = |C|\sigma(C)$. Similarly, the *expected recall* is defined as $\rho(C) = |C_i|(|C_i| - 1)\sigma(C_i)$. Based on this, the *expected F-measure* is

$$eF(C) = \frac{2}{\frac{1}{\pi(C)} + \frac{1}{\rho(C)}}. \tag{1}$$

---

[2] http://terrier.org/

3.3 Results

Our preliminary evaluation focused on the ability of clustering to identify a candidate for the cognitive overlap. We applied the well-known MRR measure using the rank of the appearance of the clusters C_{prec}, C_{pair} and C_{rep}, respectively[3]. Table 1 shows the different MRR values for the different cluster ranking methods.

Table 1. MRR values for different cluster ranking strategies

	C_{prec}	C_{pair}	C_{rep}
$arith(C)$	0.337	0.303	0.575
$eF(C)$	0.113	0.112	0.0757

Our results show that the $arith(C)$ based ranking performs better than the eF based ranking. $arith(C)$ rewards clusters with high $\Pr(R|d, r_i)$, which explains its fair C_{rep} performance. Not surprisingly eF performs slightly better when it comes to C_{prec} and C_{pair} than regarding C_{rep}, which again can be explained by the way we defined eF. But still it needs to be investigated why eF produces lower values than $arith(C)$. As a whole the $arith(C)$ scores suggest that the cluster-based polyrepresentation can be a feasible option, though it requires further examination as this paper can only be the beginning.

4 Conclusion and Future Work

In this paper we have discussed how clustering and polyrepresentation can go along together. Based on the OCF we defined a clustering approach that makes use of different information need representations as applied in polyrepresentation. Our preliminary results show that the $arith(C)$ based ranking performs better than the $eF(C)$ based ranking for this task. We take this as a motivation to continue our work, also bearing in mind that the iSearch collection in general produced rather lower evaluation values [15,18], which may explain our results as well.

A deeper analysis of the relationship of clustering and polyrepresentation will be part of our future work. Further evaluations with the iSearch collection will be performed, in particular involving a proper non-polyrepresentative baseline. We will also have a look at the polyrepresentation of documents: if $d_{r_1}, \ldots, d_{r_n}$ are different representations of a document d and q is a query, then we can apply the OCF based on $\Pr(R|d_{r_i}, q)$. A possible collection to perform these experiments is Amazon+LibraryThing [9].

---

[3] Note that in our experiments all these clusters were non-ambiguous.

References

1. Ingwersen, P.: Polyrepresentation of Information Needs and Semantic Entities, Elements of a Cognitive Theory for Information Retrieval Interaction. In: Croft, B.W., van Rijsbergen, C.J. (eds.) Proceedings SIGIR 1994, pp. 101–111. Springer, London (1994)
2. Ingwersen, P., Järvelin, K.: The turn: integration of information seeking and retrieval in context. Springer-Verlag New York, Inc., Secaucus (2005)
3. Ingwersen, P.: Cognitive perspectives of information retrieval interaction: Elements of a cognitive IR theory. The Journal of Documentation 52, 3–50 (1996)
4. Kelly, D., Fu, X.: Eliciting better information need descriptions from users of information search systems. Information Processing & Management 43(1), 30–46 (2007)
5. Diriye, A., Blandford, A., Tombros, A.: A polyrepresentational approach to interactive query expansion. In: Proceedings JCDL 2009 (2009)
6. White, R.W.: Using searcher simulations to redesign a polyrepresentative implicit feedback interface. Information Processing & Management 42(5), 1185 (2006)
7. Efron, M., Winget, M.: Query Polyrepresentation for Ranking Retrieval Systems Without Relevance Judgments. JASIST 61(6), 1081–1091 (2010)
8. Skov, M., Larsen, B., Ingwersen, P.: Inter and intra-document contexts applied in polyrepresentation for best match IR. Information Processing & Management 44(5), 1673–1683 (2008)
9. Beckers, T.: Supporting Polyrepresentation and Information Seeking Strategies. In: Proceedings of the 3rd Symposium on Future Directions in Information Access (FDIA) (April 2009)
10. Larsen, B., Ingwersen, P., Kekäläinen, J.: The polyrepresentation continuum in IR. In: Proceeding IIiX 2006, pp. 88–96. ACM, New York (2006)
11. Frommholz, I., Larsen, B., Piwowarski, B., Lalmas, M., Ingwersen, P., van Rijsbergen, K.: Supporting Polyrepresentation in a Quantum-inspired Geometrical Retrieval Framework. In: Proceedings of IIiX 2010, pp. 115–124. ACM, New Brunswick (2010)
12. Fuhr, N., Lechtenfeld, M., Stein, B., Gollub, T.: The Optimum Clustering Framework: Implementing the Cluster Hypothesis. Information Retrieval 14 (2011)
13. van Rijsbergen, C.J.: Information Retrieval. Butterworths, London (1979)
14. Lykke, M., Larsen, B., Lund, H., Ingwersen, P.: Developing a Test Collection for the Evaluation of Integrated Search. In: Gurrin, C., He, Y., Kazai, G., Kruschwitz, U., Little, S., Roelleke, T., Rüger, S., van Rijsbergen, K. (eds.) ECIR 2010. LNCS, vol. 5993, pp. 627–630. Springer, Heidelberg (2010)
15. Lioma, C., Larsen, B., Ingwersen, P.: Preliminary experiments using subjective logic for the polyrepresentation of information needs. In: Proceedings IIiX 2012, pp. 174–183. ACM Press, New York (2012)
16. Ounis, I., Amati, G., Plachouras, V., He, B., Macdonald, C., Lioma, C.: Terrier: A High Performance and Scalable Information Retrieval Platform. In: Proceedings of ACM SIGIR 2006 Workshop on Open Source Information Retrieval, OSIR 2006 (2006)
17. Raiber, F., Kurland, O.: Exploring the cluster hypothesis, and cluster-based retrieval, over the web. In: Proceedings CIKM 2012, pp. 2507–2510. ACM Press, New York (2012)
18. Schaer, P.: Der Nutzen informetrischer Analysen und nicht-textueller Dokumentattribute für das Information Retrieval in digitalen Bibliotheken. Phd thesis (2013)

Effects of Position Bias
on Click-Based Recommender Evaluation

Katja Hofmann[1], Anne Schuth[2], Alejandro Bellogín[3], and Maarten de Rijke[2]

[1] Microsoft Research
katja.hofmann@microsoft.com
[2] ISLA, University of Amsterdam
{anne.schuth,derijke}@uva.nl
[3] CWI
alejandro.bellogin@cwi.nl

Abstract. Measuring the quality of recommendations produced by a recommender system (RS) is challenging. Labels used for evaluation are typically obtained from users of a RS, by asking for *explicit* feedback, or inferring labels from *implicit* feedback. Both approaches can introduce significant biases in the evaluation process. We investigate biases that may affect labels inferred from *implicit* feedback. Implicit feedback is easy to collect but can be prone to biases, such as position bias. We examine this bias using click models, and show how bias following these models would affect the outcomes of RS evaluation. We find that evaluation based on implicit and explicit feedback can agree well, but only when the evaluation metrics are designed to take user behavior and preferences into account, stressing the importance of understanding user behavior in deployed RSs.

1 Introduction and Related Work

Recommender systems (RSs) aim to recommend to their users items of interest, such as movies, news articles, or music. Despite the success and popularity of such systems, measuring the quality of an RS is a challenge. In this paper, we examine how bias in user interactions, when used as implicit feedback, may affect RS evaluation.

Traditionally, RS evaluation has built upon *explicit* feedback, often in the form of user ratings, and error-based metrics such as Mean Absolute Error (MAE) or Root Mean Squared Error (RMSE) [9]. Recently, ranking-based metrics, such as precision, or normalized Discounted Cumulative Gain (nDCG) are being considered [1]. In both cases, ground truth typically comes in the form of user ratings for items. Creating explicit ratings for RS evaluation requires substantial user effort. In some applications, such as movie recommenders, users may be willing to expend that effort, especially when the rating process is embedded in an engaging interface. However, in applications such as news or music recommendation, explicit ratings are typically sparse and hard to obtain.

To increase the amount of data available for evaluation (or learning), researchers and practitioners have started to consider *implicit* feedback. Implicit feedback is a natural side-product of user interactions: labels are inferred based on, e.g., how many times an item has been clicked, viewed, or purchased [6, 8]. Given access to a deployed RS,

M. de Rijke et al. (Eds.): ECIR 2014, LNCS 8416, pp. 624–630, 2014.
© Springer International Publishing Switzerland 2014

collecting implicit feedback is typically easier, and the amount of data more plentiful, than for explicit feedback. However, accurately interpreting implicit feedback is non-trivial [7, 8].

In this paper we investigate how RS evaluation based on implicit feedback relates to traditional rating-based evaluation, and how evaluation outcomes may be affected by bias in user behavior. Unlike ratings, click feedback does not capture negative feedback, as it conflates items that were not clicked because they were not examined and items that were examined but not clicked because they were disliked. Interpreting user interactions as feedback therefore requires a good model of how users interact with the system and how those interactions reflect experienced system quality. Here, we specifically focus on the well-known position bias [7]. We investigate this effect using bias models that capture how user behavior is hypothesized to be affected by the rank and quality of previously viewed items. We investigate how the estimated quality of a set of standard retrieval models is affected by these models, compared to the unbiased rating-based estimates.

In our idealized experimental setup, bias in user behavior can strongly affect the inferred outcomes of evaluation. Good agreement between evaluation based on implicit and explicit information can be achieved when metrics reflect user behavior and expectations.

2 Approach

To investigate bias in RS evaluation with implicit feedback, we need a way to generate implicit feedback under varying models of bias. Here, we leverage a simulation setup that was previously proposed for assessing IR evaluation methods [5]. It simulates the interactions between an interactive system and its users, so that the behavior of the system can be tested under varying assumptions about user behavior. We simulate the interactions between a representative set of RSs and user models that capture key aspects of position bias. The RSs are first trained on a representative data set, and then "deployed" to interact with the generative user models. The models simulate interactions that reflect the assumptions under which we want to examine RS evaluation. All simulated interactions are recorded and applied to evaluate each system. Finally, we compare the evaluation outcomes to each other and to ground truth evaluations to identify cases where bias affects outcomes. Next, we detail our user models and their assumptions on user behavior.

2.1 User Models

We describe the implemented user models and how they capture user behavior and position bias. These models have been developed for the IR domain, and have been evaluated in large-scale evaluations using click log data [2–4]. We start with the simplest possible model, and build up complexity to cover assumptions that may affect evaluation outcomes based on the resulting clicks on recommended items. We phrase models in terms of clicks, but other user actions could be interpreted in the same way, e.g., watching a recommended movie, saving an item as a favorite, or rating it.

Table 1. User model specifications and definitions of examination probabilities

model	intuition	parameters	definition
examination [10]	examination probability depends only on item position	item position (a_p)	$P(E = 1|p(i)) = a_p$
browsing [3]	users give up examination after seeing many bad items	distance from previous clicked item (a_d)	$P(C = 1|0) = 1$ $P(E = 1|d) = a_d$
cascade [2, 4]	user step through a list of items from top to bottom, stop when satisfied	satisfaction with previous item (s_{i-1})	$P(E = 1|p(i) = 1) = 1$ $P(E = 1|i, s_{i-1}) = 1 - s_{i-1}$

We consider three click models, which we refer to as examination, cascade, and browsing model. All models capture the probability of a click on a given item, $P(C = 1|i)$, by decomposing this probability into examination and relevance components: $P(C = 1|i) = P(E = 1|i)P(R = 1|i)$. Here, $P(E = 1|i)$ is the probability that item i is examined by the user. This may depend on several factors, as will be shown below. The second component, $P(R = 1|i)$, captures the relevance of the item for a given user. Items that are not examined are never clicked, i.e., $E_i = 0 \Rightarrow P(C_i) = 0$.

The user models differ in how they capture items' examination probabilities (see Table 1). In the *examination model* [10], the examination probability depends on the position of an item (represented by a_p), and it is independent of the examination probabilities of all other items. The *browsing model* (based on the User Browsing Model, UBM [3]) represents examination probabilities as a function of the distance from the closest previously clicked item: users may be more motivated to continue examining items if they liked the previous items. The *cascade model* [2] (cf., the extension for multiple clicks [4]) relaxes this independence assumption, and posits a linear examination order. Users are assumed to examine one item at a time, clicking on promising ones, until a clicked item was found to be satisfactory (s_{i-1} captures the satisfaction probability of the previous item). A special case of all three click models is the *no-bias* model, which sets the examination probability of all items to 1. We use this special case as baseline.

2.2   Ground Truth and Metrics

Metrics are used in two ways. First, we choose the metrics for implicit and explicit evaluation to reflect a typical RS evaluation setup. Second, we detail how the results are compared to each other, to assess differences and similarities in how the explicit and implicit evaluation metrics judge RS performance.

We assess the following *explicit* metrics, based on explicit user-item ratings provided with our data set: (1) Precision at N (P@N), the portion of highly rated items in the top N recommended items; (2) NDCG@N, a commonly used IR metric that rewards systems for showing highly rated items as high as possible in a recommendation list. We compare these metrics for explicit ratings to the *implicit* click through rate (CTR) at N, the ratio of clicked to non-clicked items in the top N recommended items.

We compare the outcomes obtained using different metrics as follows: (a) Rank the RSs by their evaluation score and compare the resulting rankings; (b) Consider the task of predicting performance based on implicit ratings from those obtained using explicit ratings. We train linear regression models for this prediction task and compare how well the performance under each user model (in terms of CTR) can be predicted from explicit rating scores. We estimate the fit of such a model using the residual standard error (RSE, i.e., how much the predicted performance deviates from actual performance).

3 Experiments

We describe the RSs we used, the data sets and the parameter settings.

3.1 Recommender Systems

We compare the following types of recommender:(1) Two non-personalized methods serve as baselines: a random (RND) recommender and a popularity-based recommender (ItemPop) that ranks the item according to the observed popularity in the training set. (2) An item-based (IB) collaborative filtering recommender; we use Pearson's correlation as similarity between the items [9]. (3) A matrix factorization (MF) method as implemented in Mahout[1] using the Expectation Maximization algorithm. We used 100 iterations and 50 features, leaving the rest of the parameters as default (i.e., 0.005 as learning rate, 0.02 as regularization parameter, and 0.005 as random noise). (4) A user-based (UB) collaborative filtering method using 50 neighbors and Pearson's correlation as similarity metric [9]. These four types of recommender (ignoring RND) are among the best performing ones in the literature [9].

3.2 Data

We obtain the ground truth and train the recommenders on the corresponding test and training splits (in an 80–20 ratio) derived from a version of the Movielens dataset;[2] we used Movielens 1M, with 1 million ratings by $6,040$ users to $3,900$ items.

3.3 Parameter Settings

For each RS, we obtain the recommendations for all users in the test set. We simulate user interactions with these recommendations for each user model and record the generated interactions. We then compute the performance of each method using this data, and average results over 25 repetitions. We compare the resulting evaluation scores across models and systems, and also compare to ground truth scores based on the explicit ratings provided with the data sets as described in §2.2. We assume that users examine a maximum of 10 items and we compute all metrics on the top $N = 10$ item. P@10 and nDCG@10 are computed using only highly rated items (labels 4 and 5).

We instantiate the parameters of the user models (cf., §2.1) to capture a wide spectrum of possible user behavior. For the examination model we use *logarithmic* (log)

---

[1] http://mahout.apache.org
[2] http://www.grouplens.org/node/73

$(a_p = 1/\log_2(p+1))$ or *quadratic* (quad) $(a_p = 1/2^{(p-1)})$ decay, $p > 0$. For the cascade model we use, for a previous item, if it is non-clicked, $s_{i-1} = 0$; otherwise, we instantiate the satisfaction $s_{i-1}(r)$, depending on the rating r of that item as follows; (*low*): $s_{i-1}(2) = 0.2$, $s_{i-1}(3) = 0.3$, $s_{i-1}(4) = 0.4$, and $s_{i-1}(5) = 0.5$; (*high*): $s_{i-1}(2) = 0.3$, $s_{i-1}(3) = 0.5$, $s_{i-1}(4) = 0.7$, and $s_{i-1}(5) = 0.9$. For the browsing model, a_d decays with the distance from the closest click above item i with logarithmic or quadratic decay, as defined for the examination model. All models use the same instantiations of probabilities $P(R = 1|i) = c_i(r)$ for examined items i with rating r, viz. $c_i(1) = 0.0$, $c_i(2) = 0.1$, $c_i(3) = 0.2$, $c_i(4) = 0.8$, and $c_i(5) = 1.0$ (chosen to reflect the quadratic gain values of nDCG). Unrated items are not clicked.[3]

4   Results

Table 2 shows the evaluation scores for each RS under the different evaluation approaches. For rating-based evaluation, we see good agreement between system performance under nDCG and Precision. In line with the RS literature, ItemPop achieves the highest scores. IB performs lowest: it recommends a high portion of items for which no ratings are available, a well-known problem. Good agreement is also observed between rating-based precision and CTR under the no-bias model (RSE = 0.00017). Thus, when no position bias is present, precision can be used as an almost perfect predictor of CTR.

When position bias is present, agreement with explicit scores can still be high, if the assumptions about user behavior underlying the metric properly reflect user behavior. This can be seen for the *browse-log* and *exam-log* models, which both show good agreement with nDCG. Both models and nDCG assume that users are more likely to examine, and therefore benefit from, highly-ranked items. However, the decay in examination probability is weak (logarithmic). As expected, the fit between explicit metrics and CTR worsens as assumptions about user behavior deviate. This can be seen for the *exam-quad* and *browse-quad* models, where position bias is much stronger than assumed by nDCG. The RSE when predicting CTR under quadratic decay in examination probabilities is up to 3.5 times (0.00254) higher than under the log decay. In addition, two RSs, viz. MF and UB, switch ranks under these models. This means that, if explicit-rating scores were used to decide which of these models to use in production, the decision may not agree with the performance that would be observed under implicit feedback. The worst fit is observed under the cascade model, where the examination probabilities depend on whether users were satisfied with higher-ranked items. Neither the strong position bias nor the dependencies between items are captured by the rating-based metrics.

Fig. 1 examines the fit of CTR under *no-bias*, *cascade-high*, and *browse-quad*. We can see the good agreement between precision and CTR for the *no-bias* model. These setups agree in the magnitude of the scores; the residuals are close to zero for all systems. But if user behavior is best reflected by the cascade model and high satisfaction probability, precision systematically over-estimates system performance. The system for which predicted performance deviates most from the observed CTR is MF, but relatively high residuals are observed for all systems. Performance would be most strongly over-estimated in cases where the *browse-quad* model best reflects user behavior. Under this

---

[3] All our code is open source and available at `https://bitbucket.org/ilps/lerot` [11].

Table 2. Performance of systems when measured using clicks produced by several click models, compared to the rating-based equivalent. Changes in system rank with respect to rating-based evaluation, and models with the highest RSE per metric are highlighted in **bold**.

	Rating-based		Click-based (CTR)						
	nDCG	Precision	no-bias	exam-log	exam-quad	browse-log	browse-quad	cascade-low	cascade-high
ItemPop	1 (0.150)	1 (0.146)	1 (0.139)	1 (0.081)	1 (0.020)	1 (0.088)	1 (0.020)	1 (0.087)	1 (0.067)
MF	2 (0.043)	2 (0.049)	2 (0.048)	2 (0.023)	**3** (0.001)	2 (0.025)	**3** (0.001)	2 (0.039)	2 (0.034)
UB	3 (0.023)	3 (0.028)	3 (0.026)	3 (0.014)	**2** (0.003)	3 (0.015)	**2** (0.003)	3 (0.022)	3 (0.020)
RND	4 (0.005)	4 (0.006)	4 (0.006)	4 (0.003)	4 (0.000)	4 (0.003)	4 (0.000)	4 (0.006)	4 (0.006)
IB	5 (0.000)	5 (0.000)	5 (0.001)	5 (0.000)	5 (0.000)	5 (0.000)	5 (0.000)	5 (0.001)	5 (0.001)
RSE nDCG			0.00363	0.00083	0.00242	0.00070	0.00242	0.00683	**0.00711**
RSE Precision			0.00017	0.00197	0.00286	0.00254	0.00286	0.00471	**0.00550**

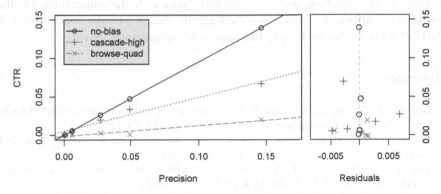

Fig. 1. Precision and CTR scores for selected user models with fitted linear model and residuals

model, the performance of MF is most strongly affected, with precision over-estimating the score (this reflects that MF is not trained specifically for ranking [6]).

Agreement between explicit rating-based evaluation and click-based evaluation depends on the degree to which assumptions about user behavior are captured by evaluation metrics. We found best agreement between precision and CTR when no position bias is present. Best agreement with nDCG was observed for the *exam-log* and *browse-log* models. Despite the small number of systems compared, we observe changes in system rankings under the *browse-quad* and *exam-quad* user models. The lowest agreement was between explicit ratings and the cascade model.

5 Conclusion

We studied the effect of position bias on RS evaluation using implicit user feedback. We modeled position bias using click models that capture key aspects of user behavior, and compared RS performance under these models to that obtained under traditional, rating-based evaluation. Good agreement between click-based and rating-based evaluation is possible if rating-based metrics properly reflect user behavior. System performance can be under- or over-estimated when user behavior strongly deviates from the assumptions underlying the metrics; this effect was strongest under steep position bias.

Our results have implications for theory and practice of rating-based and click-based RS evaluation. First, they highlight the importance of understanding user behavior in deployed systems. Only when interactions are well understood can we design metrics that accurately reflect users' perception of RS quality. A direction for future work is to investigate how bias affects RS when implicit feedback is used for training.

Acknowledgments. This research was supported by the European Community's Seventh Framework Programme (FP7/2007-2013) under grant agreement nr 288024 (LiMoSINe project), the Netherlands Organisation for Scientific Research (NWO) under project nrs 640.004.802, 727.011.005, 612.001.116, HOR-11-10, the Center for Creation, Content and Technology, the QuaMerdes project funded by the CLARIN-nl program, the TROVe project funded by the CLARIAH program, the Dutch national program COMMIT, the ESF Research Network Program ELIAS, the Elite Network Shifts project funded by the Royal Dutch Academy of Sciences, the Netherlands eScience Center under project nr 027.012.105 and the Yahoo! Faculty Research and Engagement Program.

References

[1] Bellogín, A., Castells, P., Cantador, I.: Precision-oriented evaluation of recommender systems: an algorithmic comparison. In: RecSys 2011, pp. 333–336 (2011)

[2] Craswell, N., Zoeter, O., Taylor, M., Ramsey, B.: An experimental comparison of click position-bias models. In: WSDM 2008, pp. 87–94 (2008)

[3] Dupret, G.E., Piwowarski, B.: A user browsing model to predict search engine click data from past observations. In: SIGIR 2008, pp. 331–338 (2008)

[4] Guo, F., Liu, C., Wang, Y.M.: Efficient multiple-click models in web search. In: WSDM 2009, pp. 124–131 (2009)

[5] Hofmann, K., Whiteson, S., de Rijke, M.: A probabilistic method for inferring preferences from clicks. In: CIKM 2011, pp. 249–258 (2011)

[6] Hu, Y., Koren, Y., Volinsky, C.: Collaborative filtering for implicit feedback datasets. In: ICDM 2008, pp. 263–272. IEEE Computer Society (2008)

[7] Joachims, T., Granka, L., Pan, B., Hembrooke, H., Gay, G.: Accurately interpreting click-through data as implicit feedback. In: SIGIR 2005, pp. 154–161. ACM (2005)

[8] Oard, D., Kim, J.: Implicit feedback for recommender systems. In: AAAI Workshop on Recommender Systems, pp. 81–83 (1998)

[9] Ricci, F., Rokach, L., Shapira, B., Kantor, P.B.: Recommender Systems Handbook. Springer (2011)

[10] Richardson, M., Dominowska, E., Ragno, R.: Predicting clicks: estimating the click-through rate for new ads. In: WWW 2007, pp. 521–530. ACM (2007)

[11] Schuth, A., Hofmann, K., Whiteson, S., de Rijke, M.: Lerot: an Online Learning to Rank Framework. In: LivingLab 2013, pp. 23–26. ACM (2013)

Bringing Information Retrieval
into Crowdsourcing: A Case Study

Qinmin Hu and Xiangji Huang

Information Retrieval and Knowledge Managment Research Lab
York University, Toronto, Canada
vhu@cse.yorku.ca, jhuang@yorku.ca

Abstract. We propose a novel and economic framework for bringing information retrieval into crowdsourcing. Both crowdsourcing and information retrieval achieve mutual benefits, which result in (1) workers' quality control by using the query-oriented training; (2) cost savings in money and time; and (3) better qualified feedback information. In our case study, the costs of crowdsourcing for 18,260 jobs are as low as $47.25 and as short as 5 hours in total. Furthermore, the experimental results show that information retrieval techniques greatly reduce the workloads of crowdsourcing, which is only 5% of the original work. At the other hand, crowdsourcing improves the accuracy of the information retrieval system through providing qualified feedback information.

1 Introduction and Motivation

Crowdsourcing is an online practice, which describes the act of outsourcing work to a large group of people of a community as a crowd. Nowadays, crowdsourcing has attracted growing attentions as a valuable solution to harness human abilities from a large population of workers [1]. The crowdsourcing of relevance judgements enables the evaluation of the information retrieval (IR) systems on the large-scale data sets.

The text relevance assessing task (TRAT) is a task in the TREC Crowdsourcing Track [2]. The goal of TRAT is to evaluate approaches to text relevance assessing. Different with the other tracks such as genomics and blog, we simulate playing the relevance assessing role of NIST in the TREC 8 ad-hoc track [3] with a subset of the TREC 8 queries. There are 18,260 query-document pairs to be judged under 10 queries.

The motivation of bringing the IR systems into the crowdsourcing procedure is that the cost will be very heavy if we manually ask the workers through the crowdsourcing system to judge the given 18,260 query-document pairs directly. What's more, the accuracy will be reduced, if there are more pairs judged by the workers, since the quality control of the workers is a big challenge in crowdsourcing. Therefore, we are motivated to not only train the workers by the query-oriented gold standard, but also refine the number of pairs to be manually judged using the IR techniques.

M. de Rijke et al. (Eds.): ECIR 2014, LNCS 8416, pp. 631–637, 2014.

Here we propose an IR-crowdsourcing framework in Figure 1, which obtains the benefits of both crowdsourcing and IR systems. The crowdsourcing system we use is called CrowdFlower[1]. The IR models are BM25, DFR, BM25_DFR and language model (LM) [4–8]. The feedback methods are interactive feedback, tf-idf feedback, modified pseudo feedback and proximity feedback [9]. There are only \$47.25 costed in the whole experiments and around 5 hours for both worker training and real jobs.

2 CrowdFlower

CrowdFlower uses crowdsourcing techniques to provide a wide range of enterprise solutions which process or create large amounts of data. CrowdFlower has over 50 labour channel partners, including Amazon Mechanical Turk and TrialPay; their network is composed of over 2 million Contributors from all over the world.

2.1 Customized Page Design

We design our page to catch the feedback information from the workers. First, the instructions are given for each task, especially the judgements of "relevance".

Input:
Queries and Data sets, e.g. TREC-8

Output:
A list of relevance judgements for the query-document pairs, which includes the queries, the document number, the binary judgement (0 for non-relevant and 1 for relevant), the probabilities of relevance and the run tag.

Procedure
(1) Preliminary results by the IR systems:
    (a) The IR systems output the first-round ranked documents under each query;
    (b) Obtain the top m documents as the candidates in crowdsourcing.
(2) The crowdsourcing stage:
    (a) Design the page in CrowdFlower;
    (b) Train the workers by queries;
    (c) Qualified workers manually judge the m document candidates as relevant/irrelevant;
    (d) Extract the feedback information.
(3) Evidence results by the IR systems:
    (a) Put the feedback information into the IR systems;
    (b) Documents candidates are updated.
(4) Repeat (2) and (3) if necessary.
(5) Output the list of judgements.

Note that:
(1) the IR systems generate N-rounds results based on the users' setting, in order to update the document candidates for crowdsourcing;
(2) the challenge steps at crowdsourcing are how to train the workers and how to design the page to make the workers involved and give useful feedback information.

Fig. 1. The Proposed Framework

---

[1] http://www.crowdfolower.com

"Relevance" judgement is not a particularly well-defined term. There is much previous work which shows that a very large number of factors affect whether a user will judge a document as relevant to her/his information needs [3]. Therefore, we adopt NIST's definition of relevance as that a document is relevant if it would be useful when the worker was writing a report on the subject of the given query. In particular, more useful strategies are defined as follows: (1) be consistent in all the judgements; (2) a document should be be judged as relevant or irrelevant based only on the title of the document; (3) do not rely too heavily on scanning terms which could be misleading the judgements.

Second, the query is well presented to the workers' reading and understanding, including the description and narrative of the relevance and irrelevance to the query when a document is judged. The document candidate is also shown as the typical Web page for the workers' reading. Third, the binary judgement of relevance or irrelevance and the keywords to support the workers' opinion, are mandatory to be submitted, in order to finish their jobs.

2.2   Quality Control by Query-Oriented Training

CrowdFlower stands apart from these individual networks because they offer the quality control, called Gold Standard Data, which has workers perform pre-completed tasks to determine their accuracy and trustworthiness. So we adopt the TREC 7 queries as the training queries to find the valued workers. More specifically, we do query-oriented training with bonus rewards.

First, we classify the given 10 queries in TRAT as four categories, according to their own aspects. Then, 2 similar queries for each category are carefully selected from TREC 7. Therefore, there are 8 similar queries in TREC 7, corresponding to the given 10 queries in TREC 8. Second, the gold standard data of the 8 queries of TREC 7 are put into CrowdFlower as the evaluation to find the trusted workers. Note that the trusted workers indicate ones whose TREC 7 queries' judgements are reliable. Third, we send private links of the TREC 8 query-document jobs to the trusted workers only. Particularly, the workers will get the same category jobs, not crossing over the different categories. Fourth, the trusted workers are paid by the bonus rewards, in order to encourage them to complete the real jobs better, such as extra 10 cents per job. Fifth, we ask 10 judgements from 10 workers for each job, as a peer review method.

3   Information Retrieval Stage

Here we present how the IR systems obtain the preliminary results and the evidence results in Figure 3. Four IR models as BM25, DFR, BM25_DFR and LM are adopted. Four feedback methods are applied in the systems as well. The detailed descriptions about the four IR models are shown in [4–6].

At the preliminary step, the IR systems find the most likely relevant documents as the candidates for the manually judgements in the crowdsourcing stage. Note that the candidate for crowdsourcing are the top five common documents in all four retrieved lists.

```
Input:
The top five overlapped retrieved documents, for each query.

Output:
Manually judged relevant documents and the supportive keywords.

(1) Design the page to make the workers involved and give useful feedback information;
(2) Train the workers by query-oriented training and find the qualified ones;
(3) Send the jobs to the trusted workers via the private link;
(4) Clean the workers' feedback.
```

Fig. 2. The Crowdsourcing Stage

At the evidence step, the IR systems treat the manually feedback information in four ways: (1) directly use the manually judged documents and the keywords provided by the workers in the IR systems; (2) use the manually judged documents and then adopt the proximity feedback method proposed by [9] to get the weighted feedback terms; (3) use the manually judged documents and then apply the standard pseudo feedback method [10]; (4) count the TF-IDF of the terms in the manually judged documents and extract the top 20 terms as the feedback terms. These are also our four runs in the experiments for evaluation.

4   Empirical Evaluation

Here we first present the experimental results. Then, we conduct the investigation on the mutual benefits of crowdsourcing and the IR systems, followed by the influence of quality control.

4.1   Experimental Results

We report the experimental results in Table 1. All runs' binary judgements are evaluated using the logistic average misclassification rate (LAM) and treated the adjudicated judgements as truth. The submitted probabilities of relevance are evaluated by the area under the ROC curve (AUC) and also treated the adjudicated binary judgements as truth.

4.2   Influence of Crowdsourcing

We draw a conclusion that relevance feedback from crowdsourcing greatly benefits the IR systems. As we know, a major limitation of most existing IR systems is that the retrieval results solely depend on the query and document representations. Information on most real users' needs and search context is largely ignored.

However, crowdsourcing provides an excellent solution to exploring implicit feedback information, under the quality control of workers. In order to evaluate the interactive information obtaining from crowdsourcing, we also apply the other three models, such as the pseudo feedback method, to extract the feedback

```
Input:
(1) Queries
(2) Data sets, e.g. TREC-8

Output:
Four runs of retrieved documents corresponding to four feedback methods.

Preliminary Results:
(1) Configure the queries and the TREC 8 collection into the IR systems
(2) Apply an IR model: { Output the first-round retrieval list;}
(3) Repeat (2) in four IR models: BM25, DFR, BM25_DFR, LM
(4) For each query {
      Extract the top five overlapped retrieved documents from four retrieved lists in (3);
      Note that these five retrieved documents have to be retrieved by all the four models;}
(5)The top five documents are ready for crowdsourcing.

Evidence Results:
(1) Use the manually judged documents and keywords as the relevance feedback
    into the IR systems {
      Output four lists under BM25, DFR, BM25_DFR and LM;}
(2) Use the manually judged document without keywords {
      Adopt the proximity feedback method proposed by [9],
      Get the weighted feedback terms as the relevance feedback;
      Output four lists under BM25, DFR, BM25_DFR and LM.}
(3) Use the manually judged document without keywords {
      Adopt the pseudo feedback method [10];
      Get the pseudo feedback terms as the relevance feedback;
      Output four lists under BM25, DFR, BM25_DFR and LM.}
(4) Use the manually judged document without keywords {
      Calculating the TF-IDF of each term in the relevant documents;
      Extract the top 20 terms as the relevance feedback;
      Output four lists under BM25, DFR, BM25_DFR and LM.}
```

Fig. 3. The IR Systems: (1) the crowdsourcing stage will happen between the "preliminary results" step and the "evidence results" step; (2) the documents retrieved after applying feedback methods can be the new candidates for CrowdFlower, which means the feedback information can be updated and refined

terms, with only using the judged relevant documents. Our experiments show that the best performance is generated by the proximity feedback method by [9], which exactly adopts the judged documents, but without the supportive keywords provided by the workers. An conclusion can be drawn that the workers make judgements under different understanding to "relevance". Therefore, we will give a further discussion about quality control with worker training at the following section.

4.3 Influence of Quality Control

Judgement's quality is a big challenge in crowdsourcing. Table 2 presents the samples without quality control. We can see that over 50% judgements do not make sense, where some workers only collect cents but do not make the right judgements. Under this situation, the misjudged documents will heavily hurt the feedback methods because of the incorrect "relevance" information. Consequently, the IR systems will be misled as well.

There are two advantages of the proposed query-oriented training. First, it selects valuable skilled workers from many of sources. Second, the workers get

Table 1. Performance of Four Runs Compared to the Official Median Value

Topics		Run01	Run02	Run03	Run04	Mdian		Run01	Run02	Run03	Run04	Mdian
411		0.195	0.238	0.206	0.179	0.15		0.5	0.407	0.474	0.459	0.86
416		0.236	0.279	0.187	0.221	0.16		0.471	0.398	0.465	0.48	0.85
417		0.277	0.536	0.165	0.199	0.2		0.489	0.5	0.451	0.429	0.75
420		0.478	0.47	0.334	0.351	0.17		0.462	0.443	0	0.476	0.71
427	LAM	0.191	0.26	0.184	0.183	0.18	AUC	0.402	0.363	0.464	0.415	0.73
432		0.468	0.345	0.381	0.364	0.27		0.47	0.431	0.503	0.454	0.71
438		0.282	0.422	0.245	0.273	0.26		0.443	0.448	0.463	0.457	0.78
445		0.418	0.419	0.192	0.2	0.19		0	0.403	0.509	0.48	0.83
446		0.499	0.448	0.176	0.205	0.21		0.489	0.497	0.491	0.478	0.82
447		0.063	0.077	0.111	0.104	0.08		0.498	0.486	0.534	0.523	0.76
	All	0.311	0.349	0.218	0.228	0.187	ALL	0.467	0.438	0.479	0.465	0.78

Table 2. The Sample CrowdSourcing Results Without Quality Control

unit id	id	judgement	keyword1	keyword2	keywords3	others
195005932	563650952	Relevant	National Assembly Religious Women	women priests	young women	
195005932	563662790	Relevant	LUTHERANS ... BLESSING OF ...	Times Staff Writers	RUSSELL CHANDLER JOHN DART	
195005932	563662882	Relevant	woman	religion	clergy	
195005932	563686956	Relevant	9	7	8	waw
195005932	563690807	Relevant	44	kjlk	h	
195005932	563694956	Irrelevant	f	k	j	l

the queries that they are familiar with, since we only send the private links to those whose judgements and supportive information meet the gold standard. Furthermore, the bonus cents also motivate workers and reword their work.

5 Conclusions and Future Work

Bringing IR techniques into crowdsourcing provides a win-win solution to figure out two main problems in crowdsourcing as quality control and cost savings. Our proposed approach finds the valuable workers by a query-oriented training so that the crowdsourcing tasks can be finished with much lower cost, less time duration and better quality. In our case study, there are 5 documents * (8 training queries + 10 real queries) * 10 needed judgements finished by the workers, instead of 18,260 query-document pairs. The rest of work is judged automatically by the IR systems. This explains why we only cost $47.25 with better LAM performance.

In the future, we will focus on worker behaviour analysis and worker interface design so that more qualified interactive information can be obtained.

References

1. Howe, J.: 1st edn. Crown Publishing Group, New York (2008)
2. Smucker, M.D., Kazai, G., Lease, M.: Overview of the trec 2012 crowdsourcing track. In: Proceedings of The 21st Text REtrieval Conference (2012)

3. Voorhees, E.M., Harman, D.: Overview of the eighth text retrieval conference (trec-8). In: Proceedings of The 8th Text REtrieval Conference (1999)
4. Amati, G.: Probabilistic Models for Information Retrieval Based on Divergence From Randomness. PhD thesis, Department of Computing Science, University of Glasgow (2003)
5. Beaulieu, M., Gatford, M., Huang, X., Robertson, S.E., Walker, S., Williams, P.: Okapi at TREC-5. In: Proceedings of 5th Text REtrieval Conference, pp. 143–166. NIST Special Publication (1997)
6. Ponte, J.M., Croft, W.B.: A Language Modeling Approach to Information Retrieval. In: Proceedings of SIGIR 1998, pp. 275–281 (1998)
7. Hu, Q., Huang, J.X.: Passage Extraction and Result Combination for Genomics Information Retrieval. J. Intell. Inf. Syst. 34(3), 249–274 (2010)
8. Huang, X., Hu, Q.: A Bayesian Learning Approach to Promoting Diversity in Ranking for Biomedical Information Retrieval. In: Proceedings of SIGIR 2009, pp. 307–314 (2009)
9. Miao, J., Huang, J.X., Ye, Z.: Proximity-based rocchio's model for pseudo relevance. In: Proceedings of SIGIR 2012, pp. 535–544 (2012)
10. He, J., Li, M., Li, Z., Zhang, H.-J., Tong, H., Zhang, C.: Pseudo Relevance Feedback Based on Iterative Probabilistic One-Class SVMs in Web Image Retrieval. In: Aizawa, K., Nakamura, Y., Satoh, S. (eds.) PCM 2004. LNCS, vol. 3332, pp. 213–220. Springer, Heidelberg (2004)

Quality-Based Automatic Classification
for Presentation Slides

Seongchan Kim, Wonchul Jung, Keejun Han, Jae-Gil Lee, and Mun Y. Yi

Dept. of Knowledge Service Engineering, KAIST, Korea
{sckim,wonchul.jung,keejun.han,jaegil,munyi@kaist.ac.kr}

Abstract. Computerized presentation slides have become essential for effective business meetings, classroom discussions, and even general events and occasions. With the exploding number of online resources and materials, locating the slides of high quality is a daunting challenge. In this study, we present a new, comprehensive framework of information quality developed specifically for computerized presentation slides on the basis of a user study involving 60 university students from two universities and extensive coding analysis, and explore the possibility of automatically detecting the information quality of slides. Using the classifications made by human annotators as the golden standard, we compare and evaluate the performance of alternative information quality features and dimensions. The experimental results support the validity of the proposed approach in automatically assessing and classifying the information quality of slides.

Keywords: Information Quality (IQ), Presentation Slides, Classification.

1 Introduction

Computerized presentation slides have become a popular and valuable medium for various occasions such as business meetings, academic lectures, formal presentations, and multi-purpose talks. Online services focused on presentation slides, SlideShare[1] and CourseShare[2] to name a few, offer the ability to search and share computerized presentation slides on the Internet. Millions of slides are available on the Web and the number is growing continuously. However, most of the slide service platforms suffer from the problem of discerning the quality of available slides. This problem is acute and getting worse as the number of slides is continuously increasing. Further, on most platforms anyone can upload their slides. An automated classification approach, if effective, offers several benefits: 1) users are directed to select high quality slides among a group of similar slides, and 2) the assessed quality of a slide can be integrated into the searching and ranking strategies of slide-specialized search engines. For instance, none of the currently available slide search engines (e.g., SlideShare and CourseShare) support automated slide categorization or ranking by quality.

---

[1] http://www.slideshare.net
[2] http://www.courseshare.org

M. de Rijke et al. (Eds.): ECIR 2014, LNCS 8416, pp. 638–643, 2014.
© Springer International Publishing Switzerland 2014

When issuing a query to these slide-specialized search engines, end-users will get only a list of keyword-relevant slides, with no information on slide quality. The automated classification of high quality slides is an important issue for the advancement of search engines focused on presentation slides.

In the area of information retrieval, measuring information quality (IQ) for Web documents and for Wikipedia has recently been attempted by several studies. A set of quality features for Web documents for improving retrieval performance [1, 3] have been suggested, and a different set of quality indicators for better ranking and automatic quality classification of articles on Wikipedia [2, 4] have also been reported. However, these quality indicators for Web documents and Wikipedia are inappropriate for presentation slides because they overlook the importance of the representational aspects.

To the best of our knowledge, this study is the first attempt to define quality metrics for automatic classification of presentation slides. In this study, we consider only lecture slides, as they are the most popular. The key contributions of this paper are: 1) we investigate presentation slide characteristics and propose a new, comprehensive framework of information quality developed specifically for presentation slides on the basis of a user study, and 2) we assess the validity of the identified quality features and dimensions for the task of automatic quality assessment.

2   Related Work

IQ related research work has recently been receiving considerable attention in the information retrieval research community; however, no research has yet been attempted that has considered slide quality. Several studies on the classification and ranking of Web documents [1, 3] and Wikipedia articles [2, 4] have been reported.

For Web documents, Zhou and Croft [3] devised a document quality model using *collection-document distance* and *information-to-noise ratio* to promote the quality of contents in the TREC corpus. More recently, Bendersky et al. [1] utilized various quality features related with content, layout, readability, and ease of navigation, including the number of visible terms, the average length of the terms, the fraction of anchor text on the page, and the depth of the URL path. They reported a significant improvement in the retrieval performance with ClueWeb and GOV2 corpus.

For Wikipedia, Hu et al. [4] suggested the PEERREVIEW model, which considers the review behavior. The PROBREVIEW model was proposed to extend the PEERREVIEW model with partial reviewership of contributors. These models were used to determine Wikipedia article quality with features including the number of authors per article, reviewers per article, words per article, and so on. The proposed models were evaluated and found to be effective in article ranking. Dalip et al. [2] tried to classify Wikipedia articles with the following quality indicators: 1) text features (length, structure, style, readability, etc.), 2) review features (review count, reviews per day/user, article age, etc.), and 3) network features (page rank, in/out degree, etc.). They demonstrated that these structure and style features are effective in quality categorization.

We adopt several content features such as entropy and readability [1, 2] from previous studies in our experiment. However, these features are not discriminative enough to determine slide quality, because they overlook useful features about representational aspects of slides. Thus we suggest new features including font color, font size and the number of bold words for representational clarity. To the best of our knowledge, this is the first study of automatic quality classification of slides.

3   Quality Features of Presentation Slides

In this section, we present the quality features developed in order to determine presentation slide quality. Table 1 shows the 28 quality indicators derived for the five IQ dimensions, whose definitions were previously presented in [5, 6].

Table 1. Description of extracted quality features

Dimension	Indicator	Description
Informativeness (I)	numSlides	Number of slides
	numTerms	Number of terms in the slides [1, 2]
	avgNumTerms	Number of terms per slide [1, 2]
	numImgs	Number of images [2]
	avgNumImgs	Number of images per slide [2]
	preExample	Presence of example
	numExamples	Number of examples
	preTable	Presence of table [1]
	numTables	Number of tables [1]
	preLeaObj	Presence of learning objective
Cohesiveness (C)	entropy	Entropy of texts in the slide [1]
Readability (R)	numStops	Number of stopwords [1]
	fracStops	Stopword / non-stopword ratio [1]
	avgTermLen	Average term length of texts [1]
Ease of Navigation (EN)	preTableCnts	Presence of table of contents
	preSlideNums	Presence of slide numbers
Representational Clarity (RC)	numBolds	Number of bolds
	numItalics	Number of italics
	numUnderlines	Number of underlines
	numShadows	Number of shadows
	sumHighlights	Sum of bolds, italics, underlines, and shadows
	numRichTexts	Number of styled text blocks
	numFontSizes	Number of font sizes
	avgFontSize	Average size of fonts
	numFontNames	Number of font names
	numFontColors	Number of font colors
	numLineSpaces	Number of line spaces
	avgLineSpace	Average line space

We adopt some quality features from previous studies [1, 2]; the others are inspired by our own user study, which was conducted to determine the quality criteria of presentation slides. Our user study involved 60 students, recruited from two universities in order to balance their backgrounds and individual characteristics, each of which was asked to view five slides and think aloud while comparing the given slides. The verbal statements were all recorded and transcribed. Through extensive coding analysis, we identified the criteria of IQ and determined their dimensions as shown in Table 1. The entropy of document D is computed over the individual document terms as in Equation (1).

$$-\sum_{w \in D} p_D(w) \log p_D(w), \quad \text{where} \quad p_D(w_i) = \frac{tf_{w_i,D}}{\sum_{w_j \in D} tf_{w_j,D}} \tag{1}$$

4   Experiments

We automatically conducted a preliminary classification for high, fair, and low quality lecture slides using the proposed 28 features from the five IQ dimensions.

We randomly collected 200 MS PowerPoint presentation slides from SlideShare[1] in two courses: *data mining* and *computer network*. We manually annotated the 200 slides according to quality by hiring six graduates who had completed those courses successfully. Three annotators were assigned per course; as a result, each slide was judged by three annotators. Annotators were instructed to classify a given slide into only one class out of the three (high, fair, and low), considering all dimensions of quality such as informativeness, representational clarity, and readability etc. The inter-annotator agreement among the three annotators was $\kappa = 0.67$, which is considered to indicate substantial agreement according to Fleiss kappa [7]. Finally, we obtained 178 slides that had yielded agreement on labeling quality (high, fair, and low) from more than two annotators. These 178 slides (high: 55, fair: 83, and low: 40) were used for our classification. In order to extract the proposed quality features of the slides, shown in Table 1, we used the Apache POI[3], which is a Java API for reading and writing Microsoft Office files such as Word, Excel, and PowerPoint. We extracted the features from textual contents, file metadata, layout, etc., of PowerPoint files (ppt/pptx).

The classification in three classes: high, fair, and low was conducted using 10-fold cross validation. We used *SVM* and *Logistic Regression (LR)*, which have been widely adopted for classification, in the Weka toolkit [8]. The default parameter values given in Weka were chosen for our experiment. We report on three measures: *precision (P)*, *recall (R)*, and *F-1 score (F1)* (micro-averaged).

The results of classification are as shown in Table 2. Performance was measured by adding features of each dimension. The results clearly reveal the effectiveness of the proposed features and show that there is a little difference between SVM and LR.

---

[3] http://poi.apache.org/

Table 2. Performance of classification by adding individual dimensions

Features	SVM			LR		
	P	R	F	P	R	F
RC	0.402	0.479	0.371	0.509	0.515	0.500
RC+I	0.547	0.533	0.476	0.551	0.550	0.549
RC+I+EN	0.602	0.592	0.577	0.592	0.586	0.586
RC+I+EN+R	0.619	0.604	0.588	**0.617**	**0.615**	**0.614**
All included	**0.622**	**0.609**	**0.596**	0.600	0.598	0.597

Table 3. Top 11 features by information gain (IG)

Rank	Features	IG	Dimension
1	numTables	0.112	Informativeness
2	numFontColors	0.095	Representational Clarity
3	preSlideNums	0.093	Ease of Navigation
4	numImgs	0.093	Informativeness
5	numItalics	0.088	Representational Clarity
6	numSlides	0.087	Informativeness
7	numFontNames	0.077	Representational Clarity
8	preTable	0.034	Informativeness
9	preTableCnts	0.033	Ease of Navigation
10	preExample	0.031	Informativeness
11	preLeaObj	0.003	Informativeness

SVM with all features from all dimensions achieves the best performance of 0.596 in F1. With LR, the best performance of 0.614 in F1 is achieved when C (cohesiveness) was excluded. Performance is increased when features of each dimension are added, except C with LR.

To analyze the individual feature importance, we computed information gain (IG) for each feature. The top 11 features by IG are reported in Table 3. The results show that 10 features among the proposed features are considerably discriminative for the classification. It should be noted that there is a significant drop between the 10th and 11th. Among the top 10 features, numTables is the most discriminant, followed by numFontColors and preSlideNums. The results imply that rich tables, font colors, images, italicized fonts, slides, and font faces, and existence of slide numbers, table, table of contents, and example are engaging characteristics with which high quality slides can be discerned reliably. Remarkably, numFontColors, preSlideNums, numItalics, numFontNames, and preTableCnts are distinctive and discriminative features for slides, even though these categories are not considered and used for other documents such as Web and Wikipedia in previous studies [1-4]. Furthermore, these results are contrary to previous reports, in which it has been seen that, in terms of quality measurement, readability features with stopword fraction and coverage are effective features for Web documents [1] and structure features related to the organization of the article such as sections, images, citations and links are useful for Wikipedia articles [2]. As for features about readability, they might not be effective for slides because most slides are written in a condensed form to summarize the contents.

It is clar that newly proposed features about representation in this study are effective in slides. These results also support the necessity of our study for development a different set of features for slides.

5   Concluding Remarks

In this paper, we presented a new, comprehensive framework of information quality developed specifically for computerized presentation slides and reported the performance of automatically detecting the information quality of slides using the features captured by the framework. Although the study results may be considered not so highly strong, the study supports the validity of the proposed approach in automatically assessing and classifying the information quality of presentation slides. In our future work, we need to develop further salient features from the slide layout and content to improve the overall performance while considering other IQ dimensions such as consistency, completeness and appropriateness.

Acknowledgements. "This work was supported by the National Research Foundation of Korea (NRF) grant funded by the Korea government (MEST) (No. 2011-0029185)." We thank the anonymous reviewers for the helpful comments.

References

1. Bendersky, M., Croft, W.B., Diao, Y.: Quality-biased ranking of web documents. In: Proceedings of the 4th ACM International Conference on Web Search and Data Mining, pp. 95–104. ACM, Hong Kong (2011)
2. Dalip, D.H., Gonçalves, M.A., Cristo, M., Calado, P.: Automatic quality assessment of content created collaboratively by web communities: a case study of wikipedia. In: Proceedings of the 9th ACM/IEEE-CS Joint Conference on Digital Libraries, pp. 295–304. ACM, Austin (2009)
3. Zhou, Y., Croft, W.B.: Document quality models for web ad hoc retrieval. In: Proceedings of the 14th ACM International Conference on Information and Knowledge Management, pp. 331–332. ACM, Bremen (2005)
4. Hu, M., Lim, E.-P., Sun, A., Lauw, H.W., Vuong, B.-Q.: Measuring article quality in wikipedia: models and evaluation. In: Proceedings of the 16th ACM International Conference on Information and Knowledge Management, pp. 243–252. ACM, Lisbon (2007)
5. Stvilia, B., Gasser, L., Twidale, M.B., Smith, L.C.: A framework for information quality assessment. J. Am. Soc. Inf. Sci. Tec. 58, 1720–1733 (2007)
6. Alkhattabi, M., Neagu, D., Cullen, A.: Assessing information quality of e-learning systems: a web mining approach. Computers in Human Behavior 27, 862–873 (2011)
7. Fleiss, J.L.: Measuring nominal scale agreement among many raters. Psychological Bulletin 76, 378–382 (1971)
8. Hall, M., Frank, E., Holmes, G., Pfahringer, B., Reutemann, P., Witten, I.H.: The WEKA data mining software: an update. SIGKDD Explorations Newsletter 11, 10–18 (2009)

Context of Seasonality in Web Search

Tomáš Kramár and Mária Bieliková

Faculty of Informatics and Information Technologies
Slovak University of Technology
Ilkovičova 2, 842 16 Bratislava, Slovakia
Bratislava, Slovakia
name.surname@stuba.sk

Abstract. In this paper we discuss human behavior in interaction with
information available on the Web via search. We consider seasonality as a
novel source of context for Web search and discuss the possible impact it
could have on search results quality. Seasonality is used in recommender
systems as an attribute of the recommended item that might influence
its perceived usefulness for particular user. We extend this idea to Web
search, introduce a seasonality search context, describe the challenges it
brings to Web search and discuss its applicability. We present our analysis
of AOL log that shows that the level of seasonal behavior varies.

1 Introduction

It has been recognized that Web search needs some form of information that
would help to understand the underlying intent, which is rarely expressed clearly
in the query [3]. This information is collectively referred to as a *search context*
and there are many sources which the search context can be implicitly inferred
from. In this paper we focus on a novel source of search context – context of
seasonality. The basic premise behind context of seasonality is that the interests
of a person change in intensity and those changes exhibit patterns that we can
analyze and predict. E.g., a person can be highly interested in skiing during
winter and in that case, during winter, we can boost ranking for documents
that deal with skiing. A good example of class of queries that could benefit
from such boosting are transactional queries, e.g. in case of a query in form
of a sportswear brands, the skiing equipment manufactured by the particular
brand should receive higher ranking than other equipment, because the interest
in skiing is peaking at this time.

There are many aspects of this source of context, however, before tackling
them deeper, we must first answer the important question whether the basic
premise of the seasonality context holds for the Web information space and
whether the patterns in user interest shifts exist. It has been shown that seasonality exists at the query level (e.g. a query ECIR repeats in a yearly interval) and
roughly 7% of all queries are seasonal [8]. According to [1], who analyzed a topically labeled query log, some topics exhibit global popularity peaks throughout
the day, while others remain constant. Whether seasonality exists at the level

M. de Rijke et al. (Eds.): ECIR 2014, LNCS 8416, pp. 644–649, 2014.

of interests of a single user is still an open question. Intuitively, it seems that seasonality is omnipresent, but in reality this must not be necessarily true and we have to be careful in using it to support tasks on the Web.

In this work, we analyze a search engine log for the existence of various patterns and show that the concept of seasonality in interest drifts is in reality not as straightforward as the intuitive notion. There are users who exhibit clean and predictable interest shifts, but there are also users who do not behave seasonally at all. We also discuss the benefits that the context of seasonality could bring and outline future research directions.

2 Context of Seasonality

There is an important difference between the concept of context in the area of recommender systems and in the area of personalized search. In recommender systems, the context is viewed as a set of external attributes of the environment that impact user's immediate preferences, such as the weather or location and many others. Traditionally, in Web search the context describes any information that can be used to infer the specific goal that the searcher wants to fulfill by issuing a query [6].

Although there are cases when the search context has been established explicitly [11], methods that require less cognitive load by capturing the context implicitly are more preferable. They can be characterized by the source from which they draw the information about user's need. Some of the most intensively studied sources of context in Web search are:

- Similarity between people; the underlying intent is inferred from behavior of similar searchers, who are grouped into ad hoc communities [10]. The communities are created based on the chosen similarity criteria (e.g., similar past queries, similar browsing behavior, etc.) and serve as a source of context in the personalization process. When a member of the search community issues a query, the search intent can be clarified by analyzing the preferences of other community members for the particular query [4].
- User's activity; where the context is inferred from previously entered queries and behavior on search engine results page [12]. In its simplest form every clicked search result in the predefined time window is incorporated to the context model, which represents a model of searcher's short-term interests. This means that documents matching the short-term interests can be boosted to receive higher ranking.

One of the contexts used in recommender systems is context of seasonality [7]. It is based on the similarity of a seasonal aspect of the recommended item with the current season of the year. Good examples are movies with the Christmas theme – users of the recommender are much more likely to accept such recommendation on and around Christmas, than they are at other time in the year.

Our idea of seasonality context for search is based on a similar idea. Based on our experiences, we hypothesize that the levels of interests that people have are unstable and change over time; sometimes increase, sometimes decrease, and that these changes form repeating patterns. Intuitively, there are many forms of interest drifts, e.g.:

- periodic drifts in interests that are correlated with the season of the year, e.g. winter sports or summer sport;
- drifts in interests caused by the seasonal appearance of the object the person is interested in, e.g. various seasonal produce or sports and cultural events that repeat periodically;
- drifts in interests related to switching between different tasks. In order for these drifts to be worth considering, the duration of the tasks must be sufficiently long and the tasks must repeat periodically. The most widespread task that matches these criteria is a regular job that most people have. We expect that people are changing interests when they are at work, i.e. people search for conceptually different information when they are working than when they relax.

By maintaining a seasonality context for the searcher we could have a model of interests for the given time and provide more relevant results. A search engine could detect if there is a seasonality context available for the given moment and use it to personalize the search results.

It is important to distinguish the patterns in the interest shifts. Simply looking back to one discrete moment in the history to see which interests were relevant in the past is not enough, because it is not clear which point in the history should we look at. Different interest drifts have different periodicity, which may range from hours (like in the example of work/leisure) to years. There are techniques of time series analysis [13,9] that can be applied to this problem.

Seasonality context could be problematic in situations when the active interest changes unexpectedly. This is not very probable for naturally developed interests, but more probable in situations when the interest was related to a certain task that is now complete and the user no longer has any interest in it, e.g. when a work task is completed and the employee is assigned to a different project, possibly from a different domain. Other weak point is the range of data that must be available in order to discover the repeating patterns. Despite its shortcomings, we believe that context of seasonality would bring benefits into the area of personalized search.

3   Search Engine Log Analysis

In order to answer the basic question – whether the periodic shifts in interests occur, we analyzed the publicly available query log from the AOL search engine[1]. Given that the AOL dataset spans only a period of 3 months, our goal was to find shorter periods of interest drifts and we concentrated on analyzing the existence of the task-related interest drifts related to the searcher's job.

---

[1] AOL dataset, http://zola.di.unipi.it/smalltext/datasets.html

We analyzed two different sets of disjoint periods where we expected difference in search intents:

- workdays (Monday-Friday) and weekend (Saturday or Sunday) – *working days* setup;
- working time (9:00-17:00) and leisure time (17:00-9:00), workdays strictly – *working hours* setup. The times were chosen as the typical business hours in the USA.

To characterize and compare search intents we built a topic model for each period using the period's clicked search results. The topic model is a vector of words and leverages lightweight semantics [5] in form of page keywords, page description (both provided by the page authors), title and ODP[2] topics. We extracted these lightweight semantics for each search result clicked in the particular period and added the words from these sources to the topic model of that period.

To compare the topic models of the periods we used Davies-Bouldin score [2], a metric commonly used in evaluating clustering methods. This metric awards clusters with low intra-cluster distances (high internal density) and high inter-cluster distances (well separated clusters). The lower the Davies-Bouldin score, the more tight and well-separated are the clusters indicating more tight and separated interests.

We applied this analysis to the top-100 most active users in the AOL log. Figure 1 shows the distribution of the Davies-Bouldin score over the top-100 AOL searchers in both setups (working days and working hours).

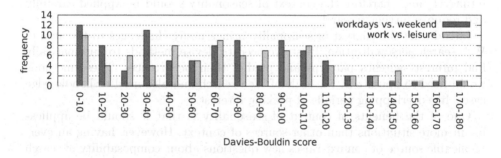

Fig. 1. Distribution of Davies-Bouldin scores across the studied dataset

We have investigated the correlation between users who exhibit the switching behavior for *working days* setup and users who exhibit the switching behavior for the *working hours* setup. We have ranked each user with the position based on the Davies-Bouldin score, i.e. in each setup, the user with lowest (best) value of Davies-Bouldin score gets ranked with 1, the runner up with 2, continuing this way all the way through the list of users.

---

[2] Open Directory Project, http://www.dmoz.org/

Using the working days setup as a baseline, we have then calculated the change in position for each of the users in working hours setup. The average positional leap in the top-100 dataset is 17 positions. However, if we look only at the top users with best Davies-Bouldin score, the average positional leap of top-5 users is 1 and the average positional leap of top-23 users is 3.17. After the 23rd position, we observe a dramatic increase in the positional leap values. This fact suggests that users who switch interests during working days and weekends are likely to switch interests during working hours and leisure hours, indicating a strict separation of work and free time.

The main discoveries from this log analysis can be summarized in the following points:

- There is no polarization in the interest drifts during the temporal periods in the two selected setups. There are some users who behave according to the intuitive notion of seasonality, i.e., they switch interests during leisure time, but roughly the same amount of users does not exhibit this behavior. In fact, the level of interest switching has uniform distribution. This indicates the need for a further research to find methods to predict the interest switching behavior of the particular user.
- Users, who switch contexts during weekends, are likely to also switch contexts during working hours and leisure time, indicating that a level of switching in one scenario can be used to predict a level of switching in different scenario.

4   Conclusions and Future Work

We have shown that not all users behave seasonally as we would have expected intuitively and therefore the context of seasonality should be applied carefully and requires further research.

Introducing the context of seasonality brings many challenges to the area of Web search personalization. First, we need to devise methods to automatically find interest switches patterns and predict their occurrences. These methods would need to operate at the scale of Web search and also need to handle edge cases, like overlapping periods of peaking interest.

One of the benefits of context of seasonality is that it should be applicable in more situations than other sources of context. However, having an ever-applicable source of context raises new questions about composability of search contexts from different sources. Is there a way that contexts coming from different sources could be combined? Or is there a way to compare contexts and always select the better one? These are the questions that should be addressed in further research by the Web search community.

Although we have looked on seasonality from point of view of a Web search, the idea is applicable to a whole range of other problems as well. Seasonality draws from the patterns in user behavior changes, and those patterns are interesting in general, not only when users are fulfilling their information needs, but also other needs, in communication, or in collaboration. We think that seasonality could be studied from other points of view, e.g. to see if there are patterns in communication styles that could be used to improve the human collaboration.

Acknowledgements. This work was partially supported by projects No. APVV-0208-10 and No. VG1/0675/11, and it is partial result of the Research and Development Operational Programme for the project University Science Park of STU Bratislava, ITMS 26240220084, co-funded by the European Regional Development Fund.

References

1. Beitzel, S.M., Jensen, E.C., Chowdhury, A., Grossman, D., Frieder, O.: Hourly analysis of a very large topically categorized web query log. In: Proc. of the 27th Annual Int. ACM SIGIR Conf. on Research and Development in Information Retrieval, SIGIR 2004, pp. 321–328. ACM, New York (2004)
2. Davies, D.L., Bouldin, D.W.: A cluster separation measure. IEEE Transactions on Pattern Analysis and Machine Intelligence 1(2), 224–227 (1979)
3. Downey, D., Dumais, S., Liebling, D., Horvitz, E.: Understanding the relationship between searchers' queries and information goals. In: Proc. of the 17th ACM Conf. on Information and Knowledge Management, CIKM 2008, pp. 449–458. ACM (2008)
4. Kramár, T., Barla, M., Bieliková, M.: Disambiguating search by leveraging a social context based on the stream of user's activity. In: De Bra, P., Kobsa, A., Chin, D. (eds.) UMAP 2010. LNCS, vol. 6075, pp. 387–392. Springer, Heidelberg (2010)
5. Kříž, J.: Keyword extraction based on implicit feedback. Bulletin of ACM Slovakia 4(2), 43–46 (2012)
6. Lawrence, S.: Context in web search. IEEE Data Eng. Bulletin 23(3), 25–32 (2000)
7. Marinho, L.B.: et al.: Improving location recommendations with temporal pattern extraction. In: Proc. of the 18th Brazilian Symposium on Multimedia and the Web, WebMedia 2012, pp. 293–296. ACM (2012)
8. Metzler, D., Jones, R., Peng, F., Zhang, R.: Improving search relevance for implicitly temporal queries. In: Proc. of the 32nd Int. ACM SIGIR Conf. on Research and Development in Information Retrieval, SIGIR 2009, pp. 700–701. ACM, New York (2009)
9. Shokouhi, M.: Detecting seasonal queries by time-series analysis. In: Proc. of the 34th Int. ACM SIGIR Conf. on Research and Development in Information Retrieval, SIGIR 2011, pp. 1171–1172. ACM, New York (2011)
10. Smyth, B.: Social and personal: communities and collaboration in adaptive web search. In: Proc. of the 1st Int. Conf. on Information Interaction in Context, IIiX, pp. 3–5. ACM (2006)
11. Smyth, B., Coyle, M., Briggs, P.: Heystaks: a real-world deployment of social search. In: Proc. of the Sixth ACM Conf. on Recommender Systems, RecSys 2012, pp. 289–292. ACM (2012)
12. White, R.W., Bennett, P.N., Dumais, S.T.: Predicting short-term interests using activity-based search context. In: Proc. of the 19th ACM Int. Conf. on Information and Knowledge Management, CIKM 2010, pp. 1009–1018. ACM (2010)
13. Zhang, Y., Jansen, B.J., Spink, A.: Time series analysis of a web search engine transaction log. Information Processing and Management 45(2), 230–245 (2009)

On the Effect of Locality
in Compressing Social Networks

Panagiotis Liakos[1], Katia Papakonstantinopoulou[1,*], and Michael Sioutis[2]

[1] University of Athens, Greece
{p.liakos,katia}@di.uoa.gr
[2] Université Lille-Nord de France, Artois, CRIL/CNRS, Lens, France
sioutis@cril.fr

Abstract. We improve the state-of-the-art method for graph compression by exploiting the locality of reference observed in social network graphs. We take advantage of certain dense parts of those graphs, which enable us to further reduce the overall space requirements. The analysis and experimental evaluation of our method confirms our observations, as our results present improvements over a wide range of social network graphs.

1 Introduction

With the arrival of the Web 2.0 era and the emerging popularity of social network sites, a number of new challenges regarding information retrieval research have been brought to the surface. Users form communities and share mass amounts of high quality information, making the effective and efficient mining of that information an important research direction for modern information retrieval. The structure of the networks studied, i.e., the graphs formed by the relationships among each network's users, is of utmost importance for fields such as user behaviour modelling, sentiment analysis, and social computing. The extraordinary pace at which social network graphs are growing has turned the focus on obtaining space-efficient in-memory representations of them. Information retrieval systems that work directly on those graphs can benefit from such representations.

Graph compression is mostly based on empirical observations of the graph structures. Concerning web graphs, a great number of links is intra-domain, and, thus, their adjacent nodes are close to each other (*locality of reference* or simply *locality*), while nodes close by tend to have similar sets of neighbours (*similarity*). In [6], these two regularities were utilized to decrease space requirements to six bits per edge. Later, Boldi and Vigna proposed a number of techniques that reduced those requirements even further [1]. *Locality* and *similarity* exist

---

* This research has been co-financed by the European Union (European Social Fund – ESF) and Greek national funds through the Operational Program "Education and Lifelong Learning" of the National Strategic Reference Framework (NSRF) – Research Funding Program: THALES. Investing in knowledge society through the European Social Fund.

M. de Rijke et al. (Eds.): ECIR 2014, LNCS 8416, pp. 650–655, 2014.
© Springer International Publishing Switzerland 2014

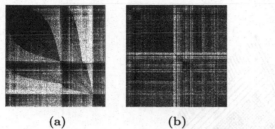

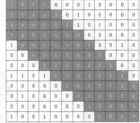

(a)	(b)	

Fig. 1. youtube-2007 before and after LLP   **Fig. 2.** An adjacency matrix

naturally in web graphs, as long as their nodes are labelled using a lexicographic order (by URL).

Although web graphs exhibit an ordering which exploits additional information – besides the graph itself – that leads to high compression rates, there is no such obvious ordering for social networks. As a consequence, research efforts have focused on testing some well-known permutations of the node labels [3] and discovering heuristics that will obtain effective node orderings [2,5]. Despite the fact that those attempts proved relatively fruitful, social networks still seem to be harder to compress than web graphs. The increased space requirements of social network graphs can be justified by the existence of a topological difference between the two kinds of graphs [5] that still is to be clarified, as well as by the fact that their compressibility characteristics have not been fully utilized.

In this paper, we concentrate on compressing social network graphs by exploiting the locality property, and build upon the state-of-the-art implementation of Boldi et al., namely, the compression framework of [1] after applying the *Layered Label Propagation* (LLP) algorithm [2] on the input graphs. Our observations not only allow for a greater compression rate – as our experimental evaluation indicates – but leave plenty of room for future exploitation as well.

2 Overview

2.1 Identifying the Dense Part of the Graph

Most compact graph representations are based either on the adjacency matrix representation [4] or on the adjacency lists representation [1] of the graph. For a given graph $G = (V, E)$ the adjacency matrix representation is preferred when G is dense, i.e., when $|E| = \Theta(|V|^2)$, while adjacency lists are preferred when G is sparse, i.e., when $|E| = O(|V|)$.

We combined the two kinds of representations after observing that social network graphs, although rather sparse in general, have a dense part around the main diagonal of the graph's adjacency matrix after the LLP algorithm [2] has been applied on them. This tendency is shown in Figure 1, where the adjacency matrix of a graph from the youtube social network is illustrated before (1a) and after (1b) the reordering of its nodes.

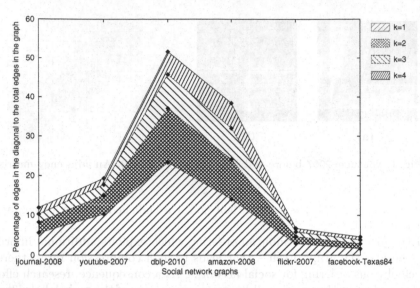

Fig. 3. Percentage of edges contained in the diagonal stripe of various social network graphs for various stripe widths

More formally, we call this dense area the *diagonal stripe*, and define it as follows: let $k \in \mathbb{Z}_+$, an edge (i, j) is in the k-diagonal stripe, iff $i - k \leq j \leq i + k$. The 3-diagonal stripe of an example adjacency matrix is illustrated in Figure 2.

In the graphs we examined experimentally, a large number of edges tends to be in the diagonal stripe, meeting our expectations regarding the locality property. Figure 3 illustrates this trend for $k \in \{1, 2, 3, 4\}$ for the graphs of our dataset, described in detail in Section 3.1.

2.2 Proposing a Hybrid Method for Graph Compression

Having identified an opportunity to compress large parts of social network graphs effectively, we propose a hybrid method, which uses a bit vector to represent the diagonal stripe and resorts to the method in [1] to address the issue of compressing the remaining edges. For the rest of this paper we will refer to our method as $BV_{\mathcal{D}}$ and to the method in [1] as BV.

Every possible pair of nodes (a, b) lying in the diagonal stripe is mapped through a simple function to the bit vector. Thus, the existence of an edge there can be verified in constant time. A big percentage of these pairs represent edges absent from the graph. However, including those pairs in our representation allows us to be aware of the position of every pair and not resort to using an index as in [4], which would not only introduce a similar space overhead, but would dramatically increase the retrieval time as well.

By using BV to compress the rest, sparse part, of the graph, we manage to provide a full graph compression framework and perform comparisons over the whole graph, not only the diagonal stripe. The computational complexity of this

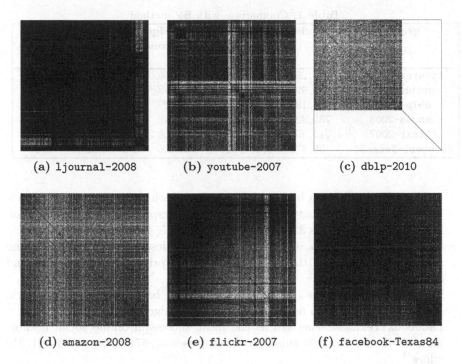

(a) ljournal-2008   (b) youtube-2007   (c) dblp-2010

(d) amazon-2008   (e) flickr-2007   (f) facebook-Texas84

Fig. 4. Visualizations of the adjacency matrices of some social network graphs

approach is approximately equal to the complexity of BV alone, as mapping the diagonal stripe to a bit vector is linear in the number of diagonal edges. Furthermore, this mapping can only decrease the query time on the compressed graph's elements, when compared with the query time of BV alone.

3 Experiments

3.1 Dataset

In order to test our approach we used a dataset of six social network graphs. Figure 4 provides an illustration of their adjacency matrices, where one can clearly see how the diagonal stripe stands out in almost all of the graphs. The origin and characteristics of our graphs are summarized in the following list:

- ljournal-2008: *LiveJournal* is a virtual community social website that started in 1999. It comprises 5, 363, 260 nodes and 79, 023, 142 edges.[1]
- youtube-2007: *Youtube* is a video-sharing website that includes a social network. It comprises 1, 138, 499 nodes and 5, 980, 886 edges.[3]

---

[1] Collected in [5], retrieved by LAW: http://law.di.unimi.it/

Table 1. Comparison with BV method

graph	# nodes	# edges	% of edges in diagonal	k	compression ratio (bits/edge)	
					BV	BV$_\mathcal{D}$
ljournal-2008	5,363,260	79,023,142	5.62%	1	11.84	11.80
youtube-2007	1,138,499	5,980,886	15.10%	2	14.18	13.79
dblp-2010	326,186	1,615,400	37.12%	2	8.63	7.76
amazon-2008	735,323	5,158,388	43.56%	5	10.77	10.56
flickr-2007	1,715,255	31,110,082	4.66%	2	9.81	9.76
facebook-Texas84	36,371	3,181,310	3.84%	3	8.82	8.80

- dblp-2010: *DBLP* is a bibliography service. Each vertex represents an author, and an edge links two vertices if the corresponding authors have collaborated. It comprises 326,186 nodes and 1,615,400 edges.[2]
- amazon-2008: *Amazon* is a symmetric graph describing similarity among books as reported by the Amazon store, comprising 735,323 nodes and 5,158,388 edges. [2]
- flickr-2007: *Flickr* is a photo-sharing website based on a social network. It comprises 1,715,255 nodes and 31,110,082 edges.[3]
- facebook-Texas84: *Facebook* is the most successful online social networking service. Its Texas84 subnetwork comprises 36,371 nodes and 3,181,310 edges.[4]

The aforementioned graphs vary in size and cover a wide range of social networking services. Thus, they form a thorough evaluation environment for our proposed method.

3.2 Compression Ratio Comparison

Table 1 shows the number of nodes and edges in each graph, the percentage of edges in the diagonal stripe, the compression ratio achieved by the BV technique [1], and the one achieved by our proposed method (BV$_\mathcal{D}$) for a given k.

As expected, the largest improvement (10%) was achieved for dblp-2010, which has the densest diagonal among all graphs in our dataset. Notable improvements were also observed for graphs youtube-2007 (3%) and amazon-2008 (2%). Surprisingly, for the other three graphs, viz., ljournal-2008, flickr-2007 and facebook-Texas84, BV$_\mathcal{D}$ also managed to surpass the performance of BV, even though the percentage of edges in their diagonal stripes is relatively small.

By outperforming BV for all the graphs in our dataset, we proved that the effect of our observations, even when utilized with a simple approach such as that of BV$_\mathcal{D}$, is very powerful on social network graphs.

---

[2] Collected by LAW: http://law.di.unimi.it/
[3] Part of the IMC 2007 datasets with LLP [2] applied on it
   (http://socialnetworks.mpi-sws.org/data-imc2007.html)
[4] The largest of the Facebook100 graphs containing friendships from 100 US universities in 2005 (https://archive.org/details/oxford-2005-facebook-matrix)

3.3   The Effect of Parameter k

Achieving a good compression ratio with $BV_\mathcal{D}$ depends heavily on choosing an appropriate width for the diagonal stripe of the given graph, defined by k. The optimal values of k for the graphs of our dataset are illustrated in Table 1.

As k increases, more and more edges are included in the diagonal stripe, which, however, becomes progressively sparser. We have found that a good selection of value for this parameter ranges between 1 and 5. The most appropriate value can only be known a posteriori, as it depends on the exact structure of the graph and does not only determine the bits per edge ratio of the diagonal part, but also the compression ratio of the subgraph compressed with BV. However, our results indicate that improvement over BV occurs for most of the values within this range; e.g., for graphs `dblp-2010` and `amazon-2008`, better results were achieved for $k \in [1, 7]$ and $k \in [1, 8]$ respectively.

4   Conclusion and Future Work

In this paper we propose a simple method for exploiting a particular property of social network graphs, namely, locality, in a more effective way than the state-of-the-art method of Boldi et al. [1,2]. Our experiments point out that our method achieves higher compression rates on a broad dataset of social network graphs, while also offering constant retrieval time for the diagonal part of the graph.

We will investigate the issue of optimizing the representation of the diagonal stripe and further decreasing the total compression ratio by using an entropy encoding algorithm, such as Huffman coding, preferably without introducing a significant access time overhead. Moreover, our intuition suggests that a rigorous study of graph reordering methods will lead to the identification of even more attractive labellings for our proposal.

Acknowledgments. We are grateful to Elias Koutsoupias and Alex Delis for discussions on aspects of this work.

References

1. Boldi, P., Vigna, S.: The WebGraph Framework I: Compression Techniques. In: WWW (2004)
2. Boldi, P., Rosa, M., Santini, M., Vigna, S.: Layered Label Propagation: A MultiResolution Coordinate-Free Ordering for Compressing Social Networks. In: WWW (2011)
3. Boldi, P., Santini, M., Vigna, S.: Permuting Web and Social Graphs. Internet Mathematics 6(3), 257–283 (2009)
4. Brisaboa, N.R., Ladra, S., Navarro, G.: k^2-trees for compact web graph representation. In: Karlgren, J., Tarhio, J., Hyyrö, H. (eds.) SPIRE 2009. LNCS, vol. 5721, pp. 18–30. Springer, Heidelberg (2009)
5. Chierichetti, F., Kumar, R., Lattanzi, S., Mitzenmacher, M., Panconesi, A., Raghavan, P.: On compressing social networks. In: KDD (2009)
6. Randall, K.H., Stata, R., Wiener, J.L., Wickremesinghe, R.G.: The Link Database: Fast Access to Graphs of the Web. In: DCC (2002)

Cross-Domain Collaborative Filtering
with Factorization Machines

Babak Loni, Yue Shi, Martha Larson, and Alan Hanjalic

Delft Multimedia Information Retrieval (D-MIR) Lab
Delft University of Technology, Netherlands
{b.loni,y.shi,m.a.larson,a.hanjalic}@tudelft.nl

Abstract. Factorization machines offer an advantage over other existing collaborative filtering approaches to recommendation. They make it
possible to work with any auxiliary information that can be encoded as
a real-valued feature vector as a supplement to the information in the
user-item matrix. We build on the assumption that different patterns
characterize the way that users interact with (i.e., rate or download)
items of a certain type (e.g., movies or books). We view interactions
with a specific type of item as constituting a particular domain and allow
interaction information from an auxiliary domain to inform recommendation in a target domain. Our proposed approach is tested on a data set
from Amazon and compared with a state-of-the-art approach that has
been proposed for Cross-Domain Collaborative Filtering. Experimental
results demonstrate that our approach, which has a lower computational
complexity, is able to achieve performance improvements.

1 Introduction

Cross-domain Collaborative Filtering (CDCF) methods exploit knowledge from
auxiliary domains (e.g., movies) containing additional user preference data to
improve recommendation on a target domain (e.g. books). While relying on a
broad scope of existing data in many cases is a key to relieving the problems of
sparse user-item data in the target domain, CDCF can also simultaneously benefits different data owners by improving quality of service in different domains.

In most CDCF approaches (e.g., [4], [5]) it is assumed that user behavior
in all domains is the same. This assumption is not always true since each user
might have different domains of interest, for example, rating items consistently
more frequently or higher in one domain than in another. In a recent work, Hu
et al. [2] argue that CDCF should consider the full *triadic* relation *user-item-
domain* to effectively exploit user preferences on items within different domains.
They represent the user-item-domain interaction with a tensor of order three
and adopt a tensor factorization model to factorize users, items and domains
into latent feature vectors. The rating of a user for an item in a domain is
calculated by element-wise product of user, item and domain latent factors. A
major problem of tensor factorization however, is that the time complexity of

M. de Rijke et al. (Eds.): ECIR 2014, LNCS 8416, pp. 656–661, 2014.

this approach is exponential as it is $O(k^m)$ where k is the number of factors and m is the number of domains.

In this paper we exploit the insight that user preferences across domains could be deployed more effectively if they are modeled separately on separate domains, and then integrated to generate a recommendation on the target domain. We therefore address the problem with *factorization machines* (FM) [6], which make such modeling possible. In addition, the FMs are more flexible than the tensor representation regarding the ways of capturing the domain-specific user preferences and could lead to more reliable recommendations. Finally, FMs are polynomial in terms of k and m, making them computationally less expensive than tensor factorization models [6].

FMs have already been applied to carry out CF in a single domain, [6,7], but have yet to be exploited to address the CDCF problem. Here we apply FMs to cross-domain recommendation in a way that allows them to incorporate user interaction patterns that are specific to particular types of items. Note that in this work, we define a domain as a type of item. The set of users is not mutually exclusive between domains, but we assume that their interaction patterns differ sufficiently to make it advantageous to model domains separately. The novel contribution of our work is to propose an extension of FMs that incorporates domains in this pattern and to demonstrate its superiority to single domain approaches and to a state-of-the-art CDCF algorithm.

2   Related Work

Cross-Domain Collaborative Filtering: An overview of CDCF approaches is available in Li [9]. Here, we restrict our discussion of related CDCF approaches to mentioning the advantages of our approach compared to the major classes of existing algorithms. *Rating pattern sharing* algorithms, exemplified by [4], groups users and items into clusters and matches cluster-level rating patterns across domains. The success of the approach depends, however, on the level of sparseness of user-item information per domain. *Latent feature sharing* approaches, exemplified by [5], transfer knowledge between domains via a common latent space and are difficult to apply when more than two domains are involved [2]. *Domain correlation* approaches, exemplified by [8], use common information (e.g., user tags) to link domains and fail when such information is lacking.

Factorization Machines: Factorization Machines (FM) [6] are general models that factorize user-item collaborative data into real valued feature vectors. Most factorization models such as Matrix Factorization can be modeled as a special case of FM [6]. Despite typical CF models where collaboration between users and items are represented by a rating matrix, in factorization machines the interaction between user and item is represented by a feature vector and the rating is considered as class label for this vector. More specifically lets assume that the data of a rating prediction problem is represented by a set S of tuples $(\mathbf{x}, y)$ where $\mathbf{x} = (x_1, \ldots, x_n) \in \mathbb{R}^n$ is a n-dimensional feature vector and y is its corresponding label. Factorization machines model all interactions between features using factorized interaction parameters. In this work we adopted a FM model with order $d = 2$ where only *pairwise* interaction between features are considered. This model can be represented as follows:

$$\hat{y}(\mathbf{x}) = w_0 + \sum_{j=1}^{n} w_j x_j + \sum_{j=1}^{n} \sum_{j'=j+1}^{n} w_{j,j'} x_j x_{j'} \tag{1}$$

where w_j are model parameters and $w_{j,j'}$ are factorized interaction parameters and are defined as $w_{j,j'} = \mathbf{v}_j . \mathbf{v}_{j'}$ where $\mathbf{v}_j$ is k-dimensional factorized vector for feature j. For a FM with n as the dimensionality of feature vectors and k as the dimensionality of factorization, the model parameters that need to be learnt are $\Theta = \{w_0, w_1, \ldots, w_n, v_{1,1}, \ldots, v_{n,k}\}$. Three learning approaches have been proposed to learn FMs [6]: Stochastic Gradient Descent (SGD), Alternating Least-Squares (ALS) and Markov Chain Monte Carlo (MCMC) method. We exploit all 3 methods in this work.

3   Cross-Domain CF with Factorization Machines

Assume we are given collaborative data of users and items in m different domains $\{\mathbb{D}_1, \ldots, \mathbb{D}_m\}$. The domains are different based on the type of items that exists in them. While rating information for a user might be very sparse in one domain (e.g. Books), he might have rich collaborative data in another domain (e.g. movies). The purpose of cross-domain CF is to transfer knowledge from different auxiliary domains to a target domains to improve rating predictions in the target domain.

To understand our approach, without loss of generality lets assume $\mathbb{D}_1$ is the target domain and $\{\mathbb{D}_2, \ldots, \mathbb{D}_m\}$ are the auxiliary domains. Also consider U_j and I_j as the set of users and items in domain $\mathbb{D}_j$. The standard rating prediction problem in the target domain $\mathbb{D}_1$ can be modeled by a target function $y : U_1 \times I_1 \to \mathbb{R}$. We represent each user-item interaction $(u, i) \in U_1 \times I_1$ with a feature vector $\mathbf{x} \in \mathbb{R}^{|U_1|+|I_1|}$ with binary variables indicating which user rated which item. In other words, if user u rated item i the feature vector $\mathbf{x}$ is represented as:

$$\mathbf{x} = (\underbrace{0, \ldots, 0, 1, 0, \ldots, 0}_{|U_1|}, \underbrace{0, \ldots, 0, 1, 0, \ldots, 0}_{|I_1|}) \tag{2}$$

where non-zero elements are corresponding to user u and item i. The feature vector $\mathbf{x}$ can also be represented by its sparse representation $\mathbf{x}(u, i) = \{(u, 1), (i, 1)\}$.

Given the feature vector $\mathbf{x}(u, i)$ in the target domain, our cross-domain CF approach extend this vector by adding collaborative information of user u from other domains. Now lets assume that $s_j(u)$ represents all items in domain $\mathbb{D}_j$ which are rated by user u. For each auxiliary domain $\mathbb{D}_j$, $j = 2, \ldots, m$, our method extend $\mathbf{x}(u, i)$ with a vector $\mathbf{z}_j(u)$ with the following sparse representation:

$$\mathbf{z}_j(u) = \{(l, \phi_j(u, l)) : l \in s_j(u)\} \tag{3}$$

where $\phi_j(u, l)$ is a domain-dependent real valued function. We define ϕ_j based on the rating of user u to item l and normalize it based on total number of items which is rated by user u in domain $\mathbb{D}_j$:

$$\phi_j(u, l) = \frac{r_j(u, l)}{|s_j(u)|} \tag{4}$$

where $r_j(u, l)$ specifies the rating of user u to item l in domain $\mathbb{D}_j$. In the above definition ϕ_j is a function of $r_j(u, l)$ which reflects rating patterns of user u in different domains. Furthermore it is normalized by considering the number of items which are rated by user in an auxiliary domain. This means that if a user is a frequent rater in an auxiliary domain, the contribution of each single rated item in this domain would be less compared to a rated item in an auxiliary domain with smaller number of user ratings. The above definition of ϕ_j prevents the model to be overwhelmed by too much information from auxiliary domains. This is one of the main advantages of factorization machines, namely to allow control of the amount of knowledge that is transferred from auxiliary domains. Note that the function ϕ_j can be also defined in other forms to reflect contribution of various domains in different ways. Based on our experiments we found the above definition of ϕ_j simple yet effective to transfer knowledge from auxiliary domains. Given the above definitions, we can now represent the extended vector $\mathbf{x}$ with the following sparse form:

$$\mathbf{x}(u, i, s_2(u), \dots, s_m(u)) = \{ \underbrace{(u, 1), (i, 1)}_{\text{target knowledge}}, \underbrace{\mathbf{z}_2(u), \dots, \mathbf{z}_m(u)}_{\text{auxiliary knowledge}} \} \tag{5}$$

The above feature vector serves as the input into the FM model in Equation (1), while the output variable y is the rating of user u to item i in the target domain. Based on our proposed feature expansion method, the FM will only need to focus on the users in the target domain, resulting in an improvement in terms of computational cost.

4   Experiments

We conducted our experiments on Amazon dataset [3] which consists of rating information of users in 4 different domains: books, music CDs, DVDs and video tapes. The dataset contains 7,593,243 ratings on the scale 1-5 provided by 1,555,170 users over 548,552 different products including 393,558 books, 103,144 music CDs, 19,828 DVDs and 26,132 VHS video tapes.

We build the training and test set in two different ways similar to [2] to be able to compare our approach with them. In the first setup, TR_{75}, 75% of data is considered as training set and the rest as test set, and in the second setup, TR_{20}, only 20% of data is considered as training set and the rest as test set.

We implemented a recommendation framework with C#[1] on top of two open source libraries for recommender systems: MyMediaLite [1] which implements

---

[1] https://github.com/babakx/delft-recommendation-framework

Table 1. Comparison of different CF methods on the Amazon dataset

Method \ Setup	TR_{75}		TR_{20}	
Target: Book	**MAE**	**RMSE**	**MAE**	**RMSE**
MF-SGD (Book)	0.62	0.86	0.89	1.14
FM-SGD (Book)	0.69	0.92	0.74	0.96
FM-ALS (Book)	0.72	0.99	0.75	1.07
FM-MCMC (Book)	0.60	0.79	0.72	0.94
FM-All-MCMC (Book)	0.60	0.79	0.76	0.99
FM-MCMC (Book, {Music, DVD, Video})	**0.46**	0.64	**0.69**	0.92
PF2-CDCF (Book, {Music, DVD, Video}) [2]	0.50	-	0.76	-
Target: Music				
FM-MCMC (Music)	0.71	0.95	0.77	1.00
FM-MCMC (Music, {Book, DVD, Video})	**0.67**	0.91	**0.74**	0.98
PF2-CDCF (Music, {Book, DVD, Video}) [2]	0.70	-	0.82	-

most common CF approaches including Matrix Factorization, and LibFM [6] which implements FM learning algorithms. We first compared FMs with matrix factorization method on two different single domains and then we compare the results of our proposed method with the state-of-the-art CDCF work [2] on the same dataset. We also compare our method with a *blind* combination of all items from all domains to show that the improvement of our results is not only due to additional training data. We used mean absolute error (MAE) and root mean square error (RMSE) as evaluation metrics in our experiments. Table 1 lists the MAE and RMSE scores on the two different setups TR_{75} and TR_{20} and based on the following approaches:

- **MF-SGD** (*D*): Matrix Factorization method using SGD learning algorithm on single domain *D*.
- **FM-X** (**D**): Factorization Machine method on single domain D based on learning algorithm X (SGD, ALS or MCMC).
- **FM-All-X** (**D**): Combining all rating data into single domain (blind combination) and testing target domain *D* by using FM with algorithm X. This approach simply increases the size of training data by including the rating data of all domains. In other words, the feature vector **x** is represented as in equation (2) and all items in different domains are treated the same.
- **FM-X** ($D_T, \{D_A\}$): Factorization Machine method on target domain D_T and auxiliary domains $\{D_A\}$ based on algorithm X.
- **PF2-CDCF**: The Cross-Domain CF method which is proposed by Hu et al. [2] on the same dataset.

Comparison of results on single domains in table 1 shows that by using MCMC learning method, FM method performs better than matrix factorization. Comparison of FM-MCMC and FM-All-MCMC methods reveals that simply including the rating data of auxiliary domains into target domain does not cause any improvement on rating prediction and it can also hurt the result since the additional data can be noisy for the target domain. The best results, FM-MCMC

(Book, {Music, DVD,Video}) and FM-MCMC (Music, {Book, DVD,Video}), are obtained using our adopted cross-domain method with MCMC learning method and are better than PF2-CDCF on the same dataset.

5   Discussion and Future Directions

In this work we adopted a model using factorization machines to exploit additional knowledge from auxiliary domains to achieve performance improvement in cross-domain CF. The success of CDCF is highly dependent on effectively transferring knowledge from auxiliary domains, which can be well exploited with FMs. A key factor of success of our approach is the ability to encode domain-specific knowledge in terms of real-valued feature vector, which became possible with FMs and which enables better exploitation of the interaction patterns in auxiliary domains. The experimental results show that our adopted method can perform better than state-of-the-art CDCF methods while it benefits from low computational cost of FMs.

In the future, we want to apply our method to more complicated CDCF scenarios particularly when the source and target domains are more heterogeneous. Another extension to our approach is to also use *contextual* information from both target and auxiliary domains to investigate whether exploiting context can result in even better CDCF performance.

Acknowledgments. This research is supported by funding from the European Commission's 7th Framework Program under grant agreement no. 610594 (CrowdRec) and PHENICX (601166).

References

1. Gantner, Z., Rendle, S., Freudenthaler, C., Schmidt-Thieme, L.: Mymedialite: a free recommender system library. In: RecSys 2011. ACM, New York (2011)
2. Hu, L., Cao, J., Xu, G., Cao, L., Gu, Z., Zhu, C.: Personalized recommendation via cross-domain triadic factorization. In: WWW 2013 (2013)
3. Leskovec, J., Adamic, L.A., Huberman, B.A.: The dynamics of viral marketing. ACM Trans. Web 1(1) (May 2007)
4. Li, B., Yang, Q., Xue, X.: Can movies and books collaborate?: cross-domain collaborative filtering for sparsity reduction. In: IJCAI 2009 (2009)
5. Pan, W., Xiang, E.W., Liu, N.N., Yang, Q.: Transfer learning in collaborative filtering for sparsity reduction. In: AAAI. AAAI Press (2010)
6. Rendle, S.: Factorization machines with libfm. ACM Trans. Intell. Syst. Technol. 3(3) (May 2012)
7. Rendle, S., Gantner, Z., Freudenthaler, C., Schmidt-Thieme, L.: Fast context-aware recommendations with factorization machines. In: SIGIR 2011. ACM (2011)
8. Shi, Y., Larson, M., Hanjalic, A.: Tags as bridges between domains: improving recommendation with tag-induced cross-domain collaborative filtering. In: Konstan, J.A., Conejo, R., Marzo, J.L., Oliver, N. (eds.) UMAP 2011. LNCS, vol. 6787, pp. 305–316. Springer, Heidelberg (2011)
9. Shi, Y., Larson, M., Hanjalic, A.: Collaborative filtering beyond the user-item matrix: A survey of the state of the art and future challenges. ACM Computing Surveys (to appear)

Improvements to Suffix Tree Clustering

Richard Elling Moe

Department of Information Science and Media Studies
University of Bergen, Norway

Abstract. We investigate document clustering through adaptation of Zamir and Etzioni's method for Suffix Tree Clustering. We modified it with substantial improvements in effectiveness and efficiency compared to the original algorithm.

1 Background

This investigation came about as part of a project with the long term goal to model the flow of news in online newspapers over time and to visualize the concentration of coverage related to various topics. An essential question is then how much overlap and recirculation there is in news production. The ability to cluster documents on the basis of having similar content would be instrumental to detecting reuse and overlap. Therefore we have explored the application of a clustering technique to a news corpus.

The corpus consisted of the 10 top stories from the front pages of 8 large online newspapers, harvested daily from December 7 to 18, 2009. A total of 960 articles were manually coded based on categories that are used by media scholars to classify news. Each article received a tag, consisting of five categories, characterizing the content of the article. For example *International-Economy-Financecrisis-Debt-Dubai*. The final preprocessing of the data was to reduce words to their ground form and to keep only certain kinds of words: nouns, verbs, adjectives and adverbs.

Zamir and Etzioni [6] demonstrate that documents can be clustered by applying the Suffix Tree method to short excerpts from them, referred to as *snippets*. This was an attractive feature in our context since such snippets may be readily available for news articles in the form of front-page matter such as headlines, captions and ingresses. For this reason we chose to adapt their use of Suffix Tree Clustering and also because they report it to outperform a number of other algorithms. More recently, Eissen et al [2] presents a more nuanced picture. They point out that the technique has some weaknesses but maintains that these have little impact when applied to shorter texts and are therefore no problem for investigation.

2 Suffix Tree Clustering

The backbone of Suffix Tree Clustering is the data structure known as a *compact trie* [5]. A trie is a tree for storing sequences of tokens. Each arc is labelled

M. de Rijke et al. (Eds.): ECIR 2014, LNCS 8416, pp. 662–667, 2014.

with a token such that a path from the root represents the sequence of labels along the path. This simple structure effectively represents sequences and their subsequences as paths whereas the branching captures shared initial sequences. Note that a path does not necessarily represent a *stored* sequence. A stored sequence will have an end-of-sequence mark attached to its final node. The trie structure can be refined for the purpose of saving space. The idea is that sections of a path containing no end-of-sequence marks and no branching can be collapsed into a single arc. The *compact* trie thus allows arcs to be labelled with sequences of tokens.

The suffix tree employed by Zamir and Etzioni is a compact trie storing all the suffixes of a set of given *phrases*, i.e. the snippets. That is, the arcs are labelled with sequences of *words*. Furthermore, the end-of-sequence mark is now the set of documents that the phrase occurs in. (In practice, the set of document ID's.)

The suffix tree forms the basis for constructing clusters of documents. Each node in the tree corresponds to a *base-cluster*. A base-cluster $\sigma : S$ is basically the set S of documents associated with the subtree rooted in the node. The *label* σ is composed of the labels along the path from the root to the node in question.

The base-clusters will be further processed to form the final clusters. That is, they are merged on grounds of being similar. Specifically, two base-clusters $\sigma : S$ and $\sigma' : S'$ are similar if and only if $\frac{|S \cap S'|}{|S|} > 0.5$ and $\frac{|S \cap S'|}{|S'|} > 0.5$.

Now consider the *similarity-graph* where the base-clusters are nodes and there is an edge between nodes if and only if they are similar. The final clusters correspond to the connected components of the similarity-graph. That is, a cluster is the union of the document-sets found in the base-clusters of a connected component. Initially the cluster is not given a designated label of its own. However, we will find use for such a label so we add one by collecting the words from the base-cluster labels and sort them by their frequencies therein.

Clearly, the construction of final clusters requires every base-cluster to be checked for similarity with every other base-cluster. This is a bottleneck in the process but Zamir and Etzioni circumvent the problem by restricting the merging to just a selection of the base-clusters. For this purpose they introduce a *score* and form the final clusters from from only the 500 highest scoring base-clusters. We refer to this limit as λ, i.e. in [6] we have $\lambda = 500$.

The score of a base-cluster $\sigma : B$ is defined to be the number $|B| \times f(\sigma)$ where the function f returns the *effective length* of σ. The effective length of a phrase is the number of words it contains that are neither too frequent, too rare nor appear in a given *stop-list* of irrelevant words. Specifically, 'too frequent' means appearing in more than a percentage θ_{max} of the (total collection of) documents whereas a word is too rare if it appears in less than θ_{min} documents. In [6] these thresholds are set to $\theta_{min} = 3$ and $\theta_{max} = 0.4$. Furthermore, f penalizes single-word phrases, is linear for phrases that are two to six words long and becomes constant for longer phrases (see [6] and [3]).

3 Modifications

In the present context we can make use of the front page matter, i.e. headline, ingress and caption, should there be a photo. So, for each news article its snippet will be the collection of such phrases. Given a snippet containing multiple phrases, each of them will be inserted into the trie separately, i.e. a suffix-tree would hold all the suffixes of each phrase in the snippet.

Our first modification is to abandon the confinement to suffixes and in stead fill the compact trie with all n-grams of snippets for a suitable n. Except from some uninteresting special cases, the number of words contained in the n-grams of a phrase is strictly fewer than the number of words in the corresponding suffixes. With fewer words to process we expect the algorithm to work faster but there is also the concern that the information held by the data then becomes impoverished. In an attempt to strike a balance, we choose n so as to maximize the number of words inserted into the trie. This is achieved by expanding a phrase of length k into its $\lceil k/2 \rceil$-grams.

Secondly, we disregard clusters where the label consists of a single word only. The presence of highly frequent words may cause texts to gravitate towards each other when clusters are formed. Even if the use of a stop-list can help reduce the impact of some very common and irrelevant words we can not blacklist every common word there is. There will inevitably be clusters cemented by the co-occurrences of a single common word. Such clusters are often large and inaccurate. A one-word cluster is not necessarily a bad cluster but it seems reasonable to assume that this is the case more often than not. Then the net effect of removing them would be positive.

Finally, we will apply a more sophisticated similarity measure. Originally, the similarity of two base-clusters is determined solely on the basis of the amount of overlap between the document-sets they are composed of. It seems likely that the decision would benefit from taking into account additional cluster characteristics such as word frequencies and label overlap.

Notation: In the following we write $\hat{s}$ to denote the set of words occurring in a sequence s, whether it be a label or an entire document.

Our approach to a new similarity measure is inspired by Chim and Deng's [1] attempt to enhance the Vector Space Document model (VSD) with suffix trees. How about doing it the other way round? Could the marriage of the two models come about by incorporating VSD concepts into the suffix tree method? For this purpose we define the *label profile* for a base-cluster $\sigma : S$ to be the function $\boldsymbol{v}$ given by

$$\boldsymbol{v}(w) = \begin{cases} \Theta(w, S^+, C) & \text{if } w \in \hat{\sigma} \\ 0 & \text{otherwise} \end{cases}$$

where Θ is the term frequency - inverse document frequency which reflects the weight of a word w in a document relative to a corpus of documents. In our case, the document in question is a concatenation S^+, of the documents contained in the base-cluster whereas the corpus is our entire collection C of articles.

Specifically, $\Theta(w, S^+, C)$ equals

$$tf(w, S^+) * \log \frac{|C|}{1 + |\{d \mid d \in C \text{ and } w \in \hat{d}\}|}$$

where the term frequency $tf(w, S^+)$ denotes how many times w occurs in S^+.

Two base-clusters $\sigma : S$ and $\sigma' : S'$ are now considered similar if, and only if, they satisfy the original similarity-measure in conjunction with having similar label profiles. Specifically that their respective profiles v and v' have a *cosine-similarity* above 0.5. That is

$$\frac{\Sigma_w(v(w) * v'(w))}{\sqrt{\Sigma_w v(w)^2} * \sqrt{\Sigma_w v'(w)^2}} \geq 0.5$$

4   Ground Truth, Precision and Recall

The manually tagged portion of our corpus can serve as ground truth for precision / recall-style investigations. Since the tags represent a human judgement as to what the document is about we think it is fair to assume that a high degree of overlap in tags will indicate overlap in content.

A *ground truth cluster* consists of all documents having identical tags, and only those documents. Thus, ground truth clusters are identifiable by tags. Assuming that C is the set of clusters generated by the algorithm and G the set of ground truth clusters, precision would measure the proportion $\frac{|C \cap G|}{|C|}$ of ground truth clusters among the clusters generated by the algorithm. Recall measures the extent to which the algorithm will recreate ground truth, i.e. $\frac{|C \cap G|}{|G|}$.

As described above, the algorithm sets the limit λ on the number of base-clusters that proceed to be merged into final clusters. This poses a serious threat to recall. Specifically, only 500 of the original 25378 base-clusters are retained. When the initial data for generating clusters is cut short by such a large amount we can not expect the algorithm to be able to recreate all the ground truth clusters. In fact, our data contains 667 ground truth clusters, while Zamir and Etzioni's original setup of the algorithm produces a total of only 159 clusters. With our corpus this has a devastating effect. A high proportion of the articles make up a ground truth cluster of its own, being the only texts on their topics. The scoring favors bigger base-clusters, leaving many singleton clusters out and causing great harm to recall.

We believe the original scoring is somewhat arbitrary and sensitive to the kind of text it is applied to. Indeed, we made the striking discovery that simply reversing the ranking induced by the score would produce far better results for our corpus. That is, the algorithm performed better on what the original scoring would hold to be the 500 *worst* base-clusters.

As our current focus is to demonstrate general improvements to the technique, as opposed to optimizing it for our specific corpus, we will explore modifications that improve the performance under two almost completely opposite scoring schemes. In contrast to Zamir and Etzioni's original ranking, referred to as $\downarrow$, we also apply the ranking $\uparrow$ obtained by slightly tweaking the $\theta_{\min}$ and $\theta_{\max}$ thresholds, to 6 and 0.5 respectively, and reversing the order.

Precision/recall studies rely on a notion of *relevance*. In our setting, a cluster is considered relevant if it matches a ground truth cluster. In case it is unreasonably strict to require a *perfect* match we incorporate some flexibility by adding a degree of perfection. A cluster C *matches ground truth with discrepancy* $5 - d$ if and only if d is the maximal number for which there is a ground-truth cluster G such that $G \subseteq C$ and d is the number of words common to all tags in $C \cup G$. Intuitively, discrepancy 0 is a perfect match while 5 means that there is no category that appears in all tags.

Relaxing the requirement for perfection by some degree of discrepancy may help to prevent the requirements for relevance from being too strict. However, it may still be so strict that it portrays the algorithm unfavorably. This is discussed in [4], which suggests an alternative measure for effectiveness. Such considerations are of less importance now since we only aim to register *improvements* in performance.

5   Evaluation

Tests have been carried out to evaluate our modifications of the algorithm. As discussed in section 4 we will attempt to neutralize possible bias from tailor-made scoring by measuring up to opposite scorings. Our benchmark is then the performances of the original algorithm under the rankings $\downarrow$ and $\uparrow$, shown in table 1.

Table 1. Benchmark: original algorithm with $\lambda = 500$

Discrepancy	Precision		Recall	
	$\downarrow$	$\uparrow$	$\downarrow$	$\uparrow$
0	0.088	0.447	0.021	0.241
≤ 1	0.157	0.450	0.039	0.244
≤ 2	0.201	0.464	0.058	0.252

Initial experiments have shown that each of our three modification will improve performance. Together they make a considerable difference. As expected the modified algorithm works faster, enabling it to process more base-clusters. In fact, λ can be raised from 500 to 800 without spending more time than the benchmark run. (About 1.5 seconds on a run-of-the-mill laptop.) Table 2 shows the overall improvements when $\lambda = 800$, i.e. the levels of precision and recall obtained using approximately the same time as in table 1.

We have already noted that recall suffers as a result of discarding base-clusters. Indeed, increasing the mass by merging more base-clusters will boost recall. For example, setting $\lambda = 8000$ increases recall under $\uparrow$ by some 160%. Unfortunately, this comes with a punishing cost in running-time. Clearly, the higher number of base-clusters flood the bottleneck of merging them.

Table 2. Modified algorithm with $\lambda = 800$

	Precision		Recall	
Discrepancy	↓	↑	↓	↑
0	0.173	0.584	0.049	0.315
≤ 1	0.213	0.584	0.063	0.315
≤ 2	0.244	0.586	0.079	0.316
Average improvement	51%	29%	77%	28%

6   Conclusion

We have modified the Suffix Tree Clustering technique with some success. Because of the confinement to a particular data set we can not claim external validity for our results. For the time being, our contribution is mainly to demonstrate an interesting potential for improvement, but also to point out directions for further work.

There are several issues that could be pursued. First, we believe that there are other varieties of similarity measures that deserves to be explored. Secondly, we observed that scoring can be sensitive to the kind of text it is applied to. We believe scoring could make more sophisticated use of cluster characteristics, such as labels, size and word-frequencies. Finally, we see that the potential for good recall values is severely hampered by the computational bottleneck of merging base-clusters. A research challenge lies in finding faster algorithms or alternative ways of forming clusters from base-clusters.

References

1. Chim, H., Deng, X.: A New Suffix Tree Similarity Measure for Document Clustering. In: Proceedings of the 16th international conference on World Wide Web, pp. 121–130. ACM, New York (2007)
2. Zu Eissen, S.M., Stein, B., Potthast, M.: The Suffix Tree Document Model Revisited. In: Tochtermann, M. (ed.) Proceedings of the I-KNOW 2005, Graz 5th International Conference on Knowledge Management Journal of Universal Computer Science, pp. 596–603 (2005)
3. Gulla, Borch, Ingvaldsen: Contextualized Clustering in Exploratory Web Search. In: do Prado, H.A., Ferneda, E. (eds.) Emerging Technologies of Text Mining: Techniques and Applications, pp. 184–207. IGI Global (2007)
4. Moe, R., Elgesem, D.: Compact trie clustering for overlap detection in news. In: Proceedings of NIK 2013 (2013)
5. Smyth, B.: Computing Patterns in Strings. Addison Wesley (2003)
6. Zamir, O., Etzioni, O.: Web Document Clustering: A Feasibility Demonstration. In: Proceedings of the 21st Annual International ACM SIGIR Conference on Research and Development in Information Retrieval, pp. 46–54. ACM, New York (1998)

Deep Learning for Character-Based Information Extraction

Yanjun Qi[1], Sujatha G. Das[2], Ronan Collobert[3], and Jason Weston[4]

[1] Department of Computer Science, University of Virginia, USA
Machine Learning Department, NEC Labs America, USA
yanjun@virginia.edu
[2] Computer Science Department, Penn State University, USA
sujatha.das@gmail.com
[3] IDIAP Research Institute, Switzerland
ronan@collobert.com
[4] Google, New York USA
jaseweston@gmail.com

Abstract. In this paper we introduce a deep neural network architecture to perform information extraction on character-based sequences, e.g. named-entity recognition on Chinese text or secondary-structure detection on protein sequences. With a task-independent architecture, the deep network relies only on simple character-based features, which obviates the need for task-specific feature engineering. The proposed discriminative framework includes three important strategies, (1) a deep learning module mapping characters to vector representations is included to capture the semantic relationship between characters; (2) abundant online sequences (unlabeled) are utilized to improve the vector representation through semi-supervised learning; and (3) the constraints of spatial dependency among output labels are modeled explicitly in the deep architecture. The experiments on four benchmark datasets have demonstrated that, the proposed architecture consistently leads to the state-of-the-art performance.

1 Introduction

In this paper, we focus on the information extraction (IE) tasks which aim to automatically extract information about pre-specified types of events, entities or relationships from unstructured or semi-structured machine-readable documents of sequences. The sequence data we focus in this paper refers to the general family of character-based natural language strings. For instance, it could be Chinese text or Japanese text in which no space exists to mark word boundary between characters. Not limited to natural languages, the target sequence could also be a protein sequence which describes the primary structure of a protein using a linear string of amino acid letters.

Several IE tasks have been identified as important for end applications such as information retrieval or text summarization. For instance, the syntactic-level part-of-speech (POS) tagging, the semantic-level named entity extraction (NER) and the word segmentation (WS) tasks are the fundamental building blocks (Table 1) for processing Chinese [6]. Similarly, the functional tagging of each amino

M. de Rijke et al. (Eds.): ECIR 2014, LNCS 8416, pp. 668–674, 2014.
© Springer International Publishing Switzerland 2014

Table 1. An example of information extraction tagging on a sample sequence of Chinese text. Three character-based IE tasks are included, (1) WS: word segmentation; (2) POS: part-of-speech tagging and (3) NER: name entity recognition.

Characters	克	林	顿	总	统	前	往	中	东
WS	B	I	E	B	E	B	E	B	E
POS	B-NR	I-NR	E-NR	B-NN	E-NN	B-VV	E-VV	B-NR	E-NR
NER	B-PER	I-PER	E-PER	O	O	O	O	B-LOC	E-LOC

acid on the protein sequences is the core endeavor of computational biology. One classic task, secondary structure (SS) prediction [7] aims to predict each amino acid's secondary structure label which provides useful intermediate information for a protein's three-dimensional structure. All these tasks could be treated as a labeling of each atomic element (e.g. Chinese character, or amino acid) in the sequence, into one of the multiple classes for the task of interest.

Differently from Romance language with words as the basic unit, Chinese text has no space to mark word boundary between characters. Tagging Chinese through assigning labels to characters (Table 1) has been shown to be a simple but effective formulation [9]. While early studies were mostly based on hand-crafted rules, recent systems have used supervised machine learning techniques such as, Hidden Markov Models (HMM), Maximum Entropy (ME) Models, Support Vector Machines (SVM), and Conditional Random Fields (CRF). The top systems from SIGHAN bakeoff competitions [6] were mostly based on the CRF and ME models. Analogous to Chinese text, the primary representation of a protein sequence includes a linear string of letters where each letter represents a certain amino acid and the string has no special letter to mark the boundary between functional segments. Many automated SS prediction methods have been described in the scientific literature [7], including for instance, neural networks, HMM and dynamic Bayesian networks. However, the above-mentioned systems mostly relied on rich sets of task-specific and hand-crafted features, which require time-consuming feature engineering and are hard to adapt to similar tasks.

Lately, "deep learning" [4] grows to bring in a lot of attentions and has won many pattern recognition contests. Our paper is motivated by a recent work from Collobert et al. [2] who demonstrated that a unified deep neural network architecture provides the state-of-art performance on multiple English Natural Language Processing (NLP) tasks using simple task-independent features. We adapt and modify this architecture to provide a unified framework for character-based IE tagging tasks by learning a hierarchy of representations given very basic inputs of character features. Three important components are employed in our framework, including *vector representation learning* of characters, semi-supervised *language modeling* and sentence-level training of output label dependencies.

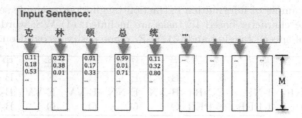

Fig. 1. An embedding module to learn vector representations for each input character

2 Method

The proposed framework provides a unified end-to-end system that, given a string of characters, it provides several layers of feature extractions and predicts labels for each character. Table 1 provides a sample sequence of Chinese text and three sample IE outputs. Features relevant to the target IE task are learned automatically by backpropagation in the deep layers of the network model.

2.1 Learning Character to Vector Representation

The deep neural network is characterized by two specialized layers: (1) a character embedding layer and (2) a segment/window feature extraction layer, followed by a series of classical neural network layers. The whole neural network architecture is displayed as Figure S1 in Supplementary [1] . The very first layer projects each character into a real-valued vector, in a M-dimensional latent embedding space (Figure 1). Here M is a hyperparameter to be chosen by the user. Within a finite character dictionary $\mathcal{D}$, each character $c_i \in \mathcal{D}$ is embedded into the feature space using a $M \times |\mathcal{D}|$ projection matrix W, such that W_{c_i}, the column vector of W at the index of c_i, is the vector representation corresponding to character c_i. Thus, using the first layer of this architecture, an input character sequence $\{c_1, c_2, \ldots c_n\}$ is transformed into a series of real valued vectors $\{W_{c_1}, W_{c_2}, \ldots W_{c_n}\}$. The parameters of the projection matrix W are learned automatically as part of the neural network training through back-propagation. The encoding weights in W are randomly initialized from a centered, uniform distribution. The resulting embedding after training is optimal for our target IE task, in the sense that it optimizes the objective cost on the training set for the specific task. Essentially this first embedding module is a special convolution neural network layer which applies the same projection W on each character in $\mathcal{D}$ regardless the position of characters in text sequences.

The second layer performs a sliding window operation on the character sequence which aggregates the output of the first layer into blocks corresponding to a fixed window of size k. To label a complete sentence, we slide the window along the sentence, labeling one character at a time. This means each character in the sequence is described through itself and its neighboring characters.

The remaining layers comprise a standard, fully connected multi-layer perceptron network with L layers of hidden units. Each hidden layer l learns to map its input to a hidden feature space (through parameter U^l), and the last layer V utilizes a softmax function to map its input to the output tag (label) space. This layer's outputs are ensured to be positive and sum to 1, thus we can interpret the outputs of the whole network as probabilities for each class [2]. Essentially the network includes the following set of parameters to be estimated, $\Theta = \{W, U^1, U^2, ..., U^{L-1}, U^L\}$.

Assuming we are given a set of training examples $\{(x_n, y_n)\}_{n=1...N}$, x_n represents a window of characters, y_n describes the label of middle character in x_n, and $f(x_n)$ represents the predicted output of x_n by the whole network. Then the deep model is trained to find the best Θ by minimizing the negative log-likelihood (NLL) loss, i.e. $E(\Theta) := \sum_{n=1}^{N} NLL_\Theta(f(x_n), y_n)$ over the training samples. $E(\Theta)$ is optimized using backpropagation through Stochastic gradient descent (SGD). In SGD, a random example (x, y) is sampled from the training set and then a gradient descent step is applied to update the network parameter Θ as: $\Theta \leftarrow \Theta - \lambda \frac{\partial E(\Theta, x, y)}{\partial \Theta}$, where λ is the learning rate hyperparameter.

2.2 Improving Representation Learning with Unlabeled Sequences

Manually labeling character-based sequences, i.e. to obtain tag label for each character, could be quite time-consuming, since it requires very detailed annotation on the character-level (Table 1). Information hiddlen inside unlabeled text has been shown to be helpful for supervised IE tagging [2]. Thus, we employ to add a so-called semi-supervised "language modeling" (LM) task [2] using unlabeled character sequences (abundant from internet, e.g. Wikipedia Chinese corpus, or Swissprot protein sequence database).

This step relies on the key observation that individual characters carry significant semantic information hidden in the unlabeled data. This translates to the basic idea that the target LM task will learn to force two character pieces with similar semantic meanings to have closer representations and two pieces with dissimilar meanings to have distant representations in a learned feature space. Intuitively, all length-k character windows from an unlabeled corpus could be labeled as positive examples, and negative fragments are generated by randomly substituting the middle character in each window. Thus, each pair of these (positive, negative) segments build up a corpus of dissimilar character pieces. A deep network, similar as the one for IE tagging, is then built to learn the representations based on the pairs of dissimilar segments from the above corpus. Again, the character embedding layer and all parameters of the subsequent network layers are automatically trained by backpropagation. Unlike supervised IE tagging using NLL loss, the LM task is trained with a margin ranking cost: $\sum_{s \in S} \sum_{c \in D} \max(0, 1 - f'(s) + f'(s^c))$ where S is the set of positive character segments, and s^c is a pseudo-negative segment where its middle character has been replaced by a random character c in the dictionary D. $f'(\cdot)$ represents the output of the deep network architecture for LM task, which indicates a function that projects its input segment to a value output. The output scalar value describes how likely a given character segment exists in the unlabeled corpus in our setup. Essentially, we are learning the network weights to rank positive character segments above synthetic negative segments. Then we utilize a popular deep learning strategy—pretraining, to connect the unsupervised LM task and the supervised tagging tasks. That is, parameter Θ' learned from LM is used as the initialization for parameter Θ for supervised IE.

2.3 Discriminative Training of Sentence-Level Label Dependency

So far, our deep framework uses a labeling-per-character strategy without exploiting the dependencies among the targeted tag labels. This assumes that the output label for each position in a character-based sequence can be predicted independently from nearby positions. Empirically, this assumption fails largely for IE tagging. For example, clearly some tags are less likely to follow another set of tags (Table 1) (e.g. O tag is less likely to follow B-PER tag for NER), and certain groups of tags are highly likely to appear as neighbors (e.g. B-PER and E-PER for NER). Thus we models the dependencies between target tags through discriminative forward training [2]. Basically it scores a whole candidate sentence by aggregating the predicted scores of each involved character and also weighting the scores by a transition matrix $[A]_{i,j}$. This parameter $[A]_{i,j}$ captures the degree of likelihood for jumping from i to j tags in successive characters. Similar to Θ, weights of A are also automatically trained by backpropagation. The sentence tagging layer with A parameter is put on the top of the basic model $f(\cdot)$, achieving a revised deep model $f''(\cdot)$. The final tagger $f''(\cdot)$ is trained to find the best tag path which optimizes each whole training sentence's output score. The prediction of the tag path is efficiently implemented through a Viterbi-like algorithm (space limit; details in [2]).

3 Experiments

Data: We demonstrate the power of deep learning architecture on four benchmark data sets for character-based IE tagging, (1), WS on Chinese tree bank (CTB) data (LDC2007T36) [8]. (2), POS on the Chinese tree bank (CTB) data [8]; (3), NER on "CITYU" data from SIGHAN3 bakeoff[6]. (4), CB513, the standard protein SS benchmark, is used [3] for the SS evaluation. All data details of the four sets (e.g. sample size, encoding of output tags, et al.) are in Supplementary [1] due to page limitation.

Metric: For WS, we compute the character-level precision, recall, and equally weighted F-measure (only F1 reported). When evaluating POS and NER, we compute overall phrase-level precision, recall, and F1-measure (only F1 reported). For the protein SS task, we use the standard "Q-score", which represents the *accuracy* evaluated at the amino acid level.

Table 2. Performance comparison on four character-based IE tagging tasks. Different combinations of the proposed strategies are compared (in percentage %). "*": for SS task, we also add the classic PSI-Blast feature in all combinations. For the 7th setup, SS is added with a multitasking strategy [7], instead of "ws+c2".

Configuration / Task	WS-Chinese	POS-Chinese	NER-Chinese	SS-Protein*
1. c1	94.73	86.74	80.61	74.5
2. c1+lm	95.57	86.93	81.79	74.8
3. c1+vit	95.38	88.41	85.81	77.6
4. c1+lm+vit	96.07	88.81	86.99	77.8
5. c1+lm+c2	95.98	88.48	83.51	∼
6. c1+lm+c2+vit	**96.62**	89.39	87.24	∼
7. c1+lm+c2+vit+ws	∼	**93.27**	88.88	**80.3***
Previous Best	95.9 [10]	91.9 [10]	**89.00** [6]	80.0 [5]
Previous Second Best	95.1 [10]	91.3 [10]	88.61 [6]	∼

Setup: The evaluation of WS and POS is carried through 10-folds cross validation. Following the same setup as (Zhang&Clark 2008) [10], we have evenly partitioned CTB into ten groups, and used nine groups for training and the rest for testing. The NER evaluation is carried through the benchmark provided train-test split. The SS evaluation follows the standard seven-folds cross validation on this set. The Torch [2] deep learning toolbox is used for coding the architecture.

Hyperparameter: Furthermore, the deep framework requires the specification of multiple hyperparameters. This includes the size k of the character window, the size h of the hidden layers, and the learning rate λ. We considered $k \in \{3, 5, 7, 9, 11, 13\}$, $h \in \{80, 100, 120, 150, 200\}$, and $\lambda \in \{0.001, 0.01, 0.03, 0.05, 0.1\}$. The parameter for the embedding layer, M, was chosen from $\{15, 30, 50, 70\}$. The hyperparameter selection was based on the cross-validation results on training. The semi-supervised language model was trained using the freely available Chinese wikipedia corpus and the swissprot protein sequence database.

Result Comparison: Table 2 summaries multiple variations of the proposed architecture and the performance comparison to the state-of-the-art systems. We systematically test the incremental combinations of multiple strategies: (1). "c1": character to vector embedding; (2). "lm": unsupervised "language model" of character. (3). "vit": sentence-level label dependency; (4). "ws": word segmentation added as discrete features [2]; (5). "c2": another embedding extraction layer for learning vector representations for character bigrams. It is clear that the final system (with the combination of all strategies) has improved over the state-of-the-art performance on WS and POS and has achieved the state-of-the-art predictive level for NER and SS tasks (detailed discussion in [1]).

Conclusion: We have proposed a deep learning framework for character-based information extraction on Chinese and protein sequences. As a flexible and robust prediction system, this architecture has achieved the state-of-art performance on four benchmark data. Our methodology could easily be adapted to additional character-based tagging tasks, such as Japanese NER.

References

1. Supplementary (December 2013), http://www.cs.cmu.edu/~qyj/zhSenna/
2. Collobert, R., Weston, J., Bottou, L., Michael, K., Kuksa: Natural language processing (almost) from scratch. JMLR 12, 2493–2537 (2011)
3. Cuff, J.A., Barton, G.J.: Evaluation and improvement of multiple sequence methods for protein secondary structure prediction. Proteins 34, 508–519 (1999)
4. Hinton, G.E., Salakhutdinov, R.R.: Reducing the dimensionality of data with neural networks. Science 313(5786), 504–507 (2006)
5. Kountouris, P., Hirst, J.D.: Prediction of backbone dihedral angles and protein secondary structure using support vector machines. BMC Bioinf. 10(437) (2009)
6. Levow, G.A.: The third international chinese language processing bakeoff: Word segmentation and named entity recognition. In: Proceedings of the Fifth SIGHAN Workshop on Chinese Language Processing, Sydney, vol. 117 (July 2006)
7. Qi, Y., Oja, M., Weston, J., Noble, W.S.: A unified multitask architecture for predicting local protein properties. PLoS One 7(3), e32235 (2012)
8. Xue, N., Xia, F., Chiou, F.D., Palmer, M.: Penn chinese treebank: Phrase structure annotation of a large corpus. Natural Language Engineering 11, 207–238 (2005)
9. Xue, N., et al.: Chinese word segmentation as character tagging. Computational Linguistics and Chinese Language Processing 8(1), 29–48 (2003)
10. Zhang, Y., Clark, S.: Joint word segmentation and pos tagging using a single perceptron. In: Proceedings of the 46th Annual Meeting of ACL, pp. 888–896 (2008)

Text-Image Topic Discovery
for Web News Data

Mingjie Qian

Department of Computer Science
University of Illinois at Urbana-Champaign, IL, USA
mqian2@illinois.edu

Abstract. We formally propose a new application problem: unsupervised text-image topic discovery. The application problem is important because almost all news articles have one picture associated. Unlike traditional topic modeling which considers text alone, the new task aims to discover heterogeneous topics from web news of multiple data types. The heterogeneous topic discovery is challenging because different media data types have different characteristics and structures, and a systematic solution that can integrate information propagation and mutual enhancement between data of different types in a principle way is not easy to obtain, especially when no supervision information is available. We propose to tackle the problem by a regularized nonnegative constrained $l_{2,1}$-norm minimization framework. We also present a new iterative algorithm to solve the optimization problem. To objectively evaluate the proposed method, we collect two real world text-image web news datasets. Experimental results show the effectiveness of the new approach.

Keywords: Text-image topic discovery.

1 Introduction

Exploring web news more efficiently and understanding them more effectively is important for absorbing information in our daily life. However, there're more and more web news articles but we have less and less time to read them. One ideal way is to automatically group the web news per their content or topics, then a user can choose a topic to read. This procedure can be done recursively so that a user could explore the news with least time. Images play an important role in news as is evident from the fact that almost all news articles have one picture associated. Thus to effectively organize news, it is important to consider both text and images. However, traditional topic modeling techniques such as LSI[4], PLSA[5], and LDA[1] are not powerful enough to handle heterogenous data because they consider only the text content. Since modern web news are usually composed of multiple data types such as text, image, and video, effective topic mining methods that can discover joint text-image topics and organize these multiple typed news data are urgently needed. The multiple typed topic discovery task is substantially different from topic modeling for a single text

M. de Rijke et al. (Eds.): ECIR 2014, LNCS 8416, pp. 675–680, 2014.
© Springer International Publishing Switzerland 2014

corpus because image is not a sequence of logical semantic units and image content is more difficult to numerically define and compute. Mining topics from heterogeneous data requires careful and insightful utilization of properties of every data types, which is not a trivial task.

Multi-view learning, which aims to learn better models to cluster data in multiple views, is a machine learning research area that can be applied for our problem, but state-of-the-art multi-view learning methods cannot do this task very well. Co-trained multi-view spectral clustering[7] iteratively uses the spectral embedding from one view to constrain the similarity graph used for the other view. However, this approach heavily relies on similarity graph for each view and completely ignores the detailed information, which may badly hurt the clustering performance due to loss of discriminative information. Also it's not straight forward to generate multi-view topic representation via this approach. [2] proposes to generalize K-means for multi-view data clustering. However, its performance tends to be dominated by the worst domain since the algorithm will assign large weight to the domain with the largest approximation error as will be demonstrated in the experiment later. There're also some heterogeneous data co-clustering work[3][8], however they require some supervision information. For example, [3] require user specified must-link and cannot-link constraint in the central type, and [8] require user preference before clustering. Besides, although heterogeneous co-clustering methods appear to be able to tackle our problem, they do not aim to explicitly learn representative and interpretable multi-view topics from heterogenous web news data. The major goal of this work is to provide an effective multi-view learning approach to discover text-image topics from web news data without any supervision.

2   Problem Formulation

In this section, we introduce the novel problem of joint text-image topic mining.

Definition 1 (Text-image Topic). *A text-image topic T is a bundle $\{V_1, V_2\}$, where V_1 is a weighted term vector or a set of weighted terms, V_2 is a set of selected images. Note that the concept of a text-image topic can be generalized as a multimedia topic which is a bundle $\{V_1, V_2, \ldots, V_M\}$ where V_1 is a weighted term vector or a set of weighted terms, V_2 is a set of selected images, V_3 is a set of selected videos, and V_M is the set of data description from the M-th media type.*

Definition 2 (Text-image Document). *A text-image document D is a general form of text document which contains 2 mutually associated "subdocuments" corresponding to text and image media types. Formally, $D = \left(d^1, d^2\right)$, where d^1 is text, d^2 is images. The joint text-image document can be further generalized to multimedia documents, which contain M mutually associated "subdocuments" corresponding to M different media types. $D = \left(d^1, d^2, \ldots, d^M\right)$, where d^1 is text, d^2 is image, d^M is the M-th media type.*

The main component of a text-image document is the text part whereas other types of data are auxiliary data. Text provides a rich context for the whole text-image document. Images with tags are not text-image documents because tags, although shown as text, do not provide a detailed description.

Definition 3 (Joint Text-Image Topic Discovery). *Joint Text-Image topic discovery is a topic mining problem which aims to mine a set of text-image topics $T_1, T_2, \ldots, T_k$ from a collection of text-image documents.*

3 Notations and Preliminaries

Throughout this paper, matrices are written as boldface capital letters and vectors are denoted as boldface lowercase letters. For matrix $\mathbf{M} = (M_{ij})$, its i-th row, j-th column are denoted by $\mathbf{m}^i$ (or $\mathbf{M}_{i:}$), $\mathbf{m}_j$ (or $\mathbf{M}_{:j}$) respectively. $\|\mathbf{M}\|_F$ is the Frobenius norm of $\mathbf{M}$ and $\mathrm{Tr}\,[\mathbf{M}]$ is the trace of $\mathbf{M}$ if $\mathbf{M}$ is square. For any matrix $\mathbf{M} \in \mathcal{R}^{r \times t}$, its $l_{2,1}$-norm is defined as $\|\mathbf{M}\|_{2,1} = \sum\limits_{i=1}^{r} \sqrt{\sum\limits_{j=1}^{p} m_{ij}^2} = \sum\limits_{i=1}^{r} \|\mathbf{m}^i\|_2$.

4 Methodology for Text-Image Topic Discovery

Existing multi-view works cannot do this task very well. E.g., [7] heavily relies on similarity graph for each view and completely ignores the detailed information, and even doesn't explicitly give topic representation; [2] assigns large weight to the domain with the largest approximation error and its performance will be dominated by the worst domain. We use matrix norm based numerical optimization in stead of probabilistic graphical model (PGM) because it is difficult to design an accurate PGM to model the structure of images. Meanwhile, outliers and noisy terms usually degrade the performance, we thus use $l_{2,1}$-norm to learn robust topics. Also, by regularizing on the image graph, two text vectors with similar topic indicators should have higher similarity computed from other media types. In this way, multiple media type information can be effectively used and mutually enhanced to get the final consistent and representative text-image topics. We thus propose a novel regularized nonnegative constrained $l_{2,1}$ norm minimization (RNL21NM) framework to tackle our task:

$$\min_{\mathbf{F} \geq 0, \mathbf{G} \geq 0} \left\| \mathbf{X} - \mathbf{G}\mathbf{F}^T \right\|_{2,1} + \lambda \mathrm{Tr}\left[\mathbf{G}^T \mathbf{L} \mathbf{G} \right] + \nu \left\| \mathbf{F} \right\|_1, \tag{1}$$

where $\mathbf{X} = [\mathbf{x}_1, \mathbf{x}_2, \ldots, \mathbf{x}_N]^T \in \mathcal{R}_+^{N \times d}$ is a nonnegative text data matrix, each row of $\mathbf{X}$ corresponds to a text subdocument, $\mathbf{F} \in \mathcal{R}_+^{d \times K}$ is a nonnegative topic basis matrix (one topic per column), K is the number of text-image topics, $\mathbf{G} \in \mathcal{R}_+^{N \times K}$ is a nonnegative topic indicator matrix (the i-th row $\mathbf{g}^i$ is the topic indicator vector for the i-th multimedia document), G_{ij} denotes the strength of the association between multimedia document D_i and text-image topic T_j; $\mathbf{L} = \sum\limits_{m=2}^{M} \alpha_m \mathbf{L}^m$ is the multimedia Laplacian matrix, where $\sum\limits_{m=2}^{M} \alpha_m = 1$, and $\mathbf{L}^m$ is the Laplacian matrix for the m-th media type given by $\mathbf{L}^m = \mathbf{D}^m - \mathbf{S}^m$ where $\mathbf{S}^m$ is a similarity matrix w.r.t. the m-th media type and $\mathbf{D}^m$ is the diagonal matrix given by $D_{ii}^m = \sum_{j=1}^N S_{ij}^m$. $\lambda > 0$ is a regularization parameter to control the topic consistency for different media types. $\nu > 0$ controls the sparseness of the topic matrix $\mathbf{F}$. We present Algorithm 1 to solve RNL21NM. The total complexity for Algorithm 1 is $\#\text{iters} \times \left(O(dNK) + \#\text{sub-iters} \times O(dK^2 + N^2K)\right)$, where N is the corpus size, and d is the vocabulary size[1].

Algorithm 1. RNL21NM

Input: $\mathbf{X}, \mathbf{L}, \lambda, \nu, \mathbf{G}_0, \mathbf{F}_0$
Output: $\mathbf{G}, \mathbf{F}$
$t \leftarrow 0$
repeat
$\quad D_{ii} \leftarrow \left\| \mathbf{x}^i - \mathbf{g}_t^i \mathbf{F}_t^T \right\|_2^{-\frac{1}{2}}$ for $i = 1, 2, \ldots, N$, otherwise $D_{ij} \leftarrow 0$ for $j \neq i$
$\quad G_{ij}^{t+1} \leftarrow \max\left(0, G_{ij}^t - \dfrac{\left[\mathbf{D}^2 \mathbf{G}_t \mathbf{F}_t^{\,T} \mathbf{F}_t - \mathbf{D}^2 \mathbf{X} \mathbf{F}_t + 2\lambda \mathbf{L} \mathbf{G}_t \right]_{ij}}{D_{ii}^2 \left[\mathbf{F}_t^{\,T} \mathbf{F}_t \right]_{jj} + 2\lambda L_{ii}} \right)$
$\quad D_{ii} \leftarrow \left\| \mathbf{x}^i - \mathbf{g}_{t+1}^i \mathbf{F}_t^T \right\|_2^{-\frac{1}{2}}$ for $i = 1, 2, \ldots, N$, otherwise $D_{ij} \leftarrow 0$ for $j \neq i$
$\quad \mathbf{F}_{:j}^{t+1} \leftarrow \max\left(0, \mathbf{F}_{:j}^t - \dfrac{\mathbf{F}_t \left[\mathbf{G}_{t+1}^T \mathbf{D}^2 \mathbf{G}_{t+1} \right]_{:j} - \left[\mathbf{X}^T \mathbf{D}^2 \mathbf{G}_{t+1} \right]_{:j} + \nu}{\left[\mathbf{G}_{t+1}^T \mathbf{D}^2 \mathbf{G}_{t+1} \right]_{jj}} \right)$
$\quad t \leftarrow t + 1$
until Convergence criterion satisfied
$\mathbf{G} \leftarrow \mathbf{G}_t$
$\mathbf{F} \leftarrow \mathbf{F}_t$

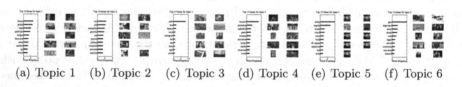

(a) Topic 1   (b) Topic 2   (c) Topic 3   (d) Topic 4   (e) Topic 5   (f) Topic 6

Fig. 1. Text-image topics discovered by RNL21NM for NPR

5   Experiments

Two widely used evaluation metrics, i.e. Normalized Mutual Information (NMI) and accuracy (ACC), are used to evaluate topic discovery performance. We crawled CNN top stories and National Public Radio (NPR) news. Titles, abstracts, and text body contents are extracted, meanwhile the associated image is stored. Text contents are stemmed and we use normalized TFIDF to represent text subdocuments. For image features, we use RGB dominant color, HSV

---

[1] All detailed derivation, proof of convergence, and complexity analysis can be found at https://sites.google.com/site/qianmingjie/

dominant color, RGB color moment, HSV color moment, RGB color histogram, HSV color histogram, four Tamura textural features, and Gabor transform. 142 text-image pairs were collected for CNN top stories from Feb. 21st to April 17th, 2011 with 8682 terms and 10 classes. 603 NPR news articles were collected from Apr. 7th to May 7th, 2013, with 17692 terms and 8 classes[2].

For single-view methods, we compare Kmeans_Text, Kmeans_Image, LDA on text, and Robust NMF (RNMF)[6] on text. For multi-view methods, we compare the state-of-the-art co-trained multi-view spectral clustering (CoTrainedMVSC)[7] and robust multi-view K-means clustering (RMKMC)[2].

We compare the best average NMI and ACC for 10 runs for all methods. For LDA[3], the best average performance is achieved by grid search with $\alpha = 50/T, \beta = 200/W$ as is suggested by the implementor, where T is the number of topics, and W is the vocabulary size. For CoTrainedMV, we use cosine similarity for text view and Gaussian normalized negative Euclidean distance for image view. For RMKMC, following its author's suggestion, we choose the best parameter γ^* by searching $log_{10}(\gamma)$ in the range from 0.1 to 2 with incremental step 0.2. For RNL21NM, we set $\alpha_2 = 1$ since only image is included. We fix ν to be 0.0001. To compute the image similarity matrix, we first calculate the negative Euclidean distance matrix and then normalize all the matrix entries into zero-one interval by Gaussian normalization. We tune λ by grid-search from $\{0, 0.2, 0.4, 0.8, 1.6, \cdots, 51.2\}$.

Table 1. Clustering Results. * means statistically significance at the 0.05 level

NMI% ± std

Dataset	Kmeans_Text	Kmeans_Image	LDA	RNMF	CoTrainedMV	RMKMC	RNL21NM
CNN	56.4 ± 7.0	22.3 ± 1.6	59.4 ± 2.5	57.7 ± 4.8	41.6 ± 1.2	32.7 ± 4.7	**68.8* ± 2.9**
NPR	26.6 ± 3.2	3.8 ± 0.4	37.7 ± 3.2	38.8 ± 4.0	17.8 ± 2.3	4.2 ± 0.4	**39.6* ± 2.5**

ACC% ± std

Dataset	Kmeans_Text	Kmeans_Image	LDA	RNMF	CoTrainedMV	RMKMC	RNL21NM
CNN	54.9 ± 6.5	25.1 ± 1.1	53.6 ± 2.7	58.0 ± 5.7	42.3 ± 5.0	37.3 ± 4.6	**68.7* ± 5.3**
NPR	39.2 ± 5.6	19.8 ± 1.8	48.4 ± 4.5	56.4 ± 7.9	37.1 ± 3.4	18.6 ± 2.3	**58.7* ± 3.3**

Results and Discussion: Due to lack of space, 6 text-image topic examples are shown in Figure 1. We then compare the average NMI and ACC in 10 runs for all methods in Table 1. The clustering results show that the proposed RNL21NM outperforms all the other methods. The results also show that image-first approach performs badly. The reason maybe that the image features are not good for capturing semantics, leading to a larger gap between image features and semantic concepts compared with text. However, image clustering performance is dramatically improved by two-stage text-first mining methods and can be further improved by our unified approach. Secondly, RNL21NM does a better job than all single text-view methods like Kmeans_Text, LDA, and RNMF, meaning

---

[2] CNN news are manually labeled. NPR provides category info for each news article in their RSS feeds.

[3] http://psiexp.ss.uci.edu/research/programs_data/toolbox.htm

that images do help improve clustering performance. We also see that state-of-the-art multi-view learning methods behave ordinarily. CoTrainedMV performs even worse than text-first K-means since it tries to find consistent topic assignments across both text and image views, but the poor quality of image features severely degrades its performance. Also, RMKMC is vulnerable to the poor feature quality in the image domain.

6 Conclusion

In this paper, we define a new concept "text-image topics" and propose a general regularized nonnegative constrained $l_{2,1}$-norm minimization framework to discover text-image topics from web news collections by using not only text data but also data of other types such as images. We propose a novel iterative algorithm to solve the optimization problem. Experimental results on the crawled CNN and NPR web news datasets validate the efficacy of the proposed approach.

Acknowledgments. This material is based upon work supported by the National Science Foundation under Grant Number CNS-1027965. I would also like to express my thankness to my advisor Prof. Chengxiang Zhai for his inspiring comments and suggestions.

References

1. Blei, D., Ng, A., Jordan, M.: Latent dirichlet allocation. The Journal of Machine Learning Research 3, 993–1022 (2003)
2. Cai, X., Nie, F., Huang, H.: Multi-view k-means clustering on big data. In: Proceedings of the 23rd International Joint Conference on Artificial Intelligence (2013)
3. Chen, Y., Wang, L., Dong, M.: Non-negative matrix factorization for semisupervised heterogeneous data coclustering. IEEE Transactions on Knowledge and Data Engineering (2010)
4. Deerwester, S., Dumais, S., Furnas, G., Landauer, T., Harshman, R.: Indexing by latent semantic analysis. Journal of the American Society for Information Science 41(6), 391–407 (1990)
5. Hofmann, T.: Probabilistic latent semantic indexing. In: Proceedings of the 22nd Annual International ACM SIGIR Conference on Research and Development in Information Retrieval, pp. 50–57. ACM (1999)
6. Kong, D., Ding, C., Huang, H.: Robust nonnegative matrix factorization using l21-norm. In: Proceedings of the 20th ACM International conference on Information and Knowledge Management, pp. 673–682. ACM (2011)
7. Kumar, A., Daumé III, H.: A co-training approach for multi-view spectral clustering. In: Proceedings of the 28th Annual International Conference on Machine Learning (2011)
8. Meng, L., Tan, A., Xu, D.: Semi-supervised heterogeneous fusion for multimedia data co-clustering. IEEE Transactions on Knowledge and Data Engineering (2013)

Detecting Event Visits in Urban Areas via Smartphone GPS Data

Richard Schaller[1], Morgan Harvey[2], and David Elsweiler[3]

[1] Computer Science, Univ. of Erlangen-Nuremberg, Germany
richard.schaller@fau.de
[2] Faculty of Informatics, University of Lugano, Switzerland
morgan.harvey@usi.ch
[3] I:IMSK, University of Regensburg, Germany
david@elsweiler.co.uk

Abstract. In geographic search tasks, where the location of the user is an important part of the task context, knowing whether or not a user has visited a location associated with a returned result could be a useful indicator of system performance. In this paper we derive and evaluate a model to estimate, based on user interaction logs, GPS information and event meta-data, the events that were visited by users of a mobile search system for an annual cultural evening where multiple events were organised across the city of Munich. Using a training / testing set derived from 111 users, our model is able to achieve high levels of accuracy, which will, in future work, allow us to explore how different ways of using the system lead to different outcomes for users.

Keywords: detection of visited locations, GPS, location-based services.

1 Introduction and Background

Evaluation of search engine performance is important, has long since been a contentious issue and recent work has lead to the insight that it is important to measure more than simply whether returned documents are relevant or not. Equally important is how the user interacted with the search system and the actual outcomes: did the user watch any of the recommended TV programmes? Were any of the found recipes actually cooked, eaten and enjoyed? Did the exploratory search to find out more about a political party change the way the searcher voted?

Of course some of these situations will be easier to measure than others - the point is that if we could glean more information about outcomes of search sessions, it would inform more about the utility of the search system than the interaction data with the system alone. In this work we try to make it possible to investigate outcomes in one particular context - geographical information needs, which represent nearly one third of all mobile search needs [2]. Such needs are dependent on location in some way. Examples include, "where is the nearest bank?" or a request for directions to a place or event. We want to establish if

M. de Rijke et al. (Eds.): ECIR 2014, LNCS 8416, pp. 681–686, 2014.

search results associated with particular locations are actually visited. Focusing
on the problem of evaluating search and recommendation systems for events, we
derive and evaluate a model to estimate, based on user interaction logs, GPS
information and event meta-data, which events were visited in the context of
the Long Night of Music - an annual cultural event organised in the city of
Munich. During one evening in May, pubs, discos, churches and museums host
diverse musical events, opening their doors to over 20,000 people. Search and
recommender systems exist for the night and it would be helpful to evaluate
which events found using such systems were visited in which situations.

2 Related Work

Geographical information is increasingly important in IR research, as demonstrated by the introduction of a TREC track dedicated to contexual suggestion
using locations [4]. There has been significant work on estimating the geographic
location of images [8]. Location information is often used in social media for
providing localised content and location-aware recommendations. Users can be
clustered based on geographic patterns [7], which can be used, for example, to
improve friend suggestion based on geographic proximity [3].

In geographical searches location is not only a vital input to search algorithms,
but could be an important component in evaluating the success of such systems.
If we can determine whether a user actually visited the place suggested or made
it to the destination they requested then this tells us something about the success
of our system i.e. the results were appealing and / or relevant enough for the
user to try them out. This applies most strongly to recommender systems for
travel or tourist situations [9] and this situation is the one we focus on in the
remainder of the paper.

Establishing points of interest (POIs) from GPS data is not straightforward
and has been investigated as a research problem [6,1]. Typical approaches tend
to apply a fixed distance radius from POIs to establish whether or not the user's
GPS signal originates from nearby and determine a visit [1]. Other approaches
additionally account for the prominence of POIs and opening hours [6].

While the methods proposed to date function well in the contexts for which
they have been applied e.g. distinguishing between activities such as shopping
and dining [6] or between POIs in different cities [1], the models are not easily transferable to other usage contexts or where the necessary granularity and
accuracy is very fine. We present a new model which accounts for similar and
additional features to those in the literature that can be used for our desired
task of establishing the outcome of search / recommendation sessions.

3 Experimental Setup

We developed an Android app with search, recommendation and browse features
to help visitors discover events of interest to them during the *Long Night*. Once
the user has chosen events of interest, the system can create a plan and guide

the user between chosen events using a map display and textual instructions [see Figure:left 3.1]. We examined user behaviour by recording all user interactions and positional data from the app. The app was available on the Google Play Store and was advertised on the official event web page. It was downloaded approximately 1300 times and 1159 users allowed us to record their interaction data. GPS positional updates were recorded if users had GPS turned on resulting in GPS data from 180 users between 5pm and 5am on the day of the event.

The main aim of this work is to determine, for each user, the sequence of events visited and the duration of time spent at each venue. We split the whole process into two parts: Extraction of the visited events (and their order) and estimation of the start and end time of each event visit. For both the human labellers and the later automatic methods we make the following assumptions: Only one event can be visited at any given time, only open events can be visited and minimum dwell time at an event is 5 minutes.

3.1 Gold Standard Dataset

The main source of information used for deciding whether a user visited a certain event or not is the recorded GPS track. We therefore concentrate further analysis on the 111 users for whom we had GPS tracks between 5pm and 5am that were sufficiently reliable to establish the events visited, as determined by manual analysis. In addition to the GPS track and accuracy of GPS signal[1] we can also consider properties of the event (such as position, name/description and opening hours) and user interactions with the event (whether the event description was viewed and read and whether the event was rated for possible tour inclusion).

We developed a tool which displays all this information on a map and allows a (high-speed) replay of the evening (see Figure:right 3.1). We instructed two

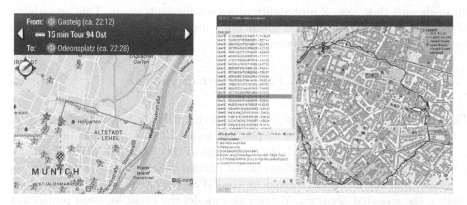

Fig. 1. Left: mobile app map view, Right: GPS track labelling interface

---

[1] As reported by the Android device. A circle around the reported location which contains the true position with 68% confidence.

human labellers to use the tool to mark the sequence of events each user most likely visited. Labellers were asked to consider all the information listed above, especially in case of multiple events at the same event location.

There are a number of aspects that make the GPS track difficult to interpret: GPS sensors are of very different quality, users may quit/restart the app or turn GPS on/off at any time, GPS does not work indoors or in streets flanked by high buildings and events are represented as a single coordinate with no information about the area of the event location given.

To compare the event sequences of both labellers we used two metrics: Jaccard-Similarity (JS) which returns a score between 0 and 1 indicating the overlap between two sets and Levenshtein-Distance (LD) which compares sequences by calculating the minimum number of edit operations (insertions/deletions) necessary to get from one sequence to the other. We compared the sequences of events from the labellers to determine what level of agreement was reached and obtained a JS of 0.818 and a LD of 1.09 event visits, suggesting generally very good overlap. Despite the manifold potential issues when working with GPS data we consider the itineraries generated by the labellers to be sufficiently accurate approximations of the true itineraries and therefore use these as a source of gold standard data for later analysis.

3.2   Experimental Systems and Their Performance

As features the model uses the same information as the labellers: GPS track and accuracy, event position and opening hours and user interactions with the event (whether the event description was read and whether the event was rated by the user). The model can also consider the popularity of the event as a prior, determined by how often all users have rated the event. For ease of computation we discretise the evening into time slots of 10 seconds. The model then calculates, for each such time slot, a probability of a visit for each event. This probability of event E having been visited given GPS measurement M and X, the true location of the user, can be formulated as:

$$P(E|M) = P(E) \int_x P(x|E)P(x|M)\ dx$$

The estimate $P(E|M)$ can be separated into the prior probability of visiting an event $P(E)$ and an integral over all possible locations $\in X$. $P(x|M)$ depends on the error of the GPS sensor which, according to the Android API documentation[2], can be modelled as a 2-dimensional normal distribution where the vector of means μ is the location received by the GPS sensor and σ is a scalar representing the reported accuracy. We can also model the event location probability $P(x|E)$ in a similar fashion, with μ being the location of the event and σ accounting for the variance in location when someone visits the event, which we assume to be equal for all events.

---

[2] developer.android.com/reference/android/location/Location.html

Thus we can solve the integral as

$$\int_x P(x|E)P(x|M)dx = \frac{1}{2\pi(\sigma_{ev}^2 + \sigma_{gps}^2)} \cdot \exp - \frac{(\mu_{ev} - \mu_{gps})^2}{2(\sigma_{ev}^2 + \sigma_{gps}^2)}$$

Note that $(\mu_{ev} - \mu_{gps})^2$ is the squared distance between the logged GPS coordinate and the event location. To factor in the extension of the building the event is located within we reduce this distance by a constant value $S_{building}$.

We calculate prior probability $P(E)$ of visiting an event based on a sum of 4 components controlled via scalar weights which must be trained. Three of these components are user dependent: The fraction of total event views for event E and whether of not event E was rated and/or selected for tour inclusion, normalised over the total number of events rated and selected by the user. The final component is the overall popularity of the event over all users, estimated as the number of times E was rated over the total number of ratings.

After calculating $P(E|M)$ over the set of all events, the event with the maximum probability is selected. Since it is possible that a user was not visiting any event at a given time we only predict a visit if $P(E|M)$ is over a threshold. Given an event visit (or no visit) for each 10 second time slot we then look at contiguous time slots with the same visited event: If the time between the first and the last time slot is longer than 5 min (our assumption of the shortest visiting time) then we consider this event to be visited during these time slots.

For evaluation we use the gold standard data described above (comprising 441 visits in total), split into 3 testing/training folds. We consider two problems: Determining a visited location and the more difficult task of determining the exact event visited (multiple events can take place at the same approximate location). We compare our model with a baseline system inspired by approaches in the literature in which a radius of 50m is set around each event[3]. An event is considered visited each time a user's GPS signal is within this radius of it for a minimum of 5 minutes. Both models were trained in order to maximise F1 score.

Table 1. Comparison of the baseline with the described model

	Location specific			Event specific		
	Precision	Recall	F1-score	Precision	Recall	F1-score
Baseline	72.7%	78.6%	75.5%	54.8%	**77.8%**	64.3%
Our model	**81.6%**	**85.6%**	**83.6%**	**71.4%**	77.1%	**74.1%**

Table 1 shows the results from both models as measured by precision, recall and F-score. The proposed model outperforms the baseline in all but one case and always in terms of precision, arguably a more important metric than recall for this task. Our model delivers particularly strong performance (relative to the baseline) for the more difficult problem of detecting which individual events were visited, achieving an F1 score improvement of over 13.2%, highlighting the extra predictive power afforded by the extra features considered.

---

[3] This value is reported by [1] and our tests also indicated this to be an optimal value.

4   Conclusions

In this paper we have explored methods of using GPS and interaction data to determine whether or not geographical search results were visited during the Long Night of Music. Using log data from a naturalistic study of a mobile search system, we evaluated the performance of two models. The first, a baseline, derived from models for similar problems in the literature was outperformed by a second, new model, accounting for additional features. This model is able to achieve high levels of accuracy, offering huge potential to understand the success of different features of the system and of different user behaviour patterns. In future work we will investigate how user behavioural strategies related to concrete outcomes for the evening, using our model to derive metrics such as the number event visits, the popularity and diversity of events visited, as well as the time spent travelling. We believe the method we have used could be tailored to be equally fruitful for similar problems such as geographical queries in mobile search.

Acknowledgments. This work was supported by the Embedded Systems Initiative (http://www.esi-anwendungszentrum.de).

References

1. Bohte, W., Maat, K.: Deriving and validating trip purposes and travel modes for multi-day gps-based travel surveys: A large-scale application in the netherlands. Transportation Research Part C: Emerging Technologies 17(3) (2009)
2. Church, K., Smyth, B.: Understanding the intent behind mobile information needs. In: Proceedings of the 14th International Conference on Intelligent User Interfaces, pp. 247–256. ACM (2009)
3. Cranshaw, J., Toch, E., Hong, J., Kittur, A., Sadeh, N.: Bridging the gap between physical location and online social networks. In: Ubicomp (2010)
4. Dean-Hall, A., Clarke, C.L., Kamps, J., Thomas, P., Voorhees, E.: Overview of the trec 2012 contextual suggestion track. In: Proceedings of TREC, vol. 12 (2012)
5. Hauff, C., Houben, G.J.: Geo-location estimation of flickr images: social web based enrichment. In: Baeza-Yates, R., de Vries, A.P., Zaragoza, H., Cambazoglu, B.B., Murdock, V., Lempel, R., Silvestri, F. (eds.) ECIR 2012. LNCS, vol. 7224, pp. 85–96. Springer, Heidelberg (2012)
6. Huang, L., Li, Q., Yue, Y.: Activity identification from gps trajectories using spatial temporal pois' attractiveness. In: Proceedings of the 2nd ACM SIGSPATIAL International Workshop on Location Based Social Networks. ACM (2010)
7. Scellato, S., Mascolo, C., Musolesi, M., Latora, V.: Distance matters: geo-social metrics for online social networks. In: Proceedings of the 3rd Conference on Online Social Networks, p. 8. USENIX Association (2010)
8. Serdyukov, P., Murdock, V., Van Zwol, R.: Placing flickr photos on a map. In: Proceedings of the 32nd International ACM SIGIR Conference on Research and Development in Information Retrieval, pp. 484–491. ACM (2009)
9. van Setten, M., Pokraev, S., Koolwaaij, J.: Context-aware recommendations in the mobile tourist application compass. In: De Bra, P.M.E., Nejdl, W. (eds.) AH 2004. LNCS, vol. 3137, pp. 235–244. Springer, Heidelberg (2004)

A Case for Hubness Removal
in High–Dimensional Multimedia Retrieval

Dominik Schnitzer[1], Arthur Flexer[1], and Nenad Tomašev[2]

[1] Austrian Research Institute for Artificial Intelligence (OFAI), Vienna, Austria
[2] Artificial Intelligence Laboratory, Jožef Stefan Institute, Ljubljana, Slovenia

Abstract. This work investigates the negative effects of hubness on multimedia retrieval systems. Because of a problem of measuring distances in
high-dimensional spaces, hub objects are close to an exceptionally large
part of the data while anti-hubs are far away from all other data points.
In the case of similarity based retrieval, hub objects are retrieved over
and over again while anti-hubs are nonexistent in the retrieval lists. We
investigate textual, image and music data and show how re-scaling methods can avoid the problem and decisively improve the overall retrieval
quality. The observations of this work suggest to make hubness analysis
an integral part when building a retrieval system.

Keywords: hubness, high dimensionality, multimedia retrieval.

1 Introduction

A number of publications have recently discussed hubness in a general machine
learning context and introduced it as a new aspect of the curse of dimensionality [12,14]. Hub objects are nearest neighbors to many other data points in
high dimensional spaces and hence are being frequently retrieved without being
semantically relevant to many of the queries. The effect is related to the phenomenon of concentration of distances and has been shown to have a negative
impact on many tasks including outlier detection [20], clustering [17] and collaborative filtering [10]. More related to multimedia retrieval, the influence of hubs
on object recognition [16] and music similarity [4] has also been investigated.
Our work studies hubness in a range of multimedia retrieval systems (text, image, music) and shows how removal of hubness via re-scaling of the distances
improves retrieval quality and overall system performance.

2 Investigation of Multimedia Retrieval Systems

To investigate the impact of hubs on multimedia retrieval systems, we build
three standard content-based multimedia retrieval systems (text, image, music).
We use the systems in a query-by-example scenario where the top–k answers are
retrieved and evaluated in terms of precision/recall with their class label. This is
one of the most common scenarios for content-based multimedia retrieval. The
three systems and eight data sets used for evaluation are described below:

M. de Rijke et al. (Eds.): ECIR 2014, LNCS 8416, pp. 687–692, 2014.

— The text retrieval system analyzes textual data by first employing stopword removal, stemming and transformation to the bag-of-words representation. Standard *tf · idf* (term frequency · inverse document frequency) weights are computed (see e.g. [2]). The word vectors are normalized to the average document length. Document vectors are compared with the cosine distance. We evaluate our text-retrieval system with the *Twitter (C1ka)* [13] and the UCI [5] *Mini Newsgroups* data sets which both show a very high extrinsic dimensionality (Twitter: 49 000, Mini Newsgroups: 8 000).

— The image retrieval system uses standard dense SIFT vectors [8] to compute spatial histograms of visual words. Each image is represented with a bag-of-visual-word (BOVW) vector [15]. To measure the similarity between two images we compute the Euclidean distance between their BOVW vectors. We evaluate our image-retrieval system with the *Caltech 101* [3], *Leeds Butterfly* [19] and *17 Flowers* [11] data sets. The extrinsic data dimensionality of the features used in this retrieval system is 36 000.

— The music retrieval system extracts MFCC features and estimates a single multivariate Gaussian representing the timbral structure of the music piece [9]. The similarity between the Gaussian representations is computed with the Jensen-Shannon divergence. This method is one of the standard methods for content based music retrieval systems. The music-retrieval system is evaluated with the *Ballroom* [6], *GTzan* [18] and *ISMIR 2004* [1] data sets. The extrinsic data dimensionality of the features is 1 275.

2.1   The Dominance of Hubs in Retrieval Lists

The described three retrieval systems are based on different kinds of data (*text, image, music*), different feature representations (*vectors, multivariate*

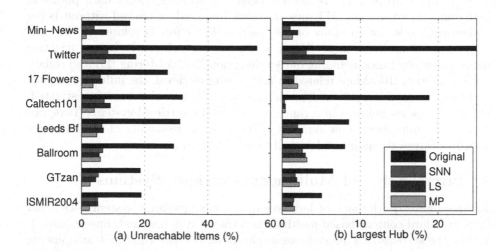

Fig. 1. At $k = 5$, (a) the number of unreachable items and (b) the size of the largest hub (as percentages of the total collection size for original and re-scaled distances)

Gaussians) and three different distance measures (*cosine, Euclidean, Jensen-Shannon*). But all systems operate with very high dimensional features (1 275 to 49 000) which should make them prone to hubness. To verify this, we analyze the k–nearest neighbor (kNN) lists of all data objects in all data sets. We compute (for $k = 5$) the percentage of items which is unreachable in our retrieval systems (do not appear in any of the kNN lists, i.e. anti-hubs) and the maximum percentage one single item appears in all kNN lists (i.e. the largest hub). In Figure 1(a) (black bars for original distance spaces) we see that on average over 20% of all objects in our system are unreachable, i.e. will never be retrieved.

Figure 1(b) (black bars for original distance spaces) shows that a single object will be present in 5% (*ISMIR 2004*) to 18% (*Caltech 101*) or over 25%

Table 1. Evaluation of multimedia retrieval systems in terms of hubness (S), precision (P), recall (R) and f–measure (F_1) for $k = 1, 5, 10$ and different distance spaces (Original, SNN, LS, MP) . Top results per line and $k = 1, 5$ or 10 in bold face.

Dataset	k	Original			SNN			LS			MP		
		1	5	10	1	5	10	1	5	10	1	5	10
Mini-	S	4.1	5.1	4.4	1.8	1.7	1.5	1.2	1.3	1.5	**0.9**	**0.6**	**0.5**
Newsgroups	P	.644	.541	.489	.597	.518	.475	.672	.571	.516	**.677**	**.576**	**.526**
text	R	.006	.027	.049	.006	.026	.047	**.007**	**.029**	.052	**.007**	**.029**	**.053**
	F_1	.013	.052	.089	.012	.049	.086	**.013**	.054	.094	**.013**	**.055**	**.096**
Twitter	S	29.2	14.6	10.7	3.9	3.7	3.2	2.3	2.9	3.1	**1.8**	**1.8**	**1.8**
(C1ka)	P	.319	.265	.245	.398	.377	.356	.478	**.473**	**.434**	**.490**	.467	**.434**
text	R	.004	.016	.028	.005	.021	.039	**.007**	**.033**	.058	**.007**	**.033**	**.059**
	F_1	.008	.029	.048	.010	.038	.067	.013	**.058**	.094	**.014**	**.058**	**.095**
17 Flowers	S	5.9	3.9	3.4	1.5	1.0	1.0	1.3	1.2	1.0	**1.0**	**0.7**	**0.5**
image	P	.520	.391	.343	.447	.367	.330	**.570**	.433	.378	.557	**.442**	**.394**
	R	.006	.024	.043	.006	.023	.041	**.007**	.027	.047	**.007**	**.028**	**.049**
	F_1	.013	.046	.076	.011	.043	.073	**.014**	.051	.084	**.014**	**.052**	**.088**
Caltech	S	80.7	50.1	34.5	1.5	1.4	1.4	1.4	1.2	1.1	**1.3**	**1.1**	**1.0**
101	P	.598	.544	.508	.591	.542	.514	**.686**	**.631**	**.587**	.672	.624	.579
image	R	.005	.021	.037	.005	.020	.037	**.007**	**.028**	**.049**	.007	.027	.048
	F_1	.011	.039	.064	.009	.038	.064	**.014**	**.052**	**.085**	.013	.051	.083
Leeds-	S	3.7	3.5	3.0	1.4	1.0	0.9	1.2	0.9	1.0	**1.0**	**0.5**	**0.4**
Butterfly	P	.584	.470	.409	.466	.394	.360	**.630**	**.529**	**.465**	.608	.522	.462
image	R	.007	.028	.050	.006	.024	.044	**.008**	**.032**	**.057**	.007	**.032**	.056
	F_1	.014	.054	.088	.011	.045	.078	**.015**	**.060**	**.101**	**.015**	**.060**	.100
Ballroom	S	3.4	2.8	2.5	1.3	0.9	0.9	1.3	1.2	1.1	**1.0**	**0.6**	**0.5**
music	P	.532	.457	.414	.520	.470	.439	.587	.505	.464	**.595**	**.512**	**.471**
	R	.006	.026	.047	.006	.026	.048	**.007**	.028	.052	**.007**	**.029**	**.053**
	F_1	.012	.049	.084	.011	.049	.087	**.013**	**.054**	.093	**.013**	**.054**	**.095**
GTzan	S	3.4	3.3	3.1	1.4	0.9	0.6	**0.9**	1.0	1.3	**0.9**	**0.4**	**0.4**
music	P	.774	.678	.599	.715	.640	.602	.796	.715	.636	**.804**	**.720**	**.652**
	R	**.008**	.034	.060	.007	.032	.060	**.008**	.036	.064	**.008**	**.036**	**.065**
	F_1	.015	.065	.109	.014	.061	.110	**.016**	.068	.116	**.016**	**.069**	**.119**
ISMIR	S	3.5	3.9	3.7	1.2	0.9	0.8	1.3	1.4	1.6	**1.0**	**0.7**	**0.4**
2004	P	.858	.808	.762	.824	.791	.762	**.914**	.846	.791	.903	**.847**	**.803**
music	R	.003	.015	.028	.003	.015	.028	**.004**	**.017**	.030	**.004**	**.017**	**.031**
	F_1	.007	.030	.053	.006	.029	.053	**.007**	**.032**	**.057**	**.007**	**.032**	**.057**

(*Twitter*) of all kNN lists. To quantify the amount of hubs/anti-hubs in a data set Radovanović et al. [12] defined a hubness measure (S^k). It measures the skewness of a histogram of each object's occurrence in the kNN lists. Positive S^k (skewed to the right) indicates that there is high hubness, i.e. there are many anti-hubs and few but very large hubs in the collection. Table 1, column *Original* records the measured hubness S for $k = 1, 5, 10$ and all data sets. The measured values range from 2.5 (*Ballroom*) to 80.7 (*Caltech 101*). With non-problematic hubness values ranging from 0 to 1, all our measurements indicate very high hubness explaining the impaired retrieval performance expressed in Figure 1.

2.2   Hubness Removal for Improved Retrieval Quality

In order to reduce hubness and its negative effects, two unsupervised methods to re-scale the high-dimensional distance spaces have been proposed [14]: Local Scaling (LS) and Mutual Proximity (MP). Both methods aim at repairing asymmetric nearest neighbor relations. The asymmetric relations are a direct consequence of the presence of hubs. A hub y is the nearest neighbor of x, but the nearest neighbor of the hub y is another point a ($a \neq x$). This is because hubs are by definition nearest neighbors to very many data points but only one data point can be the nearest neighbor to a hub. The principle of the scaling algorithms is to re-scale distances to enhance symmetry of nearest neighbors. A small distance between two objects should be returned only if their nearest neighbors concur. LS does that by computing a local statistic of the nearest neighbors to rescale the distances. MP assumes that all pairwise distances in a data set follow a certain distribution and computes the mutual probability that two points are true nearest neighbors. We use LS with a neighborhood range of $s = 10$ and MP with the empirical distribution. In addition to LS and MP we test if the shared-nearest neighbor approach (SNN) is also able to reduce hubness in our systems. We added SNN since there are a number of publications reporting its positive effects on high dimensional data [7] and there is apparent similarity to LS and MP. To use SNN first a neighborhood size (s) has to be set (in the following experiments we use $s = 50$). The new similarity between two points x and y is then the overlap in terms of their nearest neighbors (NN). Similarly to LS or MP, this enforces symmetric neighborhoods: $SNN(x, y) = |NN(x) \cap NN(y)|$.

In the previous section we have shown that all of the presented retrieval systems are strongly affected by hubs. Figure 1(a)(b) shows the impact of LS, MP and SNN in terms of largest hub size and the number of unreachable items (anti-hubs). Across all systems and data collections a pronounced decrease of unreachable items and hub sizes can be observed. Compared to the original system the number of unreachable items (Figure 1(a)) comes down from about 20% to ~5%. In 7 out of 8 data sets MP yields the best results. Using any of the methods described also leads to a significant decrease of the largest hub sizes (Figure 1(b)). The most positive effect of the scaling methods can be observed in the *Caltech 101* dataset, where the size of the largest hub comes down to around 1% (from 18.8%). In 7 out of 8 cases SNN preforms best in terms of largest hub size. Table 1 also shows the results in terms of hubness (S) at different retrieval sizes ($k = 1, 5, 10$) for LS, MP and SNN. As expected from the previous results, S also decreases with all methods to much more normal levels between 1.8 and 0.4 (with the exception of SNN and LS for data set *Twitter*: 2.3 to 3.9). MP reduces the measured hubness to the lowest values.

Besides hubness, we have also evaluated the quality impact of hubness removal on the retrieval systems in terms of precision (P), recall (R) and f–measure (F_1). The results for all three systems are shown in Table 1 (best values are in bold font). Together with lower hubness, we observe increased retrieval quality across all domains when using LS and MP. In terms of F_1, MP performs best on 2 out of 2 text, 1 out of 3 image (*17 Flowers*) and 3 out of 3 music data sets, while LS is best on 2 out 3 image (*Leeds Butterfly, Caltech 101*) data sets. While LS and MP at the same time decrease hubness and increase the overall system performance (in terms of F_1, P and R), SNN only seems to reduce hubness and its retrieval quality often decreases even below the values achieved on the original distance spaces. For example in the case of *Mini Newsgroups*, F_1 at $k = 5$ decreases from originally 0.052 to 0.049, while it increases 0.054 with LS and to 0.055 with MP.

3   Summary

We have demonstrated the effect of hubness occurring in multimedia retrieval systems across three different domains (text, image, music) operating with high dimensional data. In the similarity based kNN retrieval systems, hubness caused about one fifth of the data to be unreachable anti-hubs which are never part of any of the retrieval lists. At the same time hub objects dominate many of the retrieval lists with hubs sometimes appearing in more than 25% of all possible lists. Application of distance re-scaling methods, most notably Local Scaling and Mutual Proximity, is able to decisively reduce hubness, increase reachability and thereby enhance the diversity of the retrieval lists. At the same time we see that by removing the hubs, the retrieval quality in terms of precision/recall across all data domains and retrieval systems is also increased. Whereas most existing retrieval systems do not take hubness into account, we hope that we made a clear case for hubness removal: many of the systems are negatively affected by hubness causing lower retrieval quality and diversity which can be avoided using simple re-scaling algorithms.

Acknowledgments. This research is supported by the Austrian Science Fund (FWF): P24095.

References

1. ISMIR 2004 Music Genre Collection,
   http://ismir2004.ismir.net/genre_contest/index.htm
2. Baeza-Yates, R., Ribeiro-Neto, B.: Modern Information Retrieval. Addison Wesley (1999)

3. Fei-Fei, L., Fergus, R., Perona, P.: Learning generative visual models from few training examples: An incremental bayesian approach tested on 101 object categories. Computer Vision and Image Understanding 106(1), 59–70 (2007)
4. Flexer, A., Schnitzer, D., Schlüter, J.: A mirex meta-analysis of hubness in audio music similarity. In: Proceedings of the 13th International Society for Music Information Retrieval Conference, pp. 175–180 (2012)
5. Frank, A., Asuncion, A.: UCI machine learning repository (2010), repository located at, http://archive.ics.uci.edu/ml
6. Gouyon, F., Dixon, S., Pampalk, E., Widmer, G.: Evaluating rhythmic descriptors for musical genre classification. In: Proceedings of the 25th International Conference Audio Engineering Society Conference, pp. 196–204 (2004)
7. Houle, M.E., Kriegel, H.-P., Kröger, P., Schubert, E., Zimek, A.: Can Shared-Neighbor Distances Defeat the Curse of Dimensionality. In: Gertz, M., Ludäscher, B. (eds.) SSDBM 2010. LNCS, vol. 6187, pp. 482–500. Springer, Heidelberg (2010)
8. Lowe, D.G.: Distinctive image features from scale-invariant keypoints. International Journal of Computer Vision 60(2), 91–110 (2004)
9. Mandel, M., Ellis, D.: Song-level features and support vector machines for music classification. In: Proceedings of the 6th International Conference on Music Information Retrieval, London, UK (2005)
10. Nanopoulos, A., Radovanović, M., Ivanović, M.: How does high dimensionality affect collaborative filtering? In: Proceedings of the Third ACM Conference on Recommender Systems, pp. 293–296 (2009)
11. Nilsback, M.E., Zisserman, A.: Automated flower classification over a large number of classes. In: Proceedings of the Sixth Indian Conference on Computer Vision, Graphics & Image Processing, pp. 722–729 (2008)
12. Radovanović, M., Nanopoulos, A., Ivanović, M.: Hubs in space: Popular nearest neighbors in high-dimensional data. Journal of Machine Learning Research 11, 2487–2531 (2010)
13. Schedl, M.: On the Use of Microblogging Posts for Similarity Estimation and Artist Labeling. In: Proceedings of the 11th International Society for Music Information Retrieval Conference, Utrecht, The Netherlands (August 2010)
14. Schnitzer, D., Flexer, A., Schedl, M., Widmer, G.: Local and global scaling reduce hubs in space. Journal of Machine Learning Research 13, 2871–2902 (2012)
15. Sivic, J., Russell, B.C., Efros, A.A., Zisserman, A., Freeman, W.T.: Discovering objects and their location in images. In: Proceedings of the Tenth IEEE International Conference on Computer Vision, vol. 1, pp. 370–377 (2005)
16. Tomašev, N., Brehar, R., Mladenić, D., Nedevschi, S.: The influence of hubness on nearest-neighbor methods in object recognition. In: Proceedings of the IEEE International Conference on Intelligent Computer Communication and Processing, pp. 367–374 (2011)
17. Tomašev, N., Radovanović, M., Mladenić, D., Ivanović, M.: The role of hubness in clustering high-dimensional data. In: Huang, J.Z., Cao, L., Srivastava, J. (eds.) PAKDD 2011, Part I. LNCS, vol. 6634, pp. 183–195. Springer, Heidelberg (2011)
18. Tzanetakis, G., Cook, P.: Musical genre classification of audio signals. IEEE Transactions on Speech and Audio Processing 10(5), 293–302 (2002)
19. Wang, J., Markert, K., Everingham, M.: Learning models for object recognition from natural language descriptions (2009)
20. Zimek, A., Schubert, E., Kriegel, H.P.: A survey on unsupervised outlier detection in high-dimensional numerical data. Statistical Analysis and Data Mining 5(5), 363–387 (2012)

Multi-evidence User Group Discovery in Professional Image Search

Theodora Tsikrika[1] and Christos Diou[2]

[1] Information Technologies Institute, CERTH, Thessaloniki, Greece
[2] ECE Department, Aristotle University of Thessaloniki, Thessaloniki, Greece
theodora.tsikrika@iti.gr, diou@mug.ee.auth.gr

Abstract. This work evaluates the combination of multiple evidence for discovering groups of users with similar interests. User groups are created by analysing the search logs recorded for a sample of 149 users of a professional image search engine in conjunction with the textual and visual features of the clicked images, and evaluated by exploiting their topical classification. The results indicate that the discovered user groups are meaningful and that combining textual and visual features improves the homogeneity of the user groups compared to each individual feature.

1 Motivation

The discovery of groups of similar users is useful to information retrieval applications, such as ranking and sponsored search, that benefit from adapting their results to users' interests. Given that data sparseness and ambiguity may lead to ineffective representations of user interests, especially for users with limited history, research has suggested to leverage evidence obtained from the user group(s) the individual belongs, to either augment their personal profile, and thus potentially lead to more effective personalisation, or to use such evidence for 'groupisation', i.e., adapt retrieval the same way for all group members [8].

Previous studies have explored several features for discovering user groups, including occupational and topical evidence [8], demographic information [8,5] with a particular focus on gender [6], geographical context [1], and reading level efficiency [2]. Their results show the usefulness of such evidence for improving retrieval. Additional studies [10,11] have investigated user segmentation outside the context of a particular retrieval application with the goal to gain an understanding of how the search behaviour of specific groups differs. To this end, they analysed large scale web search logs in terms of the users' query topics and/or session characteristics, together with the users' demographic profile information augmented with U.S. census data. Their results showed that it is possible to identify distinct patterns of behaviour along different demographic features.

Our study aims at further developing an understanding of user group identification by investigating the hypothesis that the combination of multiple evidence improves the discovery of users groups with similar topical interests compared to the groupings based on individual features. To this end, we analysed the search

M. de Rijke et al. (Eds.): ECIR 2014, LNCS 8416, pp. 693–699, 2014.
© Springer International Publishing Switzerland 2014

log data collected by the commercial picture portal of a European news agency for a sample of 149 of their registered users, in conjuction with the captions, visual features, and a topical classification of the available images. Our analysis takes place in a context that is different to the search environments examined in all of the above work in at least one of the following aspects. Ours is a professional, rather than a web, environment, and furthermore it is oriented towards image, rather than text retrieval. Also, although we examine topical features for discovering user groups, similarly to [8], we consider the users' log activity rather than data collected through a user study. Moreover, to the best of our knowledge, visual features have not been previously investigated in such a context.

2   Data Acquisition and Processing

The search log data used in this study were collected by the news agency over a two year period (June 2007 – July 2009), with a three-month hiatus (October – December 2007). The sample considered in this study consists only of registered users logged into their account and was processed as follows. First, the logs were segmented into sessions, i.e., series of a single user's consecutive search actions assumed to correspond to a single information need. No intent-aware session detection was applied [3], and session boundaries were identified when the period of inactivity between two successive actions exceeded a 30-minute timeout, similarly to [11]. Next, the queries were 'lightly' normalised by converting them to lower case and removing punctuation, quotes, special characters, extraneous whitespace, URLs, and the names of major photo agencies. Furthermore, empty queries and queries consisting only of numbers or whitespace characters were removed. No stemming or stopword removal was applied at this stage. Furthermore, consecutive identical queries submitted in the same session were conflated. The final step was to further sample the logs so as to include only "active" users,

Table 1. IPTC subject codes and the IPTC distribution of clicked images

IPTC subject codes		#images	% classified images	% all images
Code	Name			
1. ACE	Arts, Culture, & Entertainment	36,413	14.3%	12.5%
2. CLJ	Crime, Law & Justice	9,768	3.8%	3.4%
3. DIS	Disaster & Accident	6,284	2.5%	2.2%
4. EBF	Economy, Business & Finance	16,148	6.4%	5.6%
5. EDU	Education	599	0.2%	0.2%
6. ENV	Environmental issue	1,819	0.7%	0.6%
7. HTH	Health	2,014	0.8%	0.7%
8. HUM	Human interest	12,528	4.9%	4.3%
9. LAB	Labour	2,006	0.8%	0.7%
10. LIF	Lifestyle & Leisure	2,411	0.9%	0.8%
11. POL	Politics	39,473	15.5%	13.6%
12. REL	Religion	2,111	0.8%	0.7%
13. SCI	Science & Technology	1,708	0.7%	0.6%
14. SOI	Social issue	1,385	0.6%	0.5%
15. SPO	Sport	110,090	43.3%	37.9%
16. WAR	Unrest, Conflicts, & War	7,608	3.0%	2.6%
17. WEA	Weather	2,093	0.8%	0.7%
		254,458	100%	87.6%

i.e., those who had issued at least 10 queries, with each followed by at least one click. Our final sample thus contains 149 users who submitted a total of 113,176 queries (55,386 unique) and clicked on a total of 519,849 images (290,593 unique).

In addition to textual captions, the content of images is also described using the top level of the International Press Telecommunications Council's (IPTC) hierarchical newscodes, i.e., the 17 IPTC *subject* codes (http://cv.iptc.org/ newscodes/subjectcode/) listed in Table 1. Out of the 290,593 unique images clicked in our sample, 254,458 (87.6%) had been manually classified by the news agency's archivists. Table 1 lists the distribution of the clicked images over the 17 IPTC subjects. Sports is by far the dominant category, indicating a slight bias in the topical interests of the sampled users, as these are reflected by their searching behaviour. Politics and cultural topics follow in almost equal measures. Economics, human interest topics, crime, and war are the next subjects of interest in descending order, with the rest following in much lower percentages.

3   User Groups

Groups of users with shared topical interests are *implicitly* formed based on the hypothesis that such users issue similar queries and/or click on similar images. To this end, clustering is performed by employing user distances that exploit textual and visual evidence associated with the users' queries and/or clicks.

The first user distance, denoted as **text**, is the Euclidean distance between term vectors that correspond to user queries and clicked image captions. Initially the text of each user's queries and the captions of clicked images are concatenated. A term vector is generated by applying stemming, but without removing stopwords, and by estimating the term weights using a *tf.idf* scheme with normalisation, using the Text to Matrix Generator (TMG) Matlab toolbox [12].

The second, denoted as **vis**, is the visual distance of the users' clicked images, defined as follows. A vector of 4,096 features was extracted from each image using the 'bag of visual words' method, with dense keypoint sampling and the color SIFT descriptor [9]. Each user u is represented by a set of clicked images $\mathbf{I}_u$ and the distance between two users u, v is computed via the Hausdorff metric [4]:

$$d_H(u,v) = \max\left\{ \max_{\mathbf{x}\in\mathbf{I}_u} \min_{\mathbf{y}\in\mathbf{I}_v} d(\mathbf{x},\mathbf{y}), \max_{\mathbf{y}\in\mathbf{I}_v} \min_{\mathbf{x}\in\mathbf{I}_u} d(\mathbf{x},\mathbf{y}) \right\} \tag{1}$$

where $d(\mathbf{x},\mathbf{y}) = \frac{1}{2}\sum_i \frac{(\mathbf{x}(i)-\mathbf{y}(i))^2}{(\mathbf{x}(i)+\mathbf{y}(i))}$ is the χ^2 distance between two visual word histograms. The distance d_H is the maximum distance found between each image of $\mathbf{I}_u$ and its closest image in $\mathbf{I}_v$. It is often used to compare sets, such as sets of points, in computer vision or computer graphics applications; here, it produces a distance between the sets of feature vectors of the images clicked by u and v.

A composite distance is then formed by the weighted sum of the two criteria, after normalising *text* and *vis*: $d_u(u,v) = w_1 d_{\text{text}}(u,v) + w_2 d_{\text{vis}}(u,v)$. We investigate all possible combinations of $w_i \in [0,1]$ at step 0.1, such that $\sum w_i = 1$.

User groupings are generated by applying agglomerative clustering with these distances and several linkage criteria (*single*, *complete*, *average*, and *weighted*), and by also varying the number of non-overlapping clusters k, as discussed next.

4 Analysis

The above **text**, **vis**, and **text_vis** (w_1/w_2) groupings are analysed so as to investigate how meaningful they are, i.e., whether group members are more similar to each other with respect to their topical interests, than to members of other groups. In particular, the inter– and intra–cluster variation in people's topical interests is examined using the following inter– and intra–cluster pairwise proximity measures: *cohesion* and *separation* [7]. Cluster cohesion is defined as the average of the pairwise proximity values of all points within the cluster, while separation as the average of the pairwise proximity values of each point within the cluster to all points in all other clusters.

In the absence of explicit ground truth, the images clicked by users are considered as implicit indicators of their topical interests. The IPTC subject distribution of the images clicked by each user can then be considered as an implicit user profile that can be used for evaluating user groupings. Cohesion and separation are then estimated by employing the *Jensen-Shannon (JS) divergence* between such distributions, and in particular its value subtracted from 1, as a pairwise user proximity measure.

The use of a ground truth dataset based on the 17 IPTC subjects motivates us to focus our analysis on clusterings with $k = 17$. Given, though, that 9 IPTC subjects are associated with less than 1% of all clicks each, it is highly likely that the topical interests of the users in our sample gravitate towards the most dominant 8 IPTC subjects. Therefore, our analysis varies k between 8 and 17.

Figure 1 shows the cohesion and separation for the groupings generated by the *weighted* linkage agglomerative clustering, which produces the best results among the different linkage criteria. For each k, *text* has higher cohesion than the whole sample, since it improves over the *average cohesion* computed over all possible user pairs. It also produces more cohesive groups than *vis*. Whereas *vis* is a rather weak source of evidence, its combination with *text* using 0.9/0.1 as weights improves significantly the cohesiveness of the groups against those generated by each individual feature. Weighing higher the contribution of *vis* is nevertheless detrimental for all other combination weights, as shown in Figure 1 for weights 0.8/0.2; the results for the remaining combinations are not shown.

Figure 1 further indicates that *text* produces user clusters that are better separated from the other clusters in the grouping, compared to the separation among clusters observed in the other groupings (the lower the separation value, the better). However, the differences in the separation values between *text* and *text_vis* (0.9/0.1) are rather insignificant, particularly for $k \geq 12$ where they are less than 0.02 (3%). Furthermore, a comparison of the cohesion values against the separation values for each of the *text* and *text_vis* groupings indicates that groups members are more similar to each user than to users in other groups.

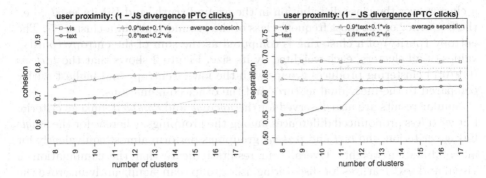

Fig. 1. Variation in group membership by comparing each grouping's intra–cluster cohesion and inter–cluster separation to the average similarity of all user pairs

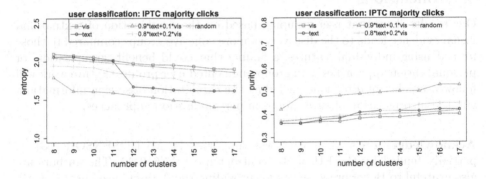

Fig. 2. Entropy and purity with respect to the *IPTC majority clicks* ground truth

Therefore, our results indicate that combining textual and visual features for discovering user groups can significantly improve the homogeneity of the formed groups compared to the groupings generated by each individual feature.

To provide further support to our hypothesis, our user groupings are evaluated against an additional implicit ground truth dataset: a user classification formed by assigning to each user the IPTC subject code of the majority of their clicked images. This corresponds to the 8 groups listed in Table 2. Given that, for personalisation purposes, it is more important to have groups that are more cohesive than well separated, our analysis focusses on measures that evaluate the extent to which a cluster contains objects of a single class: (i) the *entropy* of

Table 2. User classification based on the IPTC subject of the majority of their clicked images. All IPTC subjects that represent less than 1% of the clicked images in Table 1, apart from REL, are no longer present. WAR is also missing.

IPTC	ACE	CLJ	DIS	EBF	HUM	POL	REL	SPO	Total
# users	39	2	1	9	12	37	1	48	149
% users	26.2%	1.3%	0.7%	6.0%	8.1%	24.8%	0.7%	32.2%	100%

a cluster over the class distribution in the ground truth, and (ii) its *purity*, i.e., the frequency of the most frequent class of the ground truth in a cluster [7]. The entropy (purity) of a clustering is computed as the sum of the entropy (purity) values of all clusters, each weighted by its size. Figure 2 shows that the *text_vis* (0.9/0.1) clustering produces groups with the lowest entropy and highest purity compared to the individual features and the other combinations.

Similar results are also observed for the *average* and *complete* linkage criteria, but with less pronounced differences among the groupings, whereas for the *single* linkage criterion, the *text* and *text_vis* groupings perform almost equivalently for all combination weights. Overall, our results indicate that the combination of visual and text features for discovering user groups can significantly improve the homogeneity of the formed groups, as measured by cohesion, entropy and purity.

5   Conclusions

Our results show that user grouping based on the combination of textual and visual evidence leads to the discovery of more cohesive groups compared to those formed using individual features, a finding that could benefit 'groupisation' or personalisation approaches with group-augmented user profiles. Future work will follow a number of directions, including the exploitation of session information, and the incorporation of such groups in personalisation approaches.

Acknowledgements. This work was supported by MULTISENSOR project, partially funded by the EU, under contract no. FP7-610411. The authors are also grateful to Belga press agency for providing the datasets used in this work.

References

1. Bennett, P.N., Radlinski, F., White, R.W., Yilmaz, E.: Inferring and using location metadata to personalize web search. In: Ma, W.Y., Nie, J.Y., Baeza-Yates, R.A., Chua, T.S., Croft, W.B. (eds.) SIGIR 2011, pp. 135–144. ACM (2011)
2. Collins-Thompson, K., Bennett, P.N., White, R.W., de la Chica, S., Sontag, D.: Personalizing web search results by reading level. In: Macdonald, C., Ounis, I., Ruthven, I. (eds.) CIKM 2011, pp. 403–412. ACM (2011)
3. Gayo-Avello, D.: A survey on session detection methods in query logs and a proposal for future evaluation. Information Sciences 179(12), 1822–1843 (2009)
4. Henrikson, J.: Completeness and total boundedness of the Hausdorff metric. MIT Undergraduate Journal of Mathematics 1, 69–80 (1999)
5. Kharitonov, E., Serdyukov, P.: Demographic context in web search re-ranking. In: Wen Chen, X., Lebanon, G., Wang, H., Zaki, M.J. (eds.) CIKM 2012, pp. 2555–2558. ACM (2012)
6. Kharitonov, E., Serdyukov, P.: Gender-aware re-ranking. In: Hersh, W.R., Callan, J., Maarek, Y., Sanderson, M. (eds.) SIGIR 2012, pp. 1081–1082. ACM (2012)
7. Tan, P.-N., Steinbach, M., Kumar, V.: Introduction to Data Mining. Addison-Wesley Longman (2005)

8. Teevan, J., Morris, M.R., Bush, S.: Discovering and using groups to improve personalized search. In: Baeza-Yates, R.A., Boldi, P., Ribeiro-Neto, B.A., Cambazoglu, B.B. (eds.) WSDM 2009, pp. 15–24. ACM (2009)
9. van de Sande, K.E.A., Gevers, T., Snoek, C.G.M.: Evaluating color descriptors for object and scene recognition. IEEE Trans. PAMI 32(9), 1582–1596 (2010)
10. Weber, I., Castillo, C.: The demographics of web search. In: Crestani, F., Marchand-Maillet, S., Chen, H.-H., Efthimiadis, E.N., Savoy, J. (eds.) SIGIR 2010, pp. 523–530. ACM (2010)
11. Weber, I., Jaimes, A.: Who uses web search for what: and how. In: King, I., Nejdl, W., Li, H. (eds.) WSDM 2011, pp. 15–24. ACM (2011)
12. Zeimpekis, D., Gallopoulos, E.: TMG: A MATLAB toolbox for generating term-document matrices from text collections. In: Kogan, J., Nicholas, C., Teboulle, M. (eds.) Grouping Multidimensional Data:Recent Advances in Clustering, pp. 187–210. Springer (2006)

Analyzing Tweets to Aid Situational Awareness

Tim L.M. van Kasteren, Birte Ulrich, Vignesh Srinivasan, and Maria E. Niessen

AGT group (R&D) GmbH,
Hilpertstr. 35, 64295 Darmstadt, Germany
{tkasteren,bulrich,vsrinivasan,mniessen}@agtinternational.com
http://www.agtinternational.com

Abstract. Social media networks can be used to gather near real-time information about safety and security events. In this paper we analyze Twitter data that was captured around fifteen real world safety and security events and use a number of analytical tools to help understand the effectiveness of certain features for event detection and to study how this data can be used to aid situational awareness.

Keywords: Social Media Analytics, Situational Awareness.

1 Introduction

The popularity of social media networks provides us with a constant flow of information which can be used as a low-cost global sensing network for gathering near real-time information about safety and security events. This information can be very valuable to emergency response teams, who rely on an accurate situational awareness picture of the emergency at hand. Obtaining a more accurate situational awareness picture allows a better and faster response and results in less damage and casualties.

The microblogging service Twitter has become very popular and has been reported to sometimes spread news before traditional news channels [1,5]. However, the data obtained from Twitter is very diverse (i.e. few constraints on what users can post) and it is not well understood which information relevant to emergencies is present in the data and at which point in time.

In this paper we analyze Twitter data that was captured around fifteen real world safety and security events from four categories (accidents, natural disasters, crowd gatherings and terrorist attacks). Using a number of analytical tools we identify recurring patterns and event specific characteristics that provide a basis for creating automated classification algorithms and show the possibilities of using this data to aid situational awareness.

The remainder of this paper is organized as follows: Section 2 discusses related work. In Section 3, we present the data and in Section 4, the analysis of the data. Finally, Section 5 concludes our findings.

M. de Rijke et al. (Eds.): ECIR 2014, LNCS 8416, pp. 700–705, 2014.
© Springer International Publishing Switzerland 2014

2   Related Work

In previous work, Twitter has been analyzed to understand its usage in emergency situations. Hughes et al. provide an analysis of Twitter usage during two emergency events and two political conventions by comparing the amount of tweets per day and the relative number of reply tweets to general Twitter usage. They show a difference in statistics when an event is taking place [4]. Vieweg et al. analyze which information can be extracted from tweets to enhance the situational awareness during natural disasters. On the basis of two datasets, one for a flood and one for a wildfire, the authors show that the percentage of geo-location usage, location references, situational updates, and retweets is higher for a fast spreading wildfire than for a slow rising flood. This indicates that the features of the datasets reflect the actual emergency situation (i.e. warning, impact or recovery phase) and the type of disaster itself (unexpected fast or predictable slow event) [6].

Twitter is also used to automatically detect safety and security events from data. Walther et al. present an event detection system that uses geospatial information from tweets to automatically identify and classify events. Their approach is evaluated on events captured from real world Twitter data [7]. Sakaki et al. give an example of how semantic analysis of tweets can be used to detect natural disasters like earthquakes and typhoons. With the geo-location information of the tweets they estimate the location of the earthquake and typhoon [5]. Bouma et al. present a method for automatic anomaly detection in Twitter data using a correlation analysis of a number of variables, such as sentiment, retweets, post frequencies and other meta data [2].

In this work we analyze tweets for a large number of events from different categories to find features that consistently help detect events.

3   Data

We collected data for fifteen safety and security events from four categories: accidents (e.g. train, plane or car crash), natural disasters (e.g. flood, earthquake), crowd gatherings (e.g. festival, demonstration) and terrorist attacks (including mass shootings). Accidents and terrorist attacks are unexpected and unplanned events, natural disasters are sometimes predicted such as in the Acapulco floods and crowd gaterings are planned in case of festivals, but unplanned in the case of raids. The data was gathered by monitoring the news for reported events and collecting one to two weeks of tweets using the Twitter REST API (version 1.1). The tweets were collected using an empty query (i.e. no keyword filtering) and a geocode centered around the event with a 15 mile radius. Details of the events can be found in Table 1.

4   Analysis

The collected data was analyzed using counts, Twitter-specific, text-based, image-based and location-based tools. Our goal was to get a better understanding of how

Table 1. List of events for which Twitter data was captured

Description	Category	Tweets	Days	Dates
Antwerp Tomorrowland festival	Crowd gathering	110.161	6	24-29 July, 2013
Zurich ZuriFascht festival	Crowd gathering	41.102	4	06-09 July, 2013
Santiago train crash	Accident	156.881	7	23-29 July, 2013
Leiden factory fire	Accident	184.063	6	14-19 August, 2013
Chicago car crash	Accident	16.343	5	14-18 August, 2013
Heidelberg shooting	Terrorist	21.043	5	16-20 August, 2013
Vadodara building collapse	Accident	11.501	4	25-28 August, 2013
Cairo raids	Crowd gathering	4.641.634	11	09-19 August, 2013
Frankfurt tree on railroad	Accident	125.833	7	30 Aug.-05 Sept., 2013
Zevenaar factory fire	Accident	156.129	8	02-09 September, 2013
Washington shooting	Terrorist	997.001	3	15-17 September, 2013
Acapulco floods	Natural disaster	202.915	12	11-22 September, 2013
Nairobi mall attack	Terrorist	1.212.249	9	16-24 September, 2013
India cyclone	Natural disaster	1.684	6	07-12 October, 2013
Philipines earthquake	Natural disaster	597.043	10	08-17 October, 2013

people tweet about an event and what information can be extracted. These results help to get a better understanding of the limitations of Twitter data and help to obtain a better understanding of the effectiveness of certain features when creating event detection algorithms. We study the data collected prior to the event, during the event and after the event, to identify the potential of using this data for prediction, detection and investigation, respectively.

4.1   Counts

Our first analysis counts the number of tweets per hour. These counts can be plotted over time (Fig. 1a) and show a very clear recurrent pattern when no event is taking place, due to the common daily activities of users. This has also been reported in previous work [3].

In the majority of events we recorded data for, we see a very clear increase in the number of tweets shortly after the event takes place. The increase in the number of tweets clearly reflects the impact the event has on the public and is strongly related to the severity of the event, such as the number of people being affected by a natural disaster or the number of deaths during an accident or terrorist attack. A count-based analysis can therefore be indicative of a big event, but does not help in determining the category of the event.

4.2   Twitter-Specific Analysis

Twitter specific analysis was done by counting the number of tweets over time and getting the relative occurrence of hashtags, retweets, replies and mentions in the tweets. These measures provide indications of how the users are using the Twitter network. Hashtags generally identify a certain concept related to a message. Retweets are a Twitter specific mechanism that allows users to re-post a previously posted message and therefore help broadcast the information contained in the tweet. Replies allow users to respond to a previously posted message and are mainly used in discussions. Finally, mentions allow users to mention other users within their messages and indicate social references.

Comparing these measures during the event to before the event, we observe an increase in retweets and mentions and often a decrease in replies. For events that cause a significant increase in the total number of tweets, and thus a high social impact, this effect becomes very substantial. For the train crash in Santiago, for example, the usage of retweets rises from 22% before the event to 69% during the event, the usage of mentions grows from 35% to 72% and replies drop from 26% to 10%. The rise in retweets can be explained by the desire of people to broadcast information of severe events which they have not witnessed. The drop in replies further supports this, since people are not directing their communication to one specific receiver.

Overall, we see some significant changes in the Twitter specific measures during events, especially when events are completely unexpected like severe accidents and terrorist attacks.

4.3 Text Analysis

The text-based analysis operates on the message contained in a tweet. Basic cleaning operations were performed such as converting the text to lower case and removing punctuation marks and stop words. To analyze the resulting data we calculated the most frequent hashtags, words, bi-grams and tri-grams. The resulting most frequent words give an indication of the content of the majority of the tweets and help to identify the nature of events. Moreover, we calculated the TF-IDF of the most frequent words over time: the term frequency (TF) was determined per hour for all tweets around the time of the event while the inverse document frequency (IDF) was calculated for a week of Twitter data at the same location but weeks after the event. Hence the TF-IDF gives a normalized view of which keywords are used out of the ordinary and can be used to detect an event.

For the analyzed events, the bigger events cause twitter users to converge to a few hash tags to refer to the event and give rise to very frequent hashtags describing the event, such as *#tomorrowland* (Antwerp) or *#accidente* (Santiago train accident). However, the location of an event, especially of accidents, are mostly used as reference by Twitter users during and after the event, for example (*#santiago*, *#zevenaar*, *#leiderdorp* (Leiden fire), and *#dossenheim* (Heidelberg shooting).

Smaller events do not always show up in the frequent hashtags, while they can be found in the tweet texts and the most frequent words and n-grams. The most frequent words and n-grams can be informative from an investigation point of view, especially for events with a high social impact. For example, the most frequent words during and after the Santiago train accident all relate to the accident, but there is a shift in focus from emergency response ('donar sangre' (donate blood), 'hospital', 'extrema necesidad' (dire need)) to the effects of the accident a few days later ('necesitan psicólogos' (psychologists needed)). TF-IDF is most informative about the shift in focus of tweet content compared to normal

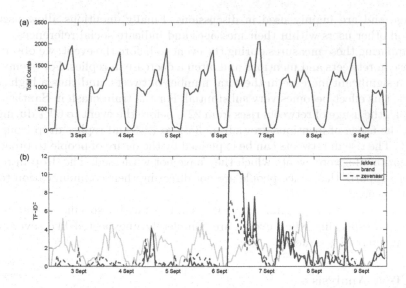

Fig. 1. (a) The total counts of tweets before, during (between approximate 6 am and 12 am on the 6th of September), and after the fire in Zevenaar and (b) the TF-IDF measure for three words: "lekker" (nice), "brand" (fire), and "Zevenaar" (the town)

situations. Figure 1 shows (a) the raw count of tweets around Zevenaar and (b) the TF-IDF of three words: 'lekker', a very common Dutch word approximately following the daily pattern of the counts, and 'brand' and 'zevenaar', which show a major peak around the time of the start of the fire that is hardly visible in the raw counts.

4.4   Image Analysis

For our image-based analysis we extracted all the URLs from the tweets and downloaded images using an automated script. We manually inspected the collected images to determine which were related to the event.

For most events no event-related images were available before the event. The exceptions were predicted natural disasters, where we found images which indicate warnings. For example, in the case of the Acapulco floods a few images show the prediction from weather institutes of storms over the region.

The percentage of event-related images shows a sharp increase during the event, in comparison with before the event. Almost half of the images collected during the event are about the event, indicating a strong interest among users to share news or information about the event. Although many images posted are duplicates, rather than unique ones, some of the images posted have a lot of potential to assist an operator in assessing an emergency situation. In particular, these images might show more information than what can be obtained from an emergency phone call. For example, in the case of Acapulco, we found images of people plundering local stores, as well as damaged and obstructed roads. Especially during the event, we see that the number of posted images related to the event is relatively high, which is the time at which the need for information is most critical.

After the event the percentage of images about the event stays high. In particular, heroic deeds (e.g. someone rescuing a child) captured in images are widely shared and retweeted. The number of images related to the event decreases with time after the event.

4.5   Location Analysis

The location-based analysis relies on the GPS coordinates that Twitter users can submit together with a tweet. On average approximately 10% of all collected tweets contain GPS coordinates. We analyzed the number of tweets as a function of the distance to the event. We see a clear increase in the number of tweets close to the location the event took place in during and after the event, for most of the events we analyzed. This indicates that events can be detected based on geo-spatial clusters of tweets.

5   Conclusion

We analyzed fifteen safety and security events from four categories using various analytical tools. Our findings show the potential of using Twitter to aid situational awareness and help understand the effectiveness of certain features for event detection. Counts, Twitter specific features, highly retweeted images and tweets with GPS coordinates are useful for detecting events with a high social impact, but lack in volume for detecting smaller events. Only the text-based feature TF-IDF is sensitive enough to detect both big and small impact events.

References

1. Becker, H., Naaman, M., Gravano, L.: Beyond trending topics: Real-world event identification on twitter. In: ICWSM (2011)
2. Bouma, H., Raaijmakers, S., Halma, A., Wedemeijer, H.: Anomaly detection for internet surveillance. In: SPIE Defense, Security, and Sensing, pp. 840807–840807. International Society for Optics and Photonics (2012)
3. Cataldi, M., Di Caro, L., Schifanella, C.: Emerging topic detection on twitter based on temporal and social terms evaluation. In: Proceedings of the Tenth International Workshop on Multimedia Data Mining, p. 4. ACM (2010)
4. Hughes, A.L., Palen, L.: Twitter adoption and use in mass convergence and emergency events. Int. Journal of Emergency Management 6(3), 248–260 (2009)
5. Sakaki, T., Okazaki, M., Matsuo, Y.: Earthquake shakes twitter users: real-time event detection by social sensors. In: Proc. of the 19th Int. Conf. on WWW (2010)
6. Vieweg, S., Hughes, A.L., Starbird, K., Palen, L.: Microblogging during two natural hazards events: what twitter contribute to situational awareness. In: Proc. of the SIGCHI Conf. on Human Factors in Comp. Sys. (2010)
7. Walther, M., Kaisser, M.: Geo-spatial event detection in the twitter stream. In: Serdyukov, P., Braslavski, P., Kuznetsov, S.O., Kamps, J., Rüger, S., Agichtein, E., Segalovich, I., Yilmaz, E. (eds.) ECIR 2013. LNCS, vol. 7814, pp. 356–367. Springer, Heidelberg (2013)

Query-Dependent Contextualization of Streaming Data

Nikos Voskarides, Daan Odijk, Manos Tsagkias,
Wouter Weerkamp, and Maarten de Rijke

ISLA, University of Amsterdam, Amsterdam, The Netherlands
{n.voskarides,d.odijk,m.tsagkias,w.weerkamp,derijke}@uva.nl

Abstract. We propose a method for linking entities in a stream of short textual documents that takes into account context both inside a document and inside the history of documents seen so far. Our method uses a generic optimization framework for combining several entity ranking functions, and we introduce a global control function to control optimization. Our results demonstrate the effectiveness of combining entity ranking functions that take into account context, which is further boosted by 6% when we use an informed global control function.

1 Introduction and Related Work

We keep track of what is happening around us through multiple dynamic sources. To keep on top of the resulting stream of information, we need useful signals. While human readers may be able to map different references as referring to the same entity, automated approaches aimed at analyzing a stream of messages might not. A key step in facilitating our understanding of streams is to move from raw text to entities. This can be achieved via a process known as *entity linking* [1, 2] that maps parts of text to entities in a knowledge base, thereby contextualizing the textual stream.

Entity linking is often performed in two steps: deciding what text to link and then where it should link to. It is common to consider entities from a general-purpose knowledge base such as Wikipedia or Freebase, since they provide sufficient coverage for most tasks and applications. For deciding what text to link *from*, the state-of-the-art performs lexical matching on the "surface forms" of entities [1, 3]. These surface forms are derived from the links created on the knowledge base, e.g., both "Barack Hussein Obama II" and "President Obama" can be used as anchor text referring to the entity "Barack Obama." Lexical matching of anchor texts can produce multiple candidate target entities that share the same surface forms (e.g., "the American President" can refer to any American president, or even the office). Deciding where to link *to* has been modeled as a learning to (re)rank problem [1, 4], where generic features represent the target entities. An alternative approach is to view this as an optimization problem [7], where the objective is to pick the best target entity for each link candidate.

When working with longer textual documents, the performance of entity linking can be enhanced by incorporating a sense of context or coherence [3, 5]. For shorter texts, context is often ignored, but even very limited textual contexts have shown to be helpful [1]. For *streaming* textual data, modeling context is not straightforward. Recent work on linking of streaming text has proposed to model context as a graph and to include it in the set of features in a learning to rerank approach [4].

M. de Rijke et al. (Eds.): ECIR 2014, LNCS 8416, pp. 706–712, 2014.
© Springer International Publishing Switzerland 2014

To contextualize streams of related pieces of short text, one needs to resolve link candidate targets in the local short text context, while benefiting from the more global history of the stream. We propose a query-dependent contextualization approach tailored for streaming text. We use a generic optimization framework that can combine information about the surface forms of candidate entities, contextual features for short pieces of text and graph-based context modeling of the stream. Our approach combines the optimization approach that was shown effective for linking entities in short text with graph-based context modeling, shown to be effective on streaming text [4, 7].

We concentrate on a filtering task where a user is monitoring a document stream, e.g., the Twitter stream, for a specific information need (e.g., query, hashtag, trending topic) and is interested in entities that are relevant to their information need. As more data enters the stream, our system should be able to become better at "understanding" the topic of the stream and, therefore, linking new tweets to the correct entities.

Our main contribution is twofold: (i) an entity linking method that extracts and links entities using context both inside a document and in the history so far, and (ii) a manually annotated dataset [1]. The main research question we aim at answering is: (RQ1) what is the entity linking effectiveness of combining entity ranking functions that take into account context both from within a document and from the history of documents seen so far compared to individual entity ranking functions and their combinations? (RQ2) what is the effect in performance when we introduce a global control function in the optimization algorithm?

2   Method

Our entity linking method starts from the text of a tweet t and uses lexical matching to extract all n-grams that are also anchors in Wikipedia articles. As each anchor a can refer to multiple Wikipedia articles (concepts), which we call candidate concepts $Cand(a)$, the challenge is to find the correct concept $c \in Cand(a)$. The set of correct concepts that correspond to anchors in t constitute the final set of entities C for t. The main problem we are trying to solve, for a given tweet t, is twofold: (i) find an optimal set of concepts C, and (ii) generate a ranking of the concepts in C. We follow [6] for generating candidate sets of C and finding an optimal set. This algorithm starts from a set of anchors and generates a reasonable set of entities (e.g., based on their popularity). It keeps a mapping M between anchors and entities, and computes a score for this set. Then it iteratively tries to maximize the set score by replacing the entities in the set in an informed way, which guarantees to arrive to a local maximum. The optimal C is obtained by $\arg\max_C \sum_{c \in C} S(c; \cdot)$, where $S(c; \cdot)$ is the score of a concept c in C, and is defined as a linear interpolation over a set of ranking functions F:

$$S(c; \cdot) = g(\cdot) \sum_{f_i \in F} w_i f_i(\cdot), \tag{1}$$

---

[1] The dataset consists of 30 topics, with a total of 69,789 tweets and an average of 2,326 tweets per topic. On average, 2.8 concepts were judged as highly relevant and 44.86 as relevant per topic. See: http://ilps.science.uva.nl/resources/query-dependent-contextualization.

where g is a global control function, f_i is a ranking function (explained below) and w_i denotes the weight for f_i. Our next step is to instantiate g, and F. We build upon previous research and consider four readily used, state-of-the-art ranking functions: (i) commonness (CMNS), (ii) tweet coherence (TC), (iii) concept graph score (CGS), (iv) link probability (LP). In particular, we define $F = \{CMNS, TC, CGS\}$ and g can be either LP or 1. Below, we describe each ranking function in turn.

Commonness (CMNS). A simple function used for tweets is to rank a candidate concept $c \in Cand(a)$ by the prior probability of c being the target for the anchor a. Then, we select the best scoring candidate concept as the concept for that anchor. CMNS is defined as follows:

$$CMNS(c, a) = \frac{|L_{a,c}|}{\sum_{c' \in Cand(a)} |L_{a,c'}|},$$

where $L_{a,c}$ is the set of links with anchor text a that map to concept c.

Tweet coherence (TC). Commoness favors popular concepts. However, this is not always desirable. Think of a tweet that mentions "python" in a programming and biology context; CMNS will return the same concept regardless. One way to tackle this problem is to leverage contextual information from within the document to select candidate concepts that are topically coherent. For example, in a tweet containing the anchors "premier league" and "spurs," it is reasonably safe to assume that the anchor "spurs" should be linked to "Tottenham Hotspur F.C." For a concept $c \in C$ we compute TC as follows:

$$TC(c, C) = \frac{1}{|C| - 1} \sum_{c' \in C, c' \neq c} sim(c, c'),$$

where $sim(c, c')$ is computed using the variant of the Normalized Google Distance proposed in [3].

Concept graph score (CGS). Another way to leverage contextual information is to consider all candidate entities found in the history of documents seen so far. We create a graph $G(V, E)$ with vertices V, and edges E where V denotes candidate concepts and an edge between two vertices, c and c', exists if it holds that $sim(c, c') > 0$ and they are not linked from the same anchor. We update the graph for every incoming tweet, and rank the nodes by the standardized degree of the node (concept). We hypothesize that concepts with a higher degree are more likely to be relevant; see also [4].

Link probability (LP). Finally, for controlling our optimization algorithm, we use the link probability function [1] for g because it can penalize concepts that are refered by n-grams that are less likely to be observed as anchors in Wikipedia. For a candidate concept, we retrieve the respective n-gram via the mapping M (see above), and compute the link probability LP for the n-gram as the number of times that an n-gram is observed as anchor over its collection frequency.

We generate a final ranking of c in C for t using (1).

3 Experimental Setup

To answer our research questions, we consider a stream of tweets related to a trending topic in Twitter. We contrast the performance of our method when it uses all possible functions with the performance of the method when it uses individual functions and combinations of them, for two global control functions: $g = 1$, and $g = LP$.

Dataset. We have crawled 700 Twitter trending topics between Sept. 3–Oct. 20, 2013. For each trending topic, we use the Twitter Search API to retrieve tweets that match the trending topic phrase, and were posted between 2 hours before and 3 hours after the topic started trending, ignoring retweets. We excluded "frivolous" topics such as "#IWishJacksonFollowedMe." For our evaluation exercise we sample 30 topics so that there is a good balance with regards to topic size, and to whether the topic is a hashtag or a phrase. As our knowledge base for linking tweets, we use an English Wikipedia dump dated March 18, 2013, with 4,070,588 articles. We filter out labels that are stop words or are raw numbers or have LP < 0.02, and entities with CMNS < 0.02 to exclude highly unlikely labels and entities.

Ground truth. Our ground truth is produced as follows. For every tweet in a topic, we extract all n-grams that are found as anchors in Wikipedia and retrieve their corresponding candidate concepts. We collect these mappings over each topic and filter out labels that appear only once in the topic stream. We assume each label that appeared in the topic stream maps to the same concept for all the tweets in that topic. Three annotators assessed each labels selecting the relevant concepts for each label. We use graded relevance for a concept: *highly relevant* if it is central to the topic (e.g., the WWE wrestler "Dustin Rhodes" for the topic "#Goldust"), *relevant* if it is related (e.g., "Vladimir Putin" for the topic "President Obama") and *non-relevant* otherwise, i.e. for labels unrelated to the topic's central subjects or not worthy of linking (e.g., "10", "lol").

This evaluation setup allows us to capture what people discuss about a topic in Twitter and is not based on pre-conceived ideas on what should be related to the topic at hand. For example, for the topic "#AfricasXtinction," which refers to the extinction of rhinoceros and elephants in Africa, the basketball player "Yao Ming" is judged as relevant, as he was involved in a campaign against the extinction.

We used 5-fold cross validation and chose the best weights of the objective function using line search over the parameters for each fold. We report on precision at N (P@N with N=1, 2, 3) at tweet level, both for highly relevant and for highly relevant or relevant. Statistical significance is tested using a two-tailed paired t-test and is marked as ▲ (or ▼) for significant differences for $\alpha = .05$.

4 Results and Analysis

In our first experiment, we set the global control function $g = 1$ and assess the effectiveness of three individual ranking functions and their combinations in the optimization framework we consider. We denote the experimental setting where we use all three ranking functions as ALL. We report our results in Table 1 (top). ALL is one of the top-performing combinations and no individual ranking function or other combination

Table 1. Performance for three individual ranking functions, i.e., commonness (CMNS), tweet coherence (TC), concept graph score (CGS) and their combinations in F, and two global control functions for g, i.e., 1 (top), and link probability (LP, bottom). ALL denotes using all three functions in F. Statistical significance is tested against ALL. Boldface marks best performance in the corresponding metric and ground truth: all relevant (relevant), only highly relevant (highly).

	P@1		P@2		P@3	
	relevant	highly	relevant	highly	relevant	highly
$g = 1$						
ALL	**0.6709**	0.4264	0.5416	0.2903	0.4481	0.2190
CMNS	0.6461	**0.4392**	0.5022▼	0.2857▼	0.3868▼	0.2110▼
TC	0.6209	0.3311▼	0.4771▼	0.2430▼	0.4012▼	0.2046▼
CGS	0.6485	0.3268▼	0.5368	0.2857	0.4475	**0.2404**
CMNS+TC	0.6624	0.4213	**0.5476**	**0.2974**	0.4501	0.2324
CMNS+CGS	0.6624	0.4213	0.5371	**0.2974**	0.4435	0.2244
TC + CGS	0.6592	0.3002▼	0.5459	0.2744▼	**0.4593**	0.2397
$g = LP$						
ALL	0.7573	0.4995	0.6044	0.3510	0.4807	0.2695
CMNS	0.7206▼	**0.5111**	0.5428▼	0.3265	0.4252▼	0.2340▼
TC	0.6801▼	0.4611▼	0.5399▼	0.3192▼	0.4420▼	0.2397▼
CGS	**0.7615**▲	0.5032	**0.6104**	**0.3593**	**0.4855**	**0.2719**
CMNS+TC	0.7289▼	0.5096	0.5640	0.3373	0.4538▼	0.2465▼
CMNS+CGS	0.7273▼	0.4980	**0.6104**	**0.3593**	**0.4855**	**0.2719**
TC + CGS	0.7557	0.4900	0.6045	0.3491	0.4817	0.2709

outperforms it in a stastically significant manner, for both all relevant and only high relevant entities. This supports our hypothesis that combining context from within the document and from the history of documents seen so far helps linking effectiveness.

There are cases where combinations marginally outperform ALL but not in a statistically significant manner. One reason for this may be due to the large intervals we used in the line search over the parameters in the training phase. Another reason can be that we ignore the decrease of an entity's CGS score over time as more nodes are added to the graph. Therefore, we are not able to achieve significant gains over the other combinations (e.g., CMNS+CGS). We plan to address this problem in future work.

In our second experiment, we set the global control function $g = LP$ and report on results in Table 1 (bottom). On average, the performance of all systems is statistically significantly better when using $g = LP$ compared to $g = 1$, with an average improvement of 6% over both highly relevant and all relevant entities. This confirms our intuition that n-grams with high LP need to be boosted during optimization, and stresses the importance of using LP in controlling the optimization function.

Our findings are similar to what we found above: no individual ranking function or combination of ranking functions significantly outperforms ALL, except for CGS at P@1. It is interesting that by controlling CGS with LP we are able to balance the problem with the CGS scores changing over time. This finding confirms the importance of modeling the stream history as a graph in the setting of Twitter topics. Future work should consider better optimization methods for combining ranking functions.

To better understand our results, we analyze the linked entities per topic. We find that our methods tend to have lower effectiveness for trending topics that are not hashtags. We looked closer at such topics and found that they are noisy, and therefore may not be centered around a particular entity, or set of entities. With regards to errors due to linking and not to noise in the stream, we found that in topic "#Goldust," for example, although the main entity is the wrestler "Dustin Rhodes," the objective function using ALL gives a higher score to the entity "WWE Raw" whereas CMNS always favors "Dustin Rhodes" because the high commonness score between the anchor "#Goldust" and the entity "Dustin Rhodes." A possible way to overcome this problem is to use a different function for optimization and ranking and take into account the number of times we have seen each entity in the stream history.

5 Conclusions

We presented an entity linking method for linking short documents in a stream of documents that uses context both from inside a document and the history of documents seen so far. Our experiments demonstrated that using context from previously seen documents leads to robust improvements in entity linking effectiveness. In terms of optimization, we introduced a global control function for a candidate entity, which improved early precision by 6% across all our experimental settings. The outcomes of our work can be used in applications which need to extract the most popular relevant entities in the stream so far. In future work, we envisage exploring other global control and ranking functions, and optimizing our method for efficiency.

Acknowledgments. This research was supported by the European Community's Seventh Framework Programme (FP7/2007-2013) under grant agreement nr 288024 (LiMoSINe project), the Netherlands Organisation for Scientific Research (NWO) under project nrs 640.004.802, 727.011.005, 612.001.116, HOR-11-10, the Center for Creation, Content and Technology (CCCT), the QuaMerdes project funded by the CLARIN-nl program, the TROVe project funded by the CLARIAH program, the Dutch national program COMMIT, the ESF Research Network Program ELIAS, the Elite Network Shifts project funded by the Royal Dutch Academy of Sciences (KNAW), the Netherlands eScience Center under project number 027.012.105 and the Yahoo! Faculty Research and Engagement Program.

References

[1] Meij, E., Weerkamp, W., de Rijke, M.: Adding semantics to microblog posts. In: WSDM 2012. ACM (2012)

[2] Mihalcea, R., Csomai, A.: Wikify!: linking documents to encyclopedic knowledge. In: CIKM 2007, pp. 233–242. ACM (2007)

[3] Milne, D., Witten, I.H.: Learning to link with Wikipedia. In: CIKM 2008, pp. 509–518. ACM (2008)

[4] Odijk, D., Meij, E., de Rijke, M.: Feeding the second screen: Semantic linking based on subtitles. In: OAIR 2013 (2013)

[5] Ratinov, L., Roth, D., Downey, D., Anderson, M.: Local and global algorithms for disambiguation to Wikipedia. In: HLT 2011, pp. 1375–1384. ACL (2011)

[6] Shen, W., Wang, J., Luo, P., Wang, M.: Liege: link entities in web lists with knowledge base. In: KDD 2012, pp. 1424–1432. ACM (2012)

[7] Shen, W., Wang, J., Luo, P., Wang, M.: Linking named entities in tweets with knowledge base via user interest modeling. In: KDD 2013, pp. 68–76. ACM (2013)

An Exploration of Tie-Breaking
for Microblog Retrieval

Yue Wang, Hao Wu, and Hui Fang

Department of Electrical & Computer Engineering,
University of Delaware, USA
{wangyue,haow,hfang}@udel.edu

Abstract. Microblog retrieval enables users to access relevant information from the huge number of tweets posted on social media. Since tweets
are different from traditional documents, existing IR models might not be
the optimal choice for this problem. Tie-breaking has been recently proposed as a new way of combining multiple retrieval signals. In this paper,
we focus on studying the potential of this approach in microblog retrieval
and propose new methods to further improve the performance. Experiment results show that these tie-breaking based methods can achieve
comparable performance with the top runs in the TREC Microblog track.

1 Introduction

Microblog has become an important information source in daily life [1]. However,
the huge number of tweets generated everyday makes it difficult for users to find
useful information. Although traditional IR models could be applied to this
domain-specific problem, they might not be the best choice because tweets are
different from regular documents in many aspects. Existing studies on microblog
retrieval have tried to overcome this limitation and incorporate new retrieval
signals such as temporal information [2–5] and quality indicators [6, 7] by either
extending existing models or using learning to rank methods.

Tie-breaking approach has been recently proposed as a new way of combining
retrieval signals [8]. The basic idea is to prioritize retrieval signals, and then
apply them one at a time to rank documents by breaking the ties created by
previously applied signals. Its advantage lies in the simplicity and flexibility of
combining multiple signals. Previous study showed the promising results on ad
hoc retrieval by combining three basic retrieval signals, i.e., document length
(DL), term frequency (TF) and inverse document frequency (IDF).

In this paper, we propose to extend the existing framework of tie-breaking to
further improve its effectiveness on microblog retrieval. Specifically, we consider
two commonly used retrieval strategies, i.e., query expansion and document expansion, and study how to incorporate them into the tie-breaking framework.
Our experiments confirm that the tie-breaking approach is more effective than
traditional retrieval models and learning to rank methods when combining multiple basic retrieval signals. Moreover, the proposed query expansion and document expansion methods can further improve the retrieval performance while

M. de Rijke et al. (Eds.): ECIR 2014, LNCS 8416, pp. 713–719, 2014.

the document expansion method is more effective. Finally, the performances of the proposed methods are comparable with that of top runs in TREC Microblog track.

2 Tie-Breaking for Microblog Retrieval

2.1 Basic Idea of Tie-Breaking

The effectiveness of a retrieval function is closely related to how it combines retrieval signals such as TF, IDF and DL. Traditional retrieval models including Okapi BM25 and pivoted normalization often combine these signals in a rather complicated way. However, it remains unclear whether such complicated combinations are necessary.

The basic idea of tie-breaking is to explore a simple yet effective way of combining multiple signals [8]. Specifically, it will first prioritize all the signals and then iteratively apply one signal at a time to rank documents by breaking the ties (i.e., the documents with the same scores) created by the previously applied signals. Figure 1 illustrates this process. Assuming that signal 1 is stronger than signal 2 and both of them are stronger than signal 3, we would then first apply signal 1 and generate a list of ranked documents. Note that the documents in the same row indicate that they receive the same scores. For example, after applying signal 1, d_5 and d_2 have the same relevance score, which is higher than the rest of documents. Similarly, d_3, d_7 and d_1 are also tied. We can then apply the next strongest signal (i.e., signal 2) to break these ties created by the signal 1, e.g., assign a higher score d_5 than d_2. This process can be repeated until all the signals are applied.

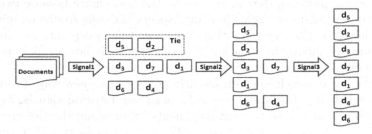

Fig. 1. Example scenario of tie-breaking

2.2 Applying Tie-Breaking to Microblog Retrieval

Experiment Setup: TREC Microblog track provides a common platform for the researchers to study the microblog retrieval problem. We conducted experiments on both the TREC Microblog 2011 and 2012 collections, i.e., **MB11** and **MB12**. We leverage the provided API [1] and crawl 10K results returned by API

---

[1] https://github.com/lintool/twitter-tools/wiki/TREC-2013-API-Specifications

Table 1. Comparing the strength of the best implementation for signals (MB11)

Retrieval signal	Implementation	MAP$_b$	MAP$_w$	MAP$_a$
TF	$\frac{c}{c+1}$	0.3906	0.2306	0.2842
IDF	$log(1 + \frac{N}{df})$	0.4969	0.2886	0.3631
DL	$\frac{1}{dl}$	0.0319	0.0220	0.0234
NOF	$log(NOF)$	0.0747	0.0562	0.0642

for each query. Due to the nature that the tweets retrieval is sensitive to the time, we build an index for each query in order to ensure that no tweets after the query time is involved. We applied the Krovetz stemming to the tweets, and no stop word is removed. The results are evaluated in terms of both mean average precision (MAP) and precision at 30 (P@30).

Methodology: The first step is to identify the retrieval signals. In additional to the three used in the previous study [8], we also consider the number of followers (NOF) because it shows the endorsement from the online community. The more follower a user has, the more influential the user could be, which further means that the content the user posted could be useful and reliable [9]. We do not consider other signals such as the number of retweets because our data sets do not contain detailed information about them.

The second step is to select the best implementation for each signal and then prioritize the signals based on their best implementation. To estimate the strength of each implementation, we would rank documents based on the implementation. Since multiple documents may receive the same score, the retrieval results would be a set of possible rankings where the ties might be broken randomly. To accurately measure the strength of an implementation, we need to look at the best, worst and expected (or average) performance of the set, e.g., MAP_b, MAP_w and MAP_a. Following the previous study [8], the selection and prioritization are done based on the same set of criteria: (1) we prefer the implementation with greater potential (i.e., larger MAP_b) and (2) we prefer the implementation with the better expected performance (i.e., larger MAP_a).

We use **MB11** as the data set to select the implementation and prioritize the signals. We tried different implementation for each signal as in the previous study [8], but only reported the performance of the best implementation for each signal in Table 1. It is clear that IDF is the strongest signal because it has the largest MAP_b and MAP_a. Among the other three signals, TF is the strongest and DL is the weakest. Thus, we should apply the signals in the following order: $IDF \oplus TF \oplus NOF \oplus DL$. This method is denoted as **Tie-breaking**.

Experiment Results: We compare the **Tie-breaking** method with three methods: (1) **Okapi**: Okapi BM25 method; (2) **IDF**: Use only IDF for term weighting; and (3) **L2R**: AdaRank with the same signals shown in Table 1. The results are shown in Table 2. All the methods are trained on **MB11**, and tested on **MB12**. Note that the results with † and * indicate improvement over **Okapi**, **IDF** is statistically significant at 0.05 level based on Wilcoxon signed-rank test.

Table 2. Performance comparison of the different retrieval methods

	Training (MB11)		Testing (MB12)	
Methods	MAP	P@30	MAP	P@30
Okapi	0.3238	0.3762	0.1922	0.3034
IDF-only	0.3548	0.4034	0.2092	0.3299
L2R	0.0748	0.0849	0.0411	0.0641
Tie-breaking	$\mathbf{0.3743}^{\dagger}$	$\mathbf{0.4204}^{\dagger}$	$\mathbf{0.2299}^{\dagger,*}$	$\mathbf{0.3684}^{\dagger,*}$

It is clear that **Tie-breaking** is more effective than the traditional retrieval functions including **Okapi** and **IDF**. Moreover, we find that it is more robust in combining simple retrieval signals than **L2R** and does not require the complicated features to work as existing learning to rank methods. The worse performance of L2R method may come from the use of simple features. In fact, when using more sophisticated features as the ones mentioned in the previous study [9], the performance of learning to rank would be comparable to our method. This further confirms that tie-breaking is a simple yet effective method of combining signals.

3 Extension of Tie-breaking Methods

So far, we have shown that tie-breaking is effective for microblog retrieval. Since tweets are often short, it is necessary to bridge the vocabulary gap between queries and documents. In this section, we explain how to extend existing tie-breaking framework to address this challenge.

3.1 Query Expansion for Tie-Breaking

Query expansion is a common way to improve the performance. One most commonly used method is pseudo relevance feedback, which extract terms from initial retrieval results to enrich the query aiming to bring more relevant query terms. To extend the tie-breaking framework for pseudo relevance feedback, we propose the following strategy. We first conduct the tie-breaking method with all the signals to get a the results. From the initial results, we choose the top N terms with the highest term frequency from the top k ties as the expansion terms. Assume there are m documents in the top k ties. The weight of the expansion term q_{exp} and that of the original query term q_{orig} are computed as:

$$weight_{q_{exp}} = \alpha \cdot \frac{\sum_{i=1}^{m} tf_i(q_{exp})}{\sum_{i=1}^{m} dl_i}, \qquad weight_{q_{orig}} = (1 - \alpha) \cdot \frac{qtf(q_{orig})}{query\_length}$$

where the $tf_i(q_{exp})$ denotes the term frequency of q_{exp} in document d_i, the dl_i denotes the length of document d_i, and the $qtf(q_{orig})$ denotes the query term frequency of the original query term q_{orig}. We then combine the original query terms and expanded query terms with the parameter α as a new query to retrieve again using tie breaking method. The α control the weight of the expansion terms. This method is denoted as **TB-PF-TB**.

3.2 Document Expansion for Tie-Breaking

Document expansion is another common way to bridge the vocabulary gap of the document and the query. As the length of the microblogs are limited, users commonly insert a URL to the tweets, which links to the webpage that describes their ideas or expressions. This feature makes it reliable to follow the links as a way of expansion for the original tweets. We propose three methods to utilize the expanded documents in the tie-breaking framework.

Merge with Original Tweets: One straightforward solution is to merge the expanded document with the original tweets. With the merged document, we could apply the **Tie-breaking** method to retrieve the tweets with the same order. We refer this method as **Merged**. It is clear this method is easy to implement. However, the effectiveness of this method relies on the quality of the expanded document. The noisy terms in the expanded document would result in the meaning of the tweet changes after we concatenate the documents.

Tie-Breaking on Two Indexes: One possible way to overcome the limitation of the **Merged** method is to build a separate index for the expanded documents, and then perform tie-breaking on each index respectively. At last we could combine the results from the two indexes into one by using a normalized score or by the ranking. The advantage of this method is that we can control the effect of the expanded document to the original tweets. To be specific, we first applied the **Tie-breaking** on the original tweets index. We then utilize the signals in the order $IDF \oplus TF \oplus DL$ in the expansion index to perform another round of retrieval. Each document will get a score from both original tweets index and expansion index. Formally, assume we have k signals and document D will get a score $S_i(D)$ with i^{th} signal s_i. Then the score for D in either original or expanded index can be computed as:

$$S(D) = \sum_i^k S_i(D) \cdot w^{k-i}$$

where the w is a weight to control the difference of each level. w is set to 100 in our experiments because the relevant score of a document would not be greater than 100. At last, we applied a normalized combine method proposed in [10] to combine the scores from these two indexes. We refer it as **Combined**.

Tie-Breaking with All Signals: Another possible way of combining information from two indexes it to extract retrieval signals from the expanded documents, and then put them together with the signals in the original index to apply **Tie-breaking**. We utilized the TF, IDF, and DL signals from the expanded index as the supplement signals to the original signals. We applied these signals in the order $IDF_{orig} \oplus TF_{exp} \oplus TF_{orig} \oplus IDF_{exp} \oplus NOF \oplus DL_{orig} \oplus DL_{exp}$ based on the preliminary results. This method is denoted as **All**.

Table 3. Performance of tie breaking with different expansion methods

Methods	Training(MB11)		Testing(MB12)	
	MAP	P@30	MAP	P@30
Tie-Breaking	0.3743	0.4204	0.2299	0.3684
TB-PF-TB	0.3811	0.4401	0.2396	0.3844
Merged	0.2957	0.4109	0.1913	0.3557
Combined	**0.4182**‡	**0.4517**‡	**0.2550**‡	0.4069‡
All	0.4028‡	0.4361	0.2532‡	**0.4201**‡

3.3 Experiment Results

We use the same experiment set up as described before and evaluate the effectiveness of the proposed expansion methods. Table 3 summarizes the performances. Similar to the previous experiment, parameters or the order of the signals are trained on MB11 and tested on MB12. The results with ‡ indicate improvement over Tie-breaking is statistically significant at 0.05 level based on Wilcoxon signed-rank test.

The results suggest that the query expansion of tie breaking can improve the performance, however, the improvement is not significant. This is probably caused by the limitation of tie-breaking method, which assumes that every query term is important and may not perform well for long queries such as the expanded ones.

Among three document expansion methods, it is less effective to merge the two indexes. And both **Combined** and **All** can improve the performance significantly. In fact, the performance improvement is similar to the improvement when using expansion methods in traditional retrieval models (e.g. the MAP of traditional query expansion is 0.3692 on MB11 and 0.2196 on MB12, and the MAP of document expansion method is 0.4019 on MB11 and 0.2447 on MB12). Moreover, the performance of the best document expansion method is comparable to the best automatic system in TREC 2011 Microblog track. Unfortunately, the overview paper for TREC 2012 Microblog track is not available, but it seems that our method could be ranked among top 3 groups based on the TREC papers about the Microblog track.

4 Conclusions and Future Work

Tie-breaking is a recently proposed new method for combining retrieval signals. In this paper, we study its potential in microblog retrieval and find that the basic tie-breaking method is more effective than traditional retrieval functions. Moreover, we proposed new methods to incorporate query expansion and document expansion into the tie-breaking framework. We find that the proposed document expansion method is more effective than the proposed query expansion method, which may reveal the limitation of tie-breaking methods for the longer queries. In the future, we plan to study how to incorporate temporal-related signals into the tie-breaking framework. Moreover, we will study how to extend the framework for longer queries.

References

1. Golovchinsky, G., Efron, M.: Making sense of twitter search. In: Proc. of CHI 2010 Workshop on Microblogging: What and How Can We Learn From It? (2010)
2. Choi, J., Croft, W.B., Kim, J.Y.: Quality models for microblog retrieval. In: Proc. of CIKM 2012 (2012)
3. Kumar, N., Carterette, B.: Time based feedback and query expansion for twitter search. In: Serdyukov, P., Braslavski, P., Kuznetsov, S.O., Kamps, J., Rüger, S., Agichtein, E., Segalovich, I., Yilmaz, E. (eds.) ECIR 2013. LNCS, vol. 7814, pp. 734–737. Springer, Heidelberg (2013)
4. Miyanishi, T., Seki, K., Uehara, K.: Combining recency and topic-dependent temporal variation for microblog search. In: Serdyukov, P., Braslavski, P., Kuznetsov, S.O., Kamps, J., Rüger, S., Agichtein, E., Segalovich, I., Yilmaz, E. (eds.) ECIR 2013. LNCS, vol. 7814, pp. 331–343. Springer, Heidelberg (2013)
5. Whiting, S., Klampanos, I.A., Jose, J.M.: Temporal pseudo-relevance feedback in microblog retrieval. In: Baeza-Yates, R., de Vries, A.P., Zaragoza, H., Cambazoglu, B.B., Murdock, V., Lempel, R., Silvestri, F. (eds.) ECIR 2012. LNCS, vol. 7224, pp. 522–526. Springer, Heidelberg (2012)
6. Massoudi, K., Tsagkias, M., de Rijke, M., Weerkamp, W.: Incorporating query expansion and quality indicators in searching microblog posts. In: Clough, P., Foley, C., Gurrin, C., Jones, G.J.F., Kraaij, W., Lee, H., Mudoch, V. (eds.) ECIR 2011. LNCS, vol. 6611, pp. 362–367. Springer, Heidelberg (2011)
7. Choi, J., Croft, W.B.: Temporal models for microblogs. In: Proc. of CIKM 2012 (2012)
8. Wu, H., Fang, H.: Tie breaker: A novel way of combining retrieval signals. In: Proc. of ICTIR 2013 (2013)
9. Han, Z., Li, X., Yang, M., Qi, H., Li, S., Zhao, T.: Hit at trec 2012 microblog track. In: TREC (2012)
10. Egozi, O., Markovitch, S., Gabrilovich, E.: Concept-based information retrieval using explicit semantic analysis. ACM TOIS, 8:1–8:34 (2011)

Efficiently Estimating Retrievability Bias

Colin Wilkie and Leif Azzopardi

University of Glasgow, 18 Lilybank Gardens, G12 8QQ, Glasgow, UK
{colin.wilkie,leif.azzopardi}@glasgow.ac.uk

Abstract. Retrievability is the measure of how easily a document can
be retrieved using a particular retrieval system. The extent to which a
retrieval system favours certain documents over others (as expressed by
their retrievability scores) determines the level of bias the system imposes on a collection. Recently it has been shown that it is possible to
tune a retrieval system by minimising the retrievability bias. However,
to perform such a retrievability analysis often requires posing millions
upon millions of queries. In this paper, we examine how many queries
are needed to obtain a reliable and useful approximation of the retrievability bias imposed by the system, and an estimate of the individual
retrievability of documents in the collection. We find that a reliable estimate of retrievability bias can be obtained, in some cases, with 90% less
queries than are typically used while estimating document retrievability
can be done with up to 60% less queries.

1 Introduction

Retrievability is a document centric evaluation measure which generates an objective score that describes the likelihood for any given document to be retrieved
by a particular retrieval system. It has been applied to a variety of different contexts. A wide range of applications exist including improving recall in retrieval
systems [5,6], improving the effectiveness of pseudo-relevance feedback [4], detecting bias towards particular organisations by search engines [1] and even relating retrievability to performance [8]. In this paper we choose not to focus on
the application of retrievability but rather, the investigation of how to accurately
estimate retrievability. We focus on estimating retrievability because in trying
to establish a complete picture of retrievability we must issue millions of queries.
This obviously, is a very time consuming and computationally intensive process.

While it is likely that a large number of queries is necessary to gain an accurate measure of the retrievability of individual documents, it may be possible
to use a reduced set of queries in order to estimate the relative retrievability
bias that the system imposes across the collection. In this paper we shall investigate the number of queries required to produce a reasonable/useful estimate
of retrievability bias and of document retrievability.

2 Background

The concept of the document-centric evaluation measure was first introduced
by Azzopardi and Vinay which they branded Retrievability [2]. This measure

M. de Rijke et al. (Eds.): ECIR 2014, LNCS 8416, pp. 720–726, 2014.
© Springer International Publishing Switzerland 2014

evaluates how likely a document is to be retrieved by a particular configuration of an IR system given the universe of potential queries. The retrievability r of a document d with respect to an IR system can be defined as:

$$r(d) \propto \sum_{q \in Q} O_q . f(k_{dq}, \{c, g\}) \tag{1}$$

where q is a query from the universe of queries Q, meaning O_q is the probability of a query being chosen. k_{dq} is the rank at which d is retrieved given q and $f(k_{dq}, \{c, g\})$ is an access function denoting how retrievable d is given q at rank cut-off c with discount factor g. To calculate retrievability, we sum the $r(d)$ across all q's in the query set Q. Obviously, it is impractical to launch every query in the universe of possible queries, as such, it is common to use a very large set of queries instead. This query set is often automatically generated bigrams [2,8]. The intuition is then the more queries that can retrieve d before the rank cut-off, the more retrievable d is. Calculating retrievability can then be performed using a number of different user models. The simplest model being a cumulative scoring model. In the cumulative model, we employ the access function $f(k_{dq}, c)$ such that $f(k_{dq}, c) = 1$ if d is retrieved in the top c documents given q otherwise $f(k_{dq}, c) = 0$. In other words, this model accumulates a score for a document so long as it is retrieved before the specified cut-off, while documents appearing after the cut-off are ignored completely which simulates a user who is willing to rigorously look at all documents up until a set point. A more complex and realistic retrievability model exists, this gravity based scoring model applies a weighting to the documents position in the ranked list, meaning as the rank approaches the cut-off, the documents contribute less score. Formally defined as $f(k_{dq}, g) = 1/k_{dq}^g$) where g is a discount factor that defines the magnitude of the penalty applied to a document given its rank position, increasing as we traverse down the ranked list. The intuition behind this model being that a document further down a ranked list is less retrievable as the users interest or attention diminishes. Therefore, a document appearing at rank 1 contributes substantially more $r(d)$ than a document at rank 10 when $c > 10$. Again, if a document appears after cut-off c, that document contributes $r(d) = 0$.

Retrievability Bias

The bias that a system imposes on the document collection can be determined by examining the distribution of $r(d)$ scores. Here, bias denotes the inequality between documents in terms of their retrievability within the collection. It is possible to visually assess the inequality / bias by plotting a Lorenz Curve [7]. In Economics and the Social Sciences, the Lorenz Curve is used to visualise the inequality in a population given their incomes.

To estimate bias we treat the retrievability of a document as its wealth. Therefore, using the Gini coefficient, we can see how skewed the distribution of retrievability is towards a particular set of documents.

In the context of retrievability, if all documents were equally retrievable then the Gini coefficient would be zero (denoting equality within the population).

On the other hand if only one document was retrievable and the rest were not then the Gini coefficient would be one (denoting total inequality). Usually, documents have some level of retrievability and thus the Gini coefficient is somewhere between one and zero. Many factors affect the retrievability bias, these include: the retrieval model/system, the parameter settings, the indexing process, the documents and collection representations/statistics, as well as how the system is used by the user (i.e. the types of queries and the number of documents that they are willing to examine). Obviously, the Gini coefficient is only a overview of the presence of bias and as such it is important to generate accurate $r(d)$ scores to examine *where* the bias lies.

Estimating Retrievability

For a given retrieval system configuration (i.e. retrieval model and parameter setting), the estimation process consists of three parts: (1) the generation and selection of a sufficiently large query set (2) issuing the queries to system and (3) calculating the retrievability of the documents, and the retrievability bias of the system.

In [2], it was pointed out that in order to obtain an exact estimate of retrievability the universe of all queries would need to be issued. However, because that is not feasible, typically, a very large set of queries is used instead (either single term queries or a subset of two word queries). Most papers that perform a retrievability analysis use two or three word queries and issue anywhere between 200,000 to 2,000,000 queries [2,6,8].

This, is obviously a very time consuming process requiring a huge amount of resources to complete. Additionally, for each configuration this would have to be repeated.

With the increasing usage of the retrievability evaluation measure, it is important to look to optimise the process to reduce the amount of time and resources required. Doing so makes retrievability a more accessbile measure and improves its viability.

One attempt has been made to improve the efficiency of retrievability. Rather than directly estimating retrievability scores for documents Bashir proposed a method for the estimation of retrievability employing techniques from machine learning [3]. Bashir extracted document features, such as normalised average term frequency, number of frequent terms, average document frequency, etc., to then estimate how likely that document is to be retrieved without resorting to posing any queries to a retrieval system. This method of determining retrievability bias is obviously far more efficient as it goes on document rankings rather than document retrievability scores. The disadvantage is that we can only gain an insight to bias but cannot understand exactly where the bias lies in terms of which individual documents are more or less favoured than others.

3   Experimental Method

We propose a set of experiments to answer some important research questions given the hypothesis; a lower bound exists, such that increasing the number of queries issued to the system provides no additional insights in terms of bias or document retrievability. Our research questions are thus:

1. Do similar trends occur when smaller sets of queries are used?
2. How many queries are needed to gain a comparable approximation of Gini?
3. How correlated are the $r(d)$ scores between varying numbers of queries?

We are ultimately interested in minimizing the amount of processing required to calculate retrievability. While previous work has managed to achieve relatively good rankings of document retrievability, the only method for effectively estimating individual document $r(d)$ remains to be in launching these huge query sets. We investigate whether we can achieve similar Gini coefficients on different sized query sets. However, it is entirely possible for similar Gini coefficients to be calculated from very differently biased sets. Therefore, we need to explore how correlated the individual document r(d) scores are to fully understand if and where this lower bound exists.

Data and Materials

Two TREC test collections: Aquaint (AQ) and Trec123 (T123) were used in these initial experiments. Each collection contains approximately one million documents and the query sets used were created by extracting bigrams from the collection and selecting those bigrams which occurred at least 20 times, giving about a quarter of a million queries for each collection. We employed four retrieval models: BM25, PL2, DPH and TF-IDF. We considered various

Table 1. AQ (Top) & T123 (Bottom): Tables of correlations between all queries and the percentage of queries stated. All correlations were found to be significant at $p < 0.05$.

AQ								
	10%	20%	30%	40%	50%	60%	70%	80%	90%
BM25	0.68	0.76	0.84	0.88	0.93	0.95	0.97	0.98	0.99
PL2	0.70	0.78	0.85	0.89	0.93	0.95	0.97	0.98	0.99
DPH	0.68	0.76	0.84	0.88	0.93	0.95	0.97	0.98	0.99
TF-IDF	0.76	0.87	0.92	0.94	0.96	0.98	0.98	0.99	1.00

T123								
	10%	20%	30%	40%	50%	60%	70%	80%	90%
BM25	0.69	0.79	0.86	0.90	0.92	0.95	0.96	0.98	0.99
PL2	0.74	0.82	0.88	0.91	0.93	0.95	0.96	0.98	0.99
DPH	0.92	0.95	0.97	0.98	0.99	0.99	0.99	1.00	1.00
TF-IDF	0.97	0.99	0.99	1.00	1.00	1.00	1.00	1.00	1.00

parameter settings for BM25 and PL2, where we performed a parameter sweep ($b = [0, 1]$ and $c = [0.1, 1, 2, \ldots, 10, 100]$) to determine which parameter value minimises the Gini coefficient as done in [8].

In [2,8] it has been observed that BM25 is the least biased model, while TF-IDF is the most biased (overly favouring long documents). We included these models in our experiments to compare how many queries are needed to find a stable estimation for different quantities of bias in models (i.e. does a more bias model require more or less queries). We also choose PL2 to compliment BM25 as these both have adjustable parameters for length normalisation. We can then investigate whether the parameter that exhibits minimum Gini changes with the number of queries issued.

4   Results and Analysis

Do similar trends occur when smaller sets of queries are used? The plots in Figure 1 show how Gini changes as the b and c parameters are varied for BM25 (left) and PL2 (right), respectively. This is shown for different numbers of queries. It is clear that the same trend is present regardless of the number of queries used. These results shows that it is possible to use substantially less queries to find the parameter value that minimise Gini. In [8], this point was correlated with high retrieval performance.

How many queries are needed to gain a comparable approximation of Gini: Figure 1 shows that regardless of how many queries are issued, the parameter setting that achieves the minimum point of Gini is found to be the same (0.7 for AQ and 0.8 for T123 on BM25). This result indicates that we can significantly reduce the amount of queries used (by up to 90% in this case) and the parameter setting at which minimum Gini occurs does not change. This means we can estimate the parameter for minimum Gini reliably with 90% less queries. These results hold for both collections used but also for PL2 where the minimum point of Gini occurs at 2 for AQ and T123 for all query sets.

Examining the plots of Figure 2 we see how Gini changes as more queries are issued on the 4 models (BM25, PL2 DPH and TF-IDF). In these plots, the minimum point of Gini discovered for BM25 and PL2 in the previous plots are used here. The first point we note is, the extremely biased model (TF-IDF) shows very little change to the Gini as the number of queries increase. This tells us that a heavily biased model reaches a stable estimation of bias in very few queries and using extremely large query sets is not necessary. The information in Table 1 backs up this claim showing that on T123, the results can converge within as few as 40% of the query set.

How correlated are the $r(d)$ scores between varying numbers of queries: Looking at the less biased models presented here we see a trend develop. If too few queries are issued (Less than 40%) the estimation of bias is not particularly accurate and provides a more biased picture than larger numbers. However between 40% and 80%, the estimation is fairly stable, giving the impression that

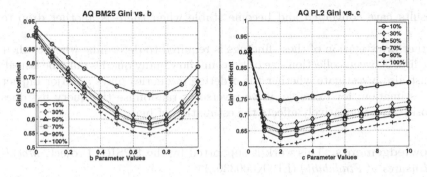

Fig. 1. AQ BM25 (Left) & PL2 (Right): Gini vs. parameter setting on BM25 and PL2

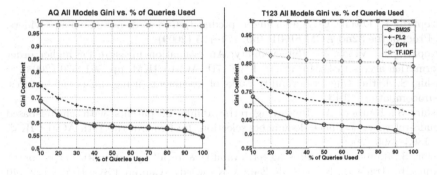

Fig. 2. AQ (Left) & T123 (Right): Gini vs. Percentage of queries used

the results have converged to a stable point but if we continue to issue more queries we see another substantial drop (> 80%) in bias. This makes estimation of bias difficult as there is often an area where it appears enough queries have been launched but an accurate estimate has not yet been reached. Table 1 shows the correlations between the volumes of queries and when all queries have been sent. We can see that high correlations exist between 40% of the queries upwards suggesting this final dip in bias is not substantially different from the stable points of previous amounts of queries.

5   Summary and Conclusions

In this paper we investigated whether or not there existed a minimum amount of queries to estimate retrievability to a highly accurate degree. We found that this minimum amount is largely dependant on the retrieval model employed. Mainly the more biased a model is, fewer queries need to be issued to reach a stable estimate of retrievability. When a model is known to be fairer we see that estimating this lower bound is very difficult as plateaus exist where it appears the results have converged and become stable when there is another drop to come.

Further investigation of these findings is required to determine whether their is some link between collection size or type and how many queries must be issued. We must also investigate whether the ordering of queries plays a major impact in the estimation; for example, if we reverse or randomise the query list does this positively or negatively affects the estimation of bias.

Acknowledgements. This work is supported by the EPSRC Project, *Models and Measures of Findability* (EP/K000330/1).

References

1. Azzopardi, L., Owens, C.: Search engine predilection towards news media providers. In: Proc. of the 32nd ACM SIGIR, pp. 774–775 (2009)
2. Azzopardi, L., Vinay, V.: Retrievability: An evaluation measure for higher order information access tasks. In: Proc. of the 17th ACM CIKM, pp. 561–570 (2008)
3. Bashir, S.: Estimating retrievability ranks of documents using document features. Neurocomputing (2013)
4. Bashir, S., Rauber, A.: Improving retrievability of patents with cluster-based pseudo-relevance feedback documents selection. In: Proc. of the 18th ACM CIKM, pp. 1863–1866 (2009)
5. Bashir, S., Rauber, A.: Improving retrievability & recall by automatic corpus partitioning. In: Trans. on Large-Scale Data & Knowledge-Centered Sys. II, pp. 122–140 (2010)
6. Bashir, S., Rauber, A.: Improving retrievability of patents in prior-art search. In: Gurrin, C., He, Y., Kazai, G., Kruschwitz, U., Little, S., Roelleke, T., Rüger, S., van Rijsbergen, K. (eds.) ECIR 2010. LNCS, vol. 5993, pp. 457–470. Springer, Heidelberg (2010)
7. Gastwirth, J.L.: The estimation of the lorenz curve and gini index. The Review of Economics and Statistics 54, 306–316 (1972)
8. Wilkie, C., Azzopardi, L.: Relating retrievability, performance and length. In: Proc. of the 36th ACM SIGIR Conference, SIGIR 2013, pp. 937–940 (2013)

A Language Modeling Approach to Personalized Search Based on Users' Microblog Behavior

Arjumand Younus[1,2], Colm O'Riordan[1], and Gabriella Pasi[2]

[1] Computational Intelligence Research Group, Information Technology,
National University of Ireland, Galway, Ireland
[2] Information Retrieval Lab, Informatics, Systems and Communication,
University of Milan Bicocca, Milan, Italy
{arjumand.younus,colm.oriordan}@nuigalway.ie, pasi@disco.unimib.it

Abstract. Personalized Web search offers a promising solution to the task of user-tailored information-seeking, and particularly in cases where the same query may represent diverse information needs. A significant component of any Web search personalization model is the means with which to model a user's interests and preferences to build what is termed as a *user profile*. This work explores the use of the Twitter microblog network as a source of *user profile* construction for Web search personalization. We propose a statistical language modeling approach taking into account various features of a user's Twitter network. The richness of the Web search personalization model leads to significant performance improvements in retrieval accuracy. Furthermore, the model is extended to include a similarity measure which further improves search engine performance.

1 Introduction and Related Work

Search engine users have diverse information needs, and it often happens that different users expect different answers to the same query [6]. In fact, given the potential of the same query to be representative of different information needs behind it, personalized Web search has emerged as a promising solution to better identify the intended information need The usual approach to the personalization process in Web search involves incorporating user's preferences into the retrieval method of the search system thereby moving from a "one size fits all" approach to the customization of search results for people with different information interests and goals.

A significant research challenge in Web search personalization is to learn about a user's interests and preferences to build what is termed as a *user profile*. The user profile is the most essential resource within the retrieval model of a personalized search system. One of the main features that can be used to differentiate between existing solutions to Web search personalization is the source used when building the user profile. Several kinds of sources have been explored by researchers in order to build a user profile, with the most popular being search and browsing histories [3,5]. However, the use of such history data may not be

M. de Rijke et al. (Eds.): ECIR 2014, LNCS 8416, pp. 727–732, 2014.

feasible given users' privacy considerations that can limit the availability of the data. Furthermore, history data are more prone to noise as previous interactions with the search system are not necessarily reflective of user needs [8]. This paper proposes microblogs as an alternative information source to build a rich user profile.

The proliferation of Web 2.0 services has created a new form of user collaboration where users engage within a social network while at the same time generating their own content, popularly known as user-generated content. Microblogs such as Twitter[1] are an immensely popular forum for such collaboration, and we show how this forum can serve as a source of information about users' preferences for the Web search personalization process. Earlier research efforts that aimed to exploit information from online social systems for personalized search rely mostly on social bookmarking and tagging systems [4,7]. However, Younus et al. [9] revealed in a user-survey based study a very low usage of social bookmarking sites as compared to other social networking tools.

A few works have considered Twitter as a source of user profile construction [1]; however, these works do not take into account features of a user's microblog network. We undertake such a direction in this work and propose a statistical language modeling approach to infer a user's profile; the proposed technique takes into account various features of a user's Twitter network thereby providing a rich model of user preferences and interests. We evaluate the proposed methods by means of a user-study and we show that retrieval performance substantially improves when using microblog behavior as a source of information about user preferences and interests for Web search personalization.

2 Methodology

This section describes the proposed personalization model in detail. We follow a strategy in which non-personalized search results returned from a search system are re-ranked with the help of the user profile to return results that are more relevant to the user [5].

2.1 Microblog Behavior Based Language Model

We adopt a statistical language model to model various aspects of Twitter behavior. Using this model, we then define our re-ranking approach.

We incorporate the *mention* and *retweet* features of Twitter within our model with the underlying intuition that those Twitterers a particular user mentions or retweets reflect, to a large extent, the user's own preferences and interests.

For the re-ranking step, we use a language modeling approach to compute the likelihood of generating a document d from a language model estimated from a user's Twitter model as follows:

$$P(u)_{lm}(d/T) = \sum_{w \in W} P(w \mid T)^{n(w,d)} \qquad (1)$$

---

[1] http://twitter.com

where w is a word in the title and snippet of a document returned by a search system (i.e., d), W the set of all the words in the title and snippet of document d, $n(w,d)$ the term frequency of w in d, and u is the user for whom we want to personalize Web search results. Here, T is used to represent the uniform mixture of the user's Twitter model as follows:

$$P(w \mid T) = \lambda_o * P(w \mid T_o) + \lambda_m * P(w \mid T_{U_m}) + \lambda_r * P(w \mid T_{U_r}) \qquad (2)$$

Let T_o denote the original tweets by the user u, T_{U_m} denotes the tweets by those Twitterers whom the user u mentions (i.e., Twitterers in set U_m) and T_{U_r} denotes the tweets by those Twitterers whom the user u retweets (i.e., Twitterers in set U_r). The individual Twitter models can be estimated as:

$$P(w \mid T_o) = \frac{1}{|T_o|} \sum_{t \in T_o} P(w \mid t) \qquad (3)$$

$$P(w \mid T_m) = \frac{1}{|U_m|} \sum_{u \in U_m} \frac{1}{|T_{u_m}|} \sum_{t \in T_{u_m}} P(w \mid t) \qquad (4)$$

$$P(w \mid T_r) = \frac{1}{|U_r|} \sum_{u \in U_r} \frac{1}{|T_{u_r}|} \sum_{t \in T_{u_r}} P(w \mid t) \qquad (5)$$

i.e., a single user's Twitter model is estimated by a mixture of his own tweets, those Twitterer's tweets whom the user mentions and those Twitterers' tweets whom the user retweets. The constituent language models for T_o, T_{U_m} and T_{U_r} are a uniform mixture of their tweets' language models employing Dirichlet prior smoothing:

$$P(w \mid t) = \frac{n(w,t) + \mu \dfrac{n(w, coll)}{|coll|}}{|t| \ | \ \mu}$$

where $n(w,.)$ denotes the frequency of word w in (.), *coll* is short for collection which refers to all tweets by user u (in case of equation (3)), all tweets by Twitterers in set U_m (in case of equation (4)) and all tweets by Twitterers in set U_r (in case of equation (5)), and $|.|$ is the overall length of the tweet or the collection.

Finally, after estimation of a user's Twitter model (using equations 2-5) we use equation (1) to re-rank the documents returned by a search system and hence, present personalized search results to the user u.

2.2 Similarity Measure between Users

In the previous section, we defined U_m as the set of users mentioned by u and U_r as the set of users whose tweets were retweeted by user u. We refine the definition of these sets to only include those users who have a sufficient similarity to the user

u. We present a network-based similarity measure which we use to decide whether or not to include a particular user in U_m and U_r. The underlying intuition behind the use of such a similarity measure is to exclude those Twitterers from the user's Twitter model who do not provide a strong indication of the user's preferences.

We calculate the similarity between the current user u and each user u_i occurring in either U_m or U_r based on the heuristic that the more people u_i follows in these sets, the more likely that user's interests overlap with the user u. Furthermore, we normalise this score by the total number of users, user u_i follows. We use the following formula to calculate the similarity score between user u and a user $u_i \in U_m$.

$$Sim(u, u_i) = \frac{follow(u_i) \cap U_m}{follow(u_i)}$$

where $follow(u_i)$ is the set of users followed by u_i.

We also calculate similarity for all users in U_r using the same approach. Finally, we retain those Twitterers in U_m and $U_r{}^2$ whose network similarity measure is above a certain threshold.

3   Experimental Evaluations

In this section we describe our experimental evaluations that demonstrate the effectiveness of our proposed approach.

1. We wish to check whether personalization through a Twitter-based user profile improves search quality over the underlying non-personalized search engine.
2. We wish to evaluate the effect of the network based similarity of section 2.2 in an attempt to study the usefulness of incorporating microblog characteristics when building a user profile for Web search personalization.

3.1   Experimental Setup

We recruited 14 active Twitter users and used their Twitter data for the purpose of experimental evaluations. We obtained the search queries, their corresponding relevance judgements and underlying corpus (i.e., search documents' collection) from a publicly available dataset called *"CiteData"* by Harpale et al. [2]. As mentioned by Harpale et al., the dataset is useful for benchmark evaluations of personalized search performance. *CiteData* comprises 81,432 academic articles and 41 queries; we asked each user who participated in our user-study to select a subset of the queries that were similar to a search query that he/she had issued at some point. Note that since the dataset comprises academic articles we recruited Twitter users who are academics with specific, personalized information needs

---

[2] These are used as part of equation (4) and equation (5) for estimation of the user's Twitter model.

for academic articles. Each user was asked to select 10 queries from the 41 queries of the dataset; of these we selected the queries that had been selected by at least three users which amounted to a total of eight unique queries. We then asked each user to mark as relevant or irrelevant 20 documents per query; we obtain these 20 documents using a BM25 non-personalized search algorithm. Note that each user in our study was asked to mark 20 documents across the queries they selected from the short-listed eight queries. Finally, we calculate the Cohen's kappa across the relevance judgements for the eight short-listed queries; for the purpose of calculating Cohen's kappa we used the relevance judgements by the graduate students of Harpale et al.'s study and the relevance judgements by the users in our study. We obtain an average Cohen's kappa value of 0.86 across all queries and all users reflecting the high reliability of the *CiteData* dataset. We perform this step of measuring inter-annotator agreement via Cohen's kappa to ensure the agreement in relevance judgements between different sets of users in the two studies (i.e., the study by Harpale et al. and ours).

Table 1. Comparison of Retrieval Performance for our Proposed Personalization Model

Chosen Algo	Measures	
	MAP	$P@10$
np	0.389	0.567
p_{ns}	0.451	0.598
p_s	0.487	0.634

3.2 Experimental Results

Once we ensure reliability of the underlying dataset and relevance judgements through the method explained in section 3.1, we evaluate the performance of our proposed personalization model using the relevance judgements of the *CiteData* dataset. As evaluation metrics, we use mean average precision (MAP) and precision at top 10 documents (P@10) which respectively measure the systems overall retrieval accuracy and its performance for those documents that are most viewed. Table 1 shows the experimental results i.e. MAP and P@10 values for the non-personalized approach (denoted as np), our personalized approach without the similarity measure of section 2.2 (denoted as p_{ns}) and our personalized approach with the similarity measure of section 2.2 (denoted as p_s)[3]. The network similarity threshold was chosen following empirical analysis; for each user a threshold equivalent to half of the maximum similarity score was found sufficient to gather a significant amount of similar users in U_m and U_r. Moreover, the parameters λ_o, λ_m, and λ_r are assigned uniform weights. The baseline non-personalized search system uses a language model approach with Dirichlet smoothing. We report the results together across the queries and judgements for all 14 users.

---

[3] We use student's t-test to verify the soundness of our evaluations and the results corresponding to p_{ns} and p_s are statistically significant with $p < 0.05$.

The results show clearly, the benefits of using Twitter data to personalize search results for users. The MAP and P@10 scores for the personalized results (p_{ns} and p_s) are superior to those achieved without personalization. Furthermore, we witness improved performance when only those Twitterers who show similarity to the active user are used in generating the model.

4 Conclusions and Future Work

The main conclusion is that exploiting evidence available from a person's microblog behaviour to allow personalization can improve the accuracy of a system. We adopt a language modeling approach and show that including similar users from the Twitter network provides the best performance. Future work will involve further analysis of the results and explore other similarity measures and sources of evidence from a user's microblog behaviour and network. We also aim to merge these sources of evidence with data available in the query and about the user's current task at hand.

References

1. Abel, F., Gao, Q., Houben, G.-J., Tao, K.: Semantic enrichment of twitter posts for user profile construction on the social web. In: Antoniou, G., Grobelnik, M., Simperl, E., Parsia, B., Plexousakis, D., De Leenheer, P., Pan, J. (eds.) ESWC 2011, Part II. LNCS, vol. 6644, pp. 375–389. Springer, Heidelberg (2011)
2. Harpale, A., Yang, Y., Gopal, S., He, D., Yue, Z.: Citedata: a new multifaceted dataset for evaluating personalized search performance. In: CIKM 2010, pp. 549–558 (2010)
3. Matthijs, N., Radlinski, F.: Personalizing web search using long term browsing history. In: WSDM 2011, pp. 25–34 (2011)
4. Noll, M.G., Meinel, C.: Web search personalization via social bookmarking and tagging. In: Aberer, K., Choi, K.-S., Noy, N., Allemang, D., Lee, K.-I., Nixon, L.J.B., Golbeck, J., Mika, P., Maynard, D., Mizoguchi, R., Schreiber, G., Cudré-Mauroux, P. (eds.) ASWC 2007 and ISWC 2007. LNCS, vol. 4825, pp. 367–380. Springer, Heidelberg (2007)
5. Tan, B., Shen, X., Zhai, C.: Mining long-term search history to improve search accuracy. In: KDD 2006, pp. 718–723 (2006)
6. Teevan, J., Dumais, S.T., Horvitz, E.: Potential for personalization. ACM Trans. Comput.-Hum. Interact. 17(1), 4:1–4:31 (2010)
7. Vallet, D., Cantador, I., Jose, J.M.: Personalizing web search with folksonomy-based user and document profiles. In: Gurrin, C., He, Y., Kazai, G., Kruschwitz, U., Little, S., Roelleke, T., Rüger, S., van Rijsbergen, K. (eds.) ECIR 2010. LNCS, vol. 5993, pp. 420–431. Springer, Heidelberg (2010)
8. Wang, Q., Jin, H.: Exploring online social activities for adaptive search personalization. In: CIKM 2010, pp. 999–1008 (2010)
9. Younus, A., O'Riordan, C., Pasi, G.: Predictors of users' willingness to personalize web search. In: Larsen, H.L., Martin-Bautista, M.J., Vila, M.A., Andreasen, T., Christiansen, H. (eds.) FQAS 2013. LNCS, vol. 8132, pp. 459–470. Springer, Heidelberg (2013)

Using Hand Gestures for Specifying Motion Queries in Sketch-Based Video Retrieval

Ihab Al Kabary and Heiko Schuldt

Databases and Information Systems Group
Department of Mathematics and Computer Science
University of Basel, Switzerland
{ihab.alkabary,heiko.schuldt}@unibas.ch

Abstract. In team sports, the analysis of a teams tactical behavior is becoming increasingly important. While this is still mostly based on the manual selection of video sequences from games, coaches and analysts increasingly demand more automated solutions to search for relevant sequences of videos, and to support this search by means of easy-to-use interfaces. In this paper, we present a novel intuitive interface for specifying sketched-based motion queries in sport videos using hand gestures. We have built the interface on top of *SportSense*, a system for interactive sports video retrieval. *SportSense* exploits spatiotemporal information incorporating various events within sport games, which is used as metadata to the actual sports videos. The interface has been designed to enable users to fully control the system and facilitate acquiring of the query object needed to perform both spatial and spatiotemporal motion queries using intuitive hand gestures.

Keywords: user interface, query-by-sketch, motion queries, gesture recognition, video retrieval.

1 Introduction

When specifying the specific tactical behavior of a sports team, coaches more and more rely on the results of an activity that is called game analysis. Currently, this is mainly a manual and tedious task as it involves the selection of relevant video snippets from sports videos based on the movements of players and the ball. However, novel technology such as on-the field cameras specifically deployed to assist in providing tracking information or using recently available light-weight wireless sensor devices explicitly designed for the sports domain allow to automatically gather this information. We have developed a novel system, named *SportSense* [1,2], that enables sketch-based motion queries in sports videos by harnessing information contained within an overlay of metadata that incorporates spatiotemporal information about various events within a game. Users can freely sketch a path showing the movement of a single player and/or the ball or a sequence of consecutive movements of one or several interacting players. Similarity search is performed and results are displayed in the form of

M. de Rijke et al. (Eds.): ECIR 2014, LNCS 8416, pp. 733–736, 2014.

Fig. 1. Screenshot of *SportSense*

video snippets ordered according to the degree in which they match the sketched motion path. Users can also interact with the system by retrospectively selecting a cascade of events, along with sketches of regions in which they happened. *SportSense* uses spatial databases to store tracking metadata in order to leverage the spatial query features needed for providing interactive response times. In this paper, we present a novel interactive interface that allows users to perform spatial as well as spatiotemporal sketch-based motion queries using hand gestures. The interface also allows users to fully control SportSense by setting query parameters, initiating the search process and browsing the results.

2   SportSense and the "Air" Interface

In the following, we present the "air" interface of *SportSense* that provides a novel intuitive user interaction for the specification of sketch-based motion queries in sports videos.

2.1   Query Formulation and Execution

The main two tasks that are performed in *SportSense* are: (i) the formulation of the query from the user-drawn sketch, (ii) this is followed by the execution of the query against the stored spatiotemporal data with the application of similarity search techniques in order to retrieve an ordered list of matching and partially matching video snippets. The system is depicted in Figure 1. More details on the building blocks of *SportSense* can be found in [1].

2.2   Hand Gesture Interaction

In order to provide interactive hand gesture interaction, SportSense exploits the recently launched LeapMotion [3] controller. It is a hardware sensor device that tracks one or several fingers (or similar items such as a pen) and converts their position(s) to coordinate input. The LeapMotion controller is analogous to a mouse, but does not require any hand contact and provides 3D instead of 2D coordinates as input. It is also able to project finger directions towards the screen to provide 2D coordinates. It uses two monochromatic IR cameras and three infrared LEDs to observe action within a hemispherical space to a distance of about one meter with reported spatial precision of about 0.01 mm. The controller recognizes the four gestures: swipe, circular, key tap and screen tap. Moreover, it provides an SDK that allows developers to implement additional gesture recognition methods using the tracking information the device captures.

Sketch-Based Query Object. In order to enable the user to sketch a spatial query or a spatiotemporal motion query by just pointing towards the screen, we need to differentiate between when users are moving their hands to reach where they want to initiate creating the query object, and when they want to practically initiate a the query object. The same applies to when they want to end the creation of the query object. After experimenting with various options, we found that the most intuitive way is for the user to initiate an "air" click event to start creating the query object, followed by moving the finger to define the area, then either returning to the start area (within a circle of radius of 25 pixel and origin being the starting point) to complete sketching the area for the spatial query or initiating another click to define the end of the motion path. Completing the query object simultaneously initiates the similarity search process. For recognizing an "air" click, the LeapMotion controller SDK already provides gesture recognition for the screen tap, but it had proven to not be very accurate in the context of drawing sketches. This is due to fact that the gesture involves having to move the tip of your finger or pen towards the screen and this slightly changes the projection of where you are pointing to, giving unintended deformations. We have developed a more intuitive Point and Focus gesture that initiates a click event when the user points and stays focused on the screen for 1.5 seconds within a circle of 25 pixel radius.

Hand Gesture Controlled GUI. The GUI contains a set of controls that allows the user to decide on which type of query (spatial or spatiotemporal) to run and another set of controls to set the query parameter settings. For Button controls, we use the same Point and Focus gesture to trigger a click event. For the Spin number selection control used to select a number from a range used for setting values such as the allowed tolerated variance of the motion query object from the exact sketch path, we initially use the Point and Focus gesture to select the control and then follow that with Set Distance gesture to define the value set to the control. We designed the Set Distance gesture to be the distance between the thumb and index fingers when placed in a near horizontal alignment.

Fig. 2. The point and focus, the set distance and the (built -in) circular gestures

The LeapMotion controller does not see fingers when placed on top of each other due to the positioning of the IR cameras below the hands. The built-in circular gesture is used to start over and clear the interface in order to initiate a new query. The hand gestures used in *SportSense* are portrayed in Figure 2.

2.3 SportSense's Air Interface in Action

For evaluating the air interface, and in particular for performing sketch-based video retrieval in a sport game using hand gestures, we have set up a test collection based on the the Manchester City Football Club (MCFC) analytics dataset [4]. This data set includes annotated metadata for events happening within a football game between Manchester City FC and Bolton Wanderers FC. Using the hand gestures provided by the air interface, users are able to capture the query object, set the query parameters, and browse the results.

3   Conclusion

We have presented a novel and intuitive user interface for enabling sketch-based video retrieval in sport videos. This interface uses hand gestures to capture the query objects, set the query parameters, and browse the resulting video snippets.

Acknowledgments. This work was supported by the Swiss National Science Foundation (Project MM-DocTable, contract no. 200020-137944/1).

References

1. Al Kabary, I., Schuldt, H.: Towards Sketch-based Motion Queries in Sports Videos. In: Proceedings of the 15th IEEE International Symposium on Multimedia (ISM 2013). IEEE, USA (2013)
2. Al Kabary, I., Schuldt, H.: SportSense: Using Motion Queries to Find Scenes in Sports Videos. In: Proceedings of the 22nd ACM International Conference on Information and Knowledge Management (CIKM 2013), pp. 2489–2492. ACM, USA (2013)
3. LeapMotion, http://www.leapmotion.com
4. ManCity Football Club, http://www.mcfc.co.uk

Page Retrievability Calculator

Leif Azzopardi, Rosanne English, Colin Wilkie, and David Maxwell

School of Computing Science
University of Glasgow, Scotland, UK
{leif.azzopardi,rosanne.english,colin.wilkie}@glasgow.ac.uk,
d.maxwell.1@research.glasgow.ac.uk

Abstract. Knowing how easily pages within a website can be retrieved using the site's search functionality provides crucial information to the site designer. If the system is not retrieving particular pages then the system or information may need to be changed to ensure that visitors to the site have the best chance of finding the relevant information. In this demo paper, we present a Page Retrievability Calculator, which estimates the retrievability of a page for a given search engine. To estimate the retrievability, instead of posing all possible queries, we focus on issuing only those likely to retrieve the page and use them to obtain an accurate approximation. We can also rank the queries associated with the page to show the site designer what queries are most likely to retrieve the pages and at what rank. With this application we can now explore how it might be possible to improve the site or content to improve the retrievability.

1 Introduction

Information Architects and Site Designers uses an array of tools to design and evaluate websites. Most of these tools are qualitative in nature, for example card sorting exercises, heuristic evaluations, and usability studies [6]. However there have been a number of attempts to develop more quantitative measures to help in designing and evaluating websites in terms of how findable they make content. For example, in [5], the authors used Information Foraging Theory to build information scent models that predicted how users would interaction with a website, while in [9,10] the authors developed measures of Navigability using Markov-Models to predict where users on a website would end up. In this paper, rather than considering how people navigate through a website, we consider how people search for information within a website and how the Retrievability of pages can be estimated and used to help Information Architects and Site Designers improve the findability of content.

Retrievability has been defined as the ease with which a document can be found using a retrieval system [2]. Retrievability has been used in a number of application areas within Information Retrieval. For example it has been used to detect bias within retrieval models [2] and search engines [1], tune retrieval models [8], analyse collections [2,4], and to improve retrieval performance [3,7]. In this demo paper we set out to develop another use for retrievability where we wish to examine the retrievability of individual pages and determine the queries that retrieve those pages.

M. de Rijke et al. (Eds.): ECIR 2014, LNCS 8416, pp. 737–741, 2014.

2 System and Method Design

To this end we have designed a command line utility that computes the retrievability of a specified URL for different search engines (called the Page Retrievability Calculator, or PRC). To compute the retrievability of a page a set of queries is first extracted from the page, or part of the page. The queries are then issued to the specified search engine. If the page is retrieved by the search engine the query receives a score based on its rank (either cumulative or gravity based, see [2]). This allows us to rank the queries according to how much they contribute to the overall retrievability of a the page. The total retrievability is computed as the sum of all the query scores. As part of the scoring process a number of components can be varied such as the search engine, the part of the page to extract query terms from, and how to select queries.

2.1 System Design

The main components of the system and how they relate to each other is described below, the source code is available on GitHub at https://github.com/leifos/ifind:

- **Page Fetcher.** This is represented by the Page Capture class. It is responsible for loading a webpage using PhantomJS and selenium. It allows us to capture the HTML of a webpage.
- **Text Extractor.** This is represented by the Position Content Extractor. This class is responsible for reading in html and removing or extracting content of divs with given ids. It is also responsible for getting a subset of the content of the html.
- **Query Extractor.** This is represented by a superclass for query generation which is responsible for extracting queries from html or text. This involves cleaning the text by a pipeline which removes features like punctuation, special characters, stop words etc. This class also calculates the number of occurrences of each term in the document for use by the query selector. There are currently three subclasses which generate single term queries, biterm queries, and 3-term queries. The biterms are generated by pairing words which are next to each other. The 3-terms are generated by grouping three terms which are next to each other.
- **Query Selector.** The query selector is responsible for calculating the probability of each query given the probability of the document and the collection. It then ranks the queries given these probabilities from most to least probable. It is then possible to get the top n queries, so that only the queries most likely to retrieve the page can be issued (instead of all of them).
- **Page Calculator.** This is responsible for calculating the score of a page given a list of queries. It generates query objects and issues them to the search engine, noting the result and calculating the cumulative and gravity based scores. It also provides a report which presents a summary of the results such as the number of queries which returned the page, and the scores.

– **Search Engines.** A number of wrappers have been implemented so that different search engines can be used. Currently, we have written wrappers for Gov.uk, Bing and SiteBing (which is Bing restricted to a particular site).

3 Demo

A site designer often wants to know how easily a page can be found, how good different search engines are, and what query terms retrieve a page. With this in mind, Table 1 presents some examples of the PRC applied to two webpages - `www.gov.uk/vehicles-you-can-drive` and `www.gov.uk/renew-adult-passport`. For each page we compare how retrievable the pages are using the site search provided by gov.uk, Bing search (via their API on Azure's Datamarket), and Bing Search but restricted to site:gov.uk (referred to as sitebing). We also provide a comparison of different methods for extracting queries, either using all the text on the page or using the text in main content div (called wrapper). Table 1 shows which engine, portion of the page, along with the number of queries issued, number of times the page was retrieved, and the cumulative and gravity based retrievability scores (where the higher is better).

Table 1. Retrievability scores for different configurations of the PRC

Run	Engine	Portion	#Q issued	Retrieved	$r_c(d)$	$r_g(d)$
\multicolumn{7}{c}{Examples 1-6 on gov.uk page vehicles-you-can-drive}						
1	gov.uk	all	150	25	22.00	10.10
2	bing	all	150	6	3	2.06
3	sitebing:govuk	all	150	33	30.00	14.03
4	gov.uk	wrapper	65	21	19.00	9.10
5	bing	wrapper	65	6	3.00	2.06
6	sitebing:govuk	wrapper	65	31	29.00	13.90
\multicolumn{7}{c}{Examples 7-12 on gov.uk page renew-adult-passport}						
7	gov.uk	all	250	161	138.00	61.39
8	bing	all	250	6	4.00	0.50
9	sitebing:govuk	all	250	101	85.00	27.35
10	gov.uk	wrapper	250	185	157.00	65.03
11	bing	wrapper	250	8	7.00	0.87
12	sitebing:govuk	wrapper	250	118	90.00	28.20

Comparing Engines: Immediately we can see that searching for the pages on the open web using Bing means that less of the queries retrieve the page resulting in substantially lower retrievability scores. When we compare the two site search variations (run 1 vs. 3), we observed that for `gov.uk/vehicle-you-can-drive`, Sitebing is more successful, retrieving the page more often (33 times vs 25 times) and so makes the page more retrievable. While for `gov.uk/renew-adult-passport` the gov.uk sitesearch is more successful at retrieving the page, retrieving the page more often and at higher ranks (as denoted by retrievability scores). It would be interesting to do a larger

740   L. Azzopardi et al.

comparison across the gov.uk domain to see which site search system performs the best, and to see if this correlates with standard effectiveness measures.

Comparing Page Extraction: Runs 4-6 and 10-12 show the results of the same comparisons. However, here we have used only part of the page to draw queries from. This is because we hypothesized that the terms in the headers, footers, and sidebars, are unlikely to be useful in retrieving the page. So we can reduce the number of queries issued to determine the retrievability of a page. When we compare the results to when the full page was used, we see that we obtain similar retrievability score (though slightly less on most occasions). For runs 7-12, we limited the number of queries issued to 250 (even though there was more possible queries). Here we see that by choosing from the main content div, we can even obtain higher retrievability scores as the queries are more closely related to the document's topic.

Top 5 Queries: Table 2 shows the top five queries which returned the page for examples 1 to 3 for page gov.uk/vehicles-you-can-drive. We can see that for Bing, while it only retrieves the page for 3 queries, the queries are pretty sensible, and are likely to be in-line with what a user might type to find information about the topic. For the site search variants we can see similar queries. However, there are also other queries that are unlikely or less likely to be issued (like "tool tells" and "old enough").

Table 2. Top 5 Queries for www.gov.uk/vehicles-you-can-drive for runs 1-3

Engine	Top Queries				
	1	2	3	4	5
gov.uk	drive different	driving transport	except otherwise	tool tells	old enough
Bing	licence drive	licence categories	categories driving	N/A	N/A
SiteBing	drive different	licences quick	adding higher	licences elsewhere	tool tells

Summary: The demo shows some of the utility of the application, but also highlights a number of areas of research and development. In future work we will examine a large subset of pages on different domains, add in new query generation methods that are more realistic, examine how to estimate the likelihood of queries to extract the best possible opens, and explore other configurations. In addition we wish to correlate the retrievability of pages with results from usability studies.

Acknowledgements. This work is supported by the EPSRC Project, *Models and Measures of Findability* (EP/K000330/1).

References

1. Azzopardi, L., Owens, C.: Search engine predilection towards news media providers. In: Proc. of the 32nd ACM SIGIR, pp. 774–775 (2009)
2. Azzopardi, L., Vinay, V.: Retrievability: An evaluation measure for higher order information access tasks. In: Proc. of the 17th ACM CIKM, pp. 561–570 (2008)

3. Bashir, S., Rauber, A.: Improving retrievability of patents with cluster-based pseudo-relevance feedback documents selection. In: Proc. of the 18th ACM CIKM, pp. 1863–1866 (2009)
4. Bashir, S., Rauber, A.: Improving retrievability of patents in prior-art search. In: Gurrin, C., He, Y., Kazai, G., Kruschwitz, U., Little, S., Roelleke, T., Rüger, S., van Rijsbergen, K. (eds.) ECIR 2010. LNCS, vol. 5993, pp. 457–470. Springer, Heidelberg (2010)
5. Chi, E.H., Pirolli, P., Chen, K., Pitkow, J.: Using information scent to model user information needs and actions and the web. In: Proc. of the SIGCHI Conference, CHI 2001, pp. 490–497. ACM (2001)
6. Morville, P.: Ambient Findability. O'Reilly Media (2005)
7. Pickens, J., Cooper, M., Golovchinsky, G.: Reverted indexing for feedback and expansion. In: Proc. of the 19th ACM CIKM, pp. 1049–1058 (2010)
8. Wilkie, C., Azzopardi, L.: Relating retrievability, performance and length. In: Proc. of the 36th ACM SIGIR Conference, SIGIR 2013, pp. 937–940 (2013)
9. Zhang, Y., Zhu, H., Greenwood, S.: Web site complexity metrics for measuring navigability. In: Proc. of the 4th QSIC, pp. 172–179 (2004)
10. Zhou, Y., Leung, H., Winoto, P.: Mnav: A markov model-based web site navigability measure. IEEE Transactions on Software Engineering 33, 869–890 (2007)

ORMA: A Semi-automatic Tool
for Online Reputation Monitoring in Twitter

Jorge Carrillo-de-Albornoz, Enrique Amigó, Damiano Spina, and Julio Gonzalo

UNED NLP & IR Group
Madrid, Spain
{jcalbornoz,enrique,damiano,julio}@lsi.uned.es
http://nlp.uned.es

Abstract. We present a semi-automatic tool that assists experts in their daily work of monitoring the reputation of entities—companies, organizations or public figures—in Twitter. The tool automatically annotates tweets for relevance (Is the tweet about the entity?), reputational polarity (Does the tweet convey positive or negative implications for the reputation of the entity?), groups tweets in topics and display topics in decreasing order of relevance from a reputational perspective. The interface helps the user to understand the content being analyzed and also to produce a manually annotated version of the data starting from the output of the automatic annotation processes. A demo is available at: http://nlp.uned.es/orma/

Keywords: Online Reputation Monitoring, Social Media, Twitter.

1 The Task of Online Reputation Monitoring in Twitter

The rise of social media brought serious concerns to companies, organizations, and public figures on what is said about them online; one of the main reasons is that negative mentions may affect businesses or careers. Monitoring online reputation has therefore become a necessity. An online reputation analyst typically has to perform at least the following tasks: given a stream of texts containing a potential mention to a company as input, (i) *filtering* out tweets that are not related to the entity of interest, (ii) determining the *polarity* (positive, neutral or negative) of the related tweets, (iii) clustering the strongly related tweets in *topics*, and (iv) assigning a relative *priority* to the clusters, in terms of whether a topic may damage the reputation of the entity.

Figure 1 describes the main steps carried out during the annotation process for Online Reputation Monitoring. The process starts selecting one of the entities assigned to the expert. In the system, each entity has a list of tweets that the expert has to annotate manually. The expert processes tweets sequentially: first, she decides whether the tweet does refer to the entity of interest or not. If the tweet is unrelated to the entity, the annotation process for the tweet finishes and the expert continues with the next tweet in the list. Otherwise, the polarity and topic annotations follow. Polarity annotation consists in deciding whether the tweet may affect positively or negatively to the reputation of the entity.

M. de Rijke et al. (Eds.): ECIR 2014, LNCS 8416, pp. 742–745, 2014.

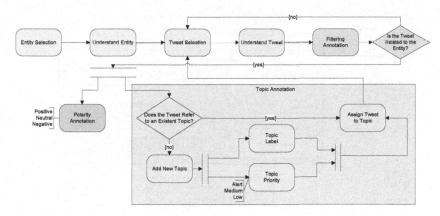

Fig. 1. Workflow of the Online Reputation Monitoring annotation process

Topic annotation consists of identifying the aspects and events related to the entity that the tweet refers to. If the tweet refers to an already identified topic, the tweet is assigned to it. Otherwise, the expert defines a new topic. A topic receives a label that summarizes what the topic is about, and it is also classified in a priority scale (Alert, Medium or Low in our tool). When the tweet is assigned to a topic, the annotation of the current tweet is finished.

2   Architecture and Implementation

Our *Online Reputation Monitoring Assistant (ORMA)* aims to assist the daily work of reputation experts, helping them to manually process the data more efficiently.

Figure 2 shows the architecture of the ORMA demo. The system is deployed into two independent elements: the *Web Client* and the *Server*. The user interface permits the user to manually label tweets about an entity of interest, following the annotation process described above. To reduce the effort of experts, the system also proposes different automatic labels for each input data, with a confidence score indicating how trustable these labels are. The input data of the demo are tweets about different entities in both English and Spanish.

The server element processes the tweets of a given entity and proposes labels for the four subtasks. The tweets are processed in the *Tweet Labeler* using the algorithms described in [3,5,4,2]. The Tweet Labeler is divided in four components, which address each of the reputation monitoring subtasks. Each component of the Tweet Labeler can be implemented using one or multiple algorithms. The labeled tweets are then analyzed with the *Confidence Score Estimator*. For each subtask, this module analyzes the output of the different algorithms and determines a confidence score representing the degree of certainty of the system for the proposed labels. This score should be considered by the experts as a threshold that determines which tweets can be labeled automatically and which tweets need to be revised. The output of this element is stored in the database that is accessed by the user interface.

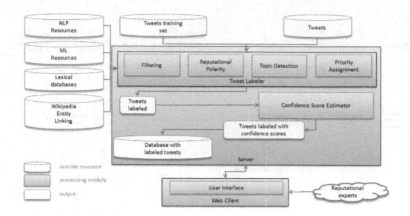

Fig. 2. Online Reputation Monitoring Assistant architecture

The *Database* stores the tweets and the proposed labels for them (both the ones assigned by the system and manually by the user) and has two salient features: first, the tweets in the database are organized at the entity level, i.e., only the tweets associated to the entity being analyzed are loaded, which allows the database to be easily deployed in a distributed environment. Second, tweets are retrieved in a lazy manner, i.e., only tweets that need to be displayed are retrieved, which makes the system more efficient and scalable.

3   Interaction Design

Once the expert is logged in, its user context is retrieved from the database, which includes the different tasks that have been assigned, the already labeled information and the work status. The system then displays the interface for labeling content.

Once the expert has selected an entity to study, the tool shows the tweets to be analyzed in the *Tweets* table. Each row in this table is a tweet containing the name of the author (which is a link to the user profile too), the text of the tweet, the date when the tweet was published and a link to the tweet. Each tweet has also associated an icon identifying if the tweet has already received some manual annotation (reputational expert icon), if all annotations are automatic (computer icon), or a question mark icon if the tweet has not been labeled yet. On the right hand side of the Tweets table, the tool shows two combo boxes (*Related* and *Tweet Polarity*) that allow the reputational experts to annotate the selected tweet with information about its relation to the entity and its polarity.

The *Topic Management* panel groups the tweets by topics and allows the experts to work at the cluster level grouping tweets strongly related in meaning. The two buttons at the bottom of the Topic Management panel, *Add* and *Edit*, are provided to create new topics or to modify the existing ones. When annotating a tweet, the expert can create a new topic and assign the tweet to it, or may assign the tweet to one of the previously created topics. To this aim, the *Tweets in Topic* table shows all tweets previously assigned to the selected topic in the Topic Management panel. Once the topic is selected, the user must assign the tweet to the topic using the *Assign Topic* button.

The *Automatic Mode* panel situated above the Tweets table allows the expert to automatically label the tweets associated to the entity being analyzed. To this aim, the interface presents four buttons, one for each of the subtasks. The panel also contains a confidence score slide bar that allows the user to determine the desired confidence threshold of the automatic labels. Pressing any of the buttons, the system loads all automatic annotations for the appropriate subtask which have a confidence score above the specified threshold —except for those tweets which have already been manually annotated—. These automatic annotations can be manually changed by the reputational expert.

4   Testing and Prototyping

An earlier version of the ORMA annotation application (which did not include the option to automatically process the data) has been tested by thirteen experts along the preparation of the RepLab 2013 test collection [1]. Over half a million annotations for 61 different entities were performed for a total workload of 21 person month. During the exercise, the application was extensively tested for robustness and user friendliness. In particular, interaction design was significantly enhanced by many GUI changes suggested by the annotators.

Acknowledgments. This research was partially supported by the European Community's FP7 Programme under grant agreement n 288024 (LiMoSINe), the Spanish Ministry of Education (FPU grant nr AP2009-0507), 8), the Spanish Ministry of Science and Innovation (Holopedia Project, TIN2010-21128-C02) and the Regional Government of Madrid and the ESF under MA2VICMR (S2009/TIC-1542).

References

1. Amigó, E., Carrillo de Albornoz, J., Chugur, I., Corujo, A., Gonzalo, J., Martín, T., Meij, E., de Rijke, M., Spina, D.: Overview of RepLab 2013: Evaluating online reputation monitoring systems. In: Forner, P., Müller, H., Paredes, R., Rosso, P., Stein, B. (eds.) CLEF 2013. LNCS, vol. 8138, pp. 333–352. Springer, Heidelberg (2013)
2. Anaya-Sánchez, H., Peñas, A., Cabaleiro, B.: UNED-READERS: Filtering Relevant Tweets using Probabilistic Signature Models. In: CLEF 2013 Labs and Workshops Notebook Papers (2013)
3. Castellanos, Á., Cigarrán, J., García-Serrano, A.: Modelling techniques for twitter contents: A step beyond classification based approaches. In: CLEF 2013 Labs and Workshops Notebook Papers (2013)
4. Peetz, M.-H., Spina, D., Gonzalo, J., de Rijke, M.: Towards an Active Learning System for Company Name Disambiguation in Microblog Streams. In: CLEF 2013 Labs and Workshops Notebook Papers (2013)
5. Spina, D., Carrillo-de-Albornoz, J., Martín, T., Amigó, E., Gonzalo, J., Giner, F.: UNED Online Reputation Monitoring Team at RepLab 2013. In: CLEF 2013 Labs and Workshops Notebook Papers (2013)

My First Search User Interface

Tatiana Gossen, Marcus Nitsche, and Andreas Nürnberger

Data and Knowledge Engineering Group, Faculty of Computer Science,
Otto von Guericke University Magdeburg, Germany
{tatiana.gossen,marcus.nitsche,andreas.nuernberger}@ovgu.de
http://www.dke.ovgu.de

Abstract. This paper describes an adaptable search user interface whose main focus group are young users. In order to address continuous and – compared to middle-age users – relatively fast changes in cognitive, fine motor skills and other abilities of young users, we developed a first prototype of an evolving search user interface (ESUI). This ESUI is able to handle changes in design requirements. It is adaptable towards individual user characteristics allowing a flexible modification of the SUI in terms of UI element properties like font size, but also UI element types and their properties. In this work, the goal of SUI adaptation is emotional support for children because positive attitudes towards the system keep them motivated.

Keywords: search user interface, adaptation, context support, human-computer interaction, information retrieval.

1 Evolving Search User Interface

Search User Interfaces (SUI) are usually optimized and adapted to general user group characteristics. For example, there exist special SUIs for children [2,3]. However, especially young users undergo continuous and – compared to middle-age users – relatively fast changes in cognitive, fine motor skills and other abilities [6]. Thus, design requirements change rapidly as well – a flexible modification of the SUI is needed. Therefore, our vision is to provide users with an *evolving* search user interface (ESUI) that adapts to individual user characteristics and allows for changes not only in properties of UI elements like color and font size, but also influences UI elements and their properties. We described the general idea and a corresponding model of an evolving search user interface in [4].

An ESUI enables personalization and therefore increases the usability of a SUI. In specific, it follows general software ergonomic principles in terms of the *ISO 9241-110*[1] like *suitability for individualization* and *conformity with user expectations*. These user interface qualities are supposed to ensure satisfaction of a user interacting with it which positively correlates with effectiveness and efficiency. Similar work [1] suggests to allow personalization of color scheme,

---

[1] http://www.iso.org/iso/home/store/
catalogue_tc/catalogue_detail.htm?csnumber=38009, accessed on 2013-02-10.

M. de Rijke et al. (Eds.): ECIR 2014, LNCS 8416, pp. 746–749, 2014.

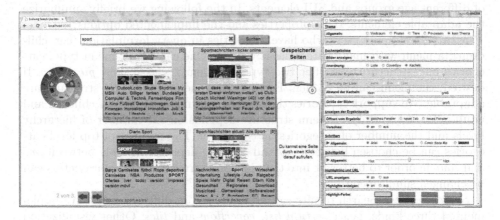

Fig. 1. Screenshot of the configuration unit (an excerpt, left) and the corresponding SUI (right): no theme, tiles result visualization, pie menu, an advanced menu structure with topics for adults, preview mode is activated, page surrogate contains URL and a thumbnail, keyword highlighting in yellow and Arial fonts in 14 pt are configured. The options of the configuration unit are described in Section 2.

search engine's name, customization in search results arrangement using drag-and-drop, and the customization of search services. In contrast, we have a strong focus on the frontend and children as a target group. We aim to customize UI elements of a SUI such as menu type, result visualization, surrogate structure, font, audio, theme and more. This is the first step towards an ESUI.

2 Design and Implementation

The proposed ESUI is an impovement of the child-centered search engine called *Knowledge Journey (KJ)* [3]. In addition to original features of KJ, the new SUI allows customization towards users' wishes. To achieve a coherent design between different variations, the positions of SUI parts are fixed. The interface consists of five groups of elements: elements for search input, menu with different categories for navigation, search result visualization, help section and storage functionality to bookmark relevant results. In order to personalize the SUI, a configuration unit that allows users to manipulate the SUI directly has been implemented (Fig. 1, right). The configuration unit contains settings for the adaptable elements and, in this way, allows to manipulate the SUI appearance. We ordered the adaptable elements according to their decreasing influence on the entire SUI from theme to audio. At this time, configuration is done under supervision with teachers or parents. Later, the configuration unit can be hidden from users and integrated in the backend resulting in an adaptive evolving system.

The implemented elements and their properties are suitable for young users or adults based on the state of the art [5]. In previous work [5], we also con-ducted a user study with children and adults to find a mapping between users of different age groups and SUI elements. In the following we briefly summarize the adaptable elements, their properties and the implemented options:

Menu. To support users who have difficulties in query formulation (e.g. children), a menu with many categories is offered. One can adapt menu type, categories and structure. We offer two types of menus, *classic* and *pie menu*. If desired, the menu can be hidden. Furthermore, menu topics and structure are adaptable. We support categories that reflect information needs of children and adults. The complexity of menu structure depends on the number of hierarchy levels and the number of categories on each level. The number of top-level categories can be choosen between *six* and *ten*, the depth is variable between *one* and *four* in order to better fit into the pie menu structure. Both parameters can be changed simultaneously.

Search results. Search results can be visualized in different ways. We implemented three kinds, i.e. a *vertical list, coverflow* and *tiles*. Other visualization types also exist, e.g. graph visualization, and can be easily added later on. One can specify the number of presented results. In addition, result list items and single tiles can be visualized differently in respect to the separation in the GUI. Three ways to adapt the separation are possible: *no separation*, separation *using lines* or *boxes*. One can also adjust the space between the boxes.

Surrogate. A document surrogate usually consists of its title, URL and a textual summary. We also added information about the search result rank. This information is especially important in case of a result visualization via tiles, where the order is not obvious. Furthermore, one can add a website thumbnail. The size of the thumbnail is continuously adaptable and is inversely proportional to the amount of text in the summary. Additional discrete adaptable surrogate features are URL address and keyword highlighting. A color can be selected to highlight the query within textual summaries.

Result page view. There are three possible ways to view a result page. One can open a page in a *new window*, a *new tab* or the *same window*. Furthermore, a preview of the result page can be turned *on* or *off*. Preview is an intermediate stage where the page is enlarged, however users can interact within the page and stay on the SUI surface.

General properties. General properties have an influence on all SUI elements. These properties are *font type* and *size, theme* and *avatars*, as well as *audio support*. The font size can be configured between 10 and 18 pt. Supported font types are *Comic Sans MS, Arial, Times New Roman* and *Impact*. One can also assign different font configurations for search input, help section, menu categories and document surrogate.

A particular feature of our SUI is the option to select a theme and a corresponding avatar. This allows individual personalization that is welcomed by users, especially by children [3]. Several options are offered: *pirate, space, animals, princess* and *no theme*. These themes fit into the search metaphor allowing especially inexperienced users to use the SUI in a more intuitive way. The theme choice influences the color scheme, background picture, background color of search results, the set of corresponding avatars and the metaphor used for the storage. For each theme there is an option to select an avatar from a predefined set of four. Another special feature of the SUI is voice-support. By hovering over UI elements, a voice explanation is played. One can turn the sound *on* or *off*, select the number of times the voice explanation to be repeated (only *the first time*, *twice* or *always*), choose between a *male, female, girl* or *boy* voice. Voice-support has several advantages. It helps users who have difficulties to read or are not familiar with SUIs at all. Frequent naming of UI elements triggers behavioral learning and thus allows users to learn faster how to use the system.

The application was developed platform- and browser-independent using HTML 5[2], Cascading Style Sheets (CSS)[3] 3 and JavaScript. We optimized our application for the *Google Chrome* Browser [7]. The prototype also works with touchscreen devices. For the backend the *Bing Search API*[4] is used with the safe search option turned on. An online demonstration of the evolving search user interface presented in this paper is available under http://www.dke-research. de/findke/en/Research/Tools+_+Demos/Knowledge+Journey+.html

We evaluated our ESUI with children and adults [5] and received a very positive user feedback about the possibility to adapt the SUI. On a five-point Likert scale from very good to very bad the children chose good (19%) or very good (81%), while 60% of the adults found it to be very good. We also obtained the mapping between the SUI options and different age groups which is important to make the SUI evolving. The direction of future research is to make the interface adaptive. Therefore, methods for automatic user group detection, e.g. based on issued queries (their topic and specific spelling errors), will be investigated.

References

1. Azzopardi, L., Dowie, D., Marshall, K.A., Glassey, R.: Mase: create your own mash-up search interface. In: Proc. of SIGIR Conf., pp. 1008–1008. ACM (2012)
2. Eickhoff, C., Azzopardi, L., Hiemstra, D., de Jong, F., de Vries, A., Dowie, D., Duarte, S., Glassey, R., Gyllstrom, K., Kruisinga, F., et al.: Emse: Initial evaluation of a child-friendly medical search system. In: IIiX Symposium (2012)
3. Gossen, T., Nitsche, M., Nürnberger, A.: Knowledge Journey: A web search interface for young users. In: Proc. of the Sixth Symposium on HCIR. ACM (2012)
4. Gossen, T., Nitsche, M., Nürnberger, A.: Evolving search user interfaces. In: Proc. of the Workshop on euroHCIR at SIGIR Conf., pp. 31–34 (2013)
5. Gossen, T., Nitsche, M., Vos, J., Nürnberger, A.: Adaptation of a search user interface towards user needs - a prototype study with children & adults. In: Proc. of the 7th Annual Symposium on HCIR. ACM (2013)
6. Kail, R.: Children and their development. Prentice Hall Upper Saddle River, NJ (2001)
7. Vos, J.: Anpassung der Nutzungsschnittstelle einer Suchmaschine an die Wünsche des Nutzers - Eine prototypische Studie. Master's thesis, Otto von Guericke University (2013)

---

[2] http://www.w3.org/TR/html-markup/spec.html, accessed on 2013-04-05.

[3] http://www.w3.org/TR/css-2010/, accessed on 2013-04-05.

[4] https://datamarket.azure.com/dataset/
5BA839F1-12CE-4CCE-BF57-A49D98D29A44, accessed on 2013-06-12.

Needle Custom Search

Recall-Oriented Search on the Web Using Semantic Annotations

Rianne Kaptein[1], Gijs Koot[1], Mirjam A.A. Huis in 't Veld[1],
and Egon L. van den Broek[2,3]

[1] TNO, Delft, The Netherlands
{rianne.kaptein,gijs.koot,mirjam.huisintveld}@tno.nl
[2] Department of Information and Computing Sciences, Utrecht University, The Netherlands
[3] Human Media Interaction group, University of Twente, The Netherlands
vandenbroek@acm.org

Abstract Web search engines are optimized for early precision, which makes it difficult to perform recall-oriented tasks using these search engines. In this article, we present our tool Needle Custom Search[1]. This tool exploits semantic annotations of Web search results and, thereby, increase the efficiency of recall-oriented search tasks. Semantic annotations, such as temporal annotations, named entities, and part-of-speech tags are used to rerank and cluster search result sets.

1 Introduction

Generally, Web searches either target homepages or are informational tasks. Both of them can be satisfied with a limited number of search results [4]. Consequently, both the search results and the interface of commercial Web search engines are optimized to excel at these tasks. In practice, only the first 10 or 20 search results are shown on the search results page. Moreover, great care is taken over the presentation of this first page; for example, results from different sources (e.g., news, images, and videos) are included [2]. Their performance can be judged using the quality of the highly ranked search results, which can be measured in terms of early precision and Normalized Discounted Cumulated Gains (NDCG) [12] (cf. [2]).

Analysis of search logs show that between 60% and 85% of searchers view solely the first search result page [4, 9]. In only 4.3% of the sessions do users look at more than three search result pages for a query. So, for most searches, commercial search engines' results are adequate. Nonetheless, there are searches in which users cannot find what they are looking for on one of the first three result pages. In this paper we look at how recall-oriented search tasks on the Web could be better supported.

Forensic investigation is an example of a recall-oriented task. Any detail on any Web page can be important in solving an investigation. Most likely, crucial information is even absent on the 10 most popular pages about the topic of investigation (e.g., a person and an event). So, an exhaustive search is needed to be able to generate an accurate picture. E-discovery is an active field of research because of its use case in the United

---

[1] The Needle Custom Search tool (NCS) can be accessed at:
http://mediaminer.nl/topic_3_context/.

M. de Rijke et al. (Eds.): ECIR 2014, LNCS 8416, pp. 750–753, 2014.
© Springer International Publishing Switzerland 2014

States, where e-discovery refers to the requirement that documents and information in electronic form stored in corporate systems are produced as evidence in litigation [7].

The goal of our Needle Custom Search tool (NCS) is to burst the filter bubble, i.e. present users with information which is not biased towards their context or search history, and including novel and diverse search results; not the typical top search results from a large Web search engine [11] (cf. [6]). The tool presented in this article is founded on the principles outlined in [5]. This article will continue with a discussion on the characteristics of recall-oriented search tasks. Subsequently, in Section 3, we present our Needle Custom Search tool. Finally, in Section 4, we present our conclusions.

2   Recall-Oriented Search

For the design of our tool, we have taken into account the following characteristics, which can be attributed to recall-oriented search tasks on the Web:

- Query expansion: The typical 2 to 3 word query used to search the Web is not sufficient to ensure the recall of all documents relevant to your search topic [6, 8]. Issuing multiple search queries containing different additions and replacements of search words increases the recall of relevant documents.
- Queries do not always have to be answered instantly. A user is likely to spend a considerable amount of time on the search results; so, often he will be willing to wait a little to get good results back. For example, in e-discovery it is more important to find all relevant documents than to get instant answers [7].
- Often searches will focus on user-generated content. Huge amounts of textual data are generated every day in social networks, blogs, and forums and via tweets. Social media has become part of mainstream e-discovery practice [3]. In contrast, more focused content can be found on (official) homepages and online encyclopedia for tasks such as entity ranking [1].

In this article, we want to investigate how we can utilize existing Web search engines for recall-oriented search. Consequently, on the one hand, we exploit engines' strengths and, on the other hand, we devise workarounds for their weaknesses. The alternative would be to develop a general Web search engine from scratch, optimized for recall-oriented search. Especially for searches in specific domains this should be considered.

3   Needle Custom Search (NCS)

Our search tool NCS uses Bing or Google's search APIs to collect Web search results. The URLs are scraped and, subsequently, the pages' textual content is extracted and saved in a database. All text is automatically annotated and, hence, is ready to be reranked and filtered or clustered. An example result page is shown in Figure 1.

Semantic annotations can help in finding a balance between the precision and the recall of the search results. They can significantly contribute to reranking and filtering out search results, which are not relevant in the particular search context [6]. Adding semantic annotations to all search results can help the user to apply a divide and conquer strategy to process the search results. Semantic annotations can be used to cluster

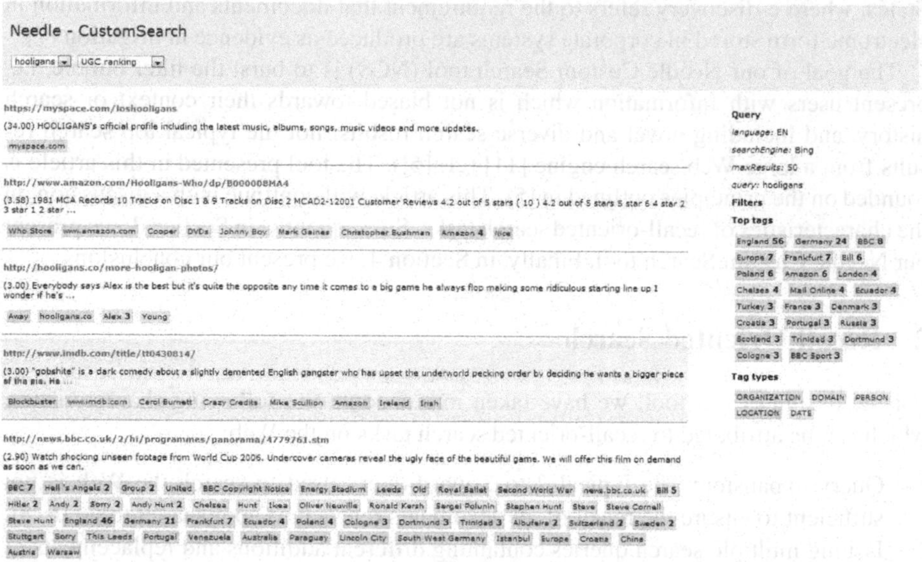

Fig. 1. The results page of Needle Custom Search (NCS) for the query 'hooligans'. The search results are ranked by user-generated content probability. For each search result the URL, a snippet, and the assigned tags are shown. Each tag type has its own color, as can be seen in the legend on the bottom right. The most frequently assigned tags are shown. Clicking on any tag will add that tag as a filter on the search results.

or rerank the search results into many dimensions. Using clustering, large number of search results can be processed more quickly by removing irrelevant clusters from your search results, and zooming in on interesting clusters.

NCS is able to make three types of annotations:

- Temporal annotations. Search results can usually be filtered using the date and time a page was last updated. However, many more date indications can be found on the Web pages itself. For example, forum post's date and time stamps.
- Entity type annotations. Named entity taggers extract entities such as persons, locations and organizations from text.
- Part-of-Speech annotations. Part-of-speech tags can tell you something about the type of language that is used. In user-generated content (e.g., a personal blog post), the distribution of part-of-speech tags will be different from those at a commercial shopping page or a Wikipedia page. NCS uses the occurrence of personal and possessive pronouns as an indicator for user-generated content.

NCS uses Heideltime to extract date and time stamps [10] and Apache's OpenNLP library[2] for entity and part-of-speech tagging.

---

[2] The Apache OpenNLP library. A machine learning based toolkit for natural language text processing; see: http://opennlp.apache.org/

The search results, retrieved via Google or Bing's API, can be ranked using:

1. The original ranking as returned by the search API; and
2. The probability that they contain user-generated content; that is, how many personal and possessive pronouns are used.

4 Conclusion

We have presented the Needle Custom Search tool (NCS; available online[1]), which supports semantic annotations to enable recall-oriented search on the Web. The aim of this tool is to help users process large numbers of search results more efficiently and prevent them from becoming captured in a filter bubble [6, 11]. For future work we would like to extend the query options to support and encourage longer and faceted queries to improve recall and to visualize the search results in a network instead of a list.

References

[1] Balog, K., Serdyukov, P., de Vries, A.P.: Overview of the TREC 2011 entity track. In: Proceedings of the Twentieth Text Retrieval Conference, TREC 2011, November 15-18. National Institute of Standards and Technology (NIST), Gaithersburg (2011)

[2] Chuklin, A., Serdyukov, P., de Rijke, M.: Using intent information to model user behavior in diversified search. In: Serdyukov, P., Braslavski, P., Kuznetsov, S.O., Kamps, J., Rüger, S., Agichtein, E., Segalovich, I., Yilmaz, E. (eds.) ECIR 2013. LNCS, vol. 7814, pp. 1–13. Springer, Heidelberg (2013)

[3] Gensler, S.: Special rules for social media discovery? Arkansas Law Review 65(7) (2012)

[4] Jansen, B.J., Spink, A.: How are we searching the World Wide Web? A comparison of nine search engine transaction logs. Information Processing & Management 42(1), 248–263 (2006)

[5] Kaptein, R., Van den Broek, E.L., Koot, G., Huis in 't Veld, M.A.A.: Recall oriented search on the web using semantic annotations. In: Sixth International Workshop on Exploiting Semantic Annotations in Information Retrieval (ESAIR 2013) (2013)

[6] Melucci, M.: Contextual search: A computational framework. Foundations and Trends in Information Retrieval 6(4-5), 257–405 (2012)

[7] Oard, D.W., Webber, W.: Information retrieval for e-discovery. Foundations and Trends in Information Retrieval 7(2-3), 99–237 (2013)

[8] Saracevic, T.: Relevance: A review of the literature and a framework for thinking on the notion in information science. Part II: Nature and manifestations of relevance. Journal of the American Society for Information Science and Technology 58(13), 1915–1933 (2007)

[9] Silverstein, C., Henzinger, M., Marais, H., Moricz, M.: Analysis of a very large web search engine query log. ACM SIGIR Forum 33(1), 6–12 (1999)

[10] Strötgen, J., Gertz, M.: Multilingual and cross-domain temporal tagging. Language Resources and Evaluation 47(2), 269–298 (2013)

[11] van der Sluis, F., van den Broek, E.L., Glassey, R.J., van Dijk, E.M.A.G., de Jong, F.M.G.: When complexity becomes interesting. Journal of the American Society for Information Science and Technology (in press, 2014)

[12] Wang, Y., Wang, L., Li, Y., He, D., Liu, T.-Y.: A theoretical analysis of NDCG type ranking measures. In: Proceedings of the 26th Conference on Learning Theory (COLT), Princeton, NJ, USA, June 12-14. JMLR: Workshop and Conference Proceedings, vol. 30, pp. 25–54. Microtome Publishing, Brookline (2013)

Khresmoi Professional: Multilingual, Multimodal Professional Medical Search

Liadh Kelly[1], Sebastian Dungs[2], Sascha Kriewel[2], Allan Hanbury[3],
Lorraine Goeuriot[1], Gareth J. F. Jones[1], Georg Langs[4], and Henning Müller[5]

[1] Dublin City University, Ireland
[2] University of Duisburg-Essen, Germany
[3] Vienna University of Technology, Austria
[4] Medical University of Vienna, Austria
[5] University of Applied Sciences Western Switzerland, Switzerland

Abstract. In this demonstration we present the Khresmoi medical search and access system. The system uses a component based architecture housed in the cloud to support target users medical information needs. This includes web systems, computer applications and mobile applications to support the multilingual and multimodal information needs of three test target groups: the general public, general practitioners (GPs) and radiologists. This demonstration presents the systems for GPs and radiologists providing multilingual text and image based (including 2D and 3D radiology images) search functionality.

1 Introduction

The Khresmoi project[1] is developing multilingual multimodal search and access systems for medical and health information. It addresses the challenges of searching medical data, including information available on the internet, as well as 2D and 3D radiology images in hospital archives. The system allows text querying in several languages, in combination with image queries. Results can be translated using a machine translation tool specifically trained on medical text. This demonstration focuses on search functionality developed for general practitioners (GPs) and radiologists.

2 Innovative Medical Search System

Components of the Khresmoi system include search, a knowledge-base, machine translation, query disambiguation and spell-checking. The modular integration of multiple software technologies in the system architecture allows for easy development of required medical search applications. These applications are innovative in several ways, including:

---

[1] http://khresmoi.eu

M. de Rijke et al. (Eds.): ECIR 2014, LNCS 8416, pp. 754–758, 2014.

Fig. 1. The Web frontend

- Multimodal search system for medical practitioners, offering multilingual support, faceted search, and several personal support components including facilities to save items and collaborate with peers.
- Large-scale image search based on visual similarity of images, supporting both 2D images (X-rays and images in publications) and 3D images (CT and MR).

These systems were developed to take into consideration users needs and requirements as determined by extensive questionnaires and analysis conducted within the Khresmoi project [1–3]. Rounds of user-centered evaluation with GPs and radiologists at both the interface component and interface system level have been used for iterative system improvement [4–6].

3   Medical Search for General Practitioners

The Khresmoi search prototype for GPs provides two user interfaces. One is a browser based web application[2] written in GWT, while the other is a Java Swing

---

[2] http://professional.khresmoi.eu

desktop application[3]. Both share a common backend service infrastructure also written in Java. A third user interface for Android devices is currently under development.

The main novelty of this search prototype is the integration in a single system of various components such as high quality machine translation, efficient information retrieval module, medically-trained summarization, collaborative components, etc.

Crawled websites with trustworthy medical information targeted at practitioners are semantically annotated using GATE technology[4] and then indexed by Mimir[5]. In addition, images from medical publications are also indexed (using ParaDISE[6]).

Figure 1 shows the web frontend which features basic functionality including free text search, and a facet explorer which enables the result set to be filtered using metadata attributes. The system gives spelling corrections and disambiguation suggestions while a user is typing a query. Result sets may include images which can be used to trigger searches for visually similar images. Users can store retrieved documents in a tray for later inspection. The personal library is a permanent storage for documents of various formats and is available to all registered users. Queries are recorded and can easily be reissued by utilizing a separate view in the interface. All components can be (un-)hidden from the perspective, re-sized and moved in the interface.

The Swing interface includes all features of the web prototype, but in addition users can issue an image search by dropping an image from their file system or browser in a special search box. The desktop client has collaborative features which enable registered users to share documents with other users or user groups. Both interfaces are fully internationalised for all Khresmoi project languages, including English, German, French, Spanish and Czech, as well as for Chinese and Vietnamese.

4   Medical Search For Radiologists

The search system for radiologists was developed based on the Swing version of the prototype interface, and shares the same technological basis as the GP system. This system enables radiologists to search and compare 2D and 3D radiology images. A use case for this system is that a radiologist faced with an unusual or unknown structure in an image can query the hospital archives for images containing a similar structure, and use the (anonymised) radiology reports associated with these images to guide the reading of the image.A demonstration of the system can be viewed[7].

---

[3] http://khresmoi.is.inf.uni-due.de/khresmoi.jnlp

[4] https://gate.ac.uk

[5] https://gate.ac.uk/mimir/

[6] ParaDISE is a new visual search engine developed in Khresmoi as a successor to the GNU Image Finding Tool (GIFT).

[7] http://youtu.be/UnPs7NSet1g

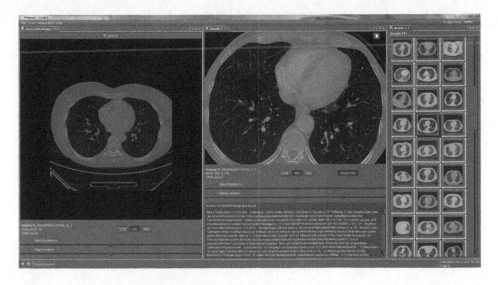

Fig. 2. Khresmoi interface for radiologists

Figure 2 presents the interface instantiated for use by radiologists. Note that this is the same interface framework shown in Figure 1, but with different tools visible. Here the query is the selected area of the image slice shown in the left panel. The images in the panel on the right are returned based on their visual similarity to the region marked in the query. In the central panel, the selected image is shown, along with the associated radiology report. For this application, only the images stored in the archives of the hospital in which the system is used are indexed. However, the possibility to do a visual search of 2D images from the medical literature is also provided.

Acknowledgments. The research leading to these results has received funding from the European Union Seventh Framework Programme (FP7/2007-2013) under grant agreement n 257528 (KHRESMOI).

References

1. Kelly, L., Goeuriot, L., Jones, G.J.F., Hanbury, A.: Considering subjects and scenarios in large-scale user-centered evaluation of a multilingual multimodal medical search system. In: CLEF (Online Working Notes/Labs/Workshop) (2012)
2. Stefanov, V., Sachs, A., Kritz, M., Samwald, M., Gschwandtner, M., Hanbury, A.: A formative evaluation of a comprehensive search system for medical professionals. In: Forner, P., Müller, H., Paredes, R., Rosso, P., Stein, B. (eds.) CLEF 2013. LNCS, vol. 8138, pp. 81–92. Springer, Heidelberg (2013)
3. Markonis, D., Holzer, M., Dungs, S., Vargas, A., Langs, G., Kriewel, S., Müller, H.: A survey on visual information search behavior and requirements of radiologists. Methods Inf. Med. 51, 539–548 (2012)

4. Samwald, M., Kritz, M., Gschwandtner, M., Stefanov, V., Hanbury, A.: Physicians searching the web for medical question answering: A european survey and local user studies. In: Proceedings of MedInfo 2013, p. 1103 (2013)
5. Kritz, M., Gschwandtner, M., Stefanov, V., Hanbury, A., Samwald, M.: Utilization and perceived problems of online medical resources and search tools among different groups of european physicians. Journal of Medical Internet Research 15 (2013)
6. Markonis, D., Baroz, F., Castaneda, R.R.D., Boyer, C., Müller, H.: User tests for assessing a medical image retrieval system: a pilot study. Stud. Health Technol. Inform. 192, 224–228 (2013)

DAIKnow: A Gamified Enterprise Bookmarking System

Michael Meder, Till Plumbaum, and Frank Hopfgartner

Technische Universität Berlin
Ernst-Reuter-Platz 7
10587 Berlin, Germany
{firstname.lastname}@tu-berlin.de

Abstract. One of the core ideas of gamification in an enterprise setting is to engage employees, i.e., to motivate them to fulfil boring, but necessary tasks. In this demo paper, we present a gamified enterprise bookmarking system which incorporates points, badges and a leaderboard. Preliminary studies indicate that these gamification methods result in an increased user engagement.

Keywords: enterprise gamification, social bookmarking system.

1 Introduction

A few years ago, one could observe a run on social bookmarking systems such as Delicious[1], i.e., online platforms that allow users to contribute, annotate and share bookmarks. Social bookmarking systems played an important role in studying techniques that are associated with the Web 2.0 such as tagging, the development of shared vocabularies (i.e., folksonomies) [1] and social networking [2]. An important domain where such systems are applied are in an enterprise scenario, i.e., they are used to provide means for employees to discover and share knowledge within the company [3] or to support innovative ideas [4]. Looking at the usage statistics of those bookmarking systems after having been deployed for a few years, one can observe a dramatic decrease of active users. Stewart et al. [5] argue that the rise and fall of such systems heavily depends on the internal corporate culture. Although the usage of such system might initially be high, its popularity can quickly erode, especially if its usage is based on a voluntary basis only and not promoted further by the management. A promising approach to increase user participation is the application of gamification methods. Deterding et al. [6] define gamification as *"the use of game design elements in non-game contexts"*, i.e., game mechanics and concepts such as leaderbords, points and badges are applied on non-gaming environments to reach specific goals. Example studies where such principles have been applied include [7–10].

In this demo paper, we introduce DAIKnow[2], an enterprise bookmarking system which got enriched with gamification elements. DAIKnow is deployed as

---

[1] http://delicious.com/
[2] http://daiknow.dai-labor.de/

M. de Rijke et al. (Eds.): ECIR 2014, LNCS 8416, pp. 759–762, 2014.

social bookmarking system of a large research institute and used to study the use of gamification principles at the workspace. In this paper, we provide an introduction into the main features of the system and provide an overview of the gamification elements that have been included.

2   Enterprise Bookmarking System

Providing similar functionalities as the Delicious system, DAIKnow allows for the creation bookmarks of websites, as well as bookmarks for files on internal file server. Given the importance of different access rights within companies, the system incorporates existing rights management. Further, it provides facilities to share bookmarks with individual users and users who used certain tags.

The system has two main views: A personal view and a general view. On the personal site, the user can see all bookmarks. Bookmarks can be either public, i.e., they are visible for everyone, or private, i.e., they only appear on the user's site. On the general site, all public items are displayed. The general view is split into a bookmark and a user view. Both bookmarks and users can be tagged and retrieved using the search box on the top right side of the screen.

Bookmarks can either be created by using a JavaScript based bookmarklet which allows users to bookmark pages those pages that are open in their web browser or by using a Windows tool.

When creating a bookmark, the system automatically suggests tags that can be used to describe the item. These suggestions are extracted from the describing text and the most commonly used tags of the user.

Besides, the system displays a "most popular" section where the most frequently used tags of the week (and of all time) are shown to enable users to understand what topic is currently important in their company.

3   Gamification Elements

In order to increase usage of the DAIKnow system, we introduced four gamification elements, namely points, badges, a leaderboard and feedback messages. These elements cover different aspects that motivate users. While points and badges are rewards that can be achieved when fulfilling different tasks, leaderboards introduce a competitive factor. Points can be gained by directly interacting with the system, e.g., by creating a new bookmark. Badges, on the other hand, are awarded after achieving certain goals such as continuously creating new bookmarks over a longer period of time. The leaderboard provides an overview of who gained the most points and badges. It is split into two parts: A monthly list, and an all time list. This split allows new users to reach the top position of the monthly list, while long term commitment is still visible, hence satisfying older players.

In the remainder of this section, we introduce the applied gamification elements in more detail.

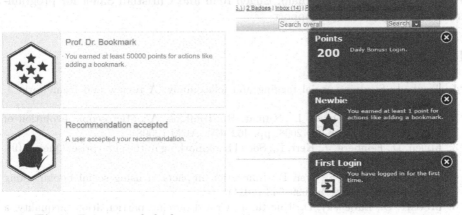

Fig. 1. Two example badges

Fig. 2. Messages appearing on the user's first log in

Points. The system provides multiple means to gain points, including daily login, adding a public/private bookmark, accepting a bookmark recommendation and many others.

Achievement Badges. Users can be awarded badges for getting points, creating bookmarks, regular usage and for providing feedback. Figure 1 depicts two example badges.

Leaderboard. The leaderboard displays the users, their points and badges in decreasing order. Two rankings are provided: A monthly leaderboard which automatically resets at the end of every month and an all-time leaderboard.

Feedback. Making sure that the user is constantly reminded of the gamification methods, the system incorporates a message system (Figure 2). These messages are displayed for ten seconds each time a user receives a reward (points or badges) and when the user increases their position on the leaderboard. In addition, the user is notified when another user copied one of their bookmarks or accepted their bookmark recommendation.

4 Summary and Outlook

In this demo, we showcase a gamified enterprise bookmarking system of a research institute with over 140 employees. User interactions are constantly logged to further study gamification methods in an enterprise environment. Preliminary studies [11] indicate the advantages of gamification principles in this scenario. The system will be used to further study these advantages and rising challenges. In the future, we intend to apply the same methods to an enterprise and desktop search scenario, e.g., to study constraints [12] that can influence search behavior.

Acknowledgments. We thank Eugen Rein and Christian Sauer for programming and graphic design support.

References

1. Trant, J.: Studying social tagging and folksonomy: A review and framework. J. Digit. Inf., 1 (2009)
2. Leskovec, J., Backstrom, L., Kumar, R., Tomkins, A.: Microscopic evolution of social networks. In: KDD 2008, pp. 462–470. ACM (2008)
3. Millen, D., Feinberg, J., Kerr, B.: Social bookmarking in the enterprise. Queue 3(9), 28–35 (2005)
4. Gray, P.H., Parise, S., Iyer, B.: Innovation impacts of using social bookmarking systems. MIS Quarterly, 629–643 (2011)
5. Stewart, O., Lubensky, D., Huerta, J.: Crowdsourcing participation inequality: a SCOUT model for the enterprise domain. In: Proceedings of the ACM SIGKDD Workshop on Human Computation, pp. 30–33 (2010)
6. Deterding, S., Dixon, D., Khaled, R., Nacke, L.: From game design elements to gamefulness: defining gamification. In: Proceeding of the 15th International Academic MindTrek Conference, pp. 9–15 (2011)
7. Dugan, C., Muller, M., Millen, D.R., Geyer, W., Beth Brownholtz, M.M.: The dogear game: a social bookmark recommender system. In: Proceedings of the 2007 International ACM Conference on Supporting Group Work, pp. 387–390 (2007)
8. Farzan, R., DiMicco, J., Millen, D.: Results from deploying a participation incentive mechanism within the enterprise. In: Proceedings of the SIGCHI Conference on Human Factors in Computing Systems, pp. 563–572 (2008)
9. Eickhoff, C., Harris, C.G., de Vries, A.P., Srinivasan, P.: Quality through flow and immersion: gamifying crowdsourced relevance assessments. In: SIGIR, pp. 871–880 (2012)
10. Poesio, M., Chamberlain, J., Kruschwitz, U., Robaldo, L., Ducceschi, L.: Phrase detectives: Utilizing collective intelligence for internet-scale language resource creation. TiiS 3(1), 3 (2013)
11. Meder, M., Plumbaum, T., Hopfgartner, F.: Perceived and actual role of gamification principles. In: UCC 2013: Proceedings of the IEEE/ACM 6th International Conference on Utility and Cloud Computing, pp. 488–493. IEEE (2013)
12. Fujikawa, K., Joho, H., Nakayama, S.: Constraint can affect human perception, behaviour, and performance of search. In: Chen, H.-H., Chowdhury, G. (eds.) ICADL 2012. LNCS, vol. 7634, pp. 39–48. Springer, Heidelberg (2012)

Search Maps
Enhancing Traceability and Overview in Collaborative Information Seeking

Dominic Stange[1] and Andreas Nürnberger[2]

[1] Volkswagen AG, Germany
dominic.stange@volkswagen.de
[2] Faculty of Computer Science, University of Magdeburg, Germany
andreas.nuernberger@ovgu.de

Abstract. We propose a search user interface that is especially designed to make individual contributions to a collaborative search task visible and traceable. The main goal is to support awareness, understanding, and sensemaking within a group working together on the same task. The support is achieved by visualizing the information seeking activities of the user group with an interactive two-dimensional Search Map. The users share the same Search Map and can actively collaborate and evolve their search topic together. The Search Map serves as a common ground and enables each user to gain a more comprehensive understanding of the domain in question by taking advantage of the shared view of the community.

1   Introduction

It is well acknowledged that designing adequate search user interfaces to support collaborative search is still an active research topic [1]. For example, interfaces should be designed that support group awareness, workspace awareness, contextual awareness, and peripheral awareness [2]. Existing collaborative search user interfaces help to store snapshots of an information seeking process like interesting websites, and search results, or queries sent to a search engine. However, they do not provide the necessary means to understand *how* the search process unfolds and *how* the different search activities of a collaborating community complement each other.

We want users to be able to share a deep understanding of each others' actions during a collaborative information seeking task. Therefore, we propose the idea of a Search Map which serves as an additional visual layer on top of existing search interfaces. This layer aims at supporting collaborative information seeking and sensemaking by making the joint search strategy of a community of users visible and traceable.

2   Related Work

Today, research in the area of collaborative information seeking has brought forth several interesting projects and tools to support a community to search

M. de Rijke et al. (Eds.): ECIR 2014, LNCS 8416, pp. 763–766, 2014.

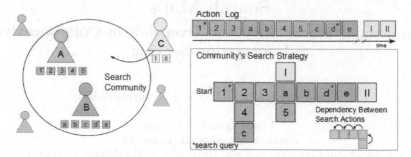

Fig. 1. A search community of three users A, B and C; their search actions represented as colored boxes; and a visualization of their joint search strategy as a list and tree

together. One of the earliest is Ariadne [3] which captures a user's search history and provides playback functionality primarily to discuss the results of a completed search task. The idea of collaborative search was later extended to also support users to search together at the same time (either co-located [4] or at remote locations [5]). More recently, approaches are made to allow for more advanced interaction capabilities, such as information collection or collaborative reporting (e.g. [6]). The challenge of collaborative sensemaking is also addressed, e.g. in [7]. Further approaches are made to enhance the understanding of what actions users contribute in a collaborative search. The authors of [8] show how a user's search actions can be (automatically) recorded using natural language descriptions of web activity, such as "user went to 'http://url.com'". This idea is closely related to ours because the descriptions work like a log which can be shared among collaborating users. In contrast to Search Maps, however, this log is one-dimensional. Therefore, the visualization of multiple search directions that originate from a single user action is very limited.

3   Concept of a Search Map

Search Maps are designed to make a community of collaborating users more aware of each others' presence and actions during the search. The motivation of a Search Map is to visualize the community's joint search strategy. For example, the left part of Fig. 1 shows a search community of three users (A)llice, (B)en, and (C)hris. Allice and Ben decide to start a search together. Later, they invite Chris to join them. The top right corner of Fig. 1 shows the order of their search actions as an action log. Allice begins with three actions: a request to a search engine (1), an opened website from one of the search results (2) and a click on a hyperlink within this website (3). Ben, then, takes this website as a starting point and follows another hyperlink to open a website (a). The search continues as pictured in the bottom right corner, which describes the dependencies between their search actions and outlines their overall search strategy. As can be seen, Allice and Ben interact with each other and complement one another's actions. By the time Chris joins them there are already a couple of directions which can

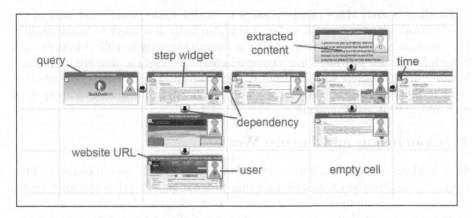

Fig. 2. Example of a Search Map as a Collaborative Search User Interface

be further extended. Search Maps are designed to reflect such interactions within a search community.

An example of a Search Map can be found in Fig. 2. The example shows the joint search strategy of the community consisting of Allice, Ben and Chris. The Search Map - like a street map - aims at improving overview, awareness, traceability, and sensemaking within the community. This is achieved by displaying all search actions as widgets in a two-dimensional grid, similar to the conceptual representation of Fig. 1.

A Search Map can be initiated by any user with access to the Search Map application which is running on a web server. Other users can be invited by sending them the URL to the currently developed Search Map. All users authenticate themselves in the application so their individual contributions can be acknowledged. Currently, a Search Map tracks four types of actions of a user as *search steps*: (a) open a website, (b) send a search query to one of a set of predefined search engines, (c) follow a hyperlink within a website to open another website, and (d) extract content from a website as a snippet. As stated above these search steps are represented as widgets. Apart from action-specific information like the URL of a website or the search query, a widget provides information about the user who performed the action, the time when it is performed (its order in the action log), and an indicator of whether the search step has been newly added since the last time the user looked at the Search Map. All search steps in a Search Map can be viewed again by clicking on them. Clicking on a query widget displays the search results. Clicking on a website widget opens the website in a separate browser window. Snippets can be viewed directly on the Search Map.

Users can add new search steps to a Search Map in any empty cell of the grid manually or by continuing an existing search step while letting the system determine where to put the corresponding widget (according to a layout strategy).

Existing Search Maps can help users to better understand what part of the information space is covered so far. They also help new users to make sense of the community's search actions and their overall strategy to collect information. Search Maps also provide a more comprehensive view of the domain in question because multiple users with different backgrounds and knowledge can contribute their individual views to the search while benefiting from each other's expertise and information seeking skills.

4   Conclusion and Future Work

Search Maps are a contribution to collaborative search user interfaces. They help to structure the information seeking process and visualize the joint search strategy of a community of collaborating users. Search Maps aim at increasing awareness, sensemaking, and traceability capabilities beyond that of existing search user interfaces. The concept of a Search Map is still evolving and the prototype application is under continuous development.

With our presentation of Search Maps we want to encourage the discussion about new types of collaborative search user interfaces and share our insights gained from an application in the automotive industry where we currently investigate its interaction capabilities in the field of technology scouting[1].

References

1. Shah, C., Marchionini, G., Kelly, D.: Learning design principles for a collaborative information seeking system. In: CHI 2009 Extended Abstracts on Human Factors in Computing Systems, CHI EA 2009, pp. 3419–3424. ACM, New York (2009)
2. Shah, C.: Effects of awareness on coordination in collaborative information seeking. JASIST 64(6), 1122–1143 (2013)
3. Twidale, M.B., Nichols, D.M.: Collaborative browsing and visualisation of the search process. In: Proceedings of ELVIRA 1996, pp. 177–182. Milton Keynes (1996)
4. Amershi, S., Morris, M.R.: Cosearch: a system for co-located collaborative web search. In: CHI 2008: Proceeding of the Twenty-Sixth Annual SIGCHI Conference on Human Factors in Computing Systems, pp. 1647–1656. ACM, New York (2008)
5. Morris, M.R., Horvitz, E.: Searchtogether: an interface for collaborative web search. In: Proceedings of the 20th Annual ACM Symposium on User Interface Software and Technology, pp. 3–12. ACM, New York (2007)
6. Shah, C.: Coagmento - a collaborative information seeking, synthesis and sensemaking framework (an integrated demo), pp. 527–528. ACM, New York (2010)
7. Paul, S.A., Morris, M.R.: Cosense: enhancing sensemaking for collaborative web search. In: Olsen Jr., D.R., Arthur, R.B., Hinckley, K., Morris, M.R., Hudson, S.E., Greenberg, S. (eds.) CHI, pp. 1771–1780. ACM (2009)
8. Wiltse, H., Nichols, J.: Playbyplay: collaborative web browsing for desktop and mobile devices. In: Olsen Jr., D.R., Arthur, R.B., Hinckley, K., Morris, M.R., Hudson, S.E., Greenberg, S. (eds.) CHI, pp. 1781–1790. ACM (2009)

---

[1] At the moment we have a community of fifteen users from different business units of the Volkswagen company using our Search Map prototype application to find and evaluate potentially interesting manufacturing technologies.

A Visual Interactive Environment
for Making Sense of Experimental Data

Marco Angelini[2], Nicola Ferro[1], Giuseppe Santucci[2], and Gianmaria Silvello[1]

[1] University of Padua, Italy
{ferro,silvello}@dei.unipd.it
[2] "La Sapienza" University of Rome, Italy
{angelini,santucci}@dis.uniroma1.it

Abstract. We present the *Visual Information Retrieval Tool for Upfront Evaluation (VIRTUE)* which is an interactive and visual system supporting two relevant phases of the experimental evaluation process: performance analysis and failure analysis.

1 Introduction

Developing and testing an *Information Retrieval (IR)* system is a challenging task, in particular when it is necessary to understand the behaviour of the system under different conditions of use in order to tune or improve it to achieve the level of effectiveness needed to meet user expectations. The complex interactions among the components of a system are often hard to trace down and to explain in the light of the obtained results. To this purpose two main activities are carried out in the context of experimental evaluation: performance analysis and failure analysis. The goal of performance analysis is to determine positive and negative aspects of the IR system under evaluation; whereas, the goal of failure analysis [6,8] is to conduct a deeper investigation for understanding the behaviour of a system determining what went well or bad.

Visual Information Retrieval Tool for Upfront Evaluation (VIRTUE) aims at reducing the effort needed to carry out both the performance and failure analyses, both at topic and experiment level, since it allows user to visually interact with and mine the experimental results. An extensive description of the background, analysis of the functionalities and evaluation of VIRTUE can be found in [4]. The running prototype of VIRTUE is available at the following URL: http://151.100.59.83:11768/Virtue/ while a video showing how the system works is available at the following URL: http://ims.dei.unipd.it/websites/ecir2014/demo_ecir2014.mp4.

This paper is organized as follows: Section 2 presents the features of VIRTUE to support performance analysis and Section 3 discusses how VIRTUE enhances failure analysis; finally, Section 4 draws some final remarks.

2 Performance Analysis

Performance analysis is one of the most consolidated activities in IR evaluation and VIRTUE allows for interactive visualization and exploration of the

M. de Rijke et al. (Eds.): ECIR 2014, LNCS 8416, pp. 767–770, 2014.

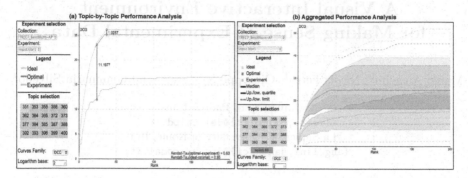

Fig. 1. The performance analysis capabilities provided by VIRTUE

experimental results, according to different metrics and parameters. It provides visual means to grasp whether the system would already have the potential to achieve the best performances or whether a new ranking strategy would be preferred. This analysis can be conducted on a topic-by-topic basis and with aggregate statistics over the whole set of topics.

In order to quantify the performances of an IR system, we adopt the (discounted) cumulated gain family of measures [7] which have proved to be especially well-suited for analyzing ranked results lists because they allow for graded relevance judgments and embed a model of the user behavior while s/he scrolls down the results list which also gives an account of her/his overall satisfaction.

We compare the result list produced by an experiment with respect to an *ideal* ranking created starting from the relevant documents in the ground-truth, which represents the best possible results that an experiment can return – this ideal ranking is what is usually used to normalize the *Discounted Cumulated Gain (DCG)* measures. In addition to what is typically done, we compare the results list with respect to an *optimal* one created with the same documents retrieved by the IR system but with a optimal ranking, i.e. a permutation of the results retrieved by the experiment aimed at maximizing its performances by sorting the retrieved documents in decreasing order of relevance. Therefore, the *ideal ranking* compares the given experiment with respect to the best results possible, i.e. considering also relevant documents not retrieved by the system, while the *optimal ranking* compares an experiment with respect to what could have been done better with the same retrieved documents.

The proposed visualization, shown in Figure 1(a), allows for interaction with these three curves, e.g. by dynamically choosing different measures in the DCG family, adjusting the discounting function, and comparing curves and their values rank by rank. Overall, this method makes it easy to determine the distance of an IR system from both its own optimal performances and the best performances possible and to get an indication about whether the system is going in the right direction or whether a completely different approach is preferable. In order to support this visual intuition, we also provide a Kendall's τ correlation analysis [9] between the three above mentioned curves [5].

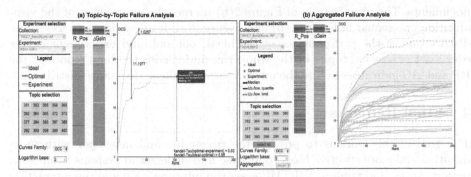

Fig. 2. The failure analysis capabilities provided by VIRTUE

VIRTUE provides also an aggregate representation based on the box-plot statistical tool showing the variability of the three DCG curves calculated either on all the topics considered by an experiment or on those selected by the user. This feature allows a user to interactively choose the topics to whose performances have to be aggregated in order to support the exploration of alternative retrieval scenarios [2]. We can see this feature in Figure 1(b).

3   Failure Analysis

For conducting failure analysis, VIRTUE exploits two indicators, called *Relative Position (RP)* and *Delta Gain (ΔG)* [3], which allow us to visually and numerically figure out the weak and strong parts of a ranking in order to quickly detect failing documents or topics and make hypotheses about how to improve them. RP quantifies the effect of misplacing relevant documents with respect to the ideal case easing the interpretation of the DCG curve, i.e. it accounts for how far a document is from its ideal position. ΔG quantifies the effect of misplacing relevant documents with respect to the ideal case in terms of the impact of the misplacement on the gain at each rank position [1].

These two indicators are paired with a visual counterpart that makes it even easier to quickly spot and inspect critical areas of the ranking. Two bars are added on the left of the visualization, as shown in Figure 2(a): one for the RP indicator and the other for the ΔG indicator. These two bars represent the ranked list of results with a box for each rank position and, by using appropriate color coding to distinguish between zero, positive and negative values and shading to represent the intensity, i.e. the absolute value of each indicator, each box represents the values of either RP or ΔG. For example, in this way, by looking at the bars and their colors a user can immediately identify non-relevant documents which have been ranked in the positions of relevant ones. Then, the visualization allows them to inspect those documents and compare them with the topic at hand in order to make a hypothesis about the causes of a failure.

The techniques described above support and ease failure analysis at the topic level and allow users to identify and guess possible causes for wrongly ranked documents. The visualization of Figure 2(b) merges the approaches of the visualizations presented in Figure 1(b) and Figure 2(a): it allows users to assess the distribution of the performances of the ideal, optimal, and experiment curves over a set of selected topics or the whole run and it adds the bars reporting the RP and ΔG indicators to ease the interpretation of the performance distribution.

4   Final Remarks

The goal of this work is to provide the researcher and developer with more intuitive and more effective tools to analyse and understand systems behaviour, performances, and failures. VIRTUE eases the interpretation and the interaction with DCG curves, allows for detecting critical areas in ranked lists, and provides an integrated way for combining topic-by-topic and aggregated analyses.

Acknowledgements. The work reported in this paper has been supported by the PROMISE network of excellence (contract n. 258191) project as a part of the 7th Framework Program of the European commission (FP7/2007-2013).

References

1. Angelini, M., Ferro, N., Järvelin, K., Keskustalo, H., Pirkola, A., Santucci, G., Silvello, G.: Cumulated Relative Position: A Metric for Ranking Evaluation. In: Catarci, T., Forner, P., Hiemstra, D., Peñas, A., Santucci, G. (eds.) CLEF 2012. LNCS, vol. 7488, pp. 112–123. Springer, Heidelberg (2012)
2. Angelini, M., Ferro, N., Santucci, G., Silvello, G.: Visual Interactive Failure Analysis: Supporting Users in Information Retrieval Evaluation. In: Proc. 4th Symposium on Information Interaction in Context (IIiX 2012), pp. 195–203. ACM Press, New York (2012)
3. Angelini, M., Ferro, N., Santucci, G., Silvello, G.: Improving Ranking Evaluation Employing Visual Analytics. In: Forner, P., Müller, H., Paredes, R., Rosso, P., Stein, B. (eds.) CLEF 2013. LNCS, vol. 8138, pp. 29–40. Springer, Heidelberg (2013)
4. Angelini, M., Ferro, N., Santucci, G., Silvello, G.: VIRTUE: A Visual Tool for Information Retrieval Performance Evaluation and Failure Analysis. Journal of Visual Languages & Computing (in print, 2014)
5. Di Buccio, E., Dussin, M., Ferro, N., Masiero, I., Santucci, G., Tino, G.: To Re-rank or to Re-query: Can Visual Analytics Solve This Dilemma? In: Forner, P., Gonzalo, J., Kekäläinen, J., Lalmas, M., de Rijke, M. (eds.) CLEF 2011. LNCS, vol. 6941, pp. 119–130. Springer, Heidelberg (2011)
6. Harman, D.K.: Some thoughts on failure analysis for noisy data. In: Proc. 2nd Workshop on Analytics for Noisy unstructured text Data (AND 2008), p. 1. ACM Press, New York (2008)
7. Järvelin, K., Kekäläinen, J.: Cumulated Gain-Based Evaluation of IR Techniques. ACM Transactions on Information Systems (TOIS) 20(4), 422–446 (2002)
8. Savoy, J.: Why do Successful Search Systems Fail for Some Topics. In: Proc. 2007 ACM Symposium on Applied Computing (SAC 2007), pp. 872–877. ACM Press, New York (2007)
9. Voorhees, E.: Evaluation by Highly Relevant Documents. In: Proc. 24th Annual International ACM SIGIR Conference on Research and Development in Information Retrieval (SIGIR 2001), pp. 74–82. ACM Press, New York (2001)

TripBuilder: A Tool for Recommending Sightseeing Tours

Igo Brilhante[2], Jose Antonio Macedo[2],
Franco Maria Nardini[1], Raffaele Perego[1], and Chiara Renso[1]

[1] ISTI-CNR, Pisa, Italy
[2] Federal University of Ceará, Fortaleza, Brasil

Abstract. We propose TRIPBUILDER, an user-friendly and interactive system for planning a time-budgeted sightseeing tour of a city on the basis of the points of interest and the patterns of movements of tourists mined from user-contributed data. The knowledge needed to build the recommendation model is entirely extracted in an unsupervised way from two popular collaborative platforms: Wikipedia[1] and Flickr[2]. TRIP-BUILDER interacts with the user by means of a friendly Web interface[3] that allows her to easily specify personal interests and time budget. The sightseeing tour proposed can be then explored and modified. We present the main components composing the system.

1 Introduction

Planning a travel itinerary is a difficult and time-consuming task for tourists approaching their destination for the first time. Different sources of information such as travel guides, maps, on-line institutional sites and travel blogs are consulted in order to devise the right blend of Points of Interest (PoIs) that best covers the subjectively interesting attractions and can be visited within the limited time planned for the travel. However, the user still need to guess how much time is needed to visit each single attraction, and to devise a *smart* strategy to schedule them moving from one attraction to the next one. Furthermore, tourist guides and even blogs, reflect the point of view of their authors, and they may result to be not authoritative sources of information when the tourist preferences diverge from the most popular flow.

These few considerations motivate TRIPBUILDER [1], an unsupervised approach that possibly overcomes the above limitations by exploiting the Wisdom-of-the-Crowds by past tourists to build a personalized plan of visit. We define the sightseeing tour generation problem as an instantiation of the Generalized Maximum Coverage (GMC) problem [2]. We model each visiting pattern by means of the PoIs and the associated Wikipedia categories, and the GMC profit function by considering PoIs popularity and the actual user preferences over the

---

[1] http://www.wikipedia.org/
[2] http://www.flickr.com/
[3] Appendix. http://www.lia.ufc.br/~igobrilhante/tripbuilder

M. de Rijke et al. (Eds.): ECIR 2014, LNCS 8416, pp. 771–774, 2014.

same Wikipedia categories. The cost function is instead built by considering the average visiting time for the PoIs in the patterns plus the time needed to move from one PoI to the next one. Our algorithm is thus able to provide visiting plans made up of actual touristy itineraries that are the most tailored to the specific preferences of the tourist TRIPBUILDER takes into account the personal preferences of the user and her temporal constraints.

In our previous work [1], we discussed the TRIPBUILDER methodology for generating personalized sightseeing tours. Here, we describe the system which implements TRIPBUILDER as a Web application. It enables the user to easily specify her preferences and ask the system for a detailed personalized tour in a given target city. Compared to other works in the literature (for example [4,3]), TRIPBUILDER solves the sightseeing tour planning problem in a completely unsupervised way and by taking user preferences into account. The touristy attractions are collected from Wikipedia while the tourist traces are mined from photo albums of Flickr.

2   TripBuilder

Time budgeted sightseeing tour generation involves four different steps (see Figure 1): i) *Data Collection*, ii) *Data Processing*, iii) *Data Storage*, and, iv) *TripBuilder Engine*.

Data Collection. It is composed by two different modules that retrieve the relevant information from Flickr and Wikipedia. In particular, the first one queries

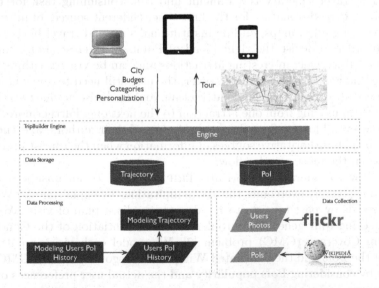

Fig. 1. Architecture of TRIPBUILDER. We outline the four modules of the system, i.e. *Data Collection, Data Processing, Data Storage* and *TripBuilder Engine*.

Flickr to retrieve the metadata (user id, timestamp, tags, geographic coordinates, etc.) of the sequences of photos taken in the given geographic area, e.g. the city bounding box. An important assumption here is that photo albums implicitly represent sightseeing itineraries within a city. To strengthen the accuracy of our method, this module retrieves only the photos having the highest geo-referencing precision. This process thus collects a large set of geo-tagged photo albums taken by different users in the given geographic area. The second module collects PoIs from Wikipedia. In particular, we assume each geo-referenced Wikipedia named entity, whose geographical coordinates falls into a given area, to be a fine-grained Point of Interest. For each PoI, we retrieve its descriptive label, its geographic coordinates as reported in the Wikipedia page, and the set of categories the PoI belongs to that are reported at the bottom of the Wikipedia page. Then, Flickr photos and Wikipedia PoIs are matched by spatial proximity according to their coordinates.

Data Processing. This module consists of different components each one manipulating part of the data previously collected to devise personalized tours and clean data. In particular, the modules here transform sequences of photos from Flickr to trajectories crossing Wikipedia PoIs to be used in the TRIPBUILDER module. Moreover, popularity and characteristics of PoIs are computed by considering the number of distinct photos taken, and the Wikipedia categories. The data obtained are then stored by means of the "Data Storage" module by using a structured database schema. This is an important point in favour of TRIP-BUILDER flexibility: different sources of information for trajectories and PoIs can be easily integrated into the system by changing/updating only the two lowest layers.

Data Storage. This component is responsible for storing, querying and indexing trajectory and PoI data. It is composed by a database management system that efficiently provides information to the "TripBuilder Engine" component. This component contains a well defined schema to enable flexibility in integrating other data sources. Geo-spatial indexes are used for searching spatial objects, such as PoIs and tourist traces, within a given region (e.g. polygon). The system also takes advantage of indexes over PoI categories and tourist traces, both represented as arrays, to efficiently retrieve relevant PoIs to the user preferences.

TripBuilder Engine. It is the core of the architecture. It receives a set of trajectories crossing a set of PoIs, a time budget, user preferences and a factor used to tune the level of personalization as input, and generates the personalized sightseeing tour. The problem of building the tour is modeled as an instance of the well-known Generalized Maximum Coverage problem [2]. Interest readers can find a detailed modeling of the problem in [1].

3   The User Interface

The user can initially select the targeted city from a list or from a photo list of the cities covered by the system. Once the city has been chosen, a map is presented jointly with a drop-down menu for which the user can set up her time budget

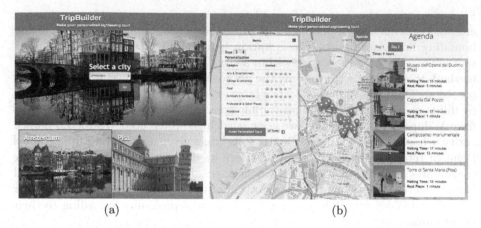

(a)   (b)

Fig. 2. A screenshot of the Web interface that lets users interact with TRIPBUILDER. The targeted city is initially selected (a). The drop-down menu on the left helps specify their preferences, the time budget and the personalization factor (b). On the right, a summary of the tour is proposed. Each PoI in the summary comes with a photo and a set of useful information (i.e., visiting time, categories, etc.).

(number of days), her preferences in terms of attractions categories (churches, bridges, museums, etc). After clicking on the "Create Personalized Tour" button, the tool returns the map of the city together with the recommended agenda. This agenda consists of distinct tours for each available day, highlighting the Points of Interests, some information and the estimated time required to visit each PoI. Indeed, the flexibility of the system allows us to easily extend the personalized sightseeing tour generation to several cities.

Acknowledgments. This work was partially supported by EU FP7 Marie Curie project SEEK (no. 295179), PRIN 2011 project ARS TECNOMEDIA, CNPQ Scholarship (no. 306806/2012-6, 211836/2013-3), CNPQ Casadinho/ PROCAD Project (no. 552578/2011-8), and CNPQ-CNR Bilateral Project (no. 490459/2011-0).

References

1. Brilhante, I., Macedo, J.A., Nardini, F.M., Perego, R., Renso, C.: Where shall we go today? planning touristic tours with tripbuilder. In: Proc. CIKM. ACM (2013)
2. Cohen, R., Katzir, L.: The generalized maximum coverage problem. Information Processing Letters 108(1), 15–22 (2008)
3. De Choudhury, M., Feldman, M., Amer-Yahia, S., Golbandi, N., Lempel, R., Yu, C.: Automatic construction of travel itineraries using social breadcrumbs. In: Proc. HT, pp. 35–44. ACM (2010)
4. Vansteenwegen, P., Souffriau, W., Berghe, G., Oudheusden, D.: The city trip planner: an expert system for tourists. Expert Systems with Applications 38(6), 6540–6546 (2011)

GTE-Cluster: A Temporal Search Interface
for Implicit Temporal Queries

Ricardo Campos[1,2,6], Gaël Dias[4], Alípio Mário Jorge[1,3], and Célia Nunes[5,6]

[1] LIAAD – INESC TEC
[2] Polytechnic Institute of Tomar, Portugal
[3] DCC – FCUP, University of Porto, Portugal
[4] HULTECH/GREYC, University of Caen Basse-Normandie, France
[5] Department of Mathematics, University of Beira Interior, Covilhã, Portugal
[6] Center of Mathematics, University of Beira Interior, Covilhã, Portugal
ricardo.campos@ipt.pt, gael.dias@unicaen.fr,
amjorge@fc.up.pt, celian@ubi.pt

Abstract. In this paper, we present GTE-Cluster an online temporal search interface which consistently allows searching for topics in a temporal perspective by clustering relevant temporal Web search results. GTE-Cluster is designed to improve user experience by augmenting *document relevance* with *temporal relevance*. The rationale is that offering the user a comprehensive temporal perspective of a topic is intuitively more informative than retrieving a result that only contains topical information. Our system does not pose any constraint in terms of language or domain, thus users can issue queries in any language ranging from business, cultural, political to musical perspective, to cite just a few. The ability to exploit this information in a temporal manner can be, from a user perspective, potentially useful for several tasks, including user query understanding or temporal clustering.

Keywords: Temporal Information Retrieval, Temporal Clustering, Implicit Temporal Queries.

1   Introduction

In recent years, a large number of temporal applications have been developed, mostly concerning Web archives (e.g. Internet Archive [1]), temporal taggers (e.g. HeidelTime [10]), temporal and spatial knowledge bases (e.g. Yago2 [5]), new forms of visualizing temporal data (e.g. SIMILE Timeline Visualization[1]), applications to track how topics evolve over time (e.g. Time Explorer [7], Google nGram viewer [8]), but also commercial services like recordedfuture.com

Despite a clear improvement of search and retrieval temporal related applications, current search engines are still mostly unaware of the temporal dimension. Indeed, in most cases, systems are limited to offer the user the chance to restrict the search to a

---

[1] http://www.simile-widgets.org/timeline/ [October 28th, 2013].

M. de Rijke et al. (Eds.): ECIR 2014, LNCS 8416, pp. 775–779, 2014.

particular time period or to simply ask him to explicitly specify a time span. If the user is not explicit in his search intents (e.g. "*los angeles earthquakes*") search engines may likely fail to present an overall historic perspective of the topic. Most search engines also provide query auto-completion suggestions to users after they start typing their query in the search box, but usually lack to include suggestions of temporal nature [1]. Similar problems can be observed for query re-formulation suggestions which are shown to the users after they submit their initial query. In both cases, query suggestions rely on past popularity of matching queries thus depending on the user's own knowledge and on the fact that some versions of the query have already been issued. While this works rather well for text suggestions it may cause some problems in case of temporal ones due to the fact that users are usually silent when it comes to explicitly express their temporal intents [9]. They are also largely unaware of the temporal dimension when the query is topically and temporally ambiguous (e.g. "*Madagascar*"). They may be able to detect the different facets of the query (country and movie), but remain alien of the fact that each one may have a different temporal nature.

The examples laid out above show that an end-to-end temporal retrieval system that consistently exploits temporal information is still to be seen. Such a retrieval system would be able to offer the user a temporal overview of the query and to provide information on its various temporal dimensions. But, it should also be able to only present the most relevant dates thus helping to improve the user satisfaction while meeting his information needs. For example, when querying "*margaret thatcher*", it would be interesting to have a few separate clusters (e.g. {1925, 1979, 1990, 2013}) highlighting the most important time periods (birth date, prime-minister period and death date) of this well-known British prime-minister.

This paper presents the GTE-Cluster temporal search interface which implements a flat temporal clustering model that groups documents at the year level. Our method is based on (1) the identification of relevant temporal expressions extracted from Web snippets and (2) a clustering methodology, where documents are grouped into the same cluster if they share a common year. The resulting clusters directly reflect groups of individual years that show a high connectivity to the text query. One such presentation of the results enables users to have a quick overview of a topic, without the need to go through an extensive list of results. As a result of our research, we publicly provide an online demo, which allows the execution of different kinds of queries, such as business (e.g. "*iPad*"), cultural (e.g. "*avatar movie*"), musical (e.g. "*Radiohead*") or natural disaster ones (e.g. "*Haiti earthquake*"). Although the main motivation of our work is focused on queries with temporal nature, the implemented prototypes allow the execution of any query including non-temporal ones.

2   System Overview

GTE-Cluster consists of five modules: Web search, Web snippet representation, temporal similarity, date filtering and temporal clustering. The demo interface receives a query from the user, fetches related Web snippets from a given search engine and applies text processing to all Web snippets. This processing task involves selecting the most relevant words/multiwords and collecting the candidate years in each Web snippet. Each candidate year is then given a temporal similarity value to the query computed by the GTE metric [3] in the temporal similarity module. We then apply a classification strategy in the date filtering module, to determine whether the candidate years are actually relevant or not to the query. Non-relevant ones will be simply discarded by the system. Each snippet is then clustered according to its associated years on the assumption that two snippets are temporally similar if they are highly related to the same set of dates. Since any Web snippet can contain several different relevant years, overlapping is allowed. The final set of clusters consists of m entities, where m is the number of relevant years. The temporally tagged m clusters are then sorted in ascending order. One of the advantages of our clustering model is that instead of considering all the temporal expressions as equally relevant as in [1], we determine which ones are more relevant to the user text query. One consequence of this is a direct impact on the quality of the retrieved clusters, as non-relevant or wrong dates are discarded. An evaluation of our approach using several performance metrics, a comparison against a well-known open-source Web snippet clustering engine and a user study demonstrates that GTE-Cluster improves the effectiveness of current approaches. Detailed results of our algorithm are available in [3,4].

3   Demo

The results of our research can be graphically explored by a demo search interface (http://wia.info.unicaen.fr/GTEAspNetFlatTempCluster_Server) made publicly available for research purposes and a video[2]. The implemented version is designed to demonstrate the current state of the demo, thus concerns of design nature where not taken into account. GTE-Cluster is designed to help users searching for information of a given topic through time without any temporal constraint. We rely on Bing Search API[3] with the *en-US* language parameter to retrieve 50 results per query. The proposed solution is computationally efficient and can easily be tested online (limited to 5000 queries per month). In response to a query submitted in a search box, GTE-Cluster displays a set of clusters generated on the fly, which can be instantly used for interactive browsing purposes. We offer two types of retrieval: one that returns only the relevant clusters (marked in blue) and one that combines relevant clusters with non-relevant ones (marked in red). Each cluster is assigned a temporal similarity value reflecting its similarity with the user query. This allows users to not only have an overall perspective of the relevance of the results, but also to evaluate the systems' effectiveness regarding decisions about the relevance of a temporal cluster. An illustration of the interface is provided in Fig. 1 for the query *"avatar movie"*.

---

[2] http://www.ccc.ipt.pt/~ricardo/software.html [October 28th, 2013].
[3] https://datamarket.azure.com/dataset/5BA839F1-12CE-4CCE-BF57-A49D98D29A44 [October 28th, 2013].

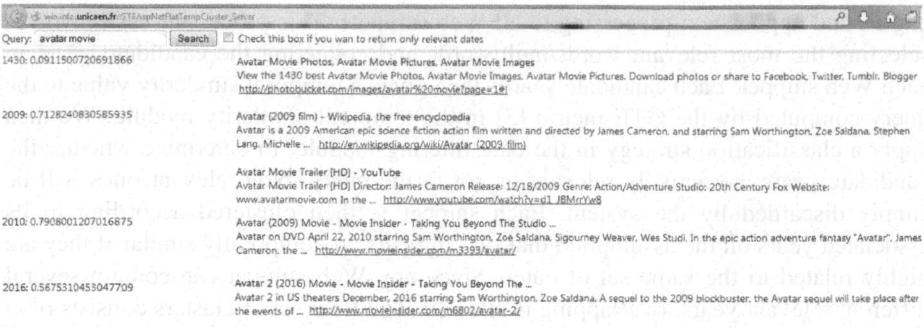

Fig. 1. GTE-Cluster interface for the query "*avatar movie*"

The values in front of the cluster reflect the similarity value computed by the GTE similarity measure. Note that clusters with a similarity value < 0.35 are considered non-relevant. In contrast, relevant clusters are marked in blue. It is worth noting that our algorithm is capable of detecting as non-relevant the clusters labeled as 1430, while detecting the most relevant ones, i.e., 2009, 2010 and 2016.

4   Conclusion and Future Work

In this paper, we presented GTE-Cluster a temporal search interface that focuses on disambiguating a text query with respect to its temporal purpose(s). We proposed a strategy for temporal clustering of Web search results, where snippets are clustered by year. We believe that the introduction of the temporal dimension will help to mitigate the limitations that users experience when their information needs include topics of a temporal nature. This is our first version approach to flat temporal clustering of search results. While we already achieved an initial stage of flat clustering by time, our proposal still lacks an approach focused on topics. We are aware that our solution is, from a clustering point of view, a straightforward algorithm. In spite of that, we believe this can open up the debate and create opportunities for future research improvements. As future research, we aim to provide an effective clustering algorithm that clusters and ranks snippets, both based on their temporal and conceptual proximities. A future approach should also consider a more elaborated mechanism in terms of ranking by applying an inter-cluster and an intra-cluster solution. This will enable to reduce the user's effort thus avoiding the need to go through all the clusters and snippets to find the most relevant one. Finally, we may use the similarity value associated to each cluster to offer an ordered set of temporal query suggestions. It would also be useful to consider re-formulations of the initial query for each identified time period as different terms may be associated to different years.

Acknowledgments. This work is financed by the ERDF – European Regional Development Fund through the COMPETE Programme (operational programme for competitiveness) and by National Funds through the FCT (Portuguese Foundation for Science and Technology) within project «FCOMP-01-0124-FEDER-037281».

References

1. Alonso, O., Gertz, M., Baeza-Yates, R.: Clustering and Exploring Search Results using Timeline Constructions. In: CIKM 2009, pp. 97–106. ACM Press (2009)
2. Campos, R., Jorge, A., Dias, G.: Using Web Snippets and Query-logs to Measure Implicit Temporal Intents in Queries. In: QRU 2011 Associated to SIGIR 2011, pp. 13–16 (2011)
3. Campos, R., Dias, G., Jorge, A.M., Nunes, C.: GTE: A Distributional Second-Order Co-Occurrence Approach to Improve the Identification of Top Relevant Dates. In: CIKM 2012, pp. 2035–2039 (2012)
4. Campos, R., Jorge, A.M., Dias, G., Nunes, C.: Disambiguating Implicit Temporal Queries by Clustering Top Relevant Dates in Web Snippets. In: WIC 2012, pp. 1–8. IEEE (2012)
5. Hoffart, J., Suchanek, F.M., Berberich, K., Lewis-Kelham, E., de Melo, G., Weikum, G.: YAGO2: Exploring and Querying World Knowledge in Time, Space, Context, and many Languages. In: WWW 2011, pp. 331–340. ACM Press (2011)
6. Kahle, B.: Preserving the Internet. Scientific American Magazine 276(3), 72–73 (1997)
7. Matthews, M., Tolchinsky, P., Blanco, R., Atserias, J., Mika, P., Zaragoza, H.: Searching through time in the New York Times. In: HCIR 2010 Workshop, pp. 41–44 (2010)
8. Michel, J.B., Shen, Y.K., Aiden, A.P., Veres, A., Gray, M.K., Pickett, J.P., Hoiberg, D., Clancy, D., Norvig, P., Orwant, J., Pinker, S., Nowak, M.A., Aiden, E.L.: Quantitative Analysis of Culture Using Millions of Digitized Books. Science 331(6014) (2011)
9. Nunes, S., Ribeiro, C., David, G.: Use of Temporal Expressions in Web Search. In: Macdonald, C., Ounis, I., Plachouras, V., Ruthven, I., White, R.W. (eds.) ECIR 2008. LNCS, vol. 4956, pp. 580–584. Springer, Heidelberg (2008)
10. Strötgen, J., Gertz, M.: HeidelTime: High Quality Rule-based Extraction and Normalization of Temporal Expressions. In: IWSE 2010 Associated to ACL 2010, pp. 321–324 (2010)

SQX-Lib: Developing a Semantic Query Expansion System in a Media Group

María Granados Buey, Ángel Luis Garrido, Sandra Escudero, Raquel Trillo,
Sergio Ilarri, and Eduardo Mena

IIS Department, University of Zaragoza, Zaragoza, Spain
{mgbuey,garrido,sandra.escudero,raqueltl,silarri,emena}@unizar.es

Abstract. Recently, there has been an exponential growth in the amount
of digital data stored in repositories. Therefore, the efficient and effective
retrieval of information from them has become a key issue. Organizations
use traditional architectures and methodologies based on classical relational databases, but these approaches do not consider the semantics of
the data or they perform complex ETL processes from relational repositories to triple repositories. Most companies do not carry out this type of
migration due to lack of time, money or knowledge.

In this paper we present a system that performs a semantic query
expansion to improve information retrieval from traditional relational
databases repositories. We have also linked it to an actual system and
we have carried out a set of tests in a real Media Group organization.
Results are very promising and show the interest of the proposal.

Keywords: information retrieval, query expansion, semantic search.

1   Introduction

Search systems in different contexts are adopting keyword-based interfaces due
to their success in traditional web search engines, such as Google or Bing. These
systems are usually based on the use of inverted indexes and different ranking
policies [1]. In the context of keyword-based search on structured sources [2],
most approaches retrieve data that exactly match the user keywords and indexed
terms. However, users often do not use the same words. So, there exist a term
mismatch problem, that reduce the retrieval effectiveness. The main problems
are the information lost (low recall) and the retrieval of non-relevant data (low
precision).

To deal with this problem, several approaches have been proposed that are
applicable in *Automatic Query Expansion (AQE)* solutions: interactive query
refinement, relevance feedback, word sense disambiguation, and search results
clustering. One of the most natural and successful techniques is to expand the
original query with other words that best capture the actual user intent, or that
simply produce a more useful query that is more likely to retrieve relevant documents. The survey presented in [3] provides a classification of existing techniques
that leverage several data sources and employ sophisticated methods for finding

M. de Rijke et al. (Eds.): ECIR 2014, LNCS 8416, pp. 780–783, 2014.
© Springer International Publishing Switzerland 2014

new features correlated with the query terms, and firmer theoretical foundations and a better understanding of the utility and limitations of AQE. In the last years, basic techniques are being used in conjunction with other mechanisms to increase their effectiveness, including the combination of methods, more active selection of information sources, and discriminative policies. AQE is currently considered a promising technique to improve the retrieval effectiveness of document ranking and it is being adopted in commercial applications, especially for desktop and intranet searches. Although AQE has received a great deal of attention and several approaches have been proposed, very little work has been done to review such studies.

Under these circumstances, we have developed a library called SQX-Lib (Semantic Query eXpansion Library) that exploits linguistic and semantic techniques to improve the quality of the keyword-based search process on the relational repositories of the Heraldo Media Group[1]. A demonstration of its operation (including a video, snapshots, etc.) can be seen at http://sid.cps.unizar.es/SEMANTICWEB/GENIE/Genie_Downloads.html. SQX-Lib is focused on automatically and semantically expanding the scope of the searches, and fine-tuning them by taking advantage of Named Entities (NE) present in the query string. For this purpose, it uses lexical resources and dictionaries, and an unsupervised disambiguation method that looks for the accurate meaning of a word.

2   System Overview

SQX-Lib can be included in a search engine or browser. The library expands and enriches semantically a given set of user keywords to improve the search process. This library can be used with any data source, as the expanded set of keywords obtained can be translated into a query in the appropriate query language of the data source, such as SQL in our case study, where relational databases are used. The semantic expansion process (Fig. 1) performs three main tasks:

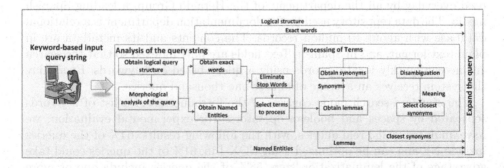

Fig. 1. Processing steps of SQX-Lib

[1] http://www.grupoheraldo.com/

1. *Analysis of the keywords of the query*: SQX-Lib analyzes the keywords in-troduced as a query. First, it obtains the logical query structure to perform an appropriate construction of the query. It processes the logical operators, parentheses and quotes that could appear in the search string. Then, SQX-Lib performs a morphological analysis of the query string using an NLP tool. Afterwards, the process carries out a second analysis of the query string in order to obtain all the NE. Moreover, it also performs the analysis of the query to obtain exact words, that is the words that have been introduced quoted. These words are searched as they have been written in the query string and they are not be processed to find their semantics. After this, SQX-Lib removes stop words. Finally, it selects the terms (common names or verbs in our case) that are going to be processed to find their semantics to enrich the input query.

2. *Processing of terms*: SQX-Lib obtains the lemmas of the terms chosen in the previous step using an NLP tool. Then, it searches the sets of synonyms for each term selected, by using the lemmas and semantic resources (dictionar-ies and lexical thesaurus). In our case, the lexical thesaurus used has been EuroWordNet [4] which is a multilingual resource for several European lan-guages. If the terms have more than one set of synonyms, SQX-Lib performs a disambiguation process to select the most appropriate according to the context, and taking into account an adaptation of the Normalized Google Distance (NGD) [5] for a document repository.

3. *Expand the query*: SQX-Lib reconstructs and expands the query from the relevant data extracted in each of the previous tasks while keeping the initial logical structure of the query.

3   Experimental Results

The aim of this work is to improve the user experience on a specific search system. The present case is a search engine that is working over the news repository used everyday by all the departments of the Heraldo Group, a leading Spanish media. The data repository used by the documentation department is a relational database with about 10 million records. The contents and its metadata are in plain text format, and in Spanish. Text fields are fully indexed so that the search engine can quickly locate records using a query based on keywords. Every day the system receives an average of almost one thousand queries.

The standard search mode consists of the introduction of a list of keywords separated by spaces and boolean operators. For experimental evaluation, we have analyzed 29353 real queries, with the following results: 64% of the queries include NE and can be optimized using SQX-Lib, 61% of the queries could take advantage of the lemmatization step, 58% of the queries include one or more common names from which the system is able to obtain synonyms and related words, and finally 13% of the queries could lead to build an incorrect query. So, we have found that about 95% of the queries made to this repository are expandable/optimized by our system.

If we analyze the results of these queries, we obtain the following conclusions: The lemmatization expansion over the query leads to an average improvement of 50% in the recall, the synonym expansion step translates into obtaining about double results on average, and the use of an automatic filter when we found a NE implies a reduction of approximately 20% of the noise. If we use all the options together, we obtain almost four times more results than using the original query, so SQX-Lib successfully minimizes the documentary silence problem. Besides, this is not done at the expense of introducing more noise.

We have carried out an opinion survey among the workers of the Heraldo Group about their use of the system improved with SQX-Lib:

- 70% of the users would include the lemmatization expansion. The rest of users would prefer that this feature be optional.
- 58% like the synonym expansion and they would include it by default.
- 88% agree about the need of a default mechanism to filter the results using the name entities embedded in the query.

In general, all the users are satisfied with the new functions although not all of them agree about whether they must be optional or applied by default. This is due to the different searching style of each department, which in turn is caused by the different types of information that they want to find. In some cases there are few records and it is preferable to use mechanisms to extend the results from the beginning, and in others the opposite occurs because the noise is initially excessive even with a simple query.

4   Demonstration

Our demo will allow to perform queries against an extract of the documentary database, it will show the expanded query, and it will compare the results of the query without expanding with the answers of the query once expanded.

Acknowledgment. This research work has been supported by the CICYT project TIN2010-21387-C02-02 and DGA-FSE. Thank you to Heraldo Group and Jorge Gracia.

References

1. Baeza-Yates, R., Ribeiro-Neto, B., et al.: Modern information retrieval, vol. 463. ACM Press, New York (1999)
2. Yu, J.X., Qin, L., Chang, L.: Keyword search in databases. Synthesis Lectures on Data Management 1(1) (2009)
3. Carpineto, C., Romano, G.: A survey of automatic query expansion in information retrieval. ACM Computing Surveys (CSUR) 44(1) (2012)
4. Vossen, P.: EuroWordNet: A multilingual database with lexical semantic networks. Kluwer Academic, Boston (1998)
5. Cilibrasi, R.L., Vitanyi, P.M.: The Google similarity distance. IEEE Transactions on Knowledge and Data Engineering 19(3), 370–383 (2007)

EntEXPO: An Interactive Search System for Entity-Bearing Queries

Xitong Liu, Peilin Yang, and Hui Fang

University of Delaware, Newark, DE, USA
{xtliu,franklyn,hfang}@udel.edu

Abstract. The paper presents EntEXPO, a search system that aims to improve the search experience for entity-bearing queries. In particular, the system exploits the entities and their relations in the context of query to identify a list of related entities and leverage them in an entity-centric query expansion method to generate more effective search results. Moreover, EntEXPO displays the related entities along with the search results to allow search users explore entity relationship to further refine the query through an interactive interface.

Keywords: entity centric, interactive search.

1 Introduction

Entities have been playing an important role in the information seeking process. Many Web search queries are related to entities. For example, Pound et al. [1] found that more than 50% of queries in a sample of Web queries are related to entities. In the meantime, significant efforts have been put to build knowledge bases such as Freebase, DBpedia and YAGO, which contain valuable information about entities and their relations. Therefore, it would be interesting to study how to leverage these knowledge bases to further improve the search experience of Web search users for entity-bearing queries.

In this paper, we describe our efforts on building a novel search system, i.e., EntEXPO, which exploits entity relationships in the context of queries. The system first identifies entities from a query and then retrieves a list of related entities based on the information from both the document collection and knowledge base. The related entities are used for (1) entity-centric query expansion [2,3], which expands original queries with the related entities to improve the search quality; and (2) entity-centric query reformulation, which allows search users to interact with the system by manually adjusting the weight for each related entity.

Our demonstration plan includes the following three parts. First, we will demonstrate the effectiveness of search by using entity-centric query expansion method. In particular, we will conduct side-by-side comparison between these results with those generated by using existing pseudo relevance feedback methods and highlight the differences. Second, we will explain that the generated entity-relation graph can help users understand the relations among entities in

M. de Rijke et al. (Eds.): ECIR 2014, LNCS 8416, pp. 784–788, 2014.

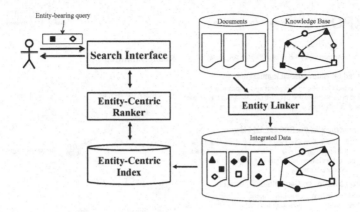

Fig. 1. System Architecture of EntEXPO

a better way and enable them to explore the information space more effectively than existing query suggestion results. Finally, we will discuss how the related entities and their relations can be used to reformulate the query by changing the weights of the related entities in the weight panel.

2 System Description

Figure 1 shows the architecture of EntEXPO, and we now describe the major components of the system in more details.

Entity Linker: Entities and their relationships exist in both the document collection and knowledge base. The information from the knowledge base is often structured, while those from the data collection is not. The main functionality of *entity linker* is to extract the entity information from the document collection using Named Entity Recognizer and link them with the corresponding entries in the knowledge base. This would allow us to leverage the entity related information in a more systematic way.

Entity-Centric Index: In addition to the traditional term-based inverted index, we also need build an entity-based inverted index on the integrated data to accelerate the process of finding documents that mention some entities. Entity-centric index is particularly useful when finding related entities for a given query.

Entity-Centric Ranker: The entity-centric ranker performs two tasks. It first retrieves a list of related entities for the query based on the entity relations in the integrated data. With the related queries, it could either automatically expand the original query using the entity-centric approach proposed in our previous studies [2,3] or allow users to reformulate the query by manually adjusting the weight for each related entity.

Search Interface: Besides the search box and document retrieval list as in the regular Web search result page, the search interface has two more components:

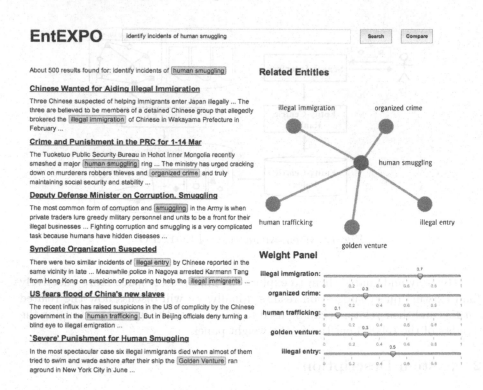

Fig. 2. Search interface of EntEXPO

related entity graph and entity weight panel. Figure 2 shows the search interface of EntEXPO. The query is "identify incidents of human smuggling". We notice that "human smuggling" is identified as a named entity and denoted in pink color in the top of left column. The right column shows the related entity graph, which visualizes the relation between the query entity and related entities[1] including "illegal immigration", "organized crime", "human trafficking", "golden venture" and "illegal entry". All the related entities are denoted in blue color in the snippets of search results. If the user clicks on one related entity in the graph, the documents which contain the related entity will be highlighted accordingly to help her understand why these documents are retrieved. If the user clicks on the bar connecting the query entity and a related entity, documents mentioning both entities will be shown in the result list to help her understand the relations between them. The entity weight panel is located beneath the related entity graph, in which each related entities is associated with a weight slider. By default each slider shows the weight suggested by EntEXPO automatically. The user can adjust the weight of each related entity individually to reformulate the query based on her inspection of document retrieval list.

---

[1] At most the top 5 related entities will be shown. If no related entity is found, the related entity graph and entity weight panel will remain blank, and search results will be retrieved based on default model without query expansion.

3   Demonstration Plan

In the demo, we use the documents from TREC 2004 Robust Track as the document collection and the English version of DBpedia 3.7 as the knowledge base. The document collection contains 528,155 documents, while the DBpedia contains 3.77 million entries with 400 million facts. Search users are allowed to type their own queries. Alternatively, they can also select a query from the official query set from TREC 2004 Robust track. The major advantage of selecting TREC queries is that users can immediately see the effectiveness of retrieval methods since the judgments for these queries are available.

We plan to demonstrate three main features of EntEXPO as follows:

- **Better retrieval effectiveness:** Users can try out their own queries and go through the search results by themselves, as shown in Figure 2. More importantly, we will also provide a comparison interface to allow users to compare the results of EntEXPO with those using existing pseudo relevance feedback methods side by side. If user selects a TREC query, the interface will display the effectiveness of both systems and highlight the difference between their search results.
- **Entity-centric information exploration:** EntEXPO displays an entity-relation graph along with the document retrieval results. The graph contains the entities from the query as well as those related entities with respect to the query. The information about these entities and relations (by clicking the nodes or edges) allows users to learn more information about the query. This feature is expected to be particularly useful for exploratory search, where users need more guidance to navigate the information spaces in order to formulate appropriate queries.
- **Interactive query formulation.** By leveraging the entity weight panel, users can reformulate the query by rewarding or penalizing the related entities. Upon receiving the new results from the entity-centric ranker, user may continue adjusting the weight to explore the parameter space iteratively, or stop when satisfying results are found. The benefits are two-fold: (1) it saves user's efforts from manually reformulating query by changing the query in the search box and conducting search once again. (2) it provides more flexibility to reformulate the query with precise control of term weighting than adding or removing terms alone.

4   Conclusion

We propose and develop a novel search system, i.e., EntEXPO, for entity-bearing queries. It exploits entity relationship to improve search accuracy as well as providing an interactive interface to explore the information space surrounded by the related entities and reformulate the query. We believe that EntEXPO could, in particular, benefit users whose information seeking tasks are exploratory and entity-related.

References

1. Pound, J., Mika, P., Zaragoza, H.: Ad-hoc Object Retrieval in the Web of Data. In: WWW, pp. 771–780 (2010)
2. Liu, X., Fang, H., Chen, F., Wang, M.: Entity Centric Query Expansion for Enterprise Search. In: CIKM, pp. 1955–1959 (2012)
3. Liu, X., Chen, F., Fang, H., Wang, M.: Exploiting Entity Relationship for Query Expansion in Enterprise Search. Information Retrieval (2013) (in press)

Column Stores as an IR Prototyping Tool

Hannes Mühleisen[1], Thaer Samar[1], Jimmy Lin[2], and Arjen P. de Vries[1]

[1] Centrum Wiskunde & Informatica, Amsterdam, The Netherlands
[2] University of Maryland, College Park, Maryland, USA
{hannes,samar}@cwi.nl, jimmylin@umd.edu, arjen@acm.org

Abstract. We make the suggestion that instead of implementing custom index structures and query evaluation algorithms, IR researchers should simply store document representations in a column-oriented relational database and write ranking models using SQL. For rapid prototyping, this is particularly advantageous since researchers can explore new ranking functions and features by simply issuing SQL queries, without needing to write imperative code. We demonstrate the feasibility of this approach by an implementation of conjunctive BM25 using MonetDB on a part of the ClueWeb12 collection.

1 Introduction

Information retrieval researchers and practitioners have long implemented specialized data structures and algorithms for document ranking [6]. Today, these techniques can be quite complex, especially with "structured queries" that span multiple nested clauses with a panoply of query operators [4]. We make the suggestion that the information retrieval community should look towards the database community for general purpose data management solutions—namely, column-oriented relational databases (or column stores). We show that one such system, MonetDB, can be used for storing postings lists and that conjunctive query evaluation with BM25 can be translated into an SQL query. We advocate this approach especially for rapid prototyping in an IR research setting.

What advantages does using a database have? We see many, beginning with a precise formal framework. Relational databases provide a formal and theoretically-sound framework in which to express any query evaluation algorithm—namely, relational calculus (or, practically, SQL). This forces IR researchers to be precise about query semantics, which may be especially useful when complex query operators are introduced in document ranking.

Second, taking advantage of relational databases yields a cleaner architecture. Almost all IR systems today are monolithic agglomerations of components for text processing, document inversion, integer compression, memory/disk management, query evaluation, feature extraction, machine learning, etc. By offloading the storage management to a relational database, we introduce a clean abstraction (via SQL) between the "low-level" components of the engine and the IR-specific components (e.g., learning to rank). This (hopefully) reduces overall system complexity and may allow different IR engines to inter-operate.

M. de Rijke et al. (Eds.): ECIR 2014, LNCS 8416, pp. 789–792, 2014.

Third, retrieval systems can benefit from advances in data management. In IR, efficiency is a relatively niche topic. In contrast, performance is a dominant preoccupation of database researchers, who make regular breakthroughs that eventually propagate to IR (for example, PForDelta [7], the best practice for index compression today, originated from database researchers). By using relational databases, IR systems can benefit from future advances "for free".

Finally, databases form a flexible rapid prototyping tool. Many IR researchers do not really care about index structures and query evaluation *per se*—they are merely means to an end, such as assessing the effectiveness of a particular ranking function. In this case, forcing researchers to design data structures and query evaluation algorithms is a burden. Using a relational database, researchers can rapidly experiment by issuing declarative SQL queries without needing to write (error-prone) imperative or object-oriented code.

There are many similarities between query evaluation in document retrieval and online analytical processing (OLAP) tasks in modern data warehouses. Both frequently involve scans, aggregations, and sorting. Thus, we believe that column-oriented databases, which excel at OLAP queries, are amenable to retrieval tasks. An overview of such databases is beyond the scope of this work, but the basic insight is to decompose relations into columns for storage and processing [2]. In this paper, we use exactly such a system, MonetDB, to illustrate document ranking on a portion of the ClueWeb12 collection.

2   Okapi BM25 in a Relational Database

All IR engines pre-process the input collection prior to indexing, and this is no different when using a relational database. For this task, we take advantage of Hadoop MapReduce to convert a document collection into a collection of relational tables. We performed tokenization and stemming using the Krovetz stemmer with a stopwords list from Terrier. The stemmed and filtered terms were mapped to integer ids and stored in a dictionary table. In the main terms table, we store all terms in a document (by term id), along with the position in which they occur. To give a concrete example, consider the document doc1 with the content "I put on my robe and wizard hat". If we assume "I", "on", "my" and "and" are stopwords, the relational tables generated from this document have the following content:

table: dict			table: terms			table: docs		
term	termid	df	termid	docid	pos	docid	name	len
1	put	1	1	1	2	1	doc1	8
2	robe	1	2	1	5			
3	wizard	1	3	1	7			
4	hat	1	4	1	8			

We have chosen Okapi BM25 to implement as an SQL query, but our approach can be easily extended to other ranking functions. Our experiments focused on conjunctive query evaluation, where the document must contain all terms;

previous work [1] has shown that this approach yields comparable end-to-end effectiveness to disjunctive query evaluation, but is substantially faster. When scoring documents based on BM25, the only score component that depends on the query–document combination is the term frequency $f(q_i, D)$. The document length $|D|$ for each document and the document frequency $n(q_i)$ for each term can be precomputed. By sorting the terms table (from above) by term id (in effect performing document inversion), we can avoid scanning the table entirely. The complete ranking function can be expressed as follows:

```
1    WITH qterms AS (SELECT termid, docid FROM terms
2        WHERE termid IN (10575, 1285, 191)),
3    subscores AS (
4    SELECT docs.docid, len, term_tf.termid,
5        tf, df, (log((45174549-df+0.5)/(df+0.5))* ((tf*(1.2+1)/(tf+1.2*(1-
6        0.75+0.75*(len/513.67)))))) AS subscore
7        FROM (SELECT termid,  docid,  COUNT(*) AS tf FROM qterms
8        GROUP BY docid, termid) AS term_tf
9        JOIN (SELECT docid FROM qterms
10           GROUP BY docid HAVING COUNT(DISTINCT termid) = 3)
11           AS cdocs ON term_tf.docid=cdocs.docid
12       JOIN docs ON term_tf.docid=docs.docid
13       JOIN dict ON term_tf.termid=dict.termid)
14   SELECT name, score FROM (SELECT docid, sum(subscore) AS score
15       FROM subscores GROUP BY docid) AS scores JOIN docs ON
16       scores.docid=docs.docid ORDER BY score DESC LIMIT 1000;
```

To explain, we map conjunctive BM25 ranking to SQL in three parts: First, we find the entries in the terms table for query terms (Lines 1 and 2).[1] In this case, the query terms have ids 10575, 1285, and 191. The second step calculates the individual scores for all term/document combinations (Lines 4-13). To express the conjunctivity in the query, we filter this intermediate result to only include term/document combinations with exactly three different term ids for each document. (Lines 9 to 11). We also collect information about document ids and lengths (Line 12) as well as the document frequency of the terms (Line 13). Then, we calculate the individual BM25 scores for each term/document combination (Lines 5 and 6) and sum the results up and sort (Lines 14 to 16). The numbers printed in bold are the only parts of the SQL query that depend on the document collection and the query terms.

---

[1] Although in theory it would be possible to "join in" the dictionary by term, in practice, performing the mapping from query terms to identifiers turns out to be preferable outside SQL.

Note that the entire ranking function is expressed as an SQL query and this approach can be extended to any scoring function that is a sum of matching query terms. Disjunctive query evaluation can be implemented by replacing inner joins with outer joins. Phrase queries are performed by simple arithmetic over term positions and relative phrase positions. In other words, IR researchers can rapidly explore different retrieval options without writing imperative code.

We have implemented the above ranking function using the open-source columnar database MonetDB [3]. Our experiments used the first segment on the first disk of ClueWeb12 ($\sim$45 million documents). We ran queries 201–250 from the TREC 2013 web track and verified that both query latency and effectiveness are comparable to existing open-source engines such as Terrier [5] and Indri [4].

3   Demonstration Description

Our demonstration allows a user to interactively compare our MonetDB solution to existing open-source search engines. Users submit queries in a web interface, which are concurrently dispatched to multiple backends. The interface displays results from each backend as soon as it completes, allowing the user to compare the performance of each search engine along with the effectiveness of the results.

Acknowledgments. This research was supported by the Netherlands Organization for Scientific Research (NWO project 640.005.001) and the Dutch national program COMMIT/.

References

1. Asadi, N., Lin, J.: Effectiveness/efficiency tradeoffs for candidate generation in multi-stage retrieval architectures. In: SIGIR, pp. 997–1000 (2013)
2. Copeland, G., Khoshafian, S.: A decomposition storage model. In: SIGMOD (1985)
3. Idreos, S., Groffen, F., Nes, N., Manegold, S., Mullender, K.S., Kersten, M.L.: MonetDB: Two decades of research in column-oriented database architectures. IEEE Data Engineering Bulletin 35(1), 40–45 (2012)
4. Metzler, D., Croft, W.B.: Combining the language model and inference network approaches to retrieval. IP&M 40(5), 735–750 (2004)
5. Ounis, I., Amati, G., Plachouras, V., He, B., Macdonald, C., Lioma, C.: Terrier: A high performance and scalable information retrieval platform. In: SIGIR Workshop on Open Source Information Retrieval (2006)
6. Zobel, J., Moffat, A.: Inverted files for text search engines. ACM Computing Surveys 38(6), 1–56 (2006)
7. Zukowski, M., Heman, S., Nes, N., Boncz, P.: Super-scalar RAM-CPU cache compression. In: ICDE, pp. 59–59 (2006)

CroSeR: Cross-language Semantic Retrieval of Open Government Data

Fedelucio Narducci[1], Matteo Palmonari[1], and Giovanni Semeraro[2]

[1] Department of Computer Science, Systems and Communication
University of Milano-Bicocca, Italy
surname@disco.unimib.it
[2] Department of Computer Science
University of Bari Aldo Moro, Italy
giovanni.semeraro@uniba.it

Abstract. CroSer (Cross-language Semantic Retrieval) is an IR system able to discover links between e-gov services described in different languages. CroSeR supports public administrators to link their own source catalogs of e-gov services described in any language to a target catalog whose services are described in English and are available in the Linked Open Data (LOD) cloud. Our system is based on a cross-language semantic matching method that i) *translates* service labels in English using a machine translation tool, ii) *extracts* a Wikipedia-based semantic representation from the translated service labels using Explicit Semantic Analysis (ESA), iii) *evaluates* the similarity between two services using their Wikipedia-based representations. The user selects a service in a source catalog and exploits the ranked list of matches suggested by CroSeR to establish a relation (of type narrower, equivalent, or broader match) with other services in the English catalog. The method is independent from the language adopted in the source catalog and it does not assume the availability of information about the services other than very short text descriptions used as service labels. CroSeR is a web application accessible via http://siti-rack.siti.disco.unimib.it:8080/croser/[1].

1 Introduction and Related Research

In the last years, many governments decided to make public their data about spending, service provision, economic indicators, and so on. These datasets are also known as Open Government Data (OGD). As of October 2013, more than 1,000,000 OGD datasets have been put online by national and local governments from more than 40 countries in 24 different languages[2]. The interconnection of

---

[1] The work presented in this paper has been partially supported by the Italian PON project PON01_00861 SMART-Services and Meta-services for smART eGovernment and PON01_00850 ASK-Health (Advanced system for the interpretations and sharing of knowledge in health care) funded by the Italian Ministry of University and Research (MIUR).

[2] http://logd.tw.rpi.edu/iogds_data_analytics

information coming from different sources has several advantages, improving the discovery and reuse of data by consumers and third-party applications and supporting rich data analysis [4]. These advantages motivate the uptake of Linked Open Data (LOD) as a paradigm for the publication of data on the Web [1] and apply also to OGD [2]. The Local Government Service List (LGSL) as part of the Electronic Service Delivery (ESD)-standards[3] is an English catalog of e-gov services already in the LOD cloud. It is one of the results of the SmartCities project[4] and is now connected with service catalogs of five european countries. LGSL counts about 1,400 distinct services. The ESD-standard lists define the semantics of public sector services. Linking service catalogs to the LOD cloud (i.e., LGSL) is a big opportunity for a large number of administrations to make their services more accessible and comparable at a cross-national level. However, discovering links between services described in different languages requires a significant human effort because of the number of service descriptions that have to be compared and because of the linguistic and cultural barriers. Automatic cross-language semantic retrieval methods can support local and national administrations in linking their service catalogs to the LOD cloud, by reducing the cost of this activity.

In this paper we present CroSeR, a system that given an e-gov service (source) in any language is able to retrieve a set of candidate services (target) from the LOD cloud that are related to the source service. The user can then establish a relation between the source and one or more target services using the SKOS *narrow*, *broader* and *equivalent* match predicates. It is worth noting that labels in different languages are not a mere translation from a language to another. For example, the German service *Briefwahl* (translated in English as *absentee ballot*), is linked to the English service *Postal voting*. We can observe that the semantics of these two services is very similar, but there are no keywords in common: a keyword-based matching is dooomed to fail.

The domain of e-gov services poses several challenges to methods proposed so far because of the poor quality of the descriptions, which often consist of the name of the service and very few other data, and the semantic heterogeneity of names refereeing to linkable service, which is due to cultural differences across countries. As our approach, also the majority of the work in the literature of cross-language ontology matching [7] reduces the problem to a monolingual task by leveraging automatic translation tools [9]. However, these systems can exploit richer representations or use training examples, and generally are deeply based on structural information that are not available in our scenario. An interesting work presented in literature applies the Explicit Semantic Analysis (ESA) to cross-language link discovery [3]. However, in this specific domain, algorithms can leverage a significant amount of text that is not available in our case. A cross-language version of ESA (CL-ESA that does not require any translation process of the input text) is proposed in [8] for cross-lingual and multilingual retrieval. CL-ESA was evaluated on multilingual documents provided with quite large textual

---

[3] http://www.esd.org.uk/esdtoolkit/
[4] http://www.smartcities.info/aim

descriptions. Another system that leverages Wikipedia for CLIR is WikiTranslate [6]. WikiTranslate performs a query translation only based on Wikipedia by representing the query in terms of Wikipedia concepts. We preliminarily evaluated these approaches in CroSeR, but results were not satisfactory, probably due to the concise descriptions available in our domain both for query (source service) and target services. The novelty and the effectiveness of the matching techniques adopted in CroSeR have been discussed in a previous paper [5]. In this system demo we show the functionalities of the system and explain features that were not included in the version described by the previous paper.

2 CroSeR: Architecture and Functionalities

CroSeR is based on the hypothesis that extracting semantic annotations from service descriptions can support effective matching methods even if the available descriptions are poor and equivalent services can be described very differently in different countries; this is in fact the case for most of the service catalogs considered in the LGSL and for most of the service catalogs provided by local administrations. We therefore assume that each service is described only by short textual description (i.e., service *labels*) which represents a high-level description of a concrete service offered by one or more providers[5].

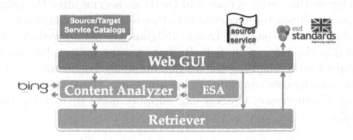

Fig. 1. CroSeR general architecture

Figure 1 depicts the general architecture of CroSeR. We can observe three main components: the *Web GUI*, the *Content Analyzer* and the *Retriever*. The user interacts with CroSeR using the *Web GUI*. The *Content Analyzer* processes the service labels and builds a semantic annotation for each service; the Content Analyzer is used to process both the source and the target catalogs; in our case the source catalog is represented by a list of services labeled in any natural language, while the target catalog is represented by the LGSL labeled in English. The *Retriever* takes a semantically annotated service (in any language in which a source service catalog is available) as input and returns a ranked list of matching services. The *Content Analyzer* uses external automatic translation tools and

---

[5] http://www.smartcities.info/files/Smart_Cities_Brief_What_is_a_service_list.pdf

Explicit Semantic Analysis (ESA) techniques to annotate services with Wikipedia concepts; however, other annotation methods can be easily plugged into the CroSeR architecture. The *Retriever* component evaluates the similarity between a query, represented by an input service, and the services in the target catalog.

3 Demonstration Summary

The first step that the CroSeR user should perform is to upload the service catalog (a set of services in a structured form) into the system (this functionality is disabled in the demo version). After that, the catalog will be semantically analyzed and indexed. The user is now able to explore the catalog just uploaded by scrolling the whole list of services or by performing a keyword-based search. Next, the user selects a source service from the catalog and CroSeR retrieves a list of *candidate target services* from the LGSL that are potentially linkable by a SKOS *narrow, broader* and *equivalent* predicates. The number of retrieved services is configurable by the user. Sometimes the potential relation between the source service and the target service is not straightforward, and cannot be effectively evaluated by looking only at the service labels. Hence, users can then select a candidate service and look at further details ("Service Info" box) directly gathered from the ESD-toolkit. Finally, users can switch on the feedback mode of CroSeR (registration and login are required) and thus the system stores the relation between the source service and the LGSL service after the selection of a candidate service from the retrieved list. Please consider that a set of catalogs already linked to the LGSL (i.e., Dutch, Belgian, German, Swedish, Norwegian) are uploaded in the demo of CroSeR. In that case the web application shows the gold standard (by highlighting the service manually connected by *owl:sameAs* statement), and the user can compare the automatic suggestion with the human annotations. Furthermore, CroSeR can support users in revising the already existing links.

References

1. Bizer, C., Heath, T., Berners-Lee, T.: Linked Data - The Story So Far. IJSWIS 5(3), 1–22 (2009)
2. Ding, L., Peristeras, V., Hausenblas, M.: Linked Open Government Data. IEEE Intelligent Systems 27(3), 11–15 (2012)
3. Knoth, P., Zilka, L., Zdrahal, Z.: Using Explicit Semantic Analysis for Cross-Lingual Link Discovery. In: 5th Int.l Workshop on Cross Lingual Information Access (2011)
4. Mendes, P.N., Jakob, M., García-Silva, A., Bizer, C.: DBpedia Spotlight: Shedding Light on the Web of Documents. In: I-SEMANTICS 2010, pp. 1–8. ACM (2011)
5. Narducci, F., Palmonari, M., Semeraro, G.: Cross-Language Semantic Retrieval and Linking of E-Gov Services. In: Alani, H., Kagal, L., Fokoue, A., Groth, P., Biemann, C., Parreira, J.X., Aroyo, L., Noy, N., Welty, C., Janowicz, K. (eds.) ISWC 2013, Part II. LNCS, vol. 8219, pp. 130–145. Springer, Heidelberg (2013)

6. Nguyen, D., Overwijk, A., Hauff, C., Trieschnigg, D.R.B., Hiemstra, D., de Jong, F.: WikiTranslate: Query Translation for Cross-Lingual Information Retrieval using only Wikipedia. In: Peters, C., Deselaers, T., Ferro, N., Gonzalo, J., Jones, G.J.F., Kurimo, M., Mandl, T., Peñas, A., Petras, V. (eds.) CLEF 2008. LNCS, vol. 5706, pp. 58–65. Springer, Heidelberg (2009)

7. Shvaiko, P., Euzenat, J.: Ontology Matching: state of the art and future challenges. IEEE Trans. Knowl. Data Eng. 25(1), 158–176 (2013)

8. Sorg, P., Cimiano, P.: Exploiting Wikipedia for Cross-lingual and Multilingual Information Retrieval. DKE 74, 26–45 (2012)

9. Spohr, D., Hollink, L., Cimiano, P.: A Machine Learning approach to Multilingual and Cross-lingual Ontology Matching. In: Aroyo, L., Welty, C., Alani, H., Taylor, J., Bernstein, A., Kagal, L., Noy, N., Blomqvist, E. (eds.) ISWC 2011, Part I. LNCS, vol. 7031, pp. 665–680. Springer, Heidelberg (2011)

Bibliometric-Enhanced Information Retrieval

Philipp Mayr[*], Andrea Scharnhorst, Birger Larsen, Philipp Schaer,
and Peter Mutschke

GESIS – Leibniz Institute for the Social Sciences,
Unter Sachsenhausen 6-8, 50667 Cologne, Germany
philipp.mayr@gesis.org

Abstract. Bibliometric techniques are not yet widely used to enhance retrieval processes in digital libraries, although they offer value-added effects for users. In this workshop we will explore how statistical modelling of scholarship, such as Bradfordizing or network analysis of coauthorship network, can improve retrieval services for specific communities, as well as for large, cross-domain collections. This workshop aims to raise awareness of the missing link between information retrieval (IR) and bibliometrics / scientometrics and to create a common ground for the incorporation of bibliometric-enhanced services into retrieval at the digital library interface.

Keywords: Bibliometrics, Informetrics, Scientometrics, Information Retrieval, Digital Libraries.

1   Background

The information retrieval (IR) and bibliometrics / scientometrics communities move more closely together with combined recent workshops like "Computational Scientometrics"[1] (held at iConference 2013 and CIKM 2013) and "Combining Bibliometrics and Information Retrieval"[2] (held in July at the ISSI conference 2013) which was organized by the authors of this workshop proposal. The ISSI workshop attracted more than 80 participants. The high interest among the bibliometricians was also generated by contributions from three "de Solla-Price"-medal winners and leading-edge bibliometricians Michel Zitt, Wolfgang Glänzel and Howard D. White. The main focus of their talks "Meso-level retrieval: field delineation and hybrid methods", "Bibliometrics-aided retrieval" and "Co-cited Author Maps, Bibliographic Retrievals, and a Viennese Author" was on the influences of IR on bibliometrics (e.g. as a tool to do better bibliometrics analyses). Two papers by Dietmar Wolfram and Birger Larsen highlighted the technical aspects of IR from a bibliometric viewpoint.

During these previous workshops it became obvious that there is a growing awareness that exploring links between bibliometric techniques and IR could be beneficial for actual both communities. They also made visible that substantial future work in

---

[1] http://www.cse.unt.edu/~ccaragea/CIKM-WS-13.html
[2] http://www.gesis.org/en/events/conferences/issiworkshop2013

M. de Rijke et al. (Eds.): ECIR 2014, LNCS 8416, pp. 798–801, 2014.
© Springer International Publishing Switzerland 2014

this direction depends on a rise in awareness in both communities. IR and bibliometrics go a long way back. Many pioneers in bibliometrics (e.g. Goffman, Brookes, Vickery), actually came from the field of IR, which is one of the traditional branches of information science. IR as a technique stays at the beginning of any scientometric exploration, and so, IR belongs to the portfolio of skills for any bibliometrician / scientometrician. However, IR and bibliometrics as special scientific fields have also grown apart over the last decades.

2   Goals, Objectives and Outcomes

Our workshop proposal aims to engage with the IR community about possible links to bibliometrics and complex network theory which also explores networks of scholarly communication. Bibliometric techniques are not yet widely used to enhance retrieval processes in digital libraries, yet they offer value-added effects for users (Mutschke et al., 2011). To give an example, recent approaches have shown the possibilities of alternative ranking methods based on citation analysis leading to an enhanced IR.

Our interests include information retrieval, information seeking, science modelling, network analysis, and digital libraries. The goal is to apply insights from bibliometrics, scientometrics, and informetrics to concrete practical problems of information retrieval and browsing.

Retrieval evaluations have shown that simple text-based retrieval methods scale up well but do not progress (Armstrong et al., 2009). Traditional retrieval has reached a high level in terms of measures like precision and recall, but scientists and scholars still face challenges present since the early days of digital libraries: mismatches between search terms and indexing terms, overload from result sets that are too large and complex, and the drawbacks of text-based relevance rankings. Therefore we will focus on statistical modelling and corresponding visualizations of the evolving science system. Such analyses have revealed not only the fundamental laws of Bradford and Lotka, but also network structures and dynamic mechanisms in scientific production (Börner et al., 2011). Statistical models of scholarly activities are increasingly used to evaluate specialties, to forecast and discover research trends, and to shape science policy (Scharnhorst et al., 2012). Their use as tools in navigating scientific information in public digital libraries is a promising but still relatively new development. We will explore how statistical modelling of scholarship (e.g. White et al., 2004) can improve retrieval services for specific communities, as well as for large, cross-domain collections. Some of these techniques are already used in working systems but not well integrated in larger scholarly IR environments.

The availability of new IR test collections that contain citation and bibliographic information like the iSearch collection (presented at the ISSI workshop by Birger Larsen, see Lykke et al., 2010) or the ACL collection (Ritchie, Teufel, and Robertson, 2006) could deliver enough ground to interest (again) the IR community in these kind of bibliographic systems. The long-term research goal is to develop and evaluate new approaches based on informetrics and bibliometrics. More specifically, we ask questions such as:

- How can we build scholarly information systems that explicitly use these approaches at the user-system interface?
- Are bibliometric-enhanced retrieval systems a value-added for scholarly work?
- How can models of science be interrelated with scholarly, task-oriented searching?
- And the other way around: Can insights from searching also improve the underlying statistical models themselves?

Although IR and scientometrics belong to one discipline, they are driven by different epistemic perspectives. In the past, experts from both sides have called for closer collaboration, but their encounters are rather ad-hoc. This workshop aims to raise awareness of the missing link between IR and bibliometrics / scientometrics and to create a common ground for the incorporation of bibliometric-enhanced services into retrieval at the digital library interface.

## 3	Format and Structure of the Workshop

The workshop will start with an inspirational keynote to kick-start thinking and discussion on the workshop topic, e.g. by one of the organizers of the ACM SIGKDD 2013 Cup, which used a large dataset from Microsoft Academic Search in an Author-Paper Identification Challenge[3]. This will be followed by paper presentations in a format found to be successful at EuroHCIR this year: Each paper is presented as a 10 minute lightning talk and discussed for 20 minutes in groups among the workshop participants followed by 1-minute pitches from each group on the main issues discussed and lessons learned. The workshop will conclude with a round-robin discussion of how to progress in enhancing IR with bibliometric methods.

## 4	Audience

The audiences (or clients) of IR and bibliometrics are different. Traditional IR serves individual information needs, and is – consequently – embedded in libraries, archives and collections alike. Scientometrics, and with it bibliometric techniques, has matured serving science policy.

We propose a half-day workshop that should bring together IR and DL researchers with an interest in bibliometric-enhanced approaches. Our interests include information retrieval, information seeking, science modelling, network analysis, and digital libraries. The goal is to apply insights from bibliometrics, scientometrics, and informetrics to concrete, practical problems of information retrieval and browsing.

The workshop is closely related to the workshop "Combining Bibliometrics and Information Retrieval" held at ISSI and tries to bring together contributions from core bibliometricians and core IR specialists but having selected those who already operate on the interface between scientometrics and IR.

---

[3] https://www.kaggle.com/c/kdd-cup-2013-author-paper-identification-challenge

5   Output

After the ISSI 2013 workshop on "Combining Bibliometrics and Information Retrieval" the workshop organizers were invited to apply for a special issue in Scientometrics. Such a dissemination serves well to account for raised awareness and contributions from the bibliometrics side and written for the bibliometrics side. We aim with the proposed workshop for a similar dissemination strategy, but now oriented towards core-IR. This way, we build a sequence of explorations, visions, results documented in scholarly discourse, and set up enough material for a sustainable bridge between bibliometrics and IR.

References

1. Armstrong, T.G., Moffat, A., Webber, W., Zobel, J.: Improvements that don't add up: ad-hoc retrieval results since 1998. In: Proceeding of the 18th ACM Conference on Information and Knowledge Management, pp. 601–610. ACM, Hong Kong (2009), doi:10.1145/1645953.1646031
2. Börner, K., Glänzel, W., Scharnhorst, A., van den Besselaar, P.: Modeling science: Studying the structure and dynamics of science. Scientometrics 89, 347–348 (2011)
3. Lykke, M., Larsen, B., Lund, H., Ingwersen, P.: Developing a Test Collection for the Evaluation of Integrated Search. In: Gurrin, C., He, Y., Kazai, G., Kruschwitz, U., Little, S., Roelleke, T., Rüger, S., van Rijsbergen, K. (eds.) ECIR 2010. LNCS, vol. 5993, pp. 627–630. Springer, Heidelberg (2010)
4. Mutschke, P., Mayr, P., Schaer, P., Sure, Y.: Science models as value-added services for scholarly information systems. Scientometrics 89(1), 349–364 (2011), doi:10.1007/s11192-011-0430-x
5. Ritchie, A., Teufel, S., Robertson, S.: Creating a Test Collection for Citation-based IR Experiments. In: Proceedings of the Main Conference on Human Language Technology Conference of the North American Chapter of the Association of Computational Linguistics, HLT-NAACL 2006, pp. 391–398. Association for Computational Linguistics, Stroudsburg (2006), doi:10.3115/1220835.1220885
6. Scharnhorst, A., Börner, K., van den Besselaar, P. (eds.): Models of Science Dynamics Encounters Between Complexity Theory and Information Sciences. Springer, Berlin (2012)
7. White, H.D., Lin, X., Buzydlowski, J.W., Chen, C.: User-controlled mapping of significant literatures. Proceedings of the National Academy of Sciences 101(suppl. 1), 5297–5302 (2004)

4th Workshop on Context-Awareness in Retrieval and Recommendation

Alan Said[1], Ernesto William De Luca[2],
Daniele Quercia[3], and Matthias Böhmer[4]

[1] CWI, The Netherlands
[2] Potsdam University of Applied Sciences, Germany
[3] Yahoo! Labs, Spain
[4] DFKI, Germany
{alansaid,dquercia}@acm.org, deluca@fh-potsdam.de,
matthias.boehmer@dfki.de

Abstract. Context-aware information is widely available in various ways and is becoming more and more important for enhancing retrieval performance and recommendation results. The current main issue to cope with is not only recommending or retrieving the most relevant items and content, but defining them ad hoc. Other relevant issues include personalizing and adapting the information and the way it is displayed to the user's current situation and interests. Ubiquitous computing further provides new means for capturing user feedback on items and providing information.

1 Introduction

The first CaRR Workshop [7] received enthusiastic feedback from the context-awareness community with a total of 15 submitted papers and more than 30 attendees at IUI 2011 in Palo Alto. A total of 9 papers were accepted for presentation at the workshop and publication in the proceedings [1]. An indication that *context-awareness in retrieval and recommendation* is a very interesting and timely topic. Among the issues discussed at the workshop were: how context-awareness has changed the goal of retrieval and recommendation, e.g. drifting from recommendations that fit to the general user interest to ad-hoc recommendations that fit to the user's context. Additional topics covered were new emerging areas where the role of retrieval and recommendation increases, e.g. recommending places to go, activities to take, or apps to use. The second time the workshop was organized, in conjunction with IUI 2012 in Lisbon, Portugal on February 14 2012 [2,6]. It received very positive feedback from the context-awareness community with a total of 10 submitted papers and more than 30 attendees. The topics covered in the workshop were discussing the roles of context in information retrieval and recommendations a choice and decision making perspective. Different work was presenting the use of context with techniques for personalization and recommendation of items. The use of context in music retrieval and user modeling was also discussed. The third edition of the CaRR

M. de Rijke et al. (Eds.): ECIR 2014, LNCS 8416, pp. 802–805, 2014.
© Springer International Publishing Switzerland 2014

Workshop [3,5] took place in Rome, Italy on February 5, 2013 and continued focusing on the integration of context for retrieval and recommendation. There were a total of 12 submissions, out of which 5 were accepted for publication. The most recent edition of the workshop[1] was organized in conjuntion to ECIR 2014 in Amsterdam, The Netherlands in February 2014. The workshop focused on context-awareness in information retrieval and web search systems.

The aim of the CaRR Workshop was to invite the community to a discussion in which we tried to find new creative ways to handle context-awareness. Furthermore, the workshop aimed at exchanging new ideas between different communities involved in research, such as HCI, machine learning, Web mining, information retrieval and recommender systems.

The proceedings of the workshop have been published in the ACM DL in the International Conference Proceedings Series [4].

2 Research Questions

The workshop was especially intended for researchers working on multidisciplinary tasks, to discuss problems and synergies. Ideas on creative and collaborative approaches for context-aware retrieval and recommendation were of special interest.

The participants were encouraged to address the following questions:

- What is context?
- Which benefits come from context-aware systems?
- In what ways can context improve the Web experience?
- How can we combine general- and user-centric context-aware technologies?
- How should context affect the way information is presented?

3 Topics

The topics of interest included, but were not limited to:

- Context-aware data mining and information retrieval
- Context-aware profiling, clustering and collaborative filtering
- Use of context-aware technologies in Web search
- Ubiquitous and context-aware computing
- Use of context-aware technologies in UI/HCI
- Context-aware advertising
- Recommendations for mobile users
- Context-awareness in portable devices
- Mobile and social search

---

[1] http://carr-workshop.org

4 Workshop Organizers

Alan Said is a postdoctoral research fellow at Centrum Wiskunde & Informatica (CWI), Amsterdam, The Netherlands. His research interests are related to evaluation of information systems. He has authored papers on national and international conferences in the fields of recommender systems, evaluation, social networks, context-awareness and other related areas. He has also chaired workshops and conferences on recommender systems and related topics.

Ernesto William De Luca is professor of information science at the Potsdam University of Applied Sciences. He is the author of more than 70 papers on national and international conferences and journals in the fields of Computational Linguistics, Information Retrieval, Adaptive Systems and other related areas. He has chaired a number of workshops on semantic personalization, context-awareness in recommender systems and related topics.

Daniele Quercia is a social media researcher at Yahoo Labs in Barcelona. Before that, he was a Horizon senior researcher at The Computer Laboratory of the University of Cambridge. He is interested in the relationship between online and offline worlds and his work has been focusing in the areas of data mining, recommender systems, computational social science, and urban informatics. He was Postdoctoral Associate at the Massachusetts Institute of Technology where he worked on social networks in a city context, and his PhD thesis at UC London was nominated for BCS Best British PhD dissertation in Computer Science.

Matthias Böhmer is a researcher at the Innovative Retail Laboratory of the German Research Center for Artificial Intelligence (DFKI) in Saarbrücken, Germany. His main research focus is understanding and supporting mobile application usage, in particular on context-aware recommender systems that suggest mobile applications. He is interested in deducing the contextual relevance of particular apps from a user's device interaction, e.g. usage patterns or icon arrangement.

5 Program Committee Members

- Luca Maria Aiello - Yahoo! Research, Spain
- Omar Alonso - Microsoft, USA
- Nicola Barbieri - Yahoo! Research, Spain
- Alejandro Bellogín, UAM, Spain
- Robin Burke - DePaul University, USA
- Pablo Castells - UAM, Spain
- Juan M. Cigarran - UNED, Spain
- Paolo Cremonesi - Politecnico di Milano, Italy
- Marco De Gemmis - University of Bari, Italy
- Ana Garcia-Serrano - UNED, Spain
- Alexandros Karatzoglou - Telefonica Research, Spain
- Martha Larson - TU Delft, The Netherlands
- Neal Lathia - University of Cambridge, UK

- Pasquale Lops - University of Bari, Italy
- Bamshad Mobasher - DePaul University, USA
- Andreas Nürnberger - OVGU, Germany
- Francesco Ricci - Free University of Bozen-Bolzano, Italy
- Yue Shi - TU Delft, The Netherlands
- Armando Stellato - University of Tor Vergata, Italy
- Domonkos Tikk - Gravity R&D, Hungary
- Arjen de Vries - CWI, The Netherlands

Acknowledgements. The organizers would like to thank all the authors for contributing to CaRR 2014 and all the members of the program committee for ascertaining the scientific quality of the workshop.

Part of the work leading to this workshop was carried out during the tenure of an ERCIM "Alain Bensoussan" Fellowship Programme. The research leading to these results has received funding from the European Union Seventh Framework Programme (FP7/2007-2013) under grant agreement no.246016.

References

1. CaRR 2011: Proceedings of the 2011 Workshop on Context-Awareness in Retrieval and Recommendation. ACM, New York (2011)
2. CaRR 2012: Proceedings of the 2nd Workshop on Context-awareness in Retrieval and Recommendation. ACM, New York (2012)
3. CaRR 2013: Proceedings of the 3rd Workshop on Context-awareness in Retrieval and Recommendation. ACM, New York (2013)
4. CaRR 2014: Proceedings of the 4th Workshop on Context-awareness in Retrieval and Recommendation. ACM, New York (2014)
5. Böhmer, M., De Luca, E.W., Said, A., Teevan, J.: 3rd workshop on context-awareness in retrieval and recommendation. In: Proceedings of the Sixth ACM International Conference on Web Search and Data Mining, WSDM 2013, pp. 789–790. ACM, New York (2013), http://doi.acm.org/10.1145/2433396.2433504
6. De Luca, E.W., Böhmer, M., Said, A., Chi, E.: 2nd workshop on context-awareness in retrieval and recommendation (carr 2012). In: Proceedings of the 2012 ACM International Conference on Intelligent User Interfaces, IUI 2012, pp. 409–412. ACM, New York (2012), http://doi.acm.org/10.1145/2166966.2167061
7. De Luca, E.W., Said, A., Böhmer, M., Michahelles, F.: Workshop on context-awareness in retrieval and recommendation. In: Proceedings of the 16th International Conference on Intelligent User Interfaces, IUI 2011, pp. 471–472. ACM, New York (2011), http://doi.acm.org/10.1145/1943403.1943506

Workshop on Gamification for Information Retrieval (GamifIR'14)

Frank Hopfgartner[1], Gabriella Kazai[2], Udo Kruschwitz[3], and Michael Meder[1]

[1] Technische Universität Berlin, Berlin, Germany
{firstname.lastname}@tu-berlin.de
[2] Microsoft Research, Cambridge, UK
v-gabkaz@microsoft.com
[3] University of Essex, Colchester, UK
udo@essex.ac.uk

Abstract. Gamification is the application of game mechanics, such as leader boards, badges or achievement points, in non-gaming environments with the aim to increase user engagement, data quality or cost effectiveness. A core aspect of gamification solutions is to infuse intrinsic motivations to participate by leveraging people's natural desires for achievement and competition. While gamification, on the one hand, is emerging as the next big thing in industry, e.g., an effective way to generate business, on the other hand, it is also becoming a major research area. However, its adoption in Information Retrieval (IR) is still in its infancy, despite the wide ranging IR tasks that may benefit from gamification techniques. These include the manual annotation of documents for IR evaluation, the participation in user studies to study interactive IR challenges, or the shift from single-user search to social search, just to mention a few.

This context provided the motivation to organise the GamifIR'14 workshop at ECIR.

1 Motivation

Many research challenges in the field of IR rely on tedious manual labour. For example, manual feedback is required to assess the relevance of documents to a given search task, to annotate documents or to evaluate interactive IR approaches. A recent trend to perform these tasks is the use of crowdsourcing techniques, i.e., relying on human computation and the "wisdom of the crowd" [1]. Although research indicates that such techniques can be useful, they fail when *motivated* users are required to perform a task for reasons other than just being paid per click or paid per minute [2].

A promising approach to increase user motivation can be achieved by employing gamification methods, i.e., game mechanics and principles such as leaderboards, points for certain activities or achievement badges are applied in non-gaming environments to reach specific goals (e.g. [3–6]).

Recently, gamification has drawn the attention of researchers from various fields, leading to workshops (e.g., [7–9]) and the eventual introduction of a new

M. de Rijke et al. (Eds.): ECIR 2014, LNCS 8416, pp. 806–809, 2014.

conference on gamification [10]. At the same time, the IR community focused on the use of crowdsourcing for relevance evaluation [11] and at searching for fun (e.g., [12]), two topics which can benefit from the use of gamification principles.

This workshop focused on the challenges and opportunities that gamification can present for the IR community. The workshop aimed to bring together researchers and practitioners from a wide range of areas including game design, information retrieval, human-computer interaction, computer games, and natural language processing. Detailed information about the workshop can be found on the workshop website at http://gamifir2014.dai-labor.de/.

2   Workshop Scope

The call for papers solicited submissions of position papers as well as novel research papers and demos addressing problems related to gamification in IR including topics such as:

- Gamification approaches in a variety of information-seeking contexts
- User engagement and motivational factors of gamification
- Player types, contests, cooperative games
- Challenges and opportunities of applying gamification in IR
- Gamification design and game mechanics
- Game-based work and crowdsourcing
- Applications and prototypes
- Evaluation of gamification techniques

3   Keynote

We were very pleased that Prof. Richard Bartle of the University of Essex (United Kingdom) agreed to provide a keynote at GamifIR'14. Bartle is one of the key researchers in the field of game design. His research lay the ground for the success of massively multiplayer online games. The gamification research community knows him best for his Player Types model [13] in which he defines different types of players that can be distinguished based on their individual behaviour.

Acknowledgements. We acknowledge the efforts of the members of the programme committee, namely:

- Omar Alonso (Microsoft Research, USA),
- Leif Azzopardi (University of Glasgow, United Kingdom),
- Yoram Bachrach (Microsoft Research, United Kingdom),
- Regina Bernhaupt (Université Paul Sabatier, France),
- Jon Chamberlain (University of Essex, United Kingdom),
- Edwin Chen (YouTube/Google, USA),
- Sebastian Deterding (Rochester Institute of Technology, USA),
- Carsten Eickhoff (ETH Zürich, Switzerland),

- Rosta Farzan (University of Pittsburgh, USA),
- Christopher G. Harris (The University Of Iowa, USA),
- Shaili Jain (Microsoft, USA),
- Hideo Joho (University of Tsukuba, Japan),
- Mounia Lalmas (Yahoo! Labs, Spain),
- Edith Law (Harvard University, USA),
- David Parkes (Harvard University, USA),
- Massimo Poesio (University of Essex, United Kingdom),
- Falk Scholer (RMIT, Australia),
- Craig Stewart (Coventry University, United Kingdom),
- Elaine Toms (University of Sheffield, United Kingdom)
- Arjen de Vries (CWI, The Netherlands),
- Albert Weichselbraun (University of Applied Sciences Chur, Switzerland),
- Lincoln C. Wood (Auckland University of Technology, New Zealand).

References

1. Alonso, O., Mizzaro, S.: Using crowdsourcing for TREC relevance assessment. Inf. Process. Manage. 48(6), 1053–1066 (2012)
2. Kazai, G., Kamps, J., Milic-Frayling, N.: An analysis of human factors and label accuracy in crowdsourcing relevance judgments. Inf. Retr. 16(2), 138–178 (2013)
3. Eickhoff, C., Harris, C.G., de Vries, A.P., Srinivasan, P.: Quality through flow and immersion: gamifying crowdsourced relevance assessments. In: SIGIR, pp. 871–880 (2012)
4. Poesio, M., Chamberlain, J., Kruschwitz, U., Robaldo, L., Ducceschi, L.: Phrase Detectives: Utilizing Collective Intelligence for Internet-Scale. ACM Trans. Interact. Intell. Syst. 3(1) (2013)
5. Meder, M., Plumbaum, T., Hopfgartner, F.: Perceived and Actual Role of Gamification Principles. In: UCC 2013: Proceedings of the IEEE/ACM 6th International Conference on Utility and Cloud Computing, pp. 488–493. IEEE (December 2013)
6. Meder, M., Plumbaum, T., Hopfgartner, F.: DAIKnow: A Gamified Enterprise Bookmarking System. In: de Rijke, M., Kenter, T., de Vries, A.P., Zhai, C., de Jong, F., Radinsky, K., Hofmann, K. (eds.) ECIR 2014. LNCS, vol. 8416, pp. 759–762. Springer, Heidelberg (2014)
7. Deterding, S., Sicart, M., Nacke, L., O'Hara, K., Dixon, D.: Gamification. using game-design elements in non-gaming contexts. In: CHI 2011 Extended Abstracts on Human Factors in Computing Systems, CHI EA 2011, pp. 2425–2428. ACM, New York (2011)
8. Deterding, S., Björk, S.L., Nacke, L.E., Dixon, D., Lawley, E.: Designing gamification: Creating gameful and playful experiences. In: CHI 2013 Extended Abstracts on Human Factors in Computing Systems, CHI EA 2013, pp. 3263–3266. ACM (2013)
9. Springer, T., Herzig, P. (eds.): Proceedings of the 2013 IEEE/ACM 6th International Conference on Utility and Cloud Computing – International Workshop on Gamification in the Cloud, CGCloud 2013 (2013)
10. Nacke, L., Harrigan, K., Randall, N. (eds.): Proceedings of the First International Conference on Gameful Design, Research, and Applications, Gamification 2013 (2013)

11. Alonso, O., Lease, M.: Crowdsourcing for information retrieval: Principles, methods, and applications. In: Proceedings of the 34th International ACM SIGIR Conference on Research and Development in Information Retrieval, SIGIR 2011, pp. 1299–1300. ACM, New York (2011)
12. Elsweiler, D., Wilson, M.L., Harvey, M. (eds.): Proceedings of the "Searching 4 Fun!" Workshop. CEUR, vol. 836 (2012)
13. Bartle, R.: Hearts, clubs, diamonds, spades: Players who suit MUDs. The Journal of Virtual Environments 1(1) (1996)) (available online)

Information Access in Smart Cities (i-ASC)

M-Dyaa Albakour[1], Craig Macdonald[1], Iadh Ounis[1], Charles L. A. Clarke[2],
and Veli Bicer[3]

[1] University of Glasgow, Glasgow, UK
{dyaa.albakour,craig.macdonald,iadh.ounis}@glasgow.ac.uk
[2] University of Waterloo, Waterloo, Canada
claclark@plg.uwaterloo.ca
[3] IBM Research Lab, Dublin, Ireland
velibice@ie.ibm.com

Abstract. Modern cities are becoming smart where a digital knowledge infrastructure is deployed by local authorities (e.g. City councils and municipalities) to better serve the information needs of their citizens, and to ensure sustainability and efficient use of power and resources. This knowledge infrastructure consists of a wide range of systems from low-level physical sensors to advanced sensing devices through social sensors. This proposed workshop will be a venue for research on digesting the city's data streams and knowledge databases in order to serve the information needs of citizens and support decision making for local authorities. Possible use cases include helping tourists to find interesting places to go or activities to do while visiting a city, or assisting journalists in reporting local incidents. Indeed, this workshop will foster the development of new information access and retrieval models that can harness effectively and efficiently the large number of heterogeneous big data streams in a city to provide a new generation of information services.

1 Background, Motivation and Themes

Cities of the 21st century do not only have the physical infrastructure of roads, buildings, and power networks, but also have the knowledge infrastructure represented with heterogeneous systems and big data platforms. These systems vary from low-level sensing devices including environmental sensors and CCTV cameras, to public databases, and social network streams. On the other hand, a recent study [1] has identified a variety of emerging information needs that citizens often have in their public urban spaces. These local information needs may be complex and are not necessarily served by existing systems such as web search engines. These include obtaining real-time information about events in the city or finding free parking spaces. The aforementioned knowledge infrastructure opens up opportunities for cities to become smarter and serve the emerging information needs of their citizens by harnessing the vast amount of diverse information stemming from the various systems within the city.

The proposed workshop would be an opportunity for researchers from the information retrieval (IR) community, and other related communities, to discuss

M. de Rijke et al. (Eds.): ECIR 2014, LNCS 8416, pp. 810–814, 2014.
© Springer International Publishing Switzerland 2014

the emerging research topic of information access in smart cities, identifying its unique challenges, opportunities and future directions. In particular, we identify four main research themes that we aim to cover:

City Data: There is a wealth of data sources that can be acquired about modern cities. However, city data sources have unique characteristics. First of all, the heterogeneity of this data poses challenges on identifying the relevant sources of information for specific needs, e.g. from different repositories in the linked data cloud [2], or otherwise fusing multiple sources of information including sensor streams [3]. Also, this data can be noisy, messy and incomplete, which adds an additional burden of cleansing, capturing the anomalies, and consolidating multiple sources. Moreover, the volume of such data can be huge and requires an efficient computing infrastructure that scales to large amount of knowledge databases and data streams. The emerging big data platforms play a significant role in providing such infrastructures.

Searching Smart Cities: Searching smart cities has unique characteristics and may not be necessarily served by existing IR models and techniques. A recent tutorial at SIGIR has highlighted some of these characteristics[1] Firstly, the information needs of the citizens can be complex and require consolidating a wide variety of city data sources [1]. Secondly, the relevance of results may depend on a number of dimensions such as geospatial, temporal and social dimensions. Furthermore, real-time search plays an important role in such environments, where it is vital to obtain the freshest results possible. New retrieval models are emerging to tackle these unique characteristics [4].

Human Interaction and Context: Providing access to information in smart cities requires suitable user interaction paradigms. The interaction paradigm may have impact on the design of the underlying retrieval models and the user interfaces. Recently, the SWIRL 2012 workshop [5] has noted that future IR systems should be able to anticipate user information needs without explicit queries, i.e. zero queries. This is particularly suitable for mobile devices, which impose a physical limitation on the user interactions. The TREC contextual suggestion track [6] introduces a typical example of a zero-query task in smart cities, where the objective is to recommend to a user interesting activities to do in a city based on contextual information of previous preferences. Developing zero-query access to information requires effective modelling and representation of contextual information about the users and the environment in order to both *anticipate* the user needs and *personalise* the results. Finally, the design of user interfaces should support the exploration and discovery of city data on emerging mobile devices to make them even more accessible.

Applications: There is a long list of domains that can benefit from services built around urban environment data. This includes transportation [3], tourism (recommendation platforms, access to heritage information), entertainment (live news),

---

[1] http://www.dublinked.ie/?q=searchinginthecityofknowledge tutorial

security and surveillance (crime prediction and prevention), crisis management [7], and urban planning.

2 Goals and Objectives

The goal of the workshop is twofold. Firstly, we aim to foster building a research community to work on the aforementioned research themes within smart cities. We envisage that this community will be inclusive of academics and practitioners not only in IR, but also in a variety of disciplines such as knowledge management, databases, machine learning, and human computer interfaces. Secondly, we aim to define a roadmap for developing information access systems in smart cities, where we identify the key challenges in each of the four themes and instantiate concrete tasks we can start working on as a community.

3 Format and Structure

We will solicit submissions of position papers (2 pages) and short technical papers (4 pages) in ACM style, from a wide range of author audience to cover the following topics within each research theme. *(1) City data* (City data acquisition and integration, Mining city databases and semantic knowledge sources, Linked data for smart cities, Fusing multiple sensor and social streams, Big data platforms); *(2) Searching smart cities* (Localised search, Searching structured city data, Real-time search over multiple data streams, Real-time filtering of multiple social and sensor streams, Searching location-based social networks); *(3) Human interaction and context* (zero-query access in smart cities, Modelling contextual information, Modelling personal and social interests, Personalisation and recommendation, Mobile interfaces, Exploration and discovery in smart cities); and *(4) Applications* (monitoring transportation, tourism accessibility, security and disaster management). The papers will be peer-reviewed by at least three members of the programme committee.

We have invited two keynote speakers from both academia and industry: Frank Kresin (Waag Society) and Pol Mac Aonghusa (IBM). Each will cover one or more of the aforementioned research themes.

We anticipate a full-day workshop structured as follows. The day starts with the first keynote speaker followed by a short presentation for each of the selected papers and then the second keynote speaker. After a break, we will hold a discussion session, where we summarise the research themes discussed and perhaps restructure the topics in these themes if necessary. The final session of the day will be a breakout session where we form a group for each research theme. Each group will discuss how they can achieve the second goal of the workshop within their theme (building the roadmap, see Section 2) and report their findings afterwards to the full workshop. The organisers will then consolidate their findings into a single report that will be submitted to the SIGIR forum, in addition to producing high quality workshop proceedings of the presented papers.

4   Committee

4.1   Programme Committee

We propose a programme committee that is balanced in terms of the contributions in the aforementioned themes. It is also balanced from the viewpoint of having both academic and industrial researchers as well as being balanced in terms of gender. The proposed names are Fernando Diaz (Microsoft Research), Jaap Kamps (University of Amsterdam), Paul Thomas (CSIRO), Daqing He (University of Pittsburgh), Omar Alonso (Microsoft Bing), Freddy Lecue (IBM Research), Cathal Gurrin (Dublin City University), and Suzan Verberne (University of Nijmegen), Raffaele Perego and Franco Maria Nardini (ISTI CNR).

4.2   Organising Committee

M-Dyaa Albakour: Dyaa is a post-doctoral researcher at the Terrier Information Retrieval in the University of Glasgow. He has been recently working on the EU SMART FP7 project which aims to build a search engine for smart cities. He has been involved in organising the annual BCS Search Solutions workshop in 2012 and 2013.

Craig Macdonald: Craig was a co-chair of CIKM 2011 with workshop remit, a past local-organiser and Industry Day Chair of ECIR, and has co-chaired workshops at WSDM and ECIR (LSDS-IR, 2012 & 2013; Diversity in Document Retrieval, 2011 & 2012). His research interests cover efficient and effective models for information retrieval in Web and social media settings, and real-world sensing with smart cities. He is a co-investigator of the EU funded SMART project.

Iadh Ounis: Iadh has authored over 100 refereed publications, and his research interests include: large-scale information retrieval; data mining; and social media retrieval. He is the principal investigator of the open source Terrier search engine. He was a track coordinator at TREC 2006-2012. He was a co-chair of ESSIR 2007, the general chair of ECIR 2008, and a general co-chair CIKM 2011.

Charles L. A. Clarke: Charles is a Professor in the School of Computer Science at the University of Waterloo, Canada. His research interests include IR, web search, and text data mining. He was a General Co-Chair for SIGIR 2003, he was a Program Co-Chair for SIGIR 2007, and he will again be a Program Co-Chair for SIGIR 2014. He has co-organized multiple workshops at SIGIR, WSDM, and ECIR, and has been the co-organizer of multiple TREC tracks.

Veli Bicer: Veli Bicer is a researcher at Smarter Cities Technology Center of IBM Research in Ireland. His research interests include semantic search, semantic data management, software engineering and statistical relational learning. He has recently co-organized the "City Data Management" workshop at CIKM 2012 and tutorial on "Searching in the City of Knowledge" at SIGIR 2013.

814   M. Albakour et al.

References

1. Kukka, H., Kostakos, V., Ojala, T., Ylipulli, J., Suopajrvi, T., Jurmu, M., Hosio, S.: This is not classified: everyday information seeking and encountering in smart urban spaces. Personal and Ubiquitous Computing 17(1), 15–27 (2013)
2. Lopez, V., Kotoulas, S., Sbodio, M.L., Stephenson, M., Gkoulalas-Divanis, A., Aonghusa, P.M.: Queriocity: A linked data platform for urban information management. In: Cudré-Mauroux, P., Heflin, J., Sirin, E., Tudorache, T., Euzenat, J., Hauswirth, M., Parreira, J.X., Hendler, J., Schreiber, G., Bernstein, A., Blomqvist, E. (eds.) ISWC 2012, Part II. LNCS, vol. 7650, pp. 148–163. Springer, Heidelberg (2012)
3. Daly, E.M., Lécué, F., Bicer, V.: Westland row why so slow? fusing social media and linked data sources for understanding real-time traffic conditions. In: Proceedings of IUI 2013, pp. 203–212 (2013)
4. Albakour, M., Macdonald, C., Ounis, I.: Identifying local events by using microblogs as social sensors. In: Proceedings of OAIR 2013, pp. 173–180 (2013)
5. Allan, J., Croft, B., Moffat, A., Sanderson, M.: Frontiers, challenges, and opportunities for information retrieval: Report from SWIRL 2012 the second strategic workshop on information retrieval in Lorne. In: ACM SIGIR Forum, vol. 46, pp. 2–32. ACM Press (2012)
6. Kamps, J., Thomas, P., Vorhees, E., Dean-Hall, A., Clarke, C.L.A.: Overview of the TREC 2012 Contextual Suggestion track. In: Proceedings of TREC 2012 (2013)
7. Maxwell, D., Raue, S., Azzopardi, L., Johnson, C., Oates, S.: Crisees: Real-time monitoring of social media streams to support crisis management. In: Baeza-Yates, R., de Vries, A.P., Zaragoza, H., Cambazoglu, B.B., Murdock, V., Lempel, R., Silvestri, F. (eds.) ECIR 2012. LNCS, vol. 7224, pp. 573–575. Springer, Heidelberg (2012)

Designing Search Usability

Tony Russell-Rose

UXLabs, London, UK
tgr@uxlabs.co.uk

Abstract. Search is not just a box and ten blue links. Search is a journey: an exploration where what we encounter along the way changes what we seek. But in order to guide people along this journey, we must understand both the art and science of user experience design.

The aim of this tutorial is to deliver a learning experience grounded in good scholarship, integrating the latest research findings with insights derived from the practical experience of designing and optimizing an extensive range of commercial search applications. It focuses on the development of transferable, practical skills that can be learnt and practiced within a half-day session.

Keywords: Site search, enterprise search, information seeking, user behaviour, information discovery, user experience design, usability.

1 Introduction

In the summer of 1804, Meriwether Lewis and William Clark embarked on an epic journey west. President Thomas Jefferson had commissioned the Corps of Discovery to explore, chart, and above all, search for a navigable water route leading to the Pacific Ocean and, by extension, commerce with Asia.

Their journey led them up the Missouri River, through the Great Plains, over the Rocky Mountains, and eventually to the Pacific coast. Over the course of 28 months and 8,000 miles, Lewis and Clark accounted for 72 native tribes, drew 140 maps, and documented more than 200 new plants and animals.

In the end, Lewis and Clark failed to find a northwest passage to Asia. Yet their journey was hardly in vain. The expedition contributed a wealth of scientific and geographic knowledge, established diplomatic relations with dozens of indigenous tribes, and explored territory never before seen by Europeans. In other words, the journey itself became more important than the destination.

The same is true when searching for information.

On the surface, search may appear to be simply a box and ten blue links—a query and a set of results. It may seem a personal rather than social activity; a brief interaction confined to a single medium. And so often, we assume that everyone knows what they are looking for in the first place.

But on closer examination, these assumptions break down. Understandably, there are times when search is simply looking up a fact or finding a particular document. But more often, search is a journey. It's an ongoing exploration where what we find

M. de Rijke et al. (Eds.): ECIR 2014, LNCS 8416, pp. 815–818, 2014.

along the way changes what we seek. It's a journey that can extend beyond a single episode, involve friends, colleagues, and even strangers, and be conducted on all manner of devices.

Our concern is with search in its broadest, most holistic sense. By investigating why and how people engage in information seeking, we learn not just about information retrieval, but how people navigate and make sense of complex digital information environments. What we learn along the way will prepare us both for the search experiences of today, and the cross-channel, information-intense experiences of tomorrow.

This tutorial explores both the art and the science of user experience design in two parts:

Part 1 focuses on theory. It sets out a conceptual framework for information seeking, investigating human characteristics, models of information seeking, the role of context, and modes of search and discovery.

Part 2 turns theory into practice. It applies the principles from Part 1 to the art and science of user experience design, from entering queries to displaying and manipulating results. It also looks briefly at faceted search and the emerging worlds of mobile and collaborative search.

2   Tutorial Content

The course comprises the following sections:

1. **Introductions and objectives**: Group introductions & ice-breaker. A brief summary of what each participant hopes to gain from the session, and what experiences they bring.

2. **Understanding search and discovery behaviour**: An overview of the key theories and models of human-information seeking behaviour, focusing on the work of Bates, Belkin, Jarvelin & Ingwersen, Marchionini, Pirolli, etc. and their relationship to more practitioner-oriented approaches, e.g. Morville, Tunkelang, etc. [1], [2], [5], [6], [7], [9], [12], [13].

3. **Varied solutions for varied contexts**: An exploration of the universal dimensions that define information-seeking behaviour, and how these translate into principles for the design of search and discovery experiences [3], [7], [10], [11].

4. **Formulating the query**: A detailed examination of the various methods by which information needs can be elicited, disambiguated and refined, and how design interventions can support this process [6], [10].

5. **Displaying results**: The counterpart to (4): an exploration of the key issues and principles underpinning the display and manipulation of search results, and how they can be used to facilitate productive user journeys [8], [10].

6. **Faceted Navigation and Search:** A review of the key principles of faceted classification and their practical implications for the design of effective faceted search applications [4], [10], [12].

7. **Conclusions and Wrap-up:** A review of the overall session, including the shared experiences of the group exercises and the contrasting findings of each. A summary of the follow-on resources and takeaways from the course and the wider HCI community.

3   Intended Audience

This intermediate tutorial is aimed at IR researchers and practitioners, information architects and search specialists interested in the designing more effective user experiences and interfaces for information retrieval and discovery. An awareness of the basic principles of user-centered design is useful (but not essential).

4   Instructor Biography

Tony Russell-Rose is founder and director of UXLabs, a research and design consultancy specializing in complex search and information access applications. Before founding UXLabs he was Manager of User Experience at Endeca and editor of the Endeca Design Pattern Library, an online resource dedicated to best practice in the design of search and discovery applications. Prior to this he was technical lead at Reuters, specializing in advanced user interfaces for information access and search. And before Reuters he was R&D manager at Canon Research Centre Europe, where he led a team developing next generation information access products and services. Earlier professional experience includes a Royal Academy of Engineering fellowship at HP Labs working on speech interfaces for mobile devices, and a Short-term Research Fellowship at BT Labs working on intelligent agents for information retrieval.

His academic qualifications include a PhD in artificial intelligence, an MSc in cognitive psychology and a first degree in engineering, majoring in human factors. He have published 70+ scientific papers on search, user experience and text analytics, and is author of "Designing the Search Experience: the Information Architecture of Discovery", published by Elsevier in 2012. He is currently vice-chair of the BCS Information Retrieval group and chair of the IEHF Human-Computer Interaction group. He also holds the position of Honorary Visiting Fellow at the Centre for Interactive Systems Research, City University, London.

References

[1] Blandford, A., Attfield, S.: Interacting with Information. Morgan & Claypool (2010)
[2] Bates, M.J.: The design of browsing and berrypicking techniques for the online search interface. Online Review 13(5), 407–424 (1989)

[3] Cool, C., Belkin, N.: A Classification of Interactions with Information. In: Bruce, H., Fidel, R., Ingwersen, P., Vakkari, P. (eds.) Emerging Frameworks and Methods. Proceedings of the Fourth International Conference on Conceptions of Library and Information Science (COLIS4), pp. 1–15. Libraries Unlimited, Greenwood Village (2002)

[4] Hearst, M.: Search User Interfaces. Cambridge University Press (2009)

[5] Jarvelin, K., Ingwersen, P.: Information seeking research needs extension towards tasks and technology. Information Research 10(1), paper 212 (2004)

[6] Morville, P., Callender, J.: Search Patterns. O'Reilly Media (2009)

[7] Marchionini, G.: Information Seeking in Electronic Environments. Cambridge University Press (1995)

[8] Nudelman, G.: Designing Search. Springer (2011)

[9] Pirolli, P., Card, S.: Information foraging. Psychological Review 106(4), 643–675 (1999)

[10] Russell-Rose, T., Tate, T.: Designing the Search Experience: the Information Architecture of Discovery. Morgan Kaufmann (2012)

[11] Ruthven, I., Kelly, D.: Interactive information seeking, behaviour and retrieval. Facet Publishing, London (2011)

[12] Tunkelang, D.: Faceted Search. Morgan & Claypool (2009)

[13] Wilson, M.L., Kules, W., Schraefel, M.C., Shneiderman, B.: From Keyword Search to Exploration: Designing Future Search Interfacesfor the Web. Foundations and Trends® in Web Science 2(1), 1–97 (2010)

Text Quantification

Fabrizio Sebastiani

Istituto di Scienza e Tecnologie dell'Informazione
Consiglio Nazionale delle Ricerche
56124 Pisa, Italy
fabrizio.sebastiani@isti.cnr.it

1   What Is Quantification?

In recent years it has been pointed out that, in a number of applications involving classification, the final goal is not determining which class (or classes) individual unlabelled data items belong to, but determining the *prevalence* (or "relative frequency") of each class in the unlabelled data. The latter task has come to be known as *quantification* [1, 3, 5–10, 15, 18, 19].

Although what we are going to discuss here applies to any type of data, in this tutorial we will mostly be interested in *text* quantification, i.e., quantification when the data items are textual documents. To see the importance of text quantification, let us examine the task of classifying textual answers returned to open-ended questions in questionnaires [4, 11, 12], and let us discuss two important such scenarios.

In the first scenario, a telecommunications company asks its current customers the question "How satisfied are you with our mobile phone services?", and wants to classify the resulting textual answers according to whether they belong or not to class MayDefectToCompetition. (Membership in this class indicates that the customer is so unhappy with the company's services that she is probably considering a switch to another company.) The company is likely interested in accurately classifying each individual customer, since it may want to call each customer that is assigned the class and offer her improved conditions.

In the second scenario, a market research agency asks respondents the question "What do you think of the recent ad campaign for product X?", and wants to classify the resulting textual answers according to whether they belong to the class LovedTheCampaign. Here, the agency is likely *not* interested in whether a specific individual belongs to the class LovedTheCampaign, but is likely interested in knowing *how many* respondents belong to it, i.e., in knowing the prevalence of the class.

In sum, while in the first scenario classification is the goal, in the second scenario the real goal is quantification. Essentially, quantification is classification evaluated at the aggregate (rather than at the individual) level. Other scenarios in which quantification is the goal may be, e.g., predicting election results by estimating the prevalence of blog posts (or tweets) supporting a given candidate or party [13], or planning the amount of human resources to allocate to different types of issues in a customer support center by estimating the prevalence of

M. de Rijke et al. (Eds.): ECIR 2014, LNCS 8416, pp. 819–822, 2014.

customer calls related to each issue [7], or supporting epidemiological research by estimating the prevalence of medical reports where a specific pathology is diagnosed [1].

2   How Should Quantification be Addressed?

In the absence of methods for estimating class prevalence more directly, the obvious method for dealing with the latter type of scenarios is *quantification via classification*, i.e., (i) classifying each unlabelled document, and then (ii) estimating class prevalence by counting the documents that have been attributed the class. However, there are two reasons why this strategy is suboptimal.

The first reason is that a good classifier may not be a good quantifier, and vice versa. To see this, one only needs to look at the definition of F_1, the standard evaluation function for binary classification, defined as

$$F_1 = \frac{2 \cdot TP}{2 \cdot TP + FP + FN} \tag{1}$$

where TP, FP and FN indicate the numbers of true positives, false positives, and false negatives, respectively. According to F_1, a binary classifier $\hat{\Phi}_1$ that returns $FP = 20$ and $FN = 20$ is worse than a classifier $\hat{\Phi}_2$ that returns, on the same test set, $FP = 0$ and $FN = 10$. However, $\hat{\Phi}_1$ is technically a better binary quantifier than $\hat{\Phi}_2$; indeed, $\hat{\Phi}_1$ is a perfect quantifier, since FP and FN are equal and thus compensate each other, so that the distribution of the test items across the class and its complement is estimated perfectly.

A second reason is that standard supervised learning algorithms are based on the assumption that the training set is drawn from the same distribution as the unlabelled data the classifier is supposed to classify. But in real-world settings this assumption is often violated, a phenomenon usually referred to as *concept drift* [17]. For instance, in a backlog of newswire stories from year 2001, the prevalence of class Terrorism in August data will likely not be the same as in September data; training on August data and testing on September data might well yield low quantification accuracy. Violations of this assumption may occur "for reasons ranging from the bias introduced by experimental design, to the irreproducibility of the testing conditions at training time" [16]. Concept drift usually comes in one of three forms [14]: (a) the class priors $p(c_i)$ may change, i.e., the one in the test set may significantly differ from the one in the training set; (b) the class-conditional distributions $p(\mathbf{x}|c_i)$ may change; (c) the posterior distribution $p(c_i|\mathbf{x})$ may change. It is the first of these three cases that poses a problem for quantification.

The previous arguments indicate that text quantification should not be considered a mere byproduct of text classification, and should be studied as a task of its own. To date, most proposed methods for solving quantification (see e.g., [2, 3, 7–10]) employ generic supervised learning methods, i.e., address quantification by elaborating on the results returned by a generic classifier. However, some approaches have been recently proposed [6, 15] that employ supervised learning methods directly addressed to quantification.

3   About This Tutorial

More research on quantification is needed, since quantification is going to be more and more important: with the advent of "big data", more and more application contexts are going to spring up in which we will simply be happy with analysing data at the aggregate (rather than at the individual) level.

The goal of this tutorial is to introduce the audience to the problem of quantification (with an emphasis on *text* quantification), to its applications, to the techniques that have been proposed for solving it, to the metrics used to evaluate them, and to the problems that are still open in the area.

References

1. Baccianella, S., Esuli, A., Sebastiani, F.: Variable-constraint classification and quantification of radiology reports under the ACR Index. Expert Systems and Applications 40(9), 3441–3449 (2013)
2. Bella, A., Ferri, C., Hernández-Orallo, J., Ramírez-Quintana, M.J.: Quantification via probability estimators. In: Proceedings of the 11th IEEE International Conference on Data Mining (ICDM 2010), pp. 737–742 (2010)
3. Bella, A., Ferri, C., Hernández-Orallo, J., Ramírez-Quintana, M.J.: Aggregative quantification for regression. Data Mining and Knowledge Discovery 28(2), 475–518 (2014)
4. Esuli, A., Sebastiani, F.: Machines that learn how to code open-ended survey data. International Journal of Market Research 52(6), 775–800 (2010)
5. Esuli, A., Sebastiani, F.: Sentiment quantification. IEEE Intelligent Systems 25(4), 72–75 (2010)
6. Esuli, A., Sebastiani, F.: Optimizing text quantifiers for multivariate loss functions. Technical Report 2013-TR-005, Istituto di Scienza e Tecnologie dell'Informazione, Consiglio Nazionale delle Ricerche, Pisa, IT (2013)
7. Forman, G.: Counting positives accurately despite inaccurate classification. In: Gama, J., Camacho, R., Brazdil, P.B., Jorge, A.M., Torgo, L. (eds.) ECML 2005. LNCS (LNAI), vol. 3720, pp. 564–575. Springer, Heidelberg (2005)
8. Forman, G.: Quantifying trends accurately despite classifier error and class imbalance. In: Proceedings of the 12th ACM International Conference on Knowledge Discovery and Data Mining (KDD 2006), Philadelphia, US, pp. 157–166 (2006)
9. Forman, G.: Quantifying counts and costs via classification. Data Mining and Knowledge Discovery 17(2), 164–206 (2008)
10. Forman, G., Kirshenbaum, E., Suermondt, J.: Pragmatic text mining: Minimizing human effort to quantify many issues in call logs. In: Proceedings of the 12th ACM International Conference on Knowledge Discovery and Data Mining (KDD 2006), Philadelphia, US, pp. 852–861 (2006)
11. Gamon, M.: Sentiment classification on customer feedback data: Noisy data, large feature vectors, and the role of linguistic analysis. In: Proceedings of the 20th International Conference on Computational Linguistics (COLING 2004), Geneva, CH, pp. 841–847 (2004)
12. Giorgetti, D., Sebastiani, F.: Automating survey coding by multiclass text categorization techniques. Journal of the American Society for Information Science and Technology 54(14), 1269–1277 (2003)

13. Hopkins, D.J., King, G.: A method of automated nonparametric content analysis for social science. American Journal of Political Science 54(1), 229–247 (2010)
14. Kelly, M.G., Hand, D.J., Adams, N.M.: The impact of changing populations on classifier performance. In: Proceedings of the 5th ACM SIGKDD International Conference on Knowledge Discovery and Data Mining (KDD 1999), San Diego, US, pp. 367–371 (1999)
15. Milli, L., Monreale, A., Rossetti, G., Giannotti, F., Pedreschi, D., Sebastiani, F.: Quantification trees. In: Proceedings of the 13th IEEE International Conference on Data Mining (ICDM 2013), Dallas, US, pp. 528–536 (2013)
16. Quiñonero-Candela, J., Sugiyama, M., Schwaighofer, A., Lawrence, N.D.: Dataset shift in machine learning. The MIT Press, Cambridge (2009)
17. Sammut, C., Harries, M.: Concept drift. In: Sammut, C., Webb, G.I. (eds.) Encyclopedia of Machine Learning, pp. 202–205. Springer, Heidelberg (2011)
18. Tang, L., Gao, H., Liu, H.: Network quantification despite biased labels. In: Proceedings of the 8th Workshop on Mining and Learning with Graphs (MLG 2010), Washington, US, pp. 147–154 (2010)
19. Xue, J.C., Weiss, G.M.: Quantification and semi-supervised classification methods for handling changes in class distribution. In: Proceedings of the 15th ACM International Conference on Knowledge Discovery and Data Mining (SIGKDD 2009), Paris, FR, pp. 897–906 (2009)

The Cluster Hypothesis in Information Retrieval

Oren Kurland

Technion, Israel Institute of Technology, Israel
kurland@ie.technion.ac.il

Abstract. The *cluster hypothesis* states that "closely associated documents tend to be relevant to the same requests" [45]. This is one of the most fundamental and influential hypotheses in the field of information retrieval and has given rise to a huge body of work. In this tutorial we will present the research topics that have emerged based on the cluster hypothesis. Specific focus will be placed on cluster-based document retrieval, the use of topic models for ad hoc IR, and the use of graph-based methods that utilize inter-document similarities. Furthermore, we will provide an in-depth survey of the suite of retrieval methods that rely, either explicitly or implicitly, on the cluster hypothesis and which are used for a variety of different tasks; e.g., query expansion, query-performance prediction, fusion and federated search, and search results diversification.

1 Tutorial Objectives

The primary objective of this tutorial is to present the cluster hypothesis and the lines of research to which it has given rise. To this end, much emphasis will be put on fundamental retrieval techniques and principles that are based on the cluster hypothesis and which have been used for a variety of IR tasks. The more specific goals of the tutorial are to provide attendees with (i) the required background to pursue research in topics that are based on the cluster hypothesis; (ii) an overview of the different tasks for which the cluster hypothesis can be leveraged; and, (iii) fundamental knowledge of the retrieval "toolkit" that was developed based on the cluster hypothesis.

2 Outline

- Introduction.
  - The cluster hypothesis [15,45].
  - A high level view of the effect of the cluster hypothesis on the IR field.
- Tests measuring the extent to which the cluster hypothesis holds [15,47,10,41,37].
- Cluster-based document retrieval.
  - Using clusters for document selection [15,4,9,10,31,44,26,34,28,35,22,24,36,23,25,38].
    * The optimal cluster problem [13,28,35,24,36,11].

M. de Rijke et al. (Eds.): ECIR 2014, LNCS 8416, pp. 823–826, 2014.

- Using clusters (or topic models) to "expand" document representations [14,40,1,34,26,42,48,51,23,20].
  - Cluster representation [36,24,39].
  - Types of clusters [15,49,43,23].
- Graph-based methods utilizing inter-document similarities [5,2,6,50,27,28,3,21].
  - Centrality-based methods [5,2,27,28,3,21].
  - Score regularization [6].
  - Cluster-based graphs [28,50,22].
  - Passage-based graphs [3,21].
- Additional IR tasks for which inter-document similarities are used.
  - Using document clusters to visualize retrieved results [13,33,32].
  - Query expansion [27,30,16].
  - Cross-lingual retrieval [8].
  - Query-performance prediction [46,7,29].
  - Fusion [18,19] and federated search [17].
  - Diversifying search results [52,12].
- Open challenges.
- Summary.

Acknowledgments. We thank the reviewers for their comments.

References

1. Azzopardi, L., Girolami, M., van Rijsbergen, K.: Topic based language models for ad hoc information retrieval. In: Proceedings of IJCNN (2004)
2. Baliński, J., Daniłowicz, C.: Re-ranking method based on inter-document distances. Information Processing and Management 41(4), 759–775 (2005)
3. Bendersky, M., Kurland, O.: Re-ranking search results using document-passage graphs. In: Proceedings of SIGIR, pp. 853–854 (2008) (poster)
4. Bruce Croft, W.: A model of cluster searching based on classification. Information Systems 5, 189–195 (1980)
5. Daniłowicz, C., Baliński, J.: Document ranking based upon Markov chains. Information Processing and Management 41(4), 759–775 (2000)
6. Diaz, F.: Regularizing ad hoc retrieval scores. In: Proceedings of CIKM, pp. 672–679 (2005)
7. Diaz, F.: Performance prediction using spatial autocorrelation. In: Proceedings of SIGIR, pp. 583–590 (2007)
8. Diaz, F.: A method for transferring retrieval scores between collections with non overlapping vocabularies. In: Proceedings of SIGIR, pp. 805–806 (2008) (poster)
9. El-Hamdouchi, A., Willett, P.: Hierarchic document clustering using Ward's method. In: Proceedings of SIGIR, pp. 149–156 (1986)
10. El-Hamdouchi, A., Willett, P.: Techniques for the measurement of clustering tendency in document retrieval systems. Journal of Information Science 13, 361–365 (1987)
11. Fuhr, N., Lechtenfeld, M., Stein, B., Gollub, T.: The optimum clustering framework: implementing the cluster hypothesis. Journal of Information Retrieval 15(2), 93–115 (2012)

12. He, J., Meij, E., de Rijke, M.: Result diversification based on query-specific cluster ranking. JASIST 62(3), 550–571 (2011)
13. Hearst, M.A., Karger, D.R., Pedersen, J.O.: Scatter/Gather as a tool for the navigation of retrieval results. In: Working Notes of the 1995 AAAI Fall Symposium on AI Applications in Knowledge Navigation and Retrieval (1995)
14. Hofmann, T.: Probabilistic latent semantic indexing. In: Proceedings of SIGIR, pp. 50–57 (1999)
15. Jardine, N., van Rijsbergen, C.J.: The use of hierarchic clustering in information retrieval. Information Storage and Retrieval 7(5), 217–240 (1971)
16. Kalmanovich, I.G., Kurland, O.: Cluster-based query expansion. In: Proceedings of SIGIR, pp. 646–647 (2009) (poster)
17. Khalaman, S., Kurland, O.: Utilizing inter-document similarities in federated search. In: Proceedings of SIGIR, pp. 1169–1170 (2012)
18. Kozorovitzky, A.K., Kurland, O.: Cluster-based fusion of retrieved lists. In: SIGIR, pp. 893–902 (2011)
19. Kozorovitzky, A.K., Kurland, O.: From "identical" to "similar": Fusing retrieved lists based on inter-document similarities. Journal of Artificial Intelligence Research (JAIR) 41, 267–296 (2011)
20. Krikon, E., Kurland, O.: A study of the integration of passage-, document-, and cluster-based information for re-ranking search results. Journal of Information Retrieval 14(6), 593–616 (2011)
21. Krikon, E., Kurland, O., Bendersky, M.: Utilizing inter-passage and inter-document similarities for re-ranking search results. ACM Transactions on Information Systems 29(1) (2010)
22. Kurland, O.: The opposite of smoothing: A language model approach to ranking query-specific document clusters. In: Proceedings of SIGIR, pp. 171–178 (2008)
23. Kurland, O.: Re-ranking search results using language models of query-specific clusters. Journal of Information Retrieval 12(4), 437–460 (2009)
24. Kurland, O., Domshlak, C.: A rank-aggregation approach to searching for optimal query-specific clusters. In: Proceedings of SIGIR, pp. 547–554 (2008)
25. Kurland, O., Krikon, E.: The opposite of smoothing: A language model approach to ranking query-specific document clusters. Journal of Artificial Intelligence Research (JAIR) 41, 367–395 (2011)
26. Kurland, O., Lee, L.: Corpus structure, language models, and ad hoc information retrieval. In: Proceedings of SIGIR, pp. 194–201 (2004)
27. Kurland, O., Lee, L.: PageRank without hyperlinks: Structural re-ranking using links induced by language models. In: Proceedings of SIGIR, pp. 306–313 (2005)
28. Kurland, O., Lee, L.: Respect my authority! HITS without hyperlinks utilizing cluster-based language models. In: Proceedings of SIGIR, pp. 83–90 (2006)
29. Kurland, O., Raiber, F., Shtok, A.: Query-performance prediction and cluster ranking: two sides of the same coin. In: Proceedings of CIKM, pp. 2459–2462 (2012)
30. Lee, K.-S., Croft, W.B., Allan, J.: A cluster-based resampling method for pseudo-relevance feedback. In: Proceedings of SIGIR, pp. 235–242 (2008)
31. Lee, K.-S., Park, Y.-C., Choi, K.-S.: Re-ranking model based on document clusters. Information Processing and Management 37(1), 1–14 (2001)
32. Leuski, A.: Evaluating document clustering for interactive information retrieval. In: Proceedings of CIKM, pp. 33–40 (2001)
33. Leouski, A., Allan, J.: Evaluating a visual navigation system for a digital library. In: Nikolaou, C., Stephanidis, C. (eds.) ECDL 1998. LNCS, vol. 1513, pp. 535–554. Springer, Heidelberg (1998)

34. Liu, X., Croft, W.B.: Cluster-based retrieval using language models. In: Proceedings of SIGIR, pp. 186–193 (2004)
35. Liu, X., Croft, W.B.: Experiments on retrieval of optimal clusters. Technical Report IR-478, Center for Intelligent Information Retrieval (CIIR), University of Massachusetts (2006)
36. Liu, X., Croft, W.B.: Evaluating text representations for retrieval of the best group of documents. In: Macdonald, C., Ounis, I., Plachouras, V., Ruthven, I., White, R.W. (eds.) ECIR 2008. LNCS, vol. 4956, pp. 454–462. Springer, Heidelberg (2008)
37. Raiber, F., Kurland, O.: Exploring the cluster hypothesis, and cluster-based retrieval, over the web. In: Proceedings of CIKM, pp. 2507–2510 (2012)
38. Raiber, F., Kurland, O.: Ranking document clusters using markov random fields. In: Proceedings of SIGIR, pp. 333–342 (2013)
39. Seo, J., Bruce Croft, W.: Geometric representations for multiple documents. In: Proceedings of SIGIR, pp. 251–258 (2010)
40. Singhal, A., Pereira, F.: Document expansion for speech retrieval. In: Proceedings of SIGIR, pp. 34–41 (1999)
41. Smucker, M.D., Allan, J.: A new measure of the cluster hypothesis. In: Proceedings of ICTIR, pp. 281–288 (2009)
42. Tao, T., Wang, X., Mei, Q., Zhai, C.: Language model information retrieval with document expansion. In: Proceedings of HLT/NAACL, pp. 407–414 (2006)
43. Tombros, A., van Rijsbergen, C.J.: Query-sensitive similarity measures for information retrieval. The Knowledge Information Systems Journal 6(5), 617–642 (2004)
44. Tombros, A., Villa, R., van Rijsbergen, C.J.: The effectiveness of query-specific hierarchic clustering in information retrieval. Information Processing and Management 38(4), 559–582 (2002)
45. van Rijsbergen, C.J.: Information Retrieval, 2nd edn. Butterworths (1979)
46. Vinay, V., Cox, I.J., Milic-Frayling, N., Wood, K.R.: On ranking the effectiveness of searches. In: Proceedings of SIGIR, pp. 398–404 (2006)
47. Voorhees, E.M.: The cluster hypothesis revisited. In: Proceedings of SIGIR, pp. 188–196 (1985)
48. Wei, X., Bruce Croft, W.: LDA-based document models for ad-hoc retrieval. In: Proceedings of SIGIR, pp. 178–185 (2006)
49. Willett, P.: Query specific automatic document classification. International Forum on Information and Documentation 10(2), 28–32 (1985)
50. Yang, L., Ji, D., Zhou, G., Nie, Y., Xiao, G.: Document re-ranking using cluster validation and label propagation. In: Proceedings of CIKM, pp. 690–697 (2006)
51. Yi, X., Allan, J.: Evaluating topic models for information retrieval. In: Proceedings of CIKM, pp. 1431–1432 (2008)
52. Zhu, X., Goldberg, A.B., Van Gael, J., Andrzejewski, D.: Improving diversity in ranking using absorbing random walks. In: Proceedings of HLT-NAACL, pp. 97–104 (2007)

Author Index

Printed in the United States
By Bookmasters